The Biochemistry and Physiology of Plant Disease

The Biochemistry and Physiology of Plant Disease

Robert N. Goodman
Zoltán Király
K. R. Wood

University of Missouri Press
Columbia, 1986

Copyright © 1986 by
The Curators of the University of Missouri
University of Missouri Press, Columbia, Missouri 65211
Printed and bound in the United States of America
All rights reserved

∞™ This paper meets the minimum requirements of
the American National Standard for Permanence of Paper
for Printed Library Materials, Z39.48, 1984.

Library of Congress Cataloging in Publication Data

Goodman, Robert N.
  The biochemistry and physiology of plant
disease.

  Includes bibliographies and indexes.
  1. Plant diseases. 2. Plants—Metabolism.
I. Király, Zoltán. II. Wood, K. R. III. Title.
IV. Title: Infectious plant disease.
SB731.G58 1985   581.2′3   84–17433
ISBN 0–8262–0349–3

Scientific advance reflects the contributions of our predecessors and current colleagues. However the work of Sir Frederick Bawden, Rothamstead Experiment Station, Professor Ernst Gäumann, Eidgenössische Techniche Hochschule, and Professor David Gottlieb, University of Illinois, in biochemical and physiological aspects of plant pathology during the past four and a half decades has been pioneering, broad-ranging, and profound. In recognition of their efforts, this book is dedicated to them.

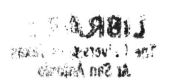

Preface

This book has had an unusually long gestation period, as if the authors were waiting for one more discovery or to include yet another piece of new information concerning biochemical and physiological aspects of plant pathology. However, science is a never-ending process of acquiring new information, hence further delay seemed unwise.

Eighteen years have passed since our earlier version of the subject appeared on 354 pages. The current effort is clearly significantly expanded, reflecting the tremendous advances that have been made by our colleagues.

The authors have used the same format as before with two exceptions; the chapter previously entitled Phenol Metabolism is now titled Secondary Metabolism, and a tenth chapter entitled Resistance has been added.

As before, each chapter, except Infection Process, Toxins, and Resistance, contains a concise modern treatment of the physiology and biochemistry of the healthy plant before describing the aberrations caused by viral, bacterial, and fungal infection.

To keep this text current, during its final stages of preparation we have added a brief section at the end of each chapter entitled Additional Comments and Citations. It is in this section that we have added recent or inadvertently omitted information of importance.

Although an effort has been made to integrate the writing styles of an American, a Hungarian, and an Englishman, personal professional concepts and biases remain. They are indeed a reflection of differences of opinion that exist, healthily, in all areas of science. We contend that time will be the final arbiter of the validity of a concept.

In preparing this text, R. N. Goodman was responsible for the material on bacteria, Z. Király for the material on fungi, and K. R. Wood for the material on viruses. The integration of the findings in these main disciplines has been considered primarily in the sections of Comparative Analysis at the end of each chapter. Collating the material from the three authors and writing the healthy plant sections were shared responsibilities. R. N. Goodman prepared and collated the materials for Chapters 4, 7, 8, and 10; Z. Király for Chapters 3, 6, and 9; and K. R. Wood for Chapters 1, 2, and 5.

The authors wish to express their gratitude to those who reviewed with generous diligence and commented upon the various chapters: M. I. Boulton, E. Brennan, B. Buchanan, M. Daly, R. F. Davis, R. Durbin, R. S. S. Fraser, M. Gidley, S. Hutcheson, N. Keen, T. Kosuge, J. Kuć, A. J. Maule, T. Moore, J. H. Newberry, R. D. Preston, L. Sequeira, R. Staples, C. Vance, L. van Loon, R. K. S. Wood, and M. Zimmerman. The comments and suggestions concerning book content and format by Dr. George Agrios are sincerely appreciated.

The authors also appreciate the opportunity afforded them by the Rockefeller Foundation to utilize the Villa Serbelloni Study and Conference Center at Bellagio, Italy, to plan and subsequently complete the writing of this book.

R. G.
Z. K.
K. W.   June 1985

Contents

Abbreviations of Virus Names

Abbreviation	Name
ALMV	alfalfa mosaic
BCMV	bean common mosaic
BCTV (CTV)	beet curly top
BGMV	bean golden mosaic
BMV	brome mosaic
BMYV	beet mild yellowing
BNYV	broccoli necrotic yellows
BPMV	bean pod mottle
BSMV	barley stripe mosaic
BWYV	beet western yellows
BYDV	barley yellow dwarf
BYMV	bean yellow mosaic
BYV	beet yellows
CaMV	cauliflower mosaic
CarMV	carnation mottle
CCMV	cowpea chlorotic mottle
CGMMV	cucumber green mottle mosaic
CLV	cherry leafroll
CMV	cucumber mosaic
CPMV	cowpea mosaic
CYMV	clover yellow mosaic
DEMV	dolichos enation mosaic (cowpea strain of tobacco mosaic)
EMV	eggplant mosaic
LNYV	lettuce necrotic yellows
MDMV	maize dwarf mosaic
MRDV	maize rough dwarf
PLRV	potato leafroll
PMTV	potato mop top
PRV	peanut rosette
PSV	peanut stunt
PVA	potato virus A
PVB	potato virus B
PVM	potato virus M

Abbreviation	Name
PVS	potato virus S
PVX	potato virus X
PVY	potato virus Y
PYFV	parsnip yellow fleck
RCMV	red clover mottle
RRV	raspberry ringspot
RTV	rice tungro
SBMV	southern bean mosaic
SCMV	sugarcane mosaic
SLRV	strawberry latent ringspot
SMV	squash mosaic
SoMV	soybean mosaic
TAMV	tomato aucuba mosaic
TAV	tomato aspermy
TBRV	tomato blackring
TBSV	tomato bushy stunt
TEV	tobacco etch
TMV	tobacco mosaic
TNDV	tobacco necrotic dwarf
TNV	tobacco necrosis
ToRSV	tomato ringspot
TRosV	turnip rosette
TRSV	tobacco ringspot
TRV	tobacco rattle
TSWV	tomato spotted wilt
TuMV	turnip mosaic
TYMV	turnip yellow mosaic
WTV	wound tumor
Satellites	
CARNA 5	CMV-associated RNA
STNV	satellite of tobacco necrosis
Viroids	
CPFV	cucumber pale fruit
PSTV	potato spindle tuber

1

The Infection Process

Much research emphasis has been given, in recent years, to the biochemical and physiological aspects of the infection process. This concentration of effort followed an extensive period of study that examined the manner in which pathogens *entered* plants. Now, it appears that researchers have embarked on a period of research that focuses on the initial *attraction* of the pathogen to its host. In other words, how does the pathogen *recognize* its host and what are the surface molecules and other "clue" compounds of host and pathogen that characterize the compatibility that may exist between them?

Viruses

Introduction

Plant viruses are normally unable to initiate infection without assistance, for, unaided, virus cannot penetrate the plant cell wall. Unlike bacteria, which cannot penetrate the cell wall either but can multiply in the intercellular spaces, viruses must, of necessity, enter the cell in order to replicate. This is achieved either through the agency of a vector (insect, nematode, or fungal zoospore, for example) or mechanically.

Here we shall be concerned with the early events in the virus infection process leading, ultimately, to virus replication, and we shall be concerned almost exclusively with the establishment of infection by mechanical transmission. Although there is still a lot to be learned about this process, we know much more about the parameters involved and the variables that influence the outcome than we do about other means of transmission (e.g., 327). Accounts of transmission by vectors can be found elsewhere (58, 205). Since the transcription and translation of the virus genome are considered in Chap. 5, in the context of nucleic acid and protein metabolism, the sequence of events outlined here stops at the uncoating stage. This is in contrast to the sections on bacterial and fungal infections, in which the infection process described includes *establishment* of the pathogen within the host tissue.

Wounding and cell penetration. When TMV was introduced into the intercellular spaces or xylem vessels of *Nicotiana glutinosa*, it did not cause infection unless the infiltrated leaves were rubbed, causing wounding or damage (48). Roberts and Price (238) have reported infection of "apparently" unwounded bean leaves with both TNV and SBMV by transport of virus to mesophyll cells through the xylem; however, under the same conditions, TMV failed to infect. Drops of virus suspension placed on a leaf surface will not penetrate and lead to infection unless the leaf is pricked with a pin or otherwise damaged. Wounding of the leaf surface may be achieved much more efficiently, however, by applying virus suspension to the leaf surface with a muslin pad, stiff brush, spatula, or even a finger. The objective appears to be to create as many small wounds as possible on the leaf surface without causing death of the epidermal cells. The efficiency of these crude inoculation procedures can, therefore, be considerably improved by the use of abrasives, either dusted onto the leaf prior to application of the virus suspension or added in defined concentration to the virus inoculum. The two in most common use are Celite (a diatomaceous earth) and Carborundum (silicon carbide).

Estimation of efficiency of infection. It will be clear already that it is essential to be able to assess the efficiency of the infection process in order to quantitate the changes in either inoculum or host. The process of virus accumulation within a plant leaf can be envisaged as taking place in several stages. Virus particles may be at least partially uncoated before cell penetration. This is followed by expression of the partially or completely uncoated genome, assembly of the replicated genome and virion components into particles, and transport of particles to adjacent cells and possibly beyond, where further replication may occur. The extent of virus replication and the nature of the resultant symptoms depend, primarily, on the genomes of virus and host but can be considerably influenced by environmental factors. We are concerned here with the establishment of infection; that is, the sequence of events leading to virus replication in the first plant cells to be infected. Ideally, the number of such infected cells indicates the efficiency of the infection process. In practice, it has been found convenient to use a host that responds hypersensitively. In this case, the result of virus replication in several adjacent cells is assessed; the initially infected cells and those adjacent become necrotic, leading to the formation of a clearly visible lesion (see Chap. 10).

This obviously involves factors other than the establishment of infection and gives a minimum estimate of efficiency, since not all cells initially infected will necessarily give rise to a visible lesion (15). Nevertheless,

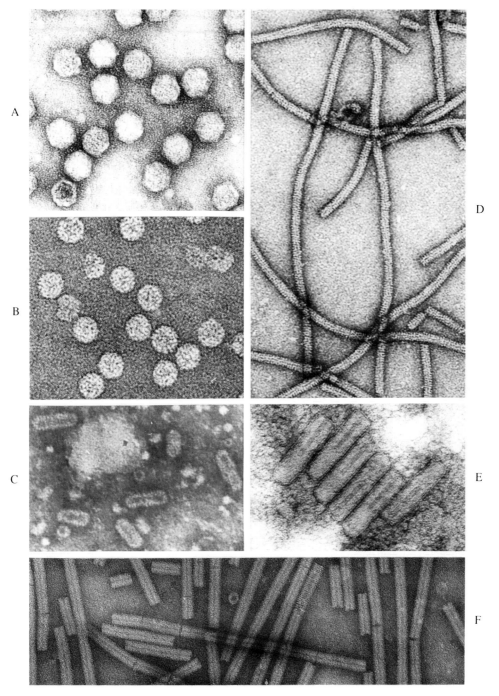

Fig. 1–1. Electron micrographs of (a) TNV; (b) CMV; (c) AMV; (d) PVY; (e) BNYV; (f) TMV. For others see references 130, 183, 200, 271 (electron micrographs courtesy of M.J. Webb).

Fig. 1–2. *N. glutinosa* leaf hair alterations after inoculation with TMV: left, the control; right, damage caused by the inoculation process (after Conti and Locci, 57).

because of the difficulties in identifying single infected cells, this procedure is used by the majority of those studying the infection process and host susceptibility to infection. If, then, the number of lesions appearing on leaves of hypersensitive host plants is taken as a measure of the efficiency of infection, use of either of these two abrasives, Celite or Carborundum, can increase efficiency a hundredfold or more. Important factors appear to be the number of abrasive particles used (37) and also their size. Conti and Locci, for example, found Carborundum "600" more efficient than "200" (57) suggesting a relationship to size and number of wounds made by the abrasive.

Trichomes and Guard Cells

A typical leaf surface may consist of $1–5 \times 10^6$ cells of various types, principally epidermal cells with some guard cells, protected by a layer of pectin, a cutinised cellulosic wall, cuticle, and, finally, on the outer surface, epicuticular wax. (The precise nature of these layers will, however, vary considerably among species.) In addition, some species possess trichomes (leaf hairs) that may be of more than one type. Abrasion of the leaf surface brings virus particles into contact with many of these cells, and the question of which cells constitute the primary site of infection has long been a subject of investigation. There is no doubt that hair cells can be inoculated and will support virus replication (28, 143, 329), though this has probably little relevance to

the process of mechanical inoculation. Leaf hairs become severely damaged (Fig. 1–2), and, though they may subsequently reappear, it is unlikely that they contribute to infection. In studies designed to assess the role of leaf hairs in the infection of *N. glutinosa* by TMV, Schlegel and his colleagues (136, 181) applied ^{14}C-labeled TMV to the leaf surface and established the distribution of radioactive label in leaf sections by autoradiography. They observed both infective and UV-inactivated virus accumulate at basal septa of broken hairs, with infective virus accumulating more in hosts than in nonhosts. It was suggested that virus particles would come into contact with cytoplasm from the broken trichomes and, thence, have access to foot cells via plasmodesmata in the basal septum. There was no indication, however, of what role, if any, these particles played in subsequent infection. An alternative explanation for virus penetration through trichomes was suggested by Hildebrand (140). When injured with a micropipette, trichomes exude a small quantity of protoplasm that is almost immediately drawn back into the cell. Virus could clearly be carried with it. When pepper plants were inoculated between leaf hairs (or leaf hairs were removed prior to inoculation), Boyle and McKinney (37) observed as many TMV lesions per unit area as when the entire leaf surface was inoculated. Hence, in this situation at least, leaf hairs seem to play a very minor role in the infection process. There are also many plant species (cowpea, for example) that do not possess

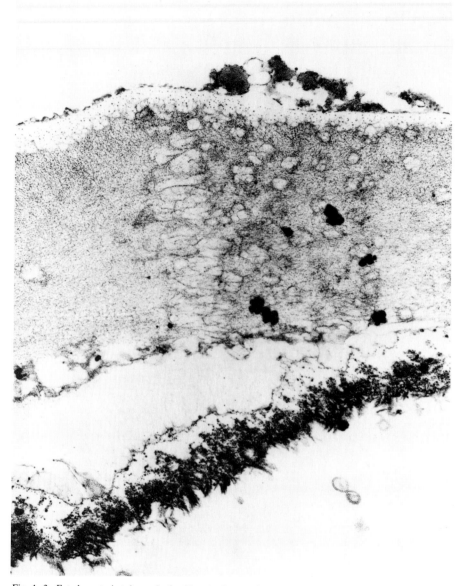

Fig. 1–3. Ectodesmata in tobacco leaf epidermis characterized by a vertical section of the loosely packed wall fibrils, ×60,000 (courtesy of deZoeten, 72).

leaf hairs. The evidence suggests that, at least in the majority of cases, trichomes present a possible but not essential route for virus entry. Nor is there evidence to implicate guard cells. Pepper plants have 20 times as many guard cells on the upper as on the lower surface, yet inoculation of either produced the same number of lesions (37).

Ectodesmata

Other sites that have been considered as likely points for the initiation of virus infection are ectodesmata, originally thought to be plasmodesmata extending through the outer wall of the epidermal cells as far as the cuticle, providing a potential link between the leaf surface and the epidermal cells. With this in mind, attempts have been made to correlate the number of ectodesmata on a leaf surface with susceptibility to infection (40, 287; Fig. 1–3). Thomas and Fulton, for example, observed that TI 245 tobacco, which produced fewer lesions on inoculation with necrotic strains of CMV or TMV than did TI 787, apparently had fewer ectodesmata and that treatments increasing the number of ectodesmata also increased the number of lesions. Though their results were consistent with the view that ectodesmata function as infection sites, others have not been quite so convinced. For example, variation in ectodesmata number failed to account for differences in susceptibility of bean to TMV (194) or the differences in susceptibility between upper and lower surfaces of *Datura stramonium* leaves to TMV (281). There is also doubt that ectodesmata do, indeed, represent pores in the outer cell wall (e.g., 252). It now seems likely (72) that ectodesmata simply represent a transient parting of the cellulose fibrils in the cell wall, and it is unlikely that they would afford direct access of virus particles to the plasmalemma. For the moment, evidence for the involvement of ectodesmata in the infection process is not substantial.

Epidermal Cells

It is likely that the primary sites of initiation of virus infection on the leaf surface are principally the epidermal cells. Hence, virus particles gain entry by the creation of small, nonfatal wounds during the process of mechanical inoculation. Each one of these cells is potentially capable of acting as an initiation site, yet even with a high virus concentration in the inoculum, the maximum number of lesions (and by inference, infection centers) attainable on a leaf of *N. glutinosa*, for example, may be on the order of 4,000. Rarely is there confluent necrosis in the initial stages of infection, which would presumably occur if all cells became infected. Thus, either there are never enough infective virus particles in the inoculum to infect all susceptible cells, or there is a practical limit either to the number of potentially infectible sites that can be created or to the number of successfully established sites that are allowed to develop into lesions. Either or both of the last two possibilities seem the more likely.

Infectible Sites

We can visualize the establishment of infection as a two-stage process; first, an infectible site is created, then this site interacts with particles in the inoculum leading to the establishment of an infection center. Usually, the two processes are almost simultaneous, with virus present during the leaf abrasion. They can, however, be separated in time by abrading the leaf surface to create an infection center and then applying virus by spraying or dipping at various times after wounding. Using the subsequent development of lesions as an indication of the number of potentially infectible sites present at the time of wounding, Furumoto and Wildman (109) established that during the inoculation of *N. glutinosa* with TMV, while many of the sites disappeared almost immediately and most were no longer infectible within the first minute, some were still present and infectible for 40 min or more. While this situation is clearly true for some virus/host combinations (140), it is not always the case (156). Jedlinski, for example (153, 154), found that for the infection of *N. rustica* with TNV, the apparent number of sites increased significantly within the first 5 min after abrasion, while for other combinations there was no change. It would appear, then, that the precise nature of the infectible sites created by wounding differs among species, and even on the same leaf surface the sites constitute a heterogeneous population. This is hardly surprising if the sites are created by the formation of many small, healable wounds. By the very nature of the process, some are going to heal quickly and rapidly render the cell inaccessible to virus particles, while others will take longer.

From studies in which virus was present at the time of creation of the infectible site, Furumoto and Mickey (107, 108) have concluded that individual cells are also heterogeneous with respect to their susceptibility to infection, while Kleczkowski had earlier proposed (176) that different areas of the leaf surface vary in susceptibility. While biochemical and physiological factors may play an additional role, this is not entirely unexpected, since the inevitable consequence of abrading an uneven leaf surface will be to wound some areas optimally for virus infection. Other areas, however, will remain relatively undamaged, or damaged too severely for formation of an infective center. Turgidity is also likely to contribute to the differential accessibility of cells to abrasive, as several reports suggest that turgid leaves are easier to infect (27, 290, 322) than less turgid ones (172, 225). This factor may also be a major contributor to the change in number of lesions appearing on leaves of plants previously inoculated on the lower leaves (105, 106).

Receptors

Under ideal circumstances, all cells on the leaf surface should be potentially infectible (107), provided that infective virus particles reach the cytoplasm. In practice, however, the number of cells infected at a given virus concentration and, consequently, the number of lesions that appear vary widely among species and depend very much on the virus/host combination and the conditions of both inoculum and plant. Although ectodesmata are possible routes of entry to which access would be provided by abrading the cuticle only, evidence points to the necessity for allowing access through the cell wall to epidermal cells. The question of whether initial interaction between virus and host involves specific receptors, either on or in the epidermal cells, remains unresolved. Homologous coat protein preparations can interfere with the establishment of infection. A mixture of TMV and coat protein, for example, produced fewer lesions on pinto bean (246), Xanthi nc tobacco (144), or *Chenopodium amaranticolor* (12, 218) than virus alone, while heterologous protein from BSMV had no effect (218). However, at low-protein concentrations, stimulation of lesion formation could be observed (144). Also, while both *Zea mays* and *Hordeum vulgare* are susceptible to BMV, but not to TMV, they were not susceptible to BMV RNA coated with TMV protein. This could be interpreted as a lack of receptors for TMV, though other explanations, such as a deficiency in uncoating, are possible (12).

It has been assumed that abrasion must damage both cuticle and cell wall of cells to be infected, exposing the plasmalemma; virus particles, possibly partially uncoated, are then either taken into the cell by pinocytosis or penetrate through small breaks in the plasmalemma. As a result of electron microscope studies on the infection of *Vigna sinensis* and *N. glutinosa* by TMV, Conti et al. (93) have proposed a slightly different mechanism. They contend that virus particles contact the plasmalemma via a transient "hyperhydration" of areas of the cell wall immediately below regions where the cuticle is damaged or reduced in thickness. Their observations support previous hypotheses that virus penetrates via output and readsorption of cellular exudates.

Protoplasts

While in vitro infection of leaf mesophyll protoplasts is contributing to the understanding of several aspects of the interaction between plant viruses and host cells (284, 312), it is doubtful whether the process of infection in vitro bears much relation to the situation that occurs in vivo. Efficient attachment to and penetration of protoplasts by virus particles is usually dependent upon the preincubation of virus with polyethylene glycol or with a polycation such as poly-L-ornithine (PLO), which probably serves two functions. If virus and protoplast have the same charge as, for example, tobacco protoplasts and TMV, then PLO forming aggregates of virus particles with a net charge opposite to that of the cell would facilitate contact between virus and cell. Secondly, the polycation may assist in virus penetration by damaging the plasmalemma sufficiently to allow virus to pass through, but not severely enough to permanently damage the cell. The alternative view is that virus aggregates are taken up by pinocytosis in a manner analogous to the penetration of animal cells by many viruses. Virus particles were first observed adhering to the plasmalemma and in pinocytotic vesicles by Cocking (54, 55) during infection of tomato fruit protoplasts with TMV, observations since confirmed with other virus/host combinations (284). However, the efficiency of cell infection appears to be the same whether performed at room (25 C) or low (2–4 C) temperature (207, 307, 330), possibly arguing against a pinocytotic mechanism. Though the situation is not yet fully resolved, it seems unlikely that pinocytosis is involved in transport of viruses across the plasmalemma either in vitro or in vivo.

Certain polycations also seem to stimulate in vivo infection (259, 295), inclusion of PLO in the inoculum enhancing the amount of TMV retained by tobacco leaves and increasing the degree of uncoating of the retained virus. It is not clear, however, whether these observations are related to the very significant stimulating effect polycations have on infection of protoplasts.

Number of Particles Required to Initiate Infection

Mechanical inoculation of a leaf surface is clearly an inefficient process due to several factors (303). First, it is unlikely that more than a small portion of the particles in an inoculum will be infective, though specific infec-

tivity will vary widely depending on the stability of the virus and the source of the inoculum. Furumoto and Wildman (110) estimated at least 10% of a TMV preparation to be infective, though for many viruses it may be less.

Several estimates of the number of particles required to establish infection have been made, though estimates vary widely depending on the virus, host, experimental technique, and expression of results. Schramm and Engler (253), for example, were able to infect systemically 50% of a batch of Samsun tobacco plants by inoculating with 5 ml of a suspension containing 10^{-16} g/ml TMV, or an average of ca. 7.5 particles/plant. To produce a lesion on *N. glutinosa*, however, 1,500–50,000 particles may be necessary (277). Using much smaller volumes of inocula, 2.5 μl, Walker and Pirone (304) were able both to achieve systemic infection of tobacco (Havana 142, TI 787) and to produce a local lesion on hypersensitive tobacco (Havana 425) with 450 particles of TMV. When, however, their results were expressed in the same way as those of Steere (277), dividing the total number of lesions on all plants of a particular culture by the total number of particles applied, then their estimates were of the same order as his. Local lesions, with consequent systemic infection, have also been obtained on Chinese cabbage by inoculation of each leaf with as few as 10 particles of TYMV contained in 0.1 μl (104), an efficiency considerably greater than that for inoculation of protoplasts in vitro. This approach has also permitted detection of virus replication in isolated protoplasts by local lesion assay. It is of interest that when single-cultured Samsun tobacco cells were inoculated by microinjection, 620 or more TMV particles were required to reproducibly infect a cell, though one out of four became infected with as few as 72 particles (128).

It would appear, therefore, that maximum efficiency and more realistic estimates of the number of particles required to initiate infection are achieved by using small volumes of concentrated inocula. Conversely, the use of large volumes may be a factor at least contributing to the estimates of 10^4–10^6 particles usually claimed to be required to produce a lesion (107). For the multicomponent viruses, it is evident that even more particles will be required to initiate infection (300, 303).

However, when one considers the total number of potentially infectible cells on the leaf surface, even 10^6 particles do not represent a vast excess of particles/cell. There are many possible explanations for more cells not becoming infected and for the majority of infective particles not leading to a successful infection. The important ones have been summarized elsewhere by Matthews (205) and are briefly as follows. First, although there may be a considerable number of infective virus particles on the leaf surface, they will not be distributed in such a way that virus reaches every susceptible cell. Second, suitable wounds for virus penetration are likely to be produced very inefficiently, so that only a small portion of the potentially infectible cells are capable of interacting with virus in a manner likely to lead to infection. Third, suitable wounds are generally short-lived, so that virus particles may not have time to reach them before they are repaired or the virus becomes inactivated. Finally, even if an infective particle contacts and penetrates a susceptible cell of a local lesion host, a *visible* lesion may not necessarily result. It is not surprising, then, that the efficiency of infection may not be very great. It seems likely, however, that for single genome viruses like TMV, one infective particle interacting with and penetrating a susceptible cell is able to initiate infection, since it is easily possible to separate individual viruses and mutants from a population by isolating single lesions obtained at high dilution (230). The single-hit (one particle/one cell) infectivity dilution curves obtained with this type of virus also lend support to the hypothesis. This does not mean, however, that other particles of the same or different strains, or even of different viruses, will necessarily be excluded from that cell. In early studies, the number of lesions on *Nicotiana sylvestris* caused by a local lesion strain of TMV (U2) was reduced by inclusion in the inoculum of a non-lesion-producing strain (U1) (246, 265, 266, 267). These results were considered to support the view that only one particle was allowed to participate in the initial infection process. However, as we shall see later, there is abundant evidence, from studies both on intact plants and on protoplasts, that two strains of the same virus, and indeed two unrelated viruses, can infect and replicate in the same cell. There is little definitive evidence for the presence of one virus excluding initial infection by a second.

In the case of viruses whose genome is in two or more segments (152, 301), it is not only possible but obligatory for more than one particle to participate in the initial infection, since the genome segments essential for infectivity are distributed among two or more particles. This requirement often, but not always, results in infectivity dilution curves that are steeper than the single-hit type and in the necessity for a large number of particles to cause a lesion (301; Fig. 1–4).

There seems little doubt that, in many cases, virus first *replicates* in epidermal cells. Shimomura (263), for example, has observed replication of TMV in lower epidermis stripped from virus-inoculated *N. glutinosa* and incubated in vitro. Ehara and Misawa (81), however, have concluded that CMV is uncoated but replicates to only a small extent, if at all, in epidermal cells of cowpea.

Factors Affecting the Establishment of Infection

As indicated at the beginning of this section, rather than attempting to assess the number of infection sites established following inoculation of plants in which the virus in question multiplies systemically, it has been found convenient to use the number of local lesions appearing on leaves of a hypersensitive host as a measure of the efficiency of the infection process. Changes in the number of lesions following modification either of the inoculum or of the host plant is then taken as a measure of change in infection efficiency or in host sus-

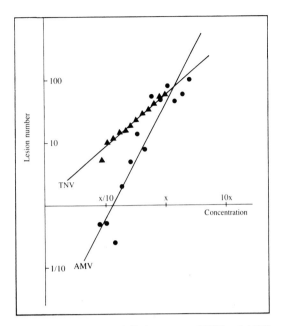

Fig. 1–4. Comparison of dilution curves of TNV and AMV 425 resulting from local lesion counts on beans (*Phaseolus vulgaris* L. var. 'Berna'). Curve values are averages of 12 half-leaves. For TNV x is 0.15 µg/ml; for AMV x is 9 µg/ml (after van Vloten-Doting et al., 301).

ceptibility. Susceptibility in this context clearly relates to establishment of infection rather than to the extent of subsequent virus replication. To produce a visible lesion, however, both virus replication and changes in host physiology are required.

The inoculum. Principally as a result of the exhaustive studies by Yarwood, it has been established that the inclusion in the inoculum of one or more of several additives, in addition to abrasives, leads in general to an increase in the number of local lesions. Of these additives, which include bentonite (323), sulphite (324, 325), sucrose (67), $MgCl_2$ (155), the most critical would appear to be phosphate (318). However, the mechanism by which phosphate affects the interaction between virus and host is not clear.

Both genetic and environmental factors contribute significantly to the outcome of the interaction between virus and host (245). Here, we shall be concerned with environmental factors, which can influence the establishment of infection to a surprising degree. They can be envisaged as affecting either the ease with which infectible sites are formed or the subsequent interaction between virus and cell and the ability of the cell to support virus replication; from the experimental data available, it is not possible to distinguish between these possibilities.

Light. Plants show a pronounced seasonal variation in susceptibility to virus infection and are often considerably more difficult to infect during the summer; there

Fig. 1–5. Increased susceptibility of a shaded tobacco leaf (left) to TNV (after Bawden and Roberts, 26, 27).

is also a diurnal variation (205). There are clearly many variable factors involved, but the two most important seem to be light intensity and temperature. Shading plants prior to inoculation, without significantly changing temperature, has been demonstrated, on many occasions, to enhance susceptibility to infection, while postinoculation darkening has the reverse effect (27, 204, 205).

Bawden and Roberts, for example (26, 27), observed an approximately tenfold increase in susceptibility of shaded tobacco and *N. glutinosa* plants to TNV and TBSV, respectively, though the effect was less pronounced for infection of *N. glutinosa* with TMV (Figs. 1–5 and 1–6). They suggested that since there did not appear to be any obvious change in the fragility of the leaves, the important factor was a change in carbohydrates, either affecting turgor (and, hence, the ease of formation of infectible sites by abrasion) or a later stage in the infection process. Subsequent experiments have supported the view that photosynthesis is important, Wiltshire (311) finding that maintaining bean plants in light in the absence of CO_2 had the same effect in their response to TNV as darkening. However, the complexity of the effect of light intensity on plant susceptibility is illustrated by the fact that a brief exposure to light before inoculation can augment the effect of darkness (135, 204, 205), while prolonged exposure can reduce susceptibility. Matthews has also reported that plants are more susceptible to inoculation in the late afternoon than before dawn (204, 205), and the age of the plant may also be important (171). Kimmins observed an increase in susceptibility of 9-to-12-day-old bean plants to TNV by preinoculation darkening, but not in susceptibility of 22-to-24-day-old plants.

It is common practice, therefore, when assaying virus suspensions, to shade plants before inoculation. The most likely effect is to alter the turgidity and possibly also the hardness of the leaf surface, thereby affecting its susceptibility to abrasion and increasing the number of infectible sites.

Temperature. The temperature at which plants are maintained prior to inoculation can also have a significant effect on their susceptibility to infection, as assessed by the appearance of local lesions. Kassanis (157), for example, found that preincubation of plants at 36 C, followed by return to normal temperature (18–23 C), increased susceptibility of several species to virus infection, as indicated in Table 1–1.

In general, the longer the treatment the more susceptible the plants become. Shimomura (264) preexposed half-leaves of *N. glutinosa* to elevated temperature (36 C) relative to control halves (25 C) prior to inoculation with TMV and observed a similar effect. He suggested, however, that it is change in temperature at the time of inoculation that is important. Increased susceptibility was also observed following transfer of plants from 30 C to 25 C and from 25 C to 17 C, though plants maintained at either 36, 30, 25 or 17 C were equally susceptible. Preinoculation cold treatment (14–16 C) of beans, *N. glutinosa*, or Xanthi nc tobacco plants had the reverse effect, decreasing susceptibility to infection by TMV.

The mechanism of the temperature effect remains obscure. As indicated above, there is a reduction in carbohydrate concentration, which could affect leaf turgidity and, hence, sensitivity to abrasion. Nevertheless, temperature induces many biochemical and physiological

Fig. 1–6. Increased susceptibility of a shaded *N. glutinosa* leaf (left) to TBSV (after Bawden and Roberts, 26, 27).

changes that may or may not be involved in determining susceptibility. As yet, causal relationships have not been established.

Postinoculation heat treatments have been applied to plants in several ways, with results depending on both the timing and the temperature of treatment. Maintaining plants at a moderately elevated temperature (e.g. 36 C) generally leads to a reduction in lesion number (157). It is not clear whether the effect is on the establishment of infection soon after inoculation or on subsequent virus replication and translocation. The effect of a short postinoculation exposure at higher temperature depends principally on the timing of the exposure. Treatment of *N. glutinosa* at 43 C immediately following inoculation with the U1 strain of TMV considerably reduced lesion numbers (315), an effect progressively decreasing in

Table 1–1. Increased susceptibility of plants preheated at 36 C (after Kassanis, 157).

Virus	Host	Number of lesions after treatment for the following times			
		0	6h	1d	2d
TMV	bean	2	4	29	46
TMV	*N. glutinosa*	32	—	98	111
TBSV	*N. glutinosa*	18	84	83	97
TSWV	tobacco	86	157	197	165

magnitude as time of treatment after inoculation increased. In this case, heat treatment apparently affected the early events in the establishment of infection. However, the appearance of visible lesions following establishment of infection is the result of the interaction between a complex series of processes, including virus multiplication, restriction by host defense mechanisms, and necrosis of infected tissue. It is conceivable that heat treatment affects all of them to some degree. The induction of lesion formation in situations in which lesions do not normally appear (TMV-, CMV-, or PVX-infected tobacco, for example; 97, 219) by short periods of hot water treatment or by cooling (220) following infection would appear to be caused by predisposition of the infected cells toward necrosis. Alternatively, treatment from about 6 h after inoculation of leaves that do normally respond with lesion formation can lead to an increase in their number and size (320, 321). In these situations, heat treatment seems to be causing expansion of lesions that, in some cases, may consist of one or a very few infected cells and thus are otherwise invisible, possibly by inactivation of host defense mechanisms, allowing virus to spread to neighboring cells and induce necrosis (314), or by induction of necrosis in cells that are already infected but not necrotic. It would seem appropriate to consider these effects in more detail in the context of the hypersensitive response (Chap. 10).

Nutrition and age. Plants growing optimally with a balanced nutrient supply are generally more susceptible to infection, producing a greater number of lesions than those retarded by either an excess or deficiency of nutrients (24). There are few indications that individual nutrients play a particularly critical role, though increased susceptibility of plants maintained at reduced potassium levels has been observed (8, 98), and calcium deficiency also had a marked effect on lesion formation in *N. glutinosa* (51). Age, too, can have a significant effect. *V. sinensis*, for example, produces few lesions if plants more than about 12 days old are inoculated with CMV, while young leaves of Xanthi nc (308) or Samsun NN (282, 283, 299) tobacco produce fewer (and generally larger) lesions than do older leaves on inoculation with TMV. These observations represent the net result of the interaction between many biochemical and physiological changes and precise explanations are still awaited. (See also Chap. 10)

Water and leaf detachment. Growth of plants with an ample water supply seems to influence their susceptibility (322). Tinsley (290) observed that several viruses produced more lesions in tobacco and *N. glutinosa* plants maintained under wet rather than dry conditions. He proposed that the former were more succulent and had thinner cuticles and were, consequently, more susceptible to abrasion. Others have also concluded that the leaf water status of darkened plants has a direct influence on susceptibility (23, 27, 172, 194). Litz and Kimmins, for example, concluded that the increased susceptibility of darkened French bean plants kept at a high relative humidity was primarily due to a more extensive

wounding of the epidermis resulting, in turn, from increased leaf turgidity (194). (It was not necessarily, however, a direct consequence of a thinner cuticle.) These conclusions would also concur with observations on the correlation between high or low leaf turgidity and induced susceptibility or resistance (106).

Water supply following inoculation clearly cannot affect the creation of infectible sites but may affect their conversion to infective centers and may, therefore, influence the establishment of infection and, possibly, also subsequent events in virus development. The effect of water supply after inoculation might be expected to be quite different from the effect of water supply before inoculation. Indeed, Yarwood (319) found that pinto bean leaves detached and wilted soon after inoculation produced more lesions than unwilted, turgid leaves. More recently, a rapid increase in leaf water deficit up to 3 h after inoculation was observed to lead to a considerable increase in TNV lesion number in French bean (cv. The Prince), an increase proportional to the water stress applied (14). The increase was observed in both attached (Fig. 1–7) and detached (Fig. 1–8) leaves. When leaves were abraded to create infectible sites and then sprayed with inoculum at various times after abrasion, the majority of potentially infectible sites had vanished within 30 min, and all had disappeared within 3 h. Increase in lesion number occasioned by wilting seemed, therefore, to be correlated with the presence of infectible sites. The increase could be accounted for by an enhancement of the water potential gradient, stimulating water movement into the leaf and, hence, also stimulating the passive movement of virions into abraded cells. More potentially infectible sites could, therefore, be converted to infective centers.

Bean leaves inoculated with TNV, then detached and

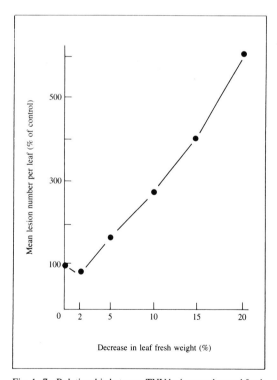

Fig. 1–7. Relationship between TNV lesion number and fresh weight loss in bean leaves (after Bailiss and Plaza-Morales, 14).

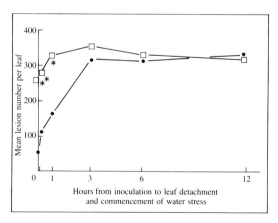

Fig. 1–8. TNV lesion numbers in leaves detached and subjected to water loss at various intervals after inoculum. (●—●) Control leaves detached, but not water stressed at the time indicated, (□—□) leaves detached and subjected to 15% fresh-weight loss at the time indicated. Values marked (*) are significantly different from control values ($P < 0.01$) (after Bailiss and Plaza-Morales, 14).

floated on water immediately after inoculation, developed fewer lesions than leaves detached later or not detached at all (23). The explanation again appeared to be one of water flux. Floating leaves on water would lead to a water flux from the abaxial to the adaxial surface, a situation promoting water movement from epidermal cells to the atmosphere. This would lead to a reduced transport of virions into abraded cells on the adaxial surface and, hence, to a reduced lesion number.

Finally, the presence or absence of water during infection can have a marked effect on the efficiency of infection. Washing or quick drying immediately after inoculation has a variety of effects depending on whether the treatments are applied separately or in conjunction, on the virus/host combination, and on the presence of other additives, particularly phosphate (318), in the inoculum (326). It may be that brief washing helps to remove inhibitors of infection (Chap. 10) from crude sap inocula that might otherwise interfere with the establishment of infection. Drying, on the other hand, concentrates virus particles, or ions that assist in the adsorption process, at the site of infection (326).

In summary, treatments that increase the fragility or turgor of leaves generally increase susceptibility, possibly by increasing the efficiency of the abrasion process and the creation of potentially infectible sites. However, the complexity of the biochemical and physiological changes induced by these treatments is such that the involvement of other as yet unspecified factors cannot be ruled out. The influence of hormones, for example, on lesion development, and hence on apparent lesion number, is discussed at length in Chap. 7.

Inhibitors of Infection

The efficiency of the infection process can be severely reduced by several compounds from many diverse sources; when present in virus inocula, their effect can be on the virus or on the host. Compounds that irreversibly inactivate viruses in vitro (acids, alcohols) are not strictly relevant to this book and will not be considered further. Compounds that, by one means or another, affect the interaction between virus and host are conventionally known as inhibitors. We include, here, compounds like bovine serum albumin, which may, for example, cause virus aggregation without inactivating the components of the aggregate. It is likely that many of the inhibitors act at the initial stages of infection, though others may interfere with one of the later stages of virus replication; some may interfere with both. Van Kammen et al. (297) have proposed an equation relating local lesion number with virus concentration and inhibitor concentration, allowing a distinction to be made between the two. However, in few cases is sufficient information available.

Inhibitors are found in extracts of many plant species and, hence, may be present in inocula prepared from infected plants, thus reducing the potential infectivity of the extract. Alternatively, they have been obtained in varying degrees of purity from uninfected plants and their inhibitory activity tested either through addition to purified virus preparations and inoculation to a local lesion host or through application to the host before or after virus inoculation. Reduction in lesion number, compared to that produced by a control inoculum without inhibitor, can give an indication of the degree of inhibition of infection at differing inhibitor concentrations. Though effective when applied to a plant as much as 24 h before virus inoculation, inhibitors are generally ineffective when added more than a few minutes after infection. If they are, then clearly they are not affecting the initial infection process, but rather one of the subsequent events in the virus replication cycle.

Generally, inhibitors permit infection of the host from which they were derived but interfere with transmission to other species (113); they may also display some virus or host specificity. It is also of interest that while these plant inhibitors may be effective in interfering with mechanical transmission, vector transmission is usually unaffected.

The majority of plant inhibitors appear to be proteins; that from pokeweed, *Phytolacca americana*, perhaps having received the most attention. It was observed that many viruses, including TMV, could be transmitted easily from pokeweed back to pokeweed but only with great difficulty to other species. The explanation is the presence in the sap of an inhibitor (78) that affects not only plant virus transmission (Table 1–2) but also the infection of mammalian cells with influenza virus (291). The compound has been shown to be a basic protein, with 116 amino acids and a molecular weight of 13,000 (316), which inhibits protein synthesis in eukaryotic cells by interfering with the elongation factor–2 catalysed translocation of aminoacyl t-RNA from the ribosome acceptor site to the donor site (114). It inhibits TMV replication in tobacco protoplasts (121), and it would appear, therefore, that it acts principally during virus replication and not in the establishment of infection.

Carnation (*Dianthus caryophyllus*) plants contain a protein with a molecular weight of ca. 10,000, that, for example, inhibits the infection of *N. glutinosa* by TMV

Table 1–2. Effect of pokeweed juice on the pathogenicity of TMV[a] (after Duggar and Armstrong, 78).

Juices constituting inoculum	Dilution	Number diseased
Diseased tobacco, dist. water	1 : 10	all (in 15 days)
Diseased tobacco, dist. water	1 : 100	all (in 11 days)
Diseased tobacco, pokeweed	1 : 5	none
Diseased tobacco, pokeweed	1 : 10	none
Diseased tobacco, pokeweed	1 : 25	none
Diseased tobacco, pokeweed	1 : 50	none
Diseased tobacco, pokeweed	1 : 75	none
Diseased tobacco, pokeweed	1 : 100	none

a. Ten plants were inoculated in each case.

(232). One explanation for its activity is that it competes with virus particles for receptor sites on the host (297), with the lysine groups playing a key role in the interaction between the protein and either virus or host cell.

The principal mode of action of the inhibitor from paprika (*Capsicum annuum*), which appears to be a flavone or related compound, in conjunction with a protein, appears to be on the host, since treatment of leaves of *Phaseolus vulgaris* on either the lower or upper surface inhibits upper surface inoculation by AMV (95). Some other inhibitors seem to be polysaccharides, the evidence suggesting that that from *Abutilon striatum* acts on virus particles (214). Alternatively, an inhibitor polysaccharide from sugar beet (*Beta vulgaris*) appears to act on the host, reducing the number of lesions produced by TMV on *N. glutinosa* and reducing the infectivity of BYMV and LMV, but not of CMV (79).

In those instances where inhibitors from plants either resistant or susceptible to a particular virus have been examined, there was no correlation between the degree of inhibition and the degree of resistance of the plant from which the inhibitor was derived (e.g., 95, 145, 268).

It is probable that tannins, phenol-oxidizing enzymes, and oxidized phenolics present in plant extracts act principally by irreversibly inactivating the virus. They may cause virus degradation or alter the coat protein in such a way that it can no longer interact appropriately with the host cell. In contrast, full infectivity is usually regained by ultracentrifugating mixtures of virus and inhibitor.

In addition to plants, inhibitors can be obtained from bacteria and insects and include enzymes, such as trypsin and ribonuclease, whose inhibitory activity is at least partially distinct from enzymic action. More comprehensive surveys of these and other plant inhibitors can be found elsewhere (21, 142, 195). We have already seen that virus coat protein can inhibit virus infection, possibly providing evidence for the presence of receptors on the host cell. However, other proteins such as bovine serum, albumin, and ovalbumin can have similar effects on TMV infectivity, and an alternative explanation is that they cause aggregation of virus particles. Polycation inhibitors such as poly-L-lysine appear to interact with viruses, while polyanions like poly-glutamic acid may interact with the host cell, both being considered to interfere with the interaction between virus and host cell (273, 295). Somewhat paradoxically, a low concentration of TMV protein or BSA can stimulate virus infection, possibly by either protecting a labile fraction of the virus or in some way interacting with the host cell and enhancing the infection process (144, 246).

The site of action of compounds, such as amino acids, nucleoside and nucleotide analogues, and antibiotics, is normally at a stage beyond the establishment of infection (see Chap. 5).

Interference and acquired resistance. The presence of a second virus either in the inoculum, in the cells to be infected, or in areas of a plant distant from that being inoculated may also interfere with the establishment or development of infection, and, if the host reacts hypersensitively, the number or size of lesions may be reduced. Although in some cases the infection process may be influenced, it is likely that many of these effects are mediated at a later stage. We felt it more logical, therefore, to consider the phenomena of *interference*, *cross-protection*, and *induced* or *acquired resistance* in the context of resistance and the hypersensitive response in Chapter 10.

Uncoating

Although it is generally agreed that the virus genome must be at least partially exposed before replication can proceed, there is very little information on the degree of uncoating required or on how or where this uncoating process occurs. The fact that uncoating does indeed take place, at least with some of the simpler viruses, has been indicated on several occasions, both directly and indirectly. In many of these experiments, estimates of the time required for uncoating have been inferred. By their very nature, however, such experiments are imprecise and cannot usually indicate the time required to remove sufficient protein from a viral RNA for that RNA to begin to be translated.

Indirect evidence relies, for example, on the sensitivity of viral RNA to inactivation by UV irradiation. Both TMV and TNV are sensitive to UV inactivation, and infection of leaves, as indicated by the number of lesions appearing on local lesion hosts such as *N. glutinosa* and French bean, can be severely curtailed if leaves are irradiated soon after inoculation with virions (159). However, the virion RNA becomes resistant to inactivation within 2 to 4 h, and thereafter irradiation has little effect on the progress of infection. This has been presumed to be due to the protective effects of association of RNA with sites within the cell where virus replication begins. Resistance could also arise from replication of the parental RNA, producing multiple copies, so that again complete restriction of infection would be difficult (e.g. 237). If, on the other hand, leaves are inoculated with viral RNA instead of virions, then resistance to UV irradiation develops within minutes.

It has been suggested that at least partial uncoating is necessary before the viral RNA can interact with the cell in such a way that it becomes resistant to UV inactivation. Hence, the difference in time between virus becoming resistant and viral RNA becoming resistant represents the time taken for uncoating to occur. Siegel et al. (267), inoculating *N. glutinosa* with TMV, found this delay to vary between strains, the uncoating process apparently taking longer with the U1 strain of TMV than with the U2 strain. This difference correlated with the greater difficulty observed in uncoating U1 TMV in vitro.

An alternative indirect approach has been to measure the time of appearance of progeny infective virus or of lesions on a local lesion host following inoculation either by virus or by viral RNA (158). Although TNV lesions began to appear on French bean leaves at about the same time after inoculation with either virus or viral RNA, the maximum *rate* of appearance was about 4 h later with virus inoculation than with RNA. The inference was that the delay in lesion appearance following virus inoculation is an indication that some degree of uncoating is required, and that it takes about 4 h.

A third approach has been to exploit the phenomenon of photoreactivation, whereby viral RNA, inactivated by exposure of the intact particle to UV irradiation, can be reactivated by exposure of the uncoated RNA to visible light. When White Burley tobacco leaves were inoculated with UV-inactivated PVX, for example, it was only after ca. 30 min that viral RNA became susceptible to reactivation, resulting in the formation of a greater number of lesions than in the absence of visible light (25). The greatest number of lesions occurred between .5 and 1.5 h, suggesting that most of the viral RNAs became susceptible within this period. Reactivation was no longer possible after ca. 3 h. Similar results were obtained with several other viruses, including CMV and TRSV, although in no case was the efficiency of photoreactivation as great as with PVX.

The interpretation of these observations was that once 30 min to 1 h had elapsed after inoculation, the RNA was sufficiently uncoated to allow access of visible light, while after 3 h, naked RNA that had not initiated infection was subject to degradation by nucleases (22). It could be inferred, therefore, that uncoating took about 1 h. To make such an inference, however, it has to be assumed that the section of RNA that has to be uncoated to allow access of visible light is the same one that needs to be exposed before translation can begin. This need not, of course, be the case.

In a more direct approach, Shaw (258, 260) produced TMV particles containing ^{14}C-labeled coat protein at the 5'- or 3'-end, by in vitro assembly of partially stripped rods and labeled coat protein. The detection of free labeled protein in inoculated leaves was then taken as evidence that uncoating had occurred. Following inoculation of Samsun tobacco leaves, 25% was released within 10 min. More was gradually lost over the next several hours, up to a maximum of ca. 50%. There was no suggestion of preferential stripping from either end. Reddi, on the other hand (237), labeled the RNA of TMV virions with ^{32}P and monitored the increase in susceptibility of the RNA to leaf ribonucleases after inoculation. He found a gradual increase in susceptibility, up to about 5% of the input RNA being degraded by 6 h after inoculation. However, detection of shorter than full-length viral RNA molecules 3.5 h after TMV inoculation of Xanthi nc tobacco seemed to indicate that most of the particles became partially uncoated (131). Kurtz-Fritsch and Hirth (184) inoculated Chinese cabbage with ^{32}P-TYMV and barley with ^{32}P-BMV, using a high ionic-strength buffer, which was claimed to remove all but "irreversibly-bound" virus particles from the leaf surface. Of the "irreversibly-bound" virus, 60% was uncoated within 10 min, and very little after this. The

two-stage process suggested by Shaw was not observed for these spherical viruses.

There are, however, problems in assessing the relevance of these somewhat disparate results to the infection process, since only a very small percentage of the particles in the inoculum participate in infection, while all particles contribute to the "direct" observations. Additionally, although it was assumed that uncoating was occurring intracellularly, presumably in epidermal cells, it is by no means certain that this is so, and observations by Kassanis and Kenten (159) point to the possibility that the majority of the uncoating observed in these experiments takes place extracellularly. TMV infiltrated into White Burley tobacco leaves, for example, lost 68% of its infectivity within 24 h, accompanied by a reduction of 66% in the number of particles detected by electron microscopy. A large portion of the particles were, therefore, uncoated and completely degraded without penetrating cells. Particles could also be uncoated on the leaf surface of a plant, *P. vulgaris* (cv. The Prince), in which TMV does not replicate. Approximately 30% of the TMV particles adsorbed by immersion in inoculum became shorter, and hence partially uncoated, within 1 h, while approximately 10% were completely uncoated. After 24 h, 80% were shorter and 50–60% completely uncoated, with electron miroscope observations suggesting uncoating preferentially from one end rather than randomly.

Nonetheless, there is evidence for the intracellular appearance of partially uncoated virus. Kiho (169, 170, 197) inoculated Xanthi nc tobacco with ^{32}P-labeled TMV and observed the formation of polysomes containing parental labeled-virus RNA, partially uncoated from the 3'-end within 2 h. These "parental" polysomes accounted for 5–8% of the RNA of adsorbed virus (170). Only a very small proportion became fully uncoated. Parental polysomes could also be obtained from a host resistant to TMV, *Brassica rapa*, suggesting that in this case, at least, the mechanism controlling resistance was operating beyond the uncoating stage (170).

There is very little information available on the more complex viruses, such as the rhabdoviruses (e.g., 101), of which LNYV is an example. The genome of these viruses is a single negative-strand RNA, which is unable to act as messenger. It seems likely that the genome remains associated with several virus proteins during the first stages of replication and that the transcription of the negative strand to positive messenger strand is accomplished by a virion-associated polymerase (102, 234).

It would appear, therefore, from the limited number of experiments reported, that a large portion of the particles in an inoculum of one of the less complicated viruses is partially uncoated soon after inoculation. There is no definitive evidence, however, to indicate the site of uncoating of those particles destined to initiate an infection, or the degree of uncoating of those particles necessary for transcription or translation to begin. Either cuticle or plasmalemma could provide the appropriate hydrophobic environment (72), and it may be that uncoating begins extracellularly and is completed, as far as required, intracellularly.

Bacteria

Introduction

Aggressive assault upon the host, characteristic of fungi, is not as readily apparent in bacterial invasion of plant tissue. Bacteria may be motile and capable of moving to the host in water, but penetration of the host plant's surface is accomplished either through natural openings (stomates, hydathodes, nectaries) or through surface wounds or breaks in fragile projections such as root hairs, trichomes, and specialized exudative glands (118).

In general, the naked bacterial cell must rapidly find an environment conducive to its proliferation or survival or it will perish; among the phytobacterial pathogens there are no spore formers. Hence, the immediate requirements are a high relative humidity and an adequate nutritional substrate. Optimal temperature and the presence of certain growth factors will certainly alter the rate of proliferation, although minimal nutrition and a high relative humidity are frequently sufficient to establish the infection.

What is the infection process as it applies to plant pathogenic bacteria? It seems logical that the infection process should include migration of the pathogen to the host, contact with host surface, penetration, and early stages of bacterial proliferation. Once the pathogen has "established" itself in the host, the infection process has terminated. This moment is conceived as the time when bacteria begin their logarithmic rate of cell division.

This infection process can be divided into three phases: migration to the host, recognition and contact, and penetration and establishment. It is apparent that bacteria in the soil can come into contact with plant roots as a consequence of root growth. Contact may also reflect locomotion by the bacteria in soil microcapillaries; in the latter instance, migration may not be random movement but rather a response to a stimulus originating from the host in the form of a gradient that can be followed by the pathogen.

Migration to the Host

Motility and aerotaxis. It has been generally assumed and recently proved (224) that flagella that provide bacteria with means for locomotion enable plant bacterial pathogens to enter the host and cause infection (Fig. 1–9). Hence, motility might be a determinant of virulence.

In studies of flagellar development in vitro of both virulent and avirulent strains of *Erwinia amylovora*, Huang (146) noted that after 10 h the virulent bacteria had their full complement of normal-size flagella. However, the avirulent bacteria had comparatively few, and these were short. After 24 h, differences between the two strains were no longer evident. Thus, if the virulent strain has an advantage due to flagella development, it is transient. Huang also noted that cells of the virulent

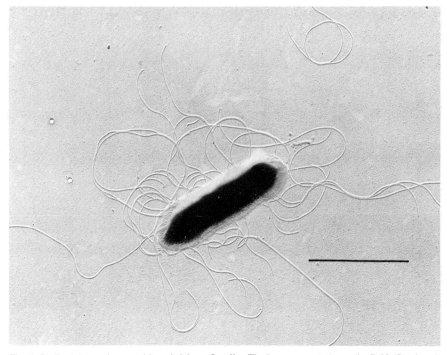

Fig. 1–9. *Erwinia amylovora* with peritrichous flagella. The bar represents 1 μm (by R.N. Goodman).

strain, observed under darkfield conditions in a hanging drop, moved much more rapidly than cells of the avirulent one. Kelman and Hruschka (163), however, have presented data that link motility with avirulence. In their experiments, 27 virulent isolates of *Pseudomonas solanacearum* grown on triphenyl tetrazolium chloride agar for 24–28 h were nonmotile. Initially, when l00% virulent isolates were grown in liquid medium and were sampled at the surface, 63% of the cells were avirulent, whereas only 2% below the surface were avirulent. Electron micrographs revealed that only 2–5% of the virulent cells were flagellated whereas 80–90% of the avirulent ones had flagella. The absence of flagella on virulent cells and their presence on avirulent *P. solanacearum* cells was also observed in scanning EM's of tomato roots by Schmit (250). In a mixture of virulent and avirulent isolates of *P. solanacearum*, the latter migrated more rapidly toward air. These same avirulent isolates, however, demonstrated no distinct migratory pattern when grown in a nitrogen atmosphere. This is the first report of aerotaxis for a plant pathogenic bacterium. It would appear, from this study, that virulent *P. solanacearum* cells migrate in or toward host tissue not necessarily with the aid of flagella and that they may follow other than an O_2 gradient.

Convincing data concerning the importance of bacterial motility to infection have been presented by Panopoulos and Schroth (224). Comparing motile and nonmotile isolates of *Pseudomonas phaseolicola*, the latter having "paralyzed" (nonfunctional) flagella, it was found that motile strains caused 12 times as many lesions as the nonmotile ones (Fig. 1–10). In this experiment, penetration was trans-stomatal. Observed speed of *P. phaseolicola* in vitro was 25 µm/sec. Hence, a bacterial cell could traverse the approximately 50 µm of the stomatal pore in Fig. 1–17a to the substomatal chamber in 2 sec. Their data also revealed that loss of motility did not signal loss of virulence, as inside the leaf inherent differences between motile and nonmotile isolates in virulence were less apparent. The invasive advantage of motile over nonmotile bacteria held for either congested or noncongested, wounded or intact leaf tissue.

Raymundo and Ries (236) have found the vigor of a motile cell is greatest at 18–23 C and that at 33 C less than 1% of the flagellated *E. amylovora* cells were motile. Weakly motile cells grown at 33 C, when incubated at 23 C, became fully motile after five generations, suggesting that temperature also controls flagella synthesis. Of primary importance, in this study, is the confirmation of Panopoulos and Schroth's observation that *E. amylovora*, like *P. phaseolicola*, is not generally motile inside plant tissue. The exception appears to be close to the wound area, i.e. the point of entry, where tissues are highly hydrated and hence water-congested. It has been suggested (224) that the motility is hampered in intercellular space because of the absence of free water. One might ask whether the increased sensitivity of water-congested tissue to bacterial infection reflects the pathogen's heightened motility in free water.

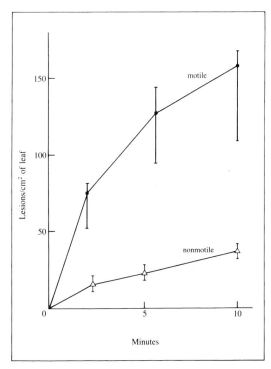

Fig. 1–10. Increase in lesion numbers in primary bean leaves by motile and nonmotile strains of *Pseudomonas phaseolicola* as a function of immersion time of the leaf in a bacterial suspension. The leaves were infiltrated with water prior to immersion (after Panopoulos and Schroth, 224).

Chemotaxis. That plant pathogenic bacteria are capable of following a chemical gradient and that this may influence infection was suggested by Chet et al. (52). They revealed that *Pseudomonas lachrymans* was attracted to substances in foliar fluids collected either as guttation or condensate. The attraction of the pathogen to solutions of amino acids, sugars, nucleotides, and vitamins was also studied. Microcapillaries containing the above nutrients were exposed for 30 min to suspensions containing 10^8 cells/ml of *P. lachrymans*, after which the number of bacteria that had migrated into the capillaries was determined. This study indicated that *P. lachrymans* was attracted to substrate from both host and nonhost plants and hence the response was nonspecific as to the source of the attractant. Guttation fluids, however, were more attractive than single amino acids, sugars, nucleotides, or vitamins. Guttation fluids with the highest concentrations of amino acids and carbohydrates were the most attractive. Of the sugars tested, ribose, arabinose, and glucose were the most attractive; of the amino acids tested, arginine and methionine were the most attractive. Raymundo and Ries (235) found chemotaxis exhibited by *E. amylovora* to be temperature (20–29 C) and pH (6–8) dependent. Positive chemotaxis was established for organic acids in apple nectar that were dicarboxylic (succinate, malonate, malate, and fumarate). Only one amino acid, aspartate, appeared attractive to *E. amylovora*. This pathogen appears to possess a single chemo-receptor site that is highly specific for 3- and 4-carbon dicarboxylic acids.

Recognition and Contact

Receptor sites. The concept of receptors on host cell surfaces for molecules on the surfaces of the bacterial cell that convey specificity to the host-parasite interaction has recently become substantive. Lippincott and Lippincott (192) described specific binding sites on the surfaces of bean leaf cells for *Agrobacterium tumefaciens*. Carborundum-induced wounds apparently exposed highly specific sites to which the crown gall bacteria must bind in order to initiate the infection process. Electron micrographs by Boger (35) and, more recently, by Matthysse et al. (206) (Fig. 1–11) are suggestive of a precise bacterial cell orientation on the surface of plant cells prior to "infusion" with bacterial DNA, i.e. the Ti plasmid. Both studies suggest that the host cell may manifest a preference for one pole of the bacterial rod over the other. This possibility draws support from the observations by Dazzo et al. (69) that the lectin trifoliin A, which exudes from clover roots and is apparently an attractant for *Rhizobium trifolii*, has a higher affinity for one of the poles of this nitrogen-fixing bacterium. Fig. 1–11 suggests that the involvement of microfibrils produced by the bacterium is instrumental in the attachment process prior to the DNA infusion. Neither polar alignment nor microfibril-dependent attachment is a certain feature of the *A. tumefaciens* infection process. Two of these studies (35, 192) presented evidence that avirulent *Agrobacterium radiobacter* cells inhibit tumor formation by competing for attachment sites in the wound area. The Lippincotts' data suggest that inhibition is not due to a simple physical blocking of wound sites, since many related bacteria are ineffective in this regard. They concluded that bacteria capable of inhibiting tumor formation have a surface quality that permits binding to those sites, which is integral to the infection process. The Lippincotts' data further suggest that the sites are discrete so that each inhibition event is due to a single inhibiting bacterium, excluding a single virulent cell from an essential receptor site. These authors also suggest that only a single site is infectible regardless of wound size. Hence, depending upon which bacterial strain reaches that site, virulent or avirulent, a tumor does or does not develop. If this is indeed the case, then one must consider the possibility that if there is a second wound site in the area, its infection is precluded by a phenomenon similar to that which usually precludes two pollen grains from fertilizing the same ovule.

Role of the bacterial cell wall. Whatley et al. (309) appear to have neatly established that the lipopolysaccharide (LPS) component of the bacterial cell envelope is responsible for attachment to the receptor (wound) site, which is essential for tumor initiation. Purified LPS, obtained by two separate procedures, prevented attachment of virulent *A. tumefaciens* cells to wound sites. It would appear, therefore, that the infection process, as accomplished by the crown gall pathogen, is at least a two-step process. The first attachment to the wound site is mediated by LPS on the surface of the bacterium and the second by infusion of bacterial DNA. The former is of chromosomal origin and the latter is a large plasmid (298). The report of Bannerjee et al. (17) characterizes more precisely the O-antigenic region of *A. tumefaciens* LPS as N-acetyl- D-galactose amine and β-D-galactose that interact with the galacturonic acid molecules of the plant cell wall during tumor initiation. Ohyama et al. (221) presented excellent evidence that binding of *A. tumefaciens* strain B6 to suspension culture cells of *Datura innoxia* L. increased gradually for 1 h and attained maximum level after 2 h. This 2-h time frame for maximum attachment was also reported for carrot suspension culture cells (206). Neither concanavalin A nor soybean lectin inhibited binding that was temperature dependent. Similarly, *Escherichia coli*, *Salmonella typhimurium*, *Rhizobium japonicum*, and *Micrococcus lysodeikticus* all failed to compete with the B6 strain for binding to *Datura* cells. Binding to the cell wall was localized and appeared similar to that described by Bogers (35), suggesting the involvement of specific sites. The binding process may be even more complicated, according to the electronmicrographic data presented by Matthysse et al. (206). Their study suggests that LPS-binding may be followed by the elaboration, *by the bacterium,* of cellulose microfibrils that anchor it to the host receptor site, thus completing a third step in the process (Fig. 1–11). Additional bacteria in the inoculum of carrot tissue culture cells are entrapped by the microfibrils, and these too produce microfibrils that secure themselves and their progeny at the site, completing step four. According to Matthysse, the bacteria in this microcolony may produce enzymes that digest pectin, initiating step five which, ostensibly, prepares the wounded surface to absorb the plasmid at the plasmalemma bacterial cell surface interface. Details concerning the final steps—the actual transfer of DNA sequences into the host cell and into the hosts' genome—remain obscured. Analogously, Cleveland and Goodman (unpublished results) have observed that *A. tumefaciens* biovar 3 isolates adsorb at high concentrations to grape suspension culture cells of the susceptible cv. Chancellor and also Seyval Blanc, whereas biovar 1 isolates do so only minimally. Mazzuchi (208) has suggested the endophytic parasitism requires a series of reciprocal recognitions between plant and pathogen, in a certain sequence.

The selectivity of rhizobial species for their specified legume hosts has also been recorded. However, from the report of Bohlool and Schmidt (36), it seemed that, in the case of soybeans, a lectin played an active role in the host-symbiont interaction. Gold and Balding (116) and Sharon and Lis (256) offer a comprehensive discussion on the chemistry and function of lectins and glycoproteins in general. What appears to be a clear case of specific recognition of plant pathogenic bacteria for their fungal host was reported by Preece and Wong

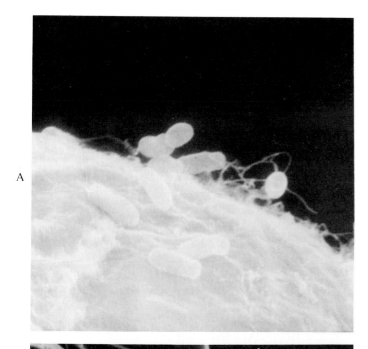

A

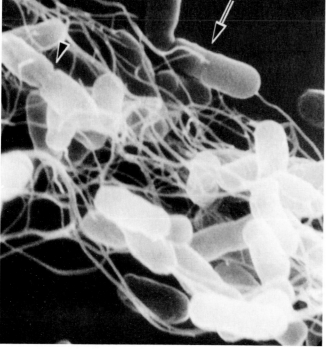

B

Fig. 1–11. The possible involvement of cellulose microfibrils produced by *Agrobacterium tumefaciens* in attachment of the bacterium to carrot callus tissue cells: (a) bacteria attached to glutaraldehyde-fixed carrot cells, thus authenticating the cellulose fibrils as bacterially produced and (b) shows the fibrils originating from sites of bacterial cells (arrows) (after Matthysse, 206).

(229). Scanning electron micrographs revealed that bacteria applied to the surfaces of a cultivated mushroom adhered rapidly and firmly. However, the pathogens of the mushroom, *Pseudomonas tolaasii* and *P. gingeri*, became attached in greater numbers (average 75% vs. 25%) than saprophyte species. In addition, the pathogenic species became attached with much greater tenac-

ity. The apparent adherence structure noted, which is probably capsular EPS, is presented in Fig. 1–12. The surge of interest in rhizobial cell surfaces and legume root lectin as possible obligatory recognition features leading to nodulation was sparked by the report by Bohlool and Schmidt (36), who presented evidence that soybean lectin bound 23 of 25 soybean-infecting strains

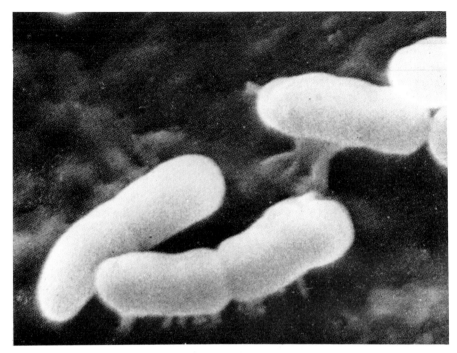

Fig. 1–12. Adherence structures produced by *Pseudomonas tolaasii*, probably EPS, on the surface of mushroom caps, *Agaricus campestris* (after Preece and Wong, 229).

while showing no affinity for 23 other strains of five species of rhizobia that do not infect soybeans. A hypothesis for this interaction features a lectin (glycoprotein) on the surface of the legumes' root and the O-antigen (LPS) on the surface of the rhizobial cell.

It seems warranted, at this point, to digress briefly in order to discuss some terminology integral to the concept of recognition. Recognition may be perceived as occurring at either the molecular or the cellular level. Molecular recognition implies that a site on one molecule, called a *ligand*, binds to a site on another molecule, called a *receptor*. Binding may be specific (i.e., sugar-lectin) or nonspecific (i.e. charge-charge). Cellular recognition, on the other hand, presupposes molecular recognition, i.e. molecules on the surface of the plant cell and the bacterial cell recognize one another and receptor and ligand bind. Secondarily, a cellular function is activated. According to Mazzucchi (208), molecular recognition is instantaneous and is a diffusion process, whereas cellular recognition implies a second reaction that may take minutes or hours to develop and is metabolic.

Both receptor and ligand are found in the literature to define lectins, and because of this disparate usage the terms convey no clear structural concept. Perhaps, as a consequence of mass, the larger of two reactants might logically be considered the receptor and the smaller the ligand. However, Burke et al. (44) have proposed the terms *cognor* and *cognon* to imply not only structure but function as well. Thus, cognor connotes the active rec-

ognizing partner, whereas cognon is the passive component that is recognized. Consequently enzyme, antibody, and lectin would be cognors, and substrates, antigens, and sugars would be cognons. Cognons could be free molecules that have been released in the environment or be a facet of the surface architecture of a cell. Hence, free cognons may participate in recognition phenomena when the cognor is some distance away, or cognons may protrude from the surface of a cell and be recognized by the cognor of a neighboring cell. Mazzucchi (208) has further defined the process of recognition with respect to the rate at which recognition proceeds. According to him, recognition may be *molecular*, which is rapid, probably instantaneous, and controlled by diffusion (of cognon to cognor). An agglutination reaction would exemplify molecular recognition (147). On the other hand, *cellular* recognition takes longer, from several minutes to hours, and presupposes molecular recognition. An increase in host cell respiration following contact with a pathogen (Fig. 3–15) or deterioration of host cell membranes during the development of HR (Fig. 10–30) is indicative of a cellular recognition reaction.

Tsien and Schmidt (293) used the term "lectin-binding polysaccharide" (LBP) for the active portion of the extracellular polysaccharide of *Rhizobium japonicum* that binds the rhizobial cell to the soybean root lectin. Apparently, LBP is not only adherent to the bacterial cell but is also diffusible from its surface (228), see Fig. 1–13a. The lectin-binding ability of *R. japonicum* cultures has been found to vary with the composition of their extracellular polysaccharide, which reflects, among other factors, genetic qualities of the isolate and age of the bacterial cell (215). Dazzo and Hubbell (68) established that clover-*R. trifolii* symbiosis reflects the attractiveness of a surface polysaccharide (cognon) of the bacterium for a multivalent lectin (cognor) in the root hair region of 24-h-old clover seedlings. Their hypothesis envisioned carbohydrate-binding sites that are anchored to the plant cell wall and appear to be specific for 2-deoxyglucose (71). Work by Jansson et al. (151) on the polysaccharide of *R. japonicum* suggests that it does not contain 2-deoxyglucose. A comprehensive review by Bauer (19) of the infection process of legumes by rhizobia suggests that the model proposed by Dazzo and Hubbell (68) is an oversimplification. Nevertheless, Bauer states that there are correlations between lectin binding and *Rhizobium* infectivity that obviate coincidence. According to Dazzo et al. (70), the sugar determinants are rather ephemeral and bind the clover root lectin, trifoliin, better when the cells are five days old than when they are either three or seven days old. Efficacy of the trifoliin receptors appears to coincide with accumulation of a fibrillar capsular polysaccharide on the surface of *R. trifolii* cells. Dazzo et al. (69) have shown that the capsular polysaccharide on the surface of *R. trifolii* is significantly altered by enzymes that are released from clover roots. The enzyme-containing root exudate alters the trifoliin-binding capsule in a way that seems to favor polar attachment of *R. trifolii* to clover root hairs. The dynamic nature of the surface of the bacterial cell suggests that the surface of the plant cell is probably similarly in flux. Hence, timing of plant cell receptivity to bacterial cell surface development is crucial to successful symbioses, as well as to infection (215). From the studies of Pueppke (231), however, it would seem that the causal relationship between lectins and rhizobia may be more apparent than real, as some mutant *R. japonicum* strains not adsorbed by the soybean lectin are, nonetheless, able to nodulate roots and fix nitrogen.

The extracellular polysaccharide (EPS) of several bacterial pathogens seems to be a determinant of virulence (13, 29, 32, 59, 229, 255) (Fig. 1–13). Clearest evidence for this is with *E. amylovora* and *P. solanacearum*, in which isolates that are devoid of EPS are completely avirulent (13, 255). That EPS appears to be toxigenic is suggested by the studies of Huang et al. (147), who reported that the EPS (amylovorin) produced by *E. amylovora* caused not only wilting but also plasmolysis of xylem parenchyma cells. Subsequently, reports by Suhayda and Goodman (280) and Goodman and White (119) confirmed that amylovorin could, indeed, cause xylem vessel occlusion accompanied by xylem parenchyma plasmolysis (Fig. 9–6). A series of reports by El Banoby and Rudolph (85, 86) and El Banoby et al. (87, 88) have indicated that EPS of a number of pathogenic pseudomonads caused water soaking in susceptible cultures but not in resistant ones. Their studies suggest that resistant cultivars produce

enzymes capable of hydrolizing EPS, precluding the development of the water-soaking symptom. A possible mode of action for EPS as a virulence factor was described by Romeiro et al. (241, 242). Their study characterized a small protein in apple seed, leaf, and stem tissue that is capable of agglutinating *E. amylovora* cells. They noted that this agglutinating activity could be reversed by EPS wherein the small highly positively charged protein precipitated the much larger negatively charged EPS polymer. Their hypothesis envisions the nonadherent toxic EPS produced by *E. amylovora* neutralizing the apple agglutinin (Fig. 10–37), thus permitting the pathogen to continue to multiply in host tissue (242).

A specific cognor on the host cell surface for the cognon and antigen on the pathogens' outer cell envelope can logically be postulated. Binding of *E. amylovora* and *P. solanacearum* to host cell surfaces seems to be related more to the elicitation of a resistance response by avirulent EPS-less isolates than to engendering pathogenesis (13, 255). In the case of bacterial pathogens, EPS appears most logically to be a suppressor of the host's defense mechanisms as it appears to physically prevent binding (228).

Is it necessary, however, for the bacterial pathogen to attach itself to the host cell in order to elicit a response? Figure 1–12 shows *P. tolaasi* firmly attached to the surface of *A. bisporus* by a pseudopod-like appendage (229) (probably EPS). Does this contact induce cellular leakage, plasmolysis, or an increase in respiratory rate?

It would seem that the case for recognition, followed by binding of the bacterial cell to its host, is not restricted to those interactions where a mutualistic association between host and bacterium is to ensue. With *A. tumefaciens* the interaction results in transfer into the host cell of a fragment of the bacterium's own genetic complement (298). Legume-rhizobial interactions appear to insure the development of nodules and a ready supply of "fixed" nitrogen. The necessity for physical binding following cell-cell contact between host and pathogen and subsequent physiological and biochemical developments are in need of additional study.

Predisposing Environmental Factors

The bacterial pathogen, although cloaked in an amorphous capsule, is without the protection of a spore wall, is sensitive to actinic rays of the sun and, hence, its survival frequently requires the rapid location of a conducive environment in order to continue development. This "friendly" environment includes a relative humidity (RH) approaching 100%, adequate nutrition and a temperature between 24 and 30 C.

Relative humidity and free water. Many studies support the contention that surface moisture is a priori necessary to effect inoculation (penetration), except when insect feeding is the vector-delivery system. Leben and coworkers (186–188) have linked surface moisture to the development of significant epiphytic populations, which may constitute an important prelude to infection.

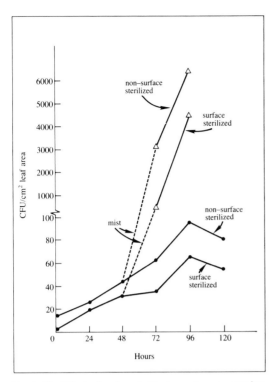

Fig. 1–14. Effects of surface sterilization and exposure to free moisture (- - -) or dry incubation (——) on resident populations (colony-forming units) of *Pseudomonas tomato* (after Schneider and Grogan, 251).

C. It is apparent that plant pathogenic bacteria can grow, albeit slowly, at 5 C. There is a rate, in each host-pathogen interaction (as yet not carefully studied), that must be achieved before infection can be established. It may be conjectured that low temperature fosters reduced bacterial replication, permitting endogenous "resistance factors" to keep pace with the feeble growth of the pathogen.

Light. The influence of light per se on infection has not been widely critically examined, although it does seem to influence the process. Smith and Kennedy (272) observed that a four-day preinoculation dark period, followed by a five-day postinoculation dark period, caused soybean varieties Acme and Harasoy, normally susceptible to races 1 and 5 of *Pseudomonas glycinea*, to develop a resistant reaction upon inoculation. When light was supplied after inoculation, however, a normal susceptible reaction developed. Light intensity over a range of 540–2,150 lux was positively correlated with the intensity of the susceptible reaction. Similarly, beans inoculated with *Xanthomonas phaseoli*, cotton with *Xanthomonas malvacearum*, and cucumbers with *P. lachrymans* all exhibited increased susceptibility when exposed to higher intensities of light. Light quality also influenced susceptibility, as both white and red light favored normal symptom expression, whereas nei-

ther blue nor green light did. Apparently, increased photosynthetic activity favored pathogenesis in these instances. This is an area of research that would benefit from further study.

Inoculum Source. Bacteria have been found in a variety of locations in the proximity of their hosts. It is obvious that soil serves as a reservoir; however, persistence there seems to be controlled by the rate at which infested debris decomposes (279). Cankers oozing viscous bacteria-laden fluid have long been recognized as sources of *P. syringae* (66, 223), and *E. amylovora* (243). Whether cankers are a consistent source of inoculum in the spring has been a subject of debate and, doubtless, depends upon such factors as root pressure, temperature, bark injury by winter sun, etc. Perhaps less obvious are the disclosures (16, 49, 50, 75, 76, 120, 160) that bacterial pathogens can persist in "apparently" healthy bud and stem tissue. The epidemiological importance of these latent "infections" is not known, nor has the cause of switchover from the latent to the active phase been precisely determined. One initiator of the "switchover" may be the induction of the meristematic activity in infected tissue. Ultrahistochemical examination of the physiological changes involved might be a worthwhile area of investigation.

Ercolani et al. (91) reported that the surfaces of hairy vetch *Vicia villosa* harbor in midwinter, under snow, populations of 10^6 cells of *P. phaseolicola* per leaf. The vetch, subsequently, provided inoculum to infect neighboring bean seedlings in late spring. Hence, for pathogens with an epiphytic capability, the importance of reservoir plants is clear. The epidemiological importance of an epiphytic potential was described by Berg (30), who noted that 12 to 64 weed species growing on Honduran banana plantations harbored *P. solanacearum.*

The importance of stomata as a source of *Xanthomonas pruni* inoculum is portrayed in Fig. 1–15. The stomata pictured is in foliar tissue that did not show symptoms until six days after the scanning electron micrograph was made.

Epiphytic populations. The literature has expanded rapidly on this subject, since it was examined by Leben (186) and Leben and Daft (187), who considered the importance of a "resident phase" for pathogens on the aerial surfaces of their host. Leben and Daft (187) have reported that large populations of pathogens could be expected to develop on leaf surfaces that remain wet for 24 h or longer at temperatures of 24–26 C. They noted in the case of *P. glycinea* on the surfaces of soybean leaves that these large populations were responsible for infection following abrasion-type injury caused by wind, rain, or hail. It is apparent that epiphytic populations constitute an inoculum potential of real significance. It may be postulated, therefore, that the so-called "resident phase" of the pathogen is required for some pathogens, in the sequence of events that are collectively termed the infection process (188).

According to Mew and Kennedy (211), the size of the

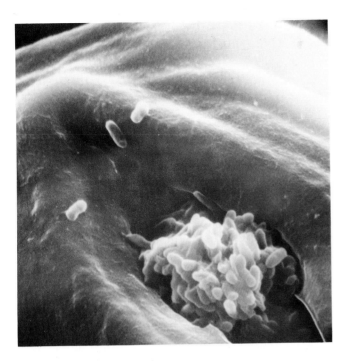

Fig. 1–15. Extrusion of *Xanthomonas pruni* from a peach leaf stomate 6 days after inoculation, yet prior to the appearance of visible symptoms of disease (after Miles et al., 212) ×5000.

epiphytic population of *P. glycinea* on soybeans closely approximates the varietal sensitivity to the pathogen. Furthermore, the population of specific races of the pathogen differing in their virulence on a given host is proportional to their aggressiveness on that host. Although Mew and Kennedy did not present direct evidence to support the Garber (111) nutritional theory of infection, foliar exudates may indeed play such a role. Crosse and Garrett (61) have reported higher populations of *Pseudomonas morsprunorum* on the leaf surfaces of susceptible cherry variety Napoleon than on resistant Roundel. Populations also increased on older trees, and this was interpreted as reflecting a denser canopy with shading from the sun and reduced drying of leaf surface moisture. The data of Chet et al. (52), which reveal the nutritional attractiveness of foliar guttation and condensates to bacteria, clearly establish leaf surfaces as a logical initial site for colonizing some hosts.

Vectors. From the foregoing discussion of inoculum source and epiphytic populations of pathogens, it would appear that the bacterial pathogen does not need a complex delivery system to insure infection. The pathogen may be delivered in a number of ways: by man, animal, insect, nematode, wind-driven rain, root abrasion as a consequence of growth, etc.

A relationship between severity of fireblight (*E. amylovora*) epiphytotics in nurseries and southerly wind-driven rains was described by Bauske (20). Lombardy poplar windbreaks protected year-old pear trees planted in nursery rows from the pathogen. Loss of trees near the center of the block was no greater that 5.5%, whereas those sections farthest from the windbreaks sustained losses of 60%. Experiments by Stevens et al.

(278) with simulated wind-driven rain substantiated this contention for *E. amylovora*. Bauske (20) linked bacteria-laden aerial strands of "ooze" previously observed by Ivanoff and Keitt (149) to the infection process and reported that simulated wind-driven rain could disseminate inoculum on pears in the form of aerial strands to blossoms of *Cotoneaster* and *Pyracantha*, causing systemic infections. The precise nature of trans-lenticellate extrusion of bacteria-laden ooze in the form of aerial strands was revealed pictorially in scanning electron micrographs by Keil and van der Zwet (161; Fig. 1–16). Strand formation was intensified by oil sprays applied to inoculated pear and apple shoots. It seems plausible that the oil injures those cells that would normally form the diaphragm of the lenticel-closing layer, causing them to become necrotic. The bacterial ooze, then, would issue through a small break in the necrosed layer, which could account for strand formation and the ridged appearance of some strands as they are extruded from the lenticel. Daft and Leben (63) have detected bacterial strands of *P. glycinea* issuing from lesions on cotyledons of germinating seeds. They concluded that *P. syringae* infection hinges on two factors, rain and wind-induced leaf injury. The epidemiological relationship of rain storms per se to fireblight epiphytotics was studied by Billing (33, 34). Her data suggest that periods of high humidity prior to and following a vigorous rainstorm tend to increase the turgescence of foliage and the severity of the epiphytotic.

Infection by insects. Perhaps the earliest account of an insect-vectored bacterial plant pathogen is that of Waite (302), who demonstrated that bees and wasps were important in disseminating the fireblight inocu-

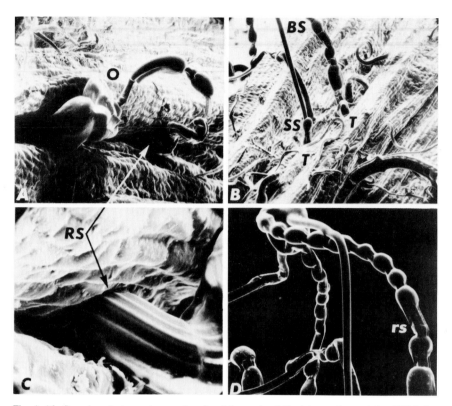

Fig. 1–16. Scanning electron micrographs of aerial strands of *Erwinia amylovora*-laden EPS being extruded from lenticels on Bartlett pear petioles. (a) Ridged strand (RS), bacterial ooze (O) being extruded from a lenticel (×150); (b) smooth strands (SS) and beaded strands (BS) near trichomes (T) on petiole surface (×90); (c) ridged strand extruded from lenticel (×750); (d) magnified view of (B) showing smooth and beaded strands, one with ridged segments (rs), all attached to a dried ooze droplet (after Keil and van der Zwet, 161).

lum. Plurad et al. (227) fed *E. amylovora* to the apple aphid, *Aphis pomi*, in apple shoot expressates through an artificial membrane. The pathogen persisted *in* the aphid for at least 72 h. However, the number of bacteria inoculated into host apple leaf tissue by a single feeding aphid did not seem to be great enough to establish a systemic infection. Perhaps the aphids' injection of bacteria into phloem rather than xylem vessels precluded extensive systemic infection.

Whether some plant pathogenic bacteria must spend at least a part of their life cycle in an insect has not been conclusively demonstrated. An early report by Petri, mentioned by Gäumann (112), suggests that the olive fly *Dacus oleae* is an obligate "vector" for *Pseudomonas savastanoi*, the olive knot pathogen. More recently, Hagen (127) has presented evidence that a symbiotic relationship exists between the fly and the bacterium. It seems that the pathogen's proteolytic enzymes assist the fly in metabolizing a protein and, apparently, the bacterium synthesizes both methionine and thiamine for the insect. It is apparent, however, that the infectivity of *P. savastanoi* is not a priori dependent upon *D. oleae*, as

the olive knot disease is common in California in absence of the fly.

Perhaps more dependent on an insect vector is *Erwinia stewartii*, which causes a wilt of corn (*Zea maize*). The pathogen is carried to its host most frequently by a flea beetle, *Chactochema pulicaria*, which harbors the bacterium during its winter hibernation (90). In fact, epiphytotics of the disease are regularly the result of mild winters during which larger populations of the vector survive. However, it has not been established that the bacterium has to spend a part of its life cycle in the flea beetle. Actually, the pathogen may also be transmitted by root-feeding larvae *Phyllophage* sp. and *Diabrotica longicornis* (148).

Penetration of Natural Openings

Stomata. Stomata may occur on upper and/or lower leaf surfaces at frequencies of 50 to 500/mm². When found on both upper and lower surfaces, stomata are more numerous on lower leaf surfaces. A corn plant may possess as many as 2×10^8 stomata; the dimensions of its stomatal orifice are 4×24 μm². Hence, stomata present a significant target for pathogenic bacteria, which are approximately 0.5–0.7 μm wide and 1.0–1.5 μm long. Whether closed stomata preclude bacterial entry was studied by Gitaitis et al. (115), who reported that sweet corn stomata that had been "closed" by abscisic acid treatment still did not exclude *Pseudomonas alboprecipitans*. When plants with hormone-"closed" stomata were compared with those having open stomata following inoculations, the disease-damaged leaf area was 20% and 35%, respectively. Their data imply that so-called closed stomata are readily penetrable by bacteria.

The motility of *P. phaseolicola*, calculated by Panopoulos and Schroth (224) to be 25 μm/s, suggests that from an inoculum of 10^6/ml, 150–375 bacteria/min can pass into the open stomates of 1.0 cm^2 of the bean leaf (Fig. 1–17a). These data indicate that random motility of individual bacteria is a decisive factor during initial stomatal invasion. However, following inoculation (by submerging) periods of 1 h, the number of bacteria inside and outside the leaf did not differ greatly. Increases in infection by motile bacteria following longer inoculation periods reflect factors other than ingress. Probably tissue water-soaking occurs that influences ultimate migration of the bacteria (53). That pathological advantage of motile over nonmotile isolates would appear to be accentuated by a period of intercellular water congestion.

Since the early experiments of Smith (270), the substomatal chamber has been an acknowledged portal of penetration for some bacterial pathogens and a site of initial proliferation. He observed this in "serial sections through very young spots on bean foliage caused by *X. phaseoli*." Colonization of the substomatal cavity was also apparent from scanning electron micrographs of Gitaitis et al. (115).

Similarly, Rolfs (240) found heavy dews and rain to favor stomatal penetration of peach leaves by *X. pruni*. Inoculum sprayed on the underside of leaves (stomata present) became infected (177 of 183 leaves became infected). Upper surfaces of 149 peach leaves (stomata absent) sprayed with the pathogen failed to develop lesions. Matthee and Daines (203) suggested that the wider stomatal aperture, promoted by nutritional factors, was partially responsible for the greater susceptibility of the cv. Sun High peach to *X. pruni* than cv. Red Haven. They contended that wider stomatal apertures favored water congestion of the substomatal chamber and neighboring intercellular space.

Lenticels. The key to lenticel penetration is the destruction of its closing layer (diaphragm), which consists of the topmost one or two rows of heavily suberized complementary cells (Fig. 1–17b). Rupture of the closing layer is fostered by relative humidities approaching 100%, which stimulate the underlying phellogen to divide rapidly, engendering an upward pressure from the large numbers of new (unsuberized) cells. Davidson (65) reported that in the potato tuber, the continual process of phellogen division and destruction of the closing

layer is at equilibrium when soil moisture conditions are optimal for plant growth. This equilibrium is accompanied by suberization of the closing layer. However, in wet soils the rate of complementary cell proliferation exceeds suberization, the diaphragm ruptures more frequently, a "pseudowound" is formed, and the bacteria not only enter but also quickly macerate the underlying tissue. Fox et al. (99, 100) studied this sequence at the ultrastructural level and reported that bacterial entry through lenticels was limited and only minimal spread of infection was detected in potato tubers held at 78% RH for periods of three weeks prior to inoculation with 10^9 cells of *E. carotovora*. Their electron micrographs revealed that the intercellular spaces of the closing layer of the complementary or "filling" tissue were blocked with suberin. This was not the case with the cells closer to the meristematic phellogen, the phelloderm; these were only lightly suberized and rather thin-walled. It was apparent from their electron micrographs that in tubers kept at or near 100% RH, the lenticels proliferate phelloderm cells profusely without suberin, and these loosely packed cells are readily macerated. It is also possible for the space between the loosely packed cells to become filled, or partially so, with water, permitting the "motility factor" of the bacteria to potentiate pathogenesis.

Leaf scars. With the important disclosure by Hewitt (137) and, subsequently, by Crosse (60) of bacterial ingress through leaf "scars," the phenomenon of leaf abscission requires brief comment at this point (Fig. 1–17c). The process, according to Osborne et al. (222), is under control of a hormone, generally acknowledged to be ethylene, which is regulated by a senescence factor (SF). In brief, SF accelerates abscission by stimulating ethylene production. Abeles and Leather (1) have reported that auxin (IAA) and abscisic acid (ABA) potentiate the process by modifying membrane permeability of organelles that enclose SF. Changes occur in two or three files of cells that are easily differentiated from neighboring cells as they are smaller, starch filled, have dense cytoplasm, and show evidence of cell division prior to abscission. The essential changes leading to abscission of leaves occur in the middle lamella and cell wall. Ethylene stimulates production of cellulase at the wall plasmalemma interface (1, 150). As a result, living cells part from one another and the tracheary and xylem cells are broken mechanically, as are epidermal cells, according to Hewitt (137). It is into these vascular elements and adjacent parenchyma that bacteria penetrate. Abscission may also be looked upon as the creation of a *pseudowound.*

Hewitt (137) noted no protective layer of "lignin-wound-gum" or suberin formed at the leaf scar of olive prior to separation. In the orange, however, Scott et al. (254) revealed that the formation of suberin increases markedly and precedes the appearance of phellogen (the meristem that produces periderm several cell layers beneath the developing abscission zone). Actually, prior or subsequent to abscission a two-component "protec-

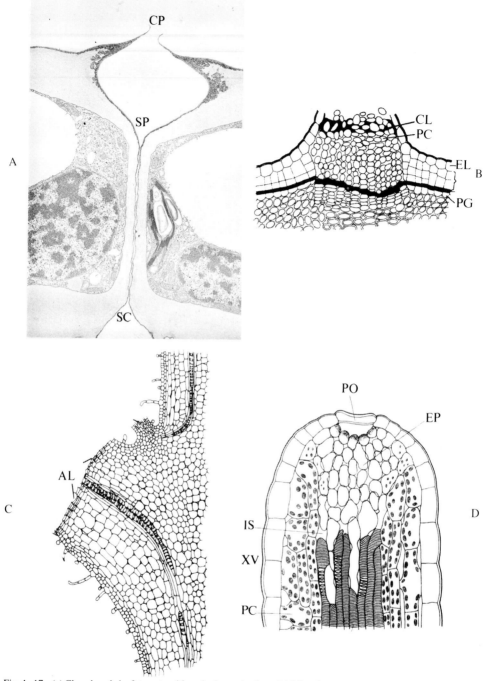

Fig. 1–17. (a) Closed apple leaf stomate with cuticular projections shielding the stomatal pore. Pore length is about 50 μm. (b) Lenticel with suberized closing layer that may rupture under elevated RH. (c) Leaf scar, site of abscissed leaf and site of suberized abscission layer. (d) Hydathode, with pore at leaf apex subtended by loosely packed parenchyma cells that abut xylem tracheids. Abbreviations: AL = abscission layer; CL = closing layer; CP = cuticular projections; EL = epidermal layer; EP = epithem; IS = intercellular space; PC = parenchyma cells; PG = phellogen; PO = pore; SC = substomatal chamber; SP = stomatal pore; XV = xylem vessel.

tive layer" is formed, the sequence being species-specific. First, the region directly behind the abscission zone, the scar or cicatrix left upon abscission, becomes impregnated with substances that have been variously referred to as suberin, lignin, or wound-gum. The precise nature of the impregnating substances and the time sequence of their formation have not been clearly established. The second component is the periderm, whose production follows cicatrization by several days. The barrier described here is now much better understood as a consequence of the biochemical studies by Kollattukudy et al. (179).

Hewitt (137) also observed that the leaf scar of olive was infectible immediately after leaf fall and that susceptibility dropped markedly during the first day. Hence, this pseudowound aged rapidly. He also noted that leaf scars maintained in high humidity produced wound gum and periderm less rapidly than those exposed to lower humidity. Histological observations revealed that penetration of P. savastanoi to a depth of five cells or more causes infection. The development of wound gum in and between the cells exposed by abscission largely reduces penetration below the requisite five-cell depth during the first 24 h after leaf drop. Hewitt also observed that bacteria entering through xylem vessels were freed into neighboring periderm when the vessels were slowly pulled apart by neoplastic growth; this could further serve to stimulate phellogen activity resulting in excessive periderm formation and tearing of nearby vessels.

In later studies, Crosse (60) reported a seasonal variation in leaf scar infection of sweet cherry by P. morsprunorum. Leaf scars exposed 7 to 8 days after delamination in November, when the greatest flush of normal abscission occurs, are much less susceptible than those exposed similarly in September. Vessel penetration monitored from 8 October to 5 November showed the average penetration dropping from 2.2 to 0.7 mm. Crosse concluded that decreased infectibility in autumn was due to shallower bacterial penetration, perhaps due to a more feeble transpiration-induced suction tension in the "freshly" exposed xylem vessels.

It is apparent that leaf drop during the course of a rainstorm could permit ingress of bacteria to xylem cells at the moment of maximum water tension. This is suggested by the data of Davis and English (66), who reported peach infection by P. syringae to be greatest when wind-driven rain gently shook off leaves. Feliciano and Daines (94) studied "spring canker" production in peach by X. pruni, and their histological examinations revealed complete suberization of the exposed abscission zone at the time of natural leaf drop. Leaf scars of this type were not infectible, whereas premature leaf removal followed immediately by inoculation resulted in infection. Delamination stimulated the development of protective forces at the leaf scar. Their data also suggest that the abscission zones for terminal leaves develop their lignosuberin protective layer later than leaves at lateral positions.

Entry of bacteria through leaf traces provides the rationale for reports of dormant bud infections (e.g., 16, 74, 77). Clear signs of bud infection, prior to bud break, have been observed in young peach and walnut foliage infected by X. pruni and X. juglandis, respectively (Goodman unpublished results).

Hydathodes and foliar trichomes. Hildebrand and MacDaniels (141) frequently associated infections of spray-inoculated, uninjured apple blossoms with bacterial penetration of nectarial stomates. The rapidity and intensity with which the blossom becomes infected suggests that the nectaries, in the case of E. amylovora, are a preferred point of ingress. Lewis and Goodman (191) have also implicated foliar secretory glandular hairs as routes of entry. As these projections are fragile, their forceful removal could expose uncutinized intercellular space to bacteria. Layne (185) clearly implicated foliar trichomes in infection of tomato Lycopersicon esculentum by Corynebacterium michiganense. He noted that younger leaves seem to have more infectible sites. This seems to be correlated with trichomes, which are abundant on upper leaf surfaces. In addition, he reported that long septate trichomes with bulbous bases were more susceptible to infection than either the short or glandular trichomes.

Schneider and Grogan (251) reported that the trichomes of tomato serve as a major habitat for the survival of P. tomato during drying conditions. A similar trend was observed by Haas and Rotem (123), when comparing survival of P. lachrymans on cucumber and potato (with trichomes) with those on glabrous pear leaves.

Although the hydathode as a site of initial penetration and infection has not been extensively studied, the anatomy of these organs as described by Haberlandt (125; Fig. l–17d) suggests that the network of comprehensive intercellular space between the loosely packed parenchyma cells under the pore is an ideal site for bacterial growth. Though the fluids may contain ions and organic substances (52, 117, 235), the osmotic concentration should be less than one atm, similar to xylem fluid. How bacteria would enter these minute xylem vessels is at present a matter of conjecture. However, Goodman and White (119) have suggested that xylem vessels in apple stem do rupture when adjacent xylem parenchyma collapse in response to the presence of rapidly multiplying E. amylovora cells. Ostensibly, collapse of parenchyma cells is due to plasmolysis caused by bacterial extracellular polysaccharide, EPS.

The wound as a rift. More and more evidence is accumulating that injury of the host is integral to the usual initial phase of the infection process. Although many bacteria cause infection by penetrating a natural opening, according to Layne (185) some of these same species seem to parasitize the host more successfully—more uniformly and intensively—through a wound site of even minute proportions, e.g. a leaf trichome basal cell. Although penetration of E. amylovora through unwounded leaf and stem surfaces, (hydathodes, lenticels, glandular trichomes) has been observed by Lewis and

Goodman (191), Crosse et al. (62) demonstrated that exposing of the vascular system by clipping the leaf apex *insured* xylem penetration by the pathogen and subsequent systemic infection (Fig. 1–18). Injury would perforce give rise to cellular leakage and enable some bacteria to follow a chemical gradient to the wound site.

Evidence presented by Leben et al. (188) strongly correlates wind-blown soil abrasion of soybean leaf surfaces with infection by *P. glycinea*. According to Daft and Leben (63), only the youngest leaves are infected following a wind-rain storm. Older leaves are infected only when, in addition to wind and rain, their surfaces are abraded. These observations suggest that the extent of injury, perhaps the number of host cells exposed, could be a factor in the process of infection. Crosse et al. (62) reported that, in wound inoculation with *E. amylovora*, resistance increased with leaf maturity. It was also apparent from this study that the percentage of shoots infected increased with an increased inoculum dose (ID). The ID_{50} could, in turn, be manipulated by altering either the inoculum concentration or the length of time the wound was exposed to the inoculum (Fig. 8–12). Hence, infectivity of a wound reflects the endogenous quality of the target cells, e.g. the number of target cells exposed, the number of pathogen cells, and the duration of host-cell exposure to the pathogen. Vakili (296) also stressed the importance of wounding tomato foliage prior to infection and that epidermal abrasion and trichome breakage facilitated infection. Daft and Leben (63) showed that soybean seedlings germinating through abrasive soil infested with *P. glycinea* developed more lesions than seedling in soil without abrasive.

Harper et al. (129) noted that "growth cracks" caused in potato tubers by excessive fertilizer applications and above-average rainfall created rifts in the protective periderm that permitted ingress of *E. carotovora*. Similarly, Thomson et al. (288) reported that sugar beet (*Beta vulgaris* L.) root cracking was accentuated by nitrogen fertilization. The incidence of this vascular necrosis caused by *E. carotovora* subsp.*betavasculorum* could be decreased by cultural practices that reduce the occurrence of cracks that serve as portals of entry.

Entrance of *P. solanacearum* into points of origin of secondary roots of tobacco was initially established by Kelman and Sequeira. These sites were ostensibly free of wounds sensu stricto. Subsequently, an ultrastructural study by Schmit (250) showed that tomato seedlings grown in soil-less culture were penetrated by *P. solanacearum* at pseudowound sites ("openings") made by secondary root emergences. Examination of the root surface by scanning electron microscopy revealed that a few hours after inoculation the inoculum was uniformly distributed over the root surface. However, after three days, the bacteria were preferentially concentrated at the origins of secondary roots.

The wound as a physiological conditioning of the host cell. Perhaps some of the most exciting research devel-opments concerning the infection process in plants by pathogenic bacteria concern the long-held concept that infection by the tumor-inducing pathogen *A. tumefaciens* required the participation of bacterial DNA. A series of studies, beginning with Klein's premature report in 1953 (177) that bacterial DNA was the long sought tumor-inducing principle (TIP), culminated with the disclosure by Van Larbecke et al. (298) in 1974 that TIP, in the form of a plasmid fragment, can be transferred to the host plant cell, thereby effecting the decisive part of the infection process.

The infection process for the pathogen may be conceived as a three-phase phenomenon. The first is wounding that, perhaps, not only exposes the infectible site but actually conditions the host cell physiologically. Phase two is the actual transfer of bacterial DNA, and phase three is the translation of the bacterial genetic information in the new host cell so that it grows in an unrestrained and undifferentiated manner. This exciting feature is described in greater detail in Chapter 5.

Establishment: Growth of the Bacteria In Vivo

Inoculum size required to establish an infection. Hildebrand (139) presented evidence that a single *A. tumefaciens* cell, preferably in a deep wound, was sufficient to induce gall formation. He contended that the environment had to be optimal for the single cell to survive, multiply, and subsequently infect host cells in the wound zone. However, increasing inoculum doses of 1, 2–10, and 50–100 resulted in increasing percentage infections of 10–60, 20–90, and 50–100, respectively. The experiments of Crosse et al. (62) confirmed Hildebrand's results. Clipping leaf apices and applying a droplet of inoculum of *E. amylovora* established the ID_{50} for systemic infection to be 38 bacterial cells. Yet, it is apparent from their data that infection could be effected with a single *E. amylovora* cell (Fig. 1–18).

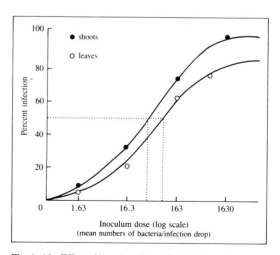

Fig. 1–18. Effect of inoculum dose of *Erwinia amylovora* at a wound site (clipped leaf tip) on infection (after Crosse et al., 62).

The sites of bacterial multiplication. Upon entry into the plant, extensive multiplication by bacteria occurs either intercellularly or, as in some vascular disease, in the xylem. Although bacteria have been shown to migrate in the phloem by Lewis and Goodman (191) and have occasionally been detected in phloem cells, they rarely multiply there (146), with the main exception being the mycoplasma-like organisms (MLOs) and the Rickettsia-like organisms (RLOs). The MLOs, being wall-less, and RLOs, having unusually complex wall structures, are able to survive the high osmotic concentration of phloem sap.

Intercellular spread. Penetration of wounds and subsequent spread through intercellular space was recorded in electron micrographs by Fox et al. (99, 100) in studies of parasitism of potato tubers by *Erwinia atroseptica.* Spread was primarily between storage parenchyma cells; however, restricted infection of xylem and phloem was also observed. In addition to causing dissolution of intercellular cement and the cell wall per se, pathogenesis results in disorganization of the cytoplasm and disruption of cellular membranes such as the tonoplast. These changes are noted two to five cells ahead of the intercellular bacterial front. Whether this derangement signals cellular leakage was not determined. It seems probable that intercellular bacterial proliferation is fostered by the leakage of both inorganic and organic nutritional substrate. It is also possible that excessive leakage of cellular fluids raises the tonicity of the intercellular environment to a level that becomes inhibitory to bacterial growth. The experiments of Shaw (261) were repeated by Goodman (unpublished data), and they indicated that molar concentrations of sucrose, mannitol, or sorbitol exceeding 0.5 M are inhibitory to growth of *E. amylovora.* This organism, however, does grow rapidly in intercellular space if bacteria gain entry through a wound (62, 280).

Intracellular growth of the pathogen. Perhaps the most revealing evidence of intracellular bacterial growth and consequences of that growth was presented by Wallis et al. (306), who clearly characterized the ultrastructural histopathology of cabbage leaf xylem infected with *Xanthomonas campestris.* Inoculation of tertiary veins of cabbage leaves resulted in intense bacterial proliferation in xylem vessels. Convincing evidence has since been presented by Goodman and White (119) and Suhayda and Goodman (280) that the initial sites of significant proliferation of *E. amylovora* are xylem vessels following entry through wounds that expose these vascular elements. The long-distance transport of *E. amylovora* and other pathogens occurs in the xylem. It is of interest to note, however, that the first xylem vessels reached by the inoculum become the site of rapid multiplication, and these appear to act as a reservoir from which the bacteria spread both laterally and distally.

The precise moment at which a pathogen becomes established is impossible to define. It is perhaps logical to consider it to be that moment when the critical bacterial mass is reached that insures the development of at least a perceptible lesion.

Fungi

Introduction

The infection process, as described in this chapter, will include (1) prior-to-entry relationships, (2) penetration, and (3) establishment of the pathogen in the host. Some *disease resistance* mechanisms are briefly described here—primarily either features of resistance that are related to the environment of the host-pathogen complex or morphological peculiarities of the host and pathogen that limit or influence the infection process. The biochemical aspects of the infection process, e.g. enzymatic degradation of the cell wall, as well as the biochemical basis for disease resistance are described in subsequent chapters.

Prior-to-Pathogen-Entry Relationships

After the fungi reach the host—whether passively as airborne fungal pathogens or by an active growth through the soil as soilborne fungi—they proceed with various activities at the surface of the plants, such as spore germination, hyphal growth, formation of appressoria, etc., that may be stimulated or inhibited by the host.

Stimuli. The chemical stimuli from the host, in most cases, seem to be not highly specific. As a rule, fungi exhibit active responses to chemical stimuli (chemotaxis), but a variety of pathogens can respond to the same chemical stimulus, and airborne spores of the rust and other fungi penetrate resistant cultivars of plants as easily as they penetrate susceptible cultivars (202). In fact, early resistance seems to develop after contact or in some instances after penetration. The spores of most of the airborne fungi will germinate satisfactorily in distilled water, an indication that host stimulus is not required for spore germination. Nevertheless, the orientation and settling of zoospores of pathogenic *Phycomycetes,* and their responses to stomata, attracted much attention (138, 244). Khew and Zentmyer (167) assumed that not only chemotaxis but also electrotaxis plays a role in causing accumulation of zoospores of *Phytophthora* species on plant roots. Recent biochemical studies have provided evidence that the host may play an important role in controlling spore germination of stem rust (*Puccinia graminis* f. sp. *tritici*) uredospores before penetration of host tissues. Mitchell and Shaw (213) reported that the host reactivates the ribosomal RNA synthesis in the germinating spores. Indeed, although earlier studies suggested that protein synthesis by uredospores is restrained, this deficiency disappears after association with a suitable host. Staples (275) found an interesting correlation between the time when the germ tube of the spore ceases to elongate and a decrease in template activity of nucleic acid preparations from spores during the later stages of germination. The amino-acid-incorporating activity of the ribosomes also declined. It seems probable that the host must stimulate continued synthesis and utilization of messenger RNA if growth of the rust fungus is to continue (233, 276).

It was thought at one time that the stimulus for forming appressoria was provided simply by a contact with the host surface. Early work of Brown (43) showed that gold-leaf and collodion membranes were able to induce appressorium formation of *Botrytis* by physical contact. Similarly, Dickinson (73) emphasized the importance of physical contact in inducing rust spores to form appressoria. But chemical stimuli, too, may occasionally be important in appressorium formation. For example, spores of *Puccinia coronata* formed appressoria on gelatin if zinc ions were present (257). A chemical stimulus is also apparent in the production of infection cushions by some soilborne pathogens such as *Rhizoctonia solani*. Kerr (165) and Kerr and Flentje (166) enclosed roots of a number of host plants in cellophane bags and then placed these bags in soil infested with various strains of *R. solani*. They found that hyphae of the strain specifically pathogenic to each host produced an appressoriumlike structure on the surface of the cellophane, an indication that the fungus responded to substances diffusing from the roots through the cellophane. Strains not pathogenic to a particular host showed no such response. Flentje (96) stressed that two distinct stimuli, a physical contact and a chemical stimulus, govern the growth of *R. solani* prior to penetration. The physical stimulus directs growth of the main hyphae, whereas, the diffusible substances (chemical stimulus) arrest hyphal growth, which in turn results in appressorium formation.

If one compares the requirements for fungal spore germination and appressorium formation, it would seem that the latter is more specific and chemical dependent. Grover (122) has suggested that appressorium formation is dependent on the balance between the stimulatory and inhibitory substances present in the infection drop.

On the other hand, the basis for the direction of germ tube growth by the bean rust fungus *Uromyces phaseoli*, and its ultimate penetration of stomates, appears to be thigmotropic, according to the definitive study by Wynn (317). When the growth of the fungus over scratches made on artificial membranes was compared with growth over ridges on leaf surfaces, both of these surface irregularities were found to be crucial to the eventual formation of appressoria. Appressoria formed preferentially either on scratches per se or on stomatal ridges.

Apparently, the physical effect of growth over ridges triggered the essential development of an appressorium. Control of the direction of growth of the germ tube is also clearly influenced by contact stimulus that tends to orient the germ tube perpendicular to the leaf's cuticular ridges. This conclusion by Wynn substantiates an earlier analogous report by Dickinson (73), suggesting that the process is not under the control of host-elaborated chemical substances. The study of Wynn (see Fig. 1–19) revealed that germ tubes growing over smooth artificial membranes were much less likely to develop appressoria. The organization of foliar ridges, whether parallel as on grasses or concentric as on beans, assures that the germ tube ultimately runs into a stomate. Wynn's data and scanning electron micrographs suggest that germ tubes are capable of two separate contact responses to distinctly different stimuli. First, the cuticular ridge pattern orients the direction of germ tube growth and, second, the stomatal lips dictate the cessation of germ tube growth and the initiation of appressorium development.

Inhibitors. There are also morphological and chemical barriers that *inhibit* or *interfere* with the penetration of plants by fungi. The moisture film on the surfaces of plant tissues contains various substances released from their underlying tissues, and Brown (42) has shown that these influence spore germination. *Mycosphaerella* blight of *Cicer arietinum* offers an interesting example. According to Hafiz (126), blight-resistant varieties of *Cicer* have more glandular hairs secreting malic acid on the leaf surface than susceptible varieties. High concentrations of malic acid inhibit spore germination and retard hyphal growth of the fungus. Similarly, resistant varieties of a number of other plants have on their leaf surfaces substances that reduce spore germination of pathogens (182, 292).

An early investigation by Walker et al. (305) demonstrated that phenolic compounds are associated with the resistance of onion varieties to *Colletotrichum circinans*. From the dead cells of outer scales, toxic phenolics diffuse to the surface and prevent spore germination. (See Chapter 10 for other examples of phenolic barriers in the waxy leaf surfaces inhibiting fungal penetration.)

Root excretions may also affect the infection process. Roots of varieties of flax resistant to *Fusarium* wilt contain a high concentration of the glucoside linamarin. In the hydrolysis of this compound, HCN is liberated at the root surface, and this can depress populations of *Fusarium* and other species in the rhizosphere (289). A similar but more specific example is seen in the *Fusarium* wilt of pea. Exudates from the roots of one pea variety reduced the germination of spores of a race of *F. oxysporum* f. sp. *pisi* nonpathogenic to this variety. This inhibitor, of unknown composition, had a far weaker effect on the pathogenic race. The substance is, thus, a selective depressor of spore germination in races of fungus to which the host variety is resistant (47).

Another example of pathogenic specificity comes from the experiments of Turner (294). Oat leaves and roots contain a fluorescent glucoside (avenacin) that inhibits the growth of several fungi. The growth of *Ophiobolus graminis*, the cause of the take-all disease of wheat, is inhibited in the sap expressed from oat roots, but the growth of *O. graminis* var. *avenae* is not. Yet the purified inhibitor was equally toxic to both strains. On the other hand, Turner has shown that a specific glucosidase was produced by var. *avenae*,

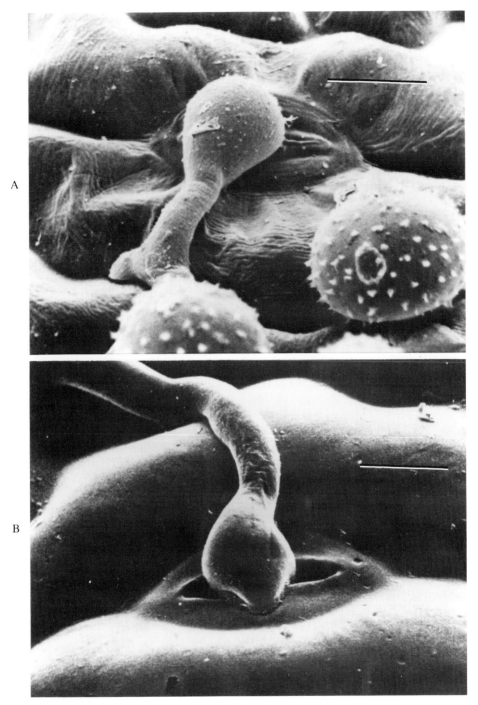

Fig. 1–19. Scanning electron micrographs of appressoria produced by *Uromyces phaseoli* over (top) closed stomate of pinto bean leaf lower surface and (bottom) open stomate on a *replica* of a pinto bean leaf lower surface (after Wynn, 317). Bar = 10 μm.

pathogenic to oats, but not by strains unable to parasitize oats. The enzyme destroys the biological activity of the inhibitor and, thereby, the fungus is able to infect the host. Thus, pathogenic specificity of *O. graminis* var. *avenae* to oats is dependent on the production of an inhibitor-inactivating enzyme.

Host *resistance* may, on the other hand, be determined by the capacity of the host to inactivate fungal toxins (112), i.e. a metabolite produced by the fungus is inactivated by the host plant. However, in the situations described above, which determine pathogenic specificity, a metabolite produced by the *host* is inactivated by the fungus.

Another fungistatic compound, tomatine (a steroidal glyco-alkaloid), was found in the expressed juice of some tomato plants. This compound is actively fungistatic toward *Fusarium oxysporum* f. sp. *lycopersici*. However, Kern (164) found no correlation between the tomatine content and wilt resistance of tomato cultivars. Arneson and Durbin (11) have shown that a leaf-infecting fungal pathogen of tomato, *Septoria lycopersici*, detoxified tomatine both in vitro and in infected leaves. A constitutive, extracellular enzyme of the pathogen is able to hydrolyze one glucose unit from the tomatine molecule, thereby inactivating it. It was presumed that this pathogen and some others, insensitive to tomatine, detoxify this compound, accounting for their success as tomato pathogens.

A quite different impediment to successful infection may be the absence or scarcity in the host of nutrients essential for the development of the pathogen. Keitt and Boone (162) isolated mutants in the apple scab fungus *Venturia inaequalis* whose requirements for vitamins, nitrogen bases, or amino acids were specific. Pathogenicity was lost in some of these mutants unless the particular requirement was supplied with the inoculum.

It is well known that several fungi that germinate in distilled water do not germinate satisfactorily when placed in water drops on the surface of leaves. It was shown by Fraser (103) that bacterial flora on the leaf surface may play a role in the restriction of fungal growth. This has been until recently a neglected field of research, and future investigations should prove fruitful.

Effect of toxins prior to pathogen entry. It is widely believed that the causal agent of milo root rot, *Periconia circinata*, produces polypeptide toxins that are able to inhibit root growth of susceptible sorghum cultivars at a concentration as low as 0.1 μg/ml prior to the pathogen entry. However, there is no direct evidence at present that the *Periconia* toxin is produced in the soil before entry of the fungus (249).

Another and better documented effect of a pathogen on its host prior to pathogen entry is that of powdery mildew on wheat. Martin and Ellingboe (201) have shown that there is an effect of inoculation with viable mildew spores on the uptake of ^{32}P by wheat leaves prior to penetration. For example, the average radioactivity in noninoculated and inoculated 1 cm-long leaf sections

was 2.4×10^5 cpm and 1.8×10^5 cpm, respectively; and the values for epidermal strips were 6.5×10^3 cpm and 3.2×10^3 cpm, respectively. Chalk dust or killed conidia did not cause a decrease in the uptake of labeled P. On the other hand, *Erysiphe graminis* f. sp. *hordei*, which is pathogenic to barley but not to wheat, reduced labeled P uptake by wheat leaves. The host may react to mildew prior to entrance of the infection peg by forming *papillae*: a matrix presumed to contain callose and other substances between the inner surface of the host wall and the plasmalemma (210). Sargent et al. (247) also point out that changes within the host cell wall and cytoplasm take place before penetration of downy mildew into lettuce. These alterations were detected below the region of contact with the appressorium. The implications of these changes remain uncertain.

Environmental factors. Environmentally conditioned susceptibility to infection, or "predisposition," is thoroughly treated by Yarwood (321) and Colhoun (56). A survey of the innumerable conflicting data in the literature precludes generalizations in this field. For example, exposing host plants to high or low temperature, high or low light intensity, or high or low soil moisture before inoculation may increase or decrease their susceptibility to several fungal pathogens. Still, high nitrogen levels in the soil or in tap solution usually favor infections caused by biotrophic fungi, such as rusts, mildews, etc. (64, 173). On the other hand, high nitrogen reduces infection elicited by necrotrophic fungi (*Alternaria*, *Septoria*, *Fusarium*, *Colletotrichum*, *Cladosporium*, *Helminthosporium*, *Botrytis*, etc.) because it increases tissue juvenility (18, 174, 198). The juvenile state of host tissues favors the growth of biotrophs, but it is unfavorable for necrotrophs or/and for the development of necrotic disease symptoms.

High phosphorus, calcium, and potassium are believed to reduce infection with several fungi (112). Even in these cases, data supporting opposing views have been recorded.

Mechanism of Penetration

Penetration of the host by the fungus occurs (a) directly through the intact surface, (b) through natural openings, and (c) through wounds. It is an *active* process.

Penetration through the cell wall was for a long time regarded as a mechanical process. The epidermis of the host tissue is covered by the cuticle, which is penetrated by an infection hypha ("peg") produced below the appressorium, a process thought to be entirely mechanical (5, 80, 274, 313). This may well be the case with *Plasmodiophora brassicae* (4). However, recent investigations by McKeen and Rimmer (210), McKeen (209), and Sargent et al. (247) brought into question the concept of mechanical penetration. For example, the infection peg of *Botrytis cinerea* is blunt and covered only by a plasma membrane; it makes a clear passage through the cuticle rather than a hole surrounded by a cuticular flange. Electron micrographs (Fig. 1–20)

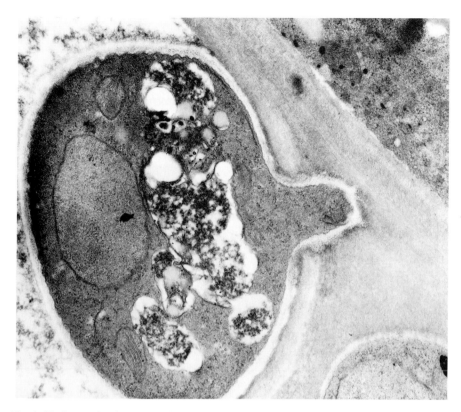

Fig. 1–20. Penetration through the thickened portion of a host cell wall by the infection peg of *Macrophomina phaseolina*. The absence of inward bending of wall laminations is suggestive of enzymatic wall degradation, ×21,400 (after Ammon et al., 9).

indicate a probable chemical degradation of cuticle at the tip of the blunt infection peg. The production of cutinase by fungi has been demonstrated satisfactorily in several instances, and it seems probable that cutinase is a regular accompaniment of the penetration process. Hence, the mechanical pressure of the infection peg of *Botrytis cinerea* is augmented by excretion of cutinase through the plasmalemma that covers the blunt tip of the peg (179, 180). (For details of the destruction of the cell wall, see Chap. 4.) Penetration of host by the barley powdery mildew fungus shows a similar pattern. Penetration of the infection peg through the epidermal wall results, in part, from "enzymatic" dissolution: the tip of the peg has no wall, is blunt, and mechanical "fracturing" was not observed under the electron microscope. It was presumed that, after dissolution of the host epidermal wall (Fig. 1–20), mechanical pressure was required to force the infection peg through the host cell wall and papilla (210).

In the case of *Olpidium brassicae*, when the entire fungal thallus is immersed in the host cytoplasm, the protoplast of the fungal thallus is extruded into the host cell through a channel in the papilla. The latter develops at the plant cell wall-plasmalemma interface (189, 190).

Non-haustoria-forming pathogens such as the fila-

mentous fungi penetrate the host directly with their hyphae. Cellulolytic and pectolytic fungal secretions may degrade the cell wall at the point of penetration. In some instances, host cells are killed by toxins of pathogens in advance of the fungus per se (5). The penetration of the cell wall of soybean root cells by *Macrophomina phaseolina* is shown in Fig. 1–20.

Fungi frequently enter plants through natural openings such as stomates. Zoospores of *Plasmopara viticola* on grape leaves are led to the stomates by a stimulus from the stomates (10). This type of penetration may also involve appressorium formation (*Puccinia, Phytophthora*) and infection hypha as in the case of direct entry through the cuticle. From the appressorium of *Puccina graminis*, a small hypha emerges and passes through the stomatal pore. Upon entering the substomatal cavity, the hypha forms a substomatal vesicle, and intercellular mycelia from this vesicle form haustoria that invaginate the protoplast.

Radioactivity from ^{14}C-labeled rust mycelia can pass to the host cells, but it is uncertain whether materials of pathogen origin are responsible for papilla formation between the host's plasmalemma and cell walls (82, 84).

The physiological mechanisms controlling the entry

of fungi through wounds are poorly understood at present. There is a large group of fungal pathogens (*Fusarium*, *Sclerotinia*, *Ceratocystis*, and many weak pathogens such as *Penicillium* and *Rhizopus*) that can penetrate only through wounds. Scheffer and Walker (248) placed tomato cuttings in a suspension of spores of *Fusarium* and were able to observe the fungus hyphae as far as 6 in above the cut surface. It is not known with certainty, however, whether spores are present in wounded, naturally infected plants.

The Colonization of the Host

After penetration, fungal pathogens may spread from the site of infection and pervade the host. Only if the fungus enters into a parasitic relationship with its host is infection successful and, hence, is pathogenesis initiated. Physiological manifestations of the disease per se are treated in the chapters that follow. In general, the protoplasm of the host cell becomes coarsely granular, and the nucleus migrates toward the infecting hyphae, later increasing in size (7). In advanced stages of disease, the nucleus begins to collapse and degenerate.

As a consequence of infection, host cell cytoplasm increases, resulting in an apparent increase in rough endoplasmic reticulum and in the concentration of ribosomes. This probably explains the increased capacity of the infected cell for protein synthesis (see also Chapter 5). Although rearrangement of the fine structure at the interface between the pathogen and its host is poorly understood at present, researchers are active in this field of investigation.

With the development of the haustorium, or the intracellular-absorbing organs of obligate or facultative parasites, the host is successfully infected and pathogenesis is initiated. It has been contended for some time (285) that penetration means getting through the host cell wall but not through its plasma membrane. Recent electron microscopic investigations have shown that, although the haustorium invaginates the host plasma membrane, the plasmalemma is, indeed, not penetrated (39, 84). The haustorium is actually isolated from the host protoplast by an extrahaustorial zone. This is the case with the rusts, powdery mildews, *Phytophthora* spp., *Peronospora* spp., and smuts and with many facultative parasites that produce intracellular-absorbing organs. In a single case, however, it has been shown that the fungus *Olpidium brassicae* actually breaks through the host plasmalemma. The entire fungal thallus of *Olpidium* is immersed in the host cytoplasm, and the pathogen is surrounded only by its own single membrane (189, 190). In another thallus-host interface (*Plasmodiophora brassicae* on cabbage), however, it was found that the host plasmalemma is invaginated and the fungus is enclosed in the host plasmalemma (310).

The extrahaustorial membrane that separates the haustorium from the host cytoplasm is probably produced by the host and not the pathogen. This membrane serves as a boundary between the host hyaloplasm and the fungus. It is spatially separated from intimate contact with the haustorial wall at the apical portion of the haustorial neck and around the haustorial head (Fig. l–21). It is continuous with the host cell plasmalemma but distinct from it both structurally and functionally.

The zone between the haustorium and the extrahaustorial membrane has long been recognized as a sacklike covering around the haustorium. Recent electron microscope and other observations show haustoria of obligate parasites (*Puccinia*, *Melampsora*, *Erysiphe*, *Peronospora*, *Albugo*) surrounded by this sacklike sheath in the host cell (38, 39, 45, 83). The sheath, encapsulation, or "zone of apposition" (84, 226) is typical neither of the host nor of the fungal cell wall and separates the haustorium from the host. It is presumed that a specific secretory process is operative in infected cells. Secretory bodies are formed in the host cytoplasm, and then move to the zone of apposition and discharge their contents. They are probably consumed by the pathogen (31, 226). However, this hypothesis has not been verified experimentally in spite of the fact that the host endoplasmic reticulum seems to be associated with the encapsulation boundary (extra-haustorial membrane). For additional detail on the relationship of haustorium, the haustorial sheath, and the invaginated host plasma membrane to rust infection, see Bracker and Littlefield (39). The non-haustoria-forming pathogens do not induce the development of membrane-bounded sheaths and encapsulations. Ehrlich and Ehrlich (84) suggested that the lack of the sheath formation may explain the rapid destruction of infected host cells by these fungi, i.e. facultative parasites. Thus, one of the functions of the encapsulation would be to exclude toxic substances from host tissues. Another hypothesis is that the encapsulation is a transit medium for materials between the host and pathogen: it is a "communication switchboard." Neither of these hypotheses has been supported experimentally. Nevertheless, the sheath around the haustorium and the extrahaustorial membrane seems to have distinctive roles in the host-parasite relation.

Host responses. One of the first visible responses of host cells to fungal penetration and colonization is the formation of *papillae* (Fig. l–22). Papilla formation is a general phenomenon that is initiated in minutes and completed within hours after fungal contact with the host cell wall. They develop between the host cell plasmalemma and the cell wall per se and have also been referred to as callosities or lignitubers (3). Papilla formation may also be regarded as a general wound-healing response of host cells and may occur before, during, or after penetration of the host cell wall by the pathogen. (See the section above on the effect of toxins on host prior to pathogen entry.) Papillae development has also been regarded as a host resistance phenomenon (46, 132, 274), because they appear to represent an initial barrier against the invading parasites. This view is well documented with host and nonhost reactions to the cowpea rust (*Uromyces phaseoli* var. *vignae*). Papillae were formed only in resistant hosts and in nonhosts like *Phaseolus vulgaris*; susceptible

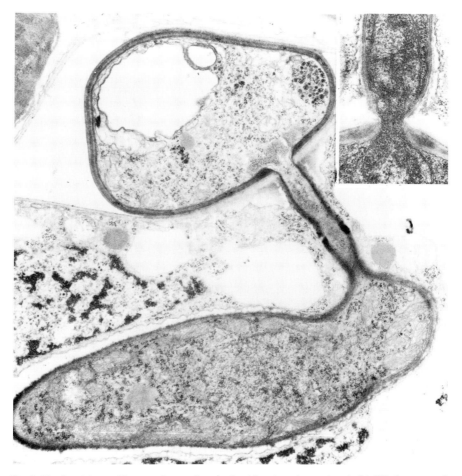

Fig. 1–21. *Puccinia sorghi* penetrating sorghum leaf and forming a haustorium (×24,600). Insert reveals magnification of haustorial head, haustorial neck, and extra haustorial membrane in cowpea rust fungus, ×84,000 (courtesy of M. Heath).

hosts never formed papillae after penetration and colonization. However, in many cases, papillae are breached by the invading pathogens, and in other host-parasite relationships papillae are equally formed in both susceptible and resistant hosts (2, 6, 39, 45, 46). Occasionally, the whole haustorium becomes encased by the cell wall deposition and is effectively walled off from the host cell ("encasement"). This is characteristic for one of the 0-type resistance reactions of hosts to rust fungi (134).

As previously mentioned, early host responses in a *compatible* host-parasite relationship may be characterized by an increase in the amount of host endoplasmic reticulum, Golgi bodies, and mitochondria. This cytoplasmic constituency is more likely to be observed in juvenile cells than in mature or damaged cells. Whether these alterations in host cytoplasm are consequences of a cytokinin-effect caused by some pathogens remains to be shown (see Chap. 7). An association between the endoplasmic reticulum and fungal haustorium

is also typical of the compatible host-pathogen complex. The host nucleus is, in many instances, in the vicinity of the haustorium and, in some instances, encircles it. However, the nucleus and the haustorium always remain separated. In most diseased plants, the first visible damage is to the chloroplasts. They usually increase in size upon infection, and the chloroplast membranes may then rupture and the thylakoids become dispersed. As a later symptom of pathogenesis, the entire membrane system of the host cell breaks down and visible senescencelike processes are initiated (262) that terminate in cell death.

Ultrastructural and histological studies of *incompatible* and *nonhost* reactions revealed diverse mechanisms of resistance to fungal infections. In the past, mainly the *hypersensitive reaction* (HR) of the host was investigated because hypersensitivity (rapid death of the penetrated host cells) seems to be the primary *visible* event in the sequence associated with host resistance to infection (199, 269). Even in this type of reaction, a

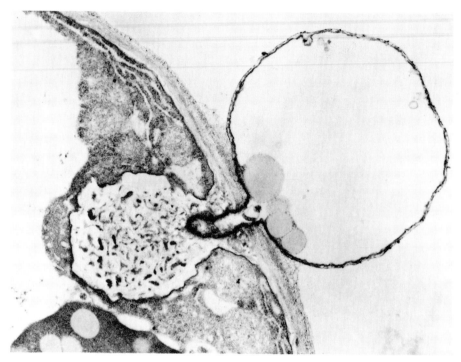

Fig. 1–22. Penetration of cabbage root hair by *Plasmodiophora brassicae* zoospore eliciting the formation of papillum, ×32,500 (after Aist and Williams, 4).

visible retardation of fungal growth and damage to the haustorium accompanies damage of the host cell, and it is difficult to demonstrate conclusively which entity is affected initially and/or primarily. Király et al. (175) and Érsek et al. (92) suggested that, in an incompatible host-pathogen complex, the pathogen is damaged initially (nonvisibly, perhaps, as an alteration of membrane permeability of the fungus), and the observed membrane damage of HR in the host is, thereby, a consequence, not the cause, of the observed resistance to infection. Indirect evidence suggests that toxic compound(s) released from the pathogen affect induction of both tissue necrosis and the phytoalexin synthesis (which are characteristics of HR). One is able to obtain this inducer of the HR even from the compatible race of *Phytophthora infestans* following severe damage to the pathogen, as may be caused by either sonication or chloroform extraction (see Table 1–3). (HR induction by bacteria is accomplished only by living bacteria.) It was suggested (92) that slight damage to the incompatible fungal parasite (e.g. nonvisible membrane alteration) would evoke the release of the inducer of HR from the pathogen, and this would be the immediate cause of the visible symptom (necrosis) of resistance.

Whether a cause or an effect relationship, these diverse ultrastructural and histological manifestations of host resistance to rust and powdery mildew infections exist. The observations suggest that hypersensitivity is not a single phenomenon (41, 89, 132, 134, 193, 196).

In several forms of plant resistance, hypersensitive necrosis does not occur at all. Host resistance is most likely a phenomenon of recognizing a nonself entity. The biochemical mechanism of the nonself recognition occurs, it would seem, at the macromolecular level, and the diverse histological and ultrastructural phenomena described are manifestations of this basic recognition.

Rohringer et al. (239) presented evidence for involvement of RNA in evoking the tissue necrosis associated with the penetration of wheat by stem rust, *Puccinia graminis*. They prepared RNA from incompatible and infected wheat leaves exhibiting a hypersensitive-type reaction. Upon injection into tissue, the RNA stimulated necrosis (hypersensitivity) at a higher rate in an incompatible host-pathogen complex than in a compatible one. They interpreted their data as demonstrating that their RNA is directly involved in the resistant reaction of wheat to stem rust. The injected RNA did not cause necrosis in the uninfected host, only in infected leaves following haustorial formation, whether in the compatible or incompatible host. In the experiments of Rohringer et al. (239), the relationship of RNA to host tissue necrosis was studied rather than the relationship to resistance per se.

Several investigations have been conducted to determine the precise time following inoculation that arrested development or death of the pathogen occurs in host tissue. Lupton (196) and, later, Ellingboe (89) and

Table 1–3. Hypersensitive reaction in potato-*Phytophthora infestans* and potato-toxin interactions (after Érsek et al., 92).

Host-pathogen/toxin interaction	Necrosis	Phytoalexin (Rishitin)
Compatible[a]	0	0
Incompatible[b]	+	+
Compatible + treatment with streptomycin or chloramphenicol	+	+
Sonicated[c]		
Compatible	+	+
Incompatible	+	+
Chloroform-treated[d]		
Compatible	+	+
Incompatible	+	+

a. Race 1 or race 1 or 1.2.3.4 of *P. infestans* on "Gülbaba" (r) or race 1.2.3.4 on "Rotkelchen" (R_1R_3) potato tuber slices.
b. Race 1 on "Rotkelchen" potato tuber slices.
c. Homogenates of mycelium of race 1 or 1.2.3.4 of *P. infestans* were sonicated. The cell-free liquids were applied to tuber slices of potato cultivar Gülbaba or Rotkelchen.
d. Cell-free liquid released by race 1 or 1.2.3.4 as a result of treatment with chloroform was applied to tuber slices of potato cultivar Gülbaba or Rotkelchen.

others showed that several different mechanisms of resistance to powdery mildew (*Erysiphe graminis*) are operative. The stages of infection by the powdery mildew fungus can be divided as follows: germination, appressorium formation, penetration of the host, haustorium development in the host cell, formation of secondary hyphae that elongate and initiate secondary appressoria and haustoria, and outgrowth of hyphae on the host surface. The formation and elongation of secondary hyphae are taken as evidence that the host and pathogen have established a compatible relationship and that pathogenesis has been initiated (Fig. 1–23.)

Ellingboe et al. (89) concluded that none of the different incompatibility- or resistance-reaction types influence either spore germination or appressorial formation. Comparative analysis of haustorial development with compatible and incompatible host-pathogen genotypes showed that a resistance mechanism is, however, expressed soon after penetration. This mechanism becomes effective about 10–12 h after inoculation. A second resistance mechanism is in operation 21–22 h after inoculation when a collapse of the young haustorium and a darkening of the cell area surrounding the haustorium is observed. A third mechanism is characterized by the collapsed haustorium without darkening of the host cell adjacent to the haustorium. This takes place 26 h after inoculation (Fig. 1–23). Some mechanisms may act later in the disease development, but experimental data for these are meager. From this study, it was concluded that the morphogenesis of the patho-

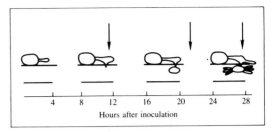

Fig. 1–23. The hour after inoculation (indicated by arrows) when the various host genes interact with the corresponding parasite genes to affect the development of *Erysiphe graminis* during primary infection (after Ellingboe, 89).

gen during the infection process was specifically inhibited according to the particular incompatible genotype involved in the host-pathogen interaction (89, 216). Littlefield (193) has shown that, in the case of flax rust, the resistance mechanism must be diverse, even in closely related host genotypes when incompatibility is controlled by the interaction of single genes in the host and the pathogen. Most of these results are from histological studies, and their biochemical bases remain to be elucidated.

Comparative Analysis of Disease Physiology—the Infection Process

The process of infection of a plant by a virus, bacterium, or fungus is characterized by an increasing order of active participation, respectively, from the three pathogens. Viruses cannot infect an intact plant unaided. Before infection can be initiated, one or more virus particles must be introduced intracellularly, either through wounds or through the agency of a vector. Bacteria, on the other hand, do not necessarily need vectors in order to reach the plant's surface. They may be carried by wind and rain and many are motile, reaching roots through the soil. Once in contact with the plant, they may penetrate the intercellular spaces via wounds or natural openings, again without assistance but in association with atmospheric (fluid) moisture. Fungal spores, carried to the plants through air or soil, may germinate on the surface and may penetrate through wounds or natural openings. These pathogens also almost always need water in which to accomplish penetration. In some cases, the elaboration of enzymes permits penetration through the intact epidermal surface. The fungal hyphae may not only penetrate the intercellular spaces but also actively penetrate the cell. Of the three pathogen groups, the fungi are the most versatile in the efforts to infect host tissue. A significant body of evidence suggests that bacteria and fungi are both stimulated and/or inhibited by plant exudates. Bacteria and fungal propagules are also capable of following diverse chemical and oxygen gradients in the solutions of the soil and the contours of aerial surfaces of the plant.

Movement to the host, contact with it, entry, and initial establishment of the pathogen are the prerequisite steps in the *infection process*. However, much current research seems to be centered on the contact and recognition aspects of the phenomenon. Although the evidence for recognition between virus and plant cell is not substantive, there is increasing evidence for recognition between bacterial lipopolysaccharides and plant cell glycoproteins.

Whereas viruses must, of necessity, replicate intracellularly, bacteria normally multiply in the intercellular spaces or, exceptionally, in the xylem. Fungi may proliferate either intra-, inter-, or extracellularly.

The comparatively small but rapidly expanding literature pertaining to MLO and RLO bacteria links their penetration of host tissue to insect vectors or asexual propagation. Their sites of initial parasitic activity internally, i.e. xylem vessels and phloem sieve tubes, reflect these delivery systems. Furthermore, their morphology, MLOs being wall-less and RLOs having an unusually complex cell wall, signifies their adaptability to fluctuating, and in phloem very high, osmotica.

References

1. Abeles, F. B., and G. R. Leather. 1971. Abscission: Control of cellulase secretion by ethylene. Planta 97:87–91.

2. Aist, J. R. 1976. Papillae and related wound plugs of plant cells. Ann. Rev. Phytopath. 14:145–63.

3. Aist, J. R., M. A. Waterman, and H. W. Israel. 1979. Papillae and penetration: Some problems, procedures and perspectives. In: Recognition and specficity in plant host-parasite interactions. J. M. Daly and I. Uritani (eds.). Jap. Sci. Soc. Press, Tokyo, pp. 85–97.

4. Aist, J. R., and P. H. Williams. 1971. The ecology and kinetics of cabbage root hair penetration by *Plasmodiophora brassicae*. Can. J. Bot. 49:2023–34.

5. Akai, S., M. Fukutomi, N. Ishida, and H. Kunoh. 1967. An anatomical approach to the mechanisms of fungal infections in plants. In: The dynamic role of molecular constituents in plant-parasite interaction. C. J. Mirocha and I. Uritani (eds.). Amer. Phytopath. Soc., St. Paul, MN, pp. 1–20.

6. Akai, S., O. Horino, M. Fukutomi, A. Nakata, H. Kunoh, and M. Shiraishi. 1971. Cell wall reaction to infection and resulting change in cell organelles. In: Morphological and biochemical events in plant-parasite interaction. S. Akai and S. Ouchi (eds.). Phytopath. Soc. Jap., Tokyo, pp. 329–67.

7. Allen, R. F. 1923. A cytological study of infection of Baart and Kanred wheats by *Puccinia graminis tritici*. J. Agr. Res. 23:131–51.

8. Allington, W. B., and E. F. Laird, Jr. 1954. The infection of *Nicotiana glutinosa* with tobacco mosaic virus as affected by potassium nutrition. Phytopathology 44:297–99.

9. Ammon, V., T. D. Wyllie, and M. F. Brown, Jr. 1974. An ultrastructural investigation of pathological alterations induced by *Macrophomina phaseolina* (Tassi) Goid in seedlings of soybean, *Glycine max* (L.) Merrill. Physiol. Plant Path. 4:1–4.

10. Arens, K. 1929. Physiologische Untersuchungen an *Plasmopara viticola*, unter besonderer Berücksichtigung der Infektionsbedingungen. Jahrb. Wiss. Bot. 70:93–157.

11. Arneson, P. A., and R. D. Durbin. 1967. Hydrolysis of tomatine by *Septoria lycopersici*: A detoxification mechanism. Phytopathology 57:1358–60.

12. Atabekov, J. G. 1975. Host specificity of plant viruses. Ann. Rev. Phytopath. 13: 127–45.

13. Ayers, A. R., S. B. Ayers, and R. N. Goodman. 1979. Extracellular polysaccharide of *Erwinia amylovora*: A correlation with virulence. Appl. Environ. Microbiol. 38:659–66.

14. Bailiss, K. W., and G. Plaza-Morales. 1980. Effects of postinoculation leaf water status on infection of French bean by tobacco necrosis virus. Physiol. Plant Path. 17:357–67.

15. Balázs, E., B. Barna, and Z. Király. 1976. Effect of kinetin on lesion development and infection sites in Xanthi-nc tobacco infected by TMV: Single-cell local lesions. Acta Phytopath. Hung. 11:1–9.

16. Baldwin, C. H., and R. N. Goodman. 1963. Prevalence of *Erwinia amylovora* in apple buds as detected by phage typing. Phytopathology 53:1299–1303.

17. Banerjee, D., M. Basu, I. Choudhury, and G. C. Chatterjee. 1981. Cell surface carbohydrates of *Agrobacterium tumefaciens* involved in adherence during crown gall tumor initiation. Biochem. Biophys. Res. Comm. 100:1384–88.

18. Barna, B., A. R. T. Sarhan, and Z. Király. 1983. The influence of nitrogen nutrition on the sensitivity of tomato plants to fusaric acid and other toxic actions. Physiol. Plant Path. 23:257–63.

19. Bauer, W. D. 1981. Infection of legumes by rhizobia. Ann. Rev. Plant Physiol. 32:407–99.

20. Bauske, R. J. 1967. Dissemination of waterborne *Erwinia amylovora* by wind in nursery plantings. Amer. Soc. Hort. Sci. 91:759–801.

21. Bawden, F. C. 1954. Inhibitors and plant viruses. Adv. Virus Res. 2:31–57.

22. Bawden, F. C. 1957. The multiplication of plant viruses. In: Nature of viruses. G. E. W. Wolstenholme and C. E. P. Millar (eds.). Little, Brown, Boston. CIBA Foundation Symposium, pp. 170–90.

23. Bawden, F. C., and B. D. Harrison. 1955. Studies on the multiplication of a tobacco necrosis virus in inoculated leaves of French-bean plants. J. Gen. Microbiol. 13: 494–508.

24. Bawden, F. C., and B. Kassanis. 1950. Some effects of host nutrition on the susceptibility of plants to infection by certain viruses. Ann. Appl. Biol. 37:46–57.

25. Bawden, F. C., and A. Kleczkowski. 1955. Studies on the ability of light to counteract the inactivating action of ultraviolet irradiation on plant viruses. J. Gen. Microbiol. 13:370–82.

26. Bawden, F. C., and F. M. Roberts. 1947. The influence of light intensity on the susceptibility of plants to certain viruses. Ann. Appl. Biol. 34:286–96.

27. Bawden, F. C., and F. M. Roberts. 1948. Photosynthesis and the predisposition of plants to infection with certain viruses. Ann. Appl. Biol. 35:418–28.

28. Benda, G. T. A. 1956. Infection of *Nicotiana glutinosa* L. following injection of two strains of tobacco mosaic virus into a single cell. Virology 2:820–27.

29. Bennett, R. A., and E. Billing. 1978. Capsulation and virulence in *Erwinia amylovora*. Ann. Appl. Biol. 89:41–45.

30. Berg, L. A. 1971. Weed hosts of the SFR strain of *Pseudomonas solanacearum*, causal organism of bacterial wilt of bananas. Phytopathology 61:1314–15.

31. Berlin, J. D., and C. C. Bowen. 1964. The host-parasite interface of *Albugo candida* on *Raphanus sativus*. Amer. J. Bot. 51:445–52.

32. Billing, E. 1960. An association between capsulation and phage sensitivity in *Erwinia amylovora*. Nature 186:819–20.

33. Billing, E. 1980. Fireblight in Kent, England in relation to weather (1955–1976). Ann. Appl. Biol. 95:341–64.

34. Billing, E. 1980. Firelight (*Erwinia amylovora*) and weather: A comparison of warning systems. Ann. App. Biol. 95:365–77.

35. Bogers, R. J. 1972. On the interaction of *Agrobacterium tumefaciens* with cells of *Kalanchoe daigremontiana*. In: Proc. Third Int. Conf. Plant Pathogenic Bact. H. P. Maas Geesteranus (ed.). Pudoc, Wageningen, Netherlands. pp. 239–50.

36. Bohlool, B. B., and E. L. Schmidt. 1974. Lectins: A possible basis for specificity in the *Rhizobium*-legume root nodule symbiosis. Science 185:269–71.

37. Boyle, L. W., and H. H. McKinney. 1938. Local virus infections in relation to leaf epidermal cells. Phytopathology 28:114–22.

38. Bracker, C. E. 1967. Ultrastructure of fungi. Ann. Rev. Phytopath. 5:343–74.

39. Bracker, C. E., and L. J. Littlefield. 1973. Structural concepts of host-pathogen interfaces. In: Fungal pathogenicity and the plant's response. R. J. W. Byrde and C. V. Cutting (eds.). Academic Press, London, and New York, pp. 159–318.

40. Brants, D. H. 1965. Relation between ectodesmata and infection of leaves by C14-labeled tobacco mosaic virus. Virology 26:554–57.

41. Brown, J. F., W. A. Shipton, and N. H. White. 1966. The relationship between hypersensitive tissue and resistance in wheat seedlings infected with *Puccinia graminis tritici*. Ann. Appl. Biol. 58:279–90.

42. Brown, W. 1922. Studies in the physiology of parasitism. VIII. On the exosmosis of nutrient substances from the host tissue into the infection drop. Ann. Bot. 36: 101–19.

43. Brown, W. 1936. The physiology of host-parasite relations. Bot. Rev. 2:236–81.

44. Burke, D., L. Mendonca-Preirato, and C. E. Ballou. 1980. Cell-cell recognition in yeast: Purification of *Hansenula wingei* 21-cell sexual agglutination factor and comparison of the factors from three genera. Proc. Nat. Acad. Sci. 77:318–22.

45. Bushnell, W. R. 1972. Physiology of fungal haustoria. Ann. Rev. Phytopath. 10: 151–76.

46. Bushnell, W. R., and S. E. Bergquist. 1975. Aggregation of host cytoplasm and the formation of papillae and haustoria in powdery mildew of barley. Phytopathology 65:310–18.

47. Buxton, E. W. 1957. Some effects of pea root exudates on physiologic races of *Fusarium oxysporum* Fr. f. *pisi* (Linf.) Snyder and Hansen. Trans. Brit. Mycol. Soc. 40:145–54.

48. Caldwell, J. 1932. Studies in the physiology of virus diseases in plants. III. Aucuba or yellow mosaic of tomato in *Nicotiana glutinosa* and other hosts. Ann. Appl. Biol. 19:144–52.

49. Calzolari, A., P. Peddles, U. Mazzucchi, P. Mori, and C. Garzena. 1982. Occurrence of *Erwinia amylovora* in buds of asymptomatic apple plants in commerce. Phytopath. Z. 103:156–62.

50. Cameron, H. R. 1970. *Pseudomonas* content of cherry trees. Phytopathology 60: 1343–46.

51. Chessin, M., and H. A. Scott. 1955. Calcium deficiency and infection of *Nicotiana glutinosa* by tobacco mosaic virus. Phytopathology 45:288–89.

52. Chet, I., Y. Zilberstein, and Y. Henis. 1973. Chemotaxis of *Pseudomonas lachrymans* to plant extracts and to water droplets from leaf surfaces of resistant and susceptible plant. Physiol. Plant Path. 3:473–79.

53. Clayton, E. E. 1936. Water soaking of leaves in relation to the development of the wildfire disease of tobacco. J. Agr. Res. 52:239–69.

54. Cocking, E. C. 1966. An electron microscope study of the initial stages of infection of isolated tomato fruit protoplasts by tobacco mosaic virus. Planta 68:206–14.

55. Cocking, E. C., and E. Pojnar. 1969. An electron microscope study of the infection of isolated tomato fruit protoplasts by tobacco mosaic virus. J. Gen. Virol. 4:305–12.

56. Colhoun, J. 1979. Predisposition by the environment. In:

Plant disease: An advanced treatise. J. G. Horsfall and E. B. Cowling (eds.). Academic Press, New York, 4:75–96.

57. Conti, G. G., and R. Locci. 1972. Leaf surface alterations of *Nicotiana glutinosa* connected with mechanical inoculation of tobacco mosaic virus. Rivista di Patologia Vegetale Ser. IV 8:85–101.

58. Corbett, M. K., and H. D. Sisler (eds.).1964. Plant virology. Univ. of Florida Press, Gainsville.

59. Corey, R. R., and M. P. Starr. 1957. Colony types of *Xanthomonas phaseoli*. J. Bact. 74:137–40.

60. Crosse, J. E. 1956. Bacterial canker of stone fruits. II. Leaf scar infection of cherry. J. Hort. Sci. 31:212–24.

61. Crosse, J. E., and C. M. E. Garrett. 1963. Studies on the bacteriophage of *Pseudomonas mors-prunorum*, *Ps. syringae* and related organisms. J. Appl. Bact. 26:159–77.

62. Crosse, J. E., R. N. Goodman, and W. H. Shaffer, Jr. 1972. Leaf damage as a predisposing factor in the infection of apple shoots by *Erwinia amylovora*. Phytopathology 62:176–82.

63. Daft, G. C., and C. Leben. 1972. Bacterial blight of soybeans: Seedling infection during and after emergence. Phytopathology 62:1167–70.

64. Daly, J. M. 1949. The influence of nitrogen source on the development of stem rust of wheat. Phytopathology 39:386–94.

65. Davidson, R. S. 1948. Factors affecting the development of bacterial soft rot of potato tuber initials. Phytopathology 38:673–87.

66. Davis, J. R., and H. English. 1969. Factors related to the development of bacterial canker in peach. Phytopathology 59:588–95.

67. Davis, R. E., and R. F. Whitcomb. 1967. Sucrose enhanced local infection of plants by viruses. Phytopathology 57:808.

68. Dazzo, F. B., and D. H. Hubbell. 1975. Cross-reactive antigens and lectin as determinants of symbiotic specificity in the *Rhizobium*-clover association. Appl. Microbiol. 30:1017–33.

69. Dazzo, F. B., G. L. Truchet, J. E. Sherwood, E. M. Hraback, and A. E. Gardiol. 1982. Alteration of the trifoliin A-binding site of *Rhizobium trifolii* 0403 by enzymes released from clover roots. Appl. Environ. Microbiol. 44:478–90.

70. Dazzo, F. B., M. R. Urbano, and W. J. Brill. 1979. Transient appearance of lectin receptors on *Rhizobium trifolii*. Current Microbiol. 2:15–20.

71. Dazzo, F. B., W. E. Yanke, and W. J. Brill. 1978. Trifoliin: A *Rhizobium* recognition protein from white clover. Biochem. Biophys. Acta 539:276–86.

72. de Zoeten, G. A. 1976. Cytology of virus infection and virus transport. In: Encyclopedia of plant physiology. New Series. R. Heitefuss and P. H. Williams (eds.).Springer-Verlag, Berlin, 4:129–49.

73. Dickinson, S. 1949. Studies in the physiology of obligate parasitism. II. The behavior of the germ-tubes of certain rusts in contact with various membranes. Ann. Bot. n.s. 13:219–36.

74. Dowler, W. M., and D. H. Petersen. 1967. Transmission of *Pseudomonas syringae* in peach trees by bud propagation. Plant Dis. Reptr. 51:666–68.

75. Dowler, W. M., and D. J. Weaver. 1975. Isolation and characterization of fluorescent pseudomonads from apparently healthy peach trees. Phytopathology 65:233–36.

76. Dueck, J., and J. B. Morand. 1975. Seasonal changes in the epiphytic population of *Erwinia amylovora* on apple and pear. Can. J. Plant Sci. 55:1007–12.

77. Dueck, J., and H. A. Quamme. 1973. Fireblight in southern Ontario in 1972. Can. Plant Dis. Surv. 53:101–4.

78. Duggar, B. M., and J. K. Armstrong. 1925. The effect of treating the virus of tobacco mosaic with the juices of various plants. Ann. Mo. Bot. Gdn. 12:359–66.

79. Ebrahim-Nesbat, F., and F. Nienhaus. 1972. Beeinflussung

von Tabakmosaik virus-und Gurkenmosaicinfektion durch hemmende Prinzipien in Pflanzenextrakten. Phytopath. Z. 73:235–50.

80. Edwards, H. H., and P. J. Allen. 1970. A fine-structure study of the primary infection process during infection of barley by *Erysiphe graminis* f. sp. *hordei*. Phytopathology 60:1504–9.

81. Ehara, Y., and T. Misawa. 1968. Studies on the infection of cucumber mosaic virus. V. The observation of epidermal cell in the local lesion. Tohoku J. Agr. Res. 19:166–72.

82. Ehrlich, H. G., and M. A. Ehrlich. 1970. Electron microscope radioautography of ^{14}C transfer from rust uredospores to wheat host cells. Phytopathology 60:1850–51.

83. Ehrlich, M. A., and H. G. Ehrlich. 1963. Electron microscopy of the sheath surrounding the haustorium of *Erysiphe graminis*. Phytopathology 53:1378–80.

84. Ehrlich, M. A., and H. G. Ehrlich. 1971. Fine structure of the host-parasite interfaces in mycoparasitism. Ann. Rev. Phytopath. 9:155–84.

85. El-Banoby, F. E., and K. Rudolph. 1979. Polysaccharide from liquid cultures of *Pseudomonas phaseolicola* which specifically induces water-soaking in bean leaves (*Phaseolus vulgaris* L.). Phytopath. Z. 95:38–50.

86. El-Banoby, F. E., and K. Rudolph. 1980. Purification of extracellular polysaccharides from *Pseudomonas phaseolicola*, which induce water-soaking in bean leaves. Physiol. Plant Path. 16:425–37.

87. El-Banoby, F. E., K. Rudolph, and A. Hutterman. 1980. Biological and physical properties of an extracellular polysaccharide from *Pseudomonas phaseolicola*, Physiol. Plant Path. 17:291–301.

88. El-Banoby, F. E., K. Rudolph, and K. Mendgen. 1981. The fate of extracellular polysacchride from *Pseudomonas phaseolicola* in leaves and leaf extracts from halo-blight susceptible and resistant bean plants (*Phaseolus vulgaris* L.) Physiol. Plant Path. 18:91–98.

89. Ellingboe, A. H. 1972. Genetics and physiology of primary infection by *Erysiphe graminis*. Phytopathology 62:401–6.

90. Elliott, C., and F. W. Poos. 1940. Seasonal development, insect vectors and host range of bacterial wilt of sweet corn. J. Agr. Res. 60:645–86.

91. Ercolani, G. L., D. J. Hagedorn, A. Kelman, and R. E. Rand. 1974. Epiphytic survival of *Pseudomonas syringae* on hairy vetch in relation to epidemiology of bacterial brown spot of bean in Wisconsin. Phytopathology 64:1330–39.

92. Érsek, T., B. Barna, and Z. Király. 1973. Hypersensitivity and the resistance of potato tuber tissues to *Phytophthora infestans*. Acta Phytopath. Hung. 8:3–12.

93. Favali, M. A., G. G. Conti, and M. Bassi. 1977. Some observations on virus-induced local lesions by transmission and scanning electron microscopy. Acta Phytopath. Hung. 12:141–50.

94. Feliciano, A., and R. H. Daines. 1970. Factors influencing ingress of *Xanthomonas pruni* through peach leaf scars and subsequent development of spring cankers. Phytopathology 60:1720–26.

95. Fischer, H., and F. Nienhaus. 1973. Virushemmende Prinzipien in Paprika pflanzen (*Capsicum annuum* L.). Phytopath. Z. 78:25–41.

96. Flentje, N. T. 1959. The physiology of penetration and infection. In: Plant pathology, problems and progress 1908–1958. C. S. Holton, G. Fischer, R. Fulton, H. Hart, and S. McCallan (eds.). Univ. of Wisconsin Press, Madison, pp. 76–87.

97. Foster, J. A., and A. F. Ross. 1975. The detection of symptomless virus-infected tissue in inoculated tobacco leaves. Phytopathology 65:600–610.

98. Foster, R. E. 1967. *Chenopodium amaranticolor* nutrition affects cucumber mosaic virus infection. Phytopathology 57:838–40.

99. Fox, R. T. V., J. G. Manners, and A. Myers. 1971. Ultrastructure of entry and spread of *Erwinia carotovora* var. *atroseptica* into potato tubers. Potato Res. 14: 61–73.

100. Fox, R. T. V., J. G. Manners, and A. Myers. 1972. Ultrastructure of tissue disintegration and host reactions in potato tubers infected by *Erwinia carotovora* var. *atroseptica*. Potato Res. 15:130–45.

101. Francki, R. I. B., E. W. Kitajima, and D. Peters. 1981. Rhabdoviruses. In: Handbook of plant virus infections: Comparative diagnosis. E. Kurstak (ed.). Elsevier/North Holland, Amsterdam, pp. 455–89.

102. Francki, R. I. B., and J. W. Randles. 1973. Some properties of lettuce necrotic yellows virus and its in vitro transcription by virion-asociated transcriptase. Virology 54:359–68.

103. Fraser, A. K. 1971. Growth restriction of pathogenic fungi on the leaf surface. In: Ecology of leaf surface micro-organisms. T. F. Preece and C. H. Dickinson (eds.). Academic Press, London and New York, pp. 529–35.

104. Fraser, L., and R. E. F. Matthews. 1980. Efficient mechanical inoculation of turnip yellow mosaic virus using small volumes of inoculum. J. Gen. Virol. 44:565–68.

105. Fraser, R. S. S. 1979. Systemic consequences of the local lesion reaction to tobacco mosaic virus in a tobacco variety lacking the N gene for hypersensitivity. Physiol. Plant Path. 14:383–94.

106. Fraser, R. S. S., S. A. R. Loughlin, and R. J. Whenham. 1979. Acquired systemic susceptibility to infection by tobacco mosaic virus in *Nicotiana glutinosa* L. J. Gen. Virol. 43:131–41.

107. Furumoto, W. A., and R. Mickey. 1967. A mathematical model for the infectivity dilution curve of tobacco mosaic virus: Theoretical considerations. Virology 32: 216–23.

108. Furumoto, W. A., and R. Mickey. 1967. A mathematical model for the infectivity dilution curve of tobacco mosaic virus: Experimental tests. Virology 32:224–33.

109. Furumoto, W. A., and S. G. Wildman. 1963. Studies on the mode of attachment of tobacco mosaic virus. Virology 20:45–52.

110. Furumoto, W. A., and S. G. Wildman. 1963. The specific infectivity of tobacco mosaic virus. Virology 20:53–61.

111. Garber, E. D. 1956. A nutrition-inhibition hypothesis of pathogenicity. Amer. Natur. 90:183–94.

112. Gäumann, E. 1950. Principles of plant infection. (Trans. W. B. Breierley.) Crosley Lockwood and Son, London.

113. Gendron, Y., and B. Kassanis. 1954. The importance of the host species in determining the action of virus inhibitors. Ann. Appl. Biol. 41:183–88.

114. Gessner, S. L., and J. D. Irvin. 1980. Inhibition of elongation factor 2— dependent translocation by the pokeweed antiviral protein and ricin. J. Biol. Chem. 255: 3251–53.

115. Gitaitis, R. D., D. A. Samuelson, and J. O. Strandberg. 1981. Scanning electron microscopy of the ingress and establishment of *Pseudomonas alboprecipitans* in sweet corn leaves. Phytopathology 71:171–75.

116. Gold, E. R., and P. Balding. 1975. Receptor-specific proteins: Plant and animal lectins. American Elsevier, New York.

117. Goodman, R. N. 1976. Physiological and cytological aspects of the bacterial infection process. In: Physiological plant pathology. R. Heitefuss and P. H. Williams (eds.). Springer-Verlag, Berlin, pp. 173–96.

118. Goodman, R. N., A. Király, and M. Zaitlin. 1967. The biochemistry and physiology of infectious plant disease. Van Nostrand Co., Princeton, NJ.

119. Goodman, R. N., and J. A. White. 1981. Xylem parenchyma plasmolysis and vessel wall disorientation are early signs of pathogenesis caused by *Erwinia amylovora*. Phytopathology 71:844–52.

120. Gowda, S. S., and R. N. Goodman. 1970. Movement and persistence of *Erwinia amylovora* in shoot, stem and root of apple. Plant Dis. Reptr. 54:576–80.

121. Grasso, S., P. Jones, and R. F. White. 1980. Inhibition of tobacco mosaic virus multiplication in tobacco protoplasts by the pokeweed inhibitor. Phytopath. Z. 98:53–58.

122. Grover, R. K. 1971. Participation of host exudate chemicals in appressorium formation by *Collectotrichum piperatum*. In: Ecology of leaf surface micro-organisms. T. F. Preece and C. H. Dickinson (eds.). Academic Press, London and New York, pp. 509–18.

123. Haas, J. H., and J. Rotem. 1976. *Pseudomonas lachrymans* absorption, survival, and infectivity following precision inoculation of leaves. Phytopathology 66:992- 97.

124. Haas, J. H., and J. Rotem. 1976. *Pseudomonas lachrymans* inoculum on infected cucumber leaves subject to dew- and rain-type wetting. Phytopathology 66:1219–23.

125. Haberlandt, G. 1914. Physiological plant anatomy. (Trans. M. Drummon.) Macmillan, London.

126. Hafiz, A. 1952. Basis of resistance in gram to *Mycosphaerella* blight. Phytopathology 42:422–24.

127. Hagen, K. S. 1966. Dependence of the olive fly, *Dacus oleae*, larvae on symbiosis with *Pseudomonas savastanoi* for the utilization of olive. Nature 209:423–24.

128. Halliwell, R. S., and W. S. Gazaway. 1975. Quantity of microinjected tobacco mosaic virus required for infection of single cultured tobacco cells. Virology 65: 583–87.

129. Harper, P. C., A. E. W. Boyd, and D. C. Graham. 1963. Growth cracking and bacterial soft rot in potato tubers. Plant Path. 12:139–42.

130. Harrison, B. D., and A. F. Murant (eds.). CMI/AAB.Descriptions of plant viruses. Commonw. Mycol. Inst. and Assn. Appl. Biol., Kew.

131. Hayashi, T. 1977. Fate of tobacco mosaic virus after entering the host cell. III. Partial uncoating. Microbiol. Immun. 21:317–24.

132. Heath, M. C. 1972. Ultrastructure of host and nonhost reaction to cowpea rust. Phytopathology 62:27–38.

133. Heath, M. C. 1974. Light and electron microscope studies of the interactions of host and non-host plants with cowpea rust—*Uromyces phaseoli* var. *vignae*. Physiol. Plant Path. 4:403–14.

134. Heath, M. C., and I. B. Heath. 1971. Ultrastructure of an immune and a susceptible reaction of cowpea leaves to rust infection. Physiol. Plant Path. 1:277–87.

135. Helms, K., and G. A. McIntyre. 1967. Light-induced susceptibility of *Phaseolus vulgaris* L. to tobacco mosaic virus infection. II. Daily variation in susceptibility. Virology 32:482–88.

136. Herridge, E. A., and D. E. Schlegel. 1962. Autoradiographic studies of tobacco mosaic virus inoculations on host and non-host species. Virology 18:517–23.

137. Hewitt, W. B. 1938. Leaf scar infection in relation to olive-knot disease. Hilgardia 12:41–71.

138. Hickman, C. J., and H. H. Ho. 1966. Behavior of zoospores in plant-pathogenic phycomycetes. Ann. Rev. Phytopath. 4:195–220.

139. Hildebrand, E. M. 1942. A micrurgical study of crown gall infection in tomato. J. Agr. Res. 65:45–59.

140. Hildebrand, E. M. 1958. The mechanism of plant virus inoculation. Phytopathology 48:262–63.(Abstr.)

141. Hildebrand, E. M., and L. H. MacDaniels. 1935. Modes of entry of *Erwinia amylovora* into the flowers of the principle pome fruits. Phytopathology 25:20. (Abstr.)

142. Hirai, T. 1977. Action of antiviral agents. In: Plant disease: An advanced treatise. J. G. Horsfall and E. B. Cowling (eds.). Academic Press, New York, 1: 285–300.

143. Hirai, T., and A. Hirai. 1964. Tobacco mosaic virus: Cytological evidence of the synthesis in the nucleus. Science 145:589–91.

144. Holoubek, V. 1964. Effect of tobacco mosaic virus protein on tobacco mosaic virus infectivity. Nature 203:499–501.

145. Hooker, W. J., and W. S. Kim. 1962. Inhibitors of potato virus X in potato leaves with different types of virus resistance. Phytopathology 52:688–93.

146. Huang, P.-Y. 1974. Ultrastructural modification by and pathogenicity of *Erwinia amylovora* in apple tissues. Ph.D. Diss., Dept. Plant Pathology, Univ. of Missouri-Columbia.

147. Huang, P.-Y., J. S. Huang, and R. N. Goodman. 1975. Resistance mechanisms of apple shoots to an avirulent strain of *Erwinia amylovora*. Physiol. Plant Path. 6: 283–87.

148. Ivanoff, S. S. 1933. Stewart's wilt disease of corn with emphasis on the life history of *Phytomonas stewarti* in relation to pathogenesis. J. Agr. Res. 47:749- 70.

149. Ivanoff, S. S., and S. W. Keitt. 1937. The occurrence of aerial strands on blossoms, fruits and shoots blighted by *Erwinia amylovora*. Phytopathology 27: 702–9.

150. Jackson, M. B., I. B. Morrow, and D. J. Osborne. 1972. Abscission and dehiscence in the squirting cucumber, *Ecballium elaterium*: Regulation by ethylene. Can. J. Bot. 50:1465–71.

151. Jansson, P.-E., B. Lundberg, and H. Ljunggren. 1979. Structural studies of the *Rhizobium trifolii* extracellular polysaccharide. Carbohyd. Res. 75:207–20.

152. Jaspars, E. M. J. 1974. Plant viruses with a multipartite genome. Adv. Virus Res. 19:37–49.

153. Jedlinski, H. 1956. Plant virus infection in relation to the interval between wounding and inoculation. Phytopathology 46:673–76.

154. Jedlinski, H. 1964. Initial infection processes by certain mechanically transmitted plant viruses. Virology 22:331–41.

155. Kado, C. I. 1963. Increase in plant virus infection by magnesium in the presence of phosphate. Nature 197:925–26.

156. Kado, C. I. 1972. Mechanical and biological inoculation principles. In: Principles and techniques in plant virology. C. I. Kado and H. O. Agrawal (eds.). Van Nostrand/Rheinhold, New York, pp. 13–31.

157. Kassanis, B. 1952. Some effects of high temperature on the susceptibility of plants to infection with viruses. Ann. Appl. Biol. 39:358–69.

158. Kassanis, B. 1960. Comparison of the early stages of infection by intact and phenol-disrupted tobacco necrosis virus. Virology 10:353–69.

159. Kassanis, B., and R. H. Kenten. 1978. Inactivation and uncoating of TMV on the surface and in the intercellular spaces of leaves. Phytopath. Z. 91:329–39.

160. Keil, H. L., and T. van der Zwet. 1972. Recovery of *Erwinia amylovora* from symptomless stems and shoots of Jonathan apple and Bartlett pear trees. Phytopathology 62:39–42.

161. Keil, H. L., and T. van der Zwet. 1972. Aerial strands of *Erwinia amylovora*: Structure and enhanced production by pesticide oil. Phytopathology 62:355–61.

162. Keitt, G. W., and D. M. Boone. 1954. Induction and inheritance of mutant characters in *Venturia inaequalis* in relation to its pathogenicity. Phytopathology 44:362–70.

163. Kelman, A., and H. Hruschka. 1973. The role of motility and aerotaxis in the selective increase of avirulent bacteria in still broth cultures of *Pseudomonas solanacearum*. J. Gen. Microbiol. 76:177–83.

164. Kern, H. 1952. Über die Beziehungen zwischen dem Alkaloidgehalt verschiedener Tomatensorten und ihrer Resistenz gegen *Fusarium lycopersici*. Phytopath. Z. 19:351–82.

165. Kerr, A. 1956. Some interactions between plant roots and pathogenic soil fungi. Aust. J. Biol. Sci. 9:45–52.

166. Kerr, A., and N. T. Flentje. 1956. Host infection in *Pellicularia filamentosa* controlled by chemical stimuli. Nature 179:204–5.

167. Khew, K. L., and G. A. Zentmyer. 1974. Electrostatic responses of zoospores of seven species of *Phytophthora*. Phytopathology 64:500–507.

168. Kiho, Y. 1970. Polysomes containing infecting viral genome in tobacco leaves infected with tobacco mosaic virus. Jap. J. Microbiol. 14:291–302.

169. Kiho, Y. 1972. Polycistronic translation of plant viral ribonucleic acid. Jap. J. Microbiol. 16:259–67.

170. Kiho, Y., H. Machida, and N. Oshima. 1972. Mechanism determining the host specificity of tobacco mosaic virus. I. Formation of polysomes containing infecting viral genome in various plants. Jap. J. Microbiol. 16:451–59.

171. Kimmins, W. C. 1967. The effect of darkening on the susceptibility of French bean to tobacco necrosis virus. Can. J. Bot. 45:543–53.

172. Kimmins, W. C., and R. E. Litz. 1967. The effect of leaf water balance on the susceptibility of French bean leaves to tobacco necrosis virus. Can. J. Bot. 45: 2115–18.

173. Király, Z. 1964. Effect of nitrogen fertilization on phenol metabolism and stem rust susceptibility of wheat. Phytopath. Z. 51:252–61.

174. Király, Z. 1976. Plant disease resistance as influenced by biochemical effects of nutrients in fertilizers. In: Fertilizer use and plant health. Proc. 12th Coll. Int. Potash Inst. Izmir., pp. 33–46.

175. Király, Z., B. Barna, and T. Érsek. 1972. Hypersensitivity as a consequence, not the cause, of plant resistance to infection. Nature 239:456–58.

176. Kleczkowski, A. 1950. Interpreting relationships between concentration of plant viruses and number of local lesions. J. Gen. Microbiol. 4:53–69.

177. Klein, R. M. 1953. The probable chemical nature of crown-gall tumor inducing principle. Amer. J. Bot. 40:597–99.

178. Knosel, D., and E. D. Garber. 1967. Pektolytische und cellulolytische enzyme bei *Xanthomonas campestris* (Pammel) Dowson. Phytopath. Z. 59:194–202.

179. Kolattukudy, P. E., and W. Köller. 1983. Fungal penetration of first line defensive barriers of plants. In: Biochemical plant pathology. J. A. Callow (ed.). Wiley, New York, pp. 79–100.

180. Köller, W., C. R. Allan, and P. E. Kolattukudy. 1982. Role of cutinase and cell wall degrading enzymes in infection of *Pisum sativum* by *Fusarium solani* f. sp. *pisi*. Physiol. Plant Path. 20:47–60.

181. Kontaxis, D. G., and D. E. Schlegel. 1962. Basal septa of broken trichomes in *Nicotiana* as possible infection sites for tobacco mosaic virus. Virology 16:244–47.

182. Kovács, A., and É. Szeoke. 1956. Die phytopathologische Bedeutung der kutikularen Exkretion. Phytopath. Z. 27:335–49.

183. Kurstak, E. (ed.). 1981. Handbook of plant virus infections. Comparative diagnosis. Elsevier/North Holland, Amsterdam.

184. Kurtz-Fritsch, C., and L. Hirth. 1972. Uncoating of two spherical plant viruses. Virology 47:385–96.

185. Layne, R. E. C. 1967. Foliar trichomes and their importance as infection sites for *Corynebacterium michiganense* on tomato. Phytopathology 57:981–85.

186. Leben, C. 1965. Epiphytic microorganisms in relation to plant disease. Ann. Rev. Phytopath. 3:209–30.

187. Leben, C., and G. C. Daft. 1967. Population variations of epiphytic bacteria. Can. J. Microbiol. 13:1151–56.

188. Leben, C., V. Rusch, and A. F. Schmitthenner. 1968. The colonization of soybean buds by *Pseudomonas glycinea* and other bacteria. Phytopathology 58:1677–81.

189. Lesemann, D. E., and W. H. Fuchs. 1970. Elektronenmikroskopische Untersuchungen über die Vorbereitung der Infektion der einzystierten Zoosporen von *Olpidium brassicae*. Arch. Mikrobiol. 71:9–19.

190. Lesemann, D. E., and W. H. Fuchs. 1970. Die Ultrastruktur des Penetrationsvorgänges von *Olpidium brassicae* an Kohlrabi-Wurzlein. Arch. Mikrobiol. 71:20–30.

191. Lewis, S., and R. N. Goodman. 1965. Mode of penetration and movement of fireblight bacteria in apple leaf and stem tissue. Phytopathology 55:719–23.

192. Lippincott, B. B., and J. A. Lippincott. 1969. Bacterial attachment to a specific wound site as an essential stage in tumor initiation by *Agrobacterium tumefaciens*. J. Bact. 97:620–28.

193. Littlefield, L. J. 1973. Histological evidence for diverse mechanisms of resistance to flax rust, *Melampsora lini* (Ehrenb.) Lev. Physiol. Plant Path. 3: 241–47.

194. Litz, R. E., and W. C. Kimmins. 1971. Interpretation of ectodesmata in relation to susceptibility to plant viruses. Can. J. Bot. 49:2011–14.

195. Loebenstein, G. 1972. Inhibition, interference and acquired resistance during infection. In: Principles and techniques in plant virology. C. I. Kado and H. O. Agrawal (eds.). Van Nostrand/Rheinhold, New York, pp. 32–61.

196. Lupton, F. G. H. 1956. Resistance mechanisms of species of *Triticum* and *Aegilops* and of amphidiploids between them to *Erysiphe graminis* D.C. Trans. Brit. Mycol. Soc. 39:51–59.

197. Machida, H., and Y. Kiho. 1970. In vivo uncoating of tobacco mosaic virus. II. Complete uncoating to TMV-RNA. Jap. J. Microbiol. 14:441–49.

198. MacKenzie, D. R. 1981. Association of potato early blight, nitrogen fertilizer rate, and potato yield. Plant Dis. 65:575–77.

199. Maclean, D. J., J. A. Sargent, I. C. Tommerup, and D. S. Ingram. 1974. Hypersensitivity as the primary event in resistance to fungal parasites. Nature 249:186–87.

200. Maramorosch, K. 1977. The atlas of insect and plant viruses. Ultrastructure in biological systems. A. J. Dalton and F. Haguenau (eds.). Academic Press, New York, Vol. 8.

201. Martin, T. J., and A. H. Ellingboe. 1973. ^{32}P uptake by wheat seedlings after inoculation and before penetration by *Erysiphe graminis* f. sp. *tritici*. 2nd Int. Cong. Plant Path., Minneapolis, MN, Sept. 5–12, #1007. (Abstr.)

202. Matta, A. 1972. Microbial penetration and immunization of uncongenial host plants. Ann. Rev. Phytopath. 9:387–410.

203. Matthee, F. N., and R. H. Daines. 1969. The influence of nutrition on susceptibility of peach foliage to water congestion and infection by *Xanthomonas pruni*. Phytopathology 59:285–87.

204. Matthews, R. E. F. 1953. Factors affecting the production of local lesions by plant viruses. I. Effect of time of day of inoculation. Ann. Appl. Biol. 40:377–83.

205. Matthews, R. E. F. 1981. Plant virology. 2d ed. Academic Press, New York.

206. Matthysse, A. G., K. V. Holmes, and R. A. G. Gurlitz. 1982. Binding of *Agrobacterium tumefaciens* to carrot protoplasts. Physiol. Plant Path. 20:27–33.

207. Maule, A. J., M. I. Boulton, and K. R. Wood. 1980. An improved method for the infection of cucumber leaf protoplasts with cucumber mosaic virus. Phytopath. Z. 97:118–26.

208. Mazzucchi, U. 1983. Recognition of bacteria by plants. In: Biochemical plant pathology. C. A. Callow (ed.). Wiley, New York, pp. 299–324.

209. McKeen, W. E. 1974. Mode of penetration of epidermal cell walls of *Vicia faba* by *Botrytis cinerea*. Phytopathology 64:461–67.

210. McKeen, W. E., and S. R. Rimmer. 1973. Initial penetration process in powdery mildew infection of susceptible barley leaves. Phytopathology 63:1049–53.

211. Mew, T. W., and B. W. Kennedy. 1971. Growth of *Pseudomonas glycinea* on the surface of soybean leaves. Phytopathology 61:715–16.

212. Miles, W. G., R. H. Daines, and J. W. Rue. 1977. Presymptomatic egress of *Xanthomonas pruni* from infected peach leaves. Phytopathology 67:895–97.

213. Mitchell, S., and M. Shaw. 1969. The nucleolus of wheat stem rust uredospores. Can. J. Bot. 47:1877–89.

214. Moraes, W. B. C., J. R. July, A. P. C. Alba, and A. R. Oliveira. 1974. The inhibitory activity of extracts of *Abutilon striatum* leaves on plant virus infection. II. Mechanism of inhibition. Phytopath. Z. 81:145–52.

215. Mort, A. J., and W. D. Bauer. 1980. Composition of the capsular and extracellular polysaccharides of *Rhizobium japonicum*. Plant Physiol. 66:158–63.

216. Mount, M. S., and R. S. Slesinski. 1971. Characterization of primary development of powdery mildew. In: Ecology of leaf surface microorganisms. T. F. Preece and C. H. Dickinson (eds.). Academic Press, London and New York, pp. 301–22.

217. Nasuno, S., and M. P. Starr. 1967. Polygalacturonic acid trans-eliminase of *Xanthomonas campestris*. Biochem. J. 104:178–85.

218. Novikov, V. K., and J. G. Atabekov. 1970. A study of the mechanisms controlling the host range of plant viruses. I. Virus-specific receptors of *Chenopodium amaranticolor*. Virology 41:101–7.

219. Ohashi, Y., and T. Shimomura. 1971. Necrotic lesion induced by heat treatment on leaves of systemic host infected with tobacco mosaic virus. Ann. Phytopath. Soc. Jap. 37:22–28.

220. Ohashi, Y., and T. Shimomura. 1971. Induction of local lesion formation on leaves systemically infected with virus by a brief heat or cold treatment. Ann. Phytopath. Soc. Jap. 37:211–14.

221. Ohyama, K., L. E. Pelcher, A. Schaeffer, and L. C. Fowke. 1979. In vitro binding of *Agrobacterium tumefaciens* to plant cells from suspension culture. Plant Physiol. 63:382–87.

222. Osborne, D. J., M. B. Jackson, and B. V. Millborrow. 1972. Physiological properties of abscission accelerator from senescent leaves. Nature 240:98–101.

223. Otta, J. D., and H. English. 1970. Epidemiology of the bacterial canker disease of French prune. Plant Dis. Reptr. 54:332–36.

224. Panopoulos, N. J., and M. N. Schroth. 1974. Role of flagellar motility in the invasion of bean leaves by *Pseudomonas phaseolicola*. Phytopathology 64:1389-97.

225. Panzer, J. D. 1957. Osmotic pressure and plant virus local lesions. Phytopathology 47:337–41.

226. Peyton, G. A., and C. C. Bowen. 1963. The host-parasite interface of *Peronospora manshurica* on *Glycine max*. Amer. J. Bot. 50:787–97.

227. Plurad, S. B., R. N. Goodman, and W. R. Enns. 1965. Persistence of *Erwinia amylovora* in the apple aphid (*Aphis pomi*, DeGeer), a probable vector. Nature 205:206.

228. Politis, D. J., and R. N. Goodman. 1980. Fine structure of extracellular polysaccharide of *Erwinia amylovora*. Appl. Environ. Microbiol. 40:596–607.

229. Preece, T. F., and W. C. Wong. 1982. Quantitative and scanning electron microscope observations on the attachment of *Pseudomonas tolasii* and other bacteria to the surface of *Agaricus bisporus*. Physiol. Plant Path. 21:251–57.

230. Price, W. C. 1964. Strains, mutation, acquired immunity and interference. In: Plant virology. M. K. Corbett and H. D. Sisler (eds.). Univ. of Florida Press, Gainesville, pp. 93–117.

231. Pueppke, S. G. 1983. Soybean lectin: Does it have an essential role in the *Rhizobium*-soybean symbiosis? In: Chemical taxonomy, molecular biology, and function of plant lectins. M. Etzzler and I. Goldstein (eds.). Alan R. Liss, New York, pp. 225–36.

232. Ragetli, H. W. J., and M. Weintraub. 1962. Purification and characteristics of a virus inhibitor from *Dianthus caryophyllus* L. II. Characterization and mode of action. Virology 18:241–48.

233. Ramakrishnan, L., and R. C. Staples. 1970. Evidence for a template RNA in resting uredospores of the bean rust fungus. Contrib. Boyce Thompson Inst. 24: 197–202.

234. Randles, J. W., and R. I. B. Francki. 1972. Infectious nucleocapsid particles of lettuce necrotic yellows virus with RNA-dependent RNA polymerase activity. Virology 50:297–300.

235. Raymundo, A. K., and S. M. Ries. 1980. Chemotaxis of *Erwinia amylovora*. Phytopathology 70:1066–69.

236. Raymundo, A. K., and S. M. Ries. 1980. Motility of *Erwinia amylovora*. Phytopathology 70:1062–65.

237. Reddi, K. K. 1966. Studies on the formation of tobacco mosaic virus ribonucleic acid. VII. Fate of tobacco mosaic virus after entering the host cell. Proc. Nat. Acad. Sci. 55:593–98.

238. Roberts, D. A., and W. C. Price. 1967. Infection of apparently uninjured leaves of bean by the viruses of tobacco necrosis and southern bean mosaic. Virology 33: 542–45.

239. Rohringer, R., N. K. Howes, W. K. Kim, and D. J. Samborski. 1974. Evidence for a gene-specific RNA determining resistance in wheat to stem rust. Nature 249:585–88.

240. Rolfs, F. M. 1915. A bacterial disease of stone fruits. Cornell Univ. Agr. Exp. Sta. Mem. 8:377–436.

241. Romeiro, R., A. Karr, and R. N. Goodman. 1981. Isolation of a factor from apple that agglutinates *Erwinia amylovora*. Plant Physiol. 68:772–77.

242. Romeiro, R., A. Karr, and R. N. Goodman. 1981. *Erwinia amylovora* cell wall receptor for apple agglutinin. Physiol. Plant Path. 19:383–90.

243. Rosen, H. R. 1938. Life span and morphology of fireblight bacteria as influenced by relative humidity, temperature and nutrition. J. Agr. Res. 56:329- 58.

244. Royle, D. J., and G. G. Thomas. 1973. Factors affecting zoospore responses towards stomata in hop mildew (*Pseudoperonospora humuli*) including some comparisons with grapevine downy mildew (*Plasmopara viticola*). Physiol. Plant Path. 3:405–17.

245. Sadasivan, T. S. 1940. A quantitative study of the interaction of viruses in plants. Ann. Appl. Biol. 27:359–67.

246. Santilli, V., J. Piacitelli, and J. H. Wu. 1961. The effect of tobacco mosaic virus protein on virus incubation period and infectivity. Virology 14:109–23.

247. Sargent, J. A., I. C. Tommerup, and D. S. Ingram. 1973. The penetration of a susceptible lettuce variety by the downy mildew fungus *Bremia lactucae* Regel. Physiol. Plant Path. 3:231–39.

248. Scheffer, R. P., and J. C. Walker. 1953. The physiology of *Fusarium* wilt of tomato. Phytopathology 43:116–25.

249. Scheffer, R. P., and O. C. Yoder. 1972. Host-specific toxins and selective toxicity. In: Phytotoxins in plant diseases. R. K. S. Wood, A. Ballio, and A. Graniti (eds.). Academic Press, New York, pp. 251–69.

250. Schmit, J. 1978. Microscopic study of early stages of infection by *Pseudomonas solanacearum* E.F.S. on "in vitro" grown tomato seedlings. Proc. 4th Int. Plant Path. Bact. Conf. (Angers), 2:841–56.

251. Schneider, R. W., and R. G. Grogan. 1977. Tomato leaf trichomes, a habitat for resident populations of *Pseudomonas tomato*. Phytopathology 67:898–902.

252. Schonherr, J., and M. J. Bukovac. 1970. Preferential polar

pathways in the cuticle and their relation to ectodesmata. Planta 92:189–201.

253. Schramm, G., and R. Engler. 1958. The latent period after infection with tobacco mosaic virus and virus nucleic acid. Nature 181:916–17.

254. Scott, R. M., M. R. Schroeder, and R. M. Turrell. 1948. Development, cell shape, suberization of internal surface, and abscission in the leaf of the Valencia orange, *Citrus sinensis*. Bot. Gaz. 109:381–411.

255. Sequeira, L., and T. L. Graham. 1977. Agglutination of avirulent strains of *Pseudomonas solanacearum* by potato lectin. Physiol. Plant Path. 11:43–54.

256. Sharon, N., and H. Lis. 1981. Glycoproteins: Research booming on long-ignored, ubiquitous compounds. Chem. Eng. News (March 30):21–28, 36–44.

257. Sharp, E. L., and F. G. Smith. 1952. The influence of pH and zinc on vesicle formation in *Puccinia coronata avenae* Corda. Phytopathology 42:581–82.

258. Shaw, J. G. 1967. In vivo removal of protein from tobacco mosaic virus after inoculation of tobacco leaves. Virology 31:665–75.

259. Shaw, J. G. 1972. Effect of poly-L-ornithine on the attachment of tobacco mosaic virus to tobacco leaves and on the uncoating of viral RNA. Virology 48: 380–85.

260. Shaw, J. G. 1973. In vivo removal of protein from tobacco mosaic virus after inoculation of tobacco leaves. III. Studies on the location on virus particles for the initial removal of protein. Virology 53:337–42.

261. Shaw, L. 1935. Intercellular humidity in relation to fireblight susceptibility in apple and pear. Cornell Univ. Agr. Exp. Sta. Mem. 181:3–40.

262. Shaw, M., and M. S. Manocha. 1965. The physiology of host-parasite relations. XV. Fine structure in rust-infected wheat leaves. Can. J. Bot. 43:1285–92.

263. Shimomura, T. 1977. The role of epidermis in local lesion formation on *Nicotiana glutinosa* leaves caused by tobacco mosaic virus. Ann. Phytopath. Soc. Jap. 43: 159–66.

264. Shimomura, T. 1977. Effect of pre-inoculation exposure of plants to high and low temperature on their susceptibility to tobacco mosaic virus. Ann. Phyto path. Soc. Jap. 43:164–74.

265. Siegel, A. 1959. Mutual exclusion of strains of tobacco mosaic virus. Virology 8:470–77.

266. Siegel, A., W. Ginoza, and S. G. Wildman. 1957. The early events of infection with tobacco mosaic virus nucleic acid. Virology 3:554–59.

267. Siegel, A., and M. Zaitlin. 1964. Infection process in plant virus diseases. Ann. Rev. Phytopath. 2:179–202.

268. Sill, W. H., Jr., and J. C. Walker. 1952. A virus inhibitor in cucumber in relation to mosaic resistance. Phytopathology 42:349–52.

269. Skipp, R. A., D. E. Harder, and D. J. Samborski. 1974. Electron microscopy studies on infection of resistant (Sr_6 gene) and susceptible near isogenic wheat lines by *Puccinia graminis* f. sp. *tritici*. Can. J. Bot. 52:2615–21.

270. Smith, E. F. 1911. Bacteria in relation to plant diseases. Carnegie Inst., Washington, DC.

271. Smith, K. M. 1972. A textbook of plant virus diseases. 3d ed. Longman, London.

272. Smith, M. A., and B. W. Kennedy. 1970. Effect of light on reactions of soybeans to *Pseudomonas glycinea*. Phytopathology 60:723–25.

273. Stahmann, M. A., and S. S. Gothoskhar. 1958. The inhibition of the infectivity of tobacco mosaic virus by some synthetic and natural polyelectrolytes. Phytopathology 48:362–65.

274. Stanbridge, B., J. L. Gay, and R. K. S. Wood. 1971. Gross and fine structural changes in *Erysiphe graminis* and barley before

and during infection. In: Ecology of leaf surface micro-organisms. T. F. Preece and C. H. Dickinson (eds.). Academic Press, New York, pp. 367–79.

275. Staples, R. C. 1968. Protein synthesis by uredospores of the bean rust fungus. In: Physiological and biochemical aspects of host-pathogen interactions. A. Fuchs and O. M. van Andel (eds.). Neth. J. Plant Path. 74(1):25–36. (suppl.)

276. Staples, R. C., Z. Yaniv, L. Ramakrishnan, and J. Lipetz. 1971. Properties of ribosomes from germinating uredospores. In: Morphological and biochemical events in plant-parasite interaction. S. Akai and S. Ouchi (eds.). Phytopath. Soc. Jap., Tokyo, pp. 59–90.

277. Steere, R. L. 1955. Concepts and problems concerning the assay of plant viruses. Phytopathology 45:196–208.

278. Stevens, F. L., W. A. Ruth, and C. S. Spooner. 1918. Pear blight wind borne. Science 48:449–50.

279. Strider, D. L. 1967. Survival studies with the tomato bacterial canker organism. Phytopathology 57:1067–71.

280. Suhayda, C. H., and R. N. Goodman. 1981. Early proliferation and migration and subsequent xylem occlusion by *Erwinia amylovora* and the fate of its extracellular polysacchride (EPS) in apple shoots. Phytopathology 71:697–707.

281. Takagi, Y. 1975. Difference in susceptibility of upper and lower surfaces of leaf to tobacco mosaic virus infection. Ann. Phytopath. Soc. Jap. 41:400–404.

282. Takahashi, T. 1972. Studies on viral pathogenesis in plant hosts. III. Leaf age dependent susceptibility to tobacco mosaic virus infection in 'Samsun NN' and 'Samsun' tobacco plants. Phytopath. Z. 75:140–55.

283. Takahashi, T. 1974. Studies on viral pathogenesis in plant hosts. VI. The rate of primary lesion growth in the leaves of 'Samsun NN' tobacco to tobacco mosaic virus. Phytopath. Z. 79:53–66.

284. Takebe, I. 1977. Protoplasts in the study of plant virus replication. In: Comprehensive virology. H. H. Fraenkel-Conrat and R. R. Wagner (eds.). Plenum Press, New York, 11:237–83.

285. Thatcher, F. S. 1942. Further studies of osmotic and permeability relations in parasitism. Can. J. Res. 20:283–311.

286. Thayer, P. L. 1965. Temperature effect on growth and pathogenicity to celery of *Pseudomonas apii* and *P. cichorii*. Phytopathology 55:1365–67.

287. Thomas, P. E., and R. W. Fulton. 1968. Correlation of ectodesmata number with nonspecific resistance to initial virus infection. Virology 34:459–69.

288. Thomson, S. V., F. J. Hills, E. D. Whitney, and M. N. Schroth. 1981. Sugar and root yield of sugar beets as affected by bacterial vascular necrosis and rot, nitrogen fertilization and plant spacing. Phytopathology 71:605–8.

289. Timonin, M. I. 1941. The interaction of higher plants and microorganisms. III. Effect of by-products of plant growth on activity of fungi and actinomycetes. Soil Sci. 52:395–413.

290. Tinsley, T. W. 1953. The effects of varying the water supply of plants on their susceptibility to infection with viruses. Ann. Appl. Biol. 40:750-60.

291. Tomlinson, J. A., V. M. Walker, T. H. Flewett, and G. R. Barclay. 1974. The inhibition of infection by cucumber mosaic virus and influenza virus by extracts from *Phytolacca americana*. J. Gen. Virol. 22:225–32.

292. Topps, J. H., and R. L. Wain. 1957. Fungistatic properties of leaf exudates. Nature 179:652–53.

293. Tsien, H. C., and E. L. Schmidt. 1980. Accumulation of soybean lectin polysaccharide during growth of *Rhizobium japonicum* as determined by hemagglutination inhibition assay. Appl. Environ. Microbiol. 39:1100–1104.

294. Turner, E. M. C. 1961. An enzymic basis for pathogenic specificity in *Ophiobolus graminis*. J. Exp. Bot. 12:169–75.

295. Tyihak, E., and E. Balázs. 1976. Antagonistic effect on TMV infectivity between poly-L-lysine and poly-L-arginine. Acta Phytopath. Hung. 11:11–16.

296. Vakili, N. G. 1967. Importance of wounds in bacterial spot (*Xanthomonas vesicatoria*) of tomatoes in the field. Phytopathology 57:1099–1103.

297. van Kammen, A., D. Noordam, and T. H. Thung. 1961. The mechanism of inhibition of infection with tobacco mosaic virus by an inhibitor from carnation sap. Virology 14:100–108.

298. Van Larbecke, N., G. Engler, M. Holsters, S. Van den Elsacker, I. Zaenen, R. A. Schilperoort, and J. Schell. 1974. Large plasmid in *Agrobacterium tumefaciens* essential for crown gall-inducing ability. Nature 252:169–70.

299. van Loon, L. C. 1976. Systemic acquired resistance, peroxidase activity and lesion size in tobacco reacting hypersensitively to tobacco mosaic virus. Physiol. Plant Path. 8: 231–42.

300. van Vloten-Doting, L., and E. M. J. Jaspars. 1977. Plant covirus systems: Three-component systems. In: Comprehensive virology. H. H. Fraenkel-Conrat and R. R. Wagner (eds.). Plenum Press, New York, 11:1–53.

301. van Vloten-Doting, L., J. Kruseman, and E. M. J. Jaspars. 1968. The biological function and mutual dependence of bottom component and top component *a* of alfalfa mosaic virus. Virology 34:728–37.

302. Waite, M. B. 1895. The cause and prevention of pear blight. Yearbook of the USDA, pp. 295–300.

303. Walker, H. L., and T. P. Pirone. 197. Particle numbers asociated with mechancal and aphid transmission of some plant viruses. Phytopathology 62:1283–88.

304. Walker, H. L., and T. P. Pirone. 1972. Number of TMV particles required to infect locally or systemically susceptible tobacco cultivars. J. Gen. Virol. 17:241–43.

305. Walker, J. C., K. P. Link, and H. R. Angell. 1929. Chemical aspects of disease resistance in the onion. Proc. Nat. Acad. Sci. 15:845–50.

306. Wallis, F. M., F. H. J. Rijkenberg, J. J. Joubert, and M. M. Martin. 1973. Ultrastructural histopathology of cabbage leaves infected with *Xanthomonas campestris*. Physiol. Plant Path. 3:371–78.

307. Watts, J. W., J. R. O. Dawson, and J. M. King. 1981. The mechanism of entry of viruses into protoplasts. In: Adhesion and microorganism pathogenicity. K. Elliott, M. O'Connor, and J. Wheeler (eds.). CIBA Sym. 80, Pitman Medical, pp. 56–71.

308. Weststeijn, E. A. 1976. Peroxidase activity in leaves of *Nicotiana tabacum* var. Xanthi nc. before and after infection with tobacco mosaic virus. Physiol. Plant Path. 8:63–71.

309. Whatley, M. H., J. S. Bodwin, B. B. Lippincott, and J. Lippincott. 1976. Role for *Agrobacterium* cell envelope lipopolysaccharide in infection site attachment. Infect. Immu. 13:1080–83.

310. Williams, P. H., and S. McNabola. 1970. Fine structure of the host-parasite interface of *Plasmodiophora brassicae* in cabbage. Phytopathology 60:1557–61.

311. Wiltshire, G. H. 1956. The effect of darkening on the susceptibility of plants to infection with viruses. I. Relation to changes in some organic acids in the French bean. Ann. Appl. Biol. 44:233–48.

312. Wood, K. R., M. I. Boulton, and A. J. Maule. 1980. Application of protoplasts in plant virus research. In: Plant cell cultures: Results and perspectives. F. Sala, B. Parisi, R. Cella, and O. Ciferri (eds.). Elsevier, Amsterdam, pp. 405–10.

313. Wood, R. K. S. 1960. Chemical ability to breach the host barriers. In: Plant pathology: An advanced treatise. J. G. Horsfall and A. E. Dimond (eds.). Academic Press, New York, 2:233–72.

314. Wu, J. H., L. M. Blakely, and J. E. Dimitman. 1969. Inactivation of a host resistance mechanism as an explanation for heat activation of TMV-infected bean leaves. Virology 37:658–66.

315. Wu, J. H., and I. Rappaport. 1961. Kinetic study of heat inactivation of tobacco mosaic virus infected centers and potentially infectible sites on *Nicotiana glutinosa*. Phytopathology 51:823–26.

316. Wyatt, S. D., and R J. Shepherd. 1969. Isolation and characterisation of a virus inhibitor from *Phytolacca americana*. Phytopathology 59:1787–94.

317. Wynn, W. K. 1976. Appressorium formation over stomates by the bean rust fungus: Response to a surface contact stimulus. Phytopathology 66:136–46.

318. Yarwood, C. E. 1952. The phosphate effect in plant virus inoculations. Phytopathology 42:137–43.

319. Yarwood, C. E. 1955. Deleterious effects of water in plant virus inoculations. Virology 1:268–85.

320. Yarwood, C. E. 1958. Heat activation of virus infections. Phytopathology 48: 39–46.

321. Yarwood, C. E. 1959. Predisposition. In: Plant pathology: An advanced treatise. J. G. Horsfall and A. E. Dimond (eds.). Academic Press, New York, 2:521–62.

322. Yarwood, C.E. 1959. Virus susceptibility increased by soaking bean leaves in water. Plant Dis. Reptr. 43:841–44.

323. Yarwood, C. E. 1966. Bentonite aids virus transmission. Virology 28:459–62.

324. Yarwood, C. E. 1968. Sequence of supplements in virus inoculations. Phytopathology 58:132–36.

325. Yarwood, C. E. 1969. Sulfite in plant virus inoculations. Virology 39:74–78.

326. Yarwood, C. E. 1973. Quick drying versus washing in virus inoculations. Phytopathology 63:72–76.

327. Yarwood, C. E., and R. W. Fulton. 1967. Mechanical transmission of plant viruses. In: Methods in virology. K. Maramorosch and H. Koprowski (eds.). Academic Press, New York, 1:237–66.

328. Yunis, H., U. Bashan, Y. Okon, and Y. Henis. 1980. Weather dependence, yield losses, and control of bacterial speck of tomato caused by *Pseudomonas tomato*. Plant Dis. 64:937–39.

329. Zech, H. 1952. Untersuchungen uber den Infectionvorgang und die Wanderung des Tabakmosaikvirus im Pflanzenkorper. Planta 40:461–514.

330. Zhuravlev, Y. N., N. F. Pisetskaya, Z. S. Yudakova, and V. G. Reifman. 1976. Uptake of labeled tobacco mosaic virus by tobacco protoplasts in the presence of metabolic inhibitors and at low temperature. Acta Virol. 20:435–38.

2

Photosynthesis

The Healthy Plant

Photosynthesis is the process whereby solar energy is converted into chemical energy, in the form of ATP and NADPH, which in turn are used in cellular biosynthetic pathways. Photosynthetic cells of higher plants use water as a hydrogen donor to reduce carbon dioxide, with release of oxygen and formation of carbohydrate, a process that is represented below in equation 1. Where the direct product of photosynthesis is glucose, molecular oxygen is formed from H_2O, rather than CO_2.

$$6\ CO_2 + 6\ H_2O \xrightarrow[\text{chloroplast}]{\text{light}} C_6H_{12}O_6 + 6\ O_2 \qquad (1)$$

Although not apparent from this simplified summary of events, there are two quite distinct and separate phases in photosynthesis, loosely referred to as the "light" and carbon-reduction reactions. During the "light" reactions, solar energy is absorbed and, through the cooperative action of systems of pigments, enzymes, and electron-carriers, is converted to ATP and NADPH, and oxygen is evolved. This chemical energy is then utilized in the carbon-reduction reactions to reduce CO_2 and form hexoses and, indirectly, many other essential organic compounds, whose biosynthesis varies with plant species, physiological status, and environmental conditions. Light is obligatory for key steps of the "light" reactions, and there is now evidence that light-generated regulatory signals are required to activate latent enzymes of the carbon-reduction reactions.

The rate of photosynthesis can be assessed by estimating either O_2 evolution, CO_2 uptake, or net assimilation rate, the accumulation of dry matter/unit leaf area/unit time. Clearly, however, account must be taken of carbohydrate catabolism, with concomitant O_2 consumption, by respiratory pathways.

The following constitutes a summary of the principal photosynthetic pathways. The reader should turn to alternative sources for more detailed information (e.g. 61).

The Chloroplast

In higher plants, photosynthesis is performed within the chloroplast, and carefully isolated organelles can also function effectively in this respect. A higher plant mesophyll cell may contain 15–50 mature chloroplasts, discoidal in shape and ca. 3–10 μm long by 1–2 μm wide. Although highly organized and complex (6, 116, 134), they may be divided into three components: the envelope, internal lamellar system, and stroma (Fig. 2–1). The *envelope* is composed of two membranes, each ca. 5 nm in thickness, separated by an intermembrane region ca. 10–20 nm wide. In common with other biological membranes (121), chloroplast membranes contain protein molecules within the lipid bilayer, and, in addition to providing structural integrity, they regulate the flow of metabolites into and out of the chloroplast. The envelope also participates in the biogenesis of the *internal lamellar* system, which consists of an extensive invagination of the inner membrane. The lamellae become paired, to form discoidal, saclike structures, the *thylakoids*, which may be stacked in groups to form *grana*. The lamellar region between the grana is the *intergranal* or *stroma lamella*. The lamellae, composed of protein and lipid, are the sites of the photochemical reactions and contain all the necessary enzymes, photosynthetic pigments, electron carriers, and cofactors (121).

Many plants of tropical origin have chloroplasts with two distinct types of lamellar organization. While the mesophyll cell chloroplasts contain regular stacks of grana, grana are absent from the bundle sheath cell chloroplasts, the internal lamellar system consisting only of parallel arrays of anastomosing tubules (82). This difference in grana content is reflected in a different emphasis in utilization of the two principal photosystems. Plants of this type, with dimorphic chloroplasts, possess an additional pathway for CO_2 utilization, the C_4 pathway (see also below), and are commonly referred to as C_4 plants (although not all C_4 plants have dimorphic chloroplasts). Both mesophyll and bundle sheath chloroplasts of these plants also have another distinctive feature uncommon in other species. Along the chloroplast periphery, the inner membrane forms a series of anastomosing channels and vesicles, which appears continuous with the inner lamellar system (82). This peripheral reticulum is not, however, a unique feature of plants with dimorphic chloroplasts; it is apparent in many plants, particularly when maintained under stress conditions.

The *stroma*, surrounding the lamellar system and enclosed within the inner envelope membrane, contains

stroma   plastoglobuli

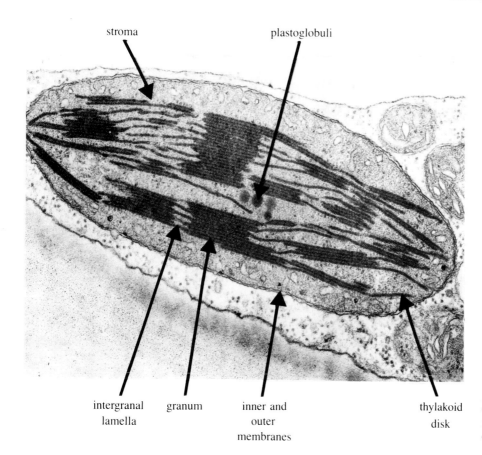

intergranal   granum   inner and   thylakoid
lamella   outer   disk
                        membranes

Fig. 2–1.
Chloroplast and
its suborganellar
components.

the enzymes and cofactors involved in the carbon-reduction reactions, of which ribulose bisphosphate carboxylase/oxygenase (ribulose-1, 5-P$_2$ carboxylase/oxygenase; "Fraction I" protein) forms a considerable portion. Dense lipid deposits (plastoglobuli) observed in the stroma are composed of lipophilic quinones (plastoquinones, vitamin K, 2-tocopherolquinone, and 2-tocophenylquinone), which serve as pools for membrane lipid formation (11). Starch granules form in illuminated chloroplasts and may be of such a size that chloroplast morphology is distorted.

Chloroplasts also contain their own DNA (which is quite different in both size and composition from nuclear DNA) and are equipped for DNA, RNA, and protein synthesis; ribosomes (70s) are also quite distinct from those of the cytoplasm (80s) (see also Chap. 5). Chloroplasts do not, however, possess the information necessary to synthesize all the proteins required for autonomy; the majority are coded for by nuclear DNA. Some of the chloroplast features referred to briefly here will be described more fully in the following sections.

Alterations in chloroplast structure during formation and aging: effect of the environment. Chloroplast development has been explored by several authors (17, 142). Early in development, the internal structure of the proplastid is minimal, consisting principally of a few vesicles continuous with the inner envelope membrane. As the chloroplast develops, many new vesicles form and fuse, arranging themselves in parallel layers. The lamellar system subsequently becomes more extensive, with development of the grana. This is a light-dependent process, closely associated with pigment production.

As the leaves senesce, the principal changes in chloroplast structure involve collapse of the grana and vesiculation of the intergranal lamellae. The grana separate, pigment is degraded, and the chloroplasts shrink. Finally, the chloroplast envelope disintegrates, and its contents are released into the cytoplasm (129). Plants under nutrient stress also exhibit significant changes in chloroplast structure and integrity. Chloroplasts of plants growing under extended conditions of deficiency in nitrogen, phosphorus, and zinc, for example, may be swollen, possess fewer grana, and have extensively vesiculated membranes (141). Inevitably, there is impairment of photosynthetic efficiency, and indeed there are many similarities with changes resulting from infection.

The Photosynthetic Process

Light reactions. The light reactions of photosynthesis (e.g. 4, 55, 65, 99) depend on the absorption of quanta of light energy by photosynthetic pigments,

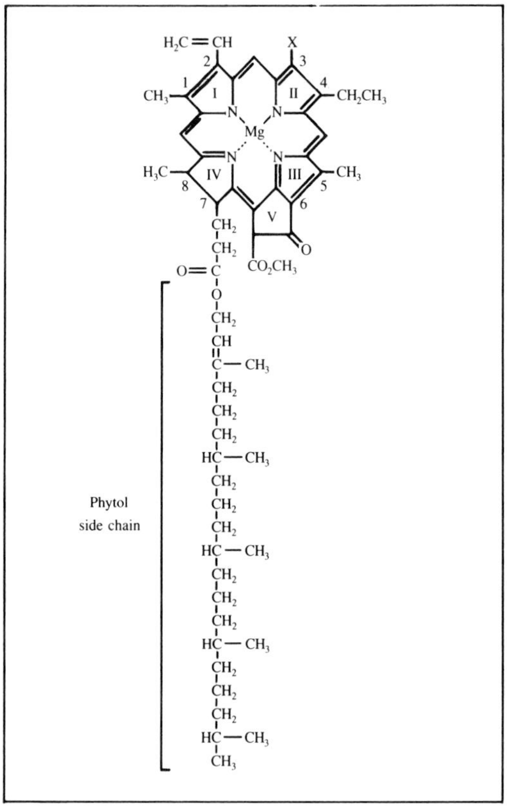

Fig. 2–2. Structures of chlorophyll a (X=CH₃) and b (X=CHO).

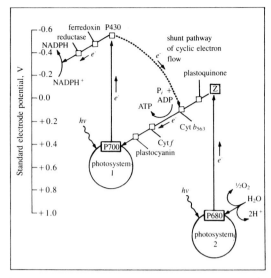

Fig. 2–3. The interaction of photosystems I and II (after Lehninger, 83).

which are then able to transfer high-energy electrons to the electron carriers of the photosystems. In doing so, they return to the ground state and are then able to again absorb light energy. The principal light-absorbing pigments are the chlorophylls, Mg^{++}-containing 5-ring structures (pheoporphyrins) with a hydrophobic phytyl side chain attached via an ester linkage to a propionic acid substituent on one of the porphyrin rings. It is the series of conjugated double bonds in the ring system that confers the ability to absorb energy. The structures of chlorophylls a and b are indicated in Fig. 2–2. It is beyond the scope of this section to detail the biosynthesis of the chlorophylls (22, 49), although the final step is relevant to some of the later sections. The final step in chlorophyll a biosynthesis involves esterification of chlorophyllide a (the ring system without the hydrophobic side chain) by phytol, a reaction catalyzed by the enzyme chlorophyllase (56). This enzyme is also able to catalyze the reverse reaction, dephytylation. Chlorophyll b may be formed from either chlorophyllide a or chlorophyll a (16). Light-absorbing pigments of lesser importance include the carotenoids, of which β-carotene and xanthophyll are commonly occurring examples.

Photosynthetic plant cells contain two types of chlorophyll. One is always chlorophyll a, and the second may or may not be chlorophyll b. These, together with other pigments, are embedded in the thylakoid membrane. Chlorophylls, for example, are located with the side chain embedded in the membrane and the light-absorbing section of the molecule on the surface, facing the stroma. In addition, molecules of chlorophylls and carotenoids are associated and interrelated in an important way. They are grouped together into two quite distinct photosystems, photosystems I and II (PSI and PSII), with PSI activated by light of longer wavelength (680–700 nm) than PSII (650 nm). Although each molecule within either a PSI or a PSII system can absorb light energy, the energy is transferred successively, often via several molecules, to a unique chlorophyll a molecule attached to a protein. This is the only one in the group capable of producing electrons and constitutes the *reaction center*. The interacting events of PSI and PSII, the so-called Z- scheme, are indicated in Fig. 2–3. The reaction centers are designated as P700 and P680, the numbers referring to the wavelengths of maximum light absorption. The purpose of the light reactions is to convert light energy (hν) to chemical energy; their net result is to transfer electrons from H_2O to $NADP^+$, the acceptor, with evolution of oxygen (equation 2), formation of 2 NADPH requiring transfer of 4 electrons.

$$2H_2O + 2NADP^+ \xrightarrow{h\nu} 2\,NADPH + 2\,H^+ + O_2 \quad (2)$$

Equation 2 is a specific, physiological example of the more general *Hill reaction*, in which electrons may be transferred from H_2O to an electron acceptor, referred to as a *Hill reagent*, under the influence of light energy (equation 3).

$$2H_2O + 2A \xrightarrow{h\nu} 2AH_2 + O_2 \qquad (3)$$

The energy required to transfer 1 electron from H_2O to $NADP^+$ is provided by the absorption of a quantum of light by each of the two photosystems. Light energy absorbed by the reaction center of PSI results in the transfer of an electron, through a series of membrane-bound electron carriers to soluble ferredoxin, a protein containing Fe and S. Finally, the electron is transferred from ferredoxin, via the flavoprotein ferredoxin-NADP reductase, to $NADP^+$. The electron lost from PSI is replaced by transfer from PSII, again through a series of electron carriers.

There is much evidence now that certain regulatory proteins play a role in the light activation of chloroplast enzymes. These proteins, thioredoxin and ferredoxin-thioredoxin reductase, are reduced in the light and, as a consequence, transmit the light signal to selected enzymes and thus modulate chloroplast enzyme activity.

In addition to transferring electrons to $NADP^+$, some of the energy released by the flow of high-energy electrons from the acceptor Z to PSI is utilized to form ATP, by ATPases, a process referred to as *photophosphorylation*. This is a noncyclic process, hence referred to as *noncyclic photophosphorylation*. However, in an alternative scheme, electrons transferred from PSI to ferredoxin can, instead of being transferred to $NADP^+$, return to PSI via a "shunt" pathway involving the electron carriers that transfer electrons from PSII to PSI. The energy thus released is used to produce additional ATP from ADP. This pathway is a cyclical one and is referred to as *cyclic photophosphorylation*. Non-cyclic photophosphorylation is accompanied by both NADPH formation and oxygen evolution, while cyclic photophosphorylation is not and provides ATP only.

It is also of interest that the H^+ produced as a result of the action of PSI and PSII is transferred through the thylakoid membrane into the intrathylakoid space. The pH gradient thus established is able to provide energy

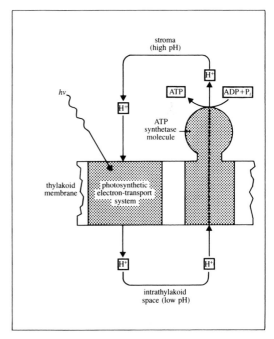

Fig. 2–4. The photosynthetic H^+ cycle (after Lehninger, 83).

for ATPase-catalyzed ATP formation, as indicated in Fig. 2–4.

Carbon-reduction reactions. The ATP and NADPH generated by the light reactions are utilized in a series of interdependent carbon-reduction reactions in which CO_2 is reduced to glucose, and thence to other carbohydrates. The principal process leading to glucose formation, largely formulated as a result of the studies of Calvin (e.g. 12), is summarized in Fig. 2–5. It is generally referred to as the *Calvin cycle*, or the reductive pentose phosphate cycle (63, 125). Six turns of the cycle are required for the formation of one molecule of

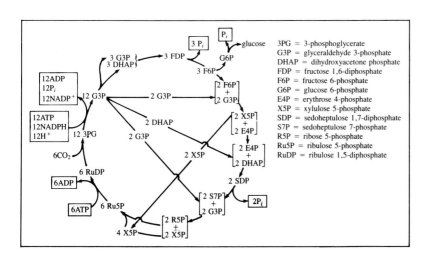

3PG = 3-phosphoglycerate
G3P = glyceraldehyde 3-phosphate
DHAP = dihydroxyacetone phosphate
FDP = fructose 1,6-diphosphate
F6P = fructose 6-phosphate
G6P = glucose 6-phosphate
E4P = erythrose 4-phosphate
X5P = xylulose 5-phosphate
SDP = sedoheptulose 1,7-diphosphate
S7P = sedoheptulose 7-phosphate
R5P = ribose 5-phosphate
Ru5P = ribulose 5-phosphate
RuDP = ribulose 1,5-diphosphate

Fig. 2–5. The photosynthetic carbon reduction, or Calvin, cycle (after Lehninger, 83).

glucose from 6 CO_2, utilizing 12 NADPH and 18 ATP. The process can be summarized by equation 4.

$$6\ CO_2 + 18\ ATP + 12\ H_2O + 12\ NADPH + 12\ H^+ \rightarrow C_6H_{12}O_6 + 18\ Pi + 18\ ADP + 12\ NADP^+ \quad (4)$$

Hence reduction of one mole of CO_2 requires utilization of two NADPH and three ATP, and, as indicated above, formation of two NADPH involves transfer of four electrons, by absorption of at least eight quanta of light, four by each of the photosystems PSI and PSII. This also produces two ATP, the other produced presumably by cyclic phosphorylation, with consumption of additional light energy. Many of the enzymes involved in this cycle are subject to various regulatory processes, including the action of light, which may both stimulate activity and lead to enhanced enzyme concentration.

The primary export product of photosynthesis, triose phosphate, can be subsequently utilized for the formation of, for example, disaccharides (including sucrose), structural polysaccharides (e.g. cellulose), or storage polysaccharides (e.g. starch).

Starch is a mixture of amylose, consisting of long chains of α-1,4-linked glucose residues and amylopectin, shorter chains of α-1,4-linked glucose residues, which are also cross-linked by α-1,6-linkages. Both are produced from glucose-6-phosphate as indicated schematically in Fig. 2–6.

Starch is utilized either by hydrolysis with amylase or by the action of phosphorylases. β-Amylase hydrolyses 2–1,4-glucose linkages at the nonreducing end of amylose/amylopectin chains. This enzyme will not, however, hydrolyse α-1,6 linkages, and to permit complete hydrolysis to glucose and maltose residues, these linkages are hydrolysed by a α-1,6-glucosidase. Amylophosphorylase, an α-1,4-glucan phosphorylase, catalyzes the phosphorolysis of glucose residues at the nonreducing ends of amylose and amylopectin to yield glucose-1-PO_4:

$$(glucose)_n + HPO_4 \rightarrow (glucose)_{n-1} + glucose\text{-}1\text{-}PO_4 \quad (5)$$

Again, α-1,6-glucosidase is required to hydrolyse α-1,6 linkages.

The initial and crucial step of the photosynthetic carbon reduction cycle is the formation of two molecules of 3-phosphoglycerate from CO_2 and ribulose-1,5-biphosphate, a reaction catalyzed by ribulose-1,5-P_2 carboxylase/oxygenase. This initial product of CO_2 fixation is a three-carbon compound; thus, plants that fix CO_2 by this pathway are referred to as C_3 plants.

There is, however, an additional pathway for photosynthetic CO_2 fixation that does not operate in C_3 plants but is utilized by many plants of tropical origin, the so-called *Hatch-Slack pathway* (33, 62, 63). In these plants, the bundle sheath cells are surrounded by mesophyll cells, and it is in these mesophyll cells, closest to the stomata and therefore the supply of CO_2, that CO_2 is first fixed to produce oxaloacetate, a four-carbon compound; the pathway is thus referred to as the C_4 pathway, and hence these plants are referred to as C_4 plants. The C_4 pathway does not replace the C_3 pathway in these plants but rather precedes it. The enzyme that catalyzes this reaction, phosphoenol pyruvate carboxylase, has a higher affinity for CO_2 than does ribulose-1,5-P_2 carboxylase/oxygenase and can, therefore, operate under conditions of low CO_2 concentration. In some plants, malic dehydrogenase catalyzes the reduction by NADPH of oxaloacetate to malate, which is then transported to neighboring bundle sheath cells, where CO_2 is regenerated with the formation of pyruvate (Fig. 2–7). This CO_2 can now enter the C_3 pathway, which operates only in the bundle sheath cells of C_4 plants, while pyruvate is returned to the mesophyll cells and used to regenerate phosphoenol pyruvate. Alternatively, mesophyll cells of other plants may contain not malic dehydrogenase but alanine aspartic transaminase. In these plants, oxaloacetate is converted to aspartate, which then acts instead of malate as a carrier of CO_2 into the bundle sheath cells.

Although the addition of this extra pathway results in the expenditure of additional energy, requiring the hydrolysis of two high-energy phosphates of ATP for phosphoenol pyruvate formation, it appears to be necessary in order to raise the concentration of CO_2 in cells performing the C_3 pathway to an optimal level. Where the plant undergoes stomatal closure to prevent water loss, its CO_2 concentration would otherwise be too low to support efficient photosynthesis.

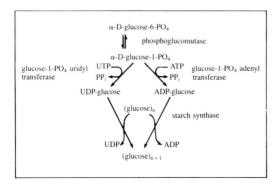

Fig. 2–6. Starch biosynthesis.

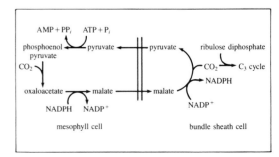

Fig. 2–7. The Hatch-Slack or C_4 pathway.

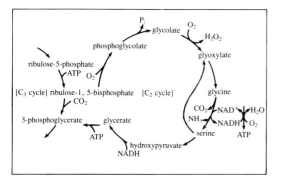

Fig. 2–8. The C_2 cycle and its integration with the C_3 cycle. The cellular location of the C_2 reactions is also indicated (modified from Lorimer and Andrews, 89).

Photorespiration and the C_2-Cycle

Although the principal pathways for the oxidation of carbohydrates produced during photosynthesis involve "dark" respiration by mitochondria, cells of photosynthetic plants possess an additional respiratory pathway that is integrated with the C_3-pathway, operates only in the light, and is referred to as the *photorespiratory carbon oxidation* cycle, or C_2 cycle (Fig. 2–8). Here, ribulose-1,5-P_2 carboxylase/oxygenase, the key enzyme of C_3 cycle, plays an alternative role, that of an oxygenase. Instead of catalyzing the reaction of ribulose-1,5-bisphosphate with CO_2 to yield two molecules of phosphoglycerate, it catalyzes the reaction with O_2 to give one molecule of phosphoglycerate and one of phosphoglycolate. Following hydrolytic phosphate release, glycolate is then transported from the chloroplast to peroxisomes and converted to glycine, via glyoxylate. Subsequent oxidative decarboxylation, in mitochondria, results in release of CO_2 and formation of serine, which is transported back to peroxisomes and, following conversion to glycerate, to the chloroplast.

The function of this cycle is, however, enigmatic. The cell does not need all the glycine and serine it provides, and energy provided by the photosynthetic light reactions appears to be consumed to no apparent purpose. Indeed, the cycle effectively reduces photosynthesis, consuming O_2 and releasing CO_2. Activity of the C_2 cycle is virtually undetectable in C_4 plants. Ribulose-1,5-P_2 carboxylase/oxygenase only functions as an oxygenase at low CO_2 or high O_2 partial pressures, so that under conditions of adequate CO_2 supply, the pathway is apparently unable to function. In C_4 plants, this adequate CO_2 supply is maintained in the bundle sheath cells by the C_4 pathway, rarely falling to a level that allows the C_2 cycle to come into operation. Consequently, ribulose-1,5-P_2 carboxylase/oxygenase is able to function more efficiently in C_4 plants than in C_3 plants, which do not possess a CO_2-concentrating mechanism.

More detailed information on the various aspects of photosynthesis outlined above can be obtained from several books and review articles (e.g. 2, 10, 43, 48, 50, 52, 53, 54, 57, 61, 83, 136).

The Influence of Virus Infection

In all interactions between host and virus that result in visible symptoms of chlorosis or necrosis, there are changes in chloroplast morphology and metabolism. It is hardly surprising, therefore, that photosynthesis and consequently growth of the infected plant are adversely affected. Several aspects of the photosynthetic process have been investigated in many virus/host combinations, usually in systemically infected tissue (26). Some of the relevant observations are detailed in the following sections.

Chloroplast Structure and Multiplicity

Reduction in the number of chloroplasts in chlorotic tissue may be a feature common to many virus infections, although reports have been relatively few. Systemic infection of sorghum and corn by MDMV or SCMV (138), of tobacco by TEV (60), and of squash by SMV (95) has, however, been observed to result in a significant reduction in chloroplast number. Much more commonly observed are chloroplast abnormalities of various kinds, including: (1) changes in external morphology and size; (2) disorganization or impaired development of lamellae; (3) membrane vesiculation; (4) increase in size and number of plastoglobuli; (5) accumulation of phytoferritin; (6) cytoplasmic invagination; (7) accumulation of starch grains; (8) aggregation (e.g. 20, 36, 81, 84, 101). Not all abnormalities are necessarily observed in any one situation, and abnormalities will clearly differ both qualitatively and quantitatively among different viruses and host plants. They are also dependent on the severity of symptoms induced by a particular strain and on the tissue that is sampled. The following examples illustrate the types of change that may be expected.

Tobacco leaves systemically infected with some strains of CMV (e.g. Prices No. 6) produce distinct yellow and green areas. Chloroplasts in yellow areas exhibit abnormalities characteristic of the time the leaf becomes infected. When chloroplasts are at an early stage of development, there is an inhibition of lamellar development (Fig. 2–9) and some organelles lack grana and contain myelinlike structures. Alternatively, when chloroplasts are at a late stage of development, only a slight inhibition of lamellar development is apparent, and many show signs of disintegration (Fig. 2–10). Proliferation of chloroplasts also appears to be inhibited (35, 67, 124).

In barley infected systemically with BSMV (20, 93), chloroplasts become rounded and swollen, aggregate together with parts of the cytoplasm trapped between adjacent organelles, and contain cytoplasmic invaginations (Fig. 2–11). In addition, several single-membrane-bound vesicles appear between the inner and outer membranes. Many of these changes resemble those observed in Chinese cabbage infected with TYMV

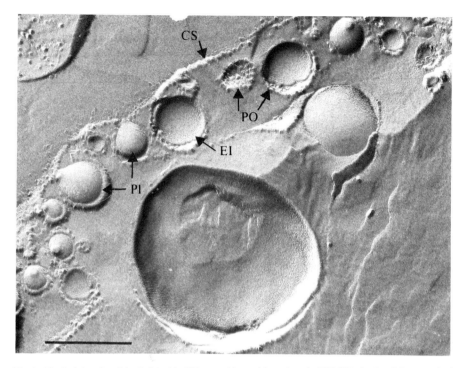

Fig. 2–12. Peripheral vesicles induced in Chinese cabbage chloroplasts by TYMV infection. Micrograph of freeze-fracture preparation with the fracture plane passing through the chloroplast. Vesicles consist of two rounded membranes; necks point toward the chloroplast surface. PI, P face of inner vesicle membrane; PO, P face of outer chloroplast membrane; EI, E face of inner chloroplast membrane; CS, chloroplast surface; bar, 250 nm (after T. Hatta et al., 64).

considered a stimulation of chlorophyllase in TMV-infected tobacco to arise as a secondary consequence of chloroplast damage, caused in this instance by an accumulation of insoluble TMV protein.

There are, however, additional observations indicating that this is not the complete story. Since the spectra of chlorophyllides are identical to those of chlorophyll, degradation of chlorophyll by chlorophyllase cannot be responsible for changes in leaf coloration. In tobacco leaves systemically infected with CMV, chlorophyllase activity decreases rather than increases (76), while in beet infected with BYV + BMYV, its activity does not change significantly (109). It is possible that in tobacco, chlorophyllase participates in the biosynthesis of chlorophyll; thus, in infected leaves, there is probably a reduction in chlorophyll synthesis, in addition to conversion to pheophytin with loss of Mg^{++}. This view is substantiated by the observation that pheophytin concentration increases in infected tissue, while that of chlorophyllide, the product of chlorophyllase action, remains negligible. There is also a substantial increase in carotenoids, particularly xanthophyll, in the yellow areas. Inhibition of synthesis was also considered to be the cause of reduced chlorophyll and carotene content of Chinese cabbage systemically infected with various strains of TYMV (25).

It is of interest that in TEV-infected tobacco (60), although the number of chloroplasts decreased, there was no change in chlorophyll content of those that remained.

Rate of Photosynthesis

Photochemical reactions. Virus infection often leads to a reduction in photochemical activity that may be greater than can be accounted for by the reduction in chlorophyll content. In chloroplasts isolated from beet infected with beet yellows, for example (130), there was a 50% reduction in Hill reaction activity, per unit weight of chlorophyll. Similarly, there was a 40% reduction in Hill reaction activity and photophosphorylation in chloroplasts from TMV-infected tobacco at nine days after infection, an effect not, however, observed when plants were supplemented with nitrogen (150).

Reduced oxygen evolution was also apparent in leaf tissue and isolated chloroplasts of MDMV-infected corn (41). However, at the time of appearance of TYMV-induced symptoms in Chinese cabbage, about one week after inoculation, chloroplasts had a higher Hill reaction activity and produced more ATP by cyclic and noncyclic photophosphorylation. It was not until three to four weeks after inoculation that photochemical activity began to decrease relative to controls (44). In tobacco

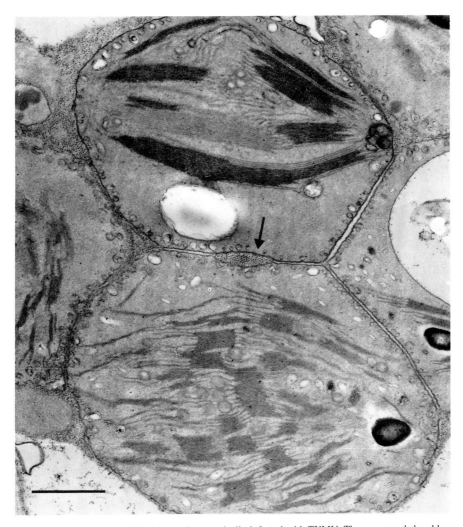

Fig. 2–13. Chloroplasts of Virginia stock systemically infected with TYMV. They are rounded and have abnormally thick grana and peripheral vesicles. Cytoplasmic pocket (→) is filled with viruslike particles; bar = 1μm (photograph courtesy of T. Hatta).

infected with TEV (68) and squash infected with SMV, Hill reaction activity and photophosphorylation apparently remained unaffected.

CO₂ assimilation. Rate of uptake of CO_2 has also been used as a measure of photosynthetic activity, and in many cases a reduction has been observed. Such instances include tobacco infected with TMV (112) or TEV (69, 113), BYDV-infected barley (73), CTV (115), and TAV-infected tomato (72). In other cases, including TMV-inoculated *N. glutinosa*, tobacco infected with PVX (114), and squash infected with SMV (95), there was no significant difference between healthy and infected tissue.

However, when CO_2 assimilation was studied in *cells* in the *early stages* of infection, a rather different picture emerged. Doke and Hirai found, for example, that in tobacco inoculated with either PVX or TMV, which produce systemic infection, or *N. glutinosa* inoculated with TMV, which produces a local lesion response, there was an early stimulation of $^{14}CO_2$ incorporation in cells in and around the site of infection (28, 29). When cells had been infected for several days, $^{14}CO_2$ assimilation declined. There were clearly areas of high and low photosynthetic activity, obscured when either large discs or whole leaves were sampled, and activity depended on the stage of virus infection.

The cause of changes in CO_2 uptake remains obscure. It has been suggested that in BYV-infected beet (59), the principal cause is an increased leaf resistance to gaseous

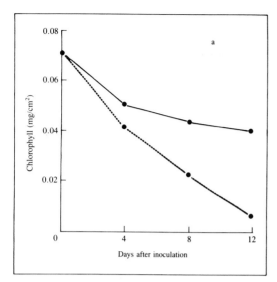

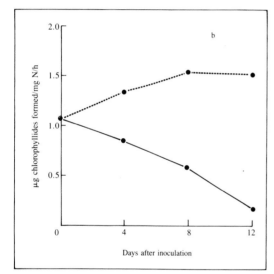

Fig. 2–14. The effect of infection by CMV on chlorophyll content (a) and chlorophyllase activity (b) of cucumber cotyledon discs; ●—● = control (after Bailiss, 8).

diffusion, probably due to differences in stomatal aperture. There was, however, no comparable effect on resistance to diffusion in TAV-infected tomato (72).

Metabolic Shifts

There is also some evidence for a shift in the utilization of carbon compounds synthesized during the early stages of photosynthesis. Hopkins and Hampton, for example (69), observing a 50% reduction in sucrose content of TEV-infected tobacco, suggested a shift toward biosynthesis of amino acids, organic acids, and glucose. Bedbrook and Matthews (see Table 2–1) found a marked increase in amino acids and organic acids, at the expense of sugars and sugar phosphates during the

Table 2–1. The distribution of ^{14}C from $^{14}CO_2$ after five minutes of photosynthesis by healthy and TYMV-infected Chinese cabbage leaf sections (after Bedbrook and Matthews, 14).

		Incorporation of ^{14}C per compound (cpm)	
		Healthy	Infected
Phosphoglyceric acid (PGA)		420	340
Sugar phosphates:	monophosphates	1,422	1,240
	diphosphates	192	80
Sugars:	sucrose	874	110
	glucose	152	136
	fructose	62	140
Phosphoenolpyruvic acid (PEP)		5	40
Organic acids:	malate	10	100
	citrate	82	422
Aminoacids:	aspartate	10	66
	alanine	192	528
	glutamate	56	54
Total cpm		3,487	3,322

phase of rapid virus multiplication in TYMV-infected Chinese cabbage. As virus replication began to decrease, however, the pattern of biosynthesis of these compounds returned to normal. There was also a preferential synthesis of amino acids, at the expense of sugars, in squash infected with SMV (95). Likewise, in TBSV- and CMV-infected *Tolmiea*, Platt et al. (119) observed a stimulation of $^{14}CO_2$ incorporation into glycine.

PEP-Carboxylase activity has been observed to increase in TYMV-infected Chinese cabbage (14) and in CMV-infected cowpea (146), although it remained unchanged in beet infected with BYV and squash infected with SMV and decreased in TSWV infected tobacco (106). These changes are summarized in Table 2–2.

Starch Metabolism

Impairment of starch metabolism is a common feature of virus infection. There may be, for example, a proliferation or enlargement of starch grains in chloroplasts in or surrounding necrotic zones and in severely chlorotic tissue (e.g. 21, 135; Fig. 2–15), although chloroplasts of healthy leaf tissue may also contain several large starch grains after a period of active photosynthesis. Starch also appears to accumulate in RTV-infected rice (40, 131) and in BYDV-infected barley (74), and in the latter case, there is also evidence for the accumulation of glycogen (107).

In some cases, accumulation can occur in infected tissue in the absence of visible symptoms. For example,

Table 2–2. Changes in enzyme activities.

Virus/host	Pep carboxylase	Aspartate amino transferase	Ribulose diphosphate carboxylase	Reference
BYV/beet	—	ND	◆	(59)
SMV/squash	—	ND	—	(95)
TYMV/Chinese cabbage	◇	◇	ND	(14)
TSWV/tobacco	◆	ND	ND	(106)
CMV/cowpea	◇	ND	ND	(146)

◇, increase; ◆, decrease; —, no change; ND, not determined.

when cucumber cotyledons are inoculated with TMV, virus is localized to cells in and around the site of infection, although there are no overt symptoms. However, if leaves are decolorized and stained with I_2/KI after a period of darkness, it is apparent that cells at and adjacent to the infection site contain much more starch than cells in surrounding tissue: both light and electron microscopy reveal a proliferation of starch grains in chloroplasts (24; Fig. 2–16). It would seem, therefore, that these cells have a reduced capacity for starch utilization.

However, both starch synthesis and degradation may be affected (26, 66). Although there is more starch in, for example, TMV-infected areas of tobacco after a period of darkness, after a period of active photosynthesis the reverse is true. Doke and Hirai (27) found starch synthesis to be inhibited in tobacco systemically infected with TMV and maintained in the light, although neither the rate of photosynthetic CO_2 fixation nor the activity of amylophosphorylase and amylase was affected. In the absence of light, starch degradation and

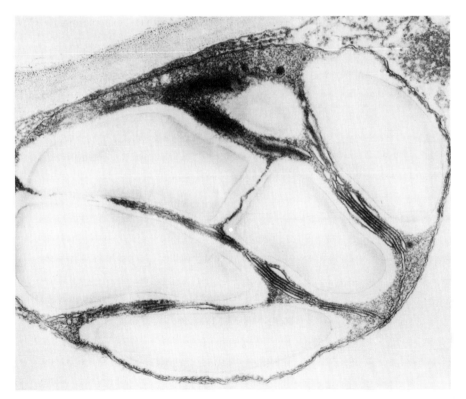

Fig. 2–15. Chloroplast in a tobacco leaf at the onset of interveinal yellowing, showing starch accumulation.

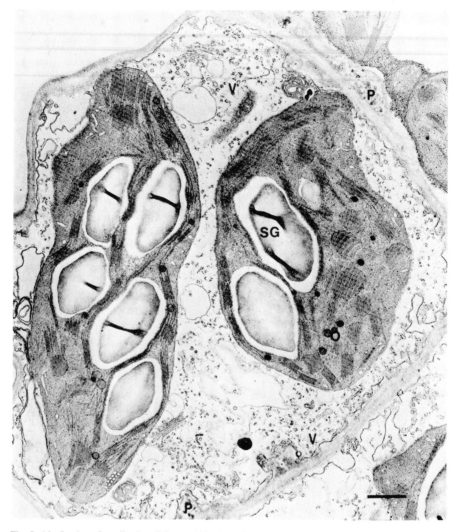

Fig. 2–16. Section of a palisade cell from within the peripheral zone of a TMV starch lesion in a cucumber cotyledon. Chloroplasts contain several starch grains (SG); P, paramural body; V, virions; O, osmiophilic material. Bar, 1 μm (after Cohen and Loebenstein, 24).

amylase activity were stimulated. Similarly, Balls and Martin (9) found amylase to be stimulated in tobacco systemically infected with a yellow strain of TMV, although infection with a common strain caused a decrease. It may be of significance that a rapid dissociation of accumulated starch grains in CYMV-infected clover could be effected by treatment with exogenous 3′,5′-cyclic adenosine monophosphate (cAMP). Dissociation was accompanied by alterations in membrane ultrastructure (137).

Carbohydrate accumulation is also a common feature of virus-infected tissue, in, for example, PLRV-infected potato, BYDV-infected barley, and BYV-infected beet; in some cases, it may be the result of impaired translocation.

It may be concluded, therefore, that virus infection leading to chlorosis or necrosis of infected leaf tissue induces in the chloroplasts of that tissue a number of structural and metabolic changes, the magnitude of which is related to the severity of the symptoms. Although there are clearly marked differences between different virus/host combinations, and effects vary with the time after infection, some general features emerge.

At the time of active virus multiplication, these changes appear to result in alterations in the metabolic pathways for the utilization of early products of photosynthesis, with a stimulation in the biosynthesis of compounds such as amino acids, essential for virus replication. Increased activity of some of the enzymes in these pathways has been observed. Although there may be a transitory stimulation of both light and dark reactions early in infection, as symptoms develop there is a reduction in chlorophyll content, a decrease in photochemical activity greater than can be accounted for by the reduction in chlorophyll content, and a reduced rate of CO_2 assimilation. There is also a decrease in chloroplast and RNA content, with a consequent decrease in number of chloroplast ribosomes (see also Chap. 5) and in chloroplast protein synthesis. In few cases, however, have observations been related to chloroplast number, so that at least a part of any observed reductions in function may be accounted for by reduction in chloroplast number, in addition to modified function.

It seems reasonable to assume that at least some of these changes associated with chloroplast degeneration and impairment of function or development are also involved in symptom expression. It has been suggested that TMV infection of tobacco leads to repression of genes for biosynthesis of chloroplast or RNA and that ultrastructural changes in the chloroplasts of tobacco infected with some strains of TMV may be the result of an accumulation of coat protein. Nevertheless, definitive evidence is lacking, and we are still some way from explaining the observed changes. Symptom severity, and hence also the extent to which chloroplast form and function are altered, differ markedly among virus strains. Symptom expression is determined by part of the viral genome, or its translation product, or by the interaction between several RNA sequences, or their translation products, and the host (see also Chap. 5). We have no idea, however, how these observed effects are mediated, and, although in many cases chloroplast changes closely resemble those occurring during senescence, it may be that the determinants differ with the situation. Much more information on the function of viral genomes, their translation products, and their interaction with the host is required.

The Influence of Bacterial Infection

The Basis for Chlorosis

Attack by a phytobacterial pathogen frequently engenders a chlorotic symptom. For example, the chlorosis that ensues following infection by a number of leaf-spotting xanthomonads and pseudomonads is suggestive of a direct effect upon chloroplasts, perhaps by toxic metabolites of the pathogen (31). As observed in virus-infected plants, yellowing of tissue should reflect an overall reduction in photosynthetic activity as well as in the rate of the process. However, the relationship of yellowing and halo formation to changes in photosynthetic processes, caused by bacterial pathogens, remains largely unknown. The reversal of chlorosis by kinetin has received some attention (38, 111, 123) and seems to be related to RNA and protein synthesis. The implications of the "kinetin effect" are discussed in detail in Chap. 7.

Additional insight regarding the chlorosis induced by toxins may be obtained from the study by Guyla and Dunleavy (58). Although the toxin produced by *Pseudomonas glycinea* is different from the *P. tabaci* toxin, their effect on the photosynthetic apparatus and upon

photosynthesis may be the same. Both pathogens cause chlorosis, and as we shall see, *P. tabaci* seems either to destroy chlorophyll or suppress its synthesis. Studying this problem in soybean infected with *P. glycinea* has the distinct advantage of permitting study of chlorosis in toxemia-affected leaves without the presence of the bacteria. Inoculation of unifoliate leaves causes first and second trifoliate leaves to become systemically chlorotic, and they are devoid of the pathogen. Five to 15 days after inoculation , the toxin-affected first trifoliate leaves were more chlorotic and stunted than the controls. This occurs to a lesser extent in the third trifoliates. The concentration of pheophytin, a chlorophyll breakdown product, revealed no significant difference between the first trifoliates of inoculated plants and controls during the 5- to 15-day period after inoculation. The chlorophyll a to b ratio, usually lower during chlorophyll breakdown, did not differ between infected and control leaf tissues.

Soybean cotyledons incubated with *P. glycinea* toxin were found to have 35% less chlorophyll. In analyses for the chlorophyll precursor aminolevulinic acid (ALA), significantly less of the precursor was found in cotyledons that had been bathed in a 100 μg/ml toxin solution. The inhibition of ALA synthesis may reflect suppressed synthesis of ALA synthetase. However, the short half-life of the enzyme has precluded the firming of this contention. It would seem that the reduction in chlorophyll precursor and the failure to detect augmented pheophytin and altered chlorophyll a to b ratios suggest that *P. glycinea*-induced chlorosis is due to suppressed synthesis rather than the breakdown of the pigment.

Early studies (18) on the chlorotic halo induced in plant leaves as a result of the action of the toxin produced by *Pseudomonas tabaci* suggested that the toxin either destroyed chlorophyll or inhibited its synthesis. Spectroscopic and chemical analyses of the chlorophyll in toxin-treated leaves revealed that concentrations of chlorophyll in chlorotic tissue were much lower than in comparable normal tissue. Since the chlorophyll extracted from toxin-treated tissue and that extracted from normal tissue were spectroscopically similar, it was concluded that the biological effect exerted by the toxin was not upon the chlorophyll molecule per se. The study by Turner presents strong evidence that the damage is to chloroplasts and is caused by ammonia (139). A detailed account of the mode of action and chemical composition of this and other chlorosis-inducing toxins is presented in Chap. 9.

Photosynthesis may also be affected in a less direct manner by the occlusion of the vascular system by a bacterial pathogen. For example, the direct relationship of *water stress* and *wilting* to *stomatal closure* and the resultant inhibition of CO_2 absorption should manifest itself in a reduction of photosynthesis.

Water Stress-Photosynthesis Relationship

Experimental evidence of an altered photosynthetic pattern in a plant invaded by a plant bacterial pathogen was provided by Beckman et al. (13). They attempted to relate the symptom of wilting in banana, induced by *Pseudomonas solanacearum*, to either vascular occlusion or a toxic effect that would influence water retention by host cells. The study was accomplished with the aid of sensitive instrumentation that continuously measured the rate and intensity of photosynthesis and respiration following inoculation. Their experiments revealed that initial fluctuations in photosynthesis and transpiration occurred six to nine days after inoculation and that normal plants deprived of water responded similarly. From previous studies, Brun (19) deduced that these fluctuations were due to a loss of turgor by the guard cells and resultant stomatal closure when water loss exceeded supply. It was possible to restore normal photosynthetic activity in leaves of infected plants when the water stress was eliminated. This was accomplished either by cutting the petioles above the vascular infection and placing them in a water reservoir or by saturating the atmosphere surrounding the leaves with water.

Beckman et al. (13) proposed, on the basis of their observation, that infected banana leaves regained transpirational and photosynthetic function if the water supply was restored, and that the final symptom of wilt was not due to a toxic effect upon either the guard cells or the chloroplasts. In fact, wilt was not demonstrated by the banana leaf until both transpiration and photosynthesis had ceased. They concluded that vascular occlusion is the primary cause of wilt in banana with *P. solanacearum*. It is also apparent that both transpiration- and photosynthesis-dependent processes may be affected well in advance of the terminal wilt symptom.

Ultrastructural Changes in the Chloroplast

Additional understanding of chlorosis, its basis, and its impact upon the metabolism of foliar parenchyma is obtained from an ultrastructural study by Goodman (45). For theoretical purposes it seemed plausible to use chlorosis induced by the *Pseudomonas tabaci* toxin as a predictive model to study changes in the photosynthetic machinery caused by other chlorosis-inducing bacterial pathogens. However, it is apparent that, although the toxin produced by *P. tabaci* differs from that of *P. phaseolicola* (see Chap. 9) and their modes of action may also vary, the chlorosis caused by the two pathogens is similar in appearance to the yellowing of tobacco leaves that occurs as a consequence of natural senescence or leaf detachment. Chlorosis in all three instances can be reversed for a short time by exogenous applications of cytokinin (38, 90). It is contended that the observed yellowing occurs as a result of a decrease in RNA and subsequently protein synthesis (111). This contention is supported by the observation that chlorosis was induced by infiltrating tobacco leaf tissue with 10 μg/ml of actinomycin-D (45). Electron micrographs of cells from tissues in early stages of chlorosis are presented in Fig. 2–17, and there are striking similarities between treatments. The major effects of all treatments whether natural or induced (pathologically, bio-chemically, or physically) include swelling of chloroplasts and reduction in both chloroplast and cytoplasmic ribosomal content. There is also pronounced decrease in chloroplast stroma and mitochondrial groundplasm (indications of reduced soluble protein). The foregoing suggest a reduction in synthesis and capacity for synthesis of nucleic acids and protein. Similar ultrastructural disorientation of chloroplasts caused by either the *P. tabaci* toxin or ammonia was reported by Durbin (31).

The plastoglobuli (plastoquinone-lipid pool which is chlorophyll free) increase in size and frequently in number as well. This observation agrees with the reports of Lichtenthaler (85, 86) on the general development and aging of chloroplasts. The globuli also lose their osmiophilic quality and become less electron dense at their periphery. It would appear from these observations that the grana are losing their lipid and light-trapping components and coalescing to form larger globules. One might infer from this a decreasing capacity for photosynthesis at least on the basis of per unit area of leaf surface. Still another feature of either induced or normal senescence is separation of the plasmalemma from the cell wall (Fig. 2–17). Even if a few plasmodesmatal connections remained intact, it would seem that mass flow of substances from cell to cell would be reduced.

Hence at the ultrastructural level chlorosis seems to be as nonspecific a symptom as it is to the unaided eye. However, the biochemically detected reduced RNA and protein synthesis and photosynthesis (70, 90, 111) are clearly predictable from ultrastructural observations (45, 47, 71, 85, 86).

The effect of a comparatively high inoculum concentration of l0⁹ cells/ml (approximately the concentration of bacteria in foliar tissue at the peak of pathogenesis) on photosynthesis was studied by Magyarosy and Buchanan (94). They compared the effects of either the compatible pathogen *Pseudomonas phaseolicola* or a saprophyte, *P. fluorescens*, on CO_2 fixation in pinto bean. They observed that the pathogen rapidly initiates a suppression of CO_2 fixation and that this intensifies over a 16-hr "infection" period until fixation is reduced to 15% of controls. At this time, chloroplast ultrastructural integrity appears to be totally destroyed. It is possible, however, that all chloroplasts are not uniformly disrupted, accounting thereby for the residual CO_2 fixation at the termination of the experiment. A comparison of the rate of CO_2 fixed when an incompatible pathogen had been used, e.g. an avirulent race of *P. phaseolicola* or perhaps *P. pisi* (91), would have been interesting. Observations made by Goodman and Plurad (47) suggest that, as a consequence of hypersensitive reaction (HR), photosynthesis would have been even more swiftly suppressed because of the rapid chloroplast degradation that ensues in tissue exposed to incompatible bacterial pathogens (Fig. 2–18). The influence of the saprophyte *P. fluorescens* on CO_2 fixation appeared in the experiments of Magyarosy and Buchanan (94) to be

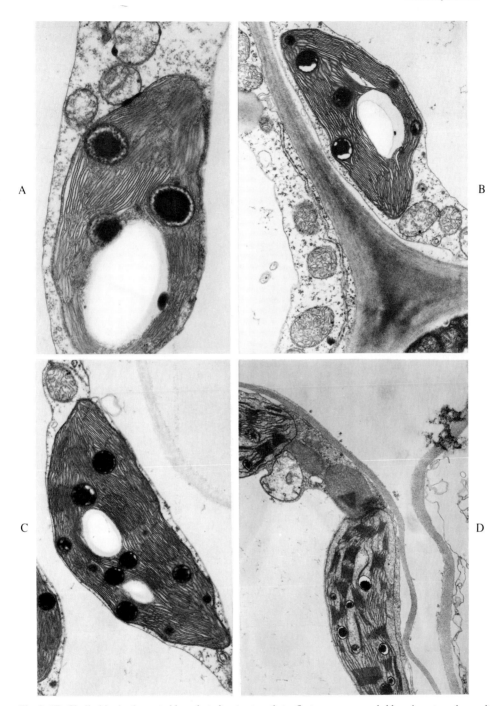

Fig. 2–17. Similarities in aberrant chloroplast ultrastructure that reflect senescence and chlorosis, yet are the result of different treatments: (a) normal senescence (yellowing); (b) senescence and yellowing caused by leaf detachment; (c) senescence and yellowing caused by treatment with actinomycin D; (d) senescence and yellowing caused by inoculation with *Pseudomonas tabaci* (after Goodman, 45).

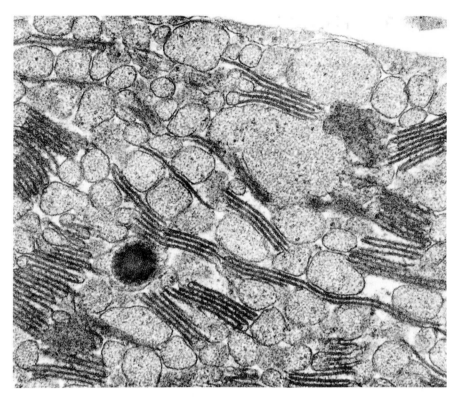

Fig. 2–18. Vesiculated grana and integranal lamellae and granal margin swelling, 7h after infiltration of 10^8 cells/ml of HR-inducing *Erwinia amylovora* into tobacco leaf tissue (after Goodman and Plurad, 47).

negligible. However, if the experiments lasted for another 3 to 4 days, a gradual but steep decline in fixation could be predicted. Experiments monitoring the influence of the saprophyte *P. fluorescens* in tobacco regularly reveal internal chlorosis in tissue infiltrated with as few as 10^6 cells/ml at the end of a four-to-six-day period (Goodman unpublished data).

Effect of HR on Chloroplasts

It has also been reported from both biochemical (70) and ultrastructural (47, 71) points of view that the chloroplast disruption in tobacco leaf tissue that ensues during the development of HR is a reflection of changes in both composition and conformation of the thylakoid proteins. That HR affects thylakoid protein composition was determined by amino acid analysis and from shifts in UV absorption spectra. Alterations in thylakoid protein were recognized by changes in solubility and the capacity to bind the phosphatide, lecithin (Table 2–3). That structural changes had occurred specifically in the protein component of chloroplast membranes was established by means of reconstitution experiments. It was noted that thylakoid protein lost its capacity to reassociate with solubilized thylakoid lipid (from control tissue) to form membranelike structures three hours after infiltration of tobacco leaf tissue with 10^8 cells/ml of the incompatible pathogen, *Erwinia amylovora* (71).

Table 2–3. Amount of phosphatide bound by chloroplast membrane proteins isolated from tobacco leaf tissues infiltrated with *Erwinia amylovora* or water (after Huang and Goodman, 70).

Structural protein source	Time after inoculation	μg of phosphorus/mg of protein	
		Initially	In final pellet[c]
H_2O[a]	20 min	0.8[d]	10.3
H_2O	3 h	0.8	11.0
H_2O	6 h	0.8	10.7
Ea[b]	20 min	0.8	10.6
Ea	3 h	0.9	6.5
Ea	6 h	0.9	5.1

a. Membrane structural protein from leaves infiltrated with H_2O.
b. Membrane structural protein from leaves infiltrated with 10^8 cells/ml *E. amylovora*.
c. Lecithin was incubated with protein (2 : 1, w/w) at pH 12.3, 26 C, for 15 min. The complex was removed by centrifugation. The pellet was washed and the amounts of protein and phosphorus were determined.
d. Average of three structural protein preparations. Three determinations were made on each preparation.

Effect of Photosynthesis on HR

According to data developed by Sasser et al. (126), suppression of photosynthesis by a dark treatment or 3-(p-chlorophenyl)-1,1-dimethyl urea (DCMU) treatment permitted an incompatible HR-inducing pathogen, *Xanthomonas phaseoli*, to multiply in pepper leaves. On the other hand, the population of *X. phaseoli* in pepper leaves kept in the light decreased steadily from 6 $\times$ 10^5 to 3 $\times$ 10^2 during the 48 h of the experiment. Of particular interest was the observation that neither the chemical nor light-dark treatments prevented the development of the HR as characterized by rapid desiccation and tissue collapse. Hence, HR-associated tissue collapse and desiccation per se did not diminish the bacterial population, but it would seem that a product of photosynthesis was at least a contributing factor. It is conceivable that the bacteria-localizing phenomenon described by Goodman et al. (46) is responsible for the HR-related bacterial suppression and that it is an active process deriving energy from a product of photosynthesis.

Support for chloroplast membrane disorientation during the development of HR has been provided by Pavlovkin and Novacky (117). Cotton cotyledons infiltrated with *P. tabaci* caused a drop in membrane potential that was equated with an irreversible change diffusion potential (permeability) of the plasmalemma of cotyledonary mesophyll cells. Their evidence suggests that membrane changes that occur during the development of HR are in the diffusion potential component of membrane potential, rather than the active energy-dependent facet (the electrogenic pump).

In Fig. 2–19, it is clear that the HR-inducing bacterium (*P. tabaci*) stimulated the electrogenic pump in the presence of air. Stimulation was heightened by fusiccocin (FC). However, in the absence of air (under N_2 anoxic conditions), the energy-dependent (ATP-requiring) pump was inactive, with or without FC.

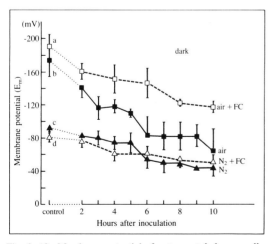

Fig. 2–19. Membrane potential of cotton cotyledonary cells undergoing HR (after Pavlovkin and Novacky, 117).

The Influence of Fungal Infection

Pathophysiological research in this field is still in an embryonic state, so what we know at present is mostly related to a general damage or aging of the "photosynthetic machinery" in the infected plant. Fungi causing diseases that terminate in necrosis seem to influence the photosynthetic process primarily by affecting the chloroplast. It would seem that chloroplast degeneration, resulting in loss of chlorophyll content and, therefore, in reduced CO_2-fixation, is the primary symptom in many diseases, particularly in those that are caused by facultative parasites (1).

Studies monitoring photosynthesis in diseases caused by biotrophs (obligate parasites) have shown that the situation is more complicated here. In addition to general injury, oat rust, for example, can reduce chloroplast rRNA content in leaves during the later stage of disease development due to a senescence effect (133). However, it was shown with barley mildew (34) that, in an early stage of disease development when neither chlorotic symptoms nor decreases in the chlorophyll content were observed, an inhibition of CO_2 uptake occurred in barley leaves. Probably, this is the result of a toxic effect of the pathogen. Indeed, Dyer and Scott (32) have demonstrated a breakdown of chloroplast polysomes in leaves as early as one day after inoculation. Since *Erysiphe graminis* f. sp. *hordei* is an ectoparasite and infects only the epidermal cells, which lack chloroplasts, the decrease in chloroplast polysomes reflects the influence of a diffusible product of the pathogen. As a result of all the above-mentioned changes, chloroplast degeneration is always associated with a decrease in the overall rate of photosynthesis. The activity of a chloroplast-associated enzyme, glycolic acid oxidase, also decreases in late stages of infection when general damage is apparent.

Light Reaction

Although loss of chlorophyll (yellowing) is a common symptom of many diseases, the *primary* or *light reactions* of photosynthesis seem not to be directly affected. For example, the primary light-mediated component of the photosynthetic process, *photosynthetic phosphorylation*, in chloroplasts isolated from rust-infected oats is not altered by infection when the results are expressed per unit of chlorophyll (149). The chlorophyll content of the leaves is reduced by infection, so the overall photosynthetic activity of the leaves is probably suppressed. Allen (3) also showed in an early study with powdery mildew (*Erysiphe graminis*)-infected wheat leaves that there is no decline in photosynthetic activity per unit of chlorophyll. On the other hand, photophosphorylation may be inhibited by toxins produced by facultative parasites (necrotrophs). For example, tentoxin, a nonspecific toxin from *Alternaria tenuis* (see Chap. 9), inhibits photophosphorylation coupling mechanisms in isolated lettuce chloroplasts but has no influence on electron transport (5). This toxin inhibits coupled electron flow (in the presence of ADP and phos-

Table 2–4. Effect of rust (*Uromyces fabae*) infection on noncyclic electron transport by *Vicia faba* chloroplasts (after Montalbini and Buchanan, 108).

Noncyclic acceptor	Days after infection						
	1	2	3	4	5	7	(average) 1–7
	Ratio of acceptor reduced photochemically by infected : healthy chloroplasts						
NADP	1.0	0.8	0.7	0.9	0.8	0.7	0.8
FeCN	1.0	0.8	0.6	0.7	0.7	0.7	0.7

phate) but does not affect basal electron flow (in the absence of ADP and phosphate) or uncoupled electron transport. Tentoxin interferes with the terminal steps of ATP synthesis, so it is an energy transfer inhibitor. It is not known whether tentoxin contributes importantly to symptom expression in the naturally infected plant. In any case, the toxin, a cyclic tetrapeptide, seems to be a valuable tool for the study of phosphorylating systems in green tissues. It has, however, no effect on electron transport or phosphorylation reactions in mitochondria. Thus, tentoxin could be used as a selective inhibitor of photophosphorylation, which would allow differentiation between photosynthetic and oxidative phosphorylation.

Inhibition of Electron Transport

Montalbini and Buchanan (108) found a toxic substance of the broad bean rust fungus to specifically inhibit the noncyclic electron transport in broad bean chloroplasts, although this observation is at variance with data from earlier studies by Wynn (149) and Allen (3). Table 2–4 shows that two days after inoculation rust infection inhibits the rate of reduction of NADP or FeCN. The latter is a nonphysiological substitute for NADP in noncyclic photophosphorylation. In other words, the noncyclic electron transport in the chloroplasts is inhibited by rust infection and to about the same extent that the formation of ATP is inhibited in noncyclic photophosphorylation. Cyclic photophosphorylation was only slightly affected, and it was also shown that the ratio of ATP formed per two electrons transferred from water to acceptor was the same in healthy or diseased chloroplasts. The latter finding indicates that a specific uncoupling of noncyclic photophosphorylation did not occur in rust infections. However, the suppression of ATP formation was mainly due to an inhibition of electron transport from water to NADP. The hypothetical existence of a toxic substance of the pathogen that behaves like DCMU or *o*-phenanthroline, which are specific inhibitors of noncyclic electron transport in chloroplasts, was suggested. However, subsequent experiments favored an alternate hypothesis. A reduction of chloroplast electron carriers (cytochromes) in infected tissue, as compared to healthy tissue, may explain the rust-induced suppres-

Table 2–5. Effect of powdery mildew infection on noncyclic photophosphorylation by isolated sugar beet chloroplasts (after Magyarosy et al., 96).

Source of chloroplasts	ATP formed	NADPH$_2$ formed	Ratio P:2e$^-$
	(μmoles/mg Chl/h)		
Healthy leaves	109	121	0.90
Infected leaves	80	64	1.25

sion of noncyclic electron transport and photophosphorylation.

Magyarosy et al. (96, 97) reported similar findings with powdery mildew (*Erysiphe polygoni*) of sugar beets. They claimed that both rust and powdery mildew fungi effect a preferential inhibition of noncyclic photophosphorylation. This inhibition leads in host leaves to a decreased rate of photosynthetic CO_2 assimilation (Fig. 2–20) and may contribute to the reduced quantum efficiency of light use (51). The inhibition of noncyclic photophosphorylation stems from an inhibition of electron transport from water to NADP (Table 2–5).

All of these changes are harmful not only to the host plant but also to biotrophic parasites, like rusts and mildew. It was shown recently by Mashaal et al. (98) that blocking photosynthesis in wheat leaves by DCMU, CMU, or by dark treatments, leads to a suppression of the development of stem rust (*Puccinia graminis*). However, this rust inhibition was successfully reversed by sucrose.

In reference to the *Hill reaction*, the available data are meager at present. Mathre (100) isolated chloroplasts from cotton leaves infected by a defoliating strain of *Verticillium albo-atrum* and found that these chloroplasts were less efficient in carrying out the Hill reaction than chloroplasts from healthy leaves. The rate of reaction was decreased in chloroplasts from both chlorotic and nonchlorotic tissues of infected leaves but less so in chloroplasts from the nonchlorotic ones. The overall rate of photosynthesis was also decreased as a consequence of the disruption of the chloroplast function.

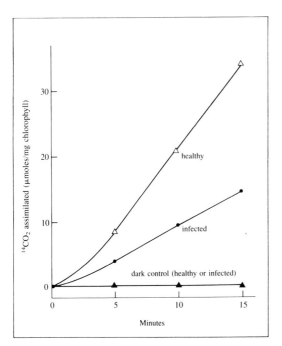

Fig. 2–20. Effect of powdery mildew infection on the rate of photosynthetic $^{14}CO_2$ assimilation by sugar beet leaf discs (after Magyarosy et al., 96).

Carbon-Reduction Reactions

The carbon-reduction reactions are usually suppressed in fungus-infected plants. Rust diseases are probably exceptions, in that CO_2 fixation is enhanced in the early stage of disease. The inhibition of photosynthesis as determined by CO_2 uptake was extensively studied by Edwards (34) in the powdery mildew disease of barley. Here, a biphasic inhibition of CO_2 fixation occurs. The first phase of inhibition seemed independent of chlorophyll content in tissues. The second, however, is due to an impairment in the photosynthetic apparatus that depends upon chlorophyll.

The first phase of inhibition of CO_2 uptake occurs prior to or during early sporulation of powdery mildew (*E. graminis* f. sp. *hordei*), when no chlorosis is seen on leaves (in other words, when there is no apparent decrease in chlorophyll content of infected tissues). Thus, the inhibition does not seem to be directly related to loss of chlorophyll. It is known that in this early stage of disease the chloroplast polysomes may be damaged in mildew-infected barley leaves (32), so the inhibition of CO_2 fixation may be related to some damage other than the loss of chlorophyll. Edwards (34) found that the situation is rather complicated in the first phase of inhibition. Actually, the inhibition of CO_2 uptake occurred only if infected leaves photosynthesized in 0.04% CO_2, a physiological concentration. On the other hand, CO_2 uptake was stimulated in infected leaves when photosynthesis proceeded at 1.0% CO_2. In this second instance, the high concentration CO_2 probably created an

Table 2–6. Effect of α-hydroxy-2-pyridine-methanesulfonic acid (HPMS) on CO_2-uptake rate of healthy and mildew-infected (50 hr) 12-day-old primary barley leaf tissue in 0.04 and 1.0% CO_2 (after Edwards, 34).

Treatment	Uptake rate	
	0.04% CO_2	1.0% CO_2
	mg CO_2/h·dm²	
Healthy	16.0	104
Healthy + HPMS	11.3	120
Infected	12.0	125
Infected + HPMS	variable	123

unnatural environmental condition. Edwards showed that a similar pattern of CO_2 uptake is seen in healthy leaves that have been treated with an inhibitor of glycolic acid oxidase. Hydroxy-pyridine-methanesulfonic acid (HPMS) in 10^{-2} M concentration inhibits photosynthesis (CO_2 uptake) of barley leaves in 0.04% CO_2 but stimulates it in 1.0% CO_2. It was concluded that the first phase of inhibition may be due to an impairment in the metabolism of glycolic acid, the major substrate for photorespiration that is similar to the action of HPMS (Table 2–6). It was shown with rust diseases that the activity of glycolic acid oxidase is strongly depressed in association with the chlorotic symptom but increased in the green islands around the infection sites (78, 79, 80). It is not currently known whether the first phase of inhibition of photosynthesis in mildew-infected barley is connected with the inhibition of glycolic acid oxidase prior to the time chlorotic symptoms develop.

In mildew-infected sugar beet leaves the decline in photosynthesis paralleled a decline in the mesophyll conductance. The latter was determined by a gas exchange procedure in which a stream of air flowed through the leaf. Under these circumstances a reduction in stomatal aperture does not influence the decline in photosynthesis (51). Infected sugar beet leaves, as well as mildewed barley leaves (144), preferentially decrease the concentration of a key enzyme of photosynthesis, ribulose-1,5-bisphosphate carboxylase/oxygenase. This protein, which can constitute 60% of total soluble leaf protein, may serve as a leaf storage protein. It is degraded during senescence (103), and it seems plausible that powdery mildew induces a senescence effect in addition to a parallel juvenescence effect (see Chap. 7).

The second phase of inhibition of CO_2 fixation takes place about six days after infection of barley with powdery mildew, when chlorotic symptoms are visible. This inhibition is directly related to chlorophyll degradation, so the impairment occurs in that part of the photosynthetic apparatus that depends upon chlorophyll.

The biphasic inhibitory pattern described for the powdery mildew disease is not valid for *rust diseases*. In the early stage of rust infection of beans, photosynthesis remains practically unchanged, and even an increase in the chlorophyll level may occur (132). Although respiration increases, the dry weight of the infected leaves increases, and this was shown to be due to enhanced dark fixation of CO_2 at the infection sites (104, 151). Similarly, a stimulation of photosynthetic CO_2 uptake in infected organs of rusted plants was detected *before* rust sporulation in bean and sunflower, if the infection intensity was not very heavy (87). Thus, in the early stage of disease, rust infection may influence the carbon cycle of photosynthesis, but this is probably not deleterious; on the contrary, it seems to be beneficial. In later stages of rust infection the rates of CO_2 fixation ($^{14}CO_2$ incorporation) decline to one-third or one-half those of healthy leaves. This inhibition is directly related to infection density and to chloroplast degradation (chlorosis), as well as to the destruction of other subcellular organelles. With lighter infection density the inhibition of $^{14}CO_2$ fixation is much lower. Degradation of subcellular organelles parallels the initiation of sporulation of the rust fungus. At this stage of disease the metabolism of the host tissues declines toward senescence, which results, for example, in reduction of chloroplast rRNA (133). Senescence associated with the late stage of rust infection of bean leaves can induce declines in both photoassimilation and photorespiration (122). However, respiration is increasing at this stage. Photorespiration does not seem to contribute to the carbon imbalance in diseased beans. Increased respiration and decreased photosynthesis are responsible for that phenomenon.

In summary, rust infection has a double effect on host tissues—a juvenescence effect in an early stage of disease and a senescence effect in the chlorotic and sporulation stage (80). This sequence is evidenced by an early increase then a late decline in CO_2 fixation.

Stimulated CO_2 Fixation in Uncolonized Leaves

It was shown (77) that healthy plants are able to compensate for the effect of removal of some leaves by increased photosynthesis. Furthermore, translocation of photosynthates from leaves remaining on the plant is also affected. Similar compensation may occur in diseased plants. Of particular interest is the finding of Livne (87) that a stimulation of CO_2 fixation in the light is characteristic of the *noninfected* leaves of heavily rusted bean plants. He found this stimulation in the uncolonized leaves under conditions that inhibited photosynthesis of the infected (colonized) leaves on the same plants. Indeed, a stimulation of CO_2 fixation in the noninfected leaves could be a source of carbohydrate for infected leaves. For data illustrating this stimulation, see Table 2–7.

The net photosynthesis in uncolonized leaves of barley infected with powdery mildew is similarly stimu-

Table 2–7. Comparison of photosynthetic activity ($^{14}CO_2$-uptake) by rust-free trifoliate leaves from healthy and rust-infected bean plants (after Livne, 87).

Days after inoculation		$^{14}CO_2$ fixed cpm·10^{-5}/g fresh wt.	Diseased / healthy
2	Healthy	14.61	1.09
	Diseased	14.81	
4	Healthy	16.30	1.05
	Diseased	17.15	
9	Healthy	15.38	1.58
	Diseased	24.32	

lated (143, 147). At the same time, photorespiration is inhibited. However, the decrease in photorespiration is not sufficiently large to account for all the increase in photosynthesis. This increase is particularly great in water-stressed plants. In measuring the apparent rate of photosynthesis of plants it is necessary to evaluate the possible changes in stomatal resistance to gas diffusion, which may influence CO_2 supply (7). It was reported by Williams and Ayres (147) that mildew inhibits stomatal opening in colonized tissues, thereby reducing transpiration rates. However, this leads to a small increase in the water potential of uncolonized tissues, particularly when plants are subjected to water stress. The stomatal opening in barley leaves increases with increasing water potential, so an increased CO_2 uptake occurs in uncolonized leaves under drying conditions. Growth regulators (e.g. cytokinins) can also increase stomatal opening, transport, and partitioning of photoassimilates that can indirectly increase CO_2 uptake in uncolonized tissues. Again, it seems unlikely that this mechanism would play a role in the net increase in photosynthesis when infected plants are not under water stress. Walters and Ayres (143) reported that the increased amount of ribulose-1,5-bisphosphate carboxylase/oxygenase protein is responsible for the major part of the stimulated CO_2 uptake of uncolonized leaves. It is noteworthy that the opposite is the case in powdery-mildew-colonized barley leaves, in which the amount of this enzyme and the activity of its carboxylase function is reduced (143).

Stomatal Apertures

Duniway and Slatyer (30) investigated the *stomatal and intracellular resistance to CO_2 diffusion* of *Fusarium*-infected tomatoes. They have shown that in a late stage of disease a strong reduction of photosynthesis of tomato leaves infected by *Fusarium oxysporium* f. sp. *lycopersici* was caused by an increase in stomatal resistance to gas diffusion and increased intracellular resistance to CO_2 uptake. Both shifts preceded the onset of water stress in a given leaf. Thus, the marked reduction in the photosynthetic capacity of in-

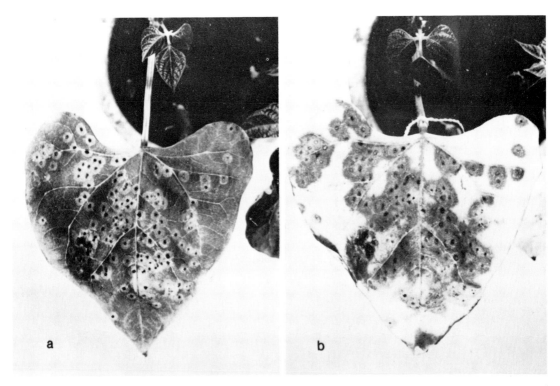

Fig. 2–21. Characteristic symptoms of rust development in pinto bean leaves kept under low light conditions. (a) early ¹ sporulation stage; (b) green island stage (after Sziráki et al., 132).

fected leaves in the late stage of disease cannot be a consequence of leaf water stress. The mechanism of increased intracellular resistance of CO_2 diffusion in diseased leaves remains unknown. However, this mechanism is associated with reductions in both dark respiration and photorespiration.

Infection of potato leaves by *Phytophthora infestans* induces *abnormally wide stomatal apertures*; this, in turn, allows photosynthesis to proceed at a higher rate at the infection sites. The biochemical processes in photosynthesis are not stimulated by infection; the high rate of photosynthesis, in this case, is related only to the maintenance of wide stomatal apertures (39).

Altered Translocation of Organic Compounds

The translocation of photosynthates and nutrients may also be influenced by infection. In diseases caused by obligate parasites, translocation is enhanced from the uninfected tissues or uninfected leaves to the infected ones, while the export from infected leaves is greatly suppressed (88, 120). This process occurs in connection with a hormonal control of translocation and with the formation of green islands (Fig. 2–21) at the infection sites particularly under low light conditions. Green islands retain chlorophyll and are photosynthetically active. Indeed, chlorophyll content of cells at the infection sites remains higher at various stages of disease than in the uninfected control cells

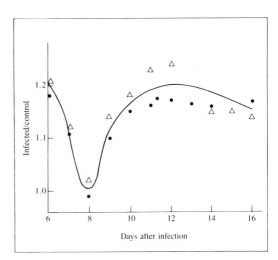

Fig. 2–22. Changes in the chlorophyll content of pinto bean leaf tissues at various stages of infection with *Uromyces phaseoli*. The open triangles and closed circles represent the results of two separate experiments. The total chlorophyll content of control tissue on the 6th day after inoculation was 25.65 nmol/100 mm² in the first, and 28.40 nmol/100 mm² in the second experiment (after Sziráki et al., 132).

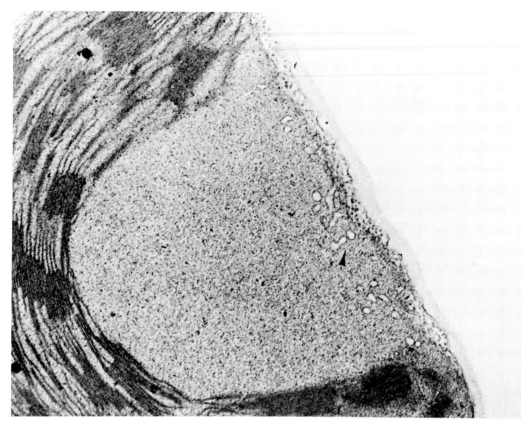

Fig. 2–23. Peripheral reticulum induced by the infection stress in a chloroplast of a green island developed under low light conditions after inoculation of pinto bean leaves with *Uromyces phaseoli* (after Sziráki et al., 132).

(Fig. 2–22), and starch tends to accumulate in these cells in greater quantities than in surrounding areas (145). It was contended that in infections by powdery mildew of wheat, the green islands result from the reformation of chlorophyll after the initial destruction upon infection. The rest of the leaf remains chlorotic (3). Chloroplasts in the green islands of rust-infected bean leaves remain in a relatively functional integrated state but contain abundant peripheral reticula (Fig. 2–23), indicating that a slight structural aberration occurs simultaneously with the juvenility effect. In green islands the activity of glycolic acid oxidase, a chloroplast-associated enzyme, increases, reaching a higher level than in healthy leaf tissues (80). All these phenomena seem to be causally related to the cytokinin production by obligate parasites or the tissues infected by them. On the other hand, in the late blight disease of potato, which is caused by a nonobligate parasite (*Phytophthora infestans*), there was no evidence for movement of photosynthetic products from uninfected to infected leaf tissue (39). Therefore, infection by *P. infestans* appears not to lead to the development of a metabolic sink comparable to that in leaves infected by rusts or powdery mildews. For a further consideration of translocation as affected by disease, see Chap. 8.

Starch Accumulation

Starch metabolism is also frequently affected by fungal parasites. Obligate parasites seem to cause starch accumulation in the diseased host chloroplasts (1) as well as in host cytoplasm of nongreen tissue (148). In rust diseases, starch content of infected leaves decreases soon after infection, then increases during or near the time of sporulation, and decreases sharply thereafter (92, 105). Fig. 2–24 shows the variation in starch content in wheat leaves infected by stripe rust (*Puccinia striiformis*).

Starch is synthesized as a result of increased glucose-1-phosphate adenyl transferase activity. MacDonald and Strobel (92) have shown that the activity of this enzyme closely parallels the starch curve during pathogenesis (Fig. 2–24).

It is an allosteric, or regulatory, enzyme that is activated by intermediates of glycolysis and inhibited by inorganic phosphate (P_i). Thus, the levels of activators and inhibitors control the rate of starch synthesis. The activity of this enzyme in rust-infected leaves not only parallels starch accumulation (Fig. 2–24 a and b), but it is also almost the reciprocal of P_i concentration (Fig. 2–24c). It was suggested that the level of P_i in diseased

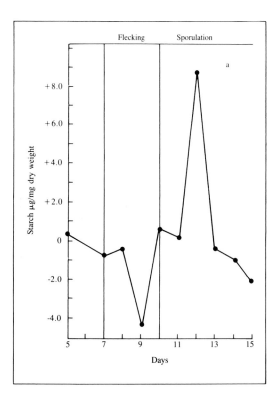

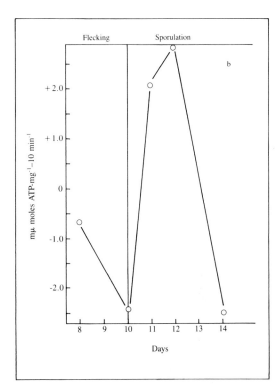

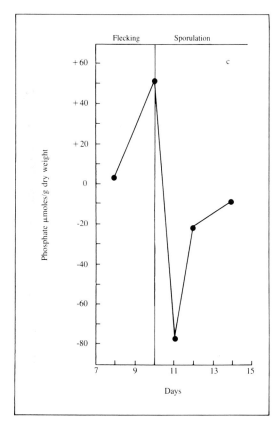

Fig. 2–24. (a) Variation in starch content in wheat leaves infected with *Puccinia striiformis* from 5 to 15 days after inoculation expressed as the difference in starch content between diseased and healthy leaves. The vertical lines indicate the division of stages of host-parasite interaction. (b) Variation in the rate of glucose-1-phosphate adenyl transferase activity in the presence of proportionate concentrations of metabolites found in healthy and *P. striiformis*-infected wheat leaves from 8 to 14 days after inoculation. The value obtained for healthy leaves was subtracted from the value obtained from diseased leaves to give each point on the graph. The data are expressed as the differences in enzyme activity between simulated metabolite preparations from diseased and healthy leaves. (c) Variation in inorganic phosphate concentration in leaves infected with *P. striiformis* from 8 to 14 days after inoculation expressed as the difference in concentration between diseased and healthy leaves. The vertical line indicates the division of stages of host-parasite interaction. The value obtained for healthy leaves was subtracted from that of diseased leaves to give each point on the graph (after MacDonald and Strobel, 92).

leaves regulates the rate of starch synthesis through glucose-1-phosphate adenyl transferase. This is an example of host-pathogen relationship, wherein the pathogen acts on a regulatory enzyme of the host, influencing an important biosynthesis. Although the significance of starch accumulation to the pathogen, if any, is not known, it is usually associated with chloroplast deterioration and occurs late in the course of pathogenesis.

Comparative Analysis of Disease Physiology—Photosynthesis

Changes in host photosynthetic behavior induced by infection with viruses, bacteria, and fungi inevitably show many points of similarity. Central to these changes are the abnormalities in chloroplast form and function that accompany chlorosis or necrosis of infected tissue. The chloroplasts may show signs of degeneration or senescence or, in some cases when young tissue becomes infected, impaired development. There is also an accompanying reduction in rRNA, with consequent reduction in ribosome function. The causes of these abnormalities remain largely unknown, although some bacteria (e.g. *Pseudomonas tabaci*) and fungi (e.g. barley mildew, broad bean rust) produce toxins that have been demonstrated to interfere with chloroplast function. Toxins may suppress chlorophyll synthesis or inhibit photophosphorylation.

Chlorosis is always accompanied by reduced chlorophyll content of the infected tissue when compared with control tissue, a reduction that leads inevitably to a reduction in the light-dependent reactions. Photosynthetic phosphorylation and Hill reaction activity are, with few exceptions, reduced. There may, however, be a reduction in light-dependent reactions greater than can be accounted for by the decrease in chlorophyll. In beets infected with beet yellows, there is a 50% reduction in Hill reaction activity per unit of chlorophyll, although photosynthetic phosphorylation is the same per unit of chlorophyll in both healthy and rust-infected oats.

Carbon dioxide assimilation may be increased or decreased, depending on the time after infection and on the tissue sampled. For example, there may be an early stimulation of CO_2 assimilation at the site of infection (in rust infections, for example), followed by a decline as symptoms progress. Areas of tissue that remain green and uninfected may have a stimulated rate of CO_2 assimilation, at least partially compensating for the reduction in infected tissue. Similar considerations apply to green areas of virus-infected tissue. Additionally, there may be pathogen-induced changes in stomatal aperture which, by altering the rate of CO_2 flow, can influence its rate of assimilation. Such is the case in, for example, bacterial wilt of banana caused by *P. solanacearum*, or *Fusarium* wilt of tomatoes.

Associated with changes in CO_2 fixation, there may be changes in activity of enzymes involved in photosynthesis, and there may also be a redirection of some

metabolic pathways, with accumulation of amino acids and organic acids at the expense of sugars and sugar phosphates. Finally, both viral and fungal infection can lead to impairment of starch metabolism, infection by viruses often resulting in depression of both synthesis and degradation.

The net effect of a disease involving chlorosis or necrosis of infected tissue is usually an ultimate reduction in photosynthetic activity of the plant, a reduction contributed to by the reduced photosynthetic capacity of infected tissue and an often reduced plant growth rate. The extent of the reduction may be a reflection of symptom severity. Although the observed changes induced by various pathogens may be similar in many respects, it must be remembered that the primary causes of these changes remain, in most instances, obscure and may, therefore, be quite diverse.

References

1. Akai, S., M. Fukutomi, N. Ishida, and H. Kunoh. 1967. An anatomical approach to mechanisms of fungal infection in plants. In:The dynamic role of molecular constituents in plant-parasite interaction. C. J. Mirocha and I. Uritani (eds.). Amer. Phytopath. Soc., St. Paul, MN, pp. 1–20.

2. Akoyunoglou, G., and J. H. Argyroudi-Akoyunoglou (eds.). 1978. Chloroplast development. Elsevier/North Holland, Amsterdam, p. 888.

3. Allen, P. J. 1942. Changes in the metabolism of wheat leaves induced by infection with powdery mildew. Amer. J. Bot. 29:425–35.

4. Arnon, D. J., H. Y. Tsujimoto, B. D. McSwain, and R. K. Chain. 1968. Separation of two photochemical systems of photosynthesis by fractionation of chloroplasts. In: Comparative biochemistry and biophysics of photosynthesis. K. Shibata, A. Takamiya, A. I. Jagendorf, and R. C. Fuller (eds). Univ. of Pennsylvania Press, Philadelphia, pp. 113–32.

5. Arntzen, C. J. 1972. Inhibition of photophosphorylation by tentoxin, a cyclic tetrapeptide. Biochim. Biophys. Acta 283:539–42.

6. Arntzen, C. J., and J. M. Briantais. 1975. Chloroplast structure and function. In: Bioenergetics of photosynthesis. Govindjee (ed.). Academic Press, New York, pp. 51–113.

7. Ayres, P. G., and J. C. Zadoks. 1979. Combined effects of powdery mildew disease and soil water level on the water relations and growth of barley. Physiol. Plant Path. 14:347–67.

8. Bailiss, K. W. 1970. Infection of cucumber cotyledons by cucumber mosaic virus and the participation of chlorophyllase in the development of the chlorotic lesions. Ann. Bot. 34:647–55.

9. Balls, A. K., and L. F. Martin. 1938. Amylase activity of mosaic tobacco. Enzymologia 5:233–38.

10. Barber, J. (ed.). 1976. The intact chloroplast. Topics in photosynthesis. Vol. 1. Elsevier, Amsterdam.

11. Barr, R., L. Mafree, and F. L. Crane. 1967. Quinone distribution in horse-chestnut chloroplasts, globules and lamellae. Amer. J. Bot. 54:365–74.

12. Bassham, J. A., and M. Calvin. 1957. The path of carbon in photosynthesis. Prentice-Hall, Englewood Cliffs, NJ.

13. Beckman, C. H., W. A. Brun, and I. W. Buddenhagen. 1962. Water relations in banana plants infected with *Pseudomonas solanacearum*. Phytopathology 52:1144–48.

14. Bedbrook, J. R., and R. E. F. Matthews. 1973. Changes in the flow of early products of photosynthetic carbon fixation associated with the replication of TYMV. Virology 53:84–91.

15. Betto, E., M. Bassi, M. A. Favali, and G. G. Conti. 1972. An electron microscope and autoradiographic study of tobacco leaves infected with the U5 strain of tobacco mosaic virus. Phytopath. Z. 75:193–201.

16. Bogorad, L. 1966. The biosynthesis of chlorophylls. In: The chlorophylls. L. P. Vernon and G. R. Seely (eds.). Academic Press, New York, pp. 481–510.

17. Bradbeer, J. W. 1981. Development of photosynthetic function during chloroplast biogenesis. In: The biochemistry of plants. Photosynthesis. M. D. Hatch and N. K. Boardman (eds.). Academic Press, New York, 8:424–72.

18. Braun, A. C. 1955. A study on the mode of action of the wildfire toxin. Phytopathology 45:659–64.

19. Brun, W. A. 1961. Photosynthesis and transpiration from upper and lower surfaces of intact banana leaves. Plant Physiol. 36:399–405.

20. Carroll, T. W. 1970. Relation of barley stripe mosaic virus to plastids. Virology 42:1015–22.

21. Carroll, T. W., and T. Kosuge. 169. Changes in structure of chloroplasts accompanying necrosis of tobacco leaves systemically infected with tobacco mosaic virus. Phytopathology 59:953–62.

22. Castelfranco, P. A., and S. I. Beale. 1981. Chlorophyll biosynthesis.In: The biochemistry of plants. Photosynthesis. M. D. Hatch and N. K. Boardman (eds.). Academic Press, New York, 8:376–421.

23. Chalcroft, J. P., and R. E. F. Matthews. 1967. Role of virus strains and leaf ontogeny in the production of mosaic patterns by turnip yellow mosaic virus. Virology 33:659–73.

24. Cohen, J., and G. Loebenstein. 1975. An electron microscope study of starch lesions in cucumber cotyledons infected with tobacco mosaic virus. Phytopathology 65:32–39.

25. Crosbie, E. S., and R. E. F. Matthews. 1974. Effects of TYMV infection on leaf growth in *Brassica pekinensis* Rupr. Physiol. Plant Path. 4:389–400.

26. Diener, T. O. 1963. Physiology of virus-infected plants. Ann. Rev. Phytopath. 1: 197–218.

27. Doke, N., and T. Hirai. 1969. Starch metabolism in tobacco leaves infected with tobacco mosaic virus. Phytopath. Z. 65:307–17.

28. Doke, N., and T. Hirai. 1970. Radioautographic studies on the photosynthetic CO_2 fixation in virus-infected leaves. Phytopathology 60:988–91.

29. Doke, N., and T. Hirai. 1970. Effects of tobacco mosaic virus infection on photosynthetic CO_2 fixation and $^{14}CO_2$ incorporation into protein in tobacco leaves. Virology 42:68–77.

30. Duniway, J. M., and R. O. Slatyer. 1971. Gas exchange studies on the transpiration and photosynthesis of tomato leaves affected by *Fusarium oxysporum* f. sp. *lycopersici*. Phytopathology 61:1377–81.

31. Durbin, R. D. 1971. Chlorosis-inducing pseudomonad toxins: Their mechanism of action and structure. In: Morphology and biochemical events in plant parasite interaction. S. Akai and S. Ouchi (eds.). Phytopath. Soc. Jap., Tokyo, pp. 369–385.

32. Dyer, T. A., and K. J. Scott. 1972. Decrease in chloroplast content of barley leaves infected with powdery mildew. Nature 236:237–38.

33. Edwards, G. E., and S. C. Huber. 1981. The C_4 pathway. In: The biochemistry of plants. Photosynthesis. M. D. Hatch and N. K. Boardman (eds.). Academic Press, New York, 8: 238–81.

34. Edwards, H. H. 1970. Biphasic inhibition of photosynthesis in powdery mildewed barley. Plant Physiol. 45:594–97.

35. Ehara, Y., and T. Misawa. 1975. Occurrence of abnormal protoplasts in tobacco leaves infected systemically with the ordinary strain of cucumber mosaic virus. Phytopath. Z. 84:233–52.

36. Esau, K. 1968. Viruses in plant hosts. Univ. of Wisconsin Press, Madison.

37. Esau, K., and J. Cronshaw. 1967. Relation of tobacco mosaic virus to the host cells. J. Cell Biol. 33:665–78.

38. Farkas, G. L., and L. Lovrekovich. 1963. Counteraction by kinetin of the toxic effect of *Pseudomonas tabaci*. Phytopath. Z. 47:391–98.

39. Farrell, G. M. 1971. Localization of photosynthetic products in potato leaves infected by *Phytophthora infestans*. Physiol. Plant Path. 1:457–67.

40. Favali, M. A., S. Pellegrini, and M. Bassi. 1975. Ultrastructural alterations induced by rice tungro virus in rice leaves. Virology 66:502–7.

41. Gates, D. W., and R. T. Gudauskas. 1969. Photosynthesis, respiration and evidence of a metabolic inhibitor in corn infected with maize dwarf mosaic virus. Phytopathology 59:575–80.

42. Gerola, F. M., M. Bassi, and G. Giussani. 1966. Some observations on the shape and localization of different viruses in experimentally infected plants and on the fine structure of the host cells. III. Turnip yellow mosaic virus in *Brassica chinensis* L. Caryologia 19:457–79.

43. Gibbs, M., and E. Latzko (eds.). 1979. Encyclopedia of plant physiology. Photosynthesis II. Springer-Verlag, Berlin, 6:578.

44. Goffeau, A., and J. M. Bové. 1965. Virus infection and photosynthesis. I. Increased photophosphorylation by chloroplasts from Chinese cabbage infected with turnip yellow mosaic virus. Virology 27:243–52.

45. Goodman, R. N. 1972. Phytotoxin-induced ultrastructural modifications of plant cells. In: Phytotoxins in plant diseases. R. K. S. Wood, A. Ballio, A. Graniti (eds.). Academic Press, New York, pp. 311–29.

46. Goodman, R. N., P.-Y. Huang, and J. A. White. 1976. Ultrastructural evidence for immobilization of an incompatible bacterium, *Pseudomonas pisi*, in tobacco leaf tissue. Phytopathology 66:754–64.

47. Goodman, R. N., and S. B. Plurad. 1971. Ultrastructural changes in tobacco undergoing the hypersensitive reaction caused by plant pathogenic bacteria. Physiol. Plant Path. l:11–15.

48. Goodwin, T. W. (ed.). 1967. Biochemistry of chloroplasts. Academic Press, New York.

49. Goodwin, T. W. (ed.). 1976. Chemistry and biochemistry of plant pigments. Vol. 1. 2d ed. Academic Press, New York.

50. Goodwin, T. W., and E. I. Mercer. 1982. Introduction to plant biochemistry. 2d ed. Pergamon Press, Oxford.

51. Gordon, T. R., and J. M. Duniway. 1982. Effects of powdery mildew infection on the efficiency of CO_2 fixation and light utilization by sugar beet leaves. Plant Physiol. 69:139–42.

52. Govindjee. 1975. Bioenergetics of photosynthesis. Academic Press, New York.

53. Govindjee (ed.). 1982. Photosynthesis. Vol. 1. Energy conversion by plants and bacteria. Academic Press, New York.

54. Govindjee (ed.). 1982. Photosynthesis. Vol. 2. Development, carbon metabolism and plant productivity. Academic Press, New York.

55. Govindjee, and R. Govindjee. 1975. Introduction to photosynthesis. In: Bioenergetics of photosynthesis. Govindjee (ed.). Academic Press, New York, pp. 1–50.

56. Granik, S. 1967. The heme and chlorophyll biosynthetic chain.In: Biochemistry of chloroplasts. T. W. Goodwin (ed.). Academic Press, New York, 2:373–410.

57. Gregory, R. P. F. 1977. Biochemistry of photosynthesis. 2d ed. Wiley, New York.

58. Guyla, T., and J.M. Dunleavy. 1979. Inhibition of chlorophyll synthesis by *Pseudomonas glycinea*. Crop Sci. 19:251–64.

59. Hall, A. E., and R. S. Loomis. 1972. An explanation for the difference in photosynthetic capabilities of healthy and beet yel-

lows virus-infected sugar beets (*Beta vulgaris* L.). Plant Physiol. 50:576–80.

60. Hampton, R. E., D. L. Hopkins, and T. G. Nye. 1966. Biochemical effects of tobacco etch virus infection on tobacco leaf tissue. I. Protein synthesis by isolated chloroplasts. Phytochemistry 5:1181–85.

61. Hatch, M. D., and N. K. Boardman (eds.). 1981. The biochemistry of plants. Vol. 8. Photosynthesis. Academic Press, New York.

62. Hatch, M. D., and C. R. Slack. 1970. The C_4-dicarboxylic acid pathway of photosynthesis. In: Progress in phytochemistry. L. Reinhold and Y. Liwschitz (eds.). Interscience, New York, 2:35–106

63. Hatch, M. D., and C. R. Slack. 1970. Photosynthetic CO_2-fixation pathways. Ann. Rev. Plant Physiol. 21:141–62.

64. Hatta, T., S. Bullivant, and R. E. F. Matthews. 1973. Fine structure of vesicles induced in chloroplasts of Chinese cabbage leaves by infection with turnip yellow mosaic virus. J. Gen. Virol. 20:37–50.

65. Hiller, R. G., and D. J. Goodchild. 1981. Thylakoid membrane and pigment organization. In: The biochemistry of plants. Photosynthesis. M. D. Hatch and N. K. Boardman (eds.). Academic Press, New York, 8:2–49.

66. Holmes, F. O. 1931. Local lesions of mosaic in *Nicotiana tabacum* L. Contrib. Boyce Thompson Inst. 3:163–72.

67. Honda, Y., and C. Matsui. 1971. Distribution of tobacco mosaic virus in etiolated tobacco leaf cells. Phytopathology 61:759–62.

68. Hopkins, D. L., and R. E. Hampton. 1969. Effects of tobacco etch virus infection upon the light reactions of photosynthesis in tobacco leaf tissue. Phytopathology 59:677–79.

69. Hopkins, D. L., and R. E. Hampton. 1969. Effects of tobacco etch virus infection upon the dark reactions of photosynthesis in tobacco leaf tissue. Phytopathology 59:1136–40.

70. Huang, J. S., and R. N. Goodman. 1972. Alterations in structural proteins from chloroplast membranes of bacterially induced hypersensitive tobacco leaves. Phytopathology 62:1428–34.

71. Huang, J. S., P.-Y. Huang, and R. N. Goodman. 1974. Ultrastructural changes in tobacco thylakoid membrane protein caused by a bacterially induced hypersensitive reaction. Physiol. Plant Path. 4:93–97.

72. Hunter, C. S., and W. E. Peat. 1973. The effect of tomato aspermy virus on photosynthesis in the young tomato plant. Physiol. Plant Path. 3:517–24.

73. Jensen, S. G. 1968. Photosynthesis, respiration and other physiological relationships in barley infected with barley yellow dwarf virus. Phytopathology 58:204–8.

74. Jensen, S. G. 1969. Composition and metabolism of barley leaves infected with barley yellow dwarf virus. Phytopathology 59:1694–98.

75. Jockusch, H. and B. Jockusch. Early cell death caused by TMV mutants with defective coat proteins. Molec. Gen. Genet. 102:204–9.

76. Kato, S. and T. Misawa. 1974. Studies on the infection and multiplication of plant viruses. VII. The breakdown of chlorophyll in tobacco leaves systemically infected with cucumber mosaic virus. Ann. Phytopath. Soc. Jap. 40:14–21.

77. Khan, A., and G. R. Sagar. 1969. Alteration of the pattern of dstribution of photosynthetic products in the tomato by manipulation of the plant. Ann. Bot. 33:753–62.

78. Király, Z. 1968. Role of cytokinins in biochemical and morphogenetic changes induced by rust infections. First Int. Congr. Plant Path. (London), p. 105. (Abstr.)

79. Király, Z., and G. L. Farkas. 1957. Decrease in glycolic acid oxidase activity in wheat leaves infected with *Puccinia graminis* var. *tritici*. Phytopathology 47:277–78.

80. Király, Z., B. I. Pozsár, and M. El Hammady. 1966. Cytokinin activity in rust-infected plants: Juvenility and senescence in diseased leaf tissues. Acta Phytopath. Hung. 1:29–37.

81. Kitajima, E. W., and A. S. Costa. 1973. Aggregates of chloroplasts in local lesions induced in *Chenopodium quinoa* willd. by turnip mosaic virus. J. Gen. Virol. 20: 413–16.

82. Laetsch, W. M. 1971. Chloroplast structural relationships in leaves of C_4 plants. In: Photosynthesis and photorespiration. M. D. Hatch, C. B. Osmond, and R. O. Slayter (eds.). Wiley Interscience, New York, pp. 323–49.

83. Lehninger, A. L. 1982. Principles of biochemistry. Worth, New York.

84. Lesemann, D. E. 1977. Virus group-specific and virus-specific cytological alterations induced by members of the tymovirus group. Phytopath. Z. 90:315–36.

85. Lichtenthaler, H. K. 1968. Plastoglobuli and the fine structure of plastids. Endeavour 102:144–49.

86. Lichtenthaler, H. K. 1969. Die plastoglobuli von spinat, ihre grosse, isolierung und lipochinonzusammensetzung. Protoplasma 68:65–77.

87. Livne, A. 1964. Photosynthesis in healthy and rust-affected plants. Plant Physiol. 39:614–21.

88. Livne, A., and J. M. Daly. 1966. Translocation in healthy and rust-affected beans. Phytopathology 56:170–75.

89. Lorimer, G. H., and T. J. Andrews. 1981. The C_2 chemo-and photo-respiratory carbon oxidation cycle. In: The biochemistry of plants. Photosynthesis. M. D. Hatch and N. K. Boardman (eds.). Academic Press, New York, 8:329–74.

90. Lovrekovich, L., Z. Klement, and G. L. Farkas. 1964. Toxic effect of *Pseudomonas tabaci* on RNA metabolism in tobacco and its counteraction by kinetin. Science 145:165.

91. Lyon, F. M., and R. K. S. Wood. 1975. Production of phaseollin, coumestrol and related compounds in bean leaves inoculated with *Pseudomonas* spp. Physiol. Plant Path. 6:117–24.

92. MacDonald, P. W., and G. A. Strobel. 1970. Adenosine diphosphate-glucose pyrophosphorylase control of starch accumulation in rust-infected wheat leaves. Plant Physiol. 46:126–35.

93. McMullen, C. R., W. S. Gardner, and G. A. Hyers. 1978. Aberrant plastids in barley leaf tissue infected with barley stripe mosaic virus. Phytopathology 68:317–25.

94. Magyarosy, A. C., and B. B. Buchanan. 1975. Effect of bacterial infiltration of photosynthesis of bean leaves. Phytopathology 65:777–80.

95. Magyarosy, A. C., B. B. Buchanan, and P. Schurmann. 1973. Effect of a systemic virus infection on chloroplast function and structure. Virology 55:426–38.

96. Magyarosy, A. C., P. Schurmann, and B. B. Buchanan. 1976. Effect of powdery mildew infection on photosynthesis by leaves and chloroplasts of sugar beets. Plant Physiol. 57:486–89.

97. Magyarosy, A. C., P. Schurmann, P. Montalbini, and B. B. Buchanan. 1977. Effect of infection by obligate parasites on photosynthesis. In: Current topics in plant pathology. Z. Király (ed.). Akadémiai Kiadó, Budapest, pp. 89–98.

98. Mashaal, S. F., B. Barna, and Z. Király. 1981. Effect of photosynthesis inhibitors on wheat stem rust development. Acta Phytopath. Hung. 16:45–48.

99. Mathis, P., and G. Paillotin. 1981. Primary processes of photosynthesis. In: The biochemistry of plants. Photosynthesis. M. D. Hatch and N. K. Boardman (eds.). Academic Press, New York, 8:98–161.

100. Mathre, D. E. 1968. Photosynthetic activities of cotton plants infected with *Verticillium albo-atrum*. Phytopathology 58:137–41.

101. Matthews, R. E. F. 1981. Plant virology. 2d ed. Academic Press, New York.

102. Matthews, R. E. F., and S. Sarkar. 1976. A light induced

structural change in chloroplasts of Chinese cabbage cells infected with turnip yellow mosaic virus. J. Gen. Virol. 33:435–46.

103. Miller, B. L., and R. C. Huffaker. 1982. Hydrolysis of ribulose–1, 5, -biphosphate carboxylase by endoproteinases from senescing barley leaves. Plant Physiol. 69: 58–62.

104. Mirocha, C. J., and P. D. Rick. 1967. Carbon dioxide fixation in the dark as a nutritional factor in parasitism. In: The dynamic role of molecular constituents in plant-parasite interaction. C. J. Mirocha and I. Uritani (eds.). Amer. Phytopath. Soc., St. Paul, MN, pp. 121–43.

105. Mirocha, C. J., and A. I. Zaki. 1966. Fluctuations in amount of starch in host plants invaded by rust and mildew fungi. Phytopathology 56:1220–24.

106. Mohamed, N. A. 1973. Some effects of systemic infection by tomato spotted wilt virus on chloroplasts of *Nicotiana tabacum* leaves. Physiol. Plant Path. 3:509–16.

107. Moline, H. E., and S. G. Jensen. 1975. Histochemical evidence for glycogen-like deposits in barley yellow dwarf virus-infected barley leaf chloroplasts. J. Vet. Res. 53:217–21.

108. Montalbini, P., and B. B. Buchanan. 1974. Effect of rust infection on photophosphorylation by isolated chloroplasts. Physiol. Plant Path. 4:191–96.

109. Montalbini, P., F. Koch, M. Burba, and E. F. Elstner. 1978. Increase in lipid-dependent carotene destruction as compared to ethylene formation and chlorophyllase activity following mixed infection of sugar beet (*Beta vulgaris* L.) with beet yellows virus and beet mild yellowing virus. Physiol. Plant Path. 12:211–23.

110. Murant, A. F., I. M. Roberts, and A. M. Hutcheson. 1975. Effects of parsnip yellow fleck virus on plant cells. J. Gen. Virol. 26:277–85.

111. Osborne, D. J. 1962. Effect of kinetin on protein and nucleic acid metabolism in *Xanthium* leaves during senescence. Plant Physiol. 37:595–602.

112. Owen, P. C. 1957. The effects of infection with tobacco mosaic virus on the photosynthesis of tobacco leaves. Ann. Appl. Biol. 45:456–61.

113. Owen, P. C. 1957. The effect of infection with tobacco etch virus on the rates of respiration and photosynthesis of tobacco leaves. Ann. Appl. Biol. 45:327–31.

114. Owen, P. C. 1958. Photosynthesis and respiration rates of leaves of *Nicotiana glutinosa* infected with tobacco mosaic virus and of *N. tabacum* infected with potato virus X. Ann. Appl. Biol. 46:198–204.

115. Panopoulos, N. J., G. Faccioli, and A. H. Gold. 1972. Kinetics of carbohydrate metabolism in curly top virus-infected tomato plants. Phytopath. Medit. 11:48–58.

116. Park, R. B. 1976. The chloroplast. In: Plant biochemistry. 3d ed. J. Bonner and J. E. Varner (eds.). Academic Press, New York, pp. 115–45.

117. Pavlovkin, J., and A. Novacky. 1984. Bacterial hypersensitivity-related alterations in cell membranes. In: Membrane transport in plants W. J. Cram, K. Janacek, R. Rybova, and K. Sigler (eds.). Academia Praha, Czechoslovakia, pp. 422–23.

118. Peterson, P. D., and H. H. McKinney. 1938. The influence of four mosaic diseases on the plastid pigments and chlorophyllase in tobacco leaves. Phytopathology 28:329–42.

119. Platt, S. G., F. Henriques, and L. Rand. 1979. Effects of virus infection on the chlorophyll content, photosynthetic rate and carbon metabolism of *Tolmiea menziesii*. Physiol. Plant Path. 15:351–65.

120. Pozsár, B. I., and Z. Király. 1966. Phloem-transport in rust-infected plants and the cytokinin-directed long-distance movement of nutrients. Phytopath. Z. 56:297–309.

121. Prebble, J. N. 1981. Mitochondria, chloroplasts and bacterial membranes. Longman, New York.

122. Raggi, V. 1980. Correlation of CO_2 compensation point

(γ) with photosynthesis and respiration and CO_2 sensitive γ in rust-affected bean leaves. Physiol. Plant Path. 16:19–24.

123. Richmond, A. E., and A. Lang. 1957. Effect of kinetin on protein content and survival of detached *Xanthium* leaves. Science 125:650–51.

124. Roberts, P. L., and K. R. Wood 1982. Effects of a severe (P6) and a mild (W) strain of cucumber mosaic virus on tobacco leaf chlorophyll, starch and cell ultrastructure. Physiol. Plant Path. 21:31–37.

125. Robinson, S. P., and D. A. Walker. 1981. Photosynthetic carbon reduction cycle. In: The biochemistry of plants. Photosynthesis. M. D. Hatch and N. K. Boardman (eds.). Academic Press, New York, 8:193–236.

126. Sasser, M., A. K. Andrews, and Z. U. Doganay. 1974. Inhibition of photosynthesis diminishes antibacterial action of pepper plants. Phytopathology 64:770–72.

127. Shalla, T. A. 1964. Assembly and aggregation of tobacco mosaic virus in tomato leaflets. J. Cell Biol. 21:253–64.

128. Shalla, T. A., L. J. Petersen, and L. Giunchedi. 1975. Partial characterization of virus-like particles in chloroplasts of plants infected with the U5 strain of TMV. Virology 66:94–105.

129. Shaw, M., and M. S. Manocha. 1965. Fine structure in detached, senescing wheat leaves. Can. J. Bot. 43:747–56.

130. Spikes, J. D., and M. Stout. 1955. Photochemical activity of chloroplasts isolated from sugar beet infected with virus yellows. Science 122:375–76.

131. Sridhar, R., P. R. Reddy, and A. Anjaneyulu. 1976. Physiology of rice tungro virus disease. Phytopath. Z. 86:136–43.

132. Sziráki, I., L. A. Mustárdy, A. Faludi-Dániel, and Z. Király. 1984. Alterations in chloroplast ultrastructure and chlorophyll content in rust infected pinto beans at different stages of disease development. Phytopathology 74:77–84.

133. Tani, T., M. Yoshikawa, and N. Naito. 1973. Effect of rust infection of oat leaves on cytoplasmic and chloroplast ribosomal ribonucleic acids. Phytopathology 63:491–94.

134. Tolbert, N. E. (ed.). 1980. The biochemistry of plants. Vol. 1. The plant cell. Academic Press, New York.

135. Tomlinson, J. A., and M. J. W. Webb. 1978. Ultrastructural changes in chloroplasts of lettuce infected with beet western yellows virus. Physiol. Plant Path. 12:13–18.

136. Trebst, A., and M. Avron. (eds.). 1977. Encyclopedia of plant physiology. Vol. 5. Photosynthesis I. Springer-Verlag, Berlin.

137. Tu, J. C. 1979. Alterations in chloroplast and cell membranes associated with cAMP-induced dissociation of starch grains in clover yellow mosaic virus infected clover. Can. J. Bot. 57:360–69.

138. Tu, J. C., R. E. Ford, and C. J. Krass. 1968. Comparisons of chloroplasts and photosynthetic rates of plants infected and not infected by maize dwarf mosaic virus. Phytopathology 58:285–88.

139. Turner, J. G. 1981. Tabtoxin, produced by *Pseudomonas tabaci*, decreases *Nicotiana tabacum* glutamine synthetase in vivo and causes accumulation of ammonia. Physiol. Plant Path. 19:57–67.

140. Ushiyama, R., and R. E. F. Matthews. 1970. The significance of chloroplast abnormalities associated with infection by turnip yellow mosaic virus. Virology 42:293–303.

141. Vesk, M., J. V. Possingham, and F. V. Mercer. 1966. The effect of mineral nutrient deficiencies on the structure of the leaf cells of tomato spinach and maize. Aust. J. Bot. 14:1–18.

142. von Wettstein, D. 1959. The formation of plastid structure. Brookhaven Symp. Biol. 11:138–59.

143. Walters, D. R., and P. G. Ayres 1983. Changes in nitrogen utilization and enzyme activities associated with CO_2 exchanges in healthy leaves of powdery mildew-infected barley. Physiol. Plant Path. 23:447–59.

144. Walters, D. R., and P. G. Ayres. 1984. Ribulose biphosp-

hate carboxylase and enzymes of CO_2 assimilation in a compatible barley/powdery mildew combination. Phytopath Z. (In press.)

145. Wang, D. 1961. The nature of starch accumulation at the rust infection site in leaves of pinto bean plants. Can. J. Bot. 39:1595–1604.

146. Welkie, G. W., S. F. Yang, and G. W. Miller. 1967. Metabolic changes induced by cucumber mosaic virus in resistant and susceptible strains of cowpea. Phytopathology 57:472–75.

147. Williams, G. M., and P. G. Ayres. 1981. Effects of powdery mildew and water stress on CO_2 exchange in uninfected leaves of barley. Plant Physiol. 68:527–30.

148. Williams, P. H., N. T. Keen, J. D. Strandberg, and S. S. McNabola. 1968. Metabolite synthesis and degradation during clubroot development in cabbage hypocotyls. Phytopathology 58:921–28.

149. Wynn, W. K., Jr. 1963. Photosynthetic phosphorylation by chloroplasts isolated from rust-infected oats. Phytopathology 53:1376–77.

150. Zaitlin, M., and A. T. Jagendorf. 1960. Photosynthetic phosphorylation and Hill Reaction activities of chloroplasts isolated from plants infected with tobacco mosaic virus. Virology 12:477–86.

151. Zaki, A. I., and C. J. Mirocha. 1965. Carbon dioxide fixation by rust-infected bean plants in the dark. Phytopathology 55:1303–8.

3

Respiration

The Healthy Plant

In healthy plant cells the function of respiration is twofold, to provide energy and carbon skeletons. Respiration, from one point of view, is the oxidation (aerobic or anaerobic breakdown) of *organic material to simple compounds* (catabolism) (Fig. 3–1). In aerobic respiration, this process takes place with the uptake of oxygen from the air. In fermentation, the degradation of organic materials is anaerobic.

One of the simplest substrates for oxidation is glucose. The overall reaction involved in the respiration of glucose is as follows:

$$C_6H_{12}O_6 + 6\ O_2 = 6\ CO_2 + 6\ H_2O$$

The respiratory process, however, can be divided into two portions (see also Fig. 3–1): (1) *The initial phase*, known as the Embden-Meyerhof-Parnas system (also known as glycolysis or fermentation), in which each glucose molecule is converted to two molecules of pyruvic acid. (2) *The terminal phase*, in which pyruvate is oxidized to CO_2 and H_2O.

The initial phase can take place in the absence of O_2 and is common both to aerobic respiration and to alcoholic or lactic fermentation (Fig. 3–2). The terminal phase is strictly aerobic. In the presence of oxygen, the full *oxidative respiration* (glycolysis + terminal phase) predominates in plant tissues. On the other hand, under anaerobic conditions, *fermentation* is carried out. Fermentation is the anaerobic degradation of pyruvate to lactic acid or alcohol. In both processes, pyruvic acid is reduced rather than oxidized.

In addition to respiration's function of providing carbon skeletons, the process of energy-liberation during respiration is of basic importance. The free energy of carbohydrates and related compounds is liberated, conserved, and transported as chemical energy. In this process chemical energy is extracted from organic nutrients. It is a characteristic of plant metabolism to extract energy also from sunlight (see photosynthetic phosphorylation in Chap. 2).

In the process of respiration, energy is not "liberated" as such but is first redistributed in molecules formed by oxidation-reduction reactions, creating bond energy. At specific oxidation steps during respiration, energy-rich groups are generated. The free energy of the cellular fuel is *conserved mostly as phosphate bond energy* of adenosine triphosphate (ATP), which is the most important energy-rich compound. The energy-rich groups generated by oxidation are capable of initiating reactions requiring energy (syntheses). Hence, the coupling of the chemically bound energy to other systems is the key reaction in biosynthetic processes of plants (Fig. 3–3).

Chemical energy from oxidation-reduction reactions of respiration is transferred to the energy-requiring reactions of syntheses mainly in the form of ATP. A second type of transfer is in the form of electrons. For example, electrons and hydrogen are required for reductions in the synthesis of some hydrogen-rich molecules (fatty acids, sterols, etc.). Energy-rich electrons produced by electron-yielding oxidative reactions are transported by such *electron-carrying coenzymes* as nicotinamide adenine dinucleotide phosphate (NADP) to the electron-requiring groups. Usually, double bonds between carbon and carbon or between carbon and oxygen are reduced to single bonds (Fig. 3–4).

Types of Oxidation

The oxidative reactions of the cell are able to "liberate" energy from the substrate molecules. Originally, the term *oxidation* referred to reactions in which oxygen combines with another substance. In modern usage, oxidation also includes processes in which hydrogen atoms or electrons are removed from the substrate (reducing agent). The hydrogens and electrons that are removed are necessarily taken up by an acceptor or oxidizing agent, which is itself reduced. Thus, every substance undergoing oxidation must be accompanied by a substance undergoing reduction.

In fact, the essential characteristic of the oxidation processes is the removal of electrons from the substance being oxidized. The substances that function in oxidation-reduction reactions have a tendency either to eject or to attract electrons. In oxidation, electrons are removed from a substance and transferred to a suitable electron acceptor, which may be oxygen or some other substance. When electrons flow from a reducing system to an oxidizing system, energy is "liberated." Thus, biological oxidation is essentially a removal of electrons from a substrate; they are then passed on through a series of acceptors until, finally, the electrons are combined with hydrogen ions and oxygen to form H_2O.

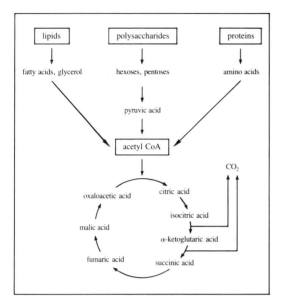

Fig. 3–1. Catabolic pathways of major nutrients.

Energy-Rich Phosphate Groups

A part of the energy liberated in the oxidation of respiratory substrates is stored as special phosphate bonds in energy-rich phosphates. The most important energy-rich compound is ATP. Actually, this and related phosphates are the sources of energy for the biosynthetic reactions of plants. Approximately 20–30 energy-rich phosphate bonds are generated in the course of the degradation of each molecule of glucose. In this way, almost half (678,000) of the total calories liberated in the complete oxidation of one molecule of glucose to CO_2 and H_2O are conserved.

The coupling of the oxidative reactions of respiration with the formation of ATP is known as *oxidative phosphorylation*. In this process, inorganic phosphate is taken up by a substrate molecule at a low energy level. In the course of the oxidation of the substrate, the "liberated" energy is, in turn, trapped by the phosphate and held at a high energy level in the oxidation product. In

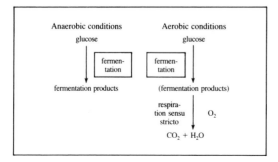

Fig. 3–2. The glycolytic (fermentation) pathway is common to both aerobic and anaerobic utilization of glucose.

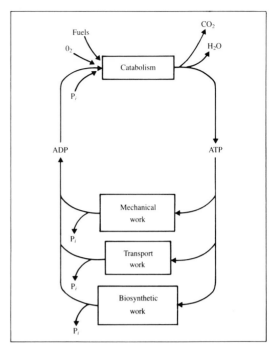

Fig. 3–3. The ATP-ADP cycle. The high-energy phosphate bonds of ATP are used in coupled reactions for carrying out energy-requiring functions; ultimately, inorganic phosphate is released. ADP is rephosphorylated to ATP during energy-yielding reactions of catabolism (after Lehninger, 64).

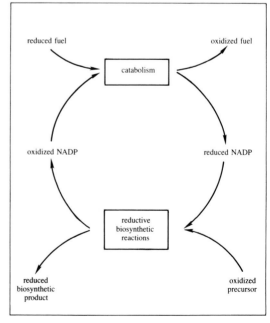

Fig. 3–4. Transfer of reducing power via the NADP cycle. Other electron-carrying coenzymes, such as flavin nucleotides, also participate in reductive biosynthesis (after Lehninger, 65).

other words, energy is redistributed in the molecule. Finally, in the presence of a transphosphorylating enzyme, energy-rich phosphate is transferred to adenosine diphosphate (ADP) with the formation of ATP. This compound then conserves the energy produced by the oxidation of the substrate (59, 64, 67).

Both the inorganic phosphate (P_i) and ADP, the latter being the final receptor for the energy released, are necessary components of the oxidative system. It is for this reason that respiration is dependent upon inorganic phosphate uptake and energy-rich phosphate production. Thus, P_i and ADP are *rate-limiting factors* of the respiration. If plant tissues utilize their energy-rich bonds for biological activity such as growth, ion accumulation, biosynthesis, etc., acceptors for energy-rich phosphate (ADP) and P_i are rapidly renewed and available for new oxidations, so respiration may proceed at a rapid rate.

$$ATP \rightarrow ADP + P_i$$

If, however, a tissue does not consume much ATP for biological work, it becomes short of ADP and P_i; consequently, the respiratory rate necessarily falls off. This is the primary control mechanism of the cell by which respiratory rate can be adjusted to energy demand (43, 62).

The Respiratory Quotient

When glucose is oxidized in respiration, one molecule of CO_2 is evolved for each molecule of O_2 taken up during the respiratory process:

$$C_6H_{12}O_6 + 6\ O_2 \rightarrow 6CO_2 + 6H_2O$$

In this case, the ratio of CO_2 evolved to O_2 consumed is 1. This ratio is known as the respiratory quotient (RQ). The RQ of plant tissues approaches unity very closely. On the other hand, when, for example, fats are oxidized, as in germinating seeds or fungus spores, more oxygen is needed for the combustion of the substrate to $CO_2 + H_2O$ because fats are poorer in oxygen and richer in hydrogen than the sugars. Hence, the RQ for the oxidation of fats is approximately 0.7. The RQ for compounds richer in oxygen than sugars is greater than 1. For example, the RQ of malic acid is 1.33:

$$C_4H_6O_5 + 3O_2 \rightarrow 4CO_2 + 3H_2O\ (CO_2/O_2 = 1.33)$$

The Initial (Anaerobic) Phase of Respiration

This phase is the so-called glycolytic phase of respiration or Embden-Meyerhof-Parnas pathway, in which hexose substrate is converted to pyruvic acid. Each molecule of hexose passes through a series of reactions yielding two molecules of pyruvic acid. The main steps of the glycolytic pathway are summarized in Fig. 3-5. A very important feature in the conversion of hexose sugar (glucose) to pyruvic acid is that phosphorylated sugars rather than the sugars themselves take part in the reactions of the pathway. For the formation of phosphorylated sugars (i.e., hexose diphosphate), each molecule

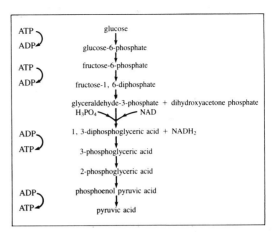

Fig. 3–5. The glycolytic phase of respiration.

of hexose used in the pathway requires two molecules of ATP. On the other hand, four molecules of ATP are produced for each one of hexose sugar utilized in the glycolytic pathway. Hence, there is a *net* synthesis of two molecules of ATP for each molecule of hexose metabolized.

The first step in the pathway consists of transfer of one phosphate group of ATP to the hexose (glucose). This reaction is catalyzed by the enzyme hexokinase. After a series of transformations, the phosphorylated hexose reacts with ATP again to form the hexose diphosphate by action of phosphofructokinase. This enzyme is rate-limiting for glycolysis because it is inhibited by high concentrations of ATP and is stimulated by ADP; therefore, it is a regulatory, or allosteric, enzyme (90). The hexose diphosphate then splits into two triose phosphate molecules. In the next steps, the oxidation of triose phosphate (glyceraldehyde phosphate) takes place. The enzyme triose phosphate dehydrogenase catalyzes this oxidative reaction in which hydrogen is removed from the substrate. Nicotinamide adenine dinucleotide (NAD), the coenzyme of the triose phosphate dehydrogenase, is reduced to nicotinamide adenine dinucleotide (NADH), phosphoglyceric acid is formed, and ATP is synthesized from ADP. Thus, we can see that this very important oxidative step is coupled with the production of an energy-rich phosphate. The detailed mechanism of this reaction was clarified by the brilliant work of Warburg and Christian (124), who demonstrated for the first time that energy produced by oxidation of an organic molecule could be conserved in the form of ATP. In aerobic respiration, the reduced NADH is reoxidized by atmospheric oxygen indirectly. On the other hand, in anaerobic fermentation NADH is reoxidized at the expense of reduction of pyruvic acid to lactic acid (lactic acid fermentation); or in alcoholic fermentation, acetaldehyde, the decarboxylated product of pyruvic acid, is reduced to alcohol with the reoxidation of NADH. In this process NAD serves as a carrier of electrons from the electron donor glyceraldehyde-3-phosphate to pyruvate or acetaldehyde. The reoxidation of NADH may proceed via anaerobic fermentation in two ways, as shown below:

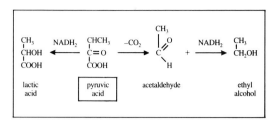

Phosphoglyceric acid, which is produced by the oxidation of triose phosphate, loses a molecule of water in the presence of the enzyme enolase, causing the energy in the molecule to be redistributed into another energy-rich phosphate group, phosphopyruvic acid. Next, the phosphopyruvic acid loses its phosphate, forming pyruvic acid, and the second molecule of ATP is formed from ADP. The formation of pyruvic acid completes the anaerobic phase of respiration. Fig. 3-5 summarizes the reactions in the glycolytic phase of respiration. Fig. 3-6 charts the fate of pyruvic acid.

Regarding the intracellular compartmentation of enzymes involved in glycolysis, it should be stressed that the entire glycolytic enzyme system is located in the *soluble* portion of cytoplasm.

The Second or Aerobic Phase of Respiration

Aerobic respiration occurs in three stages: (a) The oxidative formation of acetyl coenzyme A (CoA) from pyruvic acid (amino acids and fatty acids). (b) Degradation of acetate by the Krebs cycle to CO_2 and H atoms. The primary function of this cycle is the dehydrogenation of acetyl CoA to form four pairs of H atoms and two molecules of CO_2. Neither O_2, inorganic P, nor ATP participates in the Krebs cycle. (c) The transfer of electrons equivalent to the H atoms yielded in the Krebs cycle to molecular oxygen by electron carriers ("terminal oxidation"). This proceeds with a great decline in free energy. The liberated energy is conserved initially in reduced nucleotide form and then as ATP by the coupled oxidative phosphorylation of ADP.

Oxidative formation of acetyl CoA from pyruvate. In the first step, pyruvic acid forms an energy-rich C_2 unit

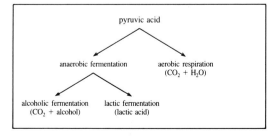

Fig. 3–6. The fate of pyruvic acid.

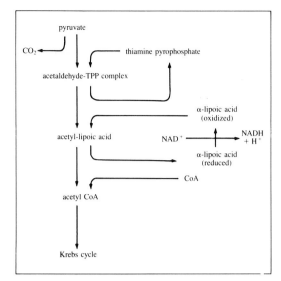

Fig. 3–7. Conversion of pyruvate to acetyl-CoA (after Bidwell, 15).

by oxidative decarboxylation. Thus, CO_2 is liberated and NADH is formed. The acetate is then combined with CoA, forming acetyl CoA (Fig. 3-7). NADH so formed is reoxidized by the cytochrome electron transport system. CoA serves as a carrier of acyl groups in a way analogous to the function of ATP as a carrier of phosphate groups. The conversion of pyruvate to acetyl CoA is mediated by the pyruvate dehydrogenase complex. This system is inhibited by ATP. Thus, the accumulation of ATP can control the production of "fuel" (acetyl CoA) for the Krebs cycle.

The Krebs cycle. The acetyl CoA formed from pyruvic acid is degraded through a cyclic series of steps known as the Krebs cycle, citric acid cycle, or tricarboxylic acid cycle. Krebs and Johnson (58) pointed out that this aerobic degradation is accomplished through the mediation of organic acids, which are widespread in plant tissues. H atoms and CO_2 are produced in the cycle. The energy of the acetyl CoA bond is used in the condensation of the "active acetate" with oxaloacetate, thus forming the six-carbon citric acid and CO_2. The citrate is decarboxylated and oxidized through the Krebs cycle, resulting in the reformation of oxaloacetic acid. Thus, with the revolution of the cycle, four pairs of hydrogen atoms are yielded by enzymatic dehydrogenation. Three pairs are used to reduce NAD and one pair to reduce the flavin adenine dinucleotide (FAD) of succinate dehydrogenase. The four pairs of hydrogen atoms become H^+ ions, and their corresponding electrons are transferred from the substrate ultimately to the molecular oxygen. Consequently, a large amount of energy is liberated and then conserved as ATP.

We can see from Fig. 3-8 that, with every oxidation as well as every oxidative decarboxylation, NAD or NADP (coenzymes of specific dehydrogenases) or FAD are reduced, forming NADH, NADPH, and FADH.

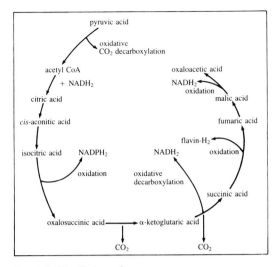

Fig. 3–8. The Krebs cycle.

Reoxidation of the reduced coenzymes and the reduced flavoprotein is carried out by other oxidative enzymes. The energy liberated in the oxidation of these cofactors is used to synthesize ATP. All the enzymes of this system are localized in the mitochondria.

In some plants and microorganisms a variation of the Krebs cycle exists (56) when the acetate must serve both as a source of energy and as a source of intermediates of carbon skeletons for biosyntheses (Fig. 3-9). Two new reactions are important in this *glyoxylate cycle,* and these result in: (a) Splitting of isocitrate to succinate and glyoxylate:

$$\text{Isocitrate} \xrightarrow{\text{isocitritase}} \text{succinate} + \text{glyoxylate,}$$

bypassing the α-ketoglutarate and thus two oxidative decarboxylations. (b) The condensation of glyoxylic

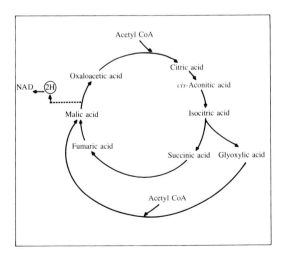

Fig. 3–9. The glyoxylate cycle.

acid with acetyl CoA, resulting in the formation of malate:

$$\text{Glyoxylate} + \text{acetyl CoA} + H_2O \xrightarrow{\text{malate synthetase}} \text{malate}$$

Thus, oxaloacetate can be regenerated both from succinate via the Krebs cycle and from glyoxylate by condensation with acetyl CoA. Two molecules of acetyl CoA are converted in this cycle: one molecule of succinate is formed and a pair of H atoms enters the respiratory chain and causes oxidative phosphorylation of ADP. The enzymes of this cycle are localized in glyoxysomes (microbodies), which possess no cytochrome system.

The pentose phosphate cycle (hexose monophosphate shunt, phosphogluconate pathway) is another pathway of glucose degradation in addition to the Krebs cycle. Its purpose is to generate reducing power (NADPH) in the extramitochondrial cytoplasm for synthesis of fatty acids and steroids, as well as to generate pentoses (D-ribose) for nucleic acid synthesis. An additional and very important function of the pentose phosphate cycle is to participate in the formation of glucose from CO_2 in the dark reaction of photosynthesis. The enzymes of this cycle are in the soluble portion of the extramitochondrial cytoplasm (5, 6, 13).

In the first step, glucose-6-phosphate is oxidized to 6-phosphogluconic acid by glucose-6-phosphate dehydrogenase, and NADP is reduced to NADPH. Phosphogluconic acid is further oxidized with the formation once again of NADPH. At the same time, it is also decarboxylated, producing CO_2 + ribulose-5-phosphate and further to ribose-5-phosphate. The latter compound can be used in the synthesis of nucleotides and RNA. The pentose phosphate pathway may end at this point, producing NADPH for reductive biosynthesis and ribose:

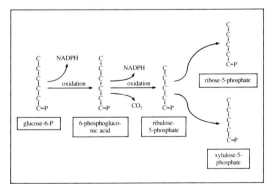

Under other conditions, the pentose phosphate pathway continues further. A sequence of further splitting and transfer reactions results in the conversion of pentose-phosphate to glyceraldehyde phosphate (which is metabolized through the glycolytic pathway) and to a hexose phosphate, which can again enter the pentose phosphate cycle:

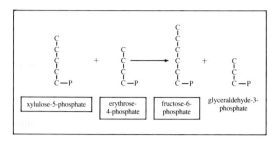

It is very likely that the pentose phosphate cycle normally rejoins the glycolytic cycle in the way outlined above.

If 6 moles of glucose-6-phosphate enter the pentose phosphate cycle they give rise to 6 moles of pentose phosphate, 12 moles of NADPH, $12H^+$, and 6 moles of CO_2. The CO_2 derives exclusively from the C-1 carbon:

$$6 \text{ hexose-P} + 12 \text{ NADP}^+ \rightarrow 6 \text{ pentose-P} + 12 \text{ NADPH} + 12 \text{ H}^+ + 6 \text{ CO}_2.$$

Furthermore, 6 moles of pentose phosphate may be converted into 4 moles of hexose phosphate plus 2 moles of triose phosphate. The 2 moles of triose phosphate may eventually combine, giving rise to a hexose unit, as was mentioned above. Hence, summarizing the situation, we can say that 6 moles of hexose phosphate enter the cycle and give rise to 5 moles of hexose phosphate and 6 moles of CO_2.

NADPH produced in the first and second steps of this cycle is reoxidized in different ways. A well-established possibility is through the synthetic processes of the cell, or eventually by the cytochrome system. However, it is not yet clear whether the oxidation of NADPH generated in the pentose phosphate cycle is coupled with oxidative phosphorylation (production of ATP). Theoretically, at least, a third way is also possible, in which oxidation of NADPH is coupled with noncytochrome systems (75). For example:

$$NADPH \rightarrow glutathione \rightarrow$$
$$ascorbate/dehydroascorbate \rightarrow ascorbic\ oxidase$$

$$NADPH \rightarrow glutathione \rightarrow ascorbate \rightarrow quinones \rightarrow$$
$$polyphenoloxidase$$

By a reversal of the pentose phosphate cycle, the conversion of pentoses to hexoses takes place. Basically, this is the pathway of carbon in photosynthesis, ribulose biphosphate being the primary acceptor of CO_2 in the carbon fixation.

Electron transfer and terminal oxidation. The respiratory substrates and intermediates do not react directly with O_2, but the pairs of electrons (plus H^+ ions) removed by their oxidation are transferred via various carriers to compounds that are capable of yielding electrons to O_2, forming H_2O or H_2O_2.

The natural electron acceptors (the coenzymes of dehydrogenase enzymes, NAD, NADP, FAD) that receive electrons originally removed from the substrate usually do not also react with O_2. The final step in the transfer is mediated by "terminal oxidases," and the end result of the terminal oxidation is the reduction of O_2 and the formation of H_2O. Through this chain of respiratory oxidations the removed electrons are finally combined with oxygen and H^+ ions.

During electron flow, much of the free energy of these electrons is broken up into three packets, and each is conserved in the form of ATP in the process of oxidative phosphorylation. Enzymes of the electron transport and oxidative phosphorylation are localized in mitochondria.

Generally speaking, in aerobic organisms the pathway of electron flow seems to occur in the following sequence: NAD → flavins → (nonheme iron proteins) → (coenzyme Q) → cytochromes → O_2. It is not yet clear whether nonheme iron proteins and coenzyme Q are true electron carriers, but it has been strongly suggested that both may be components of the respiratory chain. Fig. 3-10 is a chart of electron flow and of the entire respiratory pathway.

The oxidation of substrates or intermediates results in the reduction of coenzymes (NAD or NADP) as we have seen above. Reoxidation of the $NADH_2$ is carried out by flavoprotein enzymes. Furthermore, nonheme iron proteins, coenzyme Q, and the cytochrome enzymes (cytochrome b, c, and a) are intermediate carriers between flavoproteins and terminal oxidase. Thus, in the oxidation of reduced flavoprotein by cytochrome b, an electron is transferred to the iron atom of cytochrome b and an H^+ ion is liberated into the medium. Cytochrome b reduces cytochrome c, which, in turn, reduces cytochrome a. Cytochrome a_3 is the cytochrome oxidase (terminal oxidase), which is responsible for the reduction of molecular oxygen.

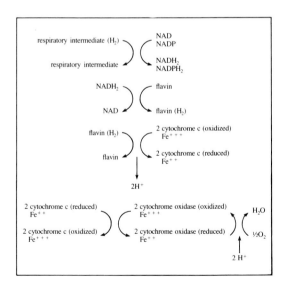

In summary, by the oxidation of NADH and NADPH, the reduction of cytochromes, and the reduction of molecular oxygen to H_2O, energy is liberated

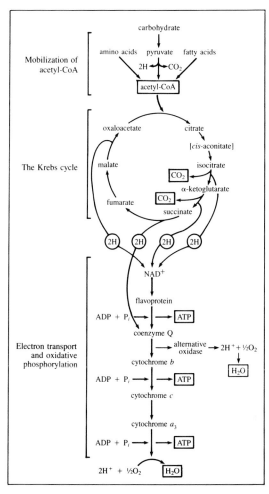

Fig. 3–10. The full respiratory pathway (after Lehninger, 65).

that is used to synthesize ATP. The electron flow from the high energy level of NADH to the low energy level of the flavin results in the liberation of energy and the formation of one mole of ATP. Similarly, ATP is formed from the reduction of cytochromes or molecular oxygen; therefore, the oxidation of one mole of NADH leads to the formation of three moles of ATP.

Finally, if one relates oxygen consumption to phosphate esterification, each oxygen atom ($\frac{1}{2}$ O_2) consumed in terminal oxidation reactions results in the esterification of three molecules of inorganic phosphate:

$$2H^+ + O \rightarrow H_2O \text{ (3 moles of ATP)}$$

Hence, the ratio of phosphate taken up to oxygen used is 3:1, or $P/O_2 = 3$. As we have seen from the Krebs cycle, four moles of reduced coenzymes (NADH and NADPH) are formed by the oxidation of one mole of pyruvic acid. Therefore, the oxidation of four moles of reduced coenzymes leads to the synthesis of 12 ATP molecules.

Other Oxidases

Earlier, other oxidases in addition to cytochrome oxidase—phenol oxidase, ascorbic acid oxidase, glycolic acid oxidase, peroxidase, catalase—were regarded as terminal oxidases. Recent investigations have not supported this view, and the exact role of these nonmitochondrial "soluble" oxidases in the metabolism of plants remains to be elucidated (16). These oxidases are organized into short nonphosphorylating electron transport chains and are capable of utilizing molecular oxygen. They may be coupled with the oxidation of NADPH, as outlined above in the description of the pentose phosphate cycle.

Phenol oxidases. These enzymes, which are copper-containing proteins, were once thought to be terminal oxidases. Later it was suggested that phenol oxidases function in vivo by transferring electrons from the phenols to a cytochrome, rather than directly to atmospheric oxygen. There is ample evidence that the oxidation products of phenols can accept electrons from various substances.

In the absence of reducing agents, the oxidation of phenols proceeds, as the result of secondary oxidations, to black or brownish-black melanin pigments. These oxidations may be responsible for the color changes observed following infection or injury of plant tissues (see Chap. 6).

Ascorbic acid oxidase. This enzyme is widely distributed in plants, and it has often been suggested that it can serve as a terminal oxidase in respiration, but this view has not been supported. The status of this copper-containing enzyme as an important intermediary in electron transport during respiration is uncertain.

Glycolic acid oxidase. Crude extracts from a variety of green tissues oxidize glycolic acid. The enzyme (a flavoprotein) is also regarded as an end oxidase that catalyzes the direct oxidation of its substrate by molecular oxygen (25). The system probably operates as follows:

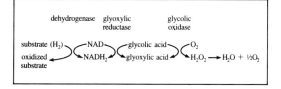

It seems possible that the glycolic oxidase-glyoxylic reductase enzyme system is, in some special cases, responsible for the final electron transfer to molecular oxygen. Glycolic acid oxidase is important in photorespiration, which is a nonmitochondrial respiration in plants. Photorespiration diverts the light-induced energy-rich power from the reduction of CO_2 into the reduction of oxygen. The substrate for photorespiration is glycolic acid, which arises from one of the intermediates of dark fixation of CO_2 during photosynthesis. Photorespiration thus prevents this reduced intermediate from being utilized for glucose formation. Photorespiration does not yield ATP; it is a wasteful process from the point of view of energy. It is characteristic of the C_3 plants; in C_4 plants photorespiration is difficult to detect.

Peroxidase and catalase. In the respiration of higher plants, peroxidase has a role in the oxidation of other metabolites induced by employing H_2O_2 as an oxidizing agent. Peroxidase is also regarded as a detoxifying agent for H_2O_2 and has the ability to oxidize phenol, as aromatic substances are attacked by it in the presence of peroxide, forming quinone and H_2O. The peroxidation of indoleacetic acid (IAA) has an important regulatory effect on IAA content in plant tissues. Catalase is an enzyme that degrades H_2O_2 into water and oxygen:

$$H_2O_2 \xrightarrow{\text{catalase}} H_2O + \tfrac{1}{2}O_2$$

Both catalase and peroxidase are distributed in the tissues of plants and contain iron porphyrins as their prosthetic groups. Their precise metabolic function in higher plants is not known.

Mechanisms Controlling Respiration

Acceptor control. In addition to the concentrations of respiratory enzymes, there are substrates, cofactors, and inhibitors that influence the actual rate of respiration. But the controlling effect of "pacemakers" is of basic importance. These pacemakers are the concentrations of ADP and P_i that must be present in the tissues in order for respiration and oxidative phosphorylation to proceed normally. Any system that promotes the breakdown of ATP will stimulate the rate of respiration by generating ADP and P_i. Substances like dinitrophenol (DNP) uncouple respiration and phosphorylation. The result is a high level of ADP. If ATP, the product of oxidative phosphorylation, is not used up for synthetic processes, a shortage in phosphate acceptor (ADP) will occur and, consequently, the rate of respiration will be lowered. Evidence indicates that a relationship exists between the concentration of phosphate acceptor (ADP) and the rate of tissue respiration ("acceptor control"). Hence, a self-regulating mechanism governs the rate of respiration in living tissues (12, 62).

Feedback control by inhibition of regulatory or allosteric enzymes. In the self-regulating multi-enzyme systems at the beginning or near the beginning of the sequence there are enzymes inhibited by the end product of the sequence; this is called "feedback inhibition." ATP itself is an end product and may function as an allosteric inhibitor.

The Pasteur effect. In the presence of oxygen, fermentation is suppressed. This phenomenon is known as the Pasteur effect. The suppressed rate of glycolysis occurs because phosphofructokinase, an allosteric glycolytic enzyme, is inhibited by ATP and citric acid, which are products of aerobic oxidation. The fermen-

tative degradation of the substrate molecules is much less efficient, from the point of view of energy yield, than is aerobic oxidation. In other words, the substrate (glucose) consumption is enormously increased in the absence of oxygen. Hence, the Pasteur effect implies a conservation of carbohydrate for the cell, since the energy yield from sugar is greater in the presence than in the absence of oxygen (118).

Repression and derepression of enzymes. Catabolism may be regulated by repression of genes responsible for the synthesis of adaptive enzymes. Under the influence of some agents, repressed genes may undergo derepression, thereby synthesizing the enzyme in question. These transcriptional control agents of DNA, i.e. of m-RNA production, that control enzyme synthesis are molecules called inducers. Antithetically, transcription may be repressed by other cell metabolites such as glucose (see Chap. 5).

The Complex Nature of Respiration in Higher Photosynthetic Cells

In addition to the normal mitochondrial "dark" respiration, one must add several newly discovered respiratory events that are typical of photosynthetic plants. Thus, the picture of respiration has become rather complex.

Cyanide-resistant respiration. This is an electron transport pathway in plant mitochondria that is coupled to little or no ATP synthesis (cf. 63). The branching from the cytochrome pathway occurs at coenzyme Q (Fig. 3-10). This "alternative" type of respiration can be specifically inhibited by salicylhydroxamic acid (SHAM). (An array of the inhibitors of the electron transport chain and their postulated site of action is presented in Fig. 3-11.)

Cyanide-resistant respiration probably acts as an overflow to drain off the energy of excess nutrients, at least in the normal, healthy plant. According to this view, respiration is related not only to energy conservation but, in some instances, to energy dissipation as well. According to another view, in those plant tissues where respiration rises dramatically due to ripening, aging, and/or wounding, the increase in cyanide-resistant respiration is coextensive with the respiration rise.

Chloroplast respiration. In chloroplasts there is an O_2-uptake reaction in which electrons from NADPH are passed via ferredoxin to O_2, producing H_2O_2. This may be a significant pathway in the light that is linked to the water-spitting system of photosynthesis. It may also occur in the dark (57); thus it can be regarded as a form of true respiration. The energy of the NADPH is not conserved. The significance of this type of respiration is not known at present.

Photorespiration. As we have seen earlier in describing the role of glycolic acid oxidase, photorespiration is a nonmitochondrial process that does not yield ATP. It introduces another set of O_2 uptake and CO_2

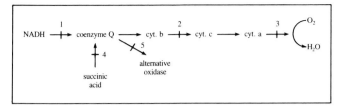

Fig. 3–11. Inhibitors in the electron transport chain in mitochondria: 1 = rotenone or piericidin A; 2 = antimycin A; 3 = CN^-, CO or sodium azide; 4 = malonic acid; 5 = salicylhydroxamic acid (SHAM).

release reactions. However, it is a wasteful process. The reducing power of glycolic acid, which is produced during photosynthesis, is utilized not for carbohydrate formation but to reduce oxygen. This type of respiration is characteristic of C_3 plants (wheat, potato, sugar beet, etc.). Its significance is that in the light it drains off excess energy in the plant. Thus it may play a regulatory role in the interactions between photosynthesis and respiration.

Summary

Breakdown of carbohydrates to smaller carbon compounds occurs by the glycolytic pathway and the pentose phosphate shunt. Subsequently, these compounds are oxidized via the Krebs cycle to $CO_2 + H_2O$. The greatest proportion of energy that becomes available to plants is from the reoxidation of coenzymes that are reduced in the Krebs cycle. During the oxidation of coenzymes, hydrogen ions and electrons are transported over various transport systems to oxygen. At the same time, energy-rich compounds (ATP) are formed from ADP and inorganic phosphate.

The only electron transport system known to be coupled with oxidative phosphorylation is that associated with cytochrome oxidase. It is generally believed that the terminal electron transport system of plants is mediated by the cytochrome system. Although the activity of other oxidases (ascorbic oxidase, polyphenoloxidase) may occasionally be very high in tissues, the in vivo function of these enzymes has not been established with certainty. Some newly discovered respiratory events, such as cyanide-resistant respiration and photorespiration, may be related to energy dissipation instead of energy conservation. They probably drain off excess energy.

Pathological Respiration Induced by Viruses

Changes in overall respiratory activity and, to a lesser extent, in the activities of individual respiratory enzymes have been investigated both in tissue in which virus replicates systemically and in which virus is confined within the area of local lesions. In either case, results must be approached with caution. First, and this applies particularly to systemically infected tissue, not only are cells infected asynchronously but many may not be infected at all. While it may be possible to exclude many of the uninfected cells by excising lesions,

this is difficult to achieve in systemically infected tissue. However, Takahashi and Hirai (112) have gone some way toward this by using stripped epidermis. In general, then, measurements are made on a heterogeneous population of cells, and results represent changes in tissue rather than in cells at a defined stage of infection. However, these considerations are not restricted to measurements of respiration but apply to any virus-induced changes and, indeed, for many purposes it may be the effect of virus infection in the whole leaf or plant that is required. Results may also be critically dependent upon the time at which tissue is sampled, age of the plant at time of infection, and environmental conditions under which the plant is maintained both before and after inoculation.

Changes in respiration of systemically infected tissue may differ considerably from comparable changes in tissue containing necrotic lesions and, with these cautionary notes in mind, they will be considered in separate sections. Additionally, it may be that a significant contribution to observed respiratory changes associated with necrosis, particularly in the later stages of infection, is attributable to increases in activity of terminal oxidases, such as the polyphenol oxidases and peroxidases, which do not contribute to the respiratory activity of healthy tissue. It has been considered more appropriate to evaluate the role of these enzymes in Chaps. 6 and 10. There are several reviews that include effects on respiration in their consideration of the physiological changes in virus-infected plants and that may be useful as guides to the early literature (11, 31, 130).

Respiration Rate

Nonhypersensitive hosts. Insofar as it is possible to generalize, inoculation of leaves of nonhypersensitive plants by viruses seems to induce a slight increase in respiration rate, as indicated by either oxygen consumption or CO_2 evolution per unit weight (Table 3–1; 14, 33, 50–52, 69, 76, 133). However, Takahashi, for example, has reported a decrease in Turkish tobacco following TMV infection (113). Owen (85, 86) similarly found the respiration rate in tobacco systemically infected with TMV to decrease slightly when results were expressed on a dry-weight basis. Yet, since virus infection caused a significant reduction in water content, an apparent increase in respiration rate emerged when results were expressed on a fresh-weight basis.

Table 3–1. Stimulation of respiration in systemic hosts.

Before symptom appearance			Coincident with symptom appearance		
Host	Virus	Ref	Host	Virus	Ref
Hordeum vulgare	BMV	22	*H. vulgare*	BYDV	50,84
			Cucumis sativus	CMV	76
			Nicotiana tabacum	PVX; TEV	33,87,88
Phaseolus vulgaris[a]	TMV	7	*P. vulgaris* cv.	AMV;	14
cv. the Prince			Bountiful	SBMV	
Vigna sinensis[a]	TMV	7	*Zea mays*	MDMV	117

a. Not clear whether stimulation is apparent before symptom appearance or not.

The studies of Jensen (50, 51) on respiration of barley leaves infected with BYDV also highlight some of the difficulties in meaningfully expressing the conclusions to be derived from these experiments. Respiration of the second foliar leaf, expressed as CO_2 release per unit fresh weight (or area), increased steadily for about two weeks after infection, then began to decline (Fig. 3–12a). Expressed on a dry-weight basis (Fig. 3–12b), however, there was little difference in the data between healthy and infected tissue throughout the three weeks of the experiment. As indicated by Jensen (50), the two apparently similar respiratory trends could be the net result of quite different physiological responses. The curve for diseased tissue arises from a rapidly increasing dry weight per unit fresh weight (Fig. 3–12c) and a slowly increasing rate of respiration per unit wet weight, while the control curve derives from a slowly increasing dry weight per unit fresh weight and a rapidly declining rate of respiration per unit fresh weight. Respiration in infected second leaves reached a maximum of 80–90% above that of healthy tissue, when expressed per unit of fresh weight (and was little different on a dry-weight basis). In the older first leaves, respiration was greater than controls when results were expressed either way, since these leaves were fully developed at the time of inoculation and the ratio of dry weight to fresh weight changed little during the course of infection with respect to controls.

The environment in which plants were maintained immediately prior to sampling is also important (51). While respiration of plants sampled after a 10-h dark period was always greater than controls, on whatever basis results were expressed (Table 3–2), respiration of infected leaves sampled after a 5-h light period was slightly higher than controls on the basis of fresh weight, but less than controls on a dry-weight basis. This effect was apparently due principally to the significant rise in the respiration rate of healthy tissue following a 5-h photoperiod. Since the accumulation of starch and carbohydrates is greater in infected than in healthy tissue, it may be assumed that infected tissue is better able to maintain a high level of respiration in the absence of photosynthesis. After a dark period, reserves of starch and carbohydrate would be low in healthy tissue, with a consequent low respiration rate. A 5-h exposure to light leads to an accumulation of photosynthetic products, allowing an enhanced respiration rate, a rate which in healthy tissue is enhanced to a greater extent than in infected leaves. It is also apparent from Jensen's work that increased respiration rates were not a direct reflection of symptom severity. Interpretation of the results is complicated by the fact that in order to achieve symptom differences, plants were maintained at different temperatures, and also measurements of respiration rate had to be made at 18C, and not the plant temperature. Plants grown at 10C eventually developed severe symptoms, while symptoms in plants grown at 27C were very mild or masked. In both cases, the respiration rate of infected tissue reached 150–160% of controls (on a fresh-weight basis). On a dry-weight basis, however, respiration of tissue from infected plants at 10C was as much as 30% below controls, while in plants at 27C it was as much as 70% greater. Unfortunately, there was no comparison of virus accumulation at the different temperatures.

Respiration was observed to increase rather more in susceptible barley plants systemically infected with BYDV than in resistant plants, though again there was no assessment of virus accumulation in the two cultivars (84).

Table 3–2. Dark respiration rates of BYDV-infected and non-inoculated barley after 10 h of darkness or 5 h of light (after Jensen, 51).

		Respiration of tissue	
	Previous		
	light	Fresh wt.	Dry wt.
Treatment	conditions	μliters O_2/min/g	
BYDV-infected	Dark	4.60	23.7
	Light	5.65	32.5
noninoculated	Dark	2.85	17.8
	Light	4.62	36.1

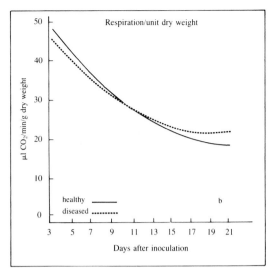

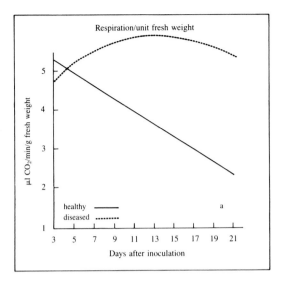

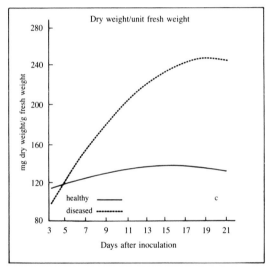

Fig. 3–12. Changes in respiration expressed either on a fresh-weight (a) or dry-weight (b) basis, and changes in dry weight/unit fresh weight (c) of healthy and BYDV-infected second leaves of Cl 666 barley (after Jensen, 51).

Merrett (77), studying respiration rates in tomato systemically infected with the aucuba strain of TMV, has compared results expressed on a fresh-weight basis with those obtained by an alternative procedure. In this instance, virus infection caused a severe stunting of the plants, so that internodes were shortened and stems were thinner. Both fresh weight and dry weight per cell were reduced, while cell number remained unchanged. Therefore, Merrett expressed his results in terms of oxygen consumed per internode, which is a reflection of respiration rate per cell. As indicated in Table 3–3, virus-infected tissue had an enhanced respiration rate when expressed per unit fresh weight but a reduced rate when compared to uninfected plants on an internode basis. The conclusion is that for systemically invaded tissue, which has a marked decrease in growth, respiration rates are more realistically compared on an organ or cell basis.

Table 3–3. The respiration rate of healthy and virus-infected tomato stem tissue expressed on either a fresh-weight or an internode (cell-number) basis (after Merrett, 77).

	Respiration rate			
	μlO_2/g fresh wt./h		μlO_2/g fresh wt./ internode/h	
Internode	Healthy	Virus-infected	Healthy	Virus-infected
Apex	84.0	150.0	11.0	6.0
3	22.1	66.8	8.0	7.0
2	53.6	56.0	70.0	13.0
1	83.3	59.0	75.0	23.0

It appears that changes in respiratory activity can take place very soon after tissue is inoculated. Owen (85, 86), for example, found an increase in CO_2 production in TMV-infected tobacco leaves as soon as one hour after infection. This increase was maintained during the period of virus synthesis before symptoms appeared, but, on appearance of systemic symptoms, respiration rate of infected tissue fell below that of healthy controls. Since in this case virus replication was only in progress for an hour, and the chain of events leading to symptom expression had barely started, interpretation of these results is difficult.

In attempts to reduce interference from uninfected cells and, therefore, obtain a more meaningful representation of changes taking place in infected cells, Takahashi and Hirai (112) inoculated the lower epidermis of tobacco leaves with TMV, stripped the epidermis at various times after inoculation, and compared the respiration rate in epidermal cells with that in buffer inoculated controls. It is evident from Fig. 3–13 that the respiration rate of infected tissue relative to controls rises quickly after inoculation and, after about two days, begins to decline, becoming the same at about six days, when virus is beginning to accumulate. It finally falls below control level as the infection progresses. If this pattern is a general one, then the importance of standardizing sampling time is clear. In this experiment, respiration could be said to be increased, decreased, or unchanged depending on the timing of the assay. Takahashi and Hirai (112) were unable to explain the early rise in respiration rate, considering it to be an unspecified host reaction to the early stages of virus infection. It has not been observed with other virus/host combinations. For example, Owen (87, 88) found that

the respiration rate in tobacco infected with TEV or PVX did not begin to increase until symptoms appeared. (When a rise did occur, it was greater than in TMV-infected tobacco, although virus accumulation was less and appeared to be a reflection of symptom expression rather than of the extent of virus multiplication.) Dwurazna and Weintraub (33), similarly, observed increases in inoculated and systemically infected tobacco leaves following infection with various strains of PVX, increases that were coincident with symptom expression and proportional to symptom severity rather than virus concentration. Few, however, have attempted to investigate respiratory rates in infected cells in the absence of uninfected ones. When whole leaf tissue is used, with cells infected asynchronously, it may be well into the later stages of infection before changes are detectable.

Even in heavily infected plants, there is little *net* increase in RNA or protein synthesis but, rather, a redirection with synthesis of virus-specific RNA and protein at the expense of cellular species. It is unlikely, then, that any observed increases are the result of responses to enhanced requirements for energy for these biosynthetic processes. Though changes may sometimes be detected prior to appearance of visible symptoms, it seems more likely that they are a response to virus-induced alterations in host metabolism and physiology, which may lead to symptom expression, rather than a response to virus multiplication per se.

Hypersensitive hosts. In virus/host combinations in which the plant responds hypersensitively with the formation of necrotic lesions, there is a much more pronounced increase in respiratory activity than in systemic hosts (Table 3–4). Increases have been detected several hours in advance of lesion appearance in *Nicotiana glutinosa* (125) or *Xanthi* tobacco (107) inoculated with

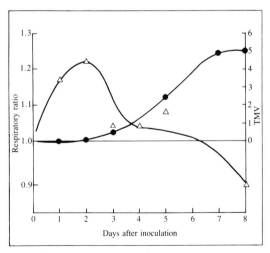

Fig. 3–13. The ratios of respiratory rates between TMV-inoculated and uninoculated tobacco leaf epidermis (open triangles) at different times following inoculation. TMV content (mg/g fresh weight, filled circles) in the inoculated epidermis was determined by serological tests (after Takahashi and Hirai, 112).

Table 3–4. Stimulation of respiration in hypersensitive hosts.

Before lesion appearance			Coincident with lesion appearance		
Host	Virus	Ref.	Host	Virus	Ref.
Gomphrena globosa	PVX	125	*G. globosa*	PVX	131
N. glutinosa	TMV	24	*N. glutinosa*	TMV	88,132
N. tabacum cv. *Xanthi*	TMV	107	*N. sylvestris*	TMV	89
P. vulgaris cv. The Prince	TNV	7	*P. vulgaris* GN59	AMV; SBMV	14
Vigna sinensis	TNV	7	*Datura stramonium*	TMV	131
			Vicia faba	CMV	131

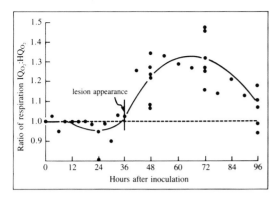

Fig. 3–14. The ratio of respiration of *N. glutinosa* half-leaves inoculated with TMV at 15C to opposite half-leaves not inoculated. Each symbol indicates values obtained in independent but comparable experiments (after Yamaguchi and Hirai, 132).

TMV and in other virus/host combinations (24, 125). Others, however, have been unable to detect changes until the appearance of lesions on TMV-inoculated *N. glutinosa* (Fig. 3–14; 88, 132) while changes were either coincident with, or appeared fractionally before, symptom appearance on tobacco inoculated with ring-spot strains of PVX (33) or *Phaseolus vulgaris* inoculated with SBMV (14).

Since virus multiplication could be detected before lesions appeared, while increase in respiratory activity, as measured by oxygen uptake, could not, Yamaguchi and Hirai (132) concluded the rise to be directly related to necrosis, rather than virus replication. Similar conclusions were reached by Parish et al. (89). Following inoculation of *N. sylvestris* with either the U2 strain of TMV, which induces lesion formation, or the U1 strain, which does not, increased oxygen uptake was detected only in U2-infected leaves. At the time measurements of respiratory activity were made, just as lesions were beginning to appear, infectivity of both strains had accumulated to approximately the same level. Weintraub et al. (125), however, considered the increase to be "correlated with virus activity other than the necrotic process," a conclusion based at least partly on the observation that necrosis induced by means other than virus infection failed to induce comparable increases in respiratory activity. In PVX-infected tobacco, however, increases in respiratory activity were correlated with symptom severity (33).

Since significant rises in polyphenol oxidases and peroxidases accompany lesion development, it has, not unnaturally, been proposed that the activities of these enzymes, particularly the polyphenoloxidases, contribute to the enhanced oxygen uptake of infected leaves. The detection of increased respiratory activity in tissue surrounding TMV lesions (125), which also had elevated polyphenoloxidase activity (121), tended to support this view, though others have failed to observe similar rises in respiratory activity (107). The evidence seems to support the conclusions reached by Weintraub

et al. (128), who consider the polyphenoloxidases to be minor contributors to elevated oxygen consumption in the early stages of lesion development, when enhanced respiratory activity is greatest, but to increase in importance as lesions develop and mitochondria degenerate. Continuing their earlier studies on *N. glutinosa* inoculated with TMV (126, 127), they also found a stimulation in mitochondrial number early in lesion formation, declining as lesions mature, a trend correlating broadly with respiratory changes. Their results, based on electron microscopic observation of leaf extracts, confirm their original data but are at variance with the conclusions of Pierpoint (92), who failed to find evidence of increase in mitochondrial content in the same host. Mitochondrial proliferation has, however, been observed in other infections in which necrosis develops (98), though not in tobacco inoculated with PVX (34). It seems premature to conclude, therefore, that a major contribution to the observed increase in respiratory activity arises from a proliferation of mitochondria.

Respiratory Pathways and Enzymes

Though few investigations have progressed beyond measurements of respiration rate, there are some indications of the metabolic changes that may be occurring following virus infection. In one of the few investigations on individual cells, Takahashi and Hirai (111) stripped the epidermis from tobacco leaves inoculated with TMV and, using cytochemical stains, compared the activities of several enzymes in cells with virus inclusions with those in apparently normal cells from the same leaf. Though the activities of some enzymes were unchanged, the activities of others, including cytochrome oxidase and some dehydrogenases, were lower in inclusion-bearing cells, leading them to conclude that there was a general decrease in synthetic activity in heavily infected cells. In potato infected by PLV and other viruses, there was a decrease in activity of several enzymes of the glycolytic pathway following infection (17, 18).

It has been suggested that at least part of the increase in respiration rate of infected tissue results from the uncoupling of oxidative phosphorylation (33). In Xanthi tobacco infected with TMV, for example, there is a decrease in the ADP/ATP ratio, though it is not clear whether this arises from an uncoupling of phosphorylation from respiration, or from a possible proliferation of mitochondria (108). There was no evidence for uncoupling in barley systemically infected with BMV (22), or in cowpea hypocotyl infected with CPMV (68).

In leaves responding with the formation of necrotic lesions, there is evidence that at least part of the increased respiration rate is due to a stimulation of carbohydrate metabolism via the pentose phosphate pathway. Solymosy and Farkas (105, 106), for example, have compared enzyme activities in the ring of yellow tissue immediately surrounding para-TMV lesions on White Burley tobacco with activities in normal, green areas. They found that while some enzymes of the Embden-Meyerhof-Parnas glycolytic pathway remain un-

Table 3–5. Effect on enzymic activities in tissue immediately beyond the necrotic lesion compared with normal green tissue (after Solymosy and Farkas, 106).

	Stimulation
Enzymes of glycolysis	
hexokinase	−
phosphohexoisomerase	−
Enzymes of hexose monophosphate shunt	
glucose-6-phosphate dehydrogenase	+
6-phosphogluconic acid dehydrogenase	+
Enzymes of the Krebs cycle	
aconitase	+
isocitric dehydrogenase	+
malic dehydrogenase	−
Other enzymes	
NADPH oxidase system	+
NADPH-dependent quinone reductase	+
NADH-dependent quinone reductase	+
polyphenol oxidase	+
cytochrome oxidase	+

changed, some enzymes of the pentose phosphate pathway are stimulated (Table 3–5).

Similar increases were observed by Dwurazna and Weintraub (34) in activity of the glucose-6-phosphate dehydrogenase and 6-phosphogluconate dehydrogenase enzymes of the pentose phosphate pathway in tobacco infected with PVX. The greatest stimulation was produced by strains inducing the most severe symptoms (Table 3–6) and at a time when necrosis was visible. However, the decrease in activity of phos-

phoriboisomerase (the third enzyme in the pathway), an observation also made by Solymosy and Farkas, remains somewhat anomalous. Though mitochondria from healthy and infected tissue had the same ability to oxidize Krebs cycle intermediates, the enhanced levels of dehydrogenase enzymes clearly indicated that infection by PVX stimulated the pentose phosphate pathway.

This observation was supported by the results of experiments using glucose-1-^{14}C and glucose-6-^{14}C to determine C_6/C_1 ratios (rate of release of $^{14}CO_2$ from glucose-6-^{14}C/rate of release of $^{14}CO_2$ from glucose-1-^{14}C), the reduced ratios of PVX-infected tissue (34) also indicating a stimulation of the pentose phosphate pathway; again, the extent of the reduction was a reflection of symptom severity. Decreases in C_6/C_1 ratios have also been observed in other virus/host combinations that resulted in necrosis (Table 3–7). In Xanthi tobacco inoculated with TMV, however, Merrett and Sunderland (78) found no evidence of a shift to the pentose phosphate pathway; C_6/C_1 ratios were similar in both healthy and infected tissue and indicated that both pathways participated in glucose catabolism. Increased glucose catabolism in infected tissue occurred by a stimulation of both glycolytic and pentose phosphate pathways.

With the exception of the observations of Dwurazna and Weintraub (34), in which the activities of enzymes of the pentose phosphate pathway were enhanced (Table 3–8) and C_6/C_1 ratios were reduced in tobacco infected with relatively mild strains of PVX, there is little evidence for a shift toward the pentose phosphate pathway in the absence of necrosis. Bell (14), for example, observed decreased C_6/C_1 ratios in *P. vulgaris* inoculated with SBMV, a combination that produced necrosis. He did not, however, find a comparable shift to the pentose phosphate pathway in the Bountiful bean variety, where necrosis did not occur. In Samsun tobacco, the

Table 3–6. The activity of enzymes of the pentose phosphate pathway in cell-free extracts of healthy leaves, and of PVX-infected tobacco leaves with well-developed symptoms (after Dwurazana and Weintraub, 34).

	Enzymic activity in moles/min/g fresh wt.				
			Potato virus X strains		
Enzyme	healthy	yellow	mild mottle	brown spot	ring-spot
glucose-6-phosphate dehydrogenase	0.09	0.125 (1.39)	0.131 (1.46)	0.156 (1.73)	0.170 (1.89)
6-phosphogluconate dehydrogenase	0.064	0.087 (1.36)	0.101 (1.58)	0.121 (1.89)	0.130 (2.03)
phosphoriboisomerase	0.056	0.052 (0.93)	0.031 (0.55)	0.040 (0.71)	0.038 (0.68)

The figures in parentheses express the ratio of infected:healthy for each strain and for each enzyme.

Table 3–7. Shift from glycolytic to pentose phosphate pathway; in hypersensitive hosts (from C_6/C_1 ratios).

Host	Virus	Stimulation	Reference
N. tabacum cv. Haranova	PVX	+	34
N. tabacum cv. Xanthi	TMV	−	78
P. vulgaris cv. GN59	SBMV	+	14
P. vulgaris cv. GN59	AMV	−	14
Datura stramonium	TMV	+	70

glycolytic pathway contributed about 80% and the pentose phosphate pathway about 20% to the total respiratory activity, a proportion that was unchanged by virus infection either during a period of active virus multiplication, 5 d after infection, or after symptom appearance 21 d after TMV infection (8). Additionally, the enzymes glucose-6-phosphate dehydrogenase, 6-phosphogluconate dehydrogenase, and phosphoriboisomerase were equally active in healthy and infected roots (10). Infection by TMV had no effect on sterol accumulation in either roots, stems, or leaves, although, if activity of enzymes of the pentose phosphate pathway influences sterol biosynthesis, an effect might have been expected had a shift toward this pathway occurred. Finally, experiments with mevalonic acid (8, 9) confirmed the view of Baur et al. that the pentose phosphate pathway is of equal importance in healthy or TMV-infected tobacco, both before and after symptom appearance. Mevalonic acid, as an inhibitor of succinic acid dehydrogenase, blocks the tricarboxylic acid cycle. If TMV infection caused a shift toward the pentose phosphate pathway, then infected plants might be expected to be more resistant to mevalonic acid treatment. In fact, there was no difference in response between healthy and infected tissue. The view of Dwurazna and Weintraub (34), therefore, that there is a "more or less quantitative shift [toward the pentose phosphate pathway] that is positively correlated with a spectrum of symptom expression of increasing severity" remains at variance with others who maintain that a shift occurs only in conjunction with necrosis.

Table 3–8. Shift from glycolytic to pentose phosphate pathway; in systemic hosts (from C_6/C_1 ratios).

Host	Virus	Stimulation	Reference
N. tabacum cv. Haranova	PVX	+	34
N. tabacum cv. Samsun	TMV	−	10
P. vulgaris cv. Bountiful	AMV; SBMV	−	14

Pathological Respiration Induced by Bacteria

A number of the studies of plant pathogenic bacteria *Agrobacterium tumefaciens*, *Pseudomonas solanacearum*, and *P. tabaci* have exposed aberrant respiratory patterns in response to infection. It should be noted, however, that the aberrant respiration observed in tissues invaded by *A. tumefaciens* should not be interpreted as indicative of the type of pathological respiration that necessarily occurs in tissue infected either by wilt-inducing bacteria such as *P. solanacearum* (48, 49) or by a leaf-necrotizing pathogen like *P. tabaci* (36). Whereas *A. tumefaciens* induces neoplastic growth, i.e. uncontrolled plant cell division with little or no differentiation (66), most other bacterial pathogens rapidly engender tissue disintegration (116) or dysfunction to varying degrees (91) accompanied by significant bacterial cell multiplication (72).

Rate and Intensity of Respiratory Change

The sensitivity of O_2-consuming metabolic events to the earliest phases of the host-pathogen interaction was demonstrated in experiments by Huang and Goodman (47) with isolated tobacco cells. Tobacco leaf cells separated by "Pectinol" were exposed to one of three bacterial species: a compatible pathogen, *Pseudomonas tabaci*; an incompatible (HR-inducing) pathogen, *P. pisi*; and a saprophyte, *P. fluorescens* (see Fig. 3–15). Oxygen consumption of the isolated leaf cells was monitored initially and then again after the leaf cells were mixed with the bacteria. The amount of O_2 consumed by the mixtures of leaf cells and bacteria was always greater than the sum of O_2 consumed by the two components. Ratios of O_2 consumed by mixtures of tobacco cells and bacteria to tobacco leaf cells alone were 2.0 or more for *P. pisi*, 1.7 for *P. tabaci*, less than 1.5 for *P. fluorescens*, and 1.0 for heat-killed bacteria. By using antimycin A, which inhibits respiration of higher plant cells but not of bacteria, it was shown that the potentiated O_2 consumption was of plant cell origin. Furthermore, the increased O_2 uptake by the mixtures of plant and bacterial cells became apparent within 20 sec of mixing. Hence, the plant cells were quickly affected by the bacteria in their environment, and they responded differentially to the three types of bacteria. The HR-inducing bacteria (possibly evoking the initial stages of the resistance reaction) caused the greatest increase in respiratory rate. The nature of the signal given the plant cells and the nature and location of the signal receptor remain unknown, and the precise biochemical segment of the respiratory pathway of the plant cell that is affected has yet to be determined. It is evident, however, that plant cells are able to recognize differences between bacterial species, and the response within 20 sec suggests that cognors and cognons of the response are on or near the surfaces of the reactants. These results are similar to those reported by Maine (73) for tobacco stem pieces being exposed to washed viable cells of *P. solanacearum*. In this instance, too, the effect was almost

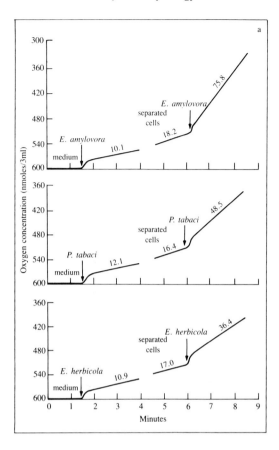

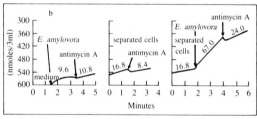

Fig. 3–15. Differential stimulation of respiration of separated tobacco leaf cells by plant pathogenic bacteria (a) *E. amylovora* (incompatible), *P. tabaci* (compatible), and *E. herbicola* (saprophyte); (b) antimycin nearly negates stimulation of respiration induced by *E. amylovora* indicating the stimulation in respiration is largely plant cell respiration (after Huang and Goodman, 47).

immediate and the increase in respiration was greater than could be accounted for by the respiration of the bacterial inoculum. Potentiated respiration during the development of HR has also been reported by Németh and Klement (82) in tobacco infiltrated with *P. pisi*, *P. tabaci*, and *P. fluorescens*. The intensity of respiration and rate of increase parallel those noted by Huang and Goodman (47). The time required for registering these changes was much greater, as the experiments were performed in tobacco leaf tissue rather than isolated

tobacco leaf cells. Németh and Klement detected maximum increases at about 5 h after inoculation, which was approximately 30 min before the development of visual symptoms of HR. It would appear, from these two sets of data, that surface component interactions between plant and bacterial cells are rapidly translated into a significantly heightened respiratory rate.

Fischer (40) compared respiratory rates of resistant and susceptible pepper fruit tissue inoculated with *Xanthomonas vesicatoria* (10^7 cell/ml). An immediate increase in O_2 consumption was detected in resistant tissue, whereas susceptible tissue did not reflect an increase in respiration until about 30 h after inoculation. This observation seemed explicable from the ATP levels detected in tissue during the 80-h course of the experiments. In resistant tissue, ATP levels fell immediately; in susceptible tissue, the decrease in ATP coincided with the delayed increase in respiration. However, it was concluded that the potentiated respiration was similar in both instances, except that it occurred more rapidly in the incompatible combination.

ATP levels were also investigated in cotton cotyledons inoculated with *Xanthomonas malvacearum* by Novacky and Ullrich-Eberius (83). In addition, measurements of transmembrane potential (E_m) were made, and it was found that the energy-dependent (ATP-dependent) portion of E_m was altered by *X. malvacearum* infection. A high E_m was maintained only in the light and not in the dark. Since the ATP level was higher in diseased than in healthy tissue and higher in dark than in light, the observed membrane depolarization could not be due to bacteria-induced inhibition of mitochondrial ATP. It was concluded that the effect of *X. malvacearum* on ATP was at the plasmalemma. Specifically, the bacteria had not affected the function of the H^+ extrusion pump of the plasmalemma. Rather, the bacteria appeared to uncouple the H^+ pump from its energy supply, in the dark. The sequence of events postulated as responsible for the uncoupling is linked to the intense bacterial growth, which required potentiated plant syntheses and ATP turnover. The demand for ATP was, apparently, satisfied in the light (by photosynthesis) but was insufficient in the dark.

A Shift from Krebs Cycle to Other "Terminal Oxidases"

Perhaps the earliest significant study of altered respiration patterns in plants infected by a phytobacterial pathogen is that of Nagy et al. (80), who compared the respiration of gall tissue on tomato (caused by *A. tumefaciens*) with contiguous healthy stem tissues. This study revealed an augmented catalase activity of approximately 160% in gall tissue (see Fig. 3–16). Oxidase activity was increased 130%, and peroxidase activity 120%. Tyrosinase was also measured, and it was noted that an extract of gall tissue destroyed half of the tyrosine in the reaction mixture, whereas an analogous extract from contiguous healthy tissue did not measurably affect tyrosine levels. Nagy et al. concluded that, since galls are in a highly active vegetative state, "it was

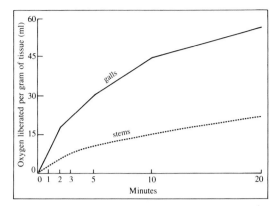

Fig. 3–16. Catalase activity of crown gall and of contiguous healthy stem tissue of tomato (after Nagy et al., 80).

not surprising to find the concentration of catalase (polyphenol oxidase), and peroxidase greater in gall than in healthy, but mature, stem tissue." Although Nagy et al. found traces of tyrosinase activity in healthy tomato tissue and in axenic cultures of the pathogen, the differences between these two and gall tissue were so great that a quantitative comparison was precluded. They concluded that a major portion of the increased tyrosinase activity reflected gall-cell tyrosinase rather than the production of this enzyme by the bacterium.

Neish and Hibbert (81) compared the rates of oxygen consumption by healthy beet root slices with slices of beet root tumors induced by *A. tumefaciens*. Their results are presented in Table 3–9 and show that the tumor respires more than twice as much O_2 as healthy tissue, and the tumor seems also to suppress the rate of respiration of the neighboring healthy cells.

Respiratory quotients for tumor and normal were 0.92 and 0.64, respectively. The low RQ of the normal tissue suggests that carbohydrates were not being completely oxidized to CO_2 and H_2O. This contention was supported by the fact that normal beet root slices formed more citric, malic, and oxalic acid than the tumor tissues. Measurement of the total carbohydrate metabo-

Table 3–9. Rates of respiration of beet tissue (after Neish and Hibbert, 81).

	Tumor tissue	Normal tissue of tumorous beet	Tissue of healthy beet
mm³ of O_2	270	70	105
adsorbed/g of	275	105	120
fresh tissue/h	280	98	115
	310	88	135
	270	80	80
	240	71	125
	—	—	75
Average value	274	86	108

lized by the two types of tissue during a 3-h span of aerobic respiration revealed that in normal tissue the CO_2 evolved accounted for 19% of the sugar metabolized. The combined oxalic, malic, and citric acid increase in tissue represented an additional 52%, or a total of 71% carbohydrate accounted for. On the other hand, 25% of the sugar metabolized by tumor tissue was accounted for as CO_2, and the fate of the remainder could not be determined. From these calculations, it is apparent that there was a qualitative difference in carbohydrate metabolism between the two types of tissue.

Of additional interest, Neish and Hibbert (81) noted that respiration of normal tissue was more sensitive to fluoride and cyanide poisoning than that of tumor tissue. This result might be interpreted as suggesting that the enolase enzyme participating in glycolysis was suppressed in the normal tissue. The insensitivity of tumor tissue respiration to fluoride, on the other hand, suggested the participation of some other respiratory pathway. This study also indicated that both normal and tumor tissues of beet root were capable of fermentation, as both produced ethanol, and the former produced lactic acid as well. However, only the normal tissue showed a decrease in carbohydrate catabolism when O_2 was supplied to the fermenting tissues. Thus, only the normal tissue elicited the classic glucose-sparing Pasteur effect (20).

The capacity of the normal tissue to convert some carbohydrate to organic acids (malic, citric, and oxalic aerobically and lactic acid under anaerobic conditions) indicated a more efficient metabolic process. In contrast, much more of the carbohydrate metabolized by tumor tissue could be accounted for as CO_2.

An analysis of this research in the light of subsequent sudies, vis-à-vis plant respiration, might permit the following interpretation. In normal tissue, oxidative phosphorylation is arrested under anaerobic conditions. The accumulation of "phosphate ester" (probably ADP) noted by Neish and Hibbert (81) in normal tissue under the influence of fluoride at least suggests this. In other words, much of the ADP produced by ATP utilization was made available for glycolysis; thus there was an increase in catabolic reactions characterized by the production of ethanol and lactic acid. However, once O_2 was made available, ADP conversion to ATP by means of *oxidative phosphorylation* ($NADH_2$ and $NADPH_2$-flavin-cytochromes) began again, and the regulation of the rate of glycolysis by aerobic events was reimposed.

That tumor tissue was metabolizing carbohydrate in a manner quite different than normal tissue was apparent under anaerobic as well as aerobic conditions. Under anaerobic conditions, respiration of tumor tissue was characterized by an alcoholic fermentation that accounted for 70% of the sugar metabolized. When O_2 was supplied to these tissues, ethanol production was inhibited (as it was in normal tissue); however, the metabolism of sugar continued at an increased rate with only 23% of it being evolved as CO_2. It would seem, therefore, that the Pasteur effect had been abolished in aerobically respiring tumor tissue.

A subsequent study by Link and Klein (66), which measured the oxygen uptake of tumors of tomato induced by auxin (IAA) and by *A. tumefaciens*, significantly increased our knowledge of respiratory processes of bacterially incited tumors. Using CO as a respiratory inhibitor and light to reverse the effect of the inhibitor, they attempted to differentiate between the portion of O_2 uptake being mediated by an iron-containing enzyme (cytochrome oxidase) and that being controlled by a copper-containing enzyme (phenol oxidase or ascorbic acid oxidase). Their data revealed that both iron- and copper-containing enzymes were involved in the uptake of O_2 in normal tissue as well as in auxin-induced and bacterially induced tumor tissue. However, it was also apparent from their experiments that in crown gall tissue there is a shift in favor of the copper-containing enzyme at the expense of the iron-containing enzyme system. They concluded that, in bacterially incited tumor tissue, terminal oxidation seems to be shifted from the cytochrome system to a copper-containing "terminal oxidase." The nature of the enzyme or enzymes is, in turn, suggested by the results of Nagy et al. (80), who found large increases in polyphenol oxidase activity, which has been shown to be activated in pathologically altered tissue (39).

Participation of the Pentose Phosphate Cycle

Later, a study by Brucker and Schmidt (21) partially corroborated the contention of a "shift" with results indicating that respiration via the Krebs cycle is reduced in tumor tissue of *Datura* and *Daucus*. They also concluded that respiration via the glycolytic pathway in these tissues was reduced, suggesting direct oxidative decarboxylation of hexoses via the *pentose phosphate cycle*.

Evidence for a potentiated activity of the pentose phosphate cycle in *A. tumefaciens*-induced tumors in beet root tissue has since been presented by Scott et al. (99). Whereas the key enzymes for catabolism of hexose via glycolysis and oxidative decarboxylation were detected in both healthy and tumor tissue, their activity levels and, hence, their degradative potentials were different. The greatest differences were observed in the pentose phosphate cycle enzymes glucose-6-phosphate dehydrogenase, 6-phosphogluconate dehydrogenase, and phosphoriboisomerase. The activities of these three enzymes were increased two to three times in tumor tissue per milligram of protein. This same tumor tissue, on the other hand, revealed small decreases in the activity of phosphohexoisomerase, 3-phosphoglycerate kinase, and phosphofructokinase, enzymes that mediate the breakdown of hexose via the glycolytic pathway.

Experiments on the dissimilation of glucose labeled in the C-1, -2, and -6 positions by normal and tumor tissues revealed C_6/C_1 ratios of less than unity, providing additional evidence that both tissues metabolize some glucose via the pentose phosphate cycle. The data concerning comparative enzyme activities suggest that tumors have a greater potential to oxidatively decarboxylate hexoses to ribose-5-phosphate than normal

tissues. It was suggested by Scott et al. (99) that enhanced oxidative decarboxylation of glucose to ribose would provide the hyperactive meristematic tissues of tumors with additional pentose phosphate and $NADPH_2$ for anabolic processes.

Additional evidence for the participation of the pentose phosphate cycle as a consequence of bacterial infection was reported by Farkas et al. (36). The activity of glucose-6-phosphate dehydrogenase in tobacco leaf tissue inoculated with *P. tabaci* was measured in the bacteria-free halo tissue and in adjacent normal green areas. The activity of this initial enzyme of the pentose phosphate cycle was approximately 250% higher in halo than in adjacent normal green tissue. In addition, the activity of this enzyme was almost 200% greater in tissues injected with culture filtrates of the pathogen than in tissues injected with water. These workers concluded that the shift from glycolysis to the pentose phosphate cycle is a nonspecific host response encountered when tissue is injured in any one of a number of ways, including pathogenic bacteria.

Origin of Potentiated Respiration . . . Host or Pathogen?

A comprehensive account of increased respiration in plant tissue affected by a vascular or necrotrophic bacterial pathogen was presented by Maine (73). In his research, a number of physiological responses of the tobacco plant to attack by *P. solanacearum* were examined to determine the origin of the respiratory increase. Maine noted that *P. solanacearum* caused significant increases in the rate of O_2 uptake in diseased tobacco tissue. The extent of these increases in susceptible and resistant tissue is presented in Fig. 3–17. The size of the bacterial population responsible for this enhanced respiration was calculated to be 2×10^8 cells/ml; and this concentration of cells in a number of synthetic media consumed, at most, 7 μl O_2/h, which

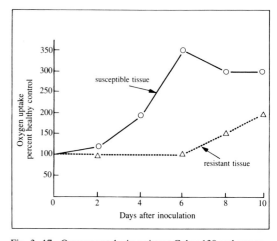

Fig. 3–17. Oxygen uptake in resistant Coker 139 and susceptible Bottom Special tobacco stem slices excised from plants following inoculation with *P. solanacearum* (after Maine, 73).

represented approximately 3% of the increased O_2 uptake observed. To minimize bacterial respiration still further in these manometric experiments, most of the bacteria were removed from xylem vessels of affected tissue by a 1/3-atm vacuum prior to measuring their respiration rates. In addition, enhanced respiratory rates were observed in leaf tissues of inoculated plants that showed evidence neither of bacterial invasion nor of wilting. From this, it is concluded that the augmented uptake of O_2 was essentially a host response to the pathogen, rather than a reflection of pathogen respiration. Essentially the same conclusion was reached in the study by Huang and Goodman (47), see Fig. 3–15.

Maine's experiments using the enzyme inhibitors sodium fluoride, dinitrophenol, O-phenanthroline, and sodium diethyldithiocarbamate had divergent effects upon the oxygen consumption of healthy and diseased stem tissues. The data were interpreted as suggesting that a noncytochrome metallocopper oxidase system was responsible for the enhanced oxygen consumption observed in diseased tissues. The oxidases with copper prosthetic groups that have most frequently been reported as being stimulated during the course of disease are polyphenol and ascorbic acid oxidase (16, 39, 44). Both were detected in tissues affected by *P. solanacearum*.

Oxidation of chlorogenic acid in infected stem slices of a susceptible tobacco variety was augmented 90%, whereas a 15% increase was recorded in infected slices from a resistant variety. From these results, it would seem that respiration in diseased susceptible and in diseased resistant tissue proceeds via different pathways. However, this difference may be more apparent than real and may, actually, reflect the degree of pathogenesis in these two tissues. It was clear, from this study, that the augmented polyphenolase activity observed in the diseased plant was due to enzyme production by the host cell, since neither bacterial cultures nor culture filtrates could catalyze the oxidation of chlorogenic acid in vitro. On the other hand, these culture filtrates, crude or dialyzed, did accelerate polyphenolase activity in previously healthy susceptible and resistant stem tissue.

Respiration Affected by Wall-Degrading Enzymes

It was also noted that if these crude or dialyzed culture filtrates were boiled, they no longer enhanced the oxidation of chlorogenic acid. Inspection of stem slices treated with the culture filtrates, following a 24-h incubation in Warburg flasks, revealed that many *pith* cells in both the susceptible and resistant tissue slices were disintegrated, although tissue slices remained intact.

When partially purified hydrolytic enzymes (possessing both polygalacturonase and cellulase activity) were prepared from culture filtrates of *P. solanacearum* and were incubated with tobacco stem slices, O_2 uptake of susceptible stem slices was increased 58%. No increase was obtained with resistant stem slices similarly treated. In addition, careful observation of the tissue slices in these studies revealed that after 24 h in the Warburg

apparatus, pith cells of *both* susceptible and resistant slices had disintegrated. Furthermore, complete separation of the vascular cylinders from cortical parenchyma occurred in many of the susceptible slices, whereas this happened less often in the resistant ones.

From the foregoing sequence of events, one might deduce that these partially purifed extracellular hydrolytic enzymes produced by *P. solanacearum* not only affect the structural integrity of the host cells but also play a direct role in the increased oxygen consumption of diseased tissues. A causal relationship between enhanced O_2 uptake and hydrolytic enzyme action is also suggested by the fact that washed viable bacterial cells of *P. solanacearum* could reproduce this effect in stem pieces almost immediately after being placed in contact with previously healthy tissue. Here, too, the increases in O_2 uptake were greater than could be accounted for by the respiration of the bacterial population. It would appear, from these experiments, that surface interactions between plant and bacterial cells quickly result in a potentiated respiratory rate.

Sequence of Increased Ascorbic Acid and Polyphenol Oxidase Activity

Although enzymatic ascorbic acid oxidation in healthy tobacco stem tissue was difficult to detect, increases in activity of this enzyme in diseased susceptible and resistant tissue were readily apparent (73). From Fig. 3-18, it is apparent that increases of 200–400% were not uncommon where pathogenesis had been initiated. In contrast with polyphenol oxidase activity, however, which continued to increase as pathogenesis progressed, ascorbic acid oxidation increases occurred early in pathogenesis. Hence, ascorbic acid oxidation did not seem to be contributing to the sustained rise in respiratory rate that could be observed during the later periods.

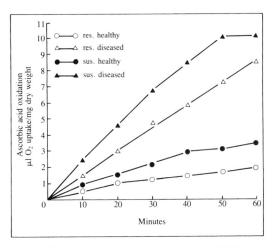

Fig. 3–18. Enzymatic ascorbic acid oxidation of 10% homogenates prepared from healthy and diseased 42-d-old resistant Coker 139 and susceptible Bottom Special stem tissues 13 d after inoculation (after Maine, 73).

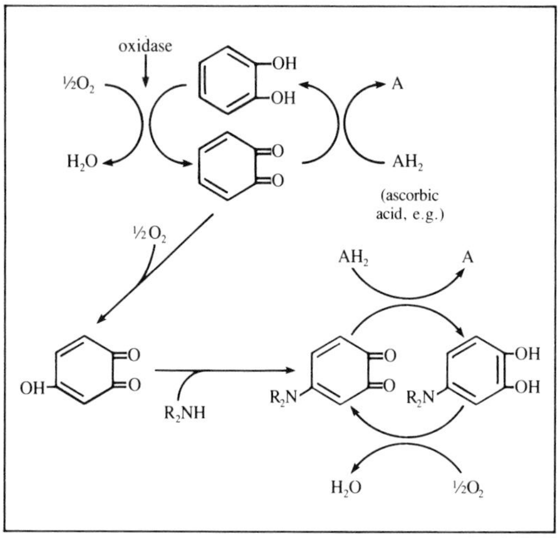

Fig. 3–19. Coupled oxidation induced by phenol oxidase with catechol as substrate. AH_2 is a reducing agent and R_2NH a secondary amine, such as dimethylamine (after Beevers, 12).

Maine (73) was also able to establish that the oxidation of polyphenols and ascorbic acid in tissues affected with *P. solanacearum* involved different enzymes. He reported that more than one enzyme was capable of catalyzing the oxidation of ascorbic acid. The one that accounted for most of the ascorbic acid oxidation was confined to the wall or particulate fraction of tobacco tissue homogenates and was not influenced by boiled homogenates of diseased tissues. The boiled homogenates were supplied, ostensibly, as a quinone source. On the other hand, the activity of the second enzyme, which was associated with cellular fluids, was at times more than doubled by the addition of boiled homogenates from diseased tissue. These findings were interpreted as indicating that ascorbic acid oxidase and a system consisting of quinone reductase and polyphenol oxidase may be responsible for the increased oxidation of ascorbic acid in diseased tissue (see Fig. 3-19).

Thus, polyphenol oxidase activity may account for a comprehensive portion of the increased oxygen consumption that occurs in diseased tissues after pathogenesis has progressed through its initial phases. The immediate responses in O_2 consumption during the first 48 h of pathogenesis may, on the other hand, signify enhanced ascorbic acid oxidation (44).

Participation of Peroxidase and Catalase in Pathological Respiration

The activities of peroxidase and catalase were studied by Rudolph and Stahmann (97) in connection with their possible role in the virulence of *Pseudomonas phaseolicola* and the susceptibility of the bean plant *Phaseolus vulgaris* to this pathogen. The general reactions of these two enzymes are

(1) $2H_2O_2 \xrightarrow{\text{catalase}} 2H_2O + O_2$

(2) $AH_2 + H_2O_2 \xrightarrow{\text{peroxidase}} 2H_2O + A$

In the second reaction, AH_2 may represent a phenolic compound, such as chlorogenic acid or caffeic acid.

Rudolph and Stahmann (97) observed that peroxidase activity remained unchanged or decreased in susceptible bean leaves infected with a virulent strain of *P. phaseolicola* and increased following infection by a less virulent one. Analogously, peroxidase activity in a resistant bean variety showed a greater increase following infection with a virulent strain than it did in a susceptible variety. Hence, it seems that heightened peroxidase activity favors resistance to *P. phaseolicola*. Indirect evidence for this contention is derived from a comparison of several isolates of *P. phaseolicola*, which showed that the virulent isolates contained a higher catalase activity than the less virulent ones. From these observations, it could appear that increased catalase activity engendered by the virulent isolates reduces the efficacy of the natural defenses of the bean plant. It was proposed that this was accomplished by a suppression of peroxidase activity, specifically by catalase destroying substrate of the peroxidase enzyme. The net result of the interaction between these two enzymes in a susceptible bean plant infected by a virulent isolate of *P. phaseolicola* seems to be a redox potential that favors maintaining phenolic compounds in their reduced and hence antimicrobially less active form, rather than as antimicrobially active quinones.

Digat (32) has reported a similar and apparently causal relationship between high catalase activity and virulence for *P. solanacearum*. Loss of this activity and of polysaccharide production appear to be involved in the loss of virulence (48, 49). In vitro experiments revealed that avirulent mutants of the pathogen are able to recover catalase activity when provided with the preformed prosthetic hemegroup. It is possible, therefore, that the avirulent mutants form the apoenzyme of catalase but not its heme coenzyme. When the avirulent mutants were grown on a heme-containing medium they regained not only their catalase activity but also their partial virulence. When these mutants were injected into tomato, they caused wilt, but much more slowly than the virulent strains. Apparently, catalase activity is but a part of the biochemical complex that controls virulence of *P. solanacearum*. Lovrekovich et al. (71) injected heat-killed *P. tabaci* into tobacco leaves and noted increased peroxidase activity, formation of new isozymes of peroxidase, and an increased resistance to subsequent inoculation with *P. tabaci*. In their experiments, increased peroxidase levels also occurred when the tobacco leaves were infiltrated with cell-free extracts of *P. tabaci* culture fluid. A companion experiment revealed that infiltration of tobacco leaf tissue with commercial peroxidase similarly resulted in increased resistance to a challenge inoculation with *P. tabaci*.

The role of peroxidase as a feature of disease development may be deduced from the experiments by Urs et al. (119, 120). They found horseradish peroxidase (HRP) at 50 μg/ml to be bactericidal in vitro to *X*.

Table 3–10. The effect of horseradish peroxidase in combination with either ascorbic acid or dehydroascorbic acid on survival of *Xanthomonas phaseoli* var. *sojense*.

Treatment number	Component	Survival (%)
1	CPB[a]	100.000
2	CPB + ascorbic acid[b]	100.000
3	CPB + dehydroascorbic acid[c]	100.000
4	CPB + H_2O_2 + KI + HRP[d]	0.090
5	CPB + H_2O_2 + KI + HRP + ascorbic acid	0.001
6	CPB + H_2O_2 + KI + HRP + dehydroascorbic acid	0.000

a. Citrate phosphate buffer.
b. 20 nmols/ml.
c. 20 nmols/ml.
d. Enzyme acitivity in units 3,625/mg protein/min (after Urs and Dunleavy, 120).

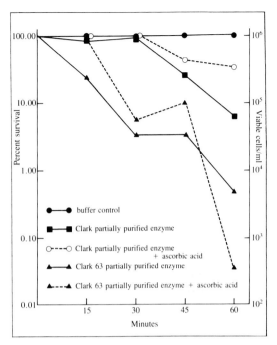

Fig. 3–20. The effect of incubation time on survival of *Xanthomonas phaseoli* var. *sojense* in the presence of partially purified and concentrated peroxidase from mitochondria of Clark and Clark 63 soybean, with or without ascorbic acid (after Urs et al., 120).

phaseoli var. *sojense* (bacterial pustule of soybean) in the presence of potassium iodide and H_2O_2. The antibacterial activity of HRP was totally suppressed by heating the enzyme at 80C for 20 min, or by leaving H_2O_2 out of the bioassay culture medium. Adding ascorbic acid (20 nmoles/ml) to the medium potentiated the activity of HRP, and dehydroascorbic acid was even more effective in this respect (see Table 3-10).

In a second study by Urs et al. (120), of the near isogenic soybean cultivars Clark, which is susceptible, and Clark 63, which is resistant to the pustule pathogen, the latter had a threefold greater mitochondria-bound peroxidase activity and a fourfold greater ascorbic acid oxidase activity than the former. Both are largely wall-bound enzymes (44). Comparative rates of O_2 uptake during the oxidation of ascorbic acid by mitochondrial fractions of Clark 63, Clark, and HRP were 3.0:1.0:0.8. In addition, the peroxidase preparation from Clark 63 had a greater antibacterial activity in the presence of H_2O_2 than did the preparation from Clark. That they may, in fact, be different isozymes of peroxidase was not determined. Partially purified peroxidase from the two cultivars revealed divergent antibacterial patterns in the presence of ascorbic acid. The partially purified Clark 63 peroxidase incubated with bacteria for one hour sustained bacterial survival at 0.5% of the original population; this was reduced to 0.03% in the presence of ascorbic acid. The Clark enzyme, on the other hand, permitted the survival of 8.0% and 50.0%, respectively, in the absence and presence of ascorbic acid (Fig. 3-20).

Urs et al. (120) have concluded that peroxidase plays an essential role in soybean resistance to bacterial pustule. However, H_2O_2 is, of course, integral to the antibacterial activity of the enzyme. Specifically, the bacteria per se produce some H_2O_2, and the high levels of ascorbic acid oxidase are able to generate additional H_2O_2 by the oxidation of ascorbic acid to dehydroascor-

bic acid. Finally, the presence of both peroxidase and H_2O_2 at elevated levels together with a phenol-like catechol constitute the antibacterial components of a peroxidase-H_2O_2-phenol system. Apparently, the components are present in more ideal quantities in Clark 63 than its near isogenic cultivar Clark.

Pathological Respiration Induced by Fungi

Respiratory Increase in Diseased Plants

A wide variety of plant species infected by pathogenic fungi exert a general reaction, which is expressed as an increase of respiratory rate (27, 38, 102, 123). Both obligate and facultative pathogens, several stresses, and chemical or mechanical injuries can induce an increase in the rate of plant respiration. Therefore, it seems that this reaction is nonspecific.

The initial increase in respiration usually occurs with the appearance of visible symptoms, rises to a maximal rate coincident with sporulation, and declines. The RQ of the diseased tissue is usually about 1.0. It has frequently been suggested that respiratory increase is due to the respiration of the host tissue and that the direct participation of the parasite in the pathological respiration is negligible. However, the extent to which fungi participate is not yet clear. For example, von Sydow and Durbin (110) have shown that the relative accumulation of C^{14}-labeled metabolites in the parasite, at least in the later stages of infection, is appreciably higher than in the host, indicating that chemical changes detected in diseased tissues may be contributed by the fungus. In barley infected with mildew (*Erysiphe graminis*), the respiratory increase is much greater at the center of mildew colonies than in the tissues adjacent to a colony. It has been suggested that the respiration of the host was stimulated by substances diffusing from the infection site and that, therefore, the respiration of the uninfected tissues near to the mildew and rust colonies is also increased (23, 101). Concerning the mechanism of the respiratory increase in diseased plants, two possibilities may be considered. First, uncoupling of oxidative phosphorylation could be the cause of the enhanced O_2 uptake. As a result of uncoupling, a high level of ADP (phosphate acceptor) accumulates, which fosters an increased rate of respiration (see The Healthy Plant, above). Second, synthetic processes stimulated in diseased plants may cause an automatic augmentation in respiration. In these instances the consumption of ATP in synthetic processes and the resultant increases in ADP lead to a higher rate of respiration.

Thus, both mechanisms are related to a lowered level of ATP and a concomitant accumulation of ADP in tissues that, in turn, favor an augmented respiration. In addition, the removal of enzyme inhibitors may be involved in the respiratory increase observed in diseased plant tissue (see Chap. 7).

The role of uncoupling was proposed as an explanation of augmented respiration of rusted wheat, first for the resistant varieties and the later stages of susceptible infections. In these cases, inorganic phosphate level increased and acid-labile phosphate (supposedly ATP) decreased, and the tissue exhibited a relative insensitivity to the uncoupler 2,4-dinitrophenol (DNP). It is well known that uncoupling agents like DNP prevent synthetic reactions, as well as the formation of ATP, and that they simultaneously accelerate respiration. According to this suggestion, infection itself exerts a DNP-like (uncoupling) effect on respiration (46, 93); consequently, diseased wheat plants are relatively insensitive to DNP. This effect was also reported for diseases caused by facultative parasites such as *Helminthosporium victoriae* (61); *Gibberella saubinetii* (122); and *Cercospora personata* (109) and for the action of T-toxin (79).

It is also possible that pathways of respiration in diseased plants are divergent from those in healthy ones and are not linked to phosphorylation (ATP formation), thereby causing an uncoupling type of respiration, i.e. increased O_2 uptake. For example, the pentose phosphate pathway is not known to be linked to oxidative phosphorylation. As will be shown later, this pathway has an important role in the pathological respiration of many plants.

A better documented explanation for the respiratory increase is that ATP is used up in stimulating metabolism, i.e. in synthetic processes such as protein, nucleic acid, carbohydrate, and aromatic synthesis. All these syntheses that require energy accelerate the breakdown

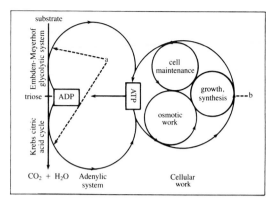

Fig. 3–21. Relationships between respiration, growth, and uncoupling agents. The uncoupling agents act essentially to dissociate the energy-producing reactions of substrate catabolism from the adenylic system (point a), thus blocking the performance of work but causing increases in enzymatic activity of the enzymes of glycolysis and the Krebs cycle. Respiratory stimulation as a result of increase in cellular work would correspond to action at point b, causing increased turnover of the adenylic system, i.e., increased ADP levels (after Daly and Sayre, 30).

of ATP and, consequently, can increase the rate of respiration by an automatic regulatory mechanism that depletes high-energy phosphate bonds. According to the results of Daly and Sayre (30), safflower infected with *Puccinia carthami* exhibited a pathologically stimulated cellular growth and a concomitant increase in tissue respiration. In supporting this abnormal growth of the diseased host cells, ATP is consumed; thus the breakdown of ATP is accelerated and respiration is enhanced.

Indeed, stimulated growth, accumulation and mobilization of metabolites, synthesis of proteins, and increase of protoplasmic streaming all demand energy, i.e. ATP. Hence, respiration is automatically regulated through the control of the level of ADP (phosphate acceptor) (see Figs. 3-3 and 3-21). It would seem that, in the early stage of disease, synthetic processes induce high rates of respiration, whereas in the late stages injury and decomposition of tissues lead to increased respiration.

One can presume that the parasitically enhanced respiration is related, at least partially, to the stimulated respiration in plants induced by wounds. In this case, the extra O_2 consumption by the damaged or wounded tissues (at late stages of infection) is caused by the activation of "unnatural" terminal oxidations (e.g. stimulated phenol oxidation). Furthermore, the wound stimulus induces protein and RNA synthesis, including the synthesis of oxidative enzymes, in plant tissues probably by derepression (26). The number of mitochondria is also increased in wounded tissues (4). This phenomenon alone may be a direct cause of the increased rate of respiration in infected plants.

Not only the respiratory rate but also the temperature of diseased tissues is higher than in healthy ones. The increase in temperature is from 0.3 to 0.7C for mildew and rust diseases (133). At present, we do not have direct evidence indicating whether respiration of infected plants is, in part at least, wasteful. Evidence that respiratory efficiency is decreased in diseased plant tissues was derived from the fact that rust-infected wheat requires more oxygen per unit of RNA synthesized in protoplasm than healthy wheat (94). In addition, it has been reported that mitochondria isolated from a victorin-treated plant exhibit little respiratory control and lower ratios of ADP to oxygen (129).

Abolition of the Pasteur Effect

As mentioned earlier, either an uncoupling of phosphorylation or an increase in the rate of synthetic reactions requiring energy raises the level of ADP and inorganic phosphorus in tissues and results in an elevated rate of respiration. Either of these events is also responsible for the abolition of the Pasteur effect. Thus, dinitrophenol as an uncoupler also abolishes the Pasteur effect. Sempio (100) was the first to stress the importance of the Pasteur effect being counteracted by infection and that this is reflected in a concomitant respiratory increase. Indeed, a counteraction of the Pasteur effect associated with the respiratory increase was demonstrated in rust-infected wheat and safflower and in sweet potato infected with *Helicobasidium mompa*. Results of the abolition of the Pasteur effect in rust-infected safflower are presented in Table 3-11, which also demonstrates an increased O_2 uptake in the diseased tissues and the lack of sensitivity of the disease-induced additional respiration to an inhibitor of glycolysis (NaF).

Since the Pasteur effect represents a more efficient utilization of glucose in respiration, a relative suppression of carbohydrate breakdown with the abolition of the Pasteur effect in diseased plants means that the aerobic breakdown of carbohydrate becomes less efficient. This loss of efficiency is believed to be due to "aerobic fermentation." Thus, respiratory efficiency is clearly decreased by the abolition of the Pasteur effect, and this result is associated with an increased rate of respiration. It will be recalled that, under conditions that favor anaerobic respiration, hexoses are not completely broken down to CO_2 and water with a maximum yield of ATP but, instead, hexose is probably degraded to alcohol.

Table 3–11. Altered respiration in safflower hypocotyls infected by *Puccinia carthami* (after Daly, 27).

	O_2 uptake		Aerobic CO_2	Anaerobic CO_2	
	μliter/g/h		μliter/g/h	μliter/g/h	Pasteur
	Control	NaF			effect
Healthy	103	52	107	66	0.62
Diseased	171	114	176	40	0.26

Table 3–12. The effect of 10 mM NaF on the respiration of resistant and susceptible tissues pretreated with victorin (after Krupka, 61).

		μliter O_2/h/g fresh weight		
Cultivar	Treatments	no NaF	NaF	Percent inhibition
Susceptible	Victorin	359	134	63
Susceptible	Control	176	63	64
Resistant	Victorin	109	38	65
Resistant	Control	102	32	69

The Role of the Embden-Meyerhof-Parnas Glycolytic System and the Krebs Cycle

The functions of these pathways have been characterized by experiments with enzyme inhibitors. The introduction of an enzyme inhibitor induces a more or less specific inhibition of particular respiratory enzymes, as well as an inhibition of the respiratory rate, and has revealed the role of the enzyme and the importance of the respiratory pathway in which it functions. For example, sodium fluoride inhibits the phosphoglyceric acid-phosphopyruvic acid reaction in glycolysis by poisoning the activity of enolase and inhibiting oxygen uptake. This reaction clearly implicates the glycolytic pathway in respiration. Sodium fluoride inhibits oxygen uptake in both healthy and toxin-treated oat plants, the latter being treated with the toxin of *Helminthosporium victoriae* (61). From this demonstration, one can deduce that stimulation of respiration by the toxin in the plant is mediated through glycolysis (see Table 3-12).

On the other hand, as was shown in Table 3-11, the rust-induced respiration was not sensitive to NaF, indicating the involvement of another pathway, the pentose phosphate cycle, in this instance.

Malonic acid, which is a competitive inhibitor of succinic dehydrogenase, is used to confirm the occurrence of the Krebs cycle in healthy plants and is a weak respiratory inhibitor in rusted wheat and mildewed *Brassica* and in groundnut tissues infected with *Cercospora* (37, 45, 54, 109). This finding suggested that a pathway other than the Krebs cycle is the main route of respiration in the above-mentioned diseased plants. On the other hand, Kristev (60) has shown that rust infection of wheat results in an increase of CoA, which is necessary for both the Krebs and the glyoxylate cycles. Actually, the process of transacetylation is stimulated in infected tissues, and the amounts of citric and malic acids are elevated in the tissues. Both acids are products of transacetylation.

Pentose Phosphate Cycle

Activation of the pentose phosphate cycle is one of the most characteristic features of the metabolism of the

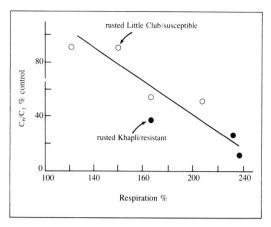

Fig. 3–22. The correlation between the C_6/C_1 ratio and oxygen uptake for rust-infected leaves of susceptible and resistant varieties. Ordinate: C_6/C_1 ratio as a percentage of the value for uninfected controls. Abscissa: oxygen uptake as a percentage of the value for uninfected controls (after Shaw and Samborski, 103).

diseased plant. If the amount of $^{14}CO_2$ respired from glucose-1-C^{14} or glucose-6-C^{14} and the C_6/C_1 ratio is calculated, the degree of participation of the pentose phosphate cycle in the respiration of a plant can be estimated. If glucose is oxidized via the pentose phosphate cycle, the C_1 carbon atom will contribute most of the evolved CO_2 and the C_6/C_1 ratio will be less than 1. In such experiments, one feeds the same amount of glucose-1-C^{14} and glucose-6-C^{14} to the tissue and determines the ratio of C_6 to C_1 evolved as $C^{14}O_2$.

The C_6/C_1 ratio of rust-infected wheat, bean, and safflower and mildew-infected barley falls sharply at the time of sporulation. Therefore, a negative correlation exists between the C_6/C_1 ratio and the respiration rate (Fig. 3-22), in which the C_6/C_1 ratio falls below unity and the rate of respiration increases. However, Daly et al. (28) have stressed that the sharp drop in the C_6/C_1 ratio is manifested only during later stages of the disease syndrome. Apparently, the observed inverse relationship between the C_6/C_1 ratio and respiration does not persist during the entire course of pathogenesis.

As might be expected, the activity of two essential enzymes of the pentose phosphate cycle, glucose-6-phosphate dehydrogenase and 6-phosphogluconate dehydrogenase, is heightened both in rust- and mildew-diseased plants, as well as in potato tubers infected with *Phytophthora infestans*. It has been determined, however, that the activation of these enzymes of the pentose phosphate cycle represents a nonspecific plant response that is also evoked by wounding, nitrogen starvation (53), excision of plant organs, virus infections, and toxin effects (35). Thus, the activation of the pentose phosphate cycle in the infected plants may also be considered a nonspecific and generally occurring alteration of the

respiratory pattern; nonetheless, it is a response by the plant to infection.

As was mentioned earlier, it has not been definitely shown that the pentose phosphate cycle is linked to oxidative phosphorylation. Hence, nonphosphorylating electron transport systems in diseased plants may be operative. Under such circumstances, an increased oxygen consumption could take place without concomitant production of ATP, and pathways not coupled with phosphorylation would cause an "uncoupling type" of respiration.

Terminal Oxidation

The main terminal oxidase in both healthy and diseased tissue appears to be cytochrome oxidase (1, 29). The NADH produced in the glycolytic pathway, as well as in the Krebs cycle, is oxidizied via the cytochrome system and is coupled with the synthesis of ATP. On the other hand, NADPH produced in the pentose phosphate pathway may be reoxidized by certain synthetic processes that are activated in diseased tissue. Theoretically, it is possible that NADPH oxidation is coupled with noncytochrome systems. For example, Mapson and Goddard (74) showed that electron transfer from NADPH to molecular O_2 can occur in a system such as:

$$NADPH \rightarrow glutathione \rightarrow ascorbic\ acid\ reductase \rightarrow ascorbic\text{-}oxidase \rightarrow O_2$$

Indeed, in rusted wheat leaves (55) and in oats treated with the toxin of *Helminthosporium victoriae* (61), ascorbic acid is oxidized at rates that increase as the O_2 uptake of tissues increases. Experiments with enzyme inhibitors suggest that activation of ascorbic acid oxidase may be important in plant diseases such as cabbage infected with *Fusarium oxysporum*, rice infected with *Pyricularia oryzae*, and others.

Polyphenoloxidase was also shown to be activated in many plant diseases such as those caused by *Fusarium* spp., *Phytophthora*, *Cochliobolus*, *Ceratocystis*, *Thielavia*, *Helicobasidium*, and others (39). An electron transport system with polyphenoloxidase functioning in vitro was also suggested (75):

$$NADPH \rightarrow glutathione \rightarrow ascorbate \rightarrow quinones \rightarrow polyphenoloxidase \rightarrow O_2$$

However, the participation of the system in vivo is questionable. Relatively little attention has been given to the investigation of coenzyme levels in diseased plants. In wheat leaves infected by leaf rust, there occurs an accumulation of NADH and NADPH that is definitely connected to high dehydrogenase activity (95).

In considering infections caused by facultative parasites, it is important to distinguish between respiratory responses occurring in the initial phase of disease and oxidative reactions occurring in the later phases of infection, i.e. at the time of tissue disintegration (necrosis). It is generally believed that, in the initial phase of plant disease, the activity of pathways normally operating in healthy plants is enhanced, and the terminal electron transport system is the cytochrome system. In later phases with degenerative changes, different "terminal" oxidases such as peroxidase, polyphenoloxidase, and ascorbic oxidase play important roles in the oxygen consumption of the diseased plant. The same is true for the hypersensitive reactions of plants against obligate parasites. These oxidases are not regarded as enzymes having any in vivo role in the respiration of healthy plant tissues; rather, their activation is conceivably connected with the disorganization of the tissues.

Pathological Respiration in Resistant Hosts

When a fungal pathogen enters a resistant host cell, a general acceleration of the metabolism of the host occurs. Oxidative metabolism will also be influenced. Generally, the change in the respiratory rate of resistant cultivars is more pronounced than in susceptible ones. It has been observed that oxygen consumption increases more rapidly in resistant plants infected either by obligate or facultative parasites. For example, an increased O_2 consumption has been registered by the potato-*Phytophthora*, the sweet potato-*Ceratocystis*, the wheat-*Puccinia*, and the barley-*Erysiphe* host-pathogen complexes. It must be pointed out, however, that acceleration of respiratory activity occurs only in an initial period of infection; later, the respiration rate gradually decreases (Fig. 3-23). It has been shown (104) that the

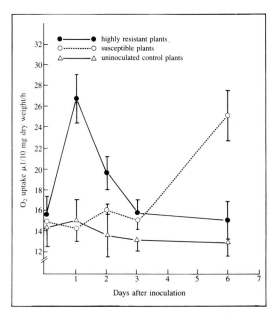

Fig. 3–23. Respiration of cultivar Sultan barley leaves resistant and susceptible to inoculation with *Erysiphe graminis* f. sp. *hordei* races 15–0 (avirulent) and 1–4 (virulent), respectively. The second leaf of each plant was used. Vertical bars represent the standard deviation (after Smedegaard-Petersen and Stølen, 104).

increased respiratory rate deprived the mildew-resistant barley of energy for growth. This led to a reduction in yield, although no symptoms appeared. Nonpathogenic leaf surface fungi may induce nonvisible reactions in the host with increased respiration and reduction in grain yield. Indeed, field experiments show that removal of fungal flora from leaf surfaces with fungicides increases the yield, although no symptoms can be seen in the untreated barley plants.

Antonelli and Daly (3) point out that the earlier and greater activation of respiration in resistant cultivars is not universal. When isogenic lines of wheat that differ in the temperature-sensitive gene for rust resistance were compared, the time of the initial responses in respiration for both resistant and susceptible infected tissues was the same. The resistant type showed a leveling in respiration rate, whereas the susceptible type continued to increase.

Of first importance was the finding that the resistance of the host can be diminished by narcotics. For example, ethanol, a potent narcotic substance, is able to suppress respiration as well as the synthesis of proteins and phenolics, and it decreases disease resistance in potato infected with *Phytophthora*. Similarly, inhibition of the phosphorylation reaction by the application of the uncoupling agent DNP is also accompanied with breaking of disease resistance (42, 114, 115). Gothoskar et al. (42) have also shown that tomato treated with DNP loses its resistance to *Fusarium* infection. Furthermore, in potatoes infected with *Phytophthora infestans*, DNP delayed the appearance of resistant reaction. These results provide evidence for the participation of oxidative metabolism in disease resistance (96). It remains to be shown which specific reactions are necessary to express the defense reaction more intensively in the untreated cells, i.e. cells in the metabolically active state.

As discussed in Chap. 6, the appearance of the hypersensitive reaction is accompanied by an accumulation and oxidation of phenolic compounds and a concomitant activation of phenol-oxidizing enzymes. Activity of polyphenoloxidase and peroxidase enzymes is known to be higher in infected resistant plants than in susceptible ones (41). In addition, Fuchs and Kotte (41) demonstrated that the resistant host became susceptible to *Phytophthora infestans* by the treatment of the potato tuber tissue with the Cu-enzyme (polyphenoloxidase) inhibitors.

Hypersensitivity, one of the most important reactions of the plant, is often accompanied by tissue browning. The appearance of necrosis is always associated with enzymatic oxidation of phenolics. Phenoloxidation products, the quinones, are usually condensed to form polyquinoid structures or to react with proteins or amino acids forming melanin substances (browning effect). The whole process is associated with activated oxidative enzymes and results in a pronounced respiratory increase.

When the plant synthesizes phytoalexins in response to infection, the cyanide-resistant pathway may also have a role in that process. Alves et al. (2) induced stress metabolite production (phytoalexins) in sliced potato tubers with an extract of *Phytophthora infestans*. Pretreatment of the slices with ethylene enhanced the tissue response to the fungal extract. When pretreatment included the specific inhibitor of cyanide-resistant respiration, salicylhydroxamic acid (SHAM), the synthesis of phytoalexins was prevented. Promoting cyanide-resistant respiration with ethylene, or inhibiting it with SHAM, either promoted or inhibited phytoalexin synthesis. Experiments by Bostock et al. (19) called attention to the fact that terpenoid phytoalexins accumulate in infected tuber tissues only after they have been held in storage. In fresh tubers, this does not occur (see Chap. 10). It is well known that the capacity for cyanide-resistant respiration increases rapidly in aged potato slices from tubers that have been held in storage. This increase parallels an increase in terpenoid phytoalexins after inoculation with *Phytophthora* (19). Thus, it would seem that a cyanide-resistant respiratory pathway is involved in phytoalexin metabolism.

Comparative Analysis of Disease Physiology—Respiration

Plants infected by fungi, bacteria, or viruses may display a common response, namely an increase of respiratory rate, which seems to be one of the most general phenomena in the physiology of diseased plants. However, this reaction is a nonspecific one, as not only plant pathogens but numerous chemicals and mechanical injury are also able to induce it.

In virus-diseased plants a slight enhancement of respiration is usually detectable. This seems to occur only in advanced stages of systemic infection when the phenomenon is observed on whole leaf tissue with cells infected asynchronously. In other instances, changes of respiratory activity may take place very soon after inoculation. Experimental results must be approached with caution, however, because wet and dry weights of plant tissue change on infection. An increase in respiratory rate expressed on a fresh-weight basis may actually reflect a decrease when expressed on a dry-weight basis.

Generally, O_2 consumption increases more rapidly in resistant hosts exhibiting necrotic symptoms than in susceptible ones, whether the disease is caused by viruses, fungi, or bacteria. Furthermore, in resistant plants respiration increases more rapidly during the earlier phases of disease, but later the respiratory rate gradually decreases. Even in those resistant cultivars that are without symptoms following fungal infection, respiration rate may increase sooner than in the susceptible ones. Isolated plant cells may respond within seconds to the presence of bacterial cells with an elevated rate of respiration. Of additional interest is the apparent differential rate response induced by incompatible, compatible, and saprophytic bacterial species. Incompatible species induced the greatest rise in respiration and saprophytes the least.

Abolition of the Pasteur effect seems to be a further characteristic response of plant tissue invaded by fungi and bacteria. As a consequence, respiratory efficiency is decreased, and this may be associated with the observed increased rate of O_2 uptake. Explanations offered for these respiratory increases have most frequently involved synthetic processes that stimulate ATP utilization or an uncoupling of oxidative phosphorylation. The precise relationship of these phenomena to the frequently reported respiratory increase is in need of additional experimentation.

Another respiratory process that alters in infectious plant disease is the pathway of glucose catabolism. The shift from the Embden-Meyerhof-Parnas sequence to the pentose phosphate pathway has been observed in tissues parasitized by fungi, viruses, and bacteria. This shift is also a nonspecific one; various types of injury, aging, and stress, as well as infection, are able to induce it. In virus infections, the pentose phosphate pathway seems to participate only in host-virus combinations resulting in local lesion, rather than in systemic infections.

It has been observed that respiration may proceed along different oxidative pathways during the course of disease development. In the initial phases of disease development caused by biotrophs, the pathway normally operative in the healthy plant, the NADH → cytochrome system, is enhanced. In the later stages of compatible interactions and during the entire course of incompatible interactions that lead to a rapid development of hypersensitive necrosis, other oxidative enzyme systems seem to be activated. These include peroxidase, polyphenol oxidase, and ascorbic acid oxidase. The function of these enzymes may be linked with the oxidation of NADPH, which is generated in the pentose phosphate cycle.

In diseases caused by necrotrophs, the production of tissue necroses also results in increased oxidative enzyme activity and a rapid enhancement of O_2 uptake. In these instances, potentiated respiration occurs during both the early and the later stages of disease development. In fact, when the disease is caused by necrotrophic fungi and bacteria, the pattern is similar in both compatible and incompatible host-pathogen combinations. However, in the incompatible combination (resistance) the whole event proceeds with greater rapidity. When necrotic symptoms are associated with phytoalexin production, the cyanide-resistant respiratory pathway may have a role in the synthesis of these "stress-associated metabolites."

It is noteworthy that resistance of the host to fungal pathogens can be diminished by oxidative enzyme inhibitors. This fact supports the contention that oxidative metabolism somehow participates in the mechanism of disease resistance.

Additional Comments and Citations

Doke in two papers (Physiol. Plant Path. 23:345–57; 23:359–67, 1983) proposed a mechanism that may have a role in inducing cell and tissue necrosis in incompatible potato/*Phytophthora* interaction. Superoxide anion (O_2^-) is elicited in incompatible, but not in compatible, interaction, and this is associated with NADPH oxidation. Superoxides are toxic, are destructive to living cells, and could be involved in rapid cell death associated with the hypersensitive necrosis. Hyphal wall components of the incompatible fungus are also able to induce O_2^- generation. Superoxide dismutase (SOD), an enzyme catalyzing the dismutation of superoxide to peroxide $(2O_2^- + 2H^+ \rightarrow H_2O_2 + O_2)$, suppressed or delayed the O_2^- formation and tissue necrosis, when applied preinfectionally. SOD is regarded as a scavanger of O_2^-.

References

1. Akazawa, T., and I. Uritani. 1955. Respiratory increase and phosphorus and nitrogen metabolism in sweet potato infected with black rot. Nature 176:1071–72.

2. Alves, L. M., E. G. Heisler, J. C. Kissinger, J. M. Patterson, and E. B. Kalan. 1979. Effects of controlled atmospheres on production of sesquiterpenoid stress metabolites by white potato tuber. Plant Physiol. 63:359–62.

3. Antonelli, E., and J. M. Daly. 1966.Decarboxylation of indoleacetic acid by near-isogenic lines of wheat resistant or susceptible to *Puccinia graminis* f. sp. *tritici*. Phytopatholology 56:610–18.

4. Ashai, T., Y. Honda, and I. Uritani. 1966. Increase of mitochondrial fraction in sweet potato root tissue after wounding or infection with *Ceratocystis fimbriata*. Plant Physiol. 41:1179–84.

5. Axelrod, B. 1967. Other pathways in carbohydrate metabolism. In: Metabolic pathways. 3d ed. D. M. Greenberg (ed.). Academic Press, New York, 1:272–308.

6. Axelrod, B., and H. Beevers. 1956. Mechanism of carbohydrate breakdown in plants. Ann. Rev. Plant Physiol. 7:267–98.

7. Bates, D. C., and S. R. Chant. 1975. The effects of virus infection on oxygen uptake and respiratory quotient of the leaves of French bean and cowpea. Physiol. Plant Path. 16:199–209.

8. Baur, J. R., R. S. Halliwell, and R. Langston. 1967. Effect of tobacco mosaic virus infection on glucose metabolism in *Nicotiana tabacum* L. var. Samsun. I. Investigations with ^{14}C labelled sugars. Virology 32:406–12.

9. Baur, J. R., R. S. Halliwell, and R. Langston. 1967. Effect of tobacco mosaic virus infection on glucose metabolism in *Nicotiana tabacum* L. var. Samsun. II. Investigation of malonic acid inhibition on respiration. Virology 32:413–15.

10. Baur, J. R., B. Richardson, R. S. Halliwell, and R. Langston. 1967. Effect of tobacco mosaic virus infection on glucose metabolism in *Nicotiana tabacum* L. var. Samsun. III. Investigation of hexosemono-phosphate shunt enzymes and steroid concentration and biosynthesis. Virology 32:580–88.

11. Bawden, F. C. 1959. Physiology of virus diseases. Ann. Rev. Plant Physiol. 10:239–56.

12. Beevers, H. 1961. Respiratory metabolism in plants. Row, Peterson and Co., Evanston, IL.

13. Beevers, H., and M. Gibbs. 1954. The direct oxidation pathway in plant respiration. Plant Physiol. 29:322–24.

14. Bell, A. A. 1964. Respiratory metabolism of *Phaseolus vulgaris* infected with alfalfa mosaic and southern bean mosaic viruses. Phytopathology 54:914–22.

15. Bidwell, R. G. S. 1974. Plant physiology. Macmillan, New York.

16. Bonner, W. D. 1957. Soluble oxidases and their functions. Ann. Rev. Plant Physiol. 8:427–52.

17. Boser, H. 1958. Einfluss pflanzlicher Virosen auf Stoffwechselfunktionen des Wirtes. II. Aktivitäten einiger Fermente des Glycolysesystems und des Citronensäurecyclus in gesunden und virosen Kartoffelblättern und knollen. Phytopath. Z. 33:197–202.

18. Boser, H. 1959. Einfluss pflanzlicher Virosen auf Stoffwechselfunktionen des Wirtes. III. P/O-Quotienten und einige Fermentaktivitäten in Knolle, Keim und Blatt viröser Kartoffelpflanzen. Phytopath. Z. 37:164–69.

19. Bostock, R. M., E. Nuckles, J. W. Henfling, and A. Kuć. 1983. Effects of potato tuber age and storage on sesquiterpenoid stress metabolic accumulation, glycoalkaloid accumulation, and response to abscisic and arachidonic acids. Phytopathology 73:435–38.

20. Brock, T. D. 1970. Biology of microorganisms. Prentice-Hall, Englewood Cliffs, NJ.

21. Brucker, W., and W. A. K. Schmidt. 1959. Zum Zuckerstoffwechsel des Kallus- und crowngall-gewebes von *Datura* und *Daucus*. Ber. Deut. Bot. Ges. 72:321–32.

22. Burroughs, R., J. A. Goss, and W. H. Sill, Jr. 1966. Alterations in respiration of barley plants infected with brome grass mosaic virus. Virology 29:580–85.

23. Bushnell, W. R., and P. J. Allen. 1962. Respiratory changes in barley leaves produced by single colonies of powdery mildew. Plant Physiol. 37:751–58.

24. Chant, S. R. 1967. Respiration rates and peroxidase activity in virus infected *Phaseolus vulgaris*. Experientia 23:676–77.

25. Clagett, C. O., N. E. Tolbert, and R. H. Burris. 1949. Oxidation of α-hydroxy acids by enzymes from plants. J. Biol. Chem. 178:977–87.

26. Click, R. E., and D. P. Hackett. 1963. The role of protein and nucleic acid synthesis in the development of respiration in potato tuber slices. Proc. Nat. Acad. Sci. 50:248–56.

27. Daly, J. M. 1967. Some metabolic consequences of infection by obligate parasites. In: The dynamic role of molecular constituents in plant-parasite interaction. C. J. Mirocha and I. Uritani (eds.). Amer. Phytopath. Soc., St. Paul, MN, pp. 144–64.

28. Daly, J. M., A. A. Bell, and L. R. Krupka. 1961. Respiratory changes during development of rust diseases. Phytopathology 51:461–71.

29. Daly, J. M., and S. G. Jensen. 1961. Evidence for functional cytochrome oxidase in tissues infected by obligate parasites. Phytopathology 51:759–65.

30. Daly, J. M., and R. M. Sayre. 1957. Relations between growth and respiratory metabolism in safflower infected by *Puccinia carthami*. Phytopathology 47:163–68.

31. Diener, T. O. 1963. Physiology of virus-infected plants. Ann. Rev. Phytopath. 1: 197–218.

32. Digat, B. 1972. The significance of catalase activity in *Pseudomonas solanacearum*. In: Proc. Third Int. Conf. Plant Pathogenic Bact. H. P. Maas Geesteranus (ed.). Pudoc, Wageningen, Netherlands, pp. 71–75.

33. Dwurazna, M. M., and M. Weintraub. 1969. Respiration of tobacco leaves infected with different strains of potato virus X. Can. J. Bot. 47:723–30.

34. Dwurazna, M. M., and M. Weintraub. 1969. The respiratory pathways of tobacco leaves infected with potato virus X. Can. J. Bot. 47:731–36.

35. Farkas, G. L., L. Dézsi, M. Horváth, K. Kisbán, and J. Udvardy. 1963/64. Common pattern of enzymatic changes in detached leaves and tissues attacked by parasites. Phytopath. Z. 49:343–54.

36. Farkas, G. L., L. Lovrekovich, and Z. Klement. 1963.

Increased activity of glucose-6-phosphate-dehydrogenase in tobacco leaves affected by *Pseudomonas tabaci*. Naturwiss. 50:22–23.

37. Farkas, G. L., and Z. Király. 1955. Studies on the respiration of wheat infected with stem rust and powdery mildew. Physiol. Plant. 8:877–87.

38. Farkas, G. L., and Z. Király. 1958. Enzymological aspects of plant diseases: I. Oxidative enzymes. Phytopath. Z. 31:251–72.

39. Farkas, G. L., and Z. Király. 1962. Role of phenolic compounds in the physiology of plant diseases and disease resistance. Phytopath. Z. 44:105–50.

40. Fischer, E. 1981. Relation between the transmembrane potential, the energy status, and the ion content in pepper fruit tissue affected by bacterial infection. Ph.D. diss., Dept. of Plant Pathology, Univ. of Missouri-Columbia.

41. Fuchs, W. H., and E. Kotte. 1954. Zur Kenntnis der Resistenz von *Solanum tuberosum* gegen *Phytophthora infestans* de By. Naturwissenschaften 41:169–70.

42. Gothoskar, S. S., R. P. Scheffer, M. A. Stahmann, and J. E. Walker. 1955. Further studies on the nature of *Fusarium* resistance in tomato. Phytopathology 45:303–6.

43. Hackett, D. 1963. Respiratory mechanism and control in higher plant tissues. In: Control mechanisms in respiration and fermentation. B. Wright (ed.). Ronald Press, New York, pp. 105–27.

44. Hallaway, M., P. D. Phethen, and J. Taggart. 1970. A critical study of the intracellular distribution of ascorbate oxidase and a comparison of the kinetics of the soluble and cell wall enzyme. Phytochemistry 9:935–44.

45. Heitefuss, R. 1957. Utersuchungen zur pathologischen Physiologie von *Peronospora parasitica* Gäum. auf *Brassica oleracea*. Ph. D. diss., Univ. of Göttingen.

46. Heitefuss, R., and W. H. Fuchs. 1963. Phosphatstoffwechsel und Sauerstoffaufnahme in Weizenkeimpflanzen nach Infektion mit *Puccinia graminis tritici*. Phytopath. Z. 46:174–98.

47. Huang, J. S., and R. N. Goodman. 1985. Recognition of pathogenic and saprophytic bacteria by tobacco leaf cells, reflected as changes in respiratory rates. Acta Phytopath. Hung. 20:7–15.

48. Husain, A., and A. Kelman. 1957. Presence of pectic and cellulolytic enzymes in tomato plants infected by *Pseudomonas solanacearum*. Phytopathology 47:111–12.

49. Husain, A., and A. Kelman. 1958. The role of pectic and cellulolytic enzymes in pathogenesis by *Pseudomonas solanacearum*. Phytopathology 48:377–86.

50. Jensen, S. G. 1968. Photosynthesis, respiration and other physiological relationships in barley infected with barley yellow dwarf virus. Phytopathology 58:204–8.

51. Jensen, S. G. 1968. Factors affecting respiration in barley yellow dwarf virus infected barley. Phytopathology 58:438–43.

52. Kato, S., and T. Misawa. 1972. Studies on the infection and the multiplication of plant viruses. VI. Analysis of metabolic changes in tobacco tissue infected with cucumber mosaic virus. Ann. Phytopath. Soc. Jap. 38:342–49.

53. Király, Z. 1964. Effect of nitrogen fertilization on phenol metabolism and stem rust susceptibility of wheat. Phytopath. Z. 51:252–61.

54. Király, Z., and G. L. Farkas. 1955. Über die parasitogen induzierte Atmungssteigerung beim Weizen. Naturwissenschaften 42:213–14.

55. Király, Z., and G. L. Farkas. 1957. On the role of ascorbic oxidase in the parasitically increased respiration of wheat. Arch. Biochem. Biophys. 66:474–85.

56. Kornberg, H. L., and H. A. Krebs. 1957. Synthesis of cell constituents from C_2- units by a modified tricarboxylic acid cycle. Nature 179:988–91.

57. Kow, Y. W., D. L. Erbes, and M. Gibbs. 1982. Chloroplast respiration. A means of supplying oxidized pyridine nucleotide for dark chloroplastic metabolism. Plant Physiol. 69:442–47.

58. Krebs, H. A., and W. A. Johnson. 1937. The role of citric acid in intermediate metabolism in animal tissues. Enzymologia 4:148–56.

59. Krebs, H. A., and H. L. Kornberg. 1957. Energy transformation in living matter. Springer-Verlag, Berlin.

60. Kristev, K. 1961. On the role of pyrophosphate metabolism in the respiratory increase of rust-infected wheat leaves. Phytopath. Z. 42:279–85.

61. Krupka, L. R. 1959. Metabolism of oats susceptible to *Helminthosporium victoriae* and victorin. Phytopathology 49:587–94.

62. Laties, G. G. 1957. Respiration and cellular work and the regulation of the respiration rate in plants. In: Survey of biological progress. B. Glass (ed.). Academic Press, New York, 3:215–99.

63. Laties, G. G. 1982. The cyanide-resistant, alternative path in higher plant respiration. Ann. Rev. Plant Physiol. 33:519–55.

64. Lehninger, A. L. 1965. Bioenergetics. 2d ed. W. A. Benjamin, New York.

65. Lehninger, A. L. 1975. Biochemistry—the molecular basis of cell structure and function. 2d ed. Worth, New York.

66. Link, G. K. K., and R. M. Klein. 1951–52. Studies on the metabolism of plant neoplasms. II. The terminal oxidase patterns of crown gall and auxin tumors of tomato. Bot. Gaz. 113:190–95.

67. Lipmann, F. 1941. Metabolic generation and utilization of phosphate bond energy. Adv. Enzymol. 1:99–162.

68. Lockhart, B. E. L., and J. S. Semancik. 1970. Respiration and nucleic acid metabolism changes in cowpea mosaic virus infected etiolated cowpea seedlings. Phytopathology 60:1852–53.

69. Loebenstein, G. 1959. Effect of a virus infection in sweet potatoes upon the host plant respiration. Virology 9:62–71.

70. Loebenstein, G., and N. Linsey. 1963. Effect of virus infection on peroxidase activity and C_6/C_1 ratios. Phytopathology 53:350. (Abstr.)

71. Lovrekovich, L., H. Lovrekovich, and M. A. Stahmann. 1968. The importance of peroxidase in the wildfire disease. Phytopathology 58:193–98.

72. Main, C. E., and J. C. Walker. 1971. Physiological responses of susceptible and resistant cucumber to *Erwinia tracheiphila*. Phytopathology 61:518–22.

73. Maine, E. C. 1960. Physiological responses in the tobacco plant to pathogenesis by *Pseudomonas solanacearum*, causal agent of bacterial wilt. Ph.D. diss., North Carolina State Univ., Raleigh.

74. Mapson, L. W., and D. R. Goddard. 1951–52. The reduction of glutathione by plant tissues. Biochem. J. 49:592–601.

75. Marrè, E. 1961. Phosphorylation in higher plants. Ann. Rev. Plant Physiol. 12: 195–218.

76. Menke, G. H., and J. C. Walker. 1963. Metabolism of resistant and susceptible cucumber varieties infected with cucumber mosaic virus. Phytopathology 53:1349–55.

77. Merrett, M. J. 1960. The respiration rate of tomato stem tissue infected by tomato aucuba mosaic virus. Ann. Bot. n. s. 24:223–31.

78. Merrett, M. J., and D. W. Sunderland. 1967. The metabolism of carbon-14 specifically labelled glucose in leaves showing tobacco mosaic virus induced local necrotic lesions. Physiol. Plant. 20:593–99.

79. Miller, R. F., and D. E. Koeppe. 1971. Southern corn leaf blight: Susceptible and resistant mitochondria. Science 173:67–69.

80. Nagy, R., A. J. Riker, and W. H. Peterson. 1938. Some physiological studies on crown gall and contiguous tissue. J. Agr. Res. 57:545–55.

81. Neish, A. C., and H. Hibbert. 1943–44. Studies on plant

tumors. Part II. Carbohydrate metabolism of normal and tumor tissue of beet root. Arch. Biochem. 3:141–57.

82. Németh, J., and Z. Klement. 1967. Changes in respiration rate of tobacco leaves infected with bacteria in relation to the hypersensitive reaction. Acta Phytopath. Hung. 2:303–8.

83. Novacky, A., and C. I. Ullrich-Eberius. 1982. Relationship between membrane potential and ATP level in *Xanthomonas campestris* pv. *malvacearum*-infected cotton cotyledons. Physiol. Plant Path. 21:237–49.

84. Orlob, G. B., and D. C. Arny. 1961. Some metabolic changes accompanying infection by barley yellow dwarf virus. Phytopathology 51:768–75.

85. Owen, P. C. 1955. The respiration of tobacco leaves in the 20-hour period following inoculation with tobacco mosaic virus. Ann. Appl. Biol. 43:114–21.

86. Owen, P. C. 1955. The respiration of tobacco leaves after systemic infection with tobacco mosaic virus. Ann. Appl. Biol. 43:265–72.

87. Owen, P. C. 1957. The effect of infection with tobacco etch virus on the rates of respiration and photosynthesis of tobacco leaves. Ann. Appl. Biol. 45:327–31.

88. Owen, P. C. 1958. Photosynthesis and respiration rates of leaves of *Nicotiana glutinosa* infected with tobacco mosaic virus and of *N. tabacum* infected with potato virus X. Ann. Appl. Biol. 46:198–204.

89. Parish, C. L., M. Zaitlin, and A. Siegel. 1965. A study of necrotic lesion formation by tobacco mosaic virus. Virology 26:413–18.

90. Passoneau, J. V., and O. H. Lowry. 1963. Phosphofructokinase and the control of the citric acid cycle. Biochem. Biophys. Res. Comm. 13:372–79.

91. Patil, S. S. 1974. Toxins produced by phytopathogenic bacteria. Ann. Rev. Phytopath. 12:259–79.

92. Pierpoint, W. S. 1968. Cytochrome oxidase and mitochondrial protein in extracts of leaves of *Nicotiana glutinosa* L. infected with tobacco mosaic virus. J. Exp. Bot. 19:264–75.

93. Pozsár, B. I., and Z. Király. 1958. Effect of rust infection on oxidative phosphorylation of wheat leaves. Nature 182:1686–87.

94. Quick, W. A. 1960. Ribonucleic acid relationships in wheat leaves infected with stem rust fungus. Master's thesis, Univ. of Saskatchewan, Canada.

95. Rohringer, R. 1964. Drifts in pyridine nucleotide levels during germination of rust spores and in wheat leaves after inoculation with leaf rust. Z. Pflanzenkr. Pflanzensch. 71:159–70.

96. Rubin, B. A., E. V. Artzichowskaja, and T. A. Proskurnikova. 1947. Oxidative changes of phenols and their relation to the resistance of potatoes against *Phytophthora infestans*. Biokhimiya 12:141–52.

97. Rudolph, K., and M. A. Stahmann. 1964. Interaction of peroxidases and catalases between *Phaseolus vulgaris* and *Pseudomonas phaseolicola* (halo blight of bean). Nature 204:474–75.

98. Russo, M., and G. P. Martelli. 1969. Cytology of *Gomphrena globosa* L. plants infected by beet mosaic virus (BMV). Phytopath. Medit. 8:65–82.

99. Scott, K. J., J. S. Craigie, and R. M. Smillie. 1964. Pathways of respiration in plant tumors. Plant Physiol. 39:323–27.

100. Sempio, C. 1950. Metabolic resistance to plant diseases. Phytopathology 40:799–819.

101. Sempio, C., and G. Barbieri. 1964. Attività respiratoria e difesa in varietà di fagiolo variamente resistenti all' *Uromyces appendiculatus*. Phytopath. Z. 50:270–82.

102. Shaw, M. 1963. The physiology and host-parasite relations of the rusts. Ann. Rev. Phytopath. 1:259–94.

103. Shaw, M., and D. J. Samborski. 1957. The physiology of host-parasite relations. III. The pattern of respiration in rusted and mildewed cereal leaves. Can. J. Bot. 35:389–407.

104. Smedegaard-Petersen, V., and O. Stølen. 1981. Effect of energy-requiring defense reactions on yield and grain quality in a powdery mildew-resistant barley cultivar. Phytopathology 71:396–99.

105. Solymosy, F., and G. L. Farkas. 1962. Simultaneous activation of pentose phosphate shunt enzymes in a virus infected local lesion host plant. Nature 195:835.

106. Solymosy, F., and G. L. Farkas. 1963. Metabolic characteristics at the enzymatic level of tobacco tissues exhibiting localised acquired resistance to viral infection. Virology 21:210–21.

107. Sunderland, D. W., and M. J. Merrett. 1965. The respiration of leaves showing necrotic local lesions following infection by tobacco mosaic virus. Ann. Appl. Biol. 56:477–84.

108. Sunderland, D. W., and M. J. Merrett. 1967. Respiration rate, adenosine diphosphate and triphosphate concentrations of leaves showing necrotic local lesions following infection by tobacco mosaic virus. Physiol. Plant. 20:368–72.

109. Swamy, R. N. 1964. Respiration in *Cercospora*-infected groundnut tissues. Part II. Effect of some enzyme inhibitors and 2, 4-dinitrophenol on respiration. Indian J. Exp. Biol. 2:193–97.

110. Sydow, B. von, and R. D. Durbin. 1962. Distribution of C^{14}-containing metabolites in wheat leaves infected with stem rust. Phytopathology 52:169–70.

111. Takahashi, T., and T. Hirai. 1963. A cytochemical study of tobacco leaf cells infected with tobacco mosaic virus. Virology 19:431–40.

112. Takahashi, T., and T. Hirai. 1964. Respiratory increase in tobacco leaf epidermis in the early stage of tobacco mosaic virus infection. Physiol. Plant. 17:63–70.

113. Takahashi, W. N. 1947. Respiration of virus infected plant tissue and effect of light on virus multiplication. Amer. J. Bot. 34:496–500.

114. Tomiyama, K. 1957. Cell-physiological studies on the resistance of potato plants to *Phytophthora infestans*. V. Effect of 2, 4-dinitrophenol upon the hypersensitive reaction of potato plant cell to infection by *P. infestans*. Ann. Phytopath. Soc. Jap. 22:75–78.

115. Tomiyama, K., R. Sakai, N. Takase, and M. Takakuwa. 1957. Physiological studies on the defence reaction of potato plant to the infection by *Phytophthora infestans*. IV. The influence of preinfectional ethanol narcosis upon the physiological reaction of potato tuber to the infection by *P. infestans*. Ann. Phytopath. Soc. Jap. 21:153–58.

116. Tribe, H. T. 1955. Studies in the physiology of parasitism. XIX. On the killing of plant cells by enzymes from *Botrytis cinerea* and *Bacterium aroideae*. Ann. Bot. 19:351–68.

117. Tu, J. C., and R. E. Ford. 1968. Effect of maize dwarf mosaic virus infection on respiration and photosynthesis of corn. Phytopathology 58:282–84.

118. Turner, J. S. 1960. Fermentation in higher plants: Its relation to respiration; the Pasteur effect. In: Encyclopedia of plant physiology. W. Ruhland (ed.). Springer-Verlag, Berlin, 12 (pt. 2):42–87.

119. Urs, N. V. R., and J. M. Dunleavy. 1974. Function of peroxidase in resistance of soybean to bacterial pustule. Crop Sci. 14:740–44.

120. Urs, N. V. R., and J. M. Dunleavy. 1974. Bactericidal activity of horseradish peroxidase on *Xanthomonas phaseoli* var. *sojensis*. Phytopathology 64:542–45.

121. van Kammen, A., and D. Brouwer. 1964. Increase of polyphenol oxidase activity by local virus infection in uninoculated parts of leaves. Virology 22:9–14.

122. Verleur, J. D. 1960. Studies on the increased respiration of

potato tuber tissue after infection with *Gibberella saubinetii* (Mont.) Sacc. Acta Bot. Neerl. 9:119–166.

123. Verleur, J.D. 1968. Regulation of carbohydrate and respiratory metabolism in fungal diseased plants. In: Biochemical regulation in diseased plants or injury. T. Hirai, Z. Hidaka, and I. Uritani (eds.). Phytopath. Soc. Jap., Tokyo, pp. 275–85.

124. Warburg, O., and Christian, W. 1939. Isolierung und Kristallisation des Proteins des oxydierenden Gärungsferments. Biochem. Z. 303:40–68.

125. Weintraub, M., W. G. Kemp, and H. W. J. Ragetli. 1960. Studies on the metabolism of leaves with localised virus infections. I. Oxygen uptake. Can. J. Microbiol. 6: 407–15.

126. Weintraub, M., and H. W. J. Ragetli. 1971. A mitochondrial disease of leaf cells infected with an apple virus. J. Ultrastruct. Res. 36:669–93.

127. Weintraub, M., H. W. J. Ragetli, and M. M. Dwurazna. 1964. Studies on the metabolism of leaves with localised virus infections. Mitochondrial activity in TMV-infected *Nicotiana glutinosa* L. Can. J. Bot. 42:541–45.

128. Weintraub, M., H. W. J. Ragetli, and E. Lo. 1972. Mitochondrial content and respiration in leaves with localized virus infections. Virology 50:841–50.

129. Wheeler, H., and P. J. Hanchey. 1966. Respiratory control: Loss in mitochondria from diseased plants. Science 154:1569–71.

130. Wynd, F. L. 1943. Metabolic phenomena associated with virus infection in plants. Bot. Rev. 9:395–465.

131. Yamaguchi, A. 1960. Increased respiration of leaves bearing necrotic local lesions. Virology 10:287–93.

132. Yamaguchi, A., and T. Hirai. 1959. The effect of local infection with tobacco mosaic virus on respiration in leaves of *N. glutinosa*. Phytopathology 49:447–49.

133. Yarwood, C. E. 1953. Heat of respiration of injured and diseased leaves. Phytopathology 43:675–81.

4

Cell Wall Composition and Metabolism

The Healthy Plant

Introduction

It is difficult to define the plant cell wall precisely from either structural or compositional points of view as it is not a *static* organelle. During growth, the wall's polymeric consitutents vary and hence its chemical properties are in constant flux. In addition the structure of some of its components has not been defined. The physical properties of cell walls are also altered as a consequence of cell metabolism (52, 143). Developmentally, the cell wall may be divided into primary wall and secondary wall, the former being thin and laid down by undifferentiated growing cells and the latter thicker and the product of cells that have stopped growing.

Configurationally, the wall may be viewed as a network of microfibrils embedded in a matrix. The microfibrils are essentially cellulose consisting of chains of glucose molecules, and the matrix is composed of hemicellulose and pectic substances. The former consists primarily of xyloglucans and the latter of homogalacturonans and rhamnogalacturonans. In the primary cell wall the cellulose microfibrils are more widely dispersed by the matrix components, whereas the cellulose fraction and lignin (a polymer of aromatic alcohols) dominate the secondary wall. The wall also contains a glycoprotein that is rich in hydroxyproline as well as serine and other amino acids. In the course of maturation the wall becomes increasingly impregnated with lignin. The external "veneer" of some cells is composed of waxes, cutin and suberin, which consist of long chained fatty acids that are liberally hydroxylated. The comprehensive cuticular veneer covering the wall of an apple leaf epidermal cell is presented in Fig. 4–1.

Cell Wall Polysaccharides

Cellulose-microfibrils. The simplest and most abundant of the major polysaccharides of the secondary cell wall is cellulose. It consists of long chains of glucose molecules linked through carbon atoms 1 and 4 by β glucosidic bonds (Fig. 4–2). Chain lengths are 1–3 μm with a degree of polymerization in the order of l0,000 or more units (10). The cellulose microfibrils in the primary cell wall are much fewer in number; they are embedded in a pectin matrix and their degree of polymerization is about 6,000 to 7,000 units (71, 139, 162). The basic biological unit of cellulose fine structure is the microfibril, which may vary in length from species to species. Its crystalline core measures approximately 8.5×4.5 nm and is surrounded by a paracrystalline cortex of hemicellulose, perhaps xyloglucan and some cellulose chains, which gives the microfibril a transverse dimension of ca. $5 \times l0$ nm. The number of glucose chains per microfibril, according to Preston (162), is about 78. On the basis of X-ray data and model studies it is believed (71) that the cellulose microfibrils are held together internally by hydrogen bonds between oxygen atoms of alternating glycosidic bonds in one glucan chain and primary hydroxyl groups at position C_6 in glycosyl residues of a neighboring chain. Adjacent chains have the same polarity and hence are parallel to each other (79). In other words, the reducing ends of these chains face in the same direction.

The microfibrillar arrangement of cellulose and its associated carbohydrate substances indicate that there are spaces (microcapillaries) between the microfibrils, and these are in the order of 20–40 nm, which varies with cell age. In the primary cell wall the capillaries are filled with hemicellulose and pectin, in the secondary wall this space also contains lignin and in some tissues suberin.

Cellulose biosynthesis and microfibril orientation. The laboratory of R. D. Preston popularized the examination of algae with cellulosic microfibrillar walls to reveal microfibril synthesis and orientation (162). Recent studies by Montezinos (146) with the alga *Oocystis spiculata* have exposed apparent enzyme complexes, which are striking examples of membrane-bound protein, that may be responsible for microfibril synthesis and orientation. It seems that the enzyme complexes (ec's) are the sites of polymerization of the glucose monomers. The developing β1–4 glucan (cellulose microfibril) traverses the periplasm (an amorphous substance between the plasmalemma and the cell wall), where it becomes coated and cross-linked with noncellulosic wall components (hemicelluloses and perhaps protein). The periplasm is derived from Golgi products via exocytosis, and it is here, ostensibly, that the three-dimensional structure of the primary wall is assembled. In *Oocystis* the ec's are paired, with each assembling one microfibril while they move in opposite directions through the fluid plasmamembrane leaving the microfibrils in place as they go. By reversing directions at the poles, a single layer of microfibrils

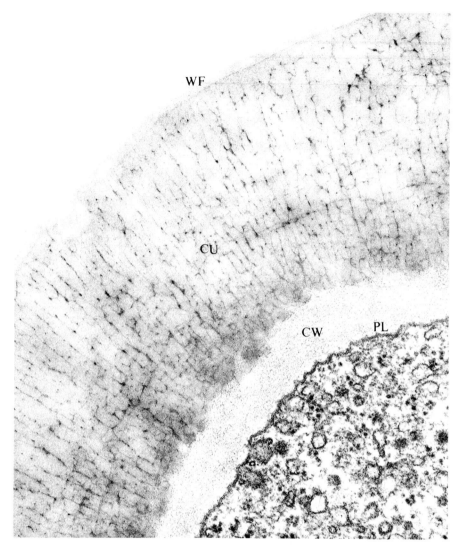

Fig. 4–1. Cuticle of apple epidermal cell showing residual wax film (WF), cuticle (CU), cell wall (CW), plasmalemma (PL), and cytoplasmic contents (courtesy P.-Y. Huang).

eventually covers the plasmamembrane surface. The ec's then dissociate from the microfibrillar layer most recently completed and reassemble in pairs to start producing fibrils at right angles to the previous layer. In higher plants, according to Vian (208), the process of synthesis and orientation of microfibrils is not clearly established; however, the fibrillogenesis takes place at the surface of the plasmalemma while hemicellulosic and pectic components are synthesized by the Golgi and delivered in vesicles to the plasmalemma (Fig. 4–3). The vesicles integrate with the plasmalemma and release the matrix components (noncellulosic) into periplasmic space. It is in this milieu of matrix-filled extracytoplasmic space that the first elements comparable to microfibrils are observed. A possible sequence for their ontogeny is that the plasmalemma-bound microfibril

synthetase complexes move in the lipid bilayer in concert with microtubules that are their ostensible supplier of cellulose precursor material. The nascent fibrils that pass through the periplasm are coated with noncellulosic polysaccharides and assume the proportions of a three-dimensionally ordered cell wall. The cellulose microfibrils in the primary wall are loosely woven in the matrix, some at random angles and others in parallel arrangement. Their distribution in the secondary wall is more complex as the alternating lamellae (Fig. 4–3) are more numerous, less fragile, and less susceptible to bond breakage. In Fig. 4–4, two strata show the typical crossed texture of the microfibrils.

Callose. Callose is mostly a β1–3 glucan and is elaborated at the surface of intact plant cells. Glucose is

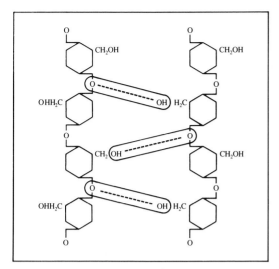

Fig. 4–2. Basic structure of the cellulose microfibril, parallel adjacent chains of β(1 → 4)-D-glucose showing hydrogen bonding between chains (after Gardner and Blackwell, 79).

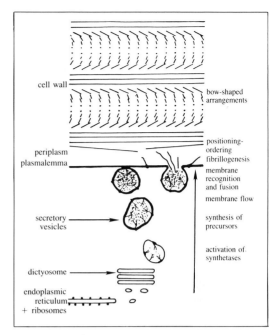

Fig. 4–3. The successive steps and events leading to the construction of a growing cell wall. A diagrammatic representation of the sequence of events and participating organelles in the synthesis of a plant cell wall. The endoplasmic reticulum generates spherosomes from which the dictyosomes develop, which in turn become secretory vesicles laden with enzymes and wall precursor materials. The vesicles (exocytes) discharge their contents into periplasmic space where wall fibrils are positioned and ordered. Alternating lamellae have been detected throughout the entire thickness of some plant cells walls wherein the bow-shaped patterns reflect small angle changes in orientation of the subunits of the lamellae (after Vian, 208).

polymerized through the participation of uridine diphosphopyridine nucleotide (UDP) glucose, and the process develops in response to cell damage or stress. Callose is generally electron transparent in normal staining procedures; however, callose deposits that form in response to infections or stress may contain materials of varying electron density. Hence they may include other components such as lignin, protein, and phospholipids.

Ultrastructural accompaniments to callose accumulation are endoplasmic reticulum dilation, haustorial sheath formation, and the so-called immune response, which results in massive and rather precisely localized callose deposition. Callose, like pectin and hemicellulose, the matrix carbohydrate polymers, may also be Golgi-derived. The enzyme involved in synthesis of callose is UDP-glucose: glucan synthetase that synthesizes β-1, 3-glucans. Potentiation of the activity of this synthetase and the production of callose is an often observed wound response. The deposition of callose and its implications are described at length by Cooper (40).

Hemicellulose (matrix). Hemicelluloses are found in the matrix of both primary and secondary plant cell walls. A comprehensive account of the major and minor components of this group has been prepared by Aspinall (10, 11). According to Northcote (154, 155), secondary thickening of the cell wall is accompanied by increased amounts of hemicellulose and cellulose. Because of the harshness of acid and basic hydrolytic fragmentation of plant cell walls, Albersheim and his coworkers (23, 105) resorted to the milder approach of enzymatic disassembly of the plant cell wall's constituents. In examining hemicelluloses, Bauer et al. (23) utilized an endo β-1-4-glucanase from the fungus *Trichoderma viride* to react with the primary cell walls of sycamore cell sus-

pension cultures. They used methylation analysis of polysaccharides released from the primary cell wall of suspension-cultured sycamore cells by endoglucanase to reveal the molecular structure of some hemicellulose polymeric components. The structure of their hemicellulose polymer is based on a repeating heptasaccharide of 4 residues of β-1–4-linked glucose and 3 residues of terminal xylose linked to carbon 6 of glucose. A fucosyl-galactose dimer is linked to one of the xylose residues (Fig. 4–5). It would appear from this study that the xyloglucan polymers are *noncovalently* linked to the cellulose fibrils. This is apparently accomplished through hydrogen-bonding every second glycosidic oxygen of the glucan chain of xyloglucan with the primary hydroxyl group at the 6 carbon position of a glucosyl residue in an adjacent cellulose chain. The glucosyl residues in the polymer are present in a celluloselike backbone, and, apparently, the xylosyl residues occur as monoxylosyl side chains linked to carbon 6 of the glycosyl residues in the glucan (hemicellulose) backbone.

Fig. 4–4. Orientation of cellulose microfibrils in algal cell wall (*Chaetamorpha*), lying as two sets of parallel microfibrils at approximately right angles to each other and as distinctly separate lamellae (courtesy of E. Frei and R.D. Preston).

The molecular weight of sycamore extracellular xyloglucan (SEPS) secreted into the culture medium by the sycamore cell suspensions is approximately 7,600, which approximates a 56-sugar residue polymer. The β-1-4-linked glucosyl backbone accounts for 25–30 of these residues. Bauer et al. (23) found the SEPS xyloglucan composition to compare favorably with xyloglucan from sycamore cell wall per se (SCW). The sugars that appeared in similar amounts included fucose, arabinose, xylose, galactose, and glucose. Mannose appeared in only one fraction of SEPS. Structural similarity exists between sycamore xyloglucan and seed xyloglucans of *Tamarindus indicus* and *Annona muricata* (116). Hence there are data that suggest a degree of universality in the plant cell walls for the hemicellulose repeating unit shown in Figure 4–5, which accounts for approximately 20% of the dry weight of the primary cell wall (see Table 4–1).

Subsequently Darvill et al. (53) revealed the presence of another hemicellulose polymer, glucuronoarabinoxylan, which approximates 5% of the mass of the primary cell wall of SEPS. The xyloglucan component of the hemicellulose fraction is estimated at 23 % of the SEPS cell wall mass. Similarity between these two polymers resides in the D-xylopyranosyl backbone; both have neutral and/or acidic side chains attached to C_2 or C_3 xylosyl residues. They are considered as structural polysaccharides that can hydrogen bond to cellulose.

Bauer et al. (23) contend that there is enough hemicellulose in the SCW to encapsulate all of the cellulose microfibrils with a monolayer of xyloglucan. In fact, they suggest that other "classical" hemicelluloses such as xylans, arabinoxylans, mannans, glucomannans, and galactoglucomannans have structures that accommodate noncovalent bonding to cellulose chains (199). This then appears to be one of their biological functions,

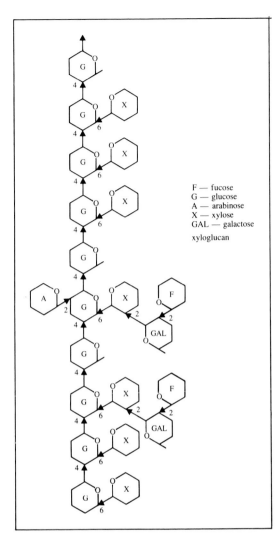

F — fucose
G — glucose
A — arabinose
X — xylose
GAL — galactose

xyloglucan

Fig. 4–5. Xyloglucan with glucose (G) backbone linked β 1–4 as in cellulose; the xylose (X) units are attached to the backbone, and both galactose (GAL) and Fucose (F) are bonded to xylose (after Bauer et al., 23).

the other being to bind *covalently* through glycosidic bonds at their reducing ends to the *pectic* polysaccharides. However, this seems to be less common than previously supposed (52, 53). Bauer et al. (23) proposed that the term *hemicellulose* be redefined to include only those plant cell wall polysaccharides that bind *noncovalently* to cellulose. Darvill et al. (53) have suggested that xyloglucan bonding to primary cell wall cellulose microfibrils prevents the latter from forming the large aggregates characteristic of secondary walls. Hemicellulose, it would appear, tends to give the primary and immature secondary wall a degree of flexibility and perhaps greater permeability.

Pectic polysaccharides (matrix). These matrix compounds are common to all higher plants and are fre-

quently characterized as middle lamella or plant cell-wall-cementing substances. The pectic polysaccharides make up approximately 35% of the primary cell wall. They are characterized by their capacity to form gels (86), a feature that may contribute to some pathological manifestations.

The pectic polysaccharides are a complex mixture of acidic and neutral polymers composed of β-1–4-linked galacturonosyl (homogalacturonan) residues in which 2-linked rhamnosyl residues are interspersed (10; Fig. 4–6). The carboxyl groups of some of the galacturonosyl residues are methyl esterified. When esterification is less than 75%, these polymers are known as pectinic acids, and those with degrees of esterification above 75% are referred to as pectins. These polymers also contain the neutral sugars D-galactose and L-arabinose as short side chains or as multiple units (galactans, arabinans, and arabino-galactans) that are usually, but not always, covalently linked to the acidic rhamnogalacturonan main chain (10; Fig. 4–6 bottom).

Rhamnogalacturonan I (RG-I) is a pectic polymer in the primary cell wall of dicots that accounts for ca. 7% of the mass of walls isolated from suspension cultured sycamore cells by an endogalacturonase purified from *Colletotrichum lindemuthianum* (144, 145). It has a molecular weight (MW) of ca. 200,000 and is composed of rhamnose, galacturonic acid, arabinose, and galactose. The backbone of the polymer contains rhamnose and galacturonic acid in a 1:1 ratio (see Fig. 4–6b). Approximately half of the rhamnose residues are 2-linked glycosidically to the C_4 of galacturonic acid. The other half of the rhamnose residues are 2–4 linked with their C_2 linked to the C_4 of galacturonic acid in the backbone polymer and their C_4 linked to a variety of side chains containing arabinose and galactose. Hence the concept of the gross structure of RG-I is consistent with the long-held view that pectic polymers contain regions rich in neutral sugars. The ratio of the total arabinose and galactose residues to 2,4-linked rhamnose is 6:1, suggesting that the average side chain length is 6 residues (53). McNeil et al. (145) originally detected 7 different side chain configurations but have since developed evidence (McNeil, personal communication) that there may be as many as 30 or 50. Hence, they have indicated that the structural complexity of RG-I suggests that it may play more than a structural role in the plant cell wall. The great diversity in the array of side chains provides at least logic for their function as regulatory molecules vis-à-vis molecular or metabolic recognition.

"Visualization" of the chemical structure of the major component of pectin as it may occur in situ has been clarified by the studies of Albersheim and his colleagues, initially by Talmadge et al. (195) and more recently McNeil et al. (143, 144) and Darvill et al. (51). They have presented a tentative structure for the rhamnogalacturonan based on analysis of sycamore cell wall fractions prepared either by an endo-α-1,4 polygalacturonase from *Colletotrichum lindemuthianum* or acid hydrolysis digestion that suggests RG-I to be the backbone of the pectic polymer.

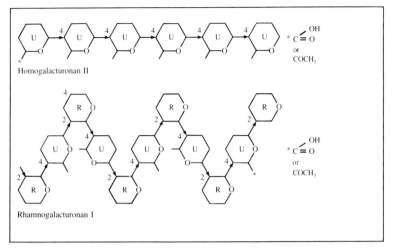

Fig. 4–6. (top) Homogalacturonan, α1 → 4 galacturonic acid constituent of the pectic polysaccharides; (bottom) Rhamnogalacturonan I shown as alternating galacturonic acid (U) and rhamnose (R) residues, which also has a variety of side chains containing arabinose and galactose (after McNeil et al., 144, 145).

Rhamnogalacturonan II (RG-II) is a pectic polymer isolated from suspension cultured sycamore cells with an endogalacturonase purified from *C. lindemuthianum* filtrates. This polymer constitutes 3–4% of the sycamore cell primary wall, and, though complex, its structure is now rather well defined. The polymer size is approximately 60 glycosyl residues; however, its diversity is attested to by its 10 different monosaccharide components.

The Albersheim group has detected two distinctly different heptasaccharides in RG-II that may comprise as much as 30–40% of the polymer. The heptasaccharides are possibly sidechains on a 4-linked galacturonic acid–rich backbone. RG-II is believed to be linked to other wall polysaccharides by 4-linked galacturonic acid chains. Spellman et al. (184) have provided the first information about the location of the rather rare 2-O-methyl fucose and apiose residues found in the cell walls of some dicots. These two sugars and the newly discovered acidic monosaccharide 3-C-carboxy-5-deoxy-L-xylose are three of the ten sugars detected in RG-II (185).

One of the heptasaccharides detected has been characterized, and its structure is shown below.

The second heptasaccharide, only partially characterized (McNeil personal communication), has the following tentative structure.

The heptasaccharide that has been completely characterized has many hydrophobic functional groups. For example, four of the residues are deoxysugars containing methyl groups instead of hydroxy-methyl groups. This oligomer has one endogenous O-methyl group and 2 O-acetyl esters. Hence the hydrophobicity of this structure lends support to the contention that interactions between carbohydrates and their receptor proteins are governed by hydrophobic bonding.

The acidic monosaccharide that was found in the first heptasaccharide, called aceric acid (185), has also been fully characterized. Its structural features are shown below.

Whereas the general picture of the pectic polysaccharides remains similar to that envisaged by Aspinall (10, 11) and Talmadge et al. (l95), more recent studies have provided important details that are related to function as well as structure (145). Clearly, these studies accentuate the complexity of the structure commonly referred to as the middle lamella and suggest the possible role of pectins as recognition molecules.

The degree to which sycamore primary cell wall pectic polymers reflect the nature of pectins in other plant species is indicated in studies of Aspinall et al. (l3) and Aspinall and Cotterell (12). They reported that the β-1,4-linked galactan of soybean seed and the arabinans of soybean, lemon peel, and mustard seed and their respective rhamnogalacturonans are similar.

Noncovalent bonding between pectic polysaccharides is also of importance. It is generally accepted that calcium ions accomplish this by linking oxygen atoms of four galacturonosyl residues distributed between two galacturonan chains. The calcium ions cross-linking the galacturonan chains confer rigidity to the cell walls. The reaction occurs specifically between calcium and unesterified polygalacturonosyl residues.

Reaction between Hydroxyproline Rich Glycoprotein and Other Wall Components

Hydroxyproline rich proteins fall into three classes: cell wall glycoproteins, soluble arabinogalactan proteins, and some lectins. We are concerned here with the hydroxyproline-rich glycoproteins of the cell wall (HPRGPCW).

The HPRGPCW per se consists primarily (95%) of six amino acids, hydroxyproline (Hyp) accounting for 46% (207). The molecule is l/3 protein and 2/3 carbohydrate with galactose comprising 3% and arabinose 97% (206). About 70% of the hyp is glycosylated with a tetra-arabinoside and about 24% with a triarabinoside. About 40% of the serines are galactosylated. Varner (207) suggested that the oligosaccharides stabilize the helical structure of HPRGPCW. This positively charged protein is a rigid rod whose peptide backbone is covered with oligosaccharides. Importantly, it accumulates in plant cell walls in response to infection, wounding, and other stress-related signals (Fig. 4–7).

The Hyp-rich protein is largely confined to the primary cell walls that are undergoing extension, and hence the name *extensin* was accorded to it by Lamport (118). The protein's role as an integral part of the primary cell wall has often been suggested as one of growth regulation. The process was envisioned as a cross-linking with itself or non-carbohydrates such as phenols. In this regard Fry (77) sought to relate dityrosine (tyrosine is sufficiently abundant in wall protein) to the coupling of proteins to wall carbohydrates in a manner shown to occur between diferulic acid and polysaccharides in the presence of peroxidase by Markewalder and Neukom (134) and Leatham et al. (123). A number of experiments with ^{14}C tyrosine have led to the detection of labeled isodityrosine (IDT), and the amount of this com-

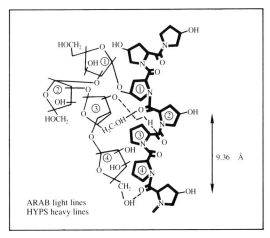

Fig. 4–7. A semiperspective drawing using a Dreiding model of a 3 residue/turn, left- handed polyhydroxyproline helix with all peptide bonds trans. The pitch is 9.36 Å. CPK models show that a β-linked tetra arabinofuranoside (linked 1–3, 1–2, 1–2, 1–4 Hyp) "nests" close to the polyhydroxyproline helix. The following hydrogen bonds are possible: No. 1. arabinose, none; No. 2. arabinose, none; No. 3. arabinose C-5 hydroxymethyl to No. 2 Hyp carboxyl; No. 3. arabinose C-2 hydroxyl to No. 3 Hyp carboxyl; No. 4 arabinose C-5 hydroxymethyl to No. 5 Hyp carboxyl. Appropriately linked arabinose residues show the H-bonds, which are not drawn in perspective (after Lamport, 120).

pound found in HPRGPCW suggests that it is the cross-linking molecule of the cell wall protein extensin (77).

A model for the manner in which extensin is actually linked to cellulose in the primary cell wall has been proposed by Lamport and Epstein (121). Extensin here implies the rodlike pentapeptide Ser-hyp-hyp-hyp-hyp, whose arabinoligosaccharides nest or wrap around and hydrogen bond with the polypeptide backbone through the C_4 hydroxyl groups of hydroxy-proline. Lamport and Epstein suggest that the primary cell wall is a weave of cellulose microfibrils penetrating an extensin mesh that is suspended in a pectin- hemicellulose gel (see Fig. 4–8). Cross-linking between two extensin chains occurs at precise intervals to accommodate the girth of the cellulose microfibril and impose the wall plasticity control that has long been sought. The xyloglucan-pectin components appear to provide the requisite slippage feature of cellulose fibrils through the restraining protein. The model obviates the necessity of covalent bonding between the interacting polymers.

A Model of a Primary Cell Wall

Albersheim (4) has collated the major disclosures concerning plant cell wall polysaccharides (10, 11, 34, 119) and proposed a model that, although indicative of the primary cell wall of suspension culture sycamore cells, seemed representative of primary cell walls of other higher plant species (10, 116, 119). The model

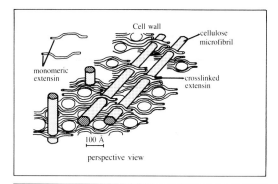

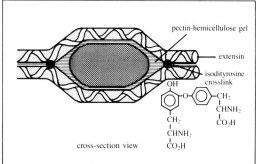

Fig. 4–8. Extensin-cellulose network, a hypothetical representation of an extensin-rich, tightly woven primary cell wall of high tensile strength. The cellulose fibers penetrate the net of extensin and the two isodityrosine cross-linked strands of extensin control both constraints and elasticity of the wall polymer and define the size and shape of a cellulose microfibril (after Lamport and Epstein, 121).

appears now to be an oversimplification of the situation. However, it provided a logical framework into which some components of the primary and the secondary wall could be integrated. The model also acknowledged the constraints of integrating polysaccharides and other encrusting substances into the more complex microfibril lattice structure of the secondary cell wall. It also provided a rationale for the sequential attack of the primary cell wall by enzymes produced by plant pathogens. McNeil and Darvill (unpublished) from Albersheim's group have put together a probable compositional summary of the cell wall's components, less its encrusting substances (Table 4–1).

Sheathing and Encrusting Substances of the Cell Wall

Almost all plant outer surfaces are cloaked in a cuticular layer of varying thickness (38, 102, 103). Included are aerial and belowground surfaces as well as surfaces of cells in contact with intercellular space and cells at natural openings such as stomata. In the latter instances the cuticle appears to become thinner when traced from the outer surface of a leaf epidermal cell or guard cell to a substomatal chamber or intercellular space (see Fig. 4–9). Martin and Juniper (138) have

written an excellent account of the cuticle's composition and function. They chose to accept the definition of Esau (63) that "the multilayered structure . . . is usually separated from the epidermal wall by a layer of pectin. This comprises an inner cuticular layer and an outer layer, the cuticle proper, made up mainly or entirely of cutin." This concept may now be modified to distinguish cutin as the chemical component of the aerial cuticle and suberin as the counterpart mantle of underground organs (113) and of certain tissues that have been wounded or naturally exposed to the atmosphere, through leaf abscission (177, 178, 183) or lenticel diaphragm rupture (63, 178).

Cuticular wax. According to Juniper (103), the wax projections on plant surfaces such as those of the bean begin to form at an early stage of development and continue throughout the period of leaf expansion (Fig. 4–l0). It is of more than passing interest to note that light intensity and duration influence the formation of these projections of surface wax. The fact that wax formation is a continuing process and not a terminal phase in the ontogeny of the leaf is also significant. We may assume, therefore, the waxes are being formed by actively metabolizing cells underlying the cuticle and that these compounds are continuously migrating toward the surface, where they eventually accumulate in greatest concentration (Fig. 4–11). The studies of Juniper (103) have not substantiated the existence of pores or galleries or plasmodesmatal-like pathways through which the surface wax is extruded. Taking the calculated plasmodesmatal distribution of Esau (63) of $36/l00 \mu m^2$ as an indication of the frequency with which one might expect the wax extrusions to be encountered on the leaf surface, the actual frequencies of wax pegs of $400/\mu m^2$ become difficult to justify. Nevertheless, there is evidence that pores do exist beneath each wax projection, and these have been observed by Hall and Donaldson (92, 93) in the leaves of *Trifolium repens*, *Brassica oleracea*, and *Poa colinsor* and in the cuticle of apple fruits. The mean diameter of these pores, seen in the electron microscope, is $6–7 \mu m$, whereas the diameters of the wax residues on the plant leaves are $38–46 \mu m$. The pores were observed on occasion to be interconnected at the surface, and apparently wax may be extruded along the length of this connection. Whether pores or galleries are necessary pathways for the seepage of waxes to aerial surfaces of plants remains to be fully substantiated.

The cuticle waxes consist of long-chain, even-numbered primary alcohols, acids, and their esters. In addition, secondary alcohols, ketones, and paraffin hydrocarbons have been detected. The cuticle waxes differ from fats in that glycerol is replaced by monohydric alcohols. These compounds have been reported to contain both odd- and even-numbered carbon-atom chains from 24 to 37. In general, the acids and alcohols consist of unbranched members of the series. The saturated acids are exemplified by lauric ($C_{11}H_{23}COOH$), palmitic ($C_{15}H_{31}COOH$), and stearic ($C_{17}H_{51}COOH$) acids and

Table 4–1. The polymers of cell walls isolated from dicots.

Cell wall component	% weight of cell wall	Quantitatively major residues	Comments
A. Cellulose	35	1,4-β-glucosyl	major structural component; forms rodlike fibers
B. Hemicelluloses			
Xyloglucan	19	1,4,6-β-glucosyl terminal-α xylosyl	hydrogen bonds to cellulose
Glucuronoarabinoxylan	5	1,4,-β-xylosyl terminal arabinosyl terminal glucuronosyl terminal 4-0-methyl glucuronosyl	major hemicellulose in monocots; believed to hydrogen bond to cellulose
C. Pectic polysaccharides			
Homogalacturonan	6	1,4-α-galacturonosyl	believed to have a structural role
Rhamnogalacturonan I	7	galacturonosyl rhamnosyl galactosyl arabinosyl	structurally complex molecule; function unknown
Rhamnogalacturonan II	3	10 different glycosyl residues	extremely complex molecule containing many unusual residues
Arabans, galactans, and arabinogalactans	18	arabinosyl	largely uncharacterized components
		galactosyl	many may be part of molecules similar to Rhamnogalacturonan I
D. Hydroxyproline-rich glycoprotein	19	hydroxyproline serine arabinosyl galactosyl	very likely a structural component of walls; may have other unknown functions

Prepared by M. McNeil and A. Darvill; Department of Chemistry, University of Colorado, Boulder.

the unsaturated ones by oleic ($C_{17}H_{33}COOH$) and linoleic ($C_{17}H_{31}COOH$) acids (115).

All of these aliphatic derivatives have but one reactive group per molecule, and as a result wax formation can involve only one molecule each of an acid and an alcohol. Polyesters may, however, be formed by esterification of hydroxy acids, such as hydroxylauric or hydroxypalmitic (juniperic) acid. These acids are joined "head to tail" to form high MW polyesters known as *estolids*.

An interesting observation was reported by Mazliak and Kader (141), who studied the metabolism of apple fruit waxes. Radioactive acetate was supplied to apple peel, and the contained waxes were extracted and saponified. Of interest from a biosynthetic point of view, the alcohols and acids of linoleic and oleic type were labeled, whereas the paraffins contained no radioactivity. It was concluded that the paraffins are synthe-

sized by a pathway other than the condensation of acetate.

Cutin. This polymer is composed of C_{16} and C_{18} families of cutin acids. The structure proposed for cutin by Kolattukudy (112, 113) appears in Fig. 4–12. The C_{16} forms predominate in fast-growing plants and consist primarily of palmitic, 16-hydroxypalmitic, l0,l6-dihydroxypalmitic acids, and forms wherein some of the hydroxyls are oxidized, e.g. 16 oxy-10 hydroxypalmitic acid. The C_{18} family is found in slower-growing plants, and representative forms are 18-hydroxyoleic acid, and 9,10,18-trihydroxystearic acid. The major esterified phenolic constituent of cutin is p-coumaric acid and occasionally ferulic acid (l73). The esterified phenols consitute ca. 1% of the cutin and appear to be covalently linked.

The biosynthesis of cutin has been studied by Kolat-

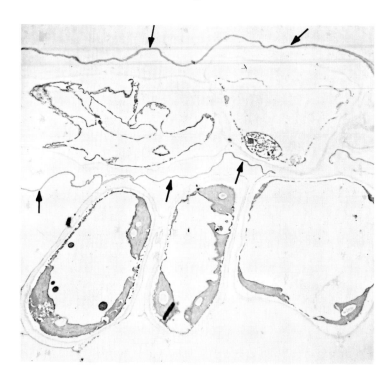

Fig. 4–9. Two epidermal cells of a Jonathan apple leaf underlain by three palisade parenchyma cells. Arrows call attention to cell surface cuticle, which is significantly thicker on the outer surfaces than on those impinging on intercellular space (after Goodman, unpublished).

tukudy and coworkers (48, 49, 112) by $1\text{-}^{14}C$ palmitic acid incorporation into *Vicia faba* leaves. The major portion of the label was indeed contained in the C_{16} triol expected to be produced with 10,16-dihydroxypalmitic acid, the major monomer of *V. faba* cutin. These results suggested that palmitic acid was first hydroxylated at the 1-position and subsequently at C-10. A proposed pathway for C_{16} cutin acids is presented in Fig. 4–13 and for the C_{18} cutin acids in Fig. 4–14. It is quite probable that, since cutin is an extracellular polymer, its synthesis occurs at an extracellular location. Incorporation of labeled monomer acids into an insoluble polymer (presumed synthesis) (48, 49, 112) was catalyzed by a particulate fraction containing cuticular membranes from excised epidermis of *V. faba*. Apparently cutin-synthesizing enzymes are present in flower petals and perhaps other tissues. Cutin synthesis seems to be roughly proportional to either leaf expansion or growth.

Suberin. There is a close chemical and functional similarity between suberin and cutin. The former is a component of the natural protective layer that develops on the surface of underground plant organs and forms when plant tissues are wounded or exposed, i.e. by rupture of lenticel closing layers or by fruit and leaf abscission (pseudowounds). Suberin has also been detected in developing tracheary walls and impregnates the Casparian strip of the endodermis. As endodermal cells mature, the entire inner surface of the cell wall may become suberized. According to Frey-Wyssling and Hausermann (72), a suberin-containing cuticle develops on the surfaces of mesophyll cells where they are exposed to intercellular air spaces (Fig. 4–19). Suberin has

been detected by Fox (70) in wounds and lenticels of potato tuber as a double layer in the cell wall per se and as an encrusting substance in the intercellular spaces of the wound or lenticel area. The lamellar structure and chemical composition of suberin in natural skin of potato, external wound periderm of potato, and potato cells lining "hollow heart" cavities was presented by Dean et al. (54). Ultrastructurally the cell walls from the three sources were similar. Isolation of suberin from the three sources revealed the components also to be similar. As with cutin, ^{14}C incorporation studies revealed that suberizing potato slices rapidly incorporate $1\text{-}^{14}C$ oleic acid and give rise to a labeled 1-hydroxyoleic and/or the corresponding dicarboxylic acid, as the major components of potato suberin (Fig. 4–14). The phenolic constituents of suberin are derived from phenylalanine much as they are in lignin formation (173). The pertinent enzymes—phenylalanine ammonia lyase and appropriate hydroxylases—are induced prior to suberization during the wound-healing process. Suberin may be generally distinguished from cutin in that the former contains large quantities of dicarboxylic acids, while the latter has them in only trace amounts (Fig. 4–12). Similarly, suberin contains very long chain (C_{20}-C_{26}) fatty acids and fatty alcohols and high phenol contents, whereas cutin contains smaller amounts of these. It is apparent, however, that cells not producing suberin can begin synthesizing this substance upon being wounded (113, 114).

Lignin. A common yet essential feature of vascular plants is their capacity to synthesize lignin. This polymer varies widely in amount with source and method of

Fig. 4–10. Electron micrograph of a shadowed carbon replica of an Alaska pea leaf surface revealing extruded wax platelets (courtesy of B.E. Juniper, Oxford University).

extraction, i.e. from 17 to 33% in wood. Its molecular weight ranges from 800 to 12,000 and it may be visualized as a three-dimensional polymer of phenyl propane units, a principal characteristic of which is its methoxyl content. A fragment of a lignin macromolecule is shown in Fig. 4–15. The importance of lignin from a plant pathological point of view has been accentuated by several studies describing lignin synthesis as an active resistance mechanism (76, 159, 172, 200).

Lignins are classified into three major groups based on their monomeric composition. For example, gymnosperm lignin is a dehydrogenation polymer of coniferyl alcohol, whereas angiosperm lignin is a mixture of coniferyl and sinapyl alcohols. Grass lignins are mixtures of sinapyl and *P*-coumaryl alcohols (89; see Fig. 4–16).

The precise site of formation of lignin precursors is not well understood; however, Clowes and Juniper (38) have speculated that precursors can move centripetally from meristematic cells to potential sites of lignifica-

tion. They suggest that the areas of lignification are "already programmed" for lignification regardless of whether the coumarin-type precursors are derived from within or outside the cells to be lignified. It is possible that phenyl propenoid precursors are produced in metabolically active cells and diffuse to polymerization sites as glycosides where glycosidases liberate the less soluble free alcohols.

The synthesis of lignin is suggested to proceed, according to Higuchi (100), in the primary walls of differentiated wood cells. The process begins in the cell corners and extends into the middle lamella as well as into primary and secondary walls.

It is of interest that as xylem differentiates and its cytoplasmic content decreases, wall lignification continues. This suggests either transport to the site of lignin perfusion or in-wall (in situ) synthesis. According to Northcote (154, 155), the hydrophobic lignin replaces water in the cell wall and encrusts both the microfibril and matrix polysaccharides, causing the wall to thicken.

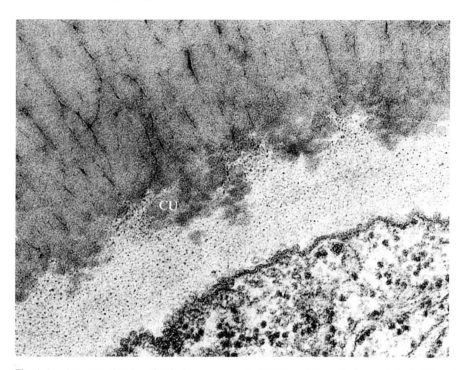

Fig. 4–11. Apparent migration of cuticular waxes or cutin (CU) through the wall of an apple leaf epidermal cell (after P.-Y. Huang, unpublished).

With the displacement of water, hydrogen bonding occurs between cellulose and hemicellulose, and covalent linkages are formed between the carbohydrates and lignin. In this regard, Markewalder and Neukom (134) reported that diferulic acid is probably esterified with both of its carboxyl groups to cell wall hemicelluloses. As a result, the wall derives great tensile strength from the microfibrils and rigidity from the lignified matrix.

The general pathway for lignin synthesis from carbohydrates proceeds through the shikimic acid pathway to the polymerization of cinnamoyl alcohols (Fig. 4–17). The switchover from general metabolism to lignification begins with the deamination of phenylalanine to cinnamic acid. Cinnamate is further converted to a number of substituted derivatives, a few of which are shown in Fig. 4–18. Deamination is catalyzed by phenylalanine ammonia lyase (PAL), the key enzyme of phenol metabolism, which is controlled by such factors as hormones and stresses of various kinds (89). The activity of PAL is followed by hydroxylases and orthomethyl transferases. In these reactions, cinnamic acid is hydroxylated in the "4" position and methoxylated in the "3" position to form ferulic acid. In the reduction of cinnamic acids to alcohols an energy source is implicit along with coenzyme A (CoA) esters and cinnamate: CoA ligases have been identified and implicated in a number of plant species (90). It is quite likely that cinnamoyl-CoA esters form in a manner similar to the activation of fatty acids. The process of activation of ferulate is shown in Fig. 4–19, and feruloyl-

CoA is an extremely reactive substrate for cinnamoyl-CoA reductase, which converts the activated moiety to the aldehyde, which is in turn converted to the alcohol by a dehydrogenase. The final step of polymerizing the coniferyl alcohol monomers to lignin occurs as a consequence of the oxidation of their phenolic hydroxyl groups. This is now firmly established as a function of peroxidase that causes the formation of phenoxyradicals shown in Fig. 4–20.

Stafford (187) has written an excellent review on the role and function of peroxidase in lignification. The enzyme apparently causes the phenoxy-free radicals to occur as monomers and as dimers, and these are further dehydrogenated and finally polymerized into the lignin macromolecule. The peroxidase enzyme is widely distributed in higher plants and is readily detected as a wall-bound enzyme in lignifying tissues. Using a pair of peroxidase enzymes from tobacco leaves and callus, Mäder and Fusl (129) have clearly established that these enzymes, like horseradish-peroxidase, catalyze the oxidation of $NADH \rightarrow NAD$ with the formation of H_2O_2. The peroxidase apparently functions in two ways, initially to provide H_2O_2 and then in the presence of cinnamyl alcohols and the primer levels of H_2O_2 to polymerize the cinnamyl alcohols to form lignin. Specifically their study revealed that coniferyl alcohol also serves a dual role in the process of lignin formation. Initially it stimulates peroxidatic oxidation of $NADH \rightarrow NAD + H_2O_2$. Then in the presence of H_2O_2, coniferyl alcohol serves as substrate for peroxidase in the formation of

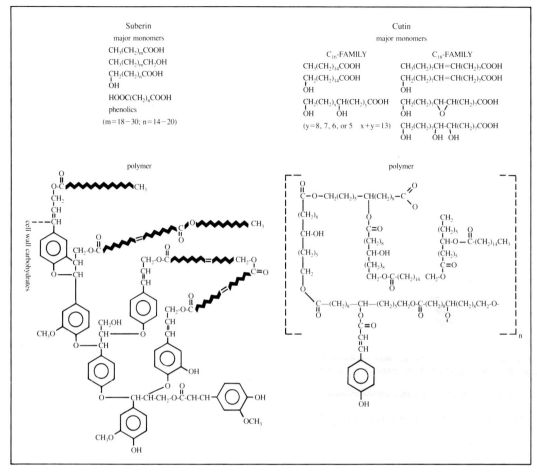

Fig. 4–12. The proposed structure of cutin and suberin (after Kolattukudy, 113).

lignin. The activity of peroxidase and the entire array of enzymes from PAL through the dehydrogenases are found wherever lignin is being synthesized. Lignin synthesis seems to proceed at greater than average rates in xylem tissue, e.g., 30 times that detected in phloem (89). Convenient tissues for these studies are found in celery stalks (89).

Wall-Degrading Enzymes

Enzymatic alteration of the cell wall by plant pathogens is perhaps more easily envisioned when one accepts, at least theoretically and perceptually, the scheme presented by Albersheim (4). The order of appearance of enzymes elaborated by the pathogen and substrate fragments supports the view of a sequential dismantling of the "macromolecule" known as the cell wall.

Rhamnogalacturonan-degrading enzymes. For many years it has been known that pectins could be degraded differentially by a series of specific enzymes (Table 4–2). Two types of enzymes disassemble the backbone of these polymers by splitting the α-1,4 bonds

of the galacturonosyl moieties. Hydrolytic cleavage is accomplished by polygalacturonases (PG) or pectinmethylgalacturonases (PMG). The former degrades

Table 4–2. Classification of pectin and pectic acid splitting enzymes.

Pectin-splitting enzymes	
Pectinmethylgalacturonases (PMG)	Pectin-lyase (PL)
1. Endo-PMG	3. Endo-PL
2. Exo-PMG	4. Exo-PL
Pectic acid-splitting enzymes	
Polygalacturonases (PG)	Pectic acid transeliminases (lyases) (PATE)
5. Endo-PG	7. Endo-PATE
6. Exo-PG	8. Exo-PATE

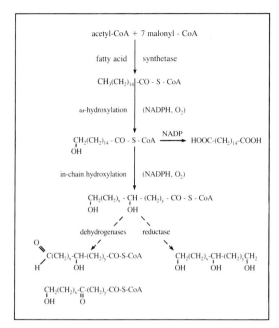

Fig. 4–13. Pathway for the biosynthesis of the C_{16} family of cutin acids. Even though CoA esters are indicated to be the substrates, the nature of the intermediates is yet to be established. Palmitoyl-CoA is probably generated by activation of the free acid, and not directly from fatty acid synthetase (after Kolattukudy, 113).

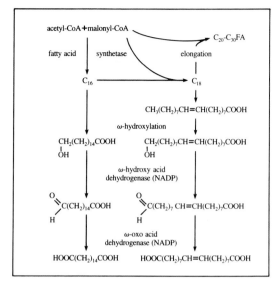

Fig. 4–14. Proposed biosynthetic pathways for the major aliphatic components of suberin (after Kolattukudy, 113).

pectic acids (nonmethyl esterified uronide chains), whereas substrate for the latter are the highly esterified pectins (homogalacturonans and rhamnogalacturonans) (Fig. 4–21a). There are additional enzymes, the trans-eliminases, now referred to as lyases, that split the glycosidic bond in a way that results in an unsaturated bond between carbons 4 and 5 of an uronide moiety (4, 5) (Fig. 4–21b). Like the hydrolases these are characterized by their substrate in that those that cleave pectin are called pectin (transeliminase) lyase (PTE) and those that cleave the acidic polymers are known as pectic acid (transeliminase) lyase (PATE). Both types of α-1,4 bond-splitting enzymes are further differentiated by the site on the polymer that they attack. Those that rupture linkages, releasing monomers, are exotypes, and those that split the polymer internally are endoenzymes. The latter are easily distinguished from the former as they much more rapidly reduce the viscosity of high-molecular-weight polymers with fewer cleavages. The exoenzymes will naturally yield a larger portion of monomers (reducing sugars) and dimers. As a rule the exolyases release unsaturated dimers (157).

Pectinmethylesterase is a third type of enzyme that alters the structure of the rhamnogalacturonide polymer. It hydrolyzes the methyl esters of α-1,4 linked galacturonosyl residues at the carboxyl of the uronide moiety exposing a uronic acid carboxyl and liberating methanol (Fig. 4–21c).

Arabinose and galactose are the neutral sugars most

frequently found covalently bound to the rhamnogalacturonan backbone (Fig. 4–5). These are split hydrolytically by exo- and endogalactanases (110) while the α-1,3 and α-1,5 linked arabans are split by exoarabinases (76). Weinstein and Albersheim (213) have purified and partially characterized wall-degrading endo- and exoarabinases and an arabinosidase from *Bacillus subtilus*. This bacterium also synthesizes a β-1–4 galactanase that degrades a β-1–4 galactan and in the process releases arabinosyl polymers with a degree of polymerization of at least 10 residues. Molecular weight (MW) determinations revealed the endoarabinase to have an MW of 32,000 whereas the arabinosidase was, as one might have anticipated, twice as large, 65,000. The smaller MW of the endoarabinase would, theoretically, enable it to penetrate between polymers to split them in endo fashion. The reader is reminded that the endoarabinase above is not of plant origin.

Hemicellulose-degrading enzymes. Bauer et al. (23) have clearly defined hydrolysis of β-1,4 linked xyloglucan in the primary wall by an endoglucanase. According to Strobel (192), β-1,4 xylans are hydrolyzed by endoxylanases, and King and Fuller (109) have reported an exoxylanase or β-xylosidase that converts xylobiose to xylose. The mannan-containing hemicelluloses, mannans, glucomannans (glucose in the mannan chain), and galactomannans (galactose side chains) are completely hydrolyzed by a series of enzymes: β-1,4 endomannase, β-mannosidase, β-glucosidase and a β-galactosidase. Reddy et al. (170) reported that *Xanthomonas alfalfa* produced at least three xylanases. Xylobiose was detected first in the reaction mixture suggesting the activity of an exoxylanase. Later xylo-oligosaccharides appeared indicating the hydrolytic activity of an endoxylanase. The

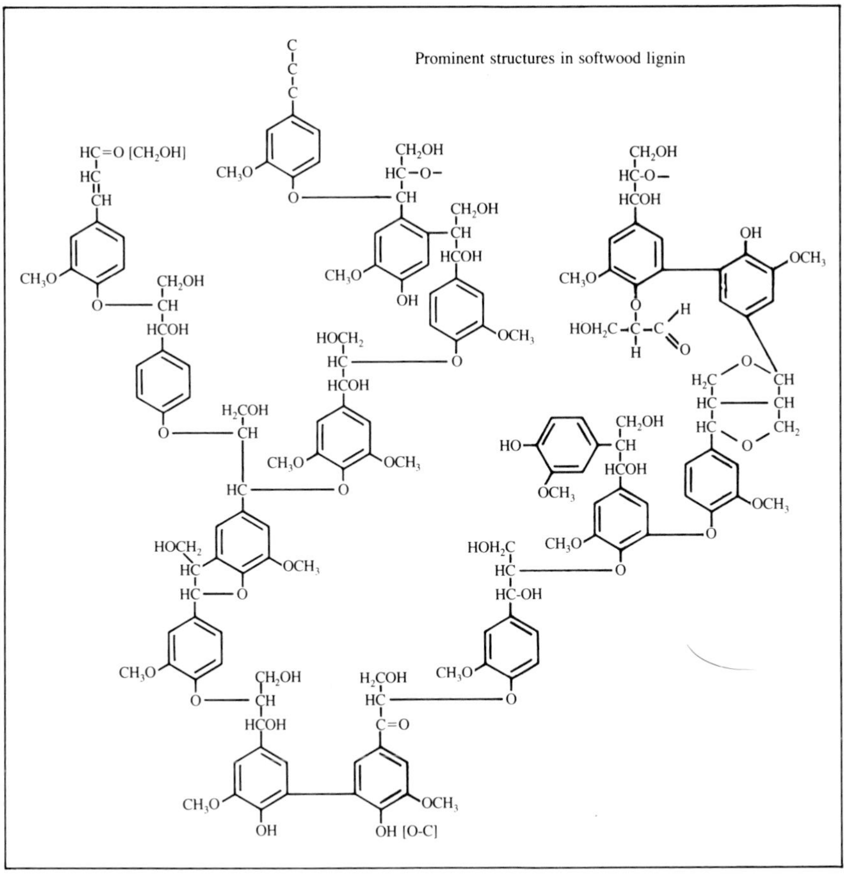

Prominent structures in softwood lignin

Fig. 4–15. Proposed model for a typical softwood lignin (from Adler, E. 1977. Lignin chemistry—past, present and future. Wood Sci. Technol. 11:169–218).

presence of the third enzyme was signaled by the appearance of xylose, which reflected the activity of a xylosidase.

Cellulose-degrading enzymes. Most enzyme-substrate relationships reveal the interaction between a soluble substrate and an enzyme that is either soluble or membrane or surface bound. The interactions of the substrate cellulose with the enzymes that degrade it are rather different, in that cellulose is an insoluble polymer and the soluble cellulases must migrate to its surface

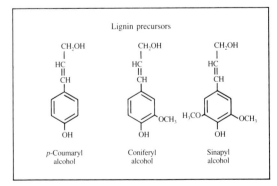

Lignin precursors

p-Coumaryl alcohol   Coniferyl alcohol   Sinapyl alcohol

Fig. 4–16. Precursor monomers of lignin.

before binding and catalytic activity ensue. Activity is measured by the detection of reducing sugars liberated from the large polymer; the lower limits of detection are fragments with fewer than seven glucose residues that are solvent in aqueous solutions.

Historically, a two-enzyme complex was believed to be involved in cellulose degradation, one producing the dimer cellobiose and a second cleaving the dimer to glucose. Later Reese et al. (171) recognized that a large number of fungal species could degrade carboxymethyl cellulose, a smaller methylated glucan. However, the native crystalline cellulose was degraded by comparatively few fungi; hence they proposed the C_1 and Cx enzyme hypothesis by which the C_1 could attack the crystalline polymer exposing it to subsequent attack by the Cx enzyme. This duo of enzymes was also augmented by β glucosidase, which could reduce resultant dimers to glucose.

Subsequently, the C_1-Cx terminology was replaced by a scheme that defined cellulose decomposition by a three-enzyme system (215; see Fig. 4–22). The first endo-1,4-β-D glucanase (EC 3.2.1.4) randomly cleaves internal glucosidic bonds within an unbroken glucan chain. The exposed nonreducing chain ends become substrates for 1,4-β-D-glucan cellobiohydrolase (EC3.2.1.9l), which cleaves the cellobiose dimers from

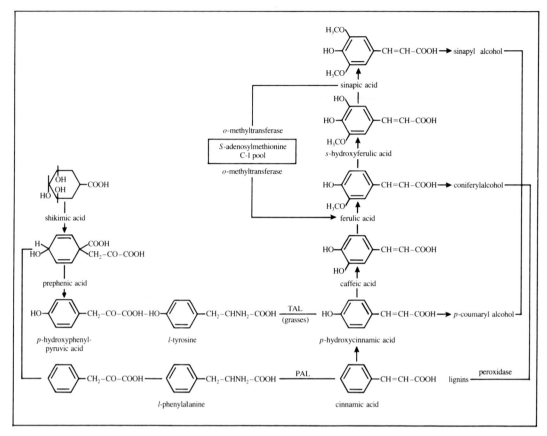

Fig. 4–17. Biosynthetic pathway of lignin from shikimic acid (after Higuchi et al., 100).

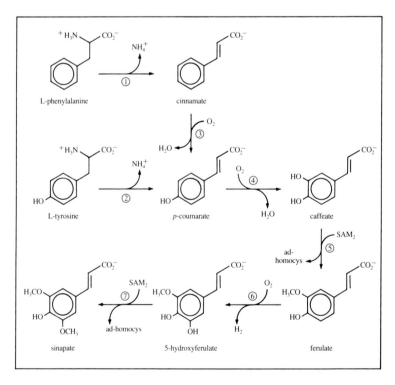

Fig. 4–18. The phenylalanine-cinnamate pathway. (1) Phenylalanine ammonia lyase; (2) tyrosine ammonia lyase; (3) cinnamate 4-hydroxylase; (4) p-coumarate 3-hydroxylase; (5, 7) catechol O-methyltransferase; (6) ferulate 5-hydroxylase (hypothetical) (after Gross et al., 89).

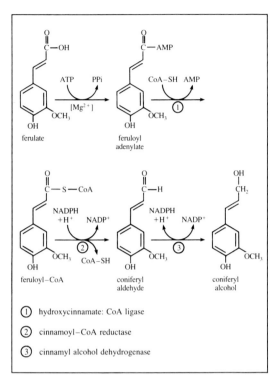

Fig. 4–19. Reduction of ferulic acid to coniferyl alcohol (after Gross et al., 90).

① hydroxycinnamate: CoA ligase

② cinnamoyl–CoA reductase

③ cinnamyl alcohol dehydrogenase

the chain releasing them into the environment. The hydrolysis of cellobiose to glucose is accomplished by β-glucosidase (EC3.2.1.2l). The activity of the endoglucanase (EG) in exposing the nonreducing chain ends for cellobiohydrolase (CBH) sets up a complementary reaction between the two enzymes that determines the rate of cellulose degradation.

Cellulose synthesized by *Acetobacter xylinum* is rapidly degraded by the enzyme complex; within 30 min products are no longer discernible in the electron microscope. *A. xylinum* cellulose is elaborated in the form of ribbons, which are aggregates of microfibrils that are called bundles. Bundles of microfibrils are clearly "splayed" within l0 min as the least extensive H-bonding occurs between bundles of fibers, and these are the first bonds to be broken during the degradation process. EG attacks glucosidic bonds within glucan chains on the surface of the bundles; the nonreducing chain ends are acted upon by CBH; and, as the bundle surfaces are eroded, H-bonds between adjacent bundles are weakened and the bundles "splay" or come apart. As the bundles come apart more surface is exposed for subsequent EG and CBH attack. The bundles are thus successively eroded and thinned by the surface degradative action of these two enzymes. The cellulase system of *Trichoderma* is estimated to contain 60% CBH and l5% EG (88) and in this approximate proportion to solubilize crystalline cellulose totally in 30 min. β- glucosidase is not essential to this process, in that solubility here does not imply total hydrolysis to monomers.

The interdependence of EG and CBH for total decrystallization of cellulose is well established. EG is capable of causing ribbon splaying by weakening H-bonding between microfibrillar surfaces; however, degradation by EG alone is slow. CBH on the other hand requires the activity of EG before it can become substantially active. By altering concentrations of the two enzymes or alternating their activity on substrate as compared with combined simultaneous exposure, the synergistic degradative effect becomes apparent.

Cowling (47) has summarized the salient structural features of cellulose microfibrils that mediate their susceptibility to enzymatic degradation. These include (l) moisture content, (2) size and diffusibility of the enzyme molecules in relation to the size and surface properties of the capillaries between the cellulose microfibrils, (3) degree of crystallinity, (4) degree of polymerization, and (5) the nature of the encrusting substances and their proximity (linkage) with cellulose, e.g. pectin, hemicellulose, lignin, suberin, etc.

Each of the above, according to Cowling, influences the accessibility of cellulose to the enzyme molecules. It is, after all, the direct physical contact between the hydroxyl groups of the cellulose chain by the enzyme molecule that is necessary to effect hydrolysis.

Protein-degrading enzymes. The presence of protein in plant cell walls has been conclusively established, as

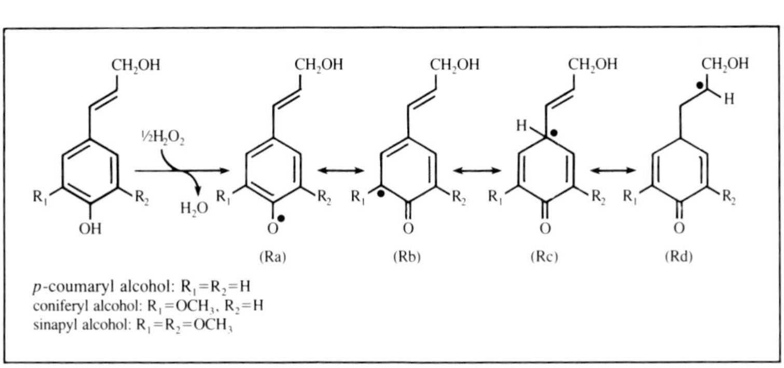

CH₂OH   CH₂OH   CH₂OH   CH₂OH   CH₂OH

p-coumaryl alcohol: $R_1 = R_2 = H$
coniferyl alcohol: $R_1 = OCH_3$, $R_2 = H$
sinapyl alcohol: $R_1 = R_2 = OCH_3$

Fig. 4–20. Formation of phenoxyradicals (•) by peroxidase (after Gross, 89).

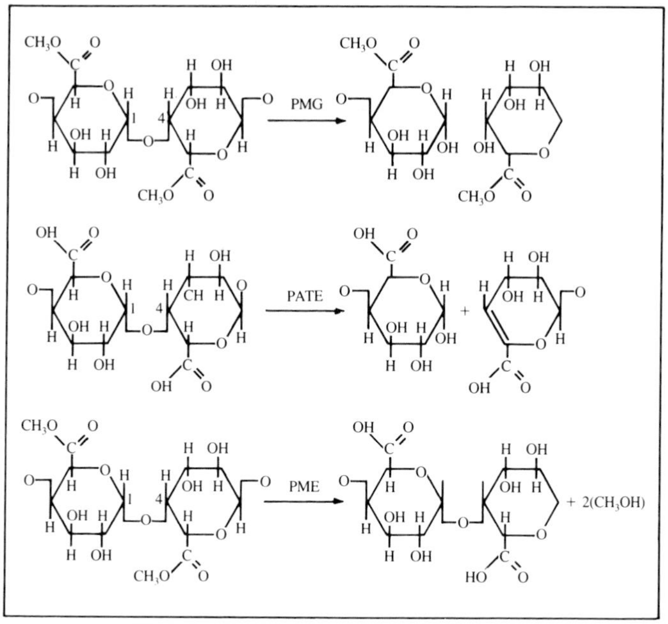

Fig. 4–21. The pectin-modifying enzymes: (top) Pectinmethylgalacturonase (PMG); (center) pectin trans-eliminase (pectin lyase)(PATE or PL); (bottom) and pectin methylesterase (PME).

has its probable structure (119) (see Fig. 4–7), although its role there is uncertain (120).

There was also the suggestion that the Hyp-rich protein extensin (NH$_2$-Ser-Hyp-Hyp-Hyp-Hyp-Ser-Hyp-Lys-CO$_2$H) (105, 119) is linked via serine to a galactose moiety of a highly branched arabinogalactan (105), together constituting a single macromolecule. The arabinogalactan may in turn be linked through galactose to the rhamnogalacturonan polymer. However, these carbohydrate-protein linkages are at present not definitively proved (119).

Release of extensin from untreated cell wall preparations by a number of protein-degrading enzymes has not been particularly successful (120). Protease and pronase treatment of sycamore cell wall preparations yielded only small amounts of Hyp-rich extensin. The recalcitrance of extensin to enzymes may be due to the large amount of hydroxyproline, which offers comparatively few enzyme-susceptible linkages, or to a carbohydrate "coating" that renders it inaccessible to proteases. Lamport has recently suggested that Hyp-rich protein may be cross-linked to itself and merely "nesting" in the larger wall polysaccharide macropolymer (120). A number of carbohydrases including pectinase and hemicellulase removed negligible amounts of extensin and a "crude" cellulase released 10–15% Hyp from the sycamore cell walls and 70–75% from tomato and tobacco cell walls (120). These cell walls had previously been boiled in water for 24 h to remove pectic substances. Pronase also liberated significant amounts of Hyp from similarly treated tomato cell walls.

Cutin-degrading enzymes. The fatty acids of cutin polymers are usually C$_{16}$ and C$_{18}$ chains with one to three hydroxyl groups. Their degradation in situ has been described by Van den Ende and Linskens (203) and ultrastructural evidence for the participation of cutinases in fungal penetration of plant hosts has been presented (42). However, the first purified depolymerase of a hydroxy fatty acid polymer (cutin) was reported by Purdy and Kolattukudy (165), who used methyl 12-hexadecanoyl-oxyoctadencanoate as a model for cutin as well as cutin prepared from apple skin. The enzyme, prepared from *Fusarium solani* f.s. *pisi*, released all of the various types of cutin monomers that were obtained by complete chemical depolymerization. Subsequently, Purdy and Kolattukudy (166, 167) isolated two cutinases from culture filtrates of *F. solani* f.s. *pisi* that had similar activities on apple cutin and similar molecular weights, 22,000 (Fig. 4–23). These cutinase isoenzymes (167) had no metal ion dependence and were not affected by thiol reagents. Both cutinases could hydrolyze cutin to oligomers and monomers. (For lignases, see Additional Comments and Citations.)

Cell Membrane Structure, Function, and Decomposition

Membrane structure. Associated with the cell wall both functionally and structurally is the cell membrane. Although the latter exhibits a greater degree of selectivity to the passage of substances, it frequently seems molded to the wall per se (Fig. 4–5). In addition, exten-

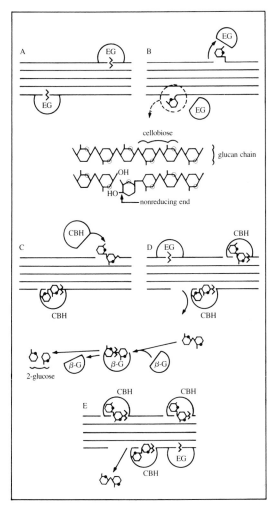

Fig. 4–22. The enzyme complex, endo-1,4-β-D-glucanase (EG); 1,4,-β-D-glucan cellobiohydrolase (CBH) and β glucosidase (β-G), that hydrolyzes cellulose polymeric chains (after White, 215).

sions of the plasmamembrane run through the wall (plasmodesmata) connecting neighboring cells.

The integrity of the cell membrane is essential to normal function, and its perturbation by plant pathogens has been recorded (135, 216). The ways in which membranes may be altered are best understood once their composition and structure is made apparent. According to Capaldi (37) and Bretscher (31) it is possible to generalize somewhat concerning the composition of cell membranes. They all contain proteins and lipids, and most also contain carbohydrates as glycolipids and glycoproteins. Although the proportions vary with species and specific organelles, proteins account for 40–50% of the membrane, lipids also about 40%, and carbohydrates 0–10%. The original bimolecular lipid leaflet proposed by Gorter and Grendel (82) and emphasized and popularized by Danielli and Davson (50)

seems to have regained popular acceptance. Accordingly, the lipid bilayer is about 4.5 nm in thickness and is the framework into and upon which the proteins are "inserted" or attached (Fig. 4–24). The lipid bilayer consists of three types of lipids— phospholipid, neutral lipid (cholesterol), and glycolipid. The proteins are of various classes and sizes but in general may be designated as *intrinsic* and *extrinsic* (Fig. 4–24). The latter have more hydrophyllic amino acids and face the cell wall, and the former contain more hydrophobic ones and are oriented toward the cytoplasm (37). The carbohydrate portion of the membrane, glycolipid or glycoprotein, is probably oriented toward the membrane surface exposed to water. In the case of the plasmalemma, the carbohydrate is located on the cell wall or "outside" of the membrane. According to Bretscher (31), the carbohydrate anchors the protein in the lipid bilayer, as both lipids and proteins are capable of lateral and rotational movement in the bilayer. He and others have popularized the concept of the fluid membrane wherein amphipathic protein and glycoprotein molecules are mobile in a sea of lipid two molecules thick with viscosity about equal to that of castor oil.

Membrane function. The integral part of cellular transport phenomena is ion transport across membranes. Ions may pass through membranes along concentration gradients (passive-transport-diffusion) or are moved actively against the concentration gradients at the expense of metabolic energy (active transport) (8, 156, 182). Both passive and active transport may lead to charge separation and the generation of electro-potential gradients, and these gradients then can serve also as a driving force without the additional expenditure of metabolic energy (approximating a Donnan equilibrium system). It is generalized, however, that most of the major nutrient ions are transported actively ("pumped") (99). Some ions such as Cl^- are moved actively inward, and some such as Na^+ are moved outward (128). The importance of outward active transport of H^+, or extrusion pump (Fig. 4–25), was investigated by Raven and Smith (169), and it is believed to play an important role in internal cell pH regulation. There is also evidence that uncharged solutes (nonelectrolytes) like sugars and amino acids are cotransported across membranes with protons (H^+) (99).

Some theories postulate carrier mechanisms (38), and a cotransport system for glucose (182) and amino acids (65) was suggested. According to this theory the same mechanism extruding H^+ outward is used by the cell to transfer these nonelectrolytes inward. The role of electropotentials in the disease process has been investigated in several instances by Novacky (156).

Membrane-degrading enzymes. Bateman and Basham (21) have neatly categorized those enzymes likely to affect the integrity of the lipid bilayer per se. Phospholipids are hydrolyzed at their acyl and phosphate ester bonds. Phospholipase A hydrolyzes one of the acylester bonds of diacylglycerophosphoryl compounds like lecithin (Fig. 4–26), yielding a fatty acid and a

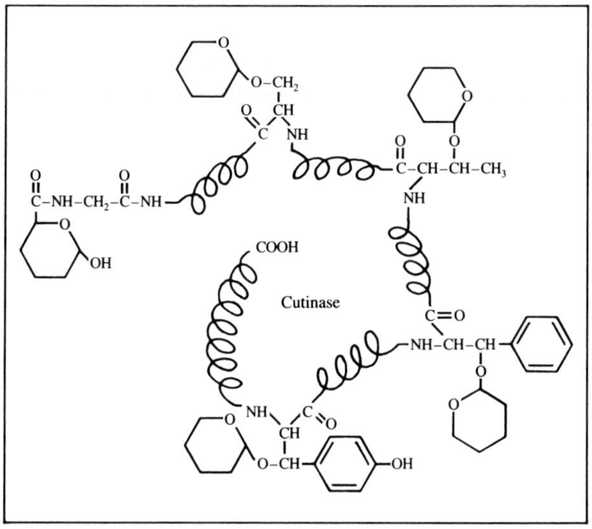

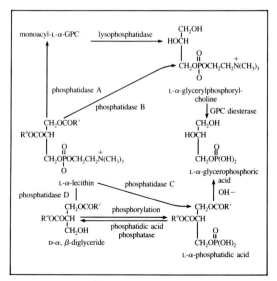

Fig. 4–23. A schematic representation of glycoprotein nature of cutinase I isolated from *F. solani pisi*. The pyranoid rings represent a carbohydrate (after Kolattukudy, 113).

lysophospholipid. Phospholipase B cleaves both of the acyl esters from the diacylglycerophosphoryl compound to yield two fatty acids and the glycerophosphoryl moiety. The third enzyme, phospholipase C, hydrolyzes the parent compound to diacylglycerol and the phosphoryl component. Phospholipase D, found only in higher plants, hydrolyzes diacylphosphoryl compounds to phosphatidic acid and the complex alcohol moiety.

Proteinases, or protein-degrading enzymes, cleave

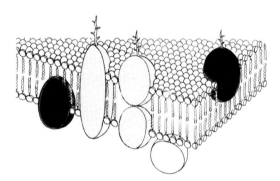

Fig. 4–24. A schematic of the structure of the plasmalemma revealing proteins permeating the lipid bilayer and projecting from either side of the membrane. The ornamentation of proteins projecting into the *outer* surface is representative of polysaccharides (after Bretscher, 31).

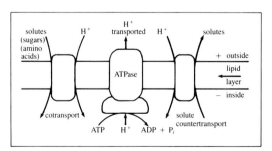

Fig. 4–25. Transport ATPase in the plasmamembrane with associated cotransport and countertransport proteins (courtesy of A. Novacky).

Fig. 4–26. The use of phosphatidases in establishing the L-α-structure and configuration of lecithin (after Bateman and Basham, 21).

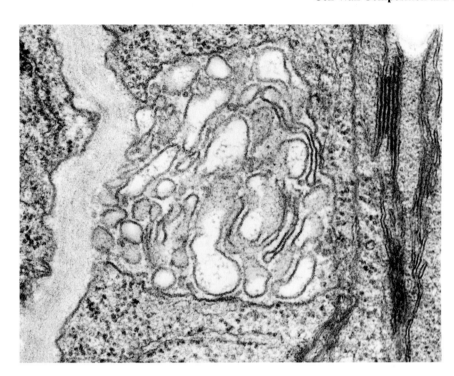

Fig. 4–27. Paramural bodies associated with CLV infection of French bean (photograph courtesy of G. P. Martelli and M. Russo).

certain exposed proteins on the surfaces of animal and microbial membranes, but in plants they seem unable to alter basic membrane structure in situ (19).

The fact that carbohydrates anchor certain proteins in the lipid bilayer at least suggests that carbohydrases may have the capacity to alter membrane integrity (174). Evidence for this concept is not at hand; however, as we shall see, such a mechanism would tend to explain how certain pectin-degrading enzymes may actually kill plant cells.

Alterations Caused by Virus Pathogens

Cell walls constitute a barrier to virus penetration (Chap. I). Viruses do not possess wall-degrading enzymes and have no mechanism for reaching the plasmalemma and initiating infection unaided. Nor do they rely on cell disruption for movement to adjacent cells (Chap. 8). Virus infection may, however, lead to significant alterations in both cell walls and the various intracellular membrane structures (136, 137, 140).

Cell Walls

Cell wall modifications involving, for example, wall thickening with deposition of callose, lignin, and other polymers are particularly pronounced when the host responds hypersensitively. These changes are discussed in the context of virus limitation to the area surrounding the infection site in Chap. l0. There are, in addition, other cell wall modifications that may accompany viral localization but are more frequently observed in systemic infections and are therefore unlikely to be involved in the processes of virus limitation. These include the formation of paramural bodies and structures that may be described as cell wall outgrowths (136).

Paramural bodies. Paramural bodies, or plasmalemmasomes, are aggregates of vesicles or membrane structures produced between the cell wall and the plasmalemma. They are found in many virus/host combinations, often in association with other wall modifications such as callose deposition or wall outgrowths and in many cases may be involved in the biosynthesis of cell wall material. A paramural body associated with systemic infection of French bean with CLV is indicated in Fig. 4–27.

Cell wall outgrowths. Cell wall outgrowths are also observed in either localized or systemic infections of several hosts, caused by many icosahedral and some bacilliform viruses. They usually take the form of narrow, elongated projections with virus particles aligned down a central channel (Fig. 4–28; 149). Their function, if any, remains speculative. Since they are often found in association with plasmodesmata, it has been suggested that they may play a role in virus transport between cells. This view, however, remains unsubstantiated (136).

Membranes

In addition to contributing to the structures described in the previous paragraphs, membranes may undergo other characteristic changes. Virus infection may, for example, lead to alterations in membrane permeability that may, in turn, influence symptom development. In hypersensitive hosts, it seems probable that impairment

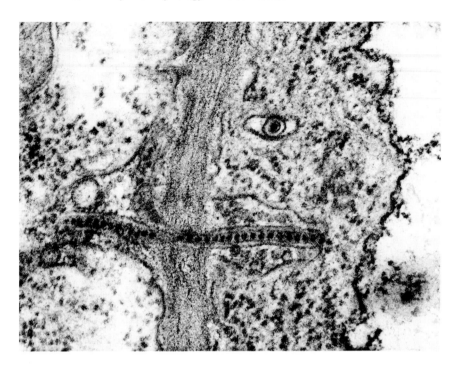

Fig. 4–28. Cell wall outgrowths associated with PYFV infection of *N. clevelandii* (after Murant et al., 149).

of membrane integrity, with consequent leakage of cell constituents, is a prelude to more extensive cell degradation. In a less severe, systemic response, extensive vesiculation of the chloroplast membrane accompanies TYMV infections of Chinese cabbage, forming structures considered to be involved in the biosynthesis of viral RNA (140), while CGMMV causes vesiculation of mitochondrial membranes in, for example, tobacco and cucumber cells (98, 193).

Much more commonly, however, virus infection leads to the formation of structures that may be collectively referred to as inclusion bodies. They may be crystalline or amorphous and, depending on virus and host, may be found in almost any part of the cell, in the nucleus or cytoplasm, or associated with other cell organelles. While crystalline inclusions are composed mainly of virus particles, amorphous inclusions may include virus particles and both virus and host-specific components, including cell membranes. They are found in the majority of viral infections; although their function remains largely unresolved, they may be involved either in biosynthesis of virus components or in virus assembly. A detailed survey of the many types of inclusions is beyond the scope of this book; they have been comprehensively reviewed elsewhere (137, 140, 175).

Alterations Caused by Bacterial Pathogens

The phytobacterial pathogen is generally considered a passive invader and is unable to develop any physical force, aside from motility, to breach the host's outer defenses. It must therefore rely exclusively upon its metabolic processes to successfully infect the host. In Chap. 1 the points of entry and the advantages that each provides to support initial multiplication have been discussed.

If the pathogen has to rely on enzyme or other metabolite activity for successful pathogenesis, an adequate supply of enzyme or metabolite is necessary. Such a supply depends on large numbers of bacteria. Since comparatively few bacterial cells are involved in an inoculation, the initial phases of the attack by a bacterial pathogen must depend on an environment that permits rapid growth supported by an adequate substrate.

Cutin and Suberin Degradation

The plant surface is the first barrier to penetration met by the bacterial cell. The aerial and subterranean parts of the plant are cuticularized, the aboveground surfaces by cutin and those below by suberin (112–15). Thus far, the literature has not provided evidence of cuticular degradation of the aerial surfaces of plants by bacteria. However, the fact that suberin degradation may occur in roots is suggested by the studies of Kelman and Sequeira (107), who concluded that *P. solanacearum* can infect roots in the absence of any external mechanical wounding of the root system, if the inoculum potential in the surrounding soil is sufficiently high. Bacterial populations were recorded in soil water surrounding infected roots of 3.2×10^8 cells/ml, which were deemed capable of infecting healthy unwounded roots. It is generally thought that penetration of roots may occur at points where secondary roots emerge. Scott and coworkers (177, 178) have reported that a mucilaginous sheath covers roots where secondary branching occurs; however this sheath may not be entirely without rifts. Kelman and Sequeira (107) also pose the possibility that the numbers of bacterial cells that they have detected may digest enzymatically the mucilaginous sheath or "other possible barriers." It is also possible that cutin layers on cell surfaces bordering such sensitive areas as substomatal chambers and intercellular space are not continuous. Rifts or interruptions in the "layer" exposing galacturonic acid polymers could provide an initial site for bacterial attack. Still another possibility is the elicitation by bacteria of *host* cuticle-degrading enzyme release. Ultrastructural studies by Goodman et al. (81) and Sequeira et al. (179) suggest that incompatible bacteria induce the separation of the cell wall surface "cuticle" from the cell wall per se.

Pectin Degradation

There now exists incontrovertible evidence of pectin-degradative activity as a prelude to successful parasitization by some pathogens. There are certain natural openings that lend themselves to rapid multiplication of bacterial plant pathogens: stomates, hydathodes, lenticels, nectaries, fissures where secondary roots emerge, etc. Bonner (27) has stated that the cell walls surrounding each of these areas that border on intercellular space retain a thin layer of pectic material covered by a veneer of cutin or suberin, and hence are susceptible to enzymatic degradation. Starr (189) has termed the pectic enzymes the "spreading factor" that enables some pathogens to move though the host via the pectic "connective tissue," creating the illusion of mass migration. Such bacterial masses were often referred to in the earlier literature of phytobacteriology as "zoogloea."

Which of the pectin-degrading enzymes are generally most effective in furthering the advance of the pathogen? For example, are the chain-splitting types more effective or more essential than the methyl-esterases? Or, put another way, what is the order of attack by these enzymes? To obtain comprehensive answers to these questions experimentally, a number of factors concerning the host, the pathogen, and the complex they form had to be determined. For example, which enzymes are produced first, how much is produced, and how rapidly are they formed? Additional controlling factors will be the pH of the environment prior to and during the interaction between host and pathogen; whether the pathogen produced the enzymes constitutively or adaptively; whether pectin-degrading enzyme inhibitors (such as phenols) are produced by the host; and others.

Perhaps the greatest block to progress in this area was the extrapolation of in vitro results to explain degradative processes in vivo. The pitfalls here were engendered by the earlier use of crude enzyme preparations that contained several enzymes rather than one and of substrates that, for example, rather than being a relatively pure pectin fraction were conglomerates of several types of pectins or other carbohydrates. Recent studies have recognized many of these pitfalls, and much effort has been expended on the purification of enzyme preparations and concomitantly their substrates.

In-Vitro Pectolytic Enzyme Production and Pathogenicity

Indicative of some of the earlier experiments that categorically linked enzymic degradation of pectin with pathogenesis are the experiments of Brown and his colleagues and students (33, 67, 87, 122). From these studies it was apparent that in vitro "macerating activity" of bacterial cultures did not necessarily reflect pathogenic activity. For example, although *Erwinia aroideae*, *E. carotovora*, and *Xanthomonas carotae* all produced culture filtrates that were pectolytic when grown in decoctions of host plants, their aggressiveness in host tissue was not assured. There were good correlations between pathogenicity and enzyme secretion for *E. aroideae* and *E. carotovora*, but the weakly parasitic *X. carotae* secreted almost as much enzyme as *E. aroideae*. From subsequent investigations it was learned that somewhat weaker parasitic activity in potato tuber tissue of *E. carotovora*, as compared with *E. aroideae*, could be attributed to the slower rate of multiplication of the former and its tendency to develop a profound acidic reaction in the culture medium. Both of these characteristics favored a reduction in enzymatic activity. Specifically, the reduced enzyme activity reflected a lowered enzyme secretion due to participation of fewer bacterial cells and an unfavorable pH, which prevented a key enzyme, pectate lyase, from performing its task with maximum efficiency (6, 7).

The experiments of Lapwood (122), also from Brown's laboratory, are particularly interesting wth respect to this matter of an a priori requirement of pectolytic activity for pathogenesis. In an examination of 39 isolates (Table 4–3), he found that there were no isolates that combined high pathogenicity with low enzyme production. Thus it appears that, although profuse pectolytic enzyme production does not necessarily insure pathogenicity, minimal enzyme production severely limits virulence.

Hydration of Host Tissues

The foregoing results imply that the enzymes themselves are important in pathogenesis, but their effective-

Table 4–3. Pathogenicity and enzymic activity of bacterial isolates (after Lapwood, 122).

| Genus | No. of isolates | Pathogenicity high | | Pathogenicity low or absent | |
		Enz. high	Enz. low	Enz. high	Enz. low
Bacterium	18	8	0	6	4
Bacillus	5	0	0	3	2
Pseudomonas	11	0	0	5	6
Unidentified	5	1	0	1	3
Totals	39	9	0	15	15

Table 4–4. Effect of water content* on amount (in grams) of rotting of potato cylinders (after Lapwood, 122).

Organism	Control	Treatment	
		Soaked	Injected and soaked
Bacterium aroideae	0.24	0.40	0.76
Pseudomonas syringae	0.0	0.54	0.92
Pseudomonas sp. No. 169	0.05	0.36	1.15
Flavobacterium sp.	0.04	0.27	0.60

*Average water contents: control, 79.1%; soaked, 83.1%; injected and soaked, 83.3%.

ness may be modified by the host or by certain environmental conditions. Prior to Lapwood's experiments, Gregg (87), also from Brown's laboratory, noted that by raising the water content of potato tubers before exposing them to culture filtrates containing the pectin-degrading enzymes, a nonpathogenic bacterium vigorously attacked the tuber tissue. This phenomenon was reinvestigated by Lapwood, and his data appear in Table 4–4. Increasing the water content of the potato tubers by soaking or by soaking and injecting (vacuum infiltration) enabled nonpathogenic species to attack potato tissue as vigorously as a bona fide pathogen. The potentiative effect of injection over and above soaking was seemingly out of proportion to the slight increase in water content attributable to vacuum infiltration. It was inferred that filling the intercellular spaces with water excluded oxygen and thus retarded the normal "wound reaction" (suberization), thereby favoring pathogenesis. Gregg (87) had suggested earlier that the water treatment may leach out certain enzyme inhibitors. Certainly dilution of inhibitory phenols would be a possibility.

An analysis of the influence of infiltrating the intercellular spaces with water leads to a number of interesting possibilities. For example, a very direct effect would be the hydration of pectin and neighboring hemicellulose and cellulose microfibrils, with a concomitant increase in the accessibility of these cell-wall and middle-lamellar components to the degrading enzymes.

A continuous water film in the network of intercellular spaces might foster more rapid movement of such flagellate species as Erwinia aroideae and E. carotovora (158). These conditions of flooding of the intercellular spaces are not unlike those responsible for symptoms of the wildfire type in tobacco foliage where inoculation during driving rainstorms causes large water-soaked lesions to form and become necrotic, as opposed to the local-lesion-halo host response that develops in tissue inoculated in the absence of water-soaked conditions.

It is possible also to propose an enzyme-inhibiting

substance that is inactivated not simply by water removal or dilution of soluble phenols but perhaps by the exclusion of oxygen, which is necessary to activate the enzyme inhibitor. This contention is supported by the fact that oxidized extracts (phenols) of a number of plant tissues are able to inhibit certain pectin depolymerizing enzymes (67, 87, 122). This inhibitor was referred to by Zucker and Hankin (222) as an "intrinsic resistance factor." In addition, they revealed a specific suberin-inhibiting mechanism that fosters maceration of potato tuber tissue by pectate lyase. Their data suggest that PAL synthesis controls suberization. The synthesis of this enzyme can be almost completely suppressed by cycloheximide with a concomitant restriction of suberin formation and potentiated susceptibility to maceration by exogenously applied purified pectate lyase.

Rate of Enzyme Production by the Pathogen as a Factor in Pathogenesis

We have presented a hypothesis for the potentiation of pectin degradation by enzymes of nonparasitic bacterial species. However, we have left unexplained why, for example, potato tissue can be parasitized by E. aroideae and not by Pseudomonas syringae, although both produce pectolytic enzymes. The answer to this paradox seems to be linked to the rapidity with which E. aroideae produces macerating enzymes, in 4 h, permitting the pathogen to overcome the host's normal wound (defense) reaction. On the other hand, P. syringae produces macerating enzymes after 10 h, by which time the defense mechanism of suberization and the incorporation of oxidized phenols in the suberized cells have proceeded far enough to prevent extensive parasitization. Evidence supplied by Lapwood (122) indicated that a temperature of 5 C for 7 d, which suppresses the growth of E. aroideae in vivo, does not proportionately interfere with suberization, but it will arrest the invasion of potato tuber tissue by the pathogen.

A rise in temperature has also been shown to foster "parasitism" by saprophytic bacteria. In a comparative study by Fernando and Stevenson (67) of potato tissue degradation by Bacillus subtilis, B. megatherium, B. mycoides and E. carotovora, it was found that at 25 C only E. carotovora vigorously attacked potato tissue. When temperatures were raised, two of the three saprophytes degraded potato tissue (see Table 4–5).

The influence of elevated temperatures on suberization and their effect upon the rate of PAL synthesis have not received sufficient attention. Thus far we have attempted to describe some of the factors that regulate the activity of the pectin-degrading or macerating enzymes in host tissue. In review these are (1) growth rate of the bacterium itself, (2) the rapidity with which the bacterium produces the macerating enzymes, (3) the production of acid by the bacterium and its effect upon enzyme activity and subsequent bacterial proliferation, (4) the host defense mechanism, e.g. the "wound response," and (5) the temperature prevailing during the postinoculation period.

Table 4–5. Effect of temperature on attack of potato tuber by saprophytic *Bacilli* (after Fernando and Stevenson, 67).

Organism	Temperature C	Number of tubers Inoculated	Number of tubers Attacked
Bacillus subtilis	37	18	18
	30	10	0
	25	10	0
B. megatherium	37	18	18
	30	10	10
	25	10	0
B. mycoides	37	18	0
	30	10	0
	25	10	0

Pectin-Degrading Enzymes In Vitro

The chemical mode of action and nature of the "macerating enzymes" were investigated in experiments conducted by Wood (217) in Brown's laboratory. These revealed that *E. aroideae*, grown on a synthetic medium, produced an enzyme (probably a mixture of enzymes) that reduced the viscosity of a high-methoxyl pectin but failed to increase significantly the reducing power of the solution containing the reaction mixture. The reaction mixture also failed to reflect the activity of pectin methylesterase. Each of these facts is significant and reveals something about the nature of the relatively crude enzyme preparation. (1) Rapid reduction in viscosity of the 0.5 percent solution of the highly methylated pectin from a value of 3.2 to 0.77 in one minute indicates the presence of an enzyme capable of splitting polymeric chains of methylesterified galacturonic acid. (2) The apparent inability of the enzyme or enzymes to increase the reducing power of the reaction solution suggests that the chains were being broken at points some distance from their terminals; hence either no monomers were being liberated or they were being rapidly metabolized. (3) Finally, the enzyme preparation was seemingly unable to de-esterify the highly methoxylated pectin substrate. The enzyme involved in the aforementioned study could have been, assuming hydrolytic splitting, an endopolymethylgalacturonase. However, Wood stated that the enzyme preparation had substantially the same effect upon demethoxylated pectins. This would suggest the presence of a second enzyme; still assuming a hydrolytic split, it could have been an endopolygalacturonase.

There is also the strong possibility that loss of viscosity was caused by pectin lyase, since the pH optimum for the enzyme preparation recorded by Wood (217) was between 8.5 and 9.0. This agrees with later studies by Albersheim et al. (6, 7), Starr and Moran (190), Nasuno and Starr (152), and Dean and Wood (55), all suggesting that transeliminative enzyme activity (pectin lyase) is more prevalent than the hydrolytic split types for bacterial plant pathogens.

A study by Kraght and Starr (117) on the pectinolytic enzymes produced by *Erwinia carotovora* utilized tests for alterations in viscosity, de-esterification and hydrolysis of terminal galacturonic acid residues to evaluate the type and intensity of the pectin-degrading activity of this soft-rot-inducing bacterium. They found that a number of *E. carotovora* isolates were capable of decreasing the viscosity of methylated pectin rapidly, and pectin methylesterase activity was also apparent since large amounts of methanol were detected in the reaction medium. Curiously, however, only a minute amount of de-esterification occurred during the initial 24 h, despite a 7,000-fold increase in bacterial population and a concomitant substantial decrease in viscosity of the pectin medium. De-methylation did occur, however, during the second 24-h period, when 66% of the methyl groups were liberated; and at the end of 72 h, hydrolysis was 94% complete.

Of additional interest was the fact that, although viscosity markedly and rapidly decreased, reducing groups could not be detected in the growing cultures supplied with pectin; this had also been noted previously by Wood (217). However, this was found to be due to the rapid utilization of galacturonic acid monomers by the bacterium. Exopolygalacturonase activity was demonstrable in either bacteria-free culture filtrates or in bacterial cultures that had been treated with 0.02 M sodium fluoride or sodium azide.

Hence the inability to demonstrate hydrolysis of terminal galacturonic acid residues in a culture filtrate obtained from a reaction of bacteria with a pectinaceous substrate need not indicate that the bacterium in question was unable to produce an exopolygalacturonase. Fig. 4–29 shows quite clearly the influence of pectin as a carbon source for *E. carotovora* on the production of exo-enzymes. It is apparent, however, that successful parasitization of host tissue by a pathogen should provide galacturonic acid residues shortly after inoculation, and hence the necessary substrate stimulus for accelerated production of pectin-degrading enzymes.

The suppression of the release of galacturonic acid (reducing groups) by glucose in Fig. 4–29 is probably an example of catabolite repression or the "glucose effect" (117). Catabolite repression has also been detected by Zucker et al. (223) for a fluorescent *Pseudomonas* sp., as specific activity of the pectin lyase it produced increased fourfold when it was grown on glycerol rather than on glucose.

Zucker et al. (223) also reported that pectate lyase synthesis by *P. marginalis* and the saprophyte *P. fluorescens* is constitutive, whereas among the *Erwinia* species it is inducible. In addition, it seemed that soft-rot-inducing pathogens produce 100 to 1000 times more macerating enzyme than other plant pathogens or saprophytes. This in vitro study revealed that extracts from a number of plants (potato, onion, carrot, and lettuce) possess enzyme regulators. Some of these stim-

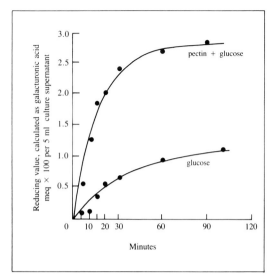

Fig. 4–29. Effect of carbon source (1% glucose or 0.5% glucose plus 0.5% pectin) in the growth medium on the production of polygalacturonase by *Erwinia carotovora* (after Kraght and Starr, 117).

ulated, whereas others severely repressed, enzyme synthesis. Whether these regulators are operative in vivo is not known. Mount et al. (148) have shown that the production of PATE by *Erwinia carotovora* was under direct control of cyclic adenosine monophosphate (CAMP; the effector of catabolite activator protein that is responsible for the regulation of all enzymes under catabolite repression). A mutant *E. carotovora*, EC 1491, which is deficient in CAMP, produced very low levels of the lyase. However, when the inducer CAMP was supplied exogenously, levels of the lyase increased significantly. Utilization of neutral sugar components of the cell wall such as L-arabinose, L-rhamnose, D-xylose, and D-galactose by *E. carotovora* is under the control of CAMP. It was concluded by Mount et al. (148) that CAMP regulates the induction of lyase and as a consequence the pathogenicity of *E. carotovora*.

Pectin-Degrading Enzymes In Vivo

From the foregoing accounts it has been established that certain bacterial species produce pectin-degrading enzymes in vitro. The culture filtrates, and in some instances *purified* enzymes, macerate plant tissue, presumably by cleaving glycosidic linkages between galacturonic acid molecules and by de-esterification of methyl groups from the uronide chains. It was also apparent that in vitro production of a macerating enzyme does not insure pathogenesis. Some questions were raised concerning the quantitative detection of pectin-degrading enzymes in infected plant tissue that could explain differences in virulence between isolates of the same genus and species.

That a causal relationship exists between pectin degradation and pathogenesis is most apparent for the soft-

rot-inducing bacteria. These bacteria are typified by *E. carotovora* and related species, and pathovars; but in addition pseudomonads, e.g. *Pseudomonas chrysanthemi* and *P. marginalis*, also express their pathogenicity in this way. Wall-degrading enzymes are also their virulence factors.

Friedman (73, 74) and Friedman and Ceponis (75) studied the degradation of celery and chicory tissue by *E. carotovora* and compared ultraviolet irradiation-induced, weakly virulent mutants with a virulent strain of the pathogen. The general symptoms were similar in potato tuber, lettuce, and carrot tissues. A "maceration" test measuring the resistance to pressure of tissue slices treated with enzymes revealed that slices treated with enzyme preparations (sterile culture filtrates) from the virulent isolates broke under weight after 32 min, whereas the enzyme from the weakly virulent mutant required 74 min. The virulent strain was found to reduce viscosity of pectin 30% and 45% in 2 and 20 min respectively, whereas the ultraviolet light-induced, weakly virulent isolates during similar time periods reduced viscosity 10% and 23%. Terminal cleavage activity, as indicated by the accumulation of galacturonic acid monomers, was 4.5 times as great with filtrates from the virulent culture as from the weakly virulent one.

From a wild-type (WT) *E. carotovora* strain, Beraha and Garber (25) obtained a streptomycin-resistant parental (P) mutant that retained its virulence. This isolate, when exposed to the mutagen N-methyl-N'-nitro-N-nitrosoguanidine (NTG), yielded a streptomycin-resistant avirulent (AV) mutant. The AV mutant was in turn exposed to NTG, and a revertant (RV) strain that retained its streptomycin resistance marker and regained its virulence (macerating activity in vivo) was obtained. The frequency of this mutant in nature was calculated to be 1 in 10^{13}.

With this array of isolates (WT, P, AV, and RV) it was possible to assess with greater reliability the role of certain enzymes in soft rot induction (in maceration). Selected data in Table 4–6 indicate that loss of virulence is marked by a reduction in polygalacturonase, pectate lyase, and pectinesterase activities. There is also the suggestion that the RV is even more competent as a tissue macerator than the WT. These data also reveal that catabolite repression of pectinesterase occurs when *E. carotovora* is grown with glucose as a carbohydrate source.

The development of basal soft rot of stem and roots of carnation reflects the pathogenic activities of two bacterial species that attack the host sequentially (28, 30). In the course of indexing carnation cuttings for *P. caryophylli*, Dickey and Nelson (58) isolated a number of bacteria that seemed nonpathogenic. However, when one of these was applied to the basal ends of carnation cuttings with *P. caryophylli*, symptom development, particularly basal soft rot, was intensified. Braithwaite and Dickey (28, 29) identified the isolate as a *Corynebacterium* sp. that is able to grow vigorously in host tissue only after *P. caryophylli* has caused membrane damage and, consequently, leakage of host cell contents

Table 4–6. Relative activity of pectinesterase (PE), polygalacturonase (PG), pectate lyase (PLv and PLc), and cellulase (Cx) in lyophilized culture extracts of *E. carotovora* and mutant strains grown in polygalacturonic acid and carboxymethylcellulose or glucose (after Beraha and Garber, 25).

		Enzyme activity[a]				
Strain	Growth	PE	PG	PLv	PLc	Cx
Polygalacturonic acid and carboxymethylcellulose						
WT	1.08	0.45	29.5	2400	0.32	545
P	0.95	0.60	30.2	3000	0.33	545
AV	1.28	0.35	13.2	429	0.18	25
RV	1.25	1.10	35.5	6000	0.35	1500
glucose						
WT	1.02	0	2.5	54	0.11	20
P	0.81	0	3.5	59	0.12	20
AV	0.56	0	0.5	13	0.09	3
RV	0.85	0	6.5	652	0.14	91

a. PE—milliequivalents of released methoxy groups per mg of enzyme powder after 71 h in a reaction mixture of 1% pectin in 0.1 M NaCl maintained at pH 7.0 at 30 C with 0.01 N NaOH. PG—μ moles of released reducing groups per ml of enzyme preparation after 24 h at 30 C during hydrolysis of the substrate containing: 0.5% polygalacturonic acid, 0.1 M NaCl, and 0.05 M sodium acetate buffer at pH 5.2. Plv—reciprocal of the time (h) required for a 50% increase in the flow rate of the reaction mixture × 100 at 30 C. Enzyme assays were made using 6 ml reactions containing: 1 ml enzyme preparation in 0.05 M tris buffer, pH 8.5, and 1.2% sodium polypectate. PLc—the change per hour in absorbancy at 547 mμ in the PLv reaction mixture, but at 2 vol substrate: 1 vol enzyme and diluted 16-fold with 0.5 N-HCl and 0.1 M thiobarbituric acid and heated before measuring absorbancy. Cx—reciprocal of time (h) required for a 50% increase in the flow rate of the reaction mixture × 100 at 30 C. Enzyme assays were made using 6 ml reactions containing: 1 ml enzyme preparation in 0.05 M tris buffer, and 1.2% carboxymethylcellulose at pH 7.0.

that support a 1,000-fold increase in the growth of the coryneform. The latter produces in vivo and in vitro, among other macerating enzymes, a pectate lyase. Maceration of carnation leaf tissue seemed causally linked to its lyase activity.

Visual evidence at the ultrastructural level that clearly links pathogenesis by *E. carotovora* to destruction of the pectinaceous middle lamella has been provided by Fox (70). Turner and Bateman (198) attempted to determine whether *E. carotovora* produced wall-degrading enzymes that could selectively or preferentially macerate the host tissue from which they were extracted. Their data in Table 4–7 reveal no specificity with respect to the host tissue in which the enzyme was produced; however, it is apparent that cucumber tissue is particularly sensitive to the macerating enzyme.

Composition of hot water-EDTA soluble (pectin) fractions from carrot, onion, and cucumber tissue was similar and readily attacked by enzymes obtained from each compatible combination. The reason for the ex-

Table 4–7. Enzymatic maceration of host and nonhost tissues by extracts of compatible host-pathogen combinations at pH 8.0[a] (after Turner and Bateman, 198).

Enzyme source[c]	Substrate	Degree of maceration[b]		
		1.5 h	4.5 h	11.5 h
E-Ca	carrot	0.0	2.0	4.0
E-Ca	cucumber	3.5	5.0	5.0
E-Ca	onion	1.0	3.0	5.0
E-Cu	carrot	2.0	5.0	5.0
E-Cu	cucumber	5.0	5.0	5.0
E-Cu	onion	2.0	4.0	5.0
P-Cu	carrot	1.0	2.0	5.0
P-Cu	cucumber	4.5	5.0	5.0
P-Cu	onion	0.5	2.0	5.0
None	carrot	0.0	0.0	0.0
None	cucumber	0.0	0.0	0.5
None	onion	0.0	0.0	0.0

a. Reaction mixtures contained 1.0 ml of 1.0% lyophilized extract and 6 tissue discs in a 5.0-ml vol containing 0.02 M tris buffer at pH 8.0. Incubation was carried out at 30 C.
b. Maceration based on a 0 to 5 index; 0 indicates no maceration and 5.0 indicates complete maceration. The reported values are means of duplicate replications.
c. Enzyme sources were the following compatible host-pathogen combinations: *E. carotovora*-carrot (E-Ca), *E. carotovora*-cucumber (E-Cu), *Pythium aphanidermatum*-cucumber (P-Cu).

treme sensitivity of cucumber tissue to the macerating enzyme remains conjectural, but the possibilities include greater intercellular space, accessibility of pectin to the enzyme, and the absence of specific enzyme inhibitors. The enzyme involved in all combinations proved to be a pectin lyase. Grounds for this conclusion are derived from the fact that cleavage of the α 1–4 linkage is eliminative, the reaction occurs at a pH of 8.0 and not at one of 4.5. Furthermore, EDTA (which complexes with Ca^{++}) entirely obliterated enzyme activity, whereas 1.5×10^{-3}M Ca^{++} increased the activity of dialyzed over undialyzed enzyme preparations.

It was shown by Garibaldi and Bateman (80) that *E. chrysanthemi* isolates EC$_{23}$ and 307 produced four and two lyases respectively in host tissue. All were potentiated by Ca^{++}, had pH optima between 8.2 and 9.8, and ranged in molecular weight from 30,000 to 36,000. Curiously, one of the four isolates from EC$_{23}$ was unable to macerate plant tissue, yet it was able to degrade pectic acid prepared from a number of plant cell wall sources. Each isolate produced exo- and endotypes, and both types could macerate tissue. In fact, from the data currently available, it would appear that only those enzymes capable of splitting the α 1–4 linkages in pectin substances significantly macerate tissue (separate cells). The isoelectric points of the four pectin lyases produced by *E. chrysanthemi* isolate EC$_{23}$ were at pH 9.4, 8.4, 7.9 and 4.6. Random cleavage of pectic acid was accomplished by all of the isozymes; however, unique to the acidic one (isoelectric point at pH 4.6) was its inability to attack *intact* plant tissue. Although its molecular size and pH optima were similar to the other three isozymes, its acidic nature may be responsible for its differential activity in vivo. It is suggested that, as a negatively charged macromolecule, becoming bound to positively charged surfaces at a distance from its substrate might interfere with its mobility. However, this interference is an in planta phenomenon that confirms frequent observations of variation in plant tissue to enzymic maceration.

Cellulose- and Hemicellulose-Degrading Enzymes in Pathogenesis

From a schematic presentation of the primary cell wall as it is envisioned by Albersheim (4), which may be an oversimplification, and from the studies of others (11, 71, 155, 162), it would seem that cellulose and the xyloglucans, glucuronoarabinoxylans, and their associated arabinogalactans (hemicelluloses) are buried beneath rhamnogalacturonans. The evidence to date suggests that wall maceration at least in its initial stages requires the degradation of the pectin and perhaps hemicellulose components in order to expose the cellulose microfibrils to enzymatic attack. The evidence to date on the relationship of cellulose and hemicellulose degradation to bacterially engendered pathogenesis is not as clearly causal as it is for the pectin polymers. However, there are data that infer an active role for cellulases and some hemicellulases in the disease process.

In a study by Kelman and Cowling (106) significant differences in alphacellulose and hemicellulose content and the degree of polymerization (DP) between healthy and *P. solanacearum*-infected tomato stem tissue provide direct evidence that this pathogen degrades cellulose during pathogenesis (Table 4–8). The participation of *P. solanacearum* cellulase in pathogenesis receives additional support from the positive correlation between virulence and cellulase activity (Table 4–9). In each of ten isolates the weakly virulent and avirulent variants displayed less cellulase activity, as measured by viscosity reduction of carobxymethylecellulose

Table 4–8. Composition of healthy tomato stem tissue and stem tissue infected by *P. solanacearum* (after Kelman and Cowling, 106).

Composition of tissue	Type of tissue	
	Healthy	Diseased
Holocellulose (%)	52.5	49.1
Alpha-cellulose (%)	32.9	23.2
Hemicellulose (%)	19.3	25.9
Degree of polymerization of holocellulose[a]	1540	990

a. Average number of anhydroglucose units per molecule of holocellulose.

Table 4–9. Relation between cellulase activity and virulence of isolates of *P. solanacearum* (after Kelman and Cowling, 106).

Host and isolate no.	Colony type[a]	Cellulase activity		
		Viscosity reduction units[b]	Glucose[c] mg/ml	Disease index[d]
Tomato	fluidal butyrous	71.4	0.68	94
25	(B₂)	0	0.29	12
Peanut	fluidal butyrous	50.0	0.90	96
31	(B₂)	0	0.10	10
Tobacco	fluidal butyrous	100.0	0.90	100
46	(B₁)	0	0.19	0
Potato	fluidal butyrous	90.9	0.70	78
51	(B₁)	26.3	0.40	0
Tomato	fluidal butyrous	125.0	0.82	96
60	(B₁)	0	0.15	0
Tobacco	fluidal butyrous	52.6	0.73	84
90	(B₂)	37.7	0.56	15
Tomato	fluidal butyrous	55.6	0.60	100
130	(B₂)	5.3	0.10	11
Pepper	fluidal butyrous	133.3	0.84	100
132	(B₁)	26.3	0.36	0
Tomato	fluidal butyrous	83.3	0.81	85
136	(B₁)	0.0	0.13	0
Dahlia	fluidal butyrous	71.4	0.64	100
227	(B₁)	0.0	0.00	0

a. Isolates 25 and 136 were obtained from Georgia and Trinidad respectively; all others from North Carolina. Colony type based on appearance on agar medium containing tetrazolium chloride. The fluidal colony type was the typical virulent wild type forming extracellular slime. Butyrous colony types were usually avirulent (B₁) or weakly virulent (B₂).
b. Viscosity reduction units = reciprocal of time for 50% reduction in efflux time (in min) of CMC × 100.
c. Reducing substances expressed as mg glucose/ml formed after incubation for 1 h at 50 C at pH 7.0 with CMC as substrate.
d. Disease index based on mean of numerical disease rating scale for 5 Rutgers tomato plants 21 d after inoculation by stem-puncture technique.

(CMC) and the formation of reducing substances from CMC (C_x or cellobiohydrolase activity), and had a much lower disease index than the related virulent types.

The study also demonstrated that the cellulase of *P. solanacearum* is constitutive, as are most bacterial cellulases, being produced on a variety of noncellulosic substrates. In addition, "S-factor" activity, which according to Simpson and Marsh (181) increases the accessibility of the cellulose microfibrils to cellulose-degrading enzymes, was detected in the culture filtrates of *P. solanacearum*. It seems plausible that their "S-factor" could be a hemicellulase or endo-1,4β-D-glucanase (215).

An apparent paradox was also revealed by this study in that *P. solanacearum* apparently possesses no β-glucosidase activity and therefore seems unable to utilize cellobiose, the end product of CMC degradation by the cellulase of the pathogen. Nevertheless, Kelman and Cowling (106) justifiably conclude that "the cellulase of *P. solanacearum* provides an enzymatic means by which the penetration of host tissues may be accomplished, even though the products of hydrolysis are not utilized (by the organism) as an energy source."

Attention is called to the ultrastructural study of Wallis and Truter (219) on pathogenesis engendered by *P. solanacearum*. Their electron microraphs seem to suggest that cell wall degradation is not a primary feature of pathogenesis. Their study suggests vessel plugging by bacterial polysaccharide to be more closely linked to symptom development (see Chap. 9).

Additional evidence of in vivo C_1 cellulolytic enzyme activity was provided by Goto and Okabe (84, 85) in their comprehensive studies on cellulolytic enzymes of phytopathogenic bacteria. They examined the degenerative activity of *Xanthomonas campestris* and *X. citri* by measuring the quantity and quality of cellulose in healthy and diseased cauliflower leaf tissue and *Citrus*. They found a marked decrease in total cellulose in the diseased tissue and were also able to distinguish decreases in an alpha fraction of cellulose in the parasitized tissue and a concomitant increase in beta and gamma cellulose fractions. An apparent cellulose and perhaps hemicellulose degradation may be inferred from electron micrographs of Wallis et al. (220) that reveal separation of spiral thickening from the primary wall of cabbage by *X. campestris*. Shredding occurs initially on the exposed wall layer and results in swelling and subsequently release of masses of fibrillar material. The shredding phenomenon has also been described in detail by Gritzali and Brown (88) as a feature of cellobiohydralase activity.

In the study by Braithwaite and Dickey (30) of a mixed inoculum, the maceration of carnation tissue was shown to be caused by the *Corynebacterium* sp. The pathogen produced both endo-PGTE and cellulase; however, maceration paralleled the lyase activity but seemed unrelated to cellulase activity.

The experiments of Beraha and Garber (25) referred to earlier seem to contradict the conclusion drawn by Turner and Bateman (198) that maceration caused by *E. carotovora* is solely a function of pectin lyase. Their data confer strong cellulase activity (C_x) upon a wild type (WT) strain of *E. carotovora* and a streptomycin resistant virulent mutant (P). An avirulent mutant (AV) exhibited a 12-fold decrease on C_x activity which, when reverted to virulence (RV), not only regained this activity but showed a 10-fold increase over the WT and P strains. These results tend to suggest that cellulase as well as pectin lyase activity are facets of virulence. However, it is possible that the two enzymes are closely linked. It should also be possible to derive specific mutants and monitor their virulence with respect to lyase and C_x activity.

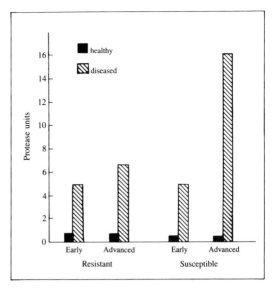

Fig. 4–30. Protease activity in extracts from healthy and diseased, resistant (22–54), and susceptible (22–102) alfalfa clones. Infected material was harvested at 4 (early) and 8 (advanced stage of disease development) d after inoculation with *Xanthomonas alfalfae* (after Reddy et al., 170).

Direct evidence for the participation of a xylanase in disease development has also been detected as a consequence of a mixed infection. Maino et al. (130) implicated an *Achromobacter* sp. in intensifying disease severity in bean caused by *P. phaseolicola*. This potentiative effect was interpreted as being due to the xylanase activity of the *Achromobacter*. *P. phaseolicola* was less able to degrade xylan and bean cell wall hemicellulose, as measured by the release of reducing sugars, than the *Achromobacter* sp. In all, they tested 41 species of plant pathogenic bacteria and three saprophytic species and concluded that xylanase production was most uncommon when corn-cob-derived xylan was used as the sole carbon source. However, the fact that some xylanase activity was produced by *P. phaseolicola* indicates that perhaps the hemicellulose source and composition closely control hemicellulase activity. As mentioned previously, Reddy et al. (170) have shown that *X. alfalfae* produces three xylan-degrading enzymes that appear to play a causal role in pathogenesis (see Fig. 4–30).

Summarizing plant cell wall maceration by pathogenic bacteria, one can say that pectin lyases play a predominant role. There is growing evidence that cellulases and hemicellulases also alter the cell wall during the wall-degrading process.

Protein and Lipid Degradation in the Plant Cell Wall and Plasmamembrane

The literature relating proteolytic and lipolytic activity to pathogenesis by bacteria is both circumstantial and meager. Friedman (73) measured the proteolytic enzyme activity of a virulent strain of *E. carotovora* and found that it liquified almost 4 times as much gelatin during a 14-d incubation period and denatured almost 2.5 times as much milk protein during an 18-d incubation period as did a weakly virulent isolate. He postulated that greater virulence could be attributed to increased proteolytic activity because of accelerated hydrolysis of host protein or heightened synthesis of bacterial protein and enzyme from the host amino acid pool. Similar conclusions were drawn by Bashan and Bateman (15) in studies of *Pseudomonas tomato* pathogenesis (see Chap. 5).

Beraha and Garber (25) observed that the virulent WT and P strains of *E. carotovora* produced phosphatidase; and Mount et al. (148) concluded that this enzyme is phosphatidase C. Beraha and Garber (25) also noted that an AV of strain P was completely incompetent with respect to the synthesis of phosphatidase. Furthermore, when the AV strain was reverted to virulence by the mutagen nitrosoguanidine, it not only regained its capability to produce phosphatidase but also produced twice as much enzyme as the virulent WT and P strains of the pathogen. Subsequently Beraha et al. (26) (Table 4–10) noted that a loss of this activity was accompanied by significant reductions in C_x and PG activity. They expressed the thought that the PT gene may "control" the elaboration of high levels of PG and particularly C_x.

The first clear evidence regarding the relationship of plant cell wall maceration to cell death was provided by Tseng and Mount (197). Their purified phosphatidase and protease preparations were unable to macerate ei-

Table 4–10. Virulence and relative enzyme activity in lyophilized culture extracts of an avirulent strain, revertants with reduced virulence, and fully virulent strains of *E. carotovora* (after Beraha et al., 26).

		Enzyme activity		
Strain[a]	Virulence[b]	PG	Cx	PT
AV 3–6	0	43.7	50	0
Rv RV 30–11	+ +	76.4	120	0
Rv RV 17–26	+ + +	80.6	162.2	0
Rv RV 4–12	+ + +	69.5	183.2	0
Rv RV 20–12	+ +	73.9	80	0
RV 8	+ + + +	118.3	2400	41.2
WT	+ + + +	100	1090.9	23.6

a. AV = avirulent; RV = revertant; Rv RV = revertant with reduced virulence; WT = wild type.
b. Scored as appearance of lesion on celery petiole 24 h after inoculation. 0 = not water soaked; no watery exudate; no lesion enlargement. + = slightly water soaked; no watery exudate; no lesion enlargement. + + = water soaked; no watery exudate; slight lesion enlargement. + + + = water soaked; some watery exudate; slight lesion enlargement. + + + + = water soaked; considerable watery exudate; lesion enlarged.

Table 4–11. Percent survival of cucumber protoplasts treated with phosphatidase C, protease, and endopolygalacturonate transeliminase (endo-PGTE) from *E. carotovora* (EC14) (after Tseng and Mount, 197).

	Percent survival of cucumber protoplasts after exposure times of				
Enzyme treatment[a]	0 h	1 h	2 h	3 h	4 h
Control[b]	100	100	100	100	100
Phosphatidase C	100	50	32	15	6
Protease	100	76	53	27	22
Endo-PGTE	100	100	100	100	100
Phosphatidase C + Protease	100	36	24	10	3
Phosphatidase C + Endo-PGTE	100	58	40	20	9
Protease + Endo-PGTE	100	68	47	42	34
Phosphatidase C + Protease + Endo-PGTE	100	47	20	17	2

a. Each reaction mixture contained either Ca^{2+} or Mg^{2+} as a cofactor; protease $(3 \times 10^{-4} \, M \, Ca^{2+})$, phosphatidase $C(3 \times 10^{-4} \, M \, Mg^{2+})$, and endo-PGTE $(3 \times 10^{-4} \, M \, Ca^{2+})$.
b. Control represents either reaction mixtures containing autoclaved enzyme with their respective cofactors or buffer alone.

ther cucumber or potato tuber disks. On the other hand, endo-PGTE completely macerated cucumber tissue in 1 h and potato in 4 h. However, there was more than just a suggestion that at least in cucumber protoplasts both phosphatidase C and protease per se could cause cell death (see Table 4–11). It would appear from these data and the results of Turner and Bateman (198) with extracts of tissue macerates produced by *E. carotovora* (Table 4–7) that cucumber tissue is exquisitely sensitive to the "enzyme complex."

Which Enzymes Kill the Host Cell?

The observation that tissue maceration (cell separation) and cell death seem to be parallel phenomena and perhaps causally related was first made by Brown (33). Subsequent investigations by Tribe (196) and Fushtey (78) with culture filtrates of *Erwinia aroideae* seemed to corroborate Brown's observation. It was Tribe (196), however, who first noted that plasmolyzed cells in tuber tissue were not killed as quickly as normal cells by the macerating culture filtrates. Tribe used KNO_3 and glucose as plasmolyzing agents and established cell death by the loss of neutral red dye from cell vacuoles following treatment of tuber tissue with partially purified culture filtrates. In one experiment $0.1 M \, KNO_3$-treated (plasmolyzed) tissue remained alive for 20 h after exposure to the filtrate, whereas nonplasmolyzed controls were dead in 20 min.

The delaying effect upon cell death by plasmolysis raised the question of whether cell death could be due to degradation of the cell wall or whether the culture filtrates exerted their effect directly upon the plasmalemma. In the latter instance it seemed plausible either that the cell membrane might be sensitive to a nonenzymatic toxic compound or that the membrane contained a component sensitive to the "macerating enzymes" in the culture filtrate.

It was apparent to Tribe (196) that one might readily determine the site of lethal action and perhaps the nature of the active factor by exposing naked protoplasts to the culture filtrates. This bold and innovative procedure yielded the following results. All protoplasts in the active enzyme preparations died in 3 h, death was recorded after 5 h for all protoplasts exposed to the heat-inactivated enzyme, and all control protoplasts survived the 5-h experiment. These results were inconclusive as they appear to implicate, in addition to the macerating enzymes, a heat stable toxic component. The nature of this "toxic" component was never established, and it may have been an artifact.

Since the experiments of Tribe (196) it has become possible to purify to homogeneity the enzymes produced in vivo by the "soft rot" pathogens. These include the pectinases, and foremost of these, for pathogenic bacteria, are the pectin lyases. In addition, these bacteria may also produce phosphatidases, proteases, and cellulases and hemicellulases.

The question we seek to answer is how plant cells are killed by bacteria that cause cell wall maceration. Let us, therefore, examine the probable target sites of these purified enzymes in order to establish the manner in which cell death is caused. Death in this instance is characterized by intense electrolyte leakage or the loss of Neutral Red dye from treated tissues, isolated cells, or naked protoplasts. The inability of isolated plant cells (with walls) and cells in tissue to plasmolyze in hypertonic solutions is also an indication of death.

The targets for the aforementioned purified enzymes are the cell wall or the cells' membranes, more specifically the plasmalemma. Basham and Bateman (14, 15) used a hypotonic solution to stretch the plasmalemma of naked protoplasts in order to expose possible pectin lyase-sensitive sites that could have been masked or obscured during the wall removal process (15). In addition, the *hypertonicity* of the plasmolyticum in which naked protoplasts are kept keeps them in a contracted condition. Nevertheless, the stretched protoplasts did not rupture or reveal any leakiness, as would have been expected had there been cryptic pectin lyase-sensitive sites on the plasmalemma (Table 4–12). The data of Stephens and Wood (191) revealed that either singly or together phosphatidase and protease ruptured naked cucumber protoplasts. These results suggest that normally the wall shields the plasmalemma from these two enzymes. The inability of either of these two enzymes to separate cells of potato disks or kill protoplasts within cell walls is indicative of their impotence in affecting walled cells. On the other hand, Stephens and Wood (191) found these protoplasts to be totally recalcitrant to the effects of pectin lyase, suggesting the absence of sensitive sites on the plasmalemma to this enzyme.

Table 4–12. Death of tobacco protoplasts subjected to osmotic stretching during exposure to endopectate lyase (after Basham and Bateman, 14).

Osmoticum molarity	% Protoplast Death[a]			
	10 min[b]		90 min[b]	
	E	B	E	B
0.63	0	0	4	0
0.51	7	7	14	6
0.40	6	17	10	30
0.29	40	47	54	61

a. Protoplast death (%) determined on counts of ca. 100 protoplasts. Differences between enzyme treatments and buffer treatments are not significant (P = 0.05) except for the 0.4 M, 90-min treatment.
b. Reaction mixtures contain ca. 6×10^3 protoplasts in 1.0 ml of 50 mM Tris-HCl buffer (pH 8.5) with mannitol osmoticum at the indicated concentrations and 1.8 units of the lyase (E) or buffer controls (B).

Tseng and Mount (197), it will be recalled, reported similar results (Table 4-11).

It has long been conjectured that in the process of wall maceration toxic products are released that alter the permeability of the plasmalemma and that these may be the basis for cell death. Basham and Bateman (15) evaluated the toxicity of peroxidase, catalase, and some low molecular weight products removed from tobacco cell walls during lytic degradation. They found no evidence that toxic products were even in part responsible for the observed cell death. Similarly, they considered the possibility that acidic enzymes with high net positive charges could deleteriously affect the plasmalemma. For example, a highly negatively charged enzyme, such as the one produced by *E. chrysanthemi* (80, 164) that had an isoelectric point at pH 4.6, might, when bound to an array of positive sites on the plasmamembrane, alter its critical spatial arrangement and its selective permeability. This concept also proved untenable.

The role of hydrogen peroxide as an adjunct to injury caused by pectin lyase has been reported (150). Apparently, H_2O_2 accumulates in both cotton (151) and cauliflower (126, 127) tissue treated with pectic enzymes. However, H_2O_2 seems not to potentiate injury to potato tissue by pectin lyase (26).

It has been proposed (218) that membrane damage in the form of plasmodesmata distortion, ostensibly due to pectin lyase-caused wall maceration, could cause cell death. Although plasmodesmata disruption probably does occur as a consequence of wall maceration, it appears not to vitally damage cells that are prepared as naked protoplasts. It should be remembered, however, that such cells are maintained in a plasmolyticum under hypertonic conditions that protect the cells from osmotic stress. In any event, it would seem that plasmodesmata have developed a process to seal and repair themselves following injury or after exposure to osmotic imbalance that leads to plasmolysis.

We believe that the evidence thus far concerning cell wall maceration and subsequent cell death (3, 15, 26, 218) suggests that the associated membrane damage caused by pectin lyase occurs *indirectly*. More specifically, pectin lyase alters the mechanical strength of the cell wall so that it is unable to support the plasmalemma, which in turn ruptures or rapidly loses its selective permeability under conditions of osmotic stress.

What are the findings that support this conclusion? (1) There are no reports of purified pectic enzymes that cause cell death without causing cell maceration (14, 15). (2) Dilutions of pectic enzymes used to treat potato disks resulted in proportional increases in time required to cause maceration and cell death (191). (3) The loss of electrolyte from potato and tobacco tissue treated with pectin lyase paralleled the course of cell wall degradation (14, 15). (4) If pectin lyase is removed from the reaction with potato disks before 50–75% of their total electrolyte is lost, the rate of electrolyte leakage is reduced fourfold (14, 15). (5) The apparent protection of plasmolyzed cells against death induced by pectin lyase may be explained by the shrunken state of the protoplast. Once pectin lyase-treated tissue (or walled cells) is removed from its plasmolyticum (the hypertonic environment that caused plasmolysis), cell death occurs swiftly and electrolyte leakage rate approximates that from free space (25, 26).

Some investigators of plant cell wall degradation are unwilling to rule out the possibility that pectin lyase directly alters plasmalemma permeability and causes cell death. Perhaps the evidence by Bretscher (31) that polysaccharides are a structural feature of the plasmalemma and other membranes and are reported to anchor proteins and enzymes in the bilipid leaflet supports this contention. Ultrastructural evidence for the presence of carbohydrates in plant cell membranes was disclosed long ago by Albersheim and Killias (5). More recently, Roseman (174) has described the presence of oligo- and polysaccharides in the membranes of both plant and animal cells.

Still another basis for the resultant plant cell death is the evidence from Albersheim's laboratory, presented by Yamazuki et al. (221), indicating that acid solubilized glucan (>1,000 MW) fragments of sycamore cell walls caused the death of suspension-cultured sycamore cells. A direct measurable effect of the fragments is the inhibition of ^{14}C leucine incorporation and thus, allegedly, interference with protein (and perhaps membrane protein) synthesis. An apparent causal relationship between the wall fragments and plant cell death caused by cell wall-degrading enzymes is the protective effect that plant cell plasmolysis exerts against either the fragments or the enzymes. Attention is called to the earlier, related study of Tribe (196), in which a toxic principle was considered at least partially responsible for plant cell death, and plasmolysis of isolated plant cells delayed the killing effect of wall-degrading enzyme preparations.

Alterations Caused by Fungal Pathogens

An important aspect in pathogenesis of fungal disease-causing organisms is their mechanism of penetrating and colonizing host tissue. In this regard the degradation of cell walls of tissues in different stages is of primary interest (20, 21). When fungi confront a plant cell wall they face a complex polymeric barrier. Because of the array of compounds and linkages, a number of specific enzymes are required for their degradation.

Decomposition of the Host Cell Wall

Many fungi produce pectic enzymes that are key factors in host cell wall degradation. These enzymes are produced both in vitro and in vivo, although the former may differ quantitatively and qualitatively from the latter. Data describing the role of cellulases, hemicellulases, proteinases, and phospholipases are less abundant, while our knowledge of the involvement of enzymes of fungi in the digestion of cutin, suberin, and lignin has recently begun to expand.

Enzymatic degradation of the cell wall has obvious significance in pathogenesis induced by necrotrophic fungi, however our knowledge of cell wall decomposition by biotrophic fungi (obligate parasites) is meager. Although Van Sumere et al. (206) reported the detection of endopolygalacturonase, cellulase, and hemicellulase during spore germination of stem rust (*Puccinia graminis* f. sp. *tritici*), the role of these enzymes in the disease symptoms is still not clear. Probably they are involved in host penetration. Our knowledge of host cell wall hydrolysis and tissue maceration derives mostly from research on necrotrophs (facultative parasites).

There are several examples where fungi have been shown to produce cell wall-degrading enzymes either constitutively or inducibly. Exogenously applied sugars regulate the production of cell wall-degrading enzymes by *Rhizoctonia solani* (212). The activity of an impressive array of specific carbohydrate-degrading enzymes can be presumed to operate in plants based on the data of Bateman (20) (Fig. 4–31), which reveal decreases in seven noncellulosic sugars and two sugar amines in the walls of bean hypocotyls infected with *Sclerotium rolfsii*. Several plant tissue components serve as stimulators of wall-degrading enzymes of pathogens. Studies with *Fusarium* spp., *Colletotrichum lindemuthianum*, and *Verticillium albo-atrum* demonstrated that cell wall-degrading enzymes are produced by these fungi in a temporal sequence. The regulation or control of synthesis of these enzymes has been described in detail by Cooper and Wood (43) in relation to pathogenesis and specificity. In particular they described the controlling elements, induction and catabolite repression. The temporal nature of enzyme synthesis is apparent from Fig. 4–32, wherein PG, PTE,

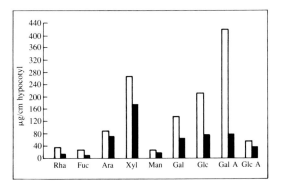

Fig. 4–31. Noncellulosic-sugar composition of bean hypocotyl cell walls 3 d after inoculating 7-d-old seedlings with *S. rolfsii*. Results are expressed as μg of each cell wall sugar constituent recovered per cm length of hypocotyl for healthy (□) and diseased (■) hypocotyls (after Bateman, 20). Rha: rhamnose; Fuc: fucose; Ara: arabinose; Xyl: xylose; Man: mannose; Gal: galactose; Glc: glucose; Gal A: galacturonic acid; Glc A: glucuronic acid.

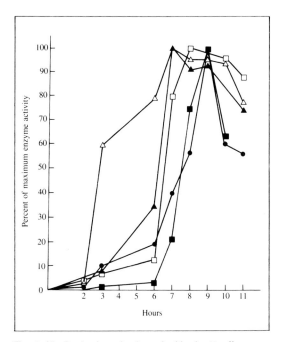

Fig. 4–32. Production of polysaccharides by *V. albo-atrum* grown at 21C on Gem susceptible cell walls. Enzymes followed by maximum activities detected are: (△—△) endopolygalacturonase, 429·2 RVU; (●—●) endo-pectin transeliminase, 9168 μmol ml^{-1} h^{-1}; (▲—▲) arabinanase, 1122 μmol ml^{-1} h^{-1}; (□—□) xylanase, 3742 μmol ml^{-1}h^{-1}; (■—■) cellulase (Cx), 684·9 RVU. pH of cultures rose from 5.5 to 8. Results are the mean of 4 replicate cultures (after Cooper and Wood, 44).

ARA, Xyl, and Cx are produced in tomato stem walls by *V. albo-atrum* over a 10-d period (41, 44). First to appear are enzymes that degrade the pectic substances, followed by those that hydrolyze the hemicelluloses, and finally the cellulose degraders are elaborated (42, 44). This pattern is observed most clearly when pathogens are cultured on cell walls of their hosts. It has been established that degradation of the rhamnogalacturonide fraction (and probably homogalacturonides), which is a key element in host cell wall structure, enhances the ability of other enzymes to hydrolyze their substrates (104, 195). Apparently, the endopolygalacturonases and lyases (transeliminases) lead the attack on the plant cell wall, and loosen its structure, exposing the other cell wall components to enzymatic hydrolysis. Thus, the endopectic enzymes function as "wall-modifying enzymes."

The correlation between in vitro production of cell wall-degrading enzymes and pathogenesis by fungi is not necessarily strong. There are great differences between infected tissues and axenic cultures (194). In axenic cultures, pectic enzymes of *Rhizoctonia solani* differ with culture age and pH . Young cultures contain mostly polygalacturonase, while older ones may reveal only pectic lyase because the pH becomes alkaline. Analogously, cell walls isolated from juvenile bean tissues are usually more sensitive to enzymatic degradation by *R. solani* than walls from mature bean tissues. It is noteworthy that deposition of ligninlike materials in cell walls after infection protects cell-wall constituents from enzymatic decomposition.

Tissue Maceration

This is a common symptom caused by enzymatic digestion of the middle lamella of the host cell wall. Enzymatic maceration has long been known (33), but the endopectic enzymes involved were characterized much later (14, 15), and it appears likely that maceration is causally related to the actual death of the affected cells. The endo-polygalacturonases and endopectate lyases are the most efficient macerating enzymes in the process. Hall and Wood (94) reported that they loosen primary wall structure, which becomes able to support the protoplast and the cell subsequently dies under osmotic stress. The process is described in detail in the previous section of this chapter. However, studies from Albersheim's laboratory (221) suggest that comparatively small glucan wall fragments may also play a role in causing pant cell death directly.

Pectic Enzymes Causing Disease Symptoms

Soft rots. The pectic enzymes of pathogens have been most widely studied in relation to soft rot induction. The infected fleshy parenchyma becomes soft and water-soaked and may remain either colorless or turn brown. Typical examples are apple rot caused by *Penicillium expansum* and *Sclerotinia fructigena*. Polygalacturonase and other macerating enzymes play an important role in pathogenesis, however environmental factors af-

Table 4–13. Effect of cations upon maceration of potato tissue by polygalacturonase from *Rhizoctonia*-infected bean tissue (after Bateman, 17).

Time (h)	Control (H_2O)	Control (PG)	PG + 5×10^{-2} M cations				
			Ba^{++}	Ca^{++}	Mg^{++}	K^+	Na^+
2.0	0.0	2.5	0.0	0.0	0.0	1.0	0.5
3.5	0.0	4.0	0.0	0.0	0.5	2.0	1.0
5.0	0.0	5.0	0.0	0.0	1.0	3.0	2.0
6.5	0.0	5.0	0.0	0.0	0.5	5.0	5.0
24.0	0.0	5.0	0.5	1.0	5.0	5.0	5.0

fect both the activity and the secretion of these enzymes. For example, increasing the water content of tissues greatly increases their susceptibility to soft rot pathogens.

In brown rot caused by *S. fructigena* the pathogen secretes polygalacturonase, which degrades the pectic substances of the cell wall, eventually killing the cells. At the same time, polyphenoloxidase acts on the host's phenols, forming oxidation products that are responsible for the developing brown color. These oxidized phenols, as well as oxidation products of leuco-anthocyanins, inactivate fungal polygalacturonase, and thus degradation of pectic substances of the wall is interfered with. Accordingly, no maceration takes place in this disease (35, 36, 39). In contrast, polyphenoloxidase is probably inhibited in the white rot caused by *P. expansum*; consequently brown discoloration does not occur. The result of this action is a substantial degradation of pectic materials of the cell wall.

Apple fruits become susceptible to soft-rot-causing fungi such as *Glomerella*, *Botryosphaeria*, and *Physalospora* after they ripen. The water-soluble pectins of mature resistant fruits are less readily utilized by the fungi as a source of carbon than are pectins of mature susceptible fruit (211). As the water-insoluble pectins and proteins decrease, during the course of ripening the fruit becomes susceptible to rot. Accompanying this process is an exchange of potassium for polyvalent cations, i.e. Ca^{++}. Thus the pectic substances of mature fruit are less extensively cross-linked by cations in the cell wall and middle lamella as complexes of pectic polymers or protein-pectic polymers. It is possible therefore that a change in pectin-protein-metal complex during the ripening process is responsible for the observed solubilization of pectin (210).

Leaf and hypocotyl lesions. In leaf-spot diseases and particularly in hypocotyl lesions caused by *R. solani*, pathogens spread only to a limited extent in host tissues. A similar picture is seen when soft-rot pathogens infect tissues of resistant plants. Bateman (17) showed that the host tissues respond to infection in a manner that makes them resistant to the action of the fungal polygalacturonase. In the early stages of *Rhizoctonia* infection of bean

hypocotyls, fungal polygalacturonase and host pectin methylesterase act together in the process of tissue destruction. In the later stages, respiration is greatly increased in diseased tissue, and this is associated with the accumulation of Ca^{++} and other ions in the infected region. This increased concentration of cations releases and activates host pectin methylesterase that had been bound to host-cell walls. Thus, the demethylated pectic materials become cross-linked by Ca^{++} and are rendered resistant to the further action of fungal polygalacturonase. The effect of various monovalent and divalent cations on pectin-degrading activity in infected tissue is shown in Table 4–13. It is evident that the activity is inhibited by the divalent cations, particularly Ba^{++} and Ca^{++}. The mechanism of this induced enzyme suppression is quite similar to that of the tomato treated with synthetic naphthalene acetic acid (NAA), an auxin. In this case, NAA promotes demethylation of pectic material, and hence Ca^{++} salt linkages form and pectic substances become more recalcitrant to the fungal (*Fusarium*) polygalacturonase (see Chap. 7). It is well known that the availability of calcium and boron and the influence of auxins affect the structure of pectic substances, thereby altering their susceptibility to pectic enzymes (46, 61).

Bateman (17, 21) made extracts from *Rhizoctonia*-infected bean hypocotyls and demonstrated the presence of a dialyzable factor that inhibits tissue maceration by polygalacturonase found in diseased tissue. This inhibitor was shown to be a divalent cation, particularly Ca^{++}. That calcium renders pectic substances resistant to hydrolysis by fungal polygalacturonase in tissue surrounding lesions was shown also in histological studies. The middle lamellae of cells adjoining lesions were intact and more difficult to hydrolyze than the middle lamellae of cells farther from lesions caused by the infection of *Rhizoctonia*. Apparently, it is by this mechanism that the host confines the fungus to a lesion of limited size in the hypocotyl tissue. On the other hand, unlike the situation in the *Rhizoctonia*-infected bean, the mobilization of Ca^{++} in lesions caused by *Colletotrichum trifolii*, the causal agent of the southern anthracnose of alfalfa, actually facilitates the fungus's invasion of the tissue by stimulating the macerating action of pectic lyase (96, 97). This work calls attention to the different pectolytic enzyme systems associated with fungal plant pathogens and perhaps a dual role for Ca^{++} in pectin metabolism of the host-pathogen relationship.

The formation of insoluble pectate salts in diseased plants represents only one of the several possibilities for the inhibition or reduction of the activity of polygalacturonase in tissues. We will describe in greater detail how oxidized phenolic compounds and tannins in fungus-infected plants may serve to inactivate the polygalacturonase of the pathogen (see Chap. 6).

Vascular wilts. Evidence has accumulated that pectic enzymes play an active part in the development of vascular wilt diseases caused by *Fusarium* and *Verticillium*. The evidence supports the concept that occlusion is caused by gels that originate from an enzyme-altered primary wall-middle lamella complex at either pit membranes or perforation plates. Gel formation at these points often precedes wilting and reduces vascular flow. The vascular browning that frequently follows probably signals intensive phenolic cross-linking of the gel (24). This impedes lateral flow through the pits and increases the effectiveness of gels as occlusions of the conducting vessels and is responsible for wilt (59). Degradation of middle lamellae of the pit membranes promotes the formation of tyloses that reduce the flow of fluids through the vessels. Furthermore, the breakdown of pectic substances releases pectin fragments that, upon passing into the vascular stream, increase its viscosity and reduce its flow rate. In addition, the oxidation of phenolics that usually occurs produces the characteristic symptom of vascular browning.

It is of considerable interest that pectic enzyme preparations can produce some of the characteristic symptoms of wilt disease in cuttings of tomato (83, 216). Furthermore, pectin methylesterase has been obtained from the vascular sap of infected plants by Waggoner and Dimond (209), and *Fusarium* readily produces pectin methylesterase, polygalacturonase, and other macerating enzymes in axenic culture. An "ideal" situation is suggested in the report of Stahmann (188), in which pectic enzyme formation in living tomato stem tissue was lower in resistant varieties than in susceptible ones even though the attacking fungi (*Fusarium* and *Verticillium*) grew equally well on both (Fig. 4-33). These reports accentuate the probable importance of pectin-degrading enzymes in vascular wilt diseases. On the other hand, McDonnell (142) and Mann (133) have presented evidence that in various UV-induced mutant strains of *Fusarium oxysporum* f. sp. *lycopersici* pathogenicity may be little related to production of pectic enzymes in vitro, since some mutants that do not produce pectolytic enzymes are also able to incite typical wilt symptoms. Both authors, however, reported that isolates deficient in PG production *in culture* exhibited reduced pathogenicity in susceptible tomato plants. Subsequently, Puhalla and Howell (163) performed similar experiments with UV-induced mutants of *V.*

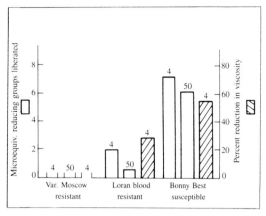

Fig. 4–33. Pectic PG formation on tomato stem tissues by two isolates (4, 50) of *Verticillium albo-atrum* (after Stahmann, 188). No enzyme formation was detected in filtrates from the culture on the highly resistant var. Moscow cultivar.

dahliae, and their data indicated that the mutants produced lower levels of pectolytic activity and caused slower symptom development. In studies by Mussell and Strand (150) and Cooper and Wood (44), purified endo-PG and both endo-PG and endo-PL (pectin lyase) were infused into cotton and tomato respectively and produced symptoms of wilt very similar to those produced by the pathogen. Mussell and Strand (152) are consequently of the opinion that the pectic enzymes produced by the wilt pathogens are causally involved in pathogenesis and some if not all of the symptoms observed in these diseases. Cooper and Wood (44, 45) have reported that the presence of endo-PL in infected vessels is responsible for profound effects on plant tissue and the lack of symptoms with heat-inactivated enzyme preparations. These studies certainly suggest a positive role for pectin-degrading enzymes. Dimond (60) suggested, quite logically, that the pits, to which the pectic enzymes are likely to be adsorbed, act as ultrafilters and actually localize enzyme activity at this highly sensitive substrate. Nevertheless, the issue is still unresolved and awaits additional data to determine the extent to which pectic enzymes and other factors may be involved in the pathogenesis of vascular wilts. The contribution of pathogen and host cell components to the formation of occluding gels bears additional research in this regard.

The Role of Cellulases and Hemicellulases

Less is known about cellulases in fungal pathogens. Much of our knowledge comes from investigations with microorganisms that decay wood and degrade textiles. In soft-rot diseases the initial attack involves a solubilization of the intermicellar pectic substances; therefore, pectic enzyme systems play a more important role than the cellulolytic systems. However, in a later stage, when an advanced disintegration of cell-wall material takes place, cellulases are also involved. Although several soft-rot pathogens (*Pythium, Rhizopus*) are actively

Table 4–14. Specificity of induction of polysaccharidases in *V. albo-atrum* by different sugars (from Cooper and Wood, 44).

Carbon source[b]	Enzyme activities (percentage of maximum[a])					
	Endo-PG	Endo-PTE	Arabinanase	β-Galactosidase	Xylanase	Cellulase (C_x)
D-Glucose	0–1.25	0	0–5.8	0.84	5.8	1.46
D-Galactose	0	0	14.4	12.90	5.2	0.37
D-Mannose	0	0	0	0.34	0	0.20
L-Arabinose	0	0	100	100	3.5	0.25
D-Xylose	0	0	3.7	0.26	100	0.08
L-Rhamnose	2.10	9.11	6.6	0	0	0.40
L-Fucose	1.38	0.16	0	0	0	0.26
D-Galacturonic acid	100	100	15.8	0	16.5	0.30
Cellobiose	3.7	1.34	—	0	—	100

a. For each enzyme, activities detected in cultures supplied with different sugars for identical periods are expressed as a percentage of the activity in cultures supplied with the specific inducer.
b. Sugars fed from diffusion capsules at linear rates of ca 2.5 to 4 mg 100 ml^{-1} h^{-1} to shake cultures.

cellulolytic in culture (186, 216), little is known about the production of cellulases by these pathogens in vivo. On the other hand, cellulase activity has been reported by Bateman (16, 18, 20) in bean hypocotyl tissue infected by *Rhizoctonia solani*. This fungus penetrates the cell walls of its host and, by destroying the native cellulose, causes a collapse of the invaded cells. The collapse results in the formation of concave lesions on the bean hypocotyl. This takes place in the later phases of pathogenesis in association with the intracellular penetration of the fungus. Similarly, *Colletrotrichum trifolii*, a pathogen of alfalfa, has been shown to invade tissue in an intracellular manner, and a high cellulolytic activity characterizes hyphal penetration of the cell wall (97).

Cellulase may also play an important role in diseases that cause wilting. *Fusarium oxysporum* and *Verticillium albo-atrum* produce cellulase in culture and degrade cellulose in vivo. By liberating large molecules from cellulose into the transpiration stream, the normal water movement is inhibited by blocking the vascular elements (57, 101).

Under natural conditions much of the cellulose in diseased plants is probably degraded by secondary invading organisms, which are not true pathogens but are strongly cellulolytic—e.g. *Trichoderma viride*.

Evidence has been accumulating that a number of fungi, particularly the wood-rotting species, produce several cellulolytic components (160, 161). Such a multiple-component cellulase system has been clearly demonstrated for the wood-destroying basidiomycete *Polyporus versicolor*. Culture filtrates of the fungus contain at least five cellulolytic components, and the molecular weights of two of the enzymes have been measured. One has a weight of 51,000 D and is a β-glucosidase that preferentially hydrolyzes oligosaccharides; the other has a weight of 11,000 D and shows the highest

activity against glucosidic bonds of polysaccharide chains.

The molecular size relationship that has been reported—the smaller enzyme splitting the long-chain polysaccharide and the larger one degrading oligosaccharides to glucose—should facilitate the degradation of cellulose in a sequential manner. The relationship supports Cowling's contention (47) that size and diffusibility of enzyme molecules in relation to the size and surface properties of the capillaries between the cellulose microfibrils determine the susceptibility of the microfibrils to enzymatic degradation.

Cellulases are rarely constitutive, as most are induced by cellobiose (132), the end product of cellulose degradation. It has been shown by Gupta and Heale (91) that cellobiose, like other low molecular weight inducers, is able to repress cellulase synthesis. Consequently, cellulase production may not occur because of catabolite repression, until the disaccharide is nearly completely metabolized.

As regards the hemicellulases, a number of pathogenic fungi have been demonstrated to produce these enzymes (*Sclerotium rolfsii, Sclerotinia sclerotiorum, Rhizoctonia solani, Fusarium* and *Diplodia* species), and the significance of degradation of hemicelluloses in pathogenesis remains unclear. Characteristics of some hemicellulases appear in Table 4-14.

The Role of Lignin Degradation

The enzymatic degradation of lignin is characteristic of saprophytic wood-rotting fungi. The nature of degradation by fungal pathogens is little understood. The ability of most plant pathogenic fungi to degrade lignin has not been investigated; still, Fiussello et al. (69) have shown that *Botrytis cinerea* has lignin-degrading activity.

Levi and Preston (124) note that all types of wood-destroying fungi remove methoxyl groups from lignin. This appears to be the main effect upon lignin of brown- and soft-rot fungi. Their experiments on the degradation of beechwood veneers by *Chaetomium globosum* revealed, among other things, that the fungus removed methoxyl groups from lignin at an increasing rate as decay ensued. They suggest that the fungus does not directly degrade lignin, but rather "modifies" it by removing methoxyl groups.

The net effect of this alteration of lignin is to expose more adjacent cellulose microfibrils to the cellulase of the attacking fungus. Hence a "lignin-modifying" or "cellulose-freeing" enzyme could, according to Levi and Preston, in some way govern the rate of cellulose decomposition in wood. It has been suggested that the enzyme involved is a transmethylase (124). (See Additional Comments and Citations.)

Cell Wall Lignification and Wall Apposition

Reaction zones (with pathogens) in host plants may contain lignin, which is regarded by many as a barrier against penetration. Induced lignification, as a consequence of infection, can make host walls more resistant to mechanical penetration and to dissolution by fungal enzymes because lignin physically shields the wall polysaccharides from fungal enzymes. The phenolic precursors of lignin can also inhibit fungi. Several infected tissues do not show induced lignification, thus the phenomenon cannot be regarded as universal. In several cases, greater concentrations of induced lignin were recovered from infected resistant host tissues than from susceptible ones. The significance of induced lignification in plant disease resistance remains a subject for additional study (9, 200).

Infected plant cells or wounded cells may respond rapidly by accumulating an amorphous, heterogenous mass of wall materials, as was shown more than 100 years ago (56). The wall depositions are also called wall appositions. Those appositions, incited by fungi, are termed *papillae*, whereas those induced by wounding are termed *wound plugs* (1, 2, 200).

The pioneering work of Fellows (66) called attention to the importance of lignin in papilla formation. When wheat was infected by *Gaeumannomyces (Ophiobolus) graminis*, the papillae contained mainly lignin as a structural component. He named the papillae lignitubers. Vance and Sherwood (201, 202) have shown that papilla formation was inhibited by the protein synthesis inhibitor cycloheximide and that induced lignification upon infection was also inhibited. They claimed that *Phalaris arundinacea* inhibits direct penetration of several alien leaf-infecting fungi by papilla formation. When the host plant was infected by *Helminthosporium avenae*, tissues had a higher lignin content and the enzymes involved in lignification were also enhanced: PAL, TAL (tyrosine ammonia lyase), hydroxycinnamate-CoA ligase, and peroxidase.

The role of papilla formation in disease resistance is not clarified as yet. Some investigations point to a possible role of lignification and papilla formation in the general (horizontal) resistance of plants to fungi (see Chaps. 1 and 10).

The Role of Cutinases

The role of cutin and cutinolytic enzymes produced by fungi in host-parasite relationship is becoming clearer (203). It has been generally supposed that an enzymatic breakdown is necessary, in addition to the mechanical penetration of the cuticle, because the cutin in the cuticle is a barrier to fungal penetration. Significant data are now available on the cutinolytic enzyme activity of some pathogenic fungi that validate this contention. The efforts of Kolattukudy et al. (112, 113, 115) provided us with a clearer understanding of the chemical composition and structure of cutin. Subsequently, Lin and Kolattukudy (125) reported the induction of cutinase by low levels of cutin monomers in an in vitro culture of *Fusarium solani* f. sp. *pisi*. Shaykh et al. (180) had previously advanced evidence that cutinase was produced by this fungus during penetration of its host. They reported that ferritin-conjugated antibodies (anticutinase I) prepared against cutinase delineates the presence of the enzyme on the surface of *Pisum sativum* cutin layer 12 h after inoculation with a conidial suspension of *F. solani* f. sp. *pisi* (see Fig. 4-34). The methods used (125) to isolate the extracellular enzyme from *F. solani* f. sp. *pisi* were also used successfully to reveal the presence of this enzyme in five additional plant pathogenic fungi (see Table 4-15). The enzymes thus far characterized by Kolattukudy and his associates have all been glycoproteins. One of their studies (131) has shown that fungal infection could be prevented by a specific inhibitor of cutinase. Hence the persistent suggestion in several ultrastructural studies of cutin disorientation (darkening) where fungi are penetrating the plant cell wall appears to have its basis in the activity of fungus-produced cutinases.

Cell Wall Components as Chemical Signals and Recognition Molecules

Cell wall degradation may exert far-reaching effects on the plant's physiology (see Chaps. 6 and 10). A large pectic polysaccharide (RG-I) that is solubilized by a fungal endopolygalacturonase from plant cell walls induces a systemic chemical signal in the plant and, as a consequence, *proteinase inhibitor* synthesis occurs in intact plant organs (176). This is a systemic action, and the inducing stimulus can be wounding, infections, or other stresses associated with cell wall perturbation. Neither the details of the mechanism nor its significance are known, but it would seem that specific cell wall fragments of pectic character provide the systemic signal for inhibitor synthesis (221).

The fragmentation of the complex plant cell wall after infection by fungi or other pathogens has been shown to liberate *elicitors of phytoalexins* (see Chap. 6). Cell wall glucans from both fungi and plant may have elicitor properties. These approximate a localized action as opposed to the above-mentioned systemic signal.

Kojima et al. (111) have reported the isolation of an

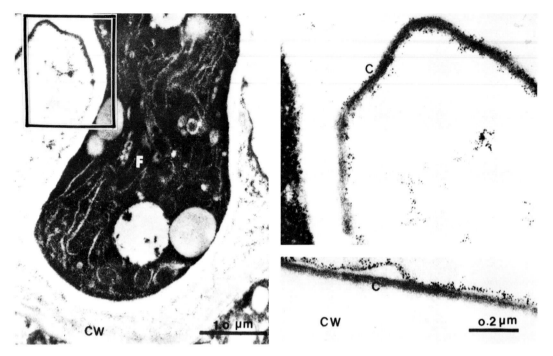

Fig. 4–34. Electron micrograph of an infected pea epicotyl (at left with bordered insert). Magnification of insert, upper right, showing ferritin granules attached to cuticle (c). In lower right electron micrograph, the granules of ferritin-conjugated anti-cutinase I clearly delineate the presence of the cutinase enzyme elaborated by *Fusarium solani* f. sp. *pisi*. The electron micrographs are of epicotyl tissue 15 h after inoculation with a conidial suspension of the pathogen (C = cutin, CW = cell wall, and F = *Fusarium*) (after Shaykh et al., 180).

agglutinin from sweet potato root cell walls that seems to be a pectic acid–like substance containing galacturonic acid as a main component. As it has a specific spore agglutinating activity in relation to *Ceratocystis* species, one may regard this factor as a recognition molecule and a determinant of specific resistance of sweet potato to black rot disease. It is not known how

Table 4–15. Molecular weight of cutinases from fungi as determined by SDS-gel electrophoresis (after Lin and Kolattukudy, 125).

Cutinase source	Molecular weight	
	Major proteins	Minor bands
F. solani f. sp. *pisi*, Cutinase I	22,000	
F. solani f. sp. *pisi*, Cutinase II	22,000	10,600 & 9,400
F. roseum culmorum	24,300	13,600 & 10,500
F. roseum sambucinum	24,800	13,800 & 12,700
U. consortiale	25,100	13,000 & 10,000
S. scabies	26,000	
H. sativum	26,300	

widespread similar agglutinins in cell walls are that influence disease resistance mechanisms.

Crude protein extracts of cell walls of several plants are capable of inhibiting a pathogen's polygalacturonases. However, it appears that they may have no role in the cultivar-specific resistance of plants to fungal infections. Perhaps they represent a general, nonspecific mechanism for resistance (68). The plant cell wall hyprich glycoproteins have been considered as disease resistance factors (64). As a consequence of fungal infections, host cell walls appear to accumulate proteins of this type that are glycosylated with arabinose and galactose. It is also possible that these glycoproteins exert a lectinlike activity, however the significance of this finding has not been established.

Vance et al. (200) have proposed that lignin precursors (phenolics) in plants can react with wall and membrane carbohydrates and proteins of pathogenic fungi, thereby activating a recognition system. The hypothesis implies that lignin precursors will complex more actively with chitinized fungal walls of *Ascomycetes* and *Basidiomycetes* than with nonchitinized walls of *Oomycetes*.

The deposition of lignin induced as a consequence of infection probably plays a role in the expressive phase of resistance (see Chap. 10). In other words, it is a secondary determinant of resistance. The primary recognition events in the determinative phase of resistance occur because protein- and carbohydrate-containing macromolecules on host and pathogen cell surfaces react with each other. This reaction leads to the agglutination, attachment, inhibition, or immobilization of the pathogen. The consequence of recognition may be, among others, the induction of lignification in the plant or in the pathogen. The induced lignification in several fungal diseases has been evaluated from the point of view of resistance by Vance et al. (200) and Ride (172).

Asada et al. (9), working with the Japanese radish (*Raphanus sativus*) infected by *Peronospora parasitica,* have shown that a lignification-inducing factor exists in the homogenate of diseased plants, which is formed also in response to injury. Another lignification phenomenon was revealed by Hammerschmidt and Kuć (95) in cucumber leaves with systemic acquired resistance. As a consequence of a prior inoculation with *Colletotrichum lagenarium,* the challenging fungus induces a rapid lignification of the epidermis. Lignification is localized around appressorium of the fungus, thus the development of *C. lagenarium* in lignified tissue is significantly restricted. Peroxidase activity is also enhanced in the resistant leaves, which may be significant because this enzyme is involved in the final step of lignin biosynthesis (cinnamyl alcohols → lignin). (See also Chaps. 6 and 10.)

Comparative Analysis of Disease Physiology—Cell Wall

Although virus pathogens do not penetrate plant cell walls directly, they are able to cause the plant cell to respond to their presence by callose deposition (wall thickening), lignification, and aromatization of the wall. The formation of paramural bodies and vesicles depicts a process for callose deposition between the plasmalemma and the plant cell wall. These phenomena suggest recognition by the plant cell and subsequent induction by viruses of metabolic processes that appear to be associated with virus limitation (localization). The poorly understood wall outgrowths, detected in plant cell walls associated with plasmodesmata, are caused by several isocosahedral and bacilliform viruses. These structures are indicative of stimulation of wall modifications by virus particles and, perhaps, by a specific protein conformation of the virus capsid.

The foregoing responses of the plant cell are also suggestive of transmembrane passage of signals from the virus to receptors on the plant cell wall, plasmalemma, or cytoplasmic membranes. In the hypersensitive host (local lesion), membrane permeability is increased and structural integrity is impaired.

Membrane vesiculation is a response of some plant cells to the presence of bacterial and fungal pathogens, as well as viruses. The formation of vesicles by the plasmalemma per se is indicative of a delivery system that is augmenting plant cell wall composition with cytoplasmic components. In instances where bacteria-induced hypersensitive reaction is in progress, it seems

that vesicles of cytoplasmic origin (perhaps Golgi) are carrying materials to the plasmalemma for release into periplasm (the space between plasmalemma and wall) or the wall itself by exocytosis.

It is apparent that both bacteria and fungi can rapidly degrade the plant cell wall in appropriate pathogen-host combinations. Degradation proceeds in a sequential fashion, and the order generally followed is cutin, pectin, hemicellulose, and cellulose. The appropriate enzymes appear to be synthesized as greater amounts of substrate are exposed. In addition, each in turn becomes crucial to the continuing process of wall degradation and host cell death. The essentiality of cutinase was clearly demonstrated by data that indicated that blockage of the enzyme precluded infection. Similar data are apparent for the pectinases and cellulases. Increasing evidence for the participation of arabinases and xylanases is being recorded. Evidence concerning the direct participation of proteases and phosphatidases in membrane disorientation remains meager.

There is good evidence that microenvironmental factors are crucial in determining the rate at which plant cell walls are degraded. For example, the level of oxygen, degree of hydration, and temperature are critical in the synthesis and activity of phenylalanine ammonia lyase, which, in turn, influences the activity of both pectin lyases and hydrolases. It would seem that the subject of catabolite repression and the role of CAMP in pectin lyase synthesis by bacterial pathogens needs additional research. The presence, synthesis, and solubility of aromatic and ionic enzyme inhibitors in the plant cell wall might also be profitably reexamined.

In relation to the control of wall-degrading enzyme synthesis by pathogens per se, it appears that incompatible pathogens are capable of causing plant wall cuticle loosening and derangement. These phenomena suggest the possible induction by the pathogens of synthesis, release, or activation of the plant's endogenous wall-altering enzymes.

Studies with mutants of *E. carotovora* have described isolates that were deficient in pectinase, cellulase, and phosphatidase. Revertants of some of these had even greater enzymatic activity than the parental strains. Studies with precisely selected mutants of bacterial and fungal pathogens might provide a better understanding of their essentiality in the cause of plant cell death by soft rot-inducing pathogens. Specifically, the role of proteases and phosphatidases must be more intensively examined. In this regard, the way in which "acid hydrolized wall fragments" cause plant cell death must also receive increased attention.

There is growing evidence that wall carbohydrates, particularly the rhamnogalacturonans, are the basis for specific recognition between the plant cell and the pathogen. However, the structural relationship between the cutin and suberin veneers and pectins on the plant cell surface remains almost totally obscured. Are there rifts in these cuticular mantles? Is there an orderly array of exposed pectin intersticed in the veneer? These are questions that will doubtlessly only be answered by microhisto- and microhistoimmuno-chemical analysis. Nevertheless, it would appear that the biochemical architecture of the plant wall projects the initial receptors for the pathogen's virulence-related ligands.

Additional Comments and Citations

Kirk, T. K. 1984. Degradation of lignin. In Microbial degradation of organic compounds. G. T. Gibson (ed.). Microbiology series vol. 13. Marcel Dekker, pp. 339–437. The review includes a concise and modern concept of composition and synthesis of the basic structure of lignin by oxidative coupling and hence a rationale for biodegradation. Assessment of microbial degradation establishes Basidiomycetes (white rot) as the most competent in this regard; secondary, slower, and less complete decomposition is effected by Ascomycetes and Fungi Imperfecti; thirdly, Basidiomycetes (brown rot) also alters lignin structure. Specifically stated is the consensus that bacteria do not play a primary role in degrading mature xylem tissues of woody plants. Enzymatic degradation involves demethylation followed by hydroxylation and degradation of the polymer, which occurs at both side chains and in aromatic nuclei that are oxidatively cleaved. Further erosion of ring cleavage fragments releases aliphatic products and frees new phenolic-hydroxylated units for sequential degradation. Lignin is degraded only at the surfaces accessible to fungal hyphae. Degradation of dimeric lignin model compounds revealed divergent cleavage processes between extracellular enzymes produced by *Fusarium solani* and *Phanaerochaete*.

References

1. Aist, J. R. 1976. Papillae and related wound plugs of plant cells. Ann. Rev. Phytopath. 14:145–63.

2. Aist, J. R., M. A. Waterman, and H. W. Israel. 1979. Papillae and penetration: Some problems, procedures and perspectives. In: Recognition and specificity in plant host-parasite interactions. J. M. Daly and I. Uritani (eds.). Jap. Sci. Soc. Press, Tokyo, pp. 85–97.

3. Alberghina, A., U. Mazzucchi, and P. Pupillo. 1973. On the effect of an endopolygalacturonate trans-eliminase on potato tissue: The influence of water potential. Phytopath. Z. 78:204–13.

4. Albersheim, P. 1975. The wall of growing plant cells. Sci. Amer. 232:80–95.

5. Albersheim, P., and U. Killias. 1963. Histochemical localization at the electron microscope level. Am. J. Bot. 50:732–45.

6. Albersheim, P., H. Neukom, and H. Deuel. 1960. Splitting of pectin chain molecules in neutral solutions. Arch. Biochem. Biophys. 90:46–51.

7. Albersheim, P., H. Neukom, and H. Deuel. 1960. Über die Bildung von ungesättigten Abbauprodukten durch ein pektinabauendes Enzym. Helvet. Chim. Acta 43:1422–26.

8. Anderson, W. P. 1972. Ion transport in the cells of higher plants. Ann. Rev. Plant Physiol. 23:51–72.

9. Asada, Y., T. Ohguchi, and I. Matsumoto. 1979. Induction of lignification in response to fungal infection. In: Recognition and specificity in plant host-parasite interactions. J. M. Daly and I. Uritani (eds.). Jap. Sci. Soc. Press, Tokyo, pp. 99–112.

10. Aspinall, G. O. 1970. Polysaccharides. Pergamon Press, New York.

11. Aspinall, G. O. 1973. Carbohydrate polymers of plant cell walls. In: Biogenesis of plant cell wall polysaccharides. F. Loewus (ed.). Academic Press, New York, pp. 95–116.

12. Aspinall, G. O., and I. W. Cotterell. 1971. Polysaccharides of soybeans. VI. Neutral polysaccharides from cotyledon meal. Can. J. Chem. 49:1019–22.

13. Aspinall, G. O., K. Hunt, and I. M. Morrison. 1967. Polysaccharides of soybeans. Part V. Acidic polysaccharides from the hulls. J. Chem. Soc. (c) 1080–86.

14. Basham, H. G., and D. F. Bateman. 1975. Killing of plant cells by pectic enzymes: The lack of direct injurious interaction between pectic enzymes or their soluble reaction products and plant cells. Phytopathology 65:141–53.

15. Basham, H. G., and D. F. Bateman. 1975. Relationship of cell death in plant tissue treated with a homogeneous endopectate lyase to cell wall degradation. Physiol. Plant Path. 5:249–62.

16. Bateman, D. F. 1963. The "macerating enzyme" of *Rhizoctonia solani*. Phytopathology 53:1178–86.

17. Bateman, D. F. 1964. An induced mechanism of tissue resistance to polygalacturonase in *Rhizoctonia*-infected hypocotyls of bean. Phytopathology 54:438–45.

18. Bateman, D. F. 1964. Cellulase and the *Rhizoctonia* disease of bean. Phytopathology 54:1372–77.

19. Bateman, D. F. 1967. Alteration of cell wall components during pathogenesis by *Rhizoctonia solani*. In: The dynamic role of molecular constituents in plant-parasite interaction. C. J. Mirocha and I. Uritani (eds.). Amer. Phytopath. Soc. St. Paul, MN, pp. 58–79.

20. Bateman, D. F. 1976. Plant cell wall hydrolysis by pathogens. In: Biochemical aspects of plant-parasite relationships. J. Friend and D. R. Threlfall (eds.). Academic Press, New York, pp. 79–103.

21. Bateman, D. F., and H. G. Basham. 1976. Degradation of plant cell walls and membranes by microbial enzymes. In: Physiological plant pathology. R. Heitefuss and P. H. Williams (eds.). Springer-Verlag, Berlin, pp. 315–55.

22. Bateman, D. F., H. D. Van Etten, H. D. English, D. J. Nevins, and P. Albersheim. 1969. Susceptibility to enzymatic degradation of cell walls from bean plants resistant and susceptible to *Rhizoctonia solani* Kuhn. Plant Physiol. 44:641–48.

23. Bauer, W. D., K. W. Talmadge, K. Keegstra, and P. Albersheim. 1973. The structure of plant cell walls. II. The hemicellulose of the walls of suspension-cultured sycamore cells. Plant Physiol. 51:174–87.

24. Beckman, C. H. 1971. The plasticizing of plant cell walls and tylose formation—a model. Physiol. Plant Path. 1:1–10.

25. Beraha, L., and E. D. Garber. 1971. Avirulence and extracellular enzymes of *Erwinia carotovora*. Phytopath. Z. 70:335–44.

26. Beraha, L., E. D. Garber, and B. A. Billeter. 1974. Enzyme profiles and virulence in mutants of *Erwinia carotovora*. Phytopath. Z. 81:15–22.

27. Bonner, J. 1950. Plant biochemistry. Academic Press, New York.

28. Braithwaite, C. W. D., and R. S. Dickey. 1970. Production of a basal soft rot symptom and maceration of carnation tissue by *Pseudomonas caryophylli* and *Corynebacterium* sp. Phytopathology 60:1040–45.

29. Braithwaite, C. W. D., and R. S. Dickey. 1971. Permeability alterations in detached carnation leaf tissue inoculated with *Pseudomonas caryophylli* and *Corynebacterium* sp. Phytopathology 61:317–21.

30. Braithwaite, C. W. D., and R. S. Dickey. 1971. Role of cellular permeability alterations and pectic and cellulolytic enzymes in maceration of carnation tissue by *Pseudomonas caryophylli* and *Corynebacterium* sp. Phytopathology 61:476–83.

31. Bretscher, M. 1973. Membrane structure: Some general principles. Science 181:622–29.

32. Brock, T. D. 1970. Biology of microorganisms. Prentice-Hall, Englewood Cliffs, NJ.

33. Brown, W. 1915. Studies in the physiology of parasitism. I. The action of *Botrytis cinerea*. Ann. Bot. 29:313–48.

34. Burke, D., P. Kaufman, M. McNeil, and P. Albersheim. 1974. The structure of plant cell walls. VI. A survey of the walls of suspension-cultured monocots. Plant Physiol. 54:109–15.

35. Byrde, R. J. W. 1963. Natural inhibitors of fungal enzymes and toxins in disease resistance. In: Perspectives of biochemical plant pathology. S. Rich (ed.). Conn. Agr. Exp. Sta. Bull. 663:31–47.

36. Byrde, R. J. W., A. H. Fielding, and A. H. Williams. 1959. The role of oxidized polyphenols in the varietal resistance of apple to brown rot. In: Phenolics in plants in health and disease. J. B. Pridham (ed.). Pergamon Press, New York, pp. 95–99.

37. Capaldi, R. A. 1974. A dynamic model of cell membranes. Sci. Amer. 230:27–33.

38. Clowes, F. A. L., and B. E. Juniper. 1968. Plant cells. Blackwell Sci. Publ., Oxford.

39. Cole, M. 1958. Oxidation products of leuco-anthocyanins as inhibitors of fungal polygalacturonase in rotting apple fruit. Nature 181:1596–97.

40. Cooper, R. M. 1982. Callose-deposit formation in radish root hairs. In: Cellulose and other natural polymer systems: Biogenesis, structure and degradation. R. M. Brown, Jr. (ed.). Plenum Press, New York, pp. 167–84.

41. Cooper, R. M. 1974. Cell wall degrading enzymes of vascular wilt fungi. Ph. D. diss., Imperial College, Univ. of London.

42. Cooper, R. M. 1977. Regulation of synthesis of cell-degrading enzymes of host-plant pathogen interactions. In: Cell wall biochemistry related to specificity in host-pathogen interactions. B. Solheim and J. Raa (eds.). Universitetsforlaget, Oslo, pp. 163-211.

43. Cooper, R. M., and R. K. S. Wood. 1973. Induction of synthesis of cell-wall degrading enzymes in vascular wilt fungi. Nature 246:309–11.

44. Cooper, R. M., and R. K. S. Wood. 1975. Regulation of synthesis of cell wall degrading enzymes by *Verticillium alboatrum* and *Fusarium oxysporum* f. sp. *lycopersici*. Physiol. Plant Path. 5:135–56.

45. Cooper, R. M., and R. K. S. Wood. 1980. Cell wall degrading enzymes of vascular wilt fungi. III. Possible involvement of endo-pectin lyase in *Verticillium* wilt of tomato. Physiol. Plant Path. 16:285–300.

46. Corden, M. E., and A. E. Dimond. 1959. The effect of growth regulating substances on disease resistance and plant growth. Phytopathology 49:68–72.

47. Cowling, E. B. 1965. Microorganisms and microbial enzyme systems as selective tools in wood anatomy. In: Cellular ultrastructure of woody plants. W. A. Cote (ed.). Syracuse Univ. Press, pp. 341–68.

48. Croteau, R., and P. E. Kolattukudy. 1974. Biosynthesis of hydroxy-fatty acid polymers. Enzymatic synthesis of cutin from monomer acids by cell free preparations from the epidermis of *Vicia faba* leaves. Biochemistry 13:3193–202.

49. Croteau, R., and P. E. Kolattukudy. 1974. Biosynthesis of pentahydroxystearic acid of cutin from linoleic acid in *Rosmarinus officinalis*. Arch. Biochem. Biophys. 162:458–70.

50. Danielli, J. F., and H. Davson. 1935. A contribution to the theory of permeability of thin films. J. Cell Comp. Physiol. 5:495–508.

51. Darvill, A. G., M. McNeil, and P. Albersheim. 1978. Structure of plant cell walls. VIII. A new pectic polysaccharide. Plant Physiol. 62:418–22.

52. Darvill, A. G., M. McNeil, P. Albersheim, and D. P. Delmer. 1980. The primary cell wall of flowering plants. In: The biochemistry of plants. P. K. Stumpf and E. E. Conn (eds.). Academic Press, New York, 1:91–162.

53. Darvill, J. E., M. McNeil, A. G. Darvill, and P. Albersheim. 1980. Structure of plant cell walls. XI. Glucuronoarabinoxylan, a second hemicellulose in the primary cell walls of suspension-cultured sycamore cells. Plant Physiol. 66:1135–39.

54. Dean, B. B., P. E. Kolattukudy, and R. W. Davis. 1977. Chemical composition and ultrastructure of suberin from hollow heart tissue of potato tubers. Plant Physiol. 59:1008–10.

55. Dean, M., and R. K. S. Wood. 1967. Cell wall degradation by a pectate transeliminase. Nature 214:408–10.

56. DeBary, A. 1863. Recherches sur le développement de quelques champignons parasites. Ann. Sci. Nat. Bot. Biol. Veg. 20:5–148.

57. Deese, D. C., and M. A. Stahmann. 1962. Pectic enzymes and cellulase formation by *Fusarium oxysporum* f. *cubense* on stem tissues from resistant and susceptible banana plants. Phytopathology 52:247–55.

58. Dickey, R. S., and P. E. Nelson. 1970. *Pseudomonas caryophylli* in carnation. IV. Unidentified bacteria isolated from carnation. Phytopathology 60:647–53.

59. Dimond, A. E. 1955. Pathogenesis in the wilt diseases. Ann. Rev. Plant Physiol. 6:329–50.

60. Dimond, A. E. 1970. Biophysics and biochemistry of the vascular wilt syndrome. Ann. Rev. Phytopath. 8:301–22.

61. Edgington, L. V., and J. C. Walker. 1958. Influence of calcium and boron nutrition on development of *Fusarium* wilt of tomato. Phytopathology 48:324–26.

62. Epstein, E. 1972. Mineral nutrition of plants: Principles and perspectives. Wiley, New York.

63. Esau, K. 1953. Plant anatomy. Wiley, New York.

64. Esquerré-Tugayé, M.-T., C. Lafitte, D. Mazau, A. Toppan, and A. Touzé. 1979. Cell surfaces in plant-microorganism interactions. II. Evidence for the accumulation of hydroxyproline-rich glycoproteins in the cell wall of diseased plants as a defense mechanism. Plant Physiol. 64:320–26.

65. Etherton, B., and G. J. Nuovo. 1974. Rapid changes in membrane potentials of out coleoptiles induced by amino acids and carbohydrates. Plant Physiol. 53:49. (Suppl.)

66. Fellows, H. 1928. Some chemical and morphological phenomena attending infection of the wheat plant by *Ophiobolus graminis*. J. Agric. Res. 37:647–61.

67. Fernando, M., and G. Stevenson. 1952. Studies on the physiology of parasitism. XVI. Effect of the condition of potato tissue as modified by temperature and water content upon attack by certain organisms and their pectinase enzymes. Ann. Bot. 16:103–14.

68. Fisher, M. L., A. J. Anderson, and P. Albersheim. 1973. Host-pathogen interactions. VI. A single plant protein effectively inhibits endopolygalacturonases secreted by *Colletotrichum lindemuthianum* and *Aspergillus niger*. Plant Physiol. 51:489–91.

69. Fiussello, N., J. Cerutiscurti, and R. Jodice. 1975. Ligninolytic activity of *Botrytis cinerea* and *Penicillium cyclopium* estimated by chromatographic methods. Allionia 20:103–7.

70. Fox, R. T. V. 1972. The ultrastructure of potato tubers injected by soft rot organism *Erwinia carotovora* var. *atroseptica*. In: Proc. Third Int. Conf. Plant Pathogenic Bact. H. P. Maas Geesteranus (ed.). Pudoc., Wageningen, Neth., pp. 95–120.

71. Frey-Wyssling, A. 1969. The ultrastructure and biogenesis of native cellulose. Fortschr. Chem. Org. Naturstoffe 27:1–30.

72. Frey-Wyssling, A., and E. Hausermann. 1960. Deutung der gestaltlosen der nektarien. Schweiz. Bot. Gesell. Ber. 70:150–62.

73. Friedman, B. A. 1962. Physiological differences between a virulent and a weakly virulent radiation-induced strain of *Erwinia carotovora*. Phytopathology 52:328–32.

74. Friedman, B. A. 1964. Carbon source and tetrazolium agar to distinguish virulence in colonies of *Erwinia carotovora*. Phytopathology 54:494–95.

75. Friedman, B. A., and M. J. Ceponis. 1959. Relationship of loss of pectolytic enzyme synthesis to avirulence of induced mutant strains of a soft rot bacterium. Phytopathology 49:227.

76. Friend, J. 1981. Plant phenolics, lignification and plant disease. Prog. Phytochem. 7:197–261. / Fuchs, A., J. A. Jobsen, and W. M. Wouts. 1965. Arabanases in phytopathogenic fungi. Nature 206:714–15.

77. Fry, S. C. 1983. Oxidative phenolic coupling reactions cross-link hydroxyproline-rich glycoprotein molecules in plant cell wall. In: Current topics in plant biochemistry and physiology. D. Randall, D. Blevins, R. Larson, and B. Rapp (eds.). Univ. of Missouri, Columbia, pp. 59–72.

78. Fushtey, S. G. 1957. Studies in the physiology of parasitism. XXIV. Further experiments on the killing of plant cells by fungal bacterial extracts. Ann. Bot. 21:273–86.

79. Gardner, K. H., and J. Blackwell. 1974. The structure of native cellulose. Biopolymers 13:1975–2001.

80. Garibaldi, A., and D. F. Bateman. 1971. Pectic enzymes produced by *Erwinia chrysanthemi* and their effects on plant tissue. Physiol. Plant Path. 1:25–40.

81. Goodman, R. N., P.-Y. Huang, and J. A. White. 1976. Ultrastructural evidence for immobilization of an incompatible bacterium, *Pseudomonas pisi* in tobacco leaf tissue. Phytopathology 66:754–64.

82. Gorter, E., and F. Grendel. 1925. On bimolecular layers of lipids on the chromocytes of the blood. J. Exp. Med. 41:439–43.

83. Gothoskar, S. S., R. P. Scheffer, J. C. Walker, and M. A. Stahmann. 1955. The role of enzymes in the development of *Fusarium* wilt of tomato. Phytopathology 45:381–87.

84. Goto, M., and N. Okabe. 1959. Studies on the cellulolytic enzymes of phylopathogenic bacteria. Part 4. On the nature of cellulose in the lesions of black rot of cauliflower (*Xanthomonas campestris*) and *Citrus* canker (*Xanthomonas citri*) Rept. Fac. Agr. Shizuoka Univ. 9:21–23.

85. Goto, M., and N. Okabe. 1960. Studies on the cellulolytic enzymes of phytopathogenic bacteria. Part 3. Some experiments on the mechanism of the decomposition of cellulose on its derivatives due to cellulases. Rept. Agr. Res. Branch Shizuoka Univ. Bull. 10:27–31.

86. Grant, G. T., E. R. Morris, D. A. Ries, P. Smith, and D. Thom. 1973. Biological interactions between polysaccharides and divalent cations. The egg-box model. FEBS Lett. 32:195–98.

87. Gregg, M. 1952. Studies on the physiology of paraitism. XVII. Enzyme secretion by strains of *Baterium carotovorum* and other pathogens in relation to parasitic vigor. Ann. Bot. 16:235–50.

88. Gritzali, M., and R. D. Brown, Jr. 1979. The cellulase system of *Trichoderma*: Relationship between purified extracellular enzymes from induced or cellulose-grown cells. In: Hydrolysis of cellulose: Mechanisms of enzymatic and acid catalysis. R. D. Brown and L. Jurasek (eds.). Adv. Chem. Series 181. Amer. Chem. Soc., Washington, DC, pp. 237–60.

89. Gross, G. G. 1978. Biosynthesis of lignin and related monomers. Rec. Adv. Phytochem. 11:141–84.

90. Gross, G. G., R. L. Mansell, and M. H. Zenk. 1975. Hydroxycinnamate: Coenzyme. A ligase from lignifying tissue of higher plants. Biochem. Physiol. Pflanzen 169:41–51.

91. Gupta, D. P., and J. B. Heale. 1971. Induction of cellulase (Cx) in *Verticillium albo-atrum*. J. Gen. Microbiol. 63:163–73.

92. Hall, D. M., and L. A. Donaldson. 1962. Secretion from pores of surface wax on plant leaves. Nature 194:1196.

93. Hall, D. M., and L. A. Donaldson. 1963. The ultrastructure of wax deposits on plant leaf surfaces. I. Growth of wax on leaves of *Trifolium repens*. J. Ultrastruc. Res. 9:259–67.

94. Hall, J. A., and R. K. S. Wood. 1973. The killing of plant cells by pectolytic enzymes. In: Fungal pathogenicity and the plant's response. R. J. W. Byrde and C. V. Cutting (eds.). Academic Press, London, pp. 19–38.

95. Hammerschmidt, R., and J. Kuć. 1982. Lignification as a mechanism for induced systemic resistance in cucumber. Physiol. Plant Path. 20:61–71.

96. Hancock, J. G., and R. L. Miller. 1965. Relative importance of polygalacturonate trans-eliminase and other pectolytic enzymes in southern anthracnose, spring black stem, and *Stemphylium* leaf spot of alfalfa. Phytopathology 55:346–55.

97. Hancock, J. G., and R. L. Miller. 1965. Association of cellulolytic, proteolytic, and xylolytic enzymes with southern anthracnose, spring black stem, and *Stemphylium* leaf spot of alfalfa. Phytopathology 55:356–60.

98. Hatta, T., and R. Ushiyama. 1973. Mitochondrial vesiculation associated with cucumber green mottle mosaic virus-infected plants. J. Gen. Virol. 21:9–17.

99. Higginbotham, N., and W. P. Anderson. 1974. Electrogenic pumps in higher plant cells. Can. J. Bot. 52:1011–21.

100. Higuchi, T. 1977. Lignin structure and morphological distribution in plant cell walls. In: Lignin biodegradation: Microbiology and potential applications. T. Kirk, T. Higuchi, and H.-M. Chang (eds.). CRC Press, Hollywood, FL, 1:1–19.

101. Husain, A., and A. E. Dimond. 1960. Role of cellulolytic enzymes in pathogenesis by *Fusarium oxysporum* f. *lycopersici*. Phytopathology 50:329–31.

102. Jeffree, C. E., E. A. Baker, and P. J. Holloway. 1976. Origins of the fine structure of plant epicuticular waxes. In: Microbiology of aerial plant surfaces. C. H. Dickinson and T. F. Preece (eds.). Academic Press, New York, pp. 119–58.

103. Juniper, B. E. 1959. The surfaces of peanuts. Endeavour 18:20–25.

104. Karr, A. L., and P. Albersheim. 1970. Polysaccharide degrading enzymes are unable to attack plant cell walls without prior action by a "wall-modifying enzyme." Plant Physiol. 46:69–80.

105. Keegstra, K., K. W. Talmadge, W. D. Bauer, and P. Albersheim. 1973. The structure of plant cell walls. III. A model of the walls of suspension-cultured sycamore cells based on the interconnections of the macromolecular components. Plant Physiol. 51:188–96.

106. Kelman, A., and E. B. Cowling. 1965. Cellulase of *Pseudomonas solanacearum* in relation to pathogenesis. Phytopathology 55:148–55.

107. Kelman, A., and L. Sequeira. 1965. Root to root spread of *Pseudomonas solanacearum* in relation to pathogenesis. Phytopathology 55:304–9.

108. King, K. W. 1966. Enzymic degradation of crystalline hydrocellulose. Biochem. Biophys. Res. Comm. 24:295–98.

109. King, N. J., and D. B. Fuller. 1968. The xylanase system of *Coniophora cerebella*. Biochem. J. 108:571–76.

110. Knee, M., and J. Friend. 1968. Extracellular "galactanase" activity from *Phytophthora infestans* (Mont.) de Bary. Phytochemistry 7:1289–91.

111. Kojima, M., K. Kawakita, and I. Uritani. 1982. Studies

on a factor in sweet potato root which agglutinates spores of *Ceratocystis fimbriata*, black rot fungus. Plant Physiol. 69:474–78.

112. Kolattukudy, P. E. 1974. Biosynthesis of a hydroxy fatty acid polymer, cutin. Identification and biosynthesis of 16-oxo–9 or 10 hydroxy-palmitic acid, a novel compound in *Vicia faba*. Biochemistry 13:1354–63.

113. Kolattukudy, P. E. 1980. Cutin, suberin, and waxes. In: The biochemistry of plants. P. K. Stumpf (ed.). Academic Press, New York, 4:571–645.

114. Kolattukudy, P. E., and V. P. Agrawal. 1974. Structure and composition of aliphatic constituents of potato tuber skin (suberin). Lipids 9:682–91.

115. Kolattukudy, P. E., R. Croteau, and L. Brown. 1974. Structure and biosynthesis of cuticular lipids. Hydroxylation of palmitic acid and decarboxylation of C_{28}, C_{30}, and C_{32} acids in *Vicia faba* flowers. Plant Physiol. 54:670–77.

116. Kooiman, P. 1961. The constitution of tamarindusamyloid. Recueil Trav. Chim Pays-Bas 80:849–65.

117. Kraght, A. J., and M. P. Starr. 1953. Pectic enzymes of *Erwinia carotovora*. Arch. Biochem. Biophys. 42:271–77.

118. Lamport, D. T. A. 1967. Hydroxyproline-O-glycosidic linkage of the plant cell wall glycoprotein extensin. Nature 216:1322–24.

119. Lamport, D. T. A. 1973. The glycopeptide linkages of extensin: O-D-galactosyl serine and O-L-arabinosyl hydroxyproline. In: Biogenesis of plant cell wall polysaccharides. F. Loewus (ed.). Academic Press, New York, pp. 149–64.

120. Lamport, D. T. A. 1980. Structure and function of plant glycoproteins. In: Biochemistry of plants. J. Preiss (ed.). Academic Press, New York, 3:501–41.

121. Lamport, D. T. A., and L. Epstein. 1983. A new model for the primary cell wall: A concatinated extensin—cellulose network. In: Current topics in plant biochemistry and physiology. D. Randall, D. Blevins, R. Larson, and B. Rapp (eds.). Univ. of Missouri, Columbia, pp. 73–83.

122. Lapwood, D. H. 1957. Studies on the physiology of parasitism. XXIII. On the parasitic vigor of certain bacteria in relation to their capacity to secrete pectolytic enzymes. Ann. Bot. 21:167–84.

123. Leatham, G. F., V. King, and M. A. Stahmann. 1980. In vitro protein polymerization by quinones or free radicals generated by plant or fungal oxidative enzymes. Phytopathology 70:1134–40.

124. Levi, M. P., and R. D. Preston. 1966. A chemical and microscopic examination of the action of the soft-rot fungus *Chaetomium globosum* on beechwood (*Fagus sylv.*). Holzforschung 19:183–90.

125. Lin, T. S., and P. E. Kolattukudy. 1980. Isolation and characterization of a cuticular polyester (cutin) hydrolyzing enzyme from phytopathogenic fungi. Physiol. Plant Path. 17:1–15.

126. Lund, B. M., and L. W. Mapson. 1970. Stimulation by *Erwinia carotovora* of the synthesis of ethylene in cauliflower tissue. Biochem. J. 119:251–63.

127. Lund, B. M. 1973. The effect of certain bacteria on ethylene production by plant tissue. In: Fungal pathogenicity and the plant's response. R. J. W. Byrde and C. V. Cutting (eds.). Academic Press, New York, pp. 69–86.

128. MacRobbie, E. A. C. 1973. Vascular ion transport. In: Ion transport in plants. W. P. Anderson (ed.). Academic Press, New York, pp. 431–446.

129. Mäder, M., and T. Fusl. 1982. Role of peroxidase in lignification of tobacco cells. II. Regulation of phenolic compounds. Plant Physiol. 70:1132–34.

130. Maino, A. L., M. N. Schroth, and N. J. Palleroni. 1974. Degradation of xylan by bacterial plant pathogens. Phytopathology 64:881–85.

131. Maitai, I. B., and P. E. Kolattukudy. 1979. Prevention of fungal infection of plants by specific inhibition of cutinase. Science 205:507–8.

132. Mandels, M., and E. T. Reese. 1960. Induction of cellulase in fungi by cellulose. J. Bact. 79:816–26.

133. Mann, B. 1962. Role of pectic enzymes in the *Fusarium* wilt syndrome of tomato. Trans. Brit. Mycol. Soc. 45:169–78.

134. Markewalder, H. K., and H. Neukom. 1976. Diferulic acid as a possible crosslink in hemicelluloses from wheat endosperm. Phytochemistry 15:836–37.

135. Marrè, E. 1980. Mechanism of action of phytotoxins affecting plasmalemma functions. Prog. Phytochem. 6:253–84.

136. Martelli, G. P. 1980. Ultrastructural aspects of possible defence reactions in virus-infected plant cells. Microbiologica 3:369–91.

137. Martelli, G. P., and M. Russo. 1977. Plant virus inclusion bodies. Adv. Virus Res. 21:175–266.

138. Martin, J. T., and B. E. Juniper. 1970. The cuticles of plants. St. Martin's Press, New York.

139. Marx-Figini, M. 1982. The control of molecular weight and molecular weight distribution in the biogenesis of cellulose. In: Cellulose and other natural polymer systems: Biogenesis, structure and degradation. R. M. Brown, Jr. (ed.). Plenum Press, New York, pp. 243–71.

140. Matthews, R. E. F. 1981. Plant Virology. 2d. ed. Academic Press, New York.

141. Mazliak, P., and J. C. Kader. 1980. Phospholipid exchange systems. In: The biochemistry of plants. Lipids: Structure and function. P. K. Stumpf (ed.). Academic Press, New York, 4:283–300.

142. McDonnell, K. 1962. Relationship of pectic enzymes and pathogenicity in the *Fusarium* wilt of tomatoes. Trans. Brit. Mycol. Soc. 45:55–62.

143. McNeil, M., A. G. Darvill, and P. Albersheim. 1978. The structural polymers of the primary cell walls of dicots. In: Progress in the chemistry of organic natural products W. Herz, H. Griesebach, and G. W. Kirby (eds.). Springer-Verlag, Vienna, 37:192–249.

144. McNeil, M., A. G. Darvill, and P. Albersheim. 1980. Structure of plant cell walls. X. Rhamnogalacturonan I, a structurally complex pectic polysaccharide in the walls of suspension-cultured sycamore cells. Plant Physiol. 66:1128–34.

145. McNeil, M., A. G. Darvill, and P. Albersheim. 1982. Structure of plant cell walls. XII. Identification of seven differently linked glycosyl residues attached to O–4 of the 2–4 linked rhamnosyl residues of rhamnogalacturonan I. Plant Physiol. 70:1586–91.

146. Montezinos, D. 1982. A cytological model of cellulose biogenesis in the alga *Oocystis apiculata*. In: Cellulose and other natural polymer systems: Biogenesis, structure and degradation. R. M. Brown, Jr. (ed.). Plenum Press, New York, pp. 3–21.

147. Mount, M. S. 1979. Regulation of endopolygalacturonate transeliminase in an adenosine 3', 5'-cyclic monophosphate-deficient mutant of *Erwinia carotovora*. Phytopathology 69:117–20.

148. Mount, M. S., D. F. Bateman, and H. G. Basham. 1970. Induction of electrolyte loss, tissue maceration, and cellular death of potato tissue by an endopolygalacturonate trans-eliminase. Phytopathology 60:924–31.

149. Murant, A. F., I. M. Roberts, and A. M. Hutcheson. 1975. Effects of parsnip yellow fleck virus on plant cells. J. Gen. Virol. 26:277–85.

150. Mussell, H. W., and L. L. Strand. 1974. Symptoms generated in intact cotton plants by purified endopolygalacturonases. Proc. Amer. Phytopath. Soc. 1:78.

151. Mussell, H. W., and L. L. Strand. 1977. Pectic enzymes: Involvement in pathogenesis and possible relevance to tolerance

and specificity. In: Cell wall biochemistry related to specificity in host-plant pathogen interactions. B. Solheim and J. Raa (eds.). Universitetsforlaget, Oslo, pp. 31–70.

152. Nasuno, S., and M. P. Starr. 1967. Polygalacturonic acid trans-eliminase of *Xanthomonas campestris*. Biochem. J. 104:178–85.

153. Norkrans, B. 1963. Degradation of cellulose. Ann. Rev. Phytopath. 1:325–50.

154. Northcote, D. H.1970. The synthesis and metabolic control of polysaccharides and lignin during the differentiation of plant cells. Essays Biochem. 6:89–137.

155. Northcote, D. H. 1972. Chemistry of the plant cell wall. Ann. Rev. Plant Physiol. 23:113–32.

156. Novacky, A. 1980. Disease-related alteration in membrane function.In: Plant membrane transport: Current and conceptual issues. R. Spanswick, W. Lucas, and J. Dainty (eds.). Elsevier/North Holland Biomedical, Amsterdam, pp. 369–80.

157. Okamoto, K., C. Hatanaka, and J. Ozawa. 1964. A saccharifying pectate *trans-* eliminase of *Erwinia aroideae*. Agr. Biol. Chem. (Tokyo) 28:331–36.

158. Panopoulos, N. J., and M. N. Schroth. 1974. Role of flagellar motility in the invasion of bean leaves by *Pseudomonas phaseolicola*. Phytopathology 64:1389- 97.

159. Pearce, R. B., and J. P. Ride. 1978. Elicitors of the lignification response of wheat. Ann. Appl. Biol. 89:304–7.

160. Petterson, G., E. B. Cowling, and J. Porath. 1963. Studies on cellulolytic enzymes: I. Isolation of low-molecular-weight cellulase from *Polyporus versicolor*. Biochem. Biopohys. Acta 67:1–8.

161. Petterson, G., and J. Porath. 1963. Studies on cellulolytic enzymes: II. Multiplicity of cellulolytic enzymes of *Polyporus versicolor*. Biochem. Biophys. Acta 67:9–15.

162. Preston, R. D. 1974. Physical biology of plant cell walls. Chapman and Hall, London.

163. Puhalla, J. E., and C. R. Howell. 1975. Significance of endopolygalacturonase activity to symptom expression of *Verticillium* wilt in cotton assessed by the use of mutants of *Verticillium dahliae*. Physiol. Plant Path. 7:147–52.

164. Pupillo, P. 1976. Pectic lyase isozymes produced by *Erwinia chrysanthemi* Burkh. et al. in polypectate broth or in Dieffenbachia leaves. Physiol. Plant Path. 9:113–20.

165. Purdy, R. E., and P. E. Kolattukudy. 1973. Depolymerization of a hydroxy fatty acid biopolymer by an extracellular enzyme from *Fusarium solani* f. *pisi*: Isolation and some properties of the enzyme. Arch. Biochem. Biophys. 159:61- 69.

166. Purdy, R. E., and P. E. Kolattukudy. 1975. Hydrolysis of plant cuticle by plant pathogens. Purification, amino acid composition, and molecular weight of two isozymes of cutinase and a nonspecific esterase from *Fusarium solani* f. *pisi*. Biochem. 14:2824–31.

167. Purdy, R. E., and P. E. Kolattukudy. 1975. Hydrolysis of plant cuticle by plant pathogens. Properties of cutinase I, cutinase II, and a nonspecific esterase isolated from *Fusarium solani* f. *pisi*. Biochem. 14:2832–40.

168. Rautela, G., and K. W. King. 1968. Significance of the crystal structure of cellulose in the production and action of cellulase. Arch. Biochem. Biophys. 123:589–601.

169. Raven, J. A., and F. A. Smith. 1976. Cytoplasmic pH regulation and electrogenic H⁺ extrusion. Curr. Adv. Plant Sci. 8:649–50.

170. Reddy, M. N., D. L. Stuteville, and E. L. Sorensen. 1974. Xylanase activity of *Xanthomonas alfalfae* in culture and during pathogenesis of bacterial leaf spot of alfalfa. Phytopathology 80:215–23.

171. Reese, E. T., R. G. H. Siu, and H. Levinson. 1966. The biological degradation of soluble cellulose derivatives and its relationship to the mechanism of cellulose hydrolysis. J. Bact. 59:485–97.

172. Ride, J. P. 1983. Cell walls and other structural barriers in defence. In: Biochemical plant pathology. J. A. Callow (ed.). Wiley, London, pp. 215–36.

173. Riley, R. G., and P. E. Kolattukudy. 1975. Evidence for covalently attached p-coumaric acid and ferulic acid in cutins and suberins. Plant Physiol. 56:650–54.

174. Roseman, S. 1975. Sugars of the cell membranes. Hosp. Prac. 61:70.

175. Rubio-Huertos, M. 1972. Inclusion bodies. In: Principles and techniques in plant virology. C. I. Kado and H. O. Agrawal (eds.). Van Nostrand/Reinhold, New York, pp. 62–75.

176. Ryan, C. A., P. Bishop, G. Pearce, A. G. Darvill, M. McNeil, and P. Albersheim. 1981. A sycamore cell wall polysaccharide and a chemically related tomato leaf polysaccharide possess similar proteinase inhibitor-inducing activities. Plant Physiol. 68:616–18.

177. Scott, F. M. 1950. Internal suberization of tissues. Bot. Gaz. 111:378–94.

178. Scott, F. M., B. G. Bystrom, and E. Bowler. 1963. Root hair, cuticle and pits. Science 140:63–64.

179. Sequeira, L., G. Gaard, and G. A. de Zoeten. 1977. Interaction of bacteria and host cell walls: Its relation to mechanisms of induced resistance. Physiol. Plant Path. 10:43–50.

180. Shaykh, M., C. Soliday, and P. E. Kolattukudy. 1977. Proof for the production of cutinase by *Fusarium solani* f. *pisi* during penetration into its host *Pisum sativum*. Plant Physiol. 60:170–72.

181. Simpson, M. E., and P. B. Marsh. 1964. Cellulose decomposition by the *Aspergilli*. U.S. Dept. Agric. Tech. Bull. 1303.

182. Slayman, C. L., and C. W. Slayman. 1974. Depolarization of the plasmamembrane of *Neurospora* during active transport of glucose: Evidence for a proton-dependent co-transport system. Proc. Nat. Acad. Sci. 71:1935–39.

183. Soliday, C. L., P. E. Kolattukudy, and R. W. Davis. 1979. Chemical and ultrastructural evidence that waxes associated with the suberin polymer constitute the major tuber (*Solanum tuberosum* L.). Planta 146:607–14.

184. Spellman, M., M. McNeil, A. Darvill, P. Albersheim, and A. Dell. 1983. Characterization of structurally complex heptasaccharide from the pectic polysaccharide rhamnogalacturonan II. Carbohyd. Res. 122:131–53.

185. Spellman, M., M. McNeil, A. Darvill, P. Albersheim, and Kim Henrick. 1983. Isolation and characterization of 3-C-carboxy–5-deoxy-L-xylose, a naturally occurring branched chain acidic monosaccharide. Carbohyd. Res. 122:115–29.

186. Srivastava, D. N., E. Echandi, and J. C. Walker. 1959. Pectolytic and cellulolytic enzymes produced by *Rhizopus stolonifer*. Phytopathology 49:145–48.

187. Stafford, H. A. 1974. Possible multienzyme complexes regulating the formation of C_6-C_3 phenolic compounds and lignins in higher plants. Rec. Adv. Phytochem. 8:53–79.

188. Stahmann, M. A. 1965. The biochemistry of proteins of the host and parasite in some plant diseases. In: Biochemische Probleme der kranken Pflanze. Tagungsberichte Nr. 74. Aschersleben, G. D. R., pp. 9–40.

189. Starr, M. P. 1961. Pectin and galacturonate metabolism of phytopathogenic bacteria. Rec. Adv. Bot. 7:632–35. Univ. of Toronto Press.

190. Starr, M. P., and F. Moran. 1962. Eliminative split of pectic substances by phytopathogenic soft-rot bacteria. Science 135:920–21.

191. Stephens, G. J., and R. K. S. Wood. 1975. Killing of protoplasts by soft-rot bacteria. Physiol. Plant Path. 5:165–81.

192. Strobel, G. A. 1963. A xylanase system produced by *Diplodia viticola*. Phytopathology 53:592–96.

193. Sugimura, Y., and R. Ushiyama. 1975. Cucumber green mottle mosaic virus infection and its bearing on cytological alterations in tobacco mesophyll protoplasts. J. Gen. Virol. 29:93–98.

194. Swinburne, T. R., and M. E. Corden. 1969. A comparison of the polygalacturonases produced in vivo and in vitro by *Penicillium expansum* Thom. J. Gen. Microbiol. 55:75–87.

195. Talmadge, K. W., K. Keegstra, W. D. Bauer, and P. Albersheim. 1973. The structure of plant cell walls. I. The macromolecular components of the walls of suspension-cultured sycamore cells with a detailed analysis of the pectic polysaccharides. Plant Physiol. 51:158–73.

196. Tribe, H. T. 1955. Studies in the physiology of parsitism. XIX. On the killing of plant cells by enzymes from *Botrytis cinerea* and *Bacterium aroideae*. Ann. Bot. 19:351–68.

197. Tseng, T. C., and M. S. Mount. 1974. Toxicity of endopolygalacturonate transeliminase, phosphatidase and protease to potato and cucumber tissue. Phytopathology 64:229–36.

198. Turner, M. T., and D. F. Bateman. 1968. Maceration of plant tissues susceptible and resistant to soft-rot pathogens by enzymes from compatible host-pathogen combinations. Phytopathology 58:1509–15.

199. Valent, B. S., and P. Albersheim. 1974. The structure of plant cell walls. V. On the binding of xyloglucan to cellulose fibers. Plant Physiol. 54:105–8.

200. Vance, C. P., T. K. Kirk, and R. T. Sherwood. 1980. Lignification as a mechanism of disease resistance. Ann. Rev. Phytopath. 18:259–88.

201. Vance, C. P., and R. T. Sherwood. 1976. Regulation of lignin formation in reed canarygrass in relation to disease resistance. Plant Physiol. 57:915–19.

202. Vance, C. P., and R. T. Sherwood. 1976. Cycloheximide treatments implicate papilla formation in resistance of reed canarygrass to fungi. Phytopathology 66:498–502.

203. Van den Ende, G., and H. F. Linskens. 1974. Cutinolytic enzymes in relation to pathogenesis. Ann. Rev. Phytopath. 12:247–58.

204. Van der Molen, G. E., C. H. Beckman, and E. Rodehurst. 1977. Vascular gelation: A general response following infection. Physiol. Plant Path. 11:95–100.

205. Van Fleet, D. S. 1961. Histochemistry and function of the endodermis. Bot. Rev. 27:165–220.

206. Van Sumere, C. F., C. Van Sumere-de Preter, and G. A. Ledingham. Cell wall- splitting enzymes of *Puccinia graminis* var. *tritici*. Can. J. Microbiol. 3:761–70.

207. Varner, J. E. 1983. Hydroxyproline-rich glycoprotein of the plant cell wall (HPRGPCW). In: Current topics in plant biochemistry and physiology. D. Randall, D. Blevins, R. Larson, and B. Rapp (eds.). Univ. of Missouri, pp. 54–58.

208. Vian, B. 1982. Organized microfibril assembly in higher plant cells. In: Cellulose and other natural polymer systems: Biogenesis, structure and degradation. R. M. Brown, Jr. (ed.). Plenum Press, New York, pp. 23–43.

209. Waggoner, P. E., and A. E. Dimond. 1955. Production and role of extracellular pectic enzymes of *Fusarium oxysporum* f. sp. *lycopersici*. Phytopathology 45:79–87.

210. Wallace, J., J. Kuć, and H. M. Draudt. 1962. Biochemical changes in the water-insoluble material of maturing apple fruit and their possible relationship to disease resistance. Phytopathology 52:1023–27.

211. Wallace, J., J. Kuć, and E. B. Williams. 1962. Production of extracellular enzymes by four pathogens of apple fruits. Phytopathology 62:1004–9.

212. Weinhold, A. R., and T. Bowman. 1974. Repression of virulence in *Rhizoctonia solani* by glucose and 3–0-methyl glucose. Phytopathology 64:985–90.

213. Weinstein, L. T., and P. Albersheim. 1979. Purification and partial characterization of a wall-degrading endoarabanase and an arabinosidase from *Bacillus subtilis*. Plant Physiol. 63:425–32.

214. Wheeler, H. 1974. Cell wall and plasmalemma modifications in diseased and injured plant tissues. Can. J. Bot. 52:1005–9.

215. White, A. R. 1982. Visualization of cellulases and cellulose degradation. In: Cellulose and other natural polymer systems: Biogenesis, structure and degradation. R. M. Brown, Jr. (ed.). Plenum Press, New York, pp. 489–509.

216. Winstead, N. N., and C. L. McCombs. 1961. Pectinolytic and cellulolytic enzyme production by *Pythium aphanidermatum*. Phytopathology 51:270–73.

217. Wood, R. K. S. 1951. Pectic enzymes produced by *Bacterium aroideae*. Nature 167:771.

218. Wood, R. K. S. 1972. The killing of plant cells by soft rot parasites. In: Phytotoxins in plant diseases. R. K. S. Wood, A. Ballio, and A. Graniti (eds.). Academic Press, London, pp. 273–88.

219. Wallis, F. M., and S. J. Truter. 1978. Histopathology of tomato plants infected with *Pseudomonas solanacearum*, with emphasis on ultrastructure. Physiol. Plant Path. 13:307–17.

220. Wallis, R. M., F. H. J. Rijkenberg, J. J. Joubert, and M. M. Martin. 1973. Ultrastructural histopathology of cabbage leaves infected with *Xanthomonas campestris*. Physiol. Plant Path. 3:371–78.

221. Yamazaki, N., S. Fry, A. Darvill, and P. Albersheim. 1983. Host pathogen interactions. XXIV. Fragments isolated from suspension-cultured sycamore cell walls inhibit the ability of the cells to incorporate ^{14}C leucine into proteins. Plant Physiol. 72:864–69.

222. Zucker, M., and L. Hankin. 1970. Physiological basis for a cycloheximide-induced soft rot of potatoes by *Pseudomonas fluorescens*. Ann. Bot. 34:1047–62.

223. Zucker, M., L. Hankin, and D. Sands. 1972. Factors governing pectate lyase synthesis in soft rot and non-soft rot bacteria. Physiol. Plant Path. 2:59–67.

5

Nucleic Acid and Protein Metabolism

The Healthy Plant

Since the first edition of this book, the increase in our knowledge of the structure and biosynthesis of nucleic acids and proteins, in both prokaryotic and eukaryotic cells, has been spectacular. A detailed treatment is clearly beyond the scope of the present volume; there are many books and reviews to which the reader may turn for further information (e.g. 34, 115, 179, 258, 272, 275, 374).

Nucleic Acids

DNA. Deoxyribonucleic acids consist of the four nucleosides—adenosine, guanosine, cytidine, and thymidine (A, G, C and T)—linked between 3'- and 5'-hydroxyl groups via phosphodiester bonds, in a sequence characteristic of each DNA species (Fig. 5–1). With the exception of those viruses whose genome consists of a single DNA strand (ssDNA), native DNA consists of two strands, in the form of a double helix; the two strands are anti-parallel and may be linear, with the 5'-terminus of one adjacent to the 3'-terminus of the second (Fig. 5–2), or circular. In eukaryotic cells, DNA is present not only in the nucleus but also in mitochondria and chloroplasts. However, the size of these DNAs, and consequently their information content, is considerably different.

Nuclear, or chromosomal, DNA (e.g. 442, 465) is associated with several very basic proteins (the histones), several nonhistone proteins, and enzymes involved in DNA and RNA synthesis, together with RNA. It forms a highly organized, compact structure, the chromatin. Haploid plant cells, containing one complete set of chromosomes, may contain 100 pg or more DNA in this form, although DNA content varies considerably, both within and between species. By comparison, an "average" bacterium may contain 0.007 pg and animal cells in the order of 1–5 pg. If all this DNA contained within the plant cell nucleus consisted of unique sequences, able to be transcribed into RNA, then the cell would contain far more information than is necessary. Clearly, this is not the case. Experiments in which the two DNA strands are separated and then allowed to reassociate, or hybridize, suggest that some sequences are unique, appearing only once in the total DNA complement of the cell; others appear many times. The former, referred to as *single-copy* or *non-*

repetitive DNA, contains almost all the sequences that are transcribed into m(messenger)RNA, which is then translated into protein. The amount of this class of DNA does not, however, seem to be determined logically. Broad bean, for example, has about seven times as much single-copy DNA as mung bean (Table 5–1). The length of single-copy sequences varies greatly among species, and plants can be broadly divided into two groups on the basis of the predominant lengths of these sequences, below or greater than ca. 1,500 nucleotide pairs. Approximately 45% of the nuclear DNA of tobacco is single copy. This would contain information for up to 10^6 polypeptides of average size, far more than the tobacco cell needs. Consequently, only a fraction of this unique DNA is transcribed into mRNA. There is evidence that in tobacco, ca. 11% of single-copy DNA, representing some 60,000 genes, may be expressed during the plant's life cycle. This would constitute about 4.6% of the plant's genome. Much more than this may, however, be transcribed, although it does not leave the nucleus and clearly does not function as mRNA. The nuclear RNA, consisting of transcripts that are processed to mRNA and those that are not, is collectively referred to as heterogeneous nuclear RNA (HnRNA).

The other type of DNA, consisting of multiple copies of the same or closely related sequences, can be subdivided into classes, depending on the frequency with which the sequences are repeated. *Satellite* DNA is a form of this *repetitive* DNA that can sometimes be physically separated from the major fraction, or "main-band," DNA by equilibrium centrifugation in neutral CsCl gradients, and which usually constitutes some 3–30% of the total. Cucumber is exceptional, containing 44% satellite. Hybridization experiments suggest that there are usually at least two types of DNA within the satellite band, with much longer repeating sequences than the short sequences usually found in mammalian satellites. These sequences are repeated thousands of times. The function of satellite DNA remains unclear. It is not chloroplast DNA, since this usually has a buoyant density in neutral CsCl more comparable to that of main-band nuclear DNA, nor is it of mitochondrial origin, since this usually constitutes no more than 1% of the total cell DNA. It is not translated into mRNA but rather may be involved in, for example, maintaining chromosome structure.

In addition to satellite, there are other classes of *re-*

150

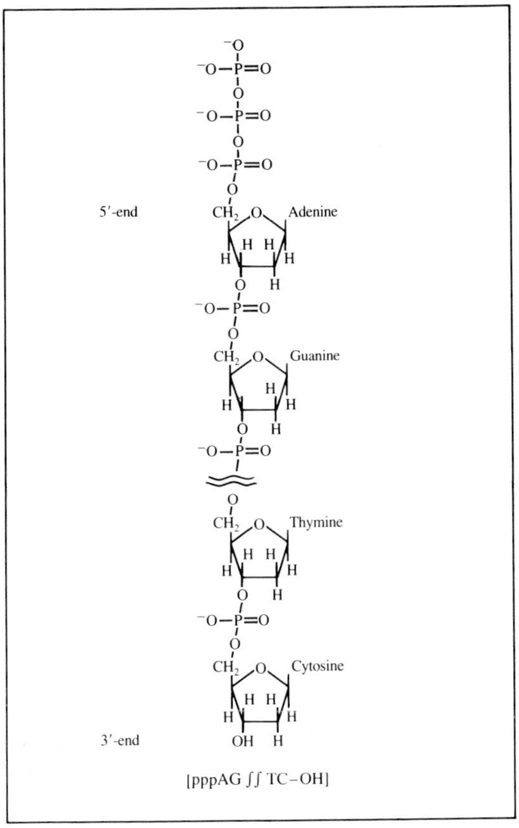

Fig. 5–1. Representation of a single strand of DNA, together with abbreviated notation.

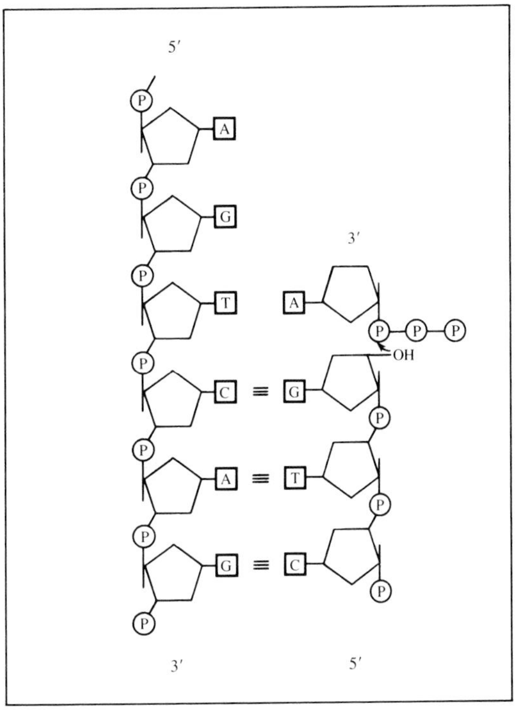

Fig. 5–2. Representation of a DNA duplex, with one strand (primer) being extended in the $5' \rightarrow 3'$ direction, using the other strand as a template.

petitive DNA that may be interspersed with single-copy sequences. The majority of nuclear DNA is of this type. Repeated units range from quite short sequences of 200–300 nucleotide pairs to units several thousands of nucleotide pairs in length, and consist of families of often widely different sequences. Again, the function of this type of DNA is unknown. Some may be genes for rRNA, 5sRNA or tRNA. Other sequences may perform a regulatory function for single-copy DNA that is transcribed into mRNA, though this type would only constitute a minor fraction.

Both *chloroplast* and *mitochondrial* DNAs (106, 439) exist predominantly as double-stranded, circular, supercoiled molecules, with sizes indicated in Table 5–2. Chloroplast DNA can usually be distinguished from "main-band" nuclear DNA by equilibrium centrifugation in neutral CsCl. The buoyant density of mitochondrial DNA is, however, often close to that of nuclear DNA (Table 5–3), so that resolution by this technique may be difficult. Lability of some chloroplast DNAs suggests that they contain some ribonucleotides. In pea, for example, there may be some 18 ribonucleotides distributed around the circular DNA. Although formerly thought of as distinct molecules with unique sequences, it is of interest that the mitochondrial and chloroplast

Table 5–1. Single-copy DNA.

Source	(Nucleotide pairs) $\times 10^9$ in haploid DNA	Single-copy DNA	
		Percentage	(nucleotide pairs) $\times 10^{-9}$
Vigna radiata (mung bean)	0.5	70	0.4
Vicia faba (broad bean)	13.0	20	2.6
Nicotiana tabacum	2.0	45	0.9

Table 5–2. Organelle DNA.

Species	MW ($\times 10^{-6}$)	Length (μm)	Copies/organelle
Mitochondrial	60–90	ca. 35–40	ca. 5
Chloroplast	70–100	ca. 40–60	10–100

Table 5–3. Buoyant densities of plant cell DNA in CsCl (g.cm.$^{-3}$).

Species	Nuclear	Chloroplast	Mitochondrial	Reference
Pea	1.695	1.698	1.705	240, 241, 242
Tobacco	1.697	1.700	—	440
Cucumber	1.692	1.701	1.706	223, 462

DNAs of maize have a sequence of 12,000 nucleotides in common (424). There is also extensive homology between chloroplast DNAs from different species (30).

DNA biosynthesis in both prokaryotic and eukaryotic cells requires the four ribonucleoside triphosphates (ATP, CTP, GTP and UTP) in addition to the four deoxyribonucleoside triphosphates (dATP, dCTP, dGTP and dTTP), and although participation in nucleic acid biosynthesis is the principal function of many of them, others, particularly ATP, also have crucial roles to play in cell metabolism. Cyclic AMP (cAMP) is not a constituent of nucleic acids but plays a central role in many regulatory processes in mammalian cells. Its role in plant cells remains to be defined (e.g. 47).

Initiation of nuclear DNA replication takes place at several points along the chromosomes (Fig. 5–3); if this were not so, replication would take several days. With the participation of several proteins (114, 404) including helix destabilizing protein (HDP), unwinding protein, and DNA-dependent ATPase, the DNA duplex separates and a short primer RNA strand is synthesized at each origin by an enzyme with RNA polymerase activity. DNA polymerase α (170, 397) then utilizes the primer RNA to synthesize a DNA strand complementary to each of the parent strands in a 5′ → 3′ direction (Fig. 5–2). Since DNA polymerase α, in common with all template-dependent nucleic acid polymerases, can only extend a chain in the 5′ → 3′ direction, one or both daughter strands is synthesized discontinuously, with repeated initiation. Ribonuclease then degrades the RNA primers, and the newly synthesized strands are joined by ligase. This is clearly a complex process and is, as yet, only understood at a superficial level in plant cells.

Replication of chloroplast DNA appears to take place bidirectionally, from two origins, in the manner indi-

cated by Tewari (439); replication of mitochondrial DNA may not be very different. The DNA polymerases of both chloroplast and mitochondrion are distinct from the DNA polymerase α of the nucleus but may be coded for by the nuclear genome.

Plant cells also contain a variety of *deoxyribonucleases*, whose precise function is uncertain. They would be expected, however, to be involved in, for example, repair and recombination.

RNA. During the process of *transcription* of DNA to RNA, one of the strands of the RNA duplex acts as template for the synthesis of a complementary RNA strand, which is extended in a 5′ → 3′ direction, antiparallel to the template DNA. Initiation takes place adjacent to sites recognized by the appropriate DNA-dependent RNA polymerase and also terminates at specific sites. As in other cells, transcription is inhibited by actinomycin D (15,388,399). Precursors are the four ribonucleoside triphosphates, ATP, CTP, GTP, and UTP, although there may be extensive modification, particularly by methylation, at later stages of RNA processing and maturation. In higher plants, there are three polymerases involved in transcription of nuclear DNA (24, 170, 171; Table 5–4), each responsible for transcription to a different class of RNA. All have a complex, multi-subunit structure (170). In addition, chloroplasts and mitochondria have polymerases distinct from those of the nucleus. That in mitochondria, for example, is structurally much simpler, with a molecular weight (MW) of ca. 64,000–65,000. Many plants also appear to have RNA-dependent RNA polymerases. The role of enzymes of this type in virus RNA biosynthesis is discussed later.

The *ribosomes* of the cytoplasm, chloroplast, and mitochondrion are also separate and distinct. Protein synthesis on cytoplasmic ribosomes is inhibited by cycloheximide (36, 284) and on those of the chloroplast by chloramphenicol (108), suggesting at least a superficial similarity between ribosomes of the chloroplast and prokaryotes. There may be some 10^7 ribosomes in a leaf cell, of which up to 40% may be in chloroplasts, with those in mitochondria constituting less than 1% of the total. Mitochondrial ribosomes also have a sedimentation coefficient (78s) similar to those of the cytoplasm, and consequently only two species, the 80s cytoplasmic and 70s chloroplast ribosomes, are normally resolved on sucrose gradients (Fig. 5–4). The subunits of the three types of ribosome also contain their own distinct RNA species (rRNA; 104, 257, Table 5–5). Electrophoretic separation on polyacrylamide or agarose

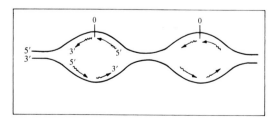

Fig. 5–3. Eukaryotic DNA replication: O, origin; ~, RNA primer; ——, DNA.

Table 5–4. Nuclear RNA polymerases.

RNA polymerase	Product
I	rRNA
II	HnRNA
III	tRNA; 5sRNA

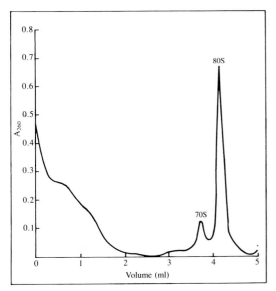

Fig. 5–4. Separation of cytoplasmic (80s) and chloroplast (70S) ribosomes from tobacco on a sucrose gradient. Sedimentation from left to right (modified from Roberts and Wood, 363).

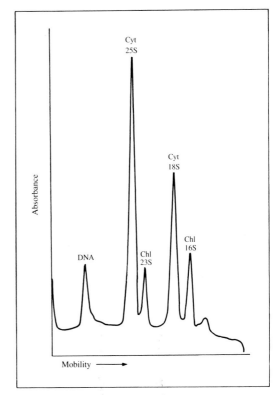

Fig. 5–5. Fractionation of high molecular weight nucleic acids from sunflowers by polyacrylamide gel electrophoresis (after Dyer and Leaver, 104).

gels normally reveals the presence of the most abundant species, 25s and 18s from cytoplasmic ribosomes and 23s and 16s from chloroplast ribosomes (Fig. 5–5). Since one species comes from each of the two ribosomal subunits, a 1:1 ratio of large to small RNA might be expected in each case. The 23s species from the chlo-

Table 5–5. Plant ribosomes and rRNA species (after Leaver, 257, and Dyer and Leaver, 104).

	Ribosome	RNA		
Source	Sedimentation coefficient(s)	Sedimentation coefficient(s)	Molecular weight (MW)	Number of nucleotides
Cytoplasm	<u>80</u>			
large subunit	60	25	1.3×10^6	3580
		5.8	5.4×10^3	157
		5	3.75×10^3	118
small subunit	40	18	0.7×10^6	1926
Chloroplast	<u>70</u>			
large subunit	50	23	1.05×10^6	2890
		5	3.94×10^3	122
		4.5	$2.1–3.3 \times 10^3$	65–103
small subunit	30	16	0.56×10^6	1541
Mitochondrion	<u>78</u>			
large subunit	60	24	$1.12–1.26 \times 10^6$	3082–3470
		5	3.88×10^3	120
small subunit	44	18.5	$0.69–0.78 \times 10^6$	1800–2146

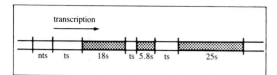

Fig. 5–6. Order of rRNA genes on nuclear DNA: ts, transcribed spacer; nts, nontranscribed spacer.

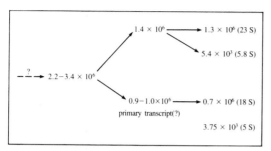

Fig. 5–7. Possible scheme for processing of cytoplasmic rRNA precursors (after Dyer and Leaver, 104).

Fig. 5–8. Possible order of chloroplast rRNA genes.

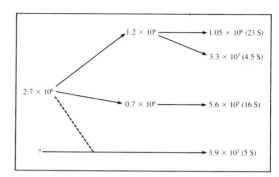

Fig. 5–9. Suggested scheme for processing of chloroplast rRNA (after Dyer and Leaver, 104).

roplast ribosome is often labile, however, so that a 1:1 ratio of 23s to 16s is not usually observed.

The order of each cytoplasmic rRNA species on the nuclear DNA is almost certainly as indicated in Fig. 5–6 (e.g. 104), a sequence that may be present in ca. 1,000–30,000 copies, grouped in a relatively few regions of the chromosome. The number of copies can vary widely, however, even within the same species, and seems to bear little relation to the rate or extent of protein synthesis required. The RNA sequence containing 18s, 5.8s, and 25s RNAs is transcribed as a single species by RNA polymerase I and processed, probably in the nucleolus, according to the scheme outlined in Fig. 5–7. The 5s species is probably synthesized separately in the nucleoplasm.

Chloroplasts contain three genes for rRNA species, contained within two large inverted repeat sections, which constitute some 4% of the chloroplast DNA. The order is as indicated in Fig. 5–8, with mature rRNA species produced by processing of larger transcripts as indicated in Fig. 5–9. The sequence of these chloroplast 23s, 16s, and 5s RNA species bears a close resemblance to those of comparable species from bacterial ribosomes.

The transfer RNAs (tRNA) of the chloroplasts and mitochondria are also distinct from those of the cytoplasm (470). They are all almost certainly produced by the processing of longer transcripts, with extensive modification of the constituent bases taking place before maturation. There is, however, comparatively little information on plant tRNAs. There are some 10 genes for each of the cytoplasmic tRNAs, while the chloroplast may have a total of 20–40 tRNA genes, distributed around the genome and occupying DNA with MW ca. 5 × 10⁶. Possibly the chloroplast contains information for all the necessary chloroplast tRNAs and does not need

to import tRNA from the cytoplasm. Again, there is little information on mitochondria.

Plant mRNAs (e.g. 104, 178) have the form outlined in Fig. 5–10. The majority have a "cap" structure (Fig. 5–11) at the 5'-terminus, which may be involved in ribosome binding and may also assist in protecting mRNA from nuclease digestion. Some plant viral mRNAs do not, however, need a cap structure. Many also have a poly-A tail, which does not appear to be obligatory for translation. All known plant messengers are monocistronic, producing one polypeptide only (e.g. 246, 247). Their biosynthesis is outlined in Fig. 5–12. RNA Polymerase II interacts with an AT-rich sequence ("TATA," or Hogness box) a few bases toward the 5'-end (upstream) from the initiation point at which transcription begins and produces a transcript to which is added a cap, by a capping enzyme, and poly-A tail, by poly-A polymerase. Any unwanted noncoding regions, or "introns," may be spliced out and "exons" retained, to give a functional mRNA. There is no evidence for the presence of poly-A tails on chloroplast mRNAs and some doubt as to their existence on mitochondrial messengers. There is no evidence for transfer of mRNAs from cytoplasm to organelle, and it would appear that mitochondria and chloroplasts synthesize all the

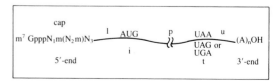

Fig. 5–10. Eukaryotic mRNA; l, leader sequence; p, protein coding region; u, untranslated sequence between p and the 3'-end; i, initiation codon; t, termination codon.

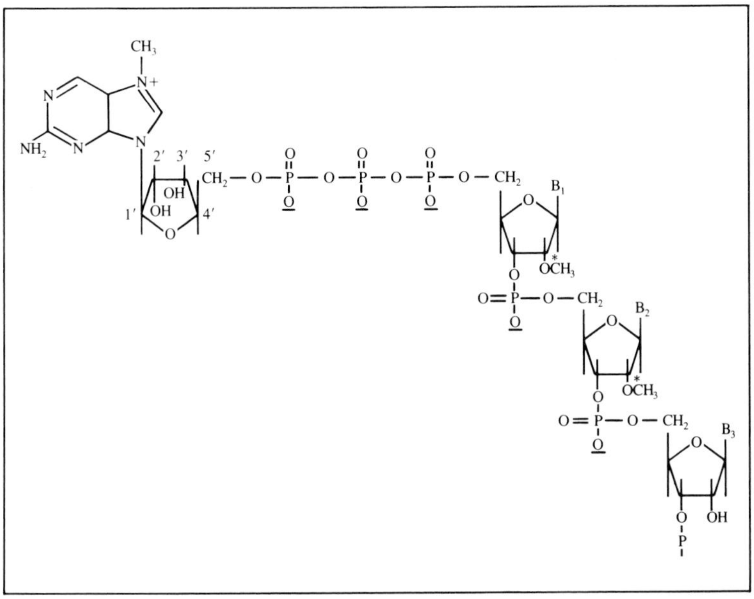

Fig. 5–11. 5′ Terminal cap structure of eukaryotic mRNAs; the 5′ position of the terminal nucleotide is linked, through triphosphate, to C5′ of guanosine, methylated at position 7. Methyl groups may or may not be present on adjacent nucleotides (*).

mRNAs they need in situ. They do, however, need a large number of proteins that have to be synthesized in the cytoplasm (e.g. 109–11), which may be synthesized with a short "presequence," or signal sequence of hydrophobic amino acids, which permits recognition of, and may also facilitate passage through, the organelle membrane. These hydrophobic sequences are subsequently removed. For example, several amino acids are cleaved from the small subunit of ribulose-1, 5-P$_2$ carboxylase after transport into the chloroplast. In total, a plant cell may contain 10^4 or more mRNA species, constituting some 5% of the total cell RNA. In tobacco, for example, polysomes from each organ system have been found to contain as much as $3.0–3.4 \times 10^7$ nucleotides of single copy transcript, representing 24,000–27,000 average-sized mRNAs (242).

Proteins

Plants contain hundreds of proteins, whose description is clearly beyond the scope of this book. The lectins (263), a group of proteins of potential significance in infection by a variety of pathogens, are referred to in Chap. 1 and at the end of this chapter. Separation by electrophoretic techniques is referred to in the virus section of this chapter.

The processes involved in the *translation* of eukaryotic mRNAs into polypeptide sequences has been largely elucidated with mammalian cells rather than plant cells. Much of our information is, therefore, inferred rather than experimental. The details of the scheme of polypeptide chain initiation and elongation outlined in Fig. 5–13 were elucidated for rabbit reticulocytes and

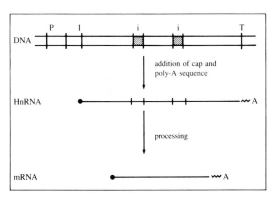

Fig. 5–12. General scheme for production of cytoplasmic mRNA. P, AT-rich promoter sequences; I, initiation; T, termination; ●, 5′-terminal cap; ⌇ A, 3′-terminal poly-A sequence; i, introns, spliced out during processing from HnRNA to mRNA.

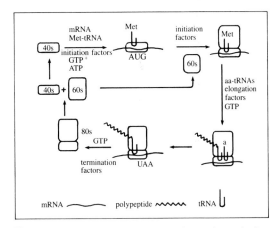

Fig. 5–13. Schematic view of eukaryotic protein synthesis.

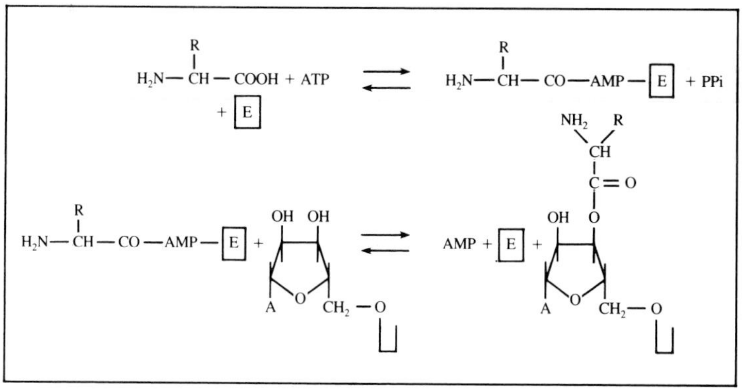

Fig. 5–14. Attachment of an amino acid to the 3′-end of t-RNA by an aminoacyl-tRNA synthetase $\boxed{E}$.

are presumed to be substantially the same, with minor differences, in the cytoplasm of plant cells (e.g. 441, 469).

Transfer RNAs are charged with the appropriate amino acid by aminoacyl-tRNA synthetases (470, Fig. 5–14), and polypeptide synthesis is initiated at the AUG initiation codon with formation of an initiation complex as indicated in Fig. 5–15. Elongation then proceeds according to the scheme outlined in Figs. 5–16 and 5–17. Several ribosomes may be attached to an mRNA at any one time, forming a polysome (Fig. 5–18), which may be intimately associated with membranes. Termination of polypeptide synthesis and release of polypep-

tide from the ribosome occurs when the ribosome reaches a termination codon, UAA, UAG or UGA, for which there is no tRNA. The overall mechanisms of protein synthesis are broadly the same in the cytoplasm and organelles, although the initiation and elongation factors differ in detail (64). The tRNA synthetases of the chloroplast and mitochondrion, although products of nuclear genes, are different from those of the cytoplasm. It is also of interest that in mitochondria from several sources, including man and yeast, codons are read differently than in the cytoplasm (387). For example, AUA, which normally codes for isoleucine, codes for methionine in mammalian mitochondria. It may be that as more DNAs are sequenced, such differences will also be found to apply to plant mitochondria.

Polypeptide production in the plant cell clearly differs both qualitatively and quantitatively in different organs and at different times during the plant's life cycle. Control may be at the level of transcription or of translation. Recent experiments on tobacco have revealed the transcription of structural genes not required for the formation of functional mRNA to transcripts that appeared in the HnRNA and did not leave the nucleus. This would suggest that regulation of some kind at a posttranscriptional level may be important. There is also evidence that, while cytoplasmic products can influence protein synthesis in organelles (109), RNA from the chloroplast can also influence cytoplasmic polypeptide production, at either transcription or translation (40).

Plants Infected by Viruses

With the exception of the viroids, whose RNA genome remains uncoated, the simplest plant viruses consist of a genome, which may be either DNA or RNA, encapsidated in multiples of one or more polypeptides. During the course of their replication cycle within the host cell, the genome is replicated, involving either DNA or RNA synthesis, and also expressed, to produce both virion and nonvirion polypeptides (see below). There is, therefore, an imposition of new nucleic acid

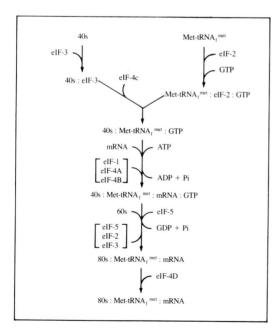

Fig. 5–15. The initiation of eukaryotic protein synthesis; there may be other factors not shown on this scheme; initiation factors remaining associated with the initiation complex are not indicated.

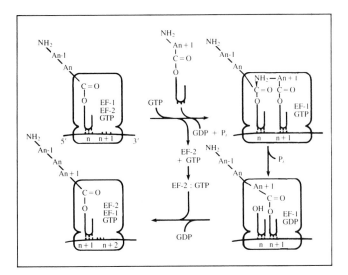

Fig. 5–16. Schematic view of the process of elongation of a polypeptide chain on an eukaryotic ribosome.

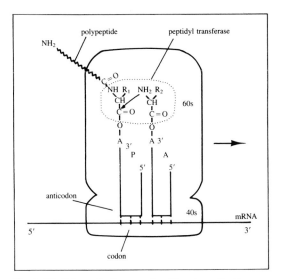

Fig. 5–17. Schematic diagram of an 80s eukaryotic ribosome with a growing polypeptide chain attached to a t-RNA occupying the P site, and t-RNA carrying the next amino acid to be added to the chain in the A site.

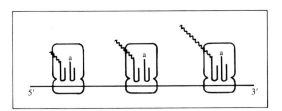

Fig. 5–18. Polysome, with several polypeptides at different stages of elongation.

and protein biosynthesis upon the host cell, which may be additional to, or partially at the expense of, normal macromolecular synthesis. This would depend on the host and virus in question. In the following sections we will assess the influence of virus infection on host nucleic acid and protein synthesis metabolism and indicate current views on the replication of some representative plant viruses.

Effect on the Host

DNA synthesis. Since the effect of virus infection on host DNA synthesis has received comparatively little attention, it is not possible to make any definitive generalizations. Using a histochemical staining technique to monitor DNA synthesis, Misawa and his colleagues (289, 290) found a stimulation in nuclear DNA synthesis following CMV infection of fully expanded tobacco leaves. The stimulation was maintained for several hours, beginning to decline some 10 h after infection. In root tips of French bean infected with TRSV, however, the rate of DNA synthesis and the number of cells capable of synthesizing DNA were reduced, at about the time the cells became invaded (8).

Although the extent of virus replication is considerable, TMV infection of Samsun tobacco plants appears to influence DNA synthesis only when small leaves are infected, changes in accumulation of DNA paralleling changes in leaf fresh weight. For example, systemic infection of leaves 1.5 cm. long with TMV reduced DNA synthesis considerably and also reduced leaf size (137). Infection of leaves that had reached maturity had no influence on DNA content; there was no evidence for degradation. Virus infection can, therefore, influence leaf development, although the basis for its effect on DNA synthesis is unknown. There is clearly scope for further investigation in both tissues and protoplasts.

RNA synthesis. The effect on host RNA synthesis has only been investigated to a significant extent with plants infected with RNA viruses. The first studies were performed on TMV-infected tobacco tissue and generally indicated a slight rise in total RNA synthesis compared to that in uninfected tissue (18, 143, 372). There was, however, no detectable rise in CMV-infected leaves (339). However, whether the virus genome is DNA or RNA, virus-specific RNA is synthesized during the virus replication cycle, in addition to host rRNA, mRNA, and tRNA. The amount of TMV RNA synthesized in an infected tobacco leaf, for example, is considerable; virus may attain a level of 10 mg/g leaf tissue, with viral RNA constituting 75% or more of the total cellular RNA. Although the majority of other viruses attain much lower concentrations than this in the leaf, the contribution of virus-specific RNA to total RNA of the infected cell is not negligible. Comprehensive assessments of the effect of virus infection on RNA metabolism would, therefore, require host and virus RNA to be separated, and the amount of each in the tissue, together with rates of synthesis and degradation, measured. Such assessments have been made with few virus/host combinations.

Fraser (136–38), Oxelfelt (321), and Hirai and Wildman (196) have all monitored the effect on tobacco RNAs of infection either by the *vulgare* strain of TMV, which causes relatively mild light green/dark green mosaic symptoms, or the *flavum* strain, which produces a much more pronounced yellowing. Using techniques of general application to the study of RNA species, total RNA was extracted from both infected and uninfected tissue, separated on polyacrylamide gels, and the relative quantities of the various RNA species estimated as peak areas from densitometer traces of the gels. Pulse labeling of tissue prior to RNA extraction with ^{32}P (136) or ^{3}H-uridine (321), coupled with estimation of radioactivity contained within the area of the gel corresponding to each RNA species, permitted estimation of the extent of de novo synthesis of virus and host RNAs during the labeling period (Fig. 5–19; 136).

Perhaps the most significant conclusion from the limited number of virus/host combinations studied to date is that the effect of virus infection on cytoplasmic rRNA (cyt rRNA) and chloroplast rRNA (ch rRNA) is quite different. The effect of TMV infection on tobacco rRNA also appears to be critically dependent upon the age of the leaf at the time of inoculation.

During the normal course of tobacco leaf development, the rate of cyt rRNA synthesis reaches a peak just before the maximum length of the leaf is reached, and then declines (Fig. 5–20). When young leaves, ca. 4 cm long, became infected systemically after inoculation of lower leaves with the *vulgare* strain of TMV, the rate of cyt rRNA synthesis declined rapidly as soon as virus replication began (Fig. 5–20) and remained at a low level. Surprisingly, leaf growth was little affected. On inoculation of older (10 cm) leaves, however, a different picture emerged. Here, three distinct phases were apparent (Fig. 5–21). The rate of cyt rRNA synthesis was

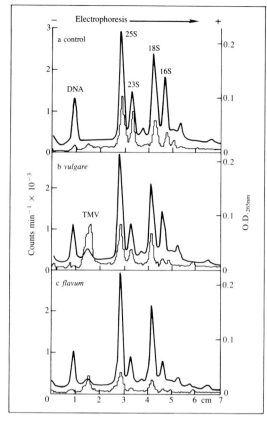

Fig. 5–19. Electrophoresis of tobacco leaf nucleic acids labeled with ^{32}P, extracted 2 d after inoculation with the b, *vulgare*, or c, *flavum*, strain of TMV. Histogram represents radioactivity present in fractions of the gel (after Fraser, 136).

stimulated with respect to that in healthy tissue one day after infection, declined to a level below that of uninfected leaves during the period of maximum virus accumulation, and then increased as the rate of virus replication declined. The decline in the rate of cyt rRNA synthesis was of the same order whether plants were inoculated with *vulgare* or *flavum*, despite the observation that *vulgare* accumulated to a much greater extent than *flavum*. This is interpreted to indicate that reduced cyt rRNA synthesis is not simply the result of competition by the virus for precursor nucleotides. If this were the case, a rather less dramatic decline in cyt rRNA synthesis might be expected in *flavum*-infected tissue, since fewer nucleotides would be required for viral RNA synthesis.

Degradation of cyt rRNA was also found to be retarded in infected tissue; its half-life was approximately twice that of healthy leaf tissue (137, 138). When, for example, tobacco leaves 13 cm long were inoculated, there was a marked reduction in the rate of loss of cyt rRNA in both *vulgare* and *flavum*-infected tissue, with the greatest reduction in tissue infected with the *vulgare* strain (Fig. 5–22).

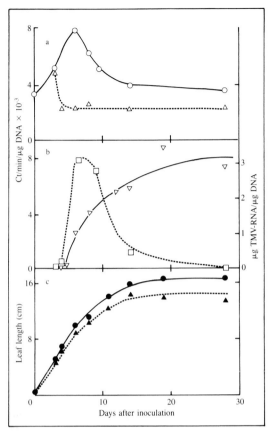

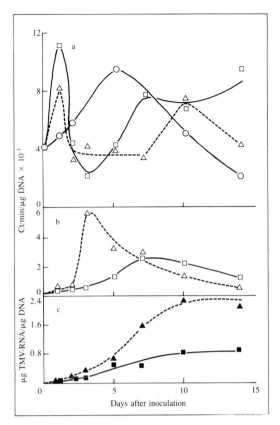

Fig. 5–20. r-RNA and TMV-RNA synthesis after secondary infection of small leaves by TMV strain *vulgare*. (a) [^{32}P] incorporation during a 5h incubation into r-RNA of healthy (○——○) and TMV-infected leaves (△——△). (b) □——□, [^{32}P] incorporation into TMV-RNA: ▽——▽ μg TMV-RNA/μg DNA. (c) Changes in leaf length with time after inoculation: ●——●, healthy leaf; ▲——▲, TMV-infected leaf (after Fraser, 138).

Fig. 5–21. r-RNA and TMV-RNA synthesis after primary inoculation of 10 cm long leaves. (a) [^{32}P] incorporation during a 5 h incubation into-RNA of healthy (○——○), TMV strain *vulgare*-infected leaves (△——△), and TMV strain *flavum*-infected leaves (□——□). (b) [^{32}P] incorporation into TMV-RNA: △——△, *vulgare*; □——□, *flavum*. (c) Changes in leaf content (μg/μg DNA) of TMV- RNA: ▲——▲, *vulgare*; ■——■, *flavum* (after Fraser, 138).

Although the mechanism by which cyt rRNA synthesis in medium-sized leaves is regulated is still unknown, the effect is to ensure conditions suitable for virus replication (138). Synthesis of cyt rRNA was stimulated before and after the rapid phase of virus multiplication and its degradation retarded, thus maintaining an adequate supply of ribosomes for virus protein synthesis. Additionally, RNA synthesis was directed toward viral RNA during the rapid phase of viral multiplication. Cyt rRNA has also been found to be stimulated during the period of TMV RNA synthesis in tomato root tissue (138). Although Reddi (353) had suggested that in TMV-infected tobacco leaves, rRNA was degraded and the products utilized for the synthesis of viral RNA, there is no evidence for this in later work (137, 321).

The effects of virus infection on ch rRNA synthesis reflected the severity of the symptoms produced. While both *flavum* and *vulgare* TMV caused a rapid inhibition of ch rRNA synthesis (136, 137, 196, 321), the *flavum* strain, producing the more severe symptoms, inhibited synthesis to the greatest extent. *Flavum* also caused an accelerated degradation of ch rRNA, an effect not observed with *vulgare*. Within 2 d of inoculation of Samsun tobacco with *flavum*, the quantities of both 23s and 16s RNA extracted from infected tissue were less than from controls (Fig. 5–19). By 4 d, both these ch rRNA species had almost disappeared, in conjunction with the appearance of low molecular weight RNA, presumably degradation products of the 23s and 16s RNA. Again, the effects were dependent upon the age of the leaf at the time of infection (137). When very small (1.5 cm) leaves became systemically infected, accumulation of ch rRNA was restricted, but not totally inhibited, by both strains; when larger leaves were inoculated, ch rRNA synthesis was quickly inhibited (Fig. 5–23).

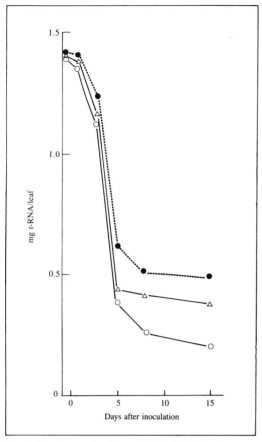

Fig. 5–22. Changes in ribosomal RNA/leaf following inoculation of 13 cm tobacco leaves with *vulgare* (△) or *flavum* (●); control (○) (after Fraser, 137).

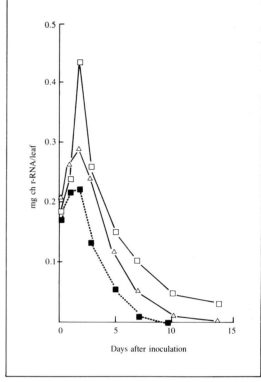

Fig. 5–23. Changes in chloroplast rRNA content/leaf following inoculation of a 10-cm tobacco leaf with *vulgare* (△) or *flavum* (■) TMV; control (□) (after Fraser, 136).

An alternative approach to extraction of RNA is to separate ribosomes by ultracentrifugation of tissue extracts through sucrose density gradients. Using this procedure, Hirai and Wildman found the U1 (or *vulgare*) strain of TMV to severely curtail incorporation of ^{32}P into 70s chloroplast ribosomes of systemically infected tobacco leaves, but not into 80s cytoplasmic ribosomes. This clearly indicated an inhibition of chloroplast rRNA synthesis, an inhibition confirmed by separation of extracted RNAs on polyacrylamide gels. The ability of isolated chloroplasts to incorporate nucleoside triphosphates into RNA was also impaired (196).

TMV infection also caused inhibition of tRNA accumulation when young tobacco leaves became systemically infected and a reduced rate of loss in older (18 cm) inoculated leaves, compared to comparable healthy ones (137, 321). The rate of loss was less in *flavum*-infected leaves, reflecting observations on cyt rRNA. It seems that in tobacco, TMV infection restricts cell development when young leaves become infected and retards senescence when older leaves are infected, effects that could be causally related to the observed effects on

RNA metabolism. As we have seen, studies have concentrated on rRNA. The *flavum* strain, for example, is able to stimulate the synthesis of cyt rRNA and retard its degradation. Precisely how these effects are mediated is unknown. It is likely, however, that effects on development also involve alterations in mRNA synthesis, although TMV infection had no effect on the polyA-containing mRNA of tobacco leaves (139) (see below).

The fact that leaves infected with the *flavum* strain develop more severe symptoms than those infected with *vulgare* may be the result of the greater inhibition of ch rRNA synthesis and its consequent effect on chloroplast protein synthesis, although there is no direct evidence for this. The possible relevance of alterations in ch rRNA and chloroplast ribosome synthesis to symptom expression is referred to in relationship to other effects on the chloroplast in Chap. 2.

Observations on other virus/host combinations have been rather less comprehensive, although the pattern is broadly the same. In Xanthi nc tobacco infected with Price's No. 6 strain of CMV, which causes a marked yellow/green mosaic, the concentration of both chloroplast and cytoplasmic ribosomes decreased at a rate greater than that observed in healthy tissue. This enhanced reduction in ribosome content was more pronounced in young, systemically infected leaves, becoming apparent when symptoms first appeared. It was not observed in tissue infected with the W strain, which caused a much less severe mosaic in Xanthi nc tobacco (363).

There was also a rapid decline in chloroplast ribosome concentration in White Burley tobacco leaves systemically infected with TSWV during the period when virus was still accumulating (291). The fact that during the same period the synthesis of ch rRNA was unaffected, and was not inhibited until virus replication began to decline, seemed to suggest a stimulation in 70s ribosome degradation. The stimulated decline in 70s ribosomes did not, however, apply to 80s cytoplasmic ribosomes, which declined in concentration at a rate comparable to healthy tissue.

In *Nicotiana glutinosa* leaves systemically infected with LNYV (349), symptoms appeared 6 d after plant inoculation, with 70s ribosome content decreasing by 7 d, until by 10 d after inoculation 70s chloroplast ribosomes could no longer be detected. As observed in TMV infection of tobacco, this ribosome loss was reflected by loss of 23s and 16s RNAs, inhibition of their synthesis, and impairment of leaf development. Although the concentration of 80s ribosomes also declined, there did not appear to be a reduction in either concentration or rate of synthesis of cyt rRNA. Unlike TMV infection of tobacco, the effects of LNYV on *N. glutinosa* were the same following infection with a mild and a severe strain.

Finally, Reid and Matthews (357) found that in TYMV-infected Chinese cabbage, the concentration of chloroplast ribosomes was much less in yellow than in green areas, while that of cytoplasmic ribosomes was little different between yellow and green areas. The more severe the symptoms, the greater the reduction. When young leaves became systemically infected, chloroplast ribosome concentration declined rapidly, as virus accumulation reached a maximum (74, 280).

We may conclude, therefore, that when young developing leaves become infected, there is a reduction in the content of chloroplast ribosomes that, in general, is correlated with symptom severity and is usually accompanied by a decrease in rate of ch rRNA synthesis.

Ribonucleases. There is always a stimulation of RNase activity following virus inoculation. Diener, for example (92), observed a stimulation in TMV-inoculated *D. stramonium* and in BPMV-inoculated *P. vulgaris*, while Wyen et al. (484) found the concentration of a relatively purine specific endoribonuclease to increase in TMV-inoculated Xanthi nc tobacco leaves. However, a significant contribution to the stimulation comes from the wounding occasioned by the inoculation process (9, 92), so that an inoculated leaf is not the best tissue in which to study virus-induced changes. The degree of stimulation appears to be correlated with lesion number and probably results from the additional cell damage caused by virus infection rather than from virus multiplication per se.

Enhanced RNase activity can also be observed in systemically infected tissue; in Chinese cabbage systemically infected with TYMV, Randles (348) found the activity of one of three host RNases to rise significantly at the time of rapid virus accumulation. Stimulation was observed only in those areas showing symptoms, where virus replication was also occurring, and not in green areas. It is not clear, however, whether these changes in RNase activity are in any way related to an enhanced degradation of ch rRNA.

Nucleotides. The available information on changes in nucleotide pools is slight. SBMV caused a slight increase in ATP in bean but otherwise had little effect (37). Tu (446) found the concentration of cyclic AMP to be reduced in clover infected with CYMV and to correlate with increased starch accumulation.

Protein synthesis. Following infection by viruses such as TMV, which attain a high concentration in infected tissue, there may be a redirection of host protein synthesis to accommodate the synthesis of large quantities of capsid protein. For viruses whose concentration remains much lower, and this may indeed be the majority, such redirection may not be necessary. Fraser found incorporation of labeled histidine into host proteins to be reduced by 50–75% following TMV infection of Samsun tobacco leaves (139), although levels returned to normal at later stages of the infection. These observations confirmed previous reports (e.g. 74, 309, 383, 475), suggesting a virus-induced redirection of host protein synthesis toward virus-specific proteins. Randles and Coleman (350) have similarly found the rate of incorporation of amino acids into host proteins to be significantly reduced in *N. glutinosa* systemically infected with LNYV within a day of the appearance of symptoms. This reduction in protein synthesis was accompanied by a reduction in the proportion of ribosomes appearing in polysomes.

Partly because of difficulties inherent in separating and identifying individual polypeptides, information on rate of synthesis, degradation, or concentration of specific proteins usually relates to enzymes or isoenzymes (e.g. peroxidases, phenylalanine ammonia lyase). The activity of many enzymes is stimulated as a result of virus infection; they are referred to in the appropriate chapters.

Fraction I protein (ribulose-1,5-bisphosphate carboxylase/oxygenase) is perhaps exceptional. It constitutes a significant proportion of the total soluble plant protein and is easily distinguishable by polyacrylamide gel electrophoresis of cell extracts or by ultracentrifugation. Its rate of synthesis and concentration in leaf tissue has been found to decline either at or soon after the time of symptom appearance in TMV- and TSWV- infected tobacco (291, 322), in LNYV-infected *N. glutinosa* (349) and in TYMV-infected Chinese cabbage (74, 280). Since one of the subunits of Fraction I protein is synthesized in the chloroplast, it may be supposed that its decline is related to the concomitant decline in 70s ribosomes. Polyacrylamide gel electrophoresis has also revealed changes in concentration of other host proteins, although they remain unidentified. Virus infection of plants that respond hypersensitively induces the biosynthesis of several "PR" or "stress" proteins that are either not present or present in very low concentrations in healthy tissue. The significance of these proteins is discussed in Chap. 10.

It is not yet clear how the changes in host protein synthesis are regulated. TMV infection of Samsun tobacco has no effect on either the concentration of poly A containing RNA or on its size distribution (139), suggesting that regulation is controlled at the translational level, with viral RNA competing more effectively for the available ribosomes than host messengers. There may, however, be significant changes in concentration of individual mRNAs within the total population, which would not be detected by the methods employed. It does not appear that reduced host protein synthesis is the result of a depleted amino acid pool (350).

Amino acids and amides. Virus infection usually seems to lead to an increase in concentration of both free amino acids and amides. There is, for example, a stimulation of free amino acids in MDMV-infected corn (280, 448) and in SoMV-infected soybean (447) and of the amides glutamine and asparagine in MRDV-infected corn (184) and CMV-infected cowpea (471). Both amino acid and amide concentrations are stimulated in, for example, PVY-infected tobacco (38), LNYV-infected *N. glutinosa* (349), and TSWV-infected tomato (70, 398). Concentration of the amino acid pipecolic acid also increases in CMV-infected cowpea (471) and in TSWV-infected tomato. Its stimulation is particularly evident in tissue responding hypersensitively.

It seems likely that increases in amino acid and amide concentration are the result of either reduced host protein synthesis or enhanced host protein degradation, or both. It must be remembered, however, that their concentration may depend on several factors and may differ at different stages of infection. It may be less than that in uninfected tissue in, for example, the early stages of TMV infection of tobacco, when virus is accumulating rapidly.

Virus-Specific Macromolecules

With a few exceptions, such as TMV, viral nucleic acids and virus-coded polypeptides constitute a small portion of the total complement of these macromolecules in the infected cell, yet the effect on cell metabolism and morphology can be dramatic. In the following sections, we examine the form and, where possible, the function of virus- specific macromolecules and indicate, where relevant, the experimental procedures applied to the study of their biosynthesis in infected tissue, particularly with respect to their discrimination from host macromolecules.

Nucleic acids. The genome of a plant virus may be either single or double-stranded RNA (ss RNA or ds RNA) or single or double-stranded DNA (ss DNA or ds DNA). However, the vast majority of known viruses have ss RNA genomes. They are usually also positive (+) stranded and able to act as messengers and hence direct polypeptide synthesis without transcription. A minority are negative (−) stranded, and must therefore produce a complementary (+) strand before translation can occur. In addition, information encoded in the viral genome may be distributed between two or more genome segments, which may be encapsidated into different particles. They are conventionally referred to as multicomponent or multipartite genome viruses. Examples of viruses representative of the various types are indicated in Table 5–6.

With the advent of more sophisticated techniques for analysis of viral genomes and their translation products, information on the structure and strategy of a rapidly growing number of plant viruses is becoming available. It is clearly beyond the scope of this book to document all this information. We do, however, indicate general features of viral genomes, their organization, and the mechanisms by which the information encoded therein appears as functional polypeptide. For a more comprehensive treatment, see several recent reviews (7, 48, 78, 180, 459).

ss (+) RNA viruses. The RNA of the ss (+) RNA viruses is constructed from the four usual ribonucleotides, without the modifications apparent in, for example, tRNA. Genome molecular weights lie between ca. 1.4 and 4.5×10^6; where the genome is divided into several segments, the total molecular weight lies between these extremes. With the advent of rapid sequencing techniques, complete sequences are becoming available (e.g. 153, 361). They often have distinctive termini.

5'-termini. Many, though not all, eukaryotic mRNAs have a "cap" structure, 7-methylguanosine linked via its 5'-OH to the 3'-OH of the penultimate base through triphosphate. Some ss (+) plant viral RNAs have a similar cap structure, although, unlike eukaryotic mRNAs, bases adjacent to the cap are not usually methylated. Capping does not appear to be obligatory for translation. TMV (497), CMV (431), and TYMV (43), for example, have caps, while STNV does not. The precise function of the cap is not clear; it may serve to protect the RNA from enzymatic degradation by phosphatases, or be involved in RNA binding to the ribosome and hence in the regulation of translation. Evidence is, however, conflicting. When a cap was added to the 5'-end of STNV, binding to ribosomes in a wheat germ in vitro translation system was approximately 2.5 times the rate of normal RNA binding (46), although there was no effect on initiation of protein synthesis (417). Removal of the cap from RNAs 3 and 4 of BMV reduced messenger activity (406). 7-Methylguanosine inhibited binding to ribosomes of capped viral RNA but not of uncapped (46). Interestingly, although all RNAs of BMV are capped, they are not equally efficient as messengers (62). If the cap is indeed involved in ribosome binding, then those mRNAs lack-

Table 5–6. Properties of viral nucleic acids.

Virus (CMI no.)[a]	Molecular weight (10^6)	% in virion	5-terminus	3-terminus
ss(+)RNA; monopartite				
Monocistronic:				
Tobamoviruses; TMV (151,184)	2.05	5	m[7] GpppGp	GCCCA-OH
Tymoviruses; TYMV (2, 214, 230)	2.0	35	m[7] GpppGp	ACCA-OH
Polycistronic: Carnation mottle (7)	1.4	20	—	—
ss(+)RNA; bipartite				
Tobraviruses; TRV (12)	2.5; 0.7, 1.0 or 1.3	5	m[7] GpppAp (short RNA)	GCCC-OH
Nepoviruses; RRV (6, 198)	2.4; 1.4	(b) M, 30; B, 43–46	Vpg	Poly (A)-OH
Comoviruses; CPMV (47, 197)	2.02; 1.37	M, 25; B, 36	Vpg-pUA	AUUpoly(A)-OH
ss(+)RNA; tripartite				
Cucumoviruses; CMV (1, 213)	1.27; 1.13; 0.82 (0.35)	18	m[7] GpppNp	CCA-OH
Bromoviruses; BMV (3, 180)	1.1; 1.0; 0.7 (0.3)	22	m[7] GpppGp	CCA-OH
Alfalfa mosaic; (46, 229)	1.04; 0.73; 0.62 (0.28)	Ta[(b)], 15.2–15.6 Tb, M; 15.5 B, 16	m[7]GpppGp	CGC-OH
ss(−)RNA				
Plant rhabdoviruses LNYV (26)	4.0	—	—	—
ds RNA				
Phytoreoviruses; WTV (34)	2.9; 2.4; 2.2; 1.8; 1.78; 1.1; 1.05; 0.83; 0.57; 0.55; 0.54; 0.032	22	—	C-OH or U-OH
Fijiviruses; MRDV (72)	2.88; 2.50; 2.35; 2.12; 1.75; 1.45; 1.25; 1.18; 1.08	—	—	—
ss DNA				
Geminiviruses; BGMV (192)	0.75	29	—	—
ds DNA				
Caulimoviruses; CaMV (24, 243)	5	17	—	—
Satellite viruses: ss(+)RNA				
CARNA 5 (−)	0.1	—	m[7] GpppGp	CCC-OH
Satellite tobacco necrosis (15)	0.4	20	ppAGU or pppAGU	CCC-OH or CCCp
Viroids				
Potato spindle tuber (−)	0.1	not encapsidated	—	—

a. Description number in Harrison and Murant (185).
b. T, M, B: 'top', 'middle', and 'bottom' components; additional data and extensive reference lists can be found in several comprehensive reviews: e.g. 180, 280.

ing a cap must clearly have an alternative structure at or near the 5'-end that can influence ribosome binding. However, although the base sequence near the 5'-end of several viruses is known, it is not yet clear precisely which regions are involved (243). Although there does not appear to be a consistent sequence between the 5'-end of the genome and the AUG triplet that initiates translation, it may be that the short UA-rich sequences ("Butler boxes") preceding initiator AUG codons in several plant viral RNAs are of significance (153). A secondary structure involving other parts of the molecule may also be required.

Other viral RNAs have a virus-coded protein linked to the 5'-end of the genome (Vpg; e.g. 78, 282). Its function is again uncertain, but in some cases, at least, it is not required for translation in vitro (63).

3'-termini. Also in common with many eukaryotic mRNAs, several viral RNAs have a poly-A sequence at the 3'-end. It can be removed from TEV (183) and BSMV (1) without apparently affecting infectivity; its role remains obscure.

In addition, the 3'-end of some plant viral RNAs possesses amino acid accepting activity, mimicking tRNA (176). The 3'-end of TYMV and other tymoviruses can be specifically charged with valine (147, 335). Type TMV accepts histidine (309, 383), although the cowpea strain accepts valine. The RNAs of BMV and BSMV can be specifically charged with tyrosine both in vitro and in barley protoplasts (265). The structures of some of these 3'-end regions, predicted from base sequences, bear a superficial resemblance to tRNA (e.g. 172) and must be able to recognize the appropriate aminoacyl tRNA synthetases. The significance of this activity is again unknown; it may be involved in RNA replication (265).

Other sites. In addition to the features referred to, the RNA of TMV, and presumably of other viruses, contains a base sequence that recognizes a group of coat protein monomers involved in the initiation of virus assembly (53), while the RNAs of AMV have sites near the 3'-end that bind coat protein. In this case, coat protein binding is a necessary prerequisite for replication (244). There must also be sequences near the 3'-end of viral RNAs able to interact with replicases to initiate synthesis of complementary (−) RNA, although, again, details are awaited.

The naked RNA genomes of these viruses, unlike those of the ss (−) RNA viruses, are able to act as mRNAs and are infective (e.g. 125, 148). In the case of the multicomponent viruses, several RNA segments may be required.

Replication. It seems likely that replication of virus ss (+) RNA occurs according to a scheme similar to that outlined in Fig. 5–24, with ss (+) RNA acting as template for the synthesis of a limited number of complementary (−) RNA strands, which in turn act as templates for the transcription of progeny (+) ss strands. For convenience, both types of replicative intermediate will be designated RI. In broad agreement with this

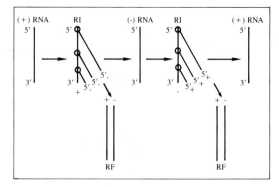

Fig. 5–24. Replication of ss (+) RNA viruses.

scheme, double-stranded RNAs have been identified from tobacco leaves inoculated or systemically infected with TMV (39, 51, 214, 305–307, 347, 407, 481), from cells isolated from TMV-infected tobacco leaves (215), and from TMV-inoculated tobacco protoplasts (5), and procedures have been developed for their isolation and separation from ss RNA (e.g. 90, 305). However, since ss RNA is much more susceptible than ds RNA to RNase, some early attempts to identify ds RNAs in infected tissue incorporated RNase into the isolation procedures, in the expectation that the ss RNAs would be destroyed. Inevitably, RI was also partially, if not completely, degraded.

The RIs indicated in Fig. 5–24 are largely single stranded and would, therefore, be expected to be susceptible to RNases. RI is, however, only partially susceptible to RNases, producing a molecule that is largely double stranded and similar in MW to the replicative form (RF). It may be that the structure of RI is as indicated in Fig. 5–25b; alternatively, the two strands may form helical regions during the isolation procedure, when replicase and possibly other RNA-associated proteins are removed. In either case, RNase would then remove only the single-stranded tails. There is, however, no evidence for the structure of RI in vivo.

Extraction of RNA species from TMV-infected tobacco cells in the absence of ribonucleases, followed by separation on polyacrylamide gels, clearly revealed the

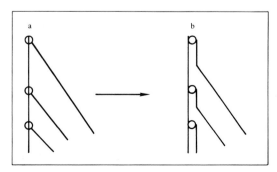

Fig. 5–25. Possible forms of the replicative intermediate (RI).

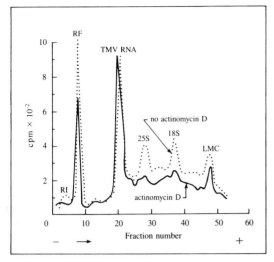

Fig. 5–26. Incorporation of ³H-uridine into the RNAs of TMV-infected tobacco leaf cells, in the presence (———) or absence (- - -) of actinomycin D. Polyacrylamide gel electrophoresis from left to right (after Jackson et al., 215).

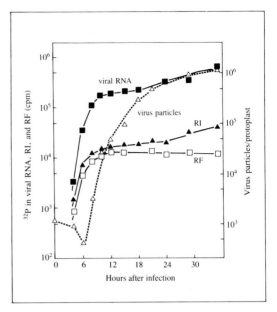

Fig. 5–27. Time course of synthesis of virus-specific RNAs and virus particle formation in TMV-infected tobacco protoplasts (after Aoki and Takebe, 5).

presence of RNA species of MW greater than that of TMV RNA, which were not present in healthy tissue (Fig. 5–26). Their synthesis also continued in the presence of actinomycin D, additional evidence for their independence from host RNA synthesis. The putative RF was resistant to RNase, while, as expected, RI yielded a molecule of MW similar to that of RF. Estimated MWs are approximately as expected, with RF ca. 3.8×10^6 (5) and RI 5×10^6 (214) or 6.7×10^6 (5), depending on the standards used in the estimation.

Information on the roles of ds forms in replication comes from pulse-chase experiments. When cells were provided with a short pulse of ³H-uridine, a large portion of the radioactive label was incorporated into ds RNAs; with a longer pulse, a greater portion was incorporated into ss (+) RNA. When the labeled uridine was followed by excess unlabeled uridine, the chase part of the experiment, then most of the label disappeared from RI and appeared in ss (+) RNA, although not all label could be chased from RF. When similar pulse chase experiments were performed with TMV-infected tobacco protoplasts (5), a significant portion of label remained in both RF and RI after chasing. Moreover, in neither case could the activity appearing in ss (+) RNA be accounted for by that lost from RF and RI. The time courses of accumulation of RF and RI in TMV-inoculated protoplasts are, however, not inconsistent with their role as intermediates in replication (Fig. 5–27).

Thus, although data are inconclusive, they suggest that RIs represent intermediates in the replication of ss (+) RNA, as indicated in Fig. 5–24, although the two types have not been distinguished. The role of RF is more elusive. It seems likely, however, that the greater proportion of RF represents terminal structures formed toward the end of (+) or (−) strand copying. It is not inconceivable that RF is formed by association between (+) and (−) strands during RNA isolation, although this does not appear to be the case with CMV (225).

In addition to the ds forms corresponding to the full-length TMV genome, shorter ds species, containing sequences corresponding to the 3'-end of the genome, have been isolated (152, 494). Their role, if any, in virus replication remains to be determined.

Information on other viruses is now becoming available, with ds RNAs revealed in extracts from protoplasts or leaf tissue infected with AMV (292), BMV (19, 264), CMV (90, 225), TNV (69), CCMV, and TYMV (33). Again, however, their role, if any, in ss (+) RNA formation awaits further study.

Replicases. In order to produce multiple copies of the infecting ss (+) RNA, an enzyme is required that can produce a complementary (+) copy of the viral RNA and then, utilizing the (−) strand as template, catalyze the synthesis of ss (+) RNA. Enzymes with this activity are referred to as RNA-dependent RNA polymerases, or replicases, and have recently been the subject of a comprehensive review (181). Although not contained within the virus particle, enzymes capable of incorporating ribonucleotides into acid-insoluble products could be detected in several hosts following virus infection and distinguished from DNA-dependent RNA polymerases by their insusceptibility to inhibition by actinomycin D. Enzyme activity was associated with either a particulate or a membrane fraction, sedimenting in the ultracentrifuge at ca. 30,000 g, or in a "soluble" fraction, the supernatant remaining after sedimentation of tissue extracts at ca. 100,000 g. Enzymes were, therefore, referred to as "membrane-bound" or "soluble."

In TMV-infected tobacco, for example, Brishammar and Juntii (44) reported the presence of a soluble enzyme in systemically infected leaves, and Zaitlin and his colleagues (41, 491) found an enzyme in particulate fractions of extracts from inoculated leaves, from which it could be released by detergent treatment. Studies on these relatively crude enzyme preparations suggested that activity of the soluble enzyme depended on the presence of added template; particle-bound enzyme was associated with an endogenous template, and its activity was stimulated by added RNA, but not wholly dependent on it. There were indications that TMV RNA was utilized more efficiently as template than other plant virus RNAs by the soluble enzyme, but unlike reports for the particle-bound enzyme, no indications existed that plant viral RNAs in general were utilized more efficiently than others. The MW of both the soluble enzyme and that solubilized from the particulate fraction was estimated to be in the order of 130,000.

Similarly, Gilliland and Symons (150) reported the presence in CMV-infected cucumber of a soluble replicase that, while dependent upon added RNA for its activity, did not appear to utilize CMV RNA in preference to others. Zabel et al. (489) found a membrane-bound replicase in CPMV-infected cowpea, associated with endogenous viral RNA template, while activity of the particle-bound enzyme from BMV-infected barley (250) was stimulated by added RNA, that of BMV and CCMV being the most effective. The particle-bound activity in TYMV-infected Chinese cabbage was associated with chloroplast membranes and appeared to have a preference for the RNA of TYMV and that of other tymoviruses (299, 346).

In the early 1970s there was, therefore, ample evidence for the presence in virus-infected tissue of replicase activity. It was generally assumed that the function of these enzymes was to replicate viral RNA and that they were coded for by the viral genome, since there was no reason to expect the host plant either to synthesize or utilize an enzyme with this specificity. However, it was becoming apparent that uninfected plants did indeed contain enzymes with replicase activity. The report by Astier-Manifacier and Cornuet (6) of replicase activity in healthy Chinese cabbage leaves was quickly followed by observations of similar activity in tobacco (101, 122, 123), and indeed it seems that replicase activity, though often at a very low level, can be detected in almost all plants. The ds RNAs found in uninfected tobacco (211) may be produced by such a replicase. It cannot be assumed, therefore, that replicase activity observed in either soluble or particulate fractions is necessarily viral in origin. These observations stimulated a reappraisal of earlier work, embodied in attempts to obtain replicase preparations, both soluble and solubilized from particulate fractions, in as pure a form as possible. Only then could meaningful comparisons between soluble and particle-bound enzymes from healthy and infected plants be made. The current situation with respect to the more important virus/host combinations is broadly as follows.

Tobacco. Uninfected tobacco plants contain a replicase enzyme that appears mainly in the soluble fraction of tissue extracts. It is template-free and therefore dependent upon exogenous RNA for its activity in vitro. It shows no specificity toward any particular viral RNA, and the products of in vitro transcription of plant viral RNAs are principally double stranded, heterogeneous, and of low MW (60, 65, 101, 122, 261).

Following infection by TMV, TNV, or AMV, the activity of this enzyme is considerably stimulated, and it may be present both in the soluble fraction and in association with membranes. Before "solubilization" and further purification, the membrane-associated enzyme is also associated with endogenous template, and its activity is usually, though not always, stimulated by addition of exogenous viral RNA. The soluble enzyme and that solubilized from its membrane-bound state, both in uninfected plants and in tissue infected with TMV and TNV, are identical by several criteria and therefore appear to be the same (209, 212). The observation that the replicase induced by TNV in cowpea is different from that induced by the same virus in tobacco, but at the same time similar to that induced in cowpea by CPMV and CCMV (210), would tend to support this view. Activity is, however, much higher in infected plants. When extensively purified, the enzyme requires added template RNA and catalyzes the synthesis of mainly ds RNA products (212). Despite earlier reports to the contrary (44, 491), there is little evidence that the enzyme uses RNA of the virus which induced it more efficiently than that of others (101, 213, 261).

Since TMV codes for a 130,000 MW polypeptide produced early in the replication cycle, it had been considered likely that this polypeptide was a subunit of the complete replicase, which associated with other polypeptide subunits contributed by the host cell to form a functional enzyme. The precedent for this hypothesis was the replicase utilized by the bacteriophage Qβ in *Escherichia coli*, which consists of one virus-coded polypeptide in association with three host subunits. This would not, however, appear to be the case. There is no evidence for the presence of this virus-coded polypeptide in purified preparations of the soluble enzyme from TMV-infected plants (369) or protoplasts (204), although the possibility of loss of subunits from membrane structures during isolation cannot be ruled out.

The soluble enzyme stimulated by AMV infection also has properties very similar to those in uninfected plants (60, 65), and it may well be that this enzyme is identical to that stimulated by TMV and TNV. Studies on temperature-sensitive mutants support the view that the replicase in AMV-infected plants does not contain a virus-coded subunit (262).

Cucumber. Infection of cucumber tissue by CMV stimulates the synthesis of a replicase found in both soluble and particulate fractions of plant extracts. Since the original reports of replicase activity in relatively crude preparations (150), both soluble and particulate enzymes have been purified extensively (149, 163, 249, 433).

The soluble enzyme appears to utilize plant virus RNAs with equal efficiency, showing no particular preference for CMV RNA. The most highly purified enzyme preparations consist of a polypeptide, MW 100,000, together with two additional ones, MW 110,000 and 35,000, in lesser and varying proportions (163). Activity appears to be associated with the MW 100,000 polypeptide. Peptide maps of the two larger polypeptides suggest that they are structurally related. The MW 110,000 polypeptide is slowly lost from enzyme preparations on storage, without affecting activity significantly, and it may be that this is the form produced in vivo. Although the enzyme stimulated by TRSV infection appeared to be slightly different from that stimulated by CMV (327), and the polypeptide cannot be found in uninfected plants (126, 433), recent work comparing enzyme preparations from cucumber infected with three strains of CMV, Q, P and T, suggests it to be of host origin. Tryptic peptide analysis indicates that none of the enzyme polypeptides corresponds to full length in vivo translation products of the viral genome. However, the possibility that virus-coded polypeptides are firmly bound within an enzyme membrane complex has not been entirely eliminated.

Cowpea. CPMV-infected tissue contains a membrane-associated replication complex that in vitro is capable of completing viral (+) strands initiated in vivo (98). This activity appears to be different from the previously reported membrane-associated replicase activity, which rises in concert with virus replication (Fig. 5–28), and is induced by infection with CPMV (99, 488). This enzyme appears to consist of a single polypeptide, MW ca. 130,000, and is apparently identical to that observed in much lower concentration in uninfected plants. There is evidence suggesting at least participation of virus-coded polypeptides in the viral replicase. The genome of CPMV consists of two RNA species, 1 and 2, each encapsidated in a separate nucleoprotein particle (B and M respectively). Both are needed for replication of complete virus particles. However, following infection of cowpea protoplasts with B component only, RNA-1 is replicated in the absence of RNA-2; RNA-2 is not replicated following infection with M component only.

Replicase activity also appears in the soluble fraction of extracts from both healthy and CCMV-infected plants (472). Activity is stimulated by exogenous viral RNA and yields mainly ds RNA. In addition, CCMV induces a membrane-bound replicase. Preparations obtained to date contain endogenous viral RNA template, which directs the synthesis in vitro of a heterogeneous population of ss RNAs, some of which comigrate as polyacrylamide gels with viral RNAs and also ds RNA.

The membrane-bound enzyme induced by TNV infection has, as indicated above, properties different from that induced by this virus in tobacco but similar to

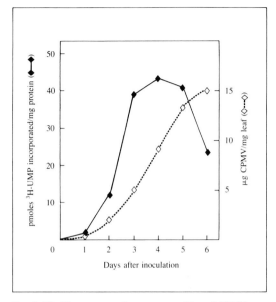

Fig. 5–28. Time course of appearance of bound CPMV replicase activity (———) and virus accumulation (- - -) in CPMV-infected cowpea leaves (after Zabel et al., 488).

those of both the enzyme from uninfected tissue and that induced by CPMV (210). Although there is some evidence, therefore, for the appearance in virus-infected tissue of replicase complexes different from those of infected tissue, the nature of the polypeptides involved remains to be determined.

Barley. The activity of a particulate-associated enzyme is stimulated in both BMV-infected plants and protoplasts (50, 175, 182, 250, 312) (Fig. 5–29). Replicase activity from infected tissue, solubilized but still relatively crude, is 50 times greater than that from healthy plants and is stimulated by exogenous template. BMV RNA is the most efficient, and products correspond in electrophoretic mobility to the RFs of the viral

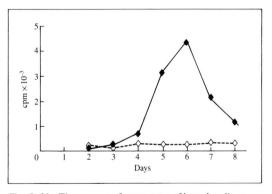

Fig. 5–29. Time course of appearance of bound replicase activity in barley: (- - -), uninfected; (———), infected with BMV (after Hadidi and Fraenkel-Conrat, 175).

RNAs. Infected plants contain a 110,000 MW polypeptide not present in uninfected tissue that may correspond to the product of in vitro translation of RNA 1 and may be part of the replicase. The evidence suggests that the replicase is either totally or partially coded for by the viral genome.

Chinese cabbage. TYMV infection induces the synthesis of a replicase that is associated with the chloroplast membrane. When solubilized and partially purified, the enzyme is dependent on exogenous template, the most efficient being RNA complementary to the viral (+) strand, i.e. (−) TYMV RNA and EMV RNA (300). It appears to consist of 4 polypeptides and to differ from host enzymes.

In summary, despite the many problems inherent in the purification of these replicases, enzymes with replicase activity have been isolated and a few have been extensively purified from several plants infected with a variety of different viruses. In most cases, however, their origin remains in doubt. Much more needs to be done on their purification, and detailed comparison of their polypeptides with virus-coded polypeptides is required. The synthesis of ds RNA species of different sizes, involving de novo synthesis of both (+) and (−) strands, depending on the template, has been demonstrated in several laboratories. There is as yet, however, little evidence for synthesis in vitro of ss (+) RNA, a necessary function of a viral replicase. With few exceptions, such as the replicase isolated by Hall and his colleagues from BMV-infected barley, they have little template specificity. We clearly still have a lot to learn about the form and function of the enzymes that produce ss (+) viral RNAs in vivo.

Actinomycin D. Although actinomycin D does not inhibit RNA-dependent RNA synthesis, it interferes with the replication of several plant viruses in both tissue and protoplasts when applied very early in the replication cycle. For example, application within a few hours of inoculation restricted replication of TMV in tobacco tissue (80, 85, 399) and of CPMV and CCMV in cowpea hypocotyls. While not found to be inhibitory to the replication of, for example, CMV in tobacco protoplasts (314), inclusion of actinomycin D in the incubation medium restricted replication of TMV in cucumber (71), TYMV in Chinese cabbage (358), PVX in tobacco (315), and AMV (303) and CPMV in cowpea (373) protoplasts. It would appear, therefore, that the replication of these, and possibly many other RNA plant viruses, requires a host-specified function at an early stage, although the nature of this function is not yet clear.

2-thiouracil. The effect of the base analogue, 2-thiouracil, on virus replication seems to differ markedly among viruses. While, for example, replication of AMV in tobacco, CMV in cucumber (277), and TNV in French bean (20) seemed little affected by 2-thiouracil treatment, and indeed replication rate of CCMV in cowpea was stimulated (83), accumulation of TMV in tobacco was reduced (20, 85). It does not, however, reduce the extent of TMV accumulation in tobacco protoplasts, but it does reduce the specific infectivity of progeny virus to 10–20% of normal. Its action may result in the formation of noninfective virus, which interferes with the spread of infection in vivo (16, 129). Application to Xanthi tobacco plants 12 h or more after infection had no effect on viral RNA or protein synthesis, although both were inhibited by treatment within 4 h. This would indicate that 2-thiouracil inhibits some early function (82).

ss (−) RNA viruses. All the known viruses of this type are enveloped and classified as rhabdoviruses. The MW of the single RNA genome is of the order of 4×10^6, and there is no reason to suggest that replication occurs by a mechanism substantially different from that indicated in Fig. 5–24, starting with (−) RNA instead of (+) RNA. However, by comparison with the ss (+) RNA viruses, little information is available either on the genome, on possible existence of RF and RI forms, or on the transcriptase involved in genome replication.

The viral ss (−) strand is unable to act as messenger, a role filled by the complementary (+) strand, so that transcription of (−) to (+) RNA is required before a replicase or replicase subunit can be produced. Consequently, (−) RNA animal viruses contain a replicase within the virion that is able to perform this function. It is of interest that one of the plant rhabdoviruses, LNYV, has been found to contain a transcriptase, that, in vitro, is able to utilize endogenous LNYV RNA as template for the synthesis of short complementary ss RNA molecules (131, 132). The fact that isolated viral ss (−) RNA is not infectious, in contrast to that of the (+) strand viruses, suggests that this enzyme is obligatory for at least the first stage of viral replication, transcription of (−) to (+) RNA.

ds RNA viruses. The ds RNA viruses have genomes consisting of either 10 (phytoreoviruses) or 12 (fijiviruses) segments, of total MW $16–19 \times 10^6$, all encapsidated within a single particle. There is evidence that during replication of ds RNA animal viruses, a (+) strand is first synthesized on each of the 10 parental ds genome segments. A complementary (−) strand is subsequently synthesized using the new (+) strand as templates to produce new ds virion RNAs. There is little information on the mode of replication of the ds plant viral RNAs. However, virions do contain a transcriptase (e.g. 32, 354) that can produce ss copies of the genome, although its role in replication is not clear. The in vitro synthesis of ds virus RNAs has not yet been demonstrated.

ss DNA viruses. Viruses with ss DNA genomes are currently encompassed within a single group, the geminiviruses (159), which, as their name suggests, have particles consisting of paired nucleocapsids. The DNA isolated from virus particles consists of circular molecules, MW 7 or 8×10^5, together with linear forms thought to derive from the circles. Cleavage of BGMV nucleic acid with restriction enzymes gives a series of fragments; when the MWs of these fragments are add-

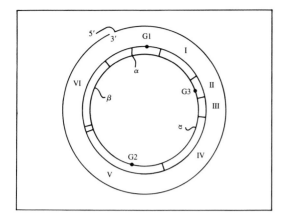

Fig. 5–30. The genome of CaMV.

strands. There is no information on the enzymes involved.

Proteins. With the exception of the viroids, plant virus genomes are encapsidated in multiple copies of one or more polypeptides, the capsid or coat proteins, coded for by the viral genome. They can be readily identified, and MWs determined, by disruption of the virion, followed by gel electrophoresis. TMV capsid protein can also be revealed in the soluble fraction of extracts from infected plants by gel electrophoresis, followed by an appropriate stain (146, 322, 458). This is, however, one of the few exceptions. Virus-coded proteins, including in many cases the capsid protein(s), may be synthesized in quantities that are very small in relation to those of the host proteins. Since the plant cell contains hundreds of proteins, separation and identification of a few virus-specific ones is difficult. In addition, some virus-coded proteins may only be synthesized during a short period of the viral replication cycle. The study of virus-coded proteins synthesized in vivo, even those of the capsid, is, therefore, by no means an easy task, and in only a few cases have such proteins been identified. Some of the procedures used, together with relevant examples, are briefly indicated below; other examples will be found under the appropriate virus.

It is possible to reveal the synthesis of virus-coded proteins in infected tissue using radioactive labeling techniques. Zaitlin and Hariharasubramanian, for example (492), employed a dual-label procedure. Proteins of uninfected tobacco leaf tissue were labeled with ^{14}C-leucine, and those of TMV-infected tissue with ^{3}H-leucine. Actinomycin D was used to suppress incorporation into host polypeptides by reducing host mRNA synthesis. The mitochondrial fractions of both healthy and infected tissue were disrupted (with SDS and mercaptoethanol), mixed together, and polypeptides separated on polyacrylamide gels in the presence of SDS. Gels were then sliced transversely, and ^{14}C or ^{3}H label incorporated into the polypeptides determined in each slice. A plot of the ratio of ^{3}H to ^{14}C against distance along the gel revealed peaks at positions in the gel corresponding to polypeptides present in infected but not in healthy tissue (Fig. 5–31). Coat protein was clearly present, together with other putative virus-coded polypeptides. A similar procedure revealed the presence of several possible virus-coded polypeptides in CMV-infected cucumber (496). The technique has limitations, however, particularly in the resolving power of the one-dimensional gels. Nevertheless, it was the first to reveal polypeptides other than capsid in infected tissue, although in the hands of others only the presence of capsid polypeptide was demonstrated (413).

There are also several difficulties inherent in the use of intact tissue. First, it is difficult to reduce incorporation of radiolabel into host polypeptides, an important consideration in single-label procedures (see below), less critical for dual labeling. Secondly, because of the asynchrony of infection, with virus replication at different stages in different cells, it is impossible to insure

ed, the sum is about twice the MW of one ss circle. It is proposed, therefore, that the genome of BGMV, and also that of several other geminivirus, consists of two ss circles, each encapsidated in a separate, paired particle. Circular ds molecules, consisting of a closed circle associated with a linear complementary strand, have been isolated from BGMV-infected tissue. DNA complementary to viral DNA is in both linear and circular forms, and it is presumed that these ds molecules are intermediates in genome replication. No details of genome replication are available, however, and it is not clear, for example, whether an RNA primer, obligatory for replication of ss DNA bacteriophages, needs to be synthesized, or which enzymes are involved.

ds DNA viruses. The known plant viruses with ds DNA genomes are closely related to the first one to be shown to contain DNA, CaMV, and all are contained within the caulimovirus group (205, 405). The genome of CaMV is typical of those examined so far consisting of a ds circle, of MW 4.5×10^6, with a discontinuity in one strand (α or 1), and two discontinuities in the other (β or 2) (Fig. 5–30), with overlaps of up to 40 bases. It may exist in various topological forms (287). The complete sequence is known (128); 5'-termini of at least some strains have free OH groups. The DNA alone is infectious. It seems probable that the genomic DNA replicates via reverse transcription of a 35s RNA transcript (173, 207).

Viroids. The genomes of the viroids, the smallest characterized agents that can produce disease, consist of small (MW $1–1.3 \times 10^5$) closed circular RNA molecules with extensive regions of intrastrand complementarity, so that they can form rigid, almost linear rods (91, 93, 95, 169). Linear molecules can also be isolated, although it is not clear whether they arise during formation of the circles or are formed from them during isolation. Although definitive information on viroid replication is still awaited, full-length RNA copies of PSTV have been isolated from infected tomato (495), suggesting that replication takes place via complementary RNA

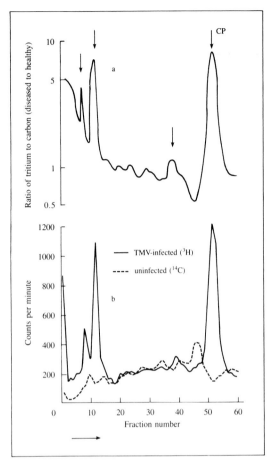

Fig. 5–31. Polyacrylamide gel electrophoresis of polypeptides from uninfected and TMV-infected tobacco leaf tissue labeled with ^{14}C-leucine or ^{3}H-leucine, respectively; a, ratio of ^{3}H to ^{14}C counts obtained from b (after Zaitlin and Hariharasubramanian, 492).

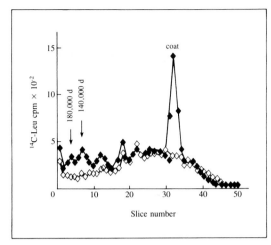

Fig. 5–32. Electrophoresis of ^{14}C-leucine labeled polypeptide from uninfected (◇) and TMV-infected (◆) tobacco protoplasts (after Sakai and Takebe, 381).

mensional gel electrophloretic techniques will also help to overcome one of its present limitations, the identification of virus-coded polypeptides in the presence of so many of host origin.

Because of the difficulties inherent in the detection and characterization of virus-coded polypeptides produced in vivo, attention has in many cases been directed

maximum incorporation of label into virus-coded polypeptides during the period of their biosynthesis or to study the timing of their synthesis. Finally, uptake and incorporation of radiolabel is inefficient.

These limitations have led to a focus on the use of infected protoplasts (e.g. 436), where infection is synchronous, host polypeptide synthesis can be reduced by, for example, UV irradiation (381), and uptake of radiolabeled precursors is efficient. Thus, the separation of extracts from TMV-infected tobacco protoplasts on polyacrylamide gels has permitted the identification of virus-coded polypeptides in addition to that of the capsid both with and without incorporation of labeled amino acid (325, 381, 408). Radiolabeled polypeptides are detected by counting label in gel slices (Fig. 5–32) or, more usually, by fluorography (Fig. 5–33). This is clearly more promising than using intact tissue. Separation of polypeptides by the more sophisticated two-di-

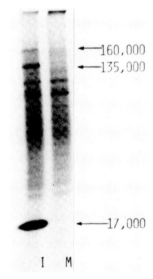

Fig. 5–33. Fluorogram of SDS-treated proteins from TMV-infected (I) and mock-infected (M) tobacco protoplasts labeled with ^{3}H-leucine 21–23 h after inoculation. Arrows indicate the position of virus-coded polypeptides (after Siegel et al., 408).

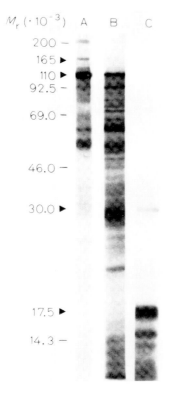

Fig. 5–34. Fluorogram of the in vitro translation products of TMV RNA in the rabbit reticulocyte lysate (a), wheat germ (b), and *E. coli* (c) systems. The position of internal MW markers and virus-coded products (▶) are indicated (after Glover and Wilson, 151).

toward alternative procedures for their detection and identification, the various in vitro translation systems. Plant virus RNAs can be translated in vitro in several systems, the two most favored being those from wheat germ and rabbit reticulocytes. The methods and their application have been comprehensively reviewed elsewhere (e.g. 77, 124).

Virus-coded polypeptides, radiolabeled in vitro, are normally characterized by gel electrophoresis (Fig. 5–34). Rarely, however, have the products of in vitro translation been detected in vivo. Until confirmation is available, therefore, the schemes for genome expression outlined below, which in many cases are largely based on results of in vitro translation, must be viewed with caution. Reservations apply particularly to those instances where more than one polypeptide is produced from the same RNA sequence, either by use of multiple initiation sites (e.g. CPMV) or readthrough of termination codons (e.g. TMV). It is not inconceivable that their formation could result from the somewhat unnatural conditions of the in vitro translation.

Organization of the Virus Genome and Its Expression

ss (+) RNA; Monopartite, Monocistronic

The genomes of viruses of this type consist of a single RNA species. However, although the majority contain information for several polypeptides, only the cistron at the 5′-end of a viral RNA is normally expressed. Internal cistrons, with initiation codons distal from the cistron at the 5′-end, are "closed." The complexities of alternative strategies for expression of viral genomes are illustrated by the following examples.

Tobacco mosaic virus. The genome of TMV is ca. 6400 nucleotides long (e.g. 198). In vitro translation, in wheat germ (49), rabbit reticulocyte lysates, or *Xenopus* oocytes (208), fails to demonstrate production of coat protein. Instead, the product is a polypeptide, MW 165,000, with a second, MW 130,000, produced in oocytes and reticulocytes. Neither contains coat protein sequences (208). Both are initiated at the same point, an AUG triplet near the 5′-end, with the larger of the two produced by readthrough of the termination codon for the smaller (Fig. 5–35; 330). The amino acid sequence of the smaller one is therefore contained within the N-terminal region of the larger one. The complete nucleotide sequence of TMV (153) indicates that both are initiated at an AUG codon 69 bases from the 5′-terminus, and that their true MWs are ca. 126,000 and 183,000. The formation of the larger of these two polypeptides represents one mechanism whereby a plant virus RNA can produce more than one polypeptide without internal initiation: readthrough. We have here two polypeptides produced from substantially the same region of the RNA, representing an efficient utilization of the information. An alternative involves translation of the message in a different reading frame. This was first observed with the bacteriophage φx 174 (e.g. 387) and subsequently with some animal viruses. It is also utilized, to a limited extent, by TMV. The sequence coding

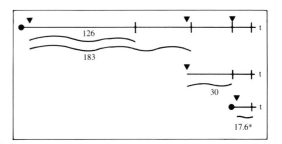

Fig. 5–35. Expression of TMV genome. There may be other subgenomic RNAs whose translation products are not yet characterized. (NOTE: In this and subsequent figures of genome expression, the following notation has been used: ■, Vpg; ●, cap; ∼ A, poly-A sequence; t, amino acid-accepting activity; @, imitiation site; +, termination site; *, coat protein. The numbers refer to the approximate molecular weights of the gene products × 10⁻³.

for a polypeptide MW 30,000 overlaps the terminal region of the 183,000 MW polypeptide sequence by five codons; this smaller polypeptide is read in a different frame.

However, neither this polypeptide nor coat protein can be translated from full-length TMV RNA, and the solution to their formation involves an alternative strategy, the formation of subgenomic RNAs, which are not essential for infectivity. Rather than being translated directly from the complete genome, coat protein is translated from a smaller RNA, synthesized in vivo (208, 409). This RNA species, MW ca. 270,000, was first observed in TMV-infected tobacco cells and referred to as LMC, or low molecular weight component (215, see Fig. 5–26). It is identical in sequence to ca. 700 nucleotides at the 3′- end of the TMV genome (22, 172) and is translated in vitro to yield coat protein. Though normally found only in vivo, in the cowpea strain of TMV this small coat protein mRNA is encapsidated and appears in purified virus preparations as a short rod (49, 194, 473). In addition, a further subgenomic RNA, I_2, coding for the MW 30,000 polypeptide, could also be found encapsidated in purified preparations of some TMV strains (23, 49, 219). This, too, appears to have a sequence in common with the 3′-end of the TMV genome, and also contains the coat protein gene. More recent investigations (152) appeared to reveal at least eight RNA species generated during the infection of tobacco by TMV, all having sequences in common with the 3′-end of the genome. At least four, however, may be electrophoretic artifacts.

It is not yet clear whether these subgenomic species are generated in vivo by transcription from $(-)$ RNAs or by cleavage of larger $(+)$ RNAs, although they can be isolated as double-stranded species analogous to RF. This would seem to favor the transcription hypothesis. In vitro translation of subgenomic RNAs suggests that TMV has a coding potential for at least six polypeptides, in addition to those already referred to, although their mRNAs have not been clearly identified with the subgenomic RNA species.

The synthesis of polypeptides MW ca. 165,000, 130,000, and 17,600 has also been observed in vivo in both infected tissue (325, 389, 492) and protoplasts (381, 408). The correspondence between the 130,000 polypeptide synthesized both in vivo and in vitro has been confirmed by chemical evidence, and there is ample evidence for the identity of the MW 17,600 coat protein. Usually, however, correspondence between polypeptides produced in vivo and in vitro rests on comparison of electrophoretic mobilities. TMV infection of tobacco protoplasts has been reported to induce the biosynthesis of up to 10 new polypeptides (204), although some, with MWs greater than the total coding capacity of the genome, would appear to be of host origin. Further studies may indicate the correspondence of others with the products of in vitro translation.

With the exception of the MW 17,600 capsid protein, the function of these polypeptides remains obscure. The MW 30,000 polypeptide appears to be involved in trans-

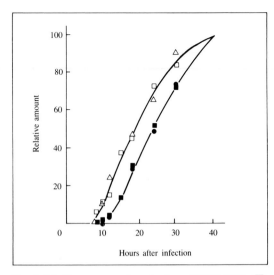

Fig. 5–36. Time course of production of TMV RNA (□) infectivity (●), MW104,000 protein (△), and coat protein (■) in TMV-infected tobacco protoplasts (after Sakai and Takebe, 381).

location, but the proposed involvement of the MW 165,000 and 130,000 polypeptides in replication, although still an attractive possibility, remains unsubstantiated.

Synchronous infection of tobacco protoplasts has permitted the timing of the synthesis of the major polypeptide and RNA species, in addition to that of the virion itself, to be determined with some precision (e.g. Fig. 5–36). The MW 165,000 and 130,000 polypeptides are synthesized early in infection, around the time of maximum rate of RNA synthesis, with the appearance of virions closely following coat protein synthesis. Virus particles could be first detected within about six hours.

Virus multiplication in cells of leaf tissue, however, is normally asynchronous, estimates of accumulation of polypeptides, RNAs, and virus particles in such tissue inevitably representing events taking place in cells in which virus replication has reached different points. Attempting to improve on the use of young, systemically infected leaves (307), in which synchrony is by no means complete, Dawson et al. (86) have applied various temperature treatments to inoculated tobacco plants. Lower leaves were inoculated and maintained in a box at 25 C, while upper leaves were allowed to penetrate the top of the box and were maintained at a low temperature, nonpermissive for complete virus replication but nevertheless permitting virus translocation to the upper leaves. On transfer to a higher, permissive temperature, virus replication, although not fully synchronous, was much more so than in untreated plants (Fig. 5–37). In this system, the maximum rate of synthesis of ds RNA was at 18–20 h and that of ss RNA at 24–26 h, with ss RNA first detectable 6–8 h after transfer to the permissive temperature. Partial synchrony

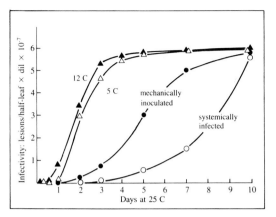

Fig. 5–37. Infectivity in tobacco leaves maintained at 12 C (▲) or 5 C (△) before becoming systemically infected with TMV at 25 C. Zero time is taken as the time of transfer to 25 C. Development of infectivity in inoculated and systemically infected leaves maintained continuously at 25 C is indicated for comparison (after Dawson et al., 86).

could also be approached by maintaining inoculated leaves at a nonpermissive temperature for several days before transfer to a higher, permissive temperature (Fig. 5–38). The delay of ca. 4 h between appearance of viral RNA and virions, observed in tobacco protoplasts, was not found here. The synchronization of infection in tissue by this kind of procedure has received comparatively little attention, and it may be that it could offer a much easier, if less precise, alternative to the use of protoplasts.

There is as yet little information on the means by

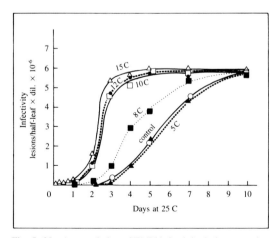

Fig. 5–38. Accumulation of TMV infectivity in leaves preincubated at restrictive temperatures. Tobacco leaves were mechanically inoculated with TMV and incubated at 15 C, 12 C, 10 C, 8 C, or 5 C for 10 d, after which they were incubated at 25 C. Control leaves were incubated at 25 C immediately after mechanical inoculation. Infectivity was determined at intervals after the shift to 25 C (after Dawson and Schlegel, 84).

which RNA and polypeptide synthesis is regulated, either qualitatively or quantitatively. It is of interest, however, that when protoplasts were prepared from Xanthi tobacco tissue in which viral RNA synthesis had virtually ceased, viral RNA synthesis resumed (121). Selective inhibition of ssRNA synthesis in TMV-infected tobacco suggests that protein synthesis is correlated with synthesis of ds rather than ss RNA (81).

The site of coat protein synthesis has been investigated by light microscopy using antibody to TMV tagged with either a fluorescent or a radioactive (^{125}I) label, or under the electron microscope using ^{125}I or ferritin-labeled antibody. However, the site of synthesis may be different from that of assembly into virus particles, and such experiments can only be convincing if virus antigen is detected very early in infection, before migration to an assembly site has taken place. This may be difficult to achieve in intact tissue. Nevertheless, Langenberg and Schlegel (256), using ^{125}I-labeled antibody as a probe for TMV coat protein in tobacco leaf tissue, concluded that protein was synthesized in the nucleus, supporting earlier observations by Hirai and Hirai (197) on tomato cells using fluorescent-labeled antibody. Experiments using ferritin-labeled antibody also indicated the presence of TMV protein in nuclei, suggesting the possibility of its synthesis there. However, there was also a significant amount in the cytoplasm that was not, apparently, assembled into rods.

Virus replication in tobacco protoplasts and callus tissue is inhibited by cycloheximide and not by chloramphenicol (381), and virus RNA can be isolated from tobacco leaf tissue as cytoplasmic polysomes (21, 415). It can be concluded, therefore, that virus-specific proteins are synthesized on cytoplasmic ribosomes, rather than in the chloroplast. Indeed, there is no convincing evidence for the involvement of the chloroplast at any stage of the replication of most TMV strains. The site of viral RNA synthesis has been investigated by several procedures, none of which is entirely convincing. In early experiments, in which UV absorption or staining was used as a guide to RNA content, a stimulation of RNA synthesis was observed in the nucleus in the early stages of infection (e.g. 460); it was concluded that viral RNA was synthesized there. It seems more likely, however, that this stimulation represented host RNA synthesis. However, using actinomycin D to suppress host RNA synthesis, Schlegel and Smith (394) observed a stimulation of uptake of radioactive precursor into RNA in the nuclei of infected tobacco discs, again suggesting the nucleus as the site of viral RNA synthesis. Radiolabel subsequently accumulated in the cytoplasm. By contrast, ds RNA and replicase have been found to be associated with cytoplasmic structures (346).

TMV RNA and coat protein can assemble in vitro into intact virions (127), an extensively studied and relatively well understood process. In essence, TMV protein, near pH 7.0, can form double discs containing two layers of 17 subunits each. This double disc can interact with the assembly region on TMV RNA and in doing so change the conformation to form a "helical lockwasher," with both RNA termini protruding from the same end. Successive discs are added to the side distant from the RNA tails and become incorporated into a helical structure into which the RNA is drawn. The rod is extended in a 5'-direction, with extension in the 3'-direction taking place at a slower rate by a mechanism that is not fully elucidated (52, 266, 280). The double disc has not yet been found in vivo, though there is no reason to believe that assembly is any different from that observed in vitro. It is of interest that where two strains multiply within the same cell, progeny may contain coat protein from both strains. Only RNA species containing the initiation site for assembly, which may differ in position among strains (144), become encapsidated into rods. RNA coding for the coat protein of the cowpea strain is encapsidated, while that of the common strain is not. In addition to coat protein, the U1 and Dahlmense strains of TMV contain approximately one copy per virion of an additional H protein, which is currently thought to consist of a branched chain fusion product between coat protein and a peptide of host or viral origin (66).

The association of labeled antibody with the cytoplasm late in infection (e.g. 316), often associated with virus particles, coupled with electron microscope visualization of particles (402), lends support to the view that particles are assembled and accumulate in the cytoplasm. Their absence from nuclei and chloroplasts rules out these organelles as assembly sites.

Evidence for the sites of synthesis of virion components, if not their assembly, is, therefore, conflicting. Despite reports to the contrary, the current view is that both RNA and protein synthesis, and also virion assembly, occur in cytoplasmic inclusions or viroplasms.

Turnip rosette virus. In common with that of the tobamoviruses, the capsid protein of TRosV is translated from a subgenomic RNA species. In addition, the genome of this virus employs a translation strategy not observed with TMV, translation into a large polypeptide that is then processed into smaller, functional polypeptides (Fig. 5–39). In vitro translation of genomic RNA yields 4 polypeptides, MW 105,000, 67,000, 35,000 and 30,000; the 67,000 and 35,000 species are produced by posttranslational processing of the 105,000 (296).

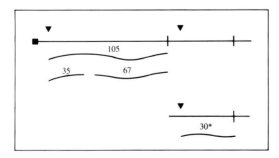

Fig. 5–39. Expression of the TRosV genome.

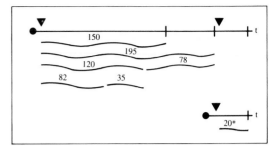

Fig. 5–40. Expression of the TYMV genome.

Turnip yellow mosaic virus. The genome of TYMV, and possibly that of other tymoviruses (the group of which TYMV is the type member), shares many features in common with that of TMV. Replication appears to proceed via ds forms, although there does not appear to be an actinomycin D sensitive stage (358). The capsid protein gene at the 3'-end of the genome is also closed, and virus preparations contain, in addition to particles with full-length RNA species and empty shells, noninfectious particles containing, in various combinations, of the order of nine subgenomic RNA species (280, 285). One of these, of MW ca. 250,000 and shown by in vitro translation to be mRNA for capsid protein, appears to be included in the majority of the particles (195, 337, 360). In vitro translation suggests that at least five are able to act as mRNAs, producing polypeptides with a common N-terminus (280). However, none of these RNA species has been observed in vivo, nor ds forms identified. Their significance remains to be determined.

The 5'-end of the TYMV genome appears, like that of TMV, to be translated into large polypeptides, the larger by readthrough of the termination codon for the smaller (Fig. 5–40). In vitro translation of the full-length genome reveals two polypeptides, MW 195,000 and 150,000, which have a common N-terminal sequence (26, 294, 490). However, in addition to the translational strategies utilized by TMV, readthrough of termination codons, and formation of subgenomic RNAs, the genome of TYMV also appears to utilize posttranslational processing of a large polypeptide. The MW 195,000 readthrough product is apparently cleaved to smaller, biologically functional polypeptides (293, 295), possibly by a proteolytic activity contained within the MW 195,000 polypeptide.

Several polypeptides have been observed in vivo in TYMV-infected protoplasts (280), although, with the exception of the MW 20,000 coat protein, there was apparently little correspondence in electrophoretic mobility between these and the in vitro products. Their characterization awaits further study.

Although there is a stimulation of RNA synthesis in the nucleus (25, 252), the association of replicase activity and ds forms of viral RNA with chloroplast membranes (345) suggests that chloroplast peripheral vesicles, a characteristic feature of TYMV-infected tissue (187, 278), are the sites of viral RNA synthesis. Since virus replication is inhibited by cycloheximide and not by chloramphenicol, it appears that capsid protein, and by inference other virus-specific polypeptides, are synthesized on 80s cytoplasmic ribosomes. The suggestion is that groups of capsid protein subunits become embedded in the outer chloroplast membrane, and particle assembly occurs at the junction between vesicle and cytoplasm (280). Empty particles appear to accumulate in the nucleus, although it is not thought that synthesis occurs there. RNA and protein do not appear to reassociate to form viral particles in vitro; although the RNA may contain a specific site for initiation of virion assembly, it has not yet been identified.

The time course of virus formation has been studied in both Chinese cabbage and *Brassica rapa* protoplasts (359, 429), with virus beginning to appear at approxiamtely 12 h after infection of protoplasts and reaching a maximum by 48 h (Fig. 5–41). Yields of 1–2 x 10^6 particles/cell were rather less than the productivity of comparable cells in inoculated leaf tissue (429).

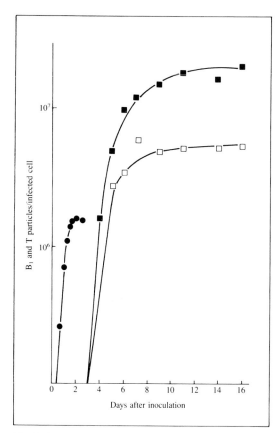

Fig. 5–41. Time course of production of TYMV empty protein shells (T, ●) and infectious nucleoprotein (B, ●) in Chinese cabbage protoplasts. Production of T (□) and B_1 (■) in cells maintained in the intact leaf is included for comparison (after Sugimura and Matthews, 429).

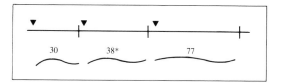

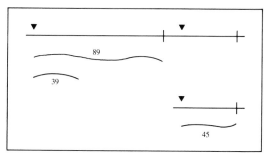

Fig. 5–42. Expression of the Car MV genome.

ss (+) RNA; Monopartite, Polycistronic

Carnation mottle virus. In contrast to those viruses that require subgenomic RNAs for complete translation, there appear to be some that can act as polycistronic messengers; several polypeptides can be translated from the complete genome by utilization of internal initiation signals. To date, only two come into this category, CarMV (382) and TNV (384).

The genome of CarMV has a MW of ca. 1.4×10^6 and appears to be translated in vitro into three polypeptides, MW 77,000, 38,000, and 30,000 (Fig. 5–42). The MW 38,000 polypeptide is that of the capsid. There is no evidence for readthrough, nor is there any information on polypeptides produced in vivo. Clearly, much more confirmatory evidence is required. More recent evidence (Harbison and Wilson, unpublished) suggests that this virus utilizes a subgenomic messenger RNA, as indicated in Fig. 5–42b

ss (+) RNA; Bipartite

Cowpea mosaic virus. Sucrose density gradient centrifugation separates purified CPMV into three components, top (T), middle (M), and bottom (B). The genome consists of two RNA species, the larger (RNA-1) encapsidated in component B and the smaller (RNA-2) in M. T consists of empty capsids.

Inoculation of cowpea protoplasts with either M + B or B alone results in the synthesis of progeny RNAs. Virus-specific RNA is not, however, produced following inoculation with M only, suggesting that replicase, or an essential subunit of it, is coded for by RNA-1 (155). Double-stranded forms of both RNAs have been observed in virus-infected cowpea leaves.

Further information on virus-coded proteins has come from the translation of the RNA species in vivo and in vitro and from the sequence of RNA-2 (461) and has led to the formulation of the scheme of genome expression outlined in Fig. 5–43 (135, 154–57). At least nine (MW 170,000, 110,000, 87,000, 84,000,

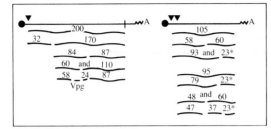

Fig. 5–43. Expression of the CPMV genome.

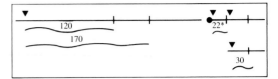

Fig. 5–44. Expression of the TRV genome.

60,000, 37,000, 32,000, 23,000, and 4,000) are produced in cowpea protoplasts inoculated with a mixture of M and B components containing both RNAs. The MW 37,000 and 23,000 species form the capsid, while the 4,000 MW species is the Vpg that becomes attached to the 5′-end of both RNA-1 and RNA-2 (421). Inoculation with B only results in synthesis of all except the MW 23,000 and 37,000 species, suggesting that these capsid proteins are coded for by RNA-2, which is contained in M. None was produced following inoculation with M, presumably because RNA-2 is unable to replicate in the absence of RNA-1. In vitro, however, RNA-2 directs the synthesis of two large polypeptides of MW 105,000 and 95,000. Incubation of these products of in vitro translation with extracts from protoplasts infected by either complete virus or B component only, with extracts from virus-infected leaves or with the in vitro translation products of RNA-1, results in cleavage into four polypeptides, with MWs 60,000, 58,000, 48,000, and 47,000. This observation suggests that one of the products of RNA-1 is a protease, involved in processing the genome primary translation products (331). Additional processing of the two large polypeptides to smaller species, MW 92,000, 79,000, and 23,000, is achieved by a second protease activity, present in infected cowpea leaves but not, apparently, in protoplasts.

It appears, therefore, that expression of the CPMV genome requires neither readthrough of termination codons nor generation of subgenomic RNAs. Instead, both RNA species are translated into large polypeptides that are processed into smaller biologically functional species. It seems likely that both virus-specific protein synthesis and virion assembly take place in the cytoplasm (318).

Construction of pseudo-recombinant viruses, containing M particles from one strain and B particles from a second, has enabled several functions to be assigned to the two RNAs (255). Capsid proteins, for example, are coded for by RNA-2, although lesion type seems to be controlled by either RNA-1 or RNA-2, or possibly both.

Tobacco rattle virus. The two genome segments of TRV are encapsidated separately into two rod-shaped particles, L (long) and S (short). Although the size of the larger of the two, RNA-1, is constant between strains, that of RNA-2 varies markedly (186). Viral RNA is replicated in tobacco protoplasts following infection with RNA-1 only, although RNA-2 is not, supporting

the view that replicase, or at least a subunit of it, is coded for by RNA-1. However, intact particles are not produced in the absence of RNA-2, which contains the coat protein gene. Evidence for the presence of at least one other gene on RNA-2 comes from the construction of pseudo-recombinants between the YS strain, which induces yellow symptoms on several hosts, and the CAM strain (364). Although the two strains have apparently identical coat proteins, symptoms are quite different and, through infections with mixtures of L + S particles from the two strains, have been shown to be controlled by RNA-2.

Polypeptides are currently thought to be produced from TRV RNAs by two mechanisms: readthrough and generation of subgenomic RNAs (Fig. 5–44). In reticulocyte lysate in vitro translation systems (332), RNA-1 of the PRN strain directs the synthesis of two polypeptides, MW 170,000 and 120,000; the longer is produced by readthrough of the termination codon for the smaller one. Although RNA-2 of PRN is reported to direct synthesis in vitro (wheat germ) of two polypeptides, coat (MW 22,000) and another (MW 30,000), only the 22,000 one is produced by CAM RNA-2 (141). The coat protein gene appears to be at the 5′-end of RNA-2, with the gene for the 30,000 polypeptide downstream and closed. This is a reversal of the more usual situation where the coat protein gene is at the 3′-end, as, for example, in TMV RNA.

The MW 30,000 polypeptide appears to be translated from a subgenomic species. Separation of RNAs from CAM-infected cells of Christie's hybrid *Nicotiana* has revealed the presence of at least five small RNAs (numbered 3–7), with MW 0.5, 0.22, 1.1, 0.37, and 0.16 × 10⁶ (31). Hybridization with radioactively labeled DNA complementary to the virion RNA suggests that RNAs 3, 4, 6 and 7 are subgenomic fragments of RNA-2. In vitro translation of RNA-3 yielded a 30,000 MW polypeptide; this species appears, therefore, to represent the 3′-end of the RNA-2. RNA-5 is a fragment of RNA-1. It is also of interest that purified preparations of the SYM strain contain two small encapsidated RNA species, MW 0.6 and 0.54 × 10⁶, which include sequences from both RNA-1 and RNA-2. The longer of the two is translated in vitro into coat protein and would, therefore, appear to be derived from RNA-2. It is suggested that the smaller, which translates to a polypeptide MW 29,000, is derived from RNA-1 (365).

There is no information on polypeptides produced in vivo, although inhibition of TRV replication by cycloheximide suggests that they are produced on 80s cytoplasmic ribosomes.

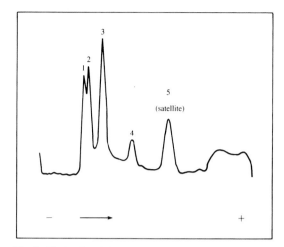

Fig. 5–45. Profile of the RNAs of CMV-W, including RNA 5, a satellite, separated by polyacrylamide gel electrophoresis.

ss (+) RNA; Tripartite

Cucumber mosaic virus. The RNA of CMV, type member of the cucumovirus group, can be separated into four species, RNAs 1–4, in order of decreasing molecular weight (Fig. 5–45). RNAs 1 and 2 are separately encapsidated, in multiples of a single polypeptide, MW 24,500, while 3 and 4 appear to be encapsidated together, to produce three types of particles. Unlike the particles of CPMV, however, which are readily separable by density gradient centrifugation, those of CMV are not (267). RNAs 1, 2, and 3 constitute the complete CMV genome (231, 268, 328) and have extensive base sequence homology at the 3'-end (432). A fifth RNA species present in many CMV strains, and encapsidated by CMV capsid protein into particles that may also contain segments of the CMV genome, has properties of a satellite; minor RNA species found on occasion appear to be degradation products of CMV genomic RNA.

The proportions of genomic RNAs isolated from purified preparations can depend on both virus strain and the host in which the virus replicates (226, 228); in addition, the proportions synthesized can depend on the host cultivar. Cucumber protoplasts, irradiated with UV light and treated with actinomycin D to inhibit host RNA synthesis, were inoculated with CMV-W and CMV RNA species synthesized at different times during the growth cycle monitored by labeling with ^{3}H-uridine prior to RNA extraction and electrophoretic separation. The proportion of RNA-2 synthesized in protoplasts from the susceptible Ashley plants was greater than that in protoplasts from resistant China-K plants (Boulton, Maule, and Wood, unpublished).

The four RNAs have been translated in vitro in wheat embryo and reticulocyte lysate systems and in toad oocytes (395, 396); the translation products produced by reticulocyte lysates are indicated in Fig. 5–46. The complete nucleotide sequence of RNA-3 of CMV-Q (166) indicates coding capacity for two polypeptides, "3A" (MW 36,700) and coat (MW 26,200), both in the same reading frame. The MWs of the products of in vitro translation are, therefore, slightly in error. The coat protein gene is contained within the 3'-end of RNA 3 but, as is often the case, is closed. Coat protein is translated from the subgenomic RNA 4 which, while not necessary for infectivity, is encapsidated along with the genomic RNAs. Although it is not clear how this RNA species is generated initially, the observation of ds forms of RNA 4 in both plants and protoplasts suggests that it may be replicated (e.g. 435).

In addition to the polypeptides illustrated in Fig. 5–46, RNA 3 directed synthesis of a MW 39,000 polypeptide in the wheat embryo system, and RNA 4, of an MW 25–29,000 polypeptide, neither of which was produced by reticulocyte lysates. They may share sequences in common with those of Fig. 5–46 and arise by initiation and termination at different points on the genome. In vivo, coat protein synthesis has been monitored in cowpea protoplasts (239), and several polypeptides have been observed in cucumber cotyledons using the dual labeling procedure (496), although their significance in relation to the products of in vitro translation is undetermined. In endeavors to determine the biological function of the three genomic RNAs, pseudo-recombinants have been constructed by mixing purified RNA species 1, 2, and 3 from several strains in different combinations (e.g. 107, 174, 255, 274, 297, 351). RNA 3, containing the coat protein gene, determined serological specificity, electrophoretic mobility, and aphid transmissibility. Symptom production, however, appears to be the result of a complex interaction between the genome of virus and host (351). In some instances, symptoms may be apparently determined by a single viral RNA species; pseudo-recombinants containing

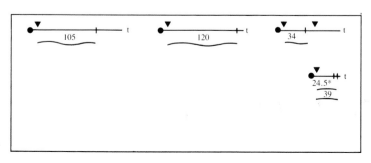

Fig. 5–46. Expression of the CMV genome.

RNA 2 from CMV-U caused necrosis in *N. ed-wardsonii*, and those with RNA 3 from CMV- K induced necrotic lesions in *V. faba*. However, the observation that pseudo-recombinants can produce symptoms atypical of either parent strain (351) suggests that, in other instances, determinants for symptom production reside in at least two species. The interaction between virus and host genome is indicated by, for example, infection of *G. globosa* and *D. stramonium* with pseudo-recmbinants containing RNAs from CMV-M. Chlorotic local lesions in *G. globosa* were produced by RNA mixtures containing RNA 3 from this virus but not by those containing RNA 2, while in *D. stramonium* lesions were produced by mixtures containing either RNA 2 or RNA 3.

In infected tissue, particles are found predominantly in the cytoplasm and less frequently in the nucleus (188, 200, 201). Inhibition of cytoplasmic ribosome function with cycloheximide or of chloroplast ribosome function with chloramphenicol inhibited virus accumulation, suggesting an obligatory role for cytoplasmic protein synthesis in virus replication (15). The association of ds RNA with vesicles associated with the tonoplast suggests that they may be the site of RNA synthesis (189). The site of assembly is uncertain.

In inoculated protoplasts, virus accumulation reaches a maximum in 24–48 h, and then declines (158, 281). In both inoculated and systemically infected leaves, however, the time course of virus accumulation is rather more complex, exhibiting a cyclic pattern of maxima and minima (276, 467, 483) (Fig. 5–47). The cause is uncertain but may be related to the presence of satellite; substantial satellite replication can affect accumulation of CMV (231).

ss (+) Satellite RNA

Cucumber mosaic virus-associated RNA 5. Separation of the RNA species of several CMV strains on

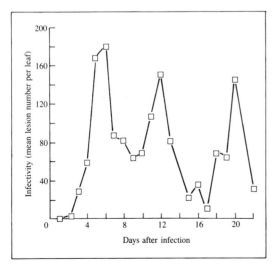

Fig. 5–47. Accumulation of CMV-W, in inoculated tobacco leaves (after Wood and Barbara, 483).

polyacrylamide or agarose gels reveals the presence of a fifth component, often representing a significant proportion of the total RNA (231, 302). This fifth component is not produced when plants are inoculated with the three CMV genomic RNAs, and it is unable to replicate autonomously. It is encapsidated by the CMV capsid protein, either in multiples or in association with CMV RNAs (231, 267). The base sequence, however, has little or no homology with that of the CMV species (165, 231), although these may be similarities in the structure at the 3′-termini (164). Double-stranded forms of this RNA are present in the population of ds CMV RNAs from infected cells, although they may be present in such high concentrations that they are unlikely to be involved in replication of the (+) RNA (231).

This fifth RNA species has, therefore, properties typical of a satellite. The first to be characterized was that isolated from CMV-S and referred to as CARNA 5. It has since been shown to have a considerable degree of sequence homology with comparable RNA species from several other CMV strains (227, 298, 361). However, not all CMV strains will act as helpers for CARNA 5. Also, while TAV and strains of CMV will support replication of the satellite from CMV-Q (165), replication of the satellite from another cucumovirus, PSV, is not supported by strains of CMV (231).

Although it seems likely that the same polymerase involved in replication of CMV RNA is also responsible for replication of satellite RNA, the participation of polypeptides coded for by satellite RNA remains a possibility. However, information on translation of satellite RNAs is limited, and the function of any virus-coded polypeptides remains unknown. The base sequence of a CARNA 5 isolated from CMV-D suggests that it could code for two polypeptides, MW 2,800 and 4,800 (361). In vitro translation in a wheat germ system reveals two, MW 3,800 and 5,200 (320). None, however, has been observed in vivo.

The extent of CARNA 5 replication appears to vary with both host and helper strain (226), replication of this and other satellites also affecting the accumulation of helper CMV genomic RNAs (228, 298, 334) (Fig. 5–48). It has also been suggested that the cyclical appearance of CMV growth curves is due to replication of satellite (231). In addition, symptoms can be modified by presence of satellite. In general, the effect is one of reduction in severity, which parallels reduction in multiplication of helper CMV. This is not always the case, however. The presence of CARNA 5, for example, leads to the production of a much more severe infection of tomatoes than that caused by CMV in its absence (230). This is not a feature common to all satellites, and it is suggested that RNA species causing necrosis in tomatoes be redesignated (n) CARNA 5, to permit distinction from others that do not cause such severe symptoms in this host (229, 298). CMV-Y contains a satellite that increases symptom severity in tobacco, producing a bright yellow mosaic instead of the more usual mild mosaic (434). Clearly, the interaction between satellite and helper is a complex one about which little is known.

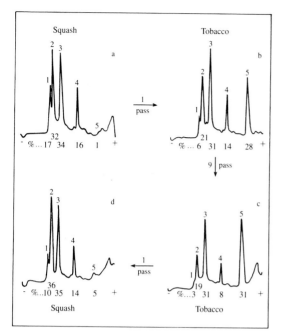

Fig. 5–48. Profiles of the RNAs of CMV + CARNA 5 separated by electrophoresis in polyacrylamide gels. RNAs were prepared from virus maintained in squash (a), then passed once in tobacco (b), maintained in tobacco for 9 subsequent passages (c), and finally passed once in squash (d) (after Kaper et al., 228).

The modification by satellite of both accumulation of helper virus and symptom expression needs further investigation.

ss (−) RNA

Lettuce necrotic yellows virus. The bacilliform particles of LNYV, one of the plant rhabdoviruses, consist of a single (−) RNA associated with at least five proteins (130–134). A glycoprotein (G) and a matrix protein (M) are associated with the viral envelope. Treatment with nonidet P40 removes the envelope to reveal three further polypeptides, N (MW 56,000), L, and NS (MW 38,000). The RNA is not infectious but requires the presence of a virion-encapsidated polypeptide for replication. Experiments monitoring incorporation of ^{3}H-uridine into de novo synthesized RNA suggest that viral RNA is synthesized in the nucleus early in infection, but later in the cytoplasm (479). Particles are probably assembled in the cytoplasm. No information exists on translation strategy or nonvirion polypeptides.

ds RNA

Wound tumor virus. Each WTV particle contains 12 double-stranded genome segments encapsidated into a nucleocapsid containing seven polypeptides, including replicase (32, 355). The coding potential of seven of the RNA species correlates well with the seven structural polypeptides. There is no information on nonvirion polypeptides, although there is a suggestion that the product of RNA-2 is involved in leafhopper transmission (354). It seems likely that virus replicates in the cytoplasm, possibly in association with viroplasms (482).

ss DNA

Bean golden mosaic virus. The genome of this virus, a typical geminivirus, is described above. There are, however, no reports on its transcription or translation. However, since these viruses are potentially useful as vehicles for introducing foreign DNA into plant cells, much more information on the strategy of their genomes can be expected in the near future. The appearance of structures containing deoxyribonucleoprotein associated with the nucleus of infected cells suggests replication within the nucleus (236).

ds DNA

Cauliflower mosaic virus. Current interest in the molecular biology of CaMV and other ds DNA-containing plant viruses has been considerably stimulated by their potential use as vehicles for transferring foreign DNA (202). The genome is encapsidated into spherical particles, with a polypeptide MW 58,000, which appears to be processed in vivo to several smaller polypeptides, MW 34,000, 37,000, and 44,000. That of MW 64,000 (205) appears to be a dimer of MW 34,000 (177).

Both linear and circular forms of the naked DNA are infective, permitting the infection of turnip (*Brassica campestris*, Just Right) leaves with either viral DNA or viral DNA into which additional DNA sequences have been inserted (145, 168, 260, 485). It may be, however, that for translocation within the leaf to occur the replicated genome needs to be encapsidated. This would, of course, place a limit on the size of foreign DNA that could be inserted into the complete CaMV genome. It is not certain whether replication and/or assembly takes place in the nucleus or at the site of the cytoplasmic inclusions or viroplasms that are a characteristic feature of CaMV-infected cells in intact tissue. However, nuclei isolated from infected turnip plants will synthesize viral DNA (4), and supercoiled viral DNA has been found in virus-infected plants as a chromatinlike nucleoprotein complex (286, 287). Also, viroplasm formation seems to be regulated independently of virus production (486), and normal viroplasms do not appear in infected protoplasts. The nucleus, therefore, seems the likely site of DNA synthesis.

The sequence of the CaMV genome reveals six potential coding regions; it appears that coat protein comes from region IV (Fig. 5–30; 76), region II codes for an aphid acquisition factor, region V for a replicase, which may be a reverse transcriptase, and region VI for inclusion body protein. Although several RNA transcripts of the α-strand of the DNA genome have been detected (102, 206), the only nonvirion polypeptide detected so far appears to be one of ca. MW 62,000, identified in both infected cells and as an in vitro translation product of a 19s mRNA isolated from infected tissue (2, 72, 73,

310). It is the major protein component of the viroplasm. Enzymes involved in DNA replication and transcription, and the aphid acquisition factor, remain uncharacterized (205).

Viroids

Potato spindle tuber viroid. The viroid genome is not encapsidated. Infective PSTV RNA appears in the nuclear fraction of extracts from infected plants (94) and can be replicated in vitro using purified nuclei as a source of enzyme. These observations support the view that viroid replication takes place in the nucleus.

Despite earlier reports of the inhibition of PSTV replication by actinomycin D (96) and of CPFV by α-amanitin under conditions in which host RNA polymerase II was inhibited (301), which suggested an involvement of DNA in viroid replication, more recent evidence favors an alternative view. Experiments by Grill and Semancik (167) suggest that replication is not sensitive to actinomycin D, while, contrary to earlier reports, tomato DNA does not appear to contain sequences complementary to PSTV RNA (493). In addition, (−) RNA complementary to the complete PSTV genome is produced in infected tomato, suggesting that replication proceeds via a complementary RNA copy (319, 495).

The nucleotide sequence of PSTV RNA does not contain an AUG initiation triplet, nor is the RNA translated in vitro (93, 94); apparently new polypeptides observed in infected tissue have been shown to be of host origin. PSTV RNA does not, therefore, appear to act as messenger, although it cannot yet be ruled out that the (−) RNA fulfills this function (279).

In the absence of detectable viroid-coded polypeptides, it seems that effects on the host are mediated via RNA, possibly interacting directly with host DNA. However, the situation both with respect to pathogenicity and mode of replication is far from resolved. It is suggested that viral (+) RNA is copied to (−) RNA by host RNA polymerase II, the (−) RNA then serving as template for transcription to (+) RNA by host RNA polymerase I.

Effect of Environment on Virus Replication

The effects on virus accumulation of changes in temperature, lighting conditions, and various contributors to plant nutrition have been observed in many virus/host combinations. Most authors have used infectivity of crude sap, from either inoculated or systemically infected leaves, as an estimate of virus accumulation. Others have purified virus before assessing infectivity or total quantity of virus either by optical density or serological methods. However, inhibitors of infection, present in crude sap, may also vary with changes in environment. Their presence may, therefore, affect the assessment of infectivity if crude sap is used (e.g. 468). In addition, the amount of virus accumulated at a given time, as assessed by these methods, is the net result of both synthesis and degradation. The majority of reports, therefore, refer to virus present at the time of assay

rather than to rate of synthesis. With these reservations in mind, the following conclusions can be drawn:

Temperature. There is usually an optimal temperature for virus accumulation. Below the optimum, virus multiplication or cell-to-cell spread is slow, although degradation is also slow. Above the optimum both rate of replication and rate of degradation may also be greater, so that, although the peak of virus accumulation may occur earlier than at the lower temperature, accumulation decreases with time as replication declines and degradation continues. For the majority of viruses, the optimum is around 20 C, although it varies between strains and hosts. Accumulation of PVX in *N. glutinosa* and tobacco (341), SMV in squash and pumpkin (*Cucurbita pepo*) (10), and TRV in tobacco (142) is optimal at 20 C or below. TMV replicates slowly in tobacco at 12 C and more rapidly at 25 C (Fig. 5–49; 86). Although the maximum concentration in tobacco discs, as assessed serologically, is reached earlier at higher temperatures (32 C), it is not so great as that at lower temperatures (24 C) (259). Eventually, a temperature is reached at which virus replication is minimal. This temperature varies between viruses, but for

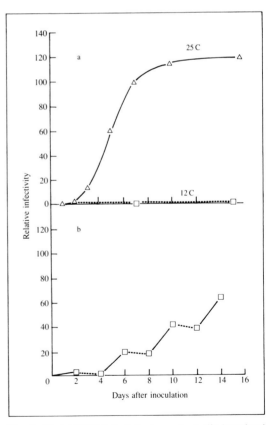

Fig. 5–49. (a) TMV infectivity in mechanically inoculated tobacco leaves maintained at 25 or 12 C. (b) TMV infectivity in mechanically inoculated tobacco leaves incubated alternately at 12 C (- - -) and 25 C (———) (after Dawson et al., 86).

TMV the upper limit is between 36 and 40 C (79, 232, 233, 487). There is little information on the nature of the functions inhibited by high temperatures, but for TMV it appears that the principal effect is to inhibit ss RNA synthesis (79). The inhibition of virus replication at elevated temperatures has found application, often in conjunction with other techniques such as callus or meristem tip culture, in the elimination of virus from infected plants.

Light. Virus accumulation is usually greatest when plants are grown, or excised tissue maintained, under a high light intensity for a relatively long period. There was, for example, more TNV in discs maintained in light than in dark conditions (20), and the concentration of TuMV in rape increased with an increasing photoperiod of between 4 and 16 h (340). However, the converse may be true for normally resistant plants. For example, although TMV multiplies to a limited extent in cowpea and cotton plants maintained in the light, dark treatment results in a substantial increase in virus accumulation (e.g. 430). It would appear that, in these cases, the effect is on virus translocation, rather than on extent of replication in infected cells.

Nutrition. Nutritional conditions favoring maximum plant growth have also been found to generally favor maximum virus accumulation, although there are exceptions. There was, for example, a close correlation between growth of tobacco plants supplied with varying concentrations of nitrogen and TMV accumulation (468). Accumulation of TuMV in *N. glutinosa* reflected host growth when supplied with nitrogen, although when phosphorus concentration varied, there was a stimulation of virus accumulation in deficient plants (342). TMV reached its highest concentration in tobacco at molybdenum concentrations optimal for plant growth (336), while in other experiments highest concentration of TMV was found in plants deficient in manganese.

It is evident that changes in plant metabolism caused by environmental changes are extremely complex, and it is not surprising that explanations for the effects on virus accumulation are not available.

Symptom formation, the visible manifestation of virus infection, may also be affected by environmental conditions. It is the result of metabolic changes that are, as yet, not understood, and changes, relative to healthy plants maintained in the same environment, are not necessarily related to the extent of virus accumulation. For example, accumulation of TSWV in tomato was greater at 20 C than at 36 C, although symptoms were more severe at 36 C (232). In extreme cases, symptom type may change completely. Hypersensitivity conferred on tobacco cultivars by the N gene from *N. glutinosa* is temperature sensitive, with necrotic local lesions produced by TMV at moderate temperatures (around 22 C) and systemic invasion produced at temperatures in excess of ca. 30 C. Genes conferring resistance in other plants may also be temperature sensitive, with the pro-

duction at higher temperatures of more virus and/or more severe symptoms (see also Chap. 10).

Plants Infected by Bacteria

Nucleic Acid and Nitrogen Metabolism in Tumor Tissue Induced by Agrobacterium tumefaciens

A review by Braun (42) has acknowledged general agreement that there are marked differences in the nitrogenous constituents between normal resting cells and actively dividing tumor cells, incited to division by the crown gall bacterium, *Agrobacterium tumefaciens*. In this regard Neish and Hibbert (304) have reported that while both normal and tumor tissues contain about the same amount of nonprotein nitrogen, tumors have three times the protein of normal tissue, and the difference in water-soluble protein is 6:1 in favor of the tumor tissue. They also noted that tumors maintain more than 60% of their Kjeldahl nitrogen in the form of protein, whereas in normal tissue this amounts to approximately 40%. These data were interpreted as indicating a greater tendency and capacity on the part of tumor tissue to synthesize protein.

According to a study by Klein (238) the greatest increases in protein and soluble nitrogen became apparent in tumor tissue late in its development, 27–35d after inoculation (Fig. 5–50), and seem to be initiated at the time when the tumor changes from primarily numerical cell increase (hyperplasia) to cell volume increase (hypertrophy). He suggested that increased protein synthesis registered in the tumor tissue at this time did not occur entirely at the expense of free amino acids and amides in close proximity to the tumor. He postulated on the basis of the stunted and nitrogen-deficient appearance of the tumorous plant that the nitrogen of the whole plant was being mobilized into the tumor.

Tumor Induction

The basis for transforming normal plant cells and tumor cells exhibiting a non-self-limiting form of growth has, as we note in Chap. 7, been resolved. It is of

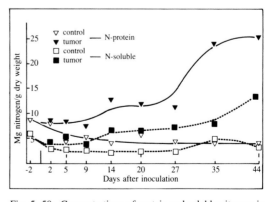

Fig. 5–50. Concentrations of protein and soluble nitrogen in control and tumorous tomato hypocotyl tissues during tumor ontogeny (after Klein, 238).

historical interest to follow an early investigative path that eventually exposed the nature of tumor inducing principle (TIP). It seemed apparent from data presented by Klein (238), who described a 30–50% increase in DNA level at approximately 48 h after inoculation, that this initial high level of DNA, occurring well in advance of cell division, represented an activity of the pretumorous cells that might be directly related to the process of tumor induction. However, any significant relationship of this early-appearing DNA to tumor induction or to TIP itself was precluded by subsequent studies from Klein's laboratory (352) and Kupila and Stern (251), which revealed similar early increases in wounded control tissue. Hence the augmented DNA synthesis occurring 1–2 d after inoculation was, apparently, a response to wounding and not causally related to bacterial activity or tumor induction. According to Braun (42), TIP stimulates in the tumor the autonomous production of IAA, myo-inositol, glutamine, and asparagine, as well as cytidylic and guanylic acids, some of the precursors for nucleoprotein and protein synthesis. This deduction is probably valid. However, these syntheses actually reflect the genetic information carried by the bacteria (457).

In 1967 Stroun et al. (428) presented evidence that seemed to support the earlier study of Klein (238). They concluded that *A. tumefaciens* DNA actually enters the plant cell nucleus and becomes integrated in the host genome. Theirs appears to be the first modern data that could explain the non-self-limiting growth of crown gall tumors. Subsequently, Chadra and Srivastava (56) reported the presence of bacteria-specific proteins in sterile tumor tissues. Hence the newly integrated bacterial DNA appeared to be coding for the synthesis of bacterial protein in the sterile tumor tissues. There were other similar substantiating reports by Schilperoort et al. (391–93).

Kado and Lurquin (221) interpreted the buoyant density data of Stroun et al. (428) as representing from 30–50% bacterial DNA into the recipient plant DNA. This rather astronomical amount of bacterial DNA seemed excessive to accomplish even the radical changes in host metabolism exemplified by tumor tissue. With highly refined DNA-DNA hybridization techniques, Kado (220) and Kado and Lurquin (221, 222) were able to detect as little as 0.01% *A. tumefaciens* DNA in tobacco callus DNA preparations. Nevertheless, they were unable to detect with confidence any bacterial DNA in the *Vinca, Phaseolus*, and tobacco tissue they treated with *A. tumefaciens* DNA.

Kado and Lurquin (222) used the estimated MW of 6×10^{12} for the *N. tabacum* tetraploid genome (410) and 2×10^9 for the genome of *A. tumefaciens* (61) and calculated that *one complete* bacterial genome per tetraploid tobacco cell would represent 0.033% of the cell's total DNA. Hence, their studies inferred that an amount less than one-third of an *A. tumefaciens* genome could be present in a tetraploid tobacco cell and escape detection.

Accepting the figure of 10,000 genes/genome for *Escherichia coli* (45) as an approximate gene number for *A. tumefaciens*, then less than 1/3 genome would still represent a significant number of genes. In conjugation it has been determined that 30–50 *E. coli* genes are transferred from a donor to recipient cell in one minute. Hence the transfer of even this small amount of DNA (3.5×10^4 base pairs) could have reflected the genetic alterations associated with the transformation of a normal plant cell to a tumor cell. It has been calculated that 23×10^4 base pairs are actually transferred from the plasmid, or about 10% of the 100–150 megadalton plasmid.

Kosuge et al. (245) provided evidence that *Pseudomonas savastanoi*, which causes gall formation on stems of olive and oleander, synthesizes IAA from tryptophan via indoleacetamide. Synthesis proceeds with the participation of the two enzymes of tryptophan mono-oxygenase (TMO) and indoleacetamide hydrolase (INH). Subsequently, Smidt and Kosuge (416) linked pathogenesis of *P. savastanoi* to IAA synthesis as IAA$^-$ mutants of the pathogen were unable to cause gall formation. The genetics of the process was clearly defined by Comai and Kosuge (67, 68), who showed that the controlling genes for IAA synthesis were extrachromosomal. The plasmid carrying the necessary information is the second largest detected in the bacterium and has a MW of 34×10^6. IAA$^-$ mutants were obtained by curing pathogenic IAA$^+$ isolates in Kings B broth and acridine orange. Acridine orange prevents inclusion of the plasmid PIAAI; as a consequence, two enzymes required for IAA synthesis, TMO and INH, are lost. Transformation of IAA$^-$ isolates with the resultant uptake of the plasmid PIAAI restored competency for the synthesis of the two enzymes, the capacity to synthesize IAA from tryptophan and to cause galls on stem tissue. Hence *P. savastonoi* is the second bacterial plant pathogen known to carry its virulence determinant on a plasmid.

Nucleic Acid Synthesis Associated with Cell Division

The second large flush of DNA synthesis observed by Klein (238) in tumorous tissue occurred ca. 14 d after inoculation and had a less controversial basis. It apparently reflected the increased DNA synthesis that accompanies cell division (chromosomal duplication). Since the DNA content of the nucleus remains more or less constant during the life of the cell, the recorded rise was simply indicative of the large number of cells per unit weight of tumor tissue (hyperplasia).

A later study by Rasch et al. (352) revealed that 3 d after inoculation prospective tumor cells were characterized by marked increases in nucleolar and nuclear volumes. Concomitant increases in RNA and protein were detected, particularly in the nucleolus and cytoplasm. The marked increase in ribonucleoprotein content of developing tumor cells was interpreted as reflecting either the activation of mechanisms of protein synthesis as a part of the transformation process or the response of prospective tumor cells to auxin being released by the bacterial cells. The stimulation of RNA and protein synthesis by auxin has been reported by Skoog (414).

A rise in RNA content was also observed in tumor tissue by Klein (238) from 27 to 35 d after inoculation and was interpreted as being causally related to the concomitant increase in protein noted in Fig. 5–50, which signaled the change in tumor tissue from numerical cell increase to cell volume increase.

Genetic Parasitism and the Specificity of A. tumefaciens *DNA expressed in Amino Acid Synthesis*

A. tumefaciens isolates (biovars 1,2,3) have, through their plasmids, developed the genetic capability to impart to their hosts the capacity to synthesize families of amino acids designated as opines (octopines, nopalines, agropines, and agrocinopines) that only the infecting strains have the ability to catabolize. Hence the infecting organism diverts a part of the metabolism of the host plant to the synthesis of amino acids that only it can use. This so-called genetic parasitism represents the ultimate form of parasitism, in which the pathogen need only be present for a short time.

It has been established that, in most instances, the tumors induced by the transferred Ti-plasmids synthesize the opine that only the inducing strain can catabolize. The amino acids coded for appear in Figure 5–51. None of these amino acids has been found in normal tissue nor, of course, have the enzymes that synthesize them.

Octopine synthetase catalyzes the reductive condensation of the pyruvate with arginine to form octopine, lysine to form lysopine, and histidine to form histopine. The enzyme has been named D(+)-lysopine dehydrogenase (LPDH), which utilizes both NADH and NADPH. Nopaline synthase is the enzyme that catalyzes the reductive condensation of either arginine or ornithine with αketoglutaric acid to form nopaline or

orniline. The biosynthesis of agropine has as yet not been worked out. A proposed structure for agropine is the condensation product of aldohexose and glutamic acid.

Influence of the Pseudomonas tabaci *Toxin and Kinetin on RNA and Protein Metabolism*

The discovery of Richmond and Lang (362) that kinetin reduced protein breakdown in detached leaves of healthy plants focused the attention of Farkas et al. (118) on this phenomenon as it influenced the effect of the wildfire toxin produced by *Pseudomonas tabaci*. Recognition of the similarity of the chlorotic halos induced by the wildfire toxin to the chlorosis normally appearing in detached leaves led these workers to use this toxic effect of phytobacterial pathogenesis to study the influence of kinetin on protein metabolism.

In their initial study, Farkas and Lovrekovich (118) were able to prevent the toxin-induced chlorosis in tobacco leaves that had previously been infiltrated with kinetin. Their second series of experiments (269) was designed to study the influence of the toxin on protein metabolism in tobacco tissue. Their data, which appear in Table 5–7, indicate that the balance between protein synthesis and breakdown is shifted to increased protein degradation by the toxin.

When half-leaves were treated with culture filtrates of *P. tabaci*, sprayed immediately with kinetin, and sprayed twice more at 24-h intervals with the hormone, there seemed to be a strong sparing effect on the degradation of protein (Table 5–8).

From these data it was clear that protein metabolism in the tobacco leaf was being adversely affected by the wildfire toxin. Convincing data were presented by Osborne (313) that suggested that both the decrease in protein level in detached leaves and the reversal of this phenomenon by kinetin were associated with changes in

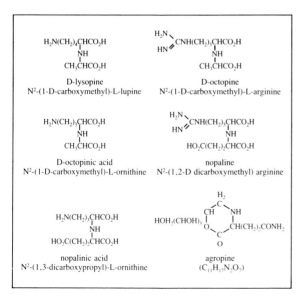

Fig. 5–51. Chemical configuration of some opines (amino acids) synthesized by enzymes coded for by the Ti plasmid.

Table 5–7. Protein, free amino acid, and ammonia content of tobacco half-leaves treated withh toxin-containing culture filtrate of *Pseudomonas tabaci* and with nutrient medium, respectively (after Farkas and Lovrekovich, 118).

Exp. no.	Protein		Amino-N		Ammonia	
	mg/g C	fr. wt. T	µg/g C	fr. wt. T	µg/g C	fr. wt. T
1	14.0	7.3	51	87	24	570
2	9.0	6.6	26	120	10	460
3	9.7	6.6	40	105	17	510

C: tissues treated with Czapek-Dox nutrient medium
T: tissues treated with toxin-containing culture filtrate

RNA concentration. Since the RNA:protein ratio is, according to Osborne, a constant, the data of Farkas and Lovrekovich (118) and the observations by Richmond and Lang (362) suggest that alterations in protein metabolism may actually reflect a primary change in RNA metabolism (protein synthesis) in leaves that are either detached or affected by wildfire toxin. This hypothesis was studied by Lovrekovich et al. (270), who found that tobacco tissue treated with the toxin contained 40% less RNA than the controls and that the treatment of tobacco leaves with 10^{-4} M kinetin following exposure of the tissue to the toxin resulted in a 36% greater RNA content than was detected in control tissues. Hence RNA levels per se were reduced in tobacco tissue by the wildfire toxin, and this effect could be at least partially offset by kinetin. The hypothesis that ribonuclease controls RNA content in the plant and that this regulatory process is interfered with by the toxin was also tested; the results indicated that the toxin potentiated ribonuclease activity 44%. Thus, the effect of the wildfire toxin could be due to an augmentation of ribonuclease activity, which in turn "damaged" RNA metabolism. Curiously, kinetin applications were not able to reverse the increase in ribonuclease activity.

Table 5–8. Effect of kinetin on the protein content of tobacco leaves treated with toxin-containing culture filtrates of *Pseudomonas tabaci* (after Farkas and Lovrekovich, 118).

Exp. no.	Protein mg/g fresh weight half leaves sprayed with	
	Water	10^{-5} M kinetin
1	5.1	9.9
2	4.3	9.9
3	6.5	12.0

Erwinia amylovora *DNA as a Protectant against Infections.*

Protection of plants against bacterial infection with whole bacteria or bacterial components is treated in detail in Chap. 10. However, the perplexing experiments performed by McIntyre et al. (283) concerning protection afforded by DNA should be commented upon here as well. The protective effect was even detected at sites removed from the point of DNA application. The authenticity and purity of DNA extracted from *Erwinia amylovora* by either sonication or the "Marmur" procedure were established satisfactorily.

In essence, DNA (and not RNA) extracted from either virulent or avirulent *E. amylovora* absorbed by Bartlett pear seedlings that had their radicles excised or given as a localized injection afforded protection against infection. A challenge inoculum of 10^2 or 10^3 virulent *E. amylovora* cells was most effectively protected against when an 18-h interval separated the protective treatment and the challenge. The effect of *E. amylovora* DNA appeared to be specific, as neither unrelated bacterial DNA nor salmon sperm DNA, a series of proteins or nucleic acid components, were competent in providing protection. The efficacy of the nonsheared (Marmur) and consequently larger fragments of DNA was greater than that of the sheared (sonicated) preparation. Of additional significance was the observation that the protective effect of DNA was destroyed following its exposure to DNase. A definitive explanation for the phenomenon has not been advanced.

Amino Acid Metabolism as Altered by the Halo-inducing Pseudomonads and Blight Bacterium Pseudomonas phaseolicola

A full analysis of alterations in the amino acid and amide pool as a result of pathogenesis was conducted by Patel and Walker (324). Their experiments were conducted with bean plants susceptible and resistant to the halo blight bacterium, *Pseudomonas phaseolicola*. Infection of the first trifoliate leaf was characterized in the resistant variety by small brown lesions. In general, the amino acid and amide content was not appreciably altered by this infection. On the other hand, the susceptible variety showed water-soaked lesions after 2–3 d that increased in size, forming conspicuous halos in 7–9 d. In addition, the second and third trifoliate leaves showed a severe systemic chlorosis.

Chromatographic analysis of healthy and inoculated leaves of the susceptible variety revealed 22 ninhydrin-positive compounds; of these, 15 were identified as amino acids and amides. Selected data showing concentration differences of amino compounds in healthy, infected, and chlorotic leaf tissue may be seen in Table 5–9. Of particular interest is the fact that ornithine and histidine, which were not detectable in healthy tissue, appeared to accumulate in significant proportions in both infected and chlorotic tissue.

Alterations in amino acids in inoculated and chlorotic tissue were determined by amino acid analyzer, and

Table 5–9. Free amino acid and amide concentrations as estimated by paper chromatography in halo blight-infected leaves from susceptible bean plants (after Patel and Walker, 324).

| Amino compound | Concentration indices | | | |
| | 1st trifoliate leaves | | 2nd trifoliate leaves | |
	Healthy	Infected	Healthy	Chlorotic
Glycine	0.75	1.50	0.75	1.25
Serine	0.05	1.75	1.00	1.75
Glutamine	1.25	2.00	1.50	2.50
Asparagine	0.75	1.25	1.00	1.50
Lysine	Trace	1.00	0.25	2.25
Histidine	0.0	2.00	0.0	2.25
Ornithine	0.0	3.50	0.0	6.00

again there was a large accumulation of ornithine, histidine and methionine in inoculated leaf tissue. In addition more than five times as much ornithine accumulated in the uninoculated chlorotic leaves (Table 5–10).

Patel and Walker also compared the amino acid metabolism in the chlorotic halo with the adjacent green tissue. A number of amino acids were higher in the halo area than in the apparently healthy green tissue, including histidine, methionine, and serine. In this analysis the accumulation of ornithine was also striking, more than 20 times as much being detected in the halo tissue.

Waitz and Schwartz (464) showed that a toxin produced by *P. phaseolicola* was responsible for the chlorosis induced by the pathogen. They demonstrated that the trifoliate leaves above infected primary leaves could become chlorotic and yet be completely devoid of bacteria. It would seem, therefore, that the toxin is by itself responsible for the altered amino acid metabolism. This contention seems plausible from the work of Rudolph and Stahmann (377), in which the total protein of susceptible bean foliage was found to decrease only

Table 5–10. Free amino acid concentrations as estimated by amino acid analyzer from inoculated and chlorotic leaves from susceptible bean plants (after Patel and Walker, 324).

| Amino compound | Concentration as percent of healthy control leaves | |
	Inoculated	Chlorotic
Ornithine	4,178	25,315
Histidine	664	1,717
Methionine	454	2,801

slightly; however, this decrease was more pronounced in the chlorotic secondary trifoliate leaves. The toxin produced by *P. phaseolicola* and its mode of action will be discussed in greater detail in Chap. 9. According to Patil (326) the toxin induced an allosteric competitive inhibition of the enzyme that converts ornithine to citrulline; hence interference with its activity resulted in ornithine accumulation and effected a block in the "ornithine cycle."

The relationship of the accumulation of ornithine and histidine, α amino acids not incorporated into protein, to the observed chlorosis can only be conjectured upon. Ornithine in the ornithine cycle regulates the removal of ammonium ions, and it is known that ammonia tends to accumulate in tissues undergoing protein degradation (see Table 5–7). Thus ornithine may be forming as a result of the amination of glutamic acid. Thus the reaction

$$\text{Glutamic acid} + NADH + NH_2 \rightarrow \text{ornithine} + NAD^+$$

may be proceeding at an accelerated rate; however, its conversion to citrulline is blocked. Patel and Walker (324) offer as evidence for this sequence the fact that citrulline was not detected in either infected or healthy tissue.

Further evidence for this conclusion was subsequently provided by Rudolph and Stahmann (377), who found that L-ornithine fed to healthy bean shoots was quickly metabolized and led to the accumulation of arginine. Significantly, arginine did not accumulate in the halo blight-infected leaves. It would appear, therefore, that the ornithine cycle was indeed blocked in bean plants infected by strains of *P. phaseolicola* capable of producing the halo-inducing toxin. Crude toxin preparations were inhibitory to the carbamoylphosphate: L-ornithine carbamoyltransferase (EC 2.1.3.3.), the enzyme that catalyzes the carbamoylation of ornithine to citrulline. The toxin does not inhibit other enzymes of the urea cycle (326). From this it is clear why ornithine increased in infected tissue and why the level of arginine decreased in chlorotic tissues.

Ammonia Accumulation in Bacterially Infected Tissue

Durbin (103, 254) studied the question of ammonia (NH$_3$) accumulation in tobacco leaves infected by *P. tabaci* and the relationship to the pathogen's toxin, tabtoxin. Four hours after the exposure of tobacco leaves to purifed tabtoxin, degradation of chloroplast structure was extensive and two- to fourfold increases in NH$_3$ were detected. It was concluded that the increases in NH$_3$ were *not* due to the degradation of plant protein. Hence, if the toxin does not inhibit glutamine synthestase (as previously thought, 412), what is responsible for the buildup of NH$_3$? The rapid disintegration of chloroplasts in tobacco leaf cells upon exposure to NH$_3$ vapors is presented in Fig. 5–52 (160).

This question has been answered by Turner (449), who has shown that the NH$_3$ accumulating in tobacco leaves treated either with tabtoxin or *P. tabaci* can be

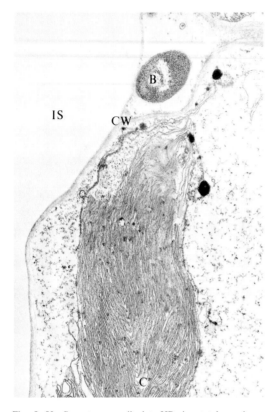

Fig. 5–52. Symptoms ascribed to HR, i.e. total membrane destruction of plant cell in contact with incompatible bacterium 6 h after inoculation. Initially believed to be due to the release of ammonia as a gas (NH_3). However, this does not occur until 24 h after inoculation. Membrane damage may be due to an accumulation of an ammonium ($NH4^+$) ion. B, bacterium; C, chloroplast; CW, cell wall; IS, intercellular space (after Goodman, 160).

quantitatively accounted for by the loss of glutamine synthetase activity. This study also revealed that the earliest detectable physiological change in tabtoxin-treated tissue was the loss of glutamine synthetase activity followed by an increase in NH_3, the enzymes' substrate. The role proposed by Turner for tabtoxin in pathogenesis engendered by *P. tabaci* is to make NH_3 available to the pathogen and reduce the host's de novo protein synthesis as glutamine levels are depleted.

Subsequently, Turner and Debbage (450) reported that NH_3 production in tabtoxin-treated plants was reduced immediately when the tissue was darkened. Darkening resulted in a cessation of NH_3 accumulation and the development of chlorosis and necrosis. This linked the source of the accumulating NH_3 to the photorespiratory-N cycle (235). In the evolution of this cycle NH_3 is formed during the conversion of two molecules of glycine into one of serine. The NH_3 is utilized by glutamine synthetase, which is eventually incorporated as an amino group into intermediates of the pho-

torespiratory carbon cycle leading back to glycine. The flow of carbon through this cycle requires ATP and NADPH produced during photosynthesis. According to Cullimore and Sims (75), the NH_3 is derived from cell protein only during photorespiration. Hence, as photorespiration is halted, so is the liberation of NH_3. Clearly, glutamine synthetase is a key enzyme in the plant nitrogen cycle, which is in turn driven by the photorespiratory C-cycle and coupled to protein catabolism. Thus, in summary, the inhibition of glutamine synthetase by tabtoxin blocks the N-cycle causing the rapid accumulation of NH_3 formed during protein catabolism.

The accumulation of NH_3 during pathogenesis has also been related to symptom development in cotton inoculated with *Xanthomonas malvacearum* (308, 451), apple shoots inoculated with *Erwinia amylovora* (271), oats infected with *Pseudomonas coronafaciens*, and to the rapid destruction of plant tissue that accompanies the development of the hypersensitive reactions (HR)(160). In the latter instance, the data seemed inconclusive in that membrane damage and plant cell collapse attributed to NH_3 accumulation were shown to occur at least 40 h before NH_3 was detected in appreciable amounts. Ullrich and Novacky (308, 451) studied membrane potentials in cotton cotyledons inoculated with the compatible pathogen *X. malvacearum* or tobacco inoculated with the HR-inducing incompatible pathogen *P. pisi* and reported the accumulation of NH_3. The accumulation was three to four times greater in young protein-rich tobacco leaves than in older ones. When the inoculated tissues were maintained in the dark their membrane potentials (E_m) were reduced to the diffusion potential levels (-95mV), apparently inactivating the membrane's active or ion pump component. Illuminating the system raised E_m slightly beyond control values of -185mV.

Ullrich and Novacky (451) have contended that in bacteria-inoculated tissues NH_3 originates from a potentiated protein catabolism. However, in the light NH_3 is partly reassimilated via the chloroplasts' glutamine synthetase-glutamate synthase cycle, which maintains E_m at normal or higher levels. In the dark, in spite of adequate ATP levels, normal E_m cannot be sustained. They suggested that NH_3 alkalinizes the cytoplasm in the dark and, as a consequence, the H^+/ATPase pump in the plasmalemma that regulates E_m is inactivated. Thus, both pathogenesis and HR seem to result in impaired plant membrane function which, in turn, causes NH_3 accumulation.

Convincing data concerning the necrogenic activity of NH_3 have been reported by Bashan et al. (17), leaving little doubt that the necrotic spots on tomato leaves and fruits caused by *P. tomato* reflect the activity of NH_3. Filtrates of *P. tomato* grown on either tomato leaf extracts or yeast peptone, but not synthetic or NO_3-glucose media, contained a necrosis-inducing component.

The necrogenic factor was dialyzable and not adsorbed by charcoal. When this fraction was eluted from a Sephadex G-25 column, the active fraction was found to contain proline, glutamic acid, tyrosine, phenylalanine, arginine, and ammonia. The eluate contained neither proteins nor peptides. Ammonia was present at 100 μ/ml, whereas the amino acids could be detected at 1–4 μ/ml levels. Only NH_3 in the active fraction could cause necrosis; when NH_3 was evaporated from this fraction, at pH 12, its necrogenic capability was eliminated. It was also possible to detect increasing levels of NH_3 in tomato leaf tissue 20 h after inoculation with *P. tomato*, and this increase continued until necrosis became apparent.

The data of Bashan et al. (17), and of several others, emphasize the relationship of increasing pH as NH_3 accumulates to toxic levels. Their data support earlier reports (160) that neither the ammonium (NH_4) ion nor a broad spectrum of amino acids is necrogenic at 100 μ/ml, whereas NH_3 is. It would appear from the data in Fig. 5–53 that NH_3 continues to accumulate for 100 h after inoculation. Twenty hours later, the pH reaches 8.0 and the NH_3 volatilizes and, in turn, damages the plasmalemma causing ion leakage (160, 271).

The accumulation of NH_3, during the course of pathogenesis, is suggestive of protein degradation. The obvious questions that arise are whether proteases or deaminases are requisite features of pathogenesis and if they contribute to the virulence of the bacterial pathogen. It is, of course, possible that protein degradation is a secondary event that signals earlier metabolic or physiological dysfunction.

Proteolysis as a Consequence of Pathogenesis

An increase in a limited number of amino acids in tobacco plants infected with *P. solanacearum* was reported by Pegg and Sequeira (329). Notably, aromatic amino acid synthesis, e.g. L-trytophan, L-tyrosine, L-phenylalanine, rose sharply. Of these L- phenylalanine showed the earliest significant increase. At 36 h after inoculation (pre-symptom period) the above amino acids registered 500–600% increases over healthy control plants. This increase, according to the authors, reflected de novo synthesis rather than proteolytic activity. Sustained sharp increases in these and other amino acids as well as ammonia at 96–144 h (intense symptom expression) may have signified protein hydrolysis.

Clear evidence of increased proteolytic activity as a consequence of infection has rarely been reported. Friedman (140) noted that virulent strains of *E. carotovora* have a greater proteolytic competence than avirulent mutants. Subsequent studies by Tseng and Mount (445) and Stephens and Wood (423) (see Table 5–11) also suggest that proteases may play a role *not* in tissue maceration caused by *E. carotovora* but in cell death per se, possibly by rearranging the membrane structure of protoplasts.

Bashan and Okon (personal communication) have also presented evidence that *P. tomato* synthesizes a protease constitutively during its logarithmic growth at 30 C that has a pH optimum of 7.0. The addition of proteinaceous substrate to the medium increased the

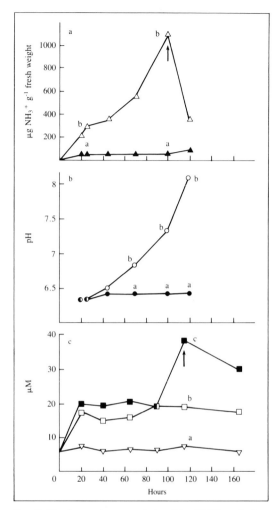

Fig. 5–53. Ammonia accumulation (a), pH (b), and electrolyte leakage (c) symptoms appearance in tomato plants infected with *Pseudomonas tomato* (arrow) and *P. fluorescens*. △, ammonia accumulation with *P. tomato*; ○, pH of leaves infected with *P. tomato*; ▲, ammonia accumulated with *P fluorescens*; ●, pH of leaves infected with *P. fluorescens*; ■, electrolyte leakage of leaves infected with *P. tomato*; □, electrolyte leakage of leaves infected with *P. fluorescens*; ▽, electrolyte leakage of untreated leaves. Points in the graph are mean of 2 experiments each carried out in 6 replicates. Different letters in the graph show significant differences at $P = 0.05$ (Bashan et al., 17).

specific activity of the enzyme. Proteolytic activity was also higher in diseased, susceptible plants than in resistant ones, and maximum activity was detected at the later stages of infection. This finding could indicate differences in population levels in resistant and susceptible hosts and does not necessarily implicate protease as a virulence factor. Closer analysis of the proteolytic activity revealed an array of six isozymes that were of either host or pathogen origin and isozymes peculiar to

Table 5–11. Effect of crude extract and enzymes on plasmolyzed protoplasts in carrot callus cells (after Stephens and Wood, 423).

Time (h)	PTE	PTE+ CMC-ase	Pr+Ph	PTE+ Pr+Ph	PTE+CMC-ase +Pr+Ph	Crude extract	Water
			Protoplasts alive (% of water control)				
1	95	101	89	66	63	53	100
2	93	104	82	57	62	48	100
3	98	102	73	54	56	46	100
4	90	100	63	47	48	42	100
5	93	95	67	47	52	43	100
6	93	95	62	46	47	36	100
10	100	98	61	58	58	30	100
20	96	94	60	54	44	24	100
24	100	102	74	68	53	20	100

Pr, proteinase; Ph, phosphatidase; CMC-ase, cellulase assayed on carboxymethylcellulose (CMC); PTE, pectin trans-eliminase

the host-pathogen interaction. Proteolytic activity was most intense around zones of developing necrosis and resulted in a decrease in soluble leaf proteins and an increase in asparagine. Bashan and Okon suggested that proteolytic activity may play a role in symptom development in the bacterial speck of disease of tomato. However, whether proteases were virulence determinants was not conclusively established. As noted earlier by Pegg and Sequeira (329), increases in free amino acids are detected following inoculation of susceptible plants with *P. solanacearum*. Probably the amino acids induce the production of deaminative enzymes that degrade asparagine and glutamine, as well as amino acids, to ammonia and carbon skeletons. The accumulation of ammonia alkalinizes the tissues, and when the pH reaches 8.0 the ammonium ion is converted to gaseous ammonia that could be responsible for the observed necrosis.

Reddy et al. (356) have detected a potential protease activity in alfalfa infected by *Xanthomonas alfalfae*. The enzyme produced by this leaf-spotting pathogen is active in vitro, releasing Congo Red from congocoll, and hydrolyzed gelatin, casein, and clotted milk. The amount of protease extracted from healthy and infected-resistant and susceptible alfalfa clones is presented in Fig. 4–30. Protease activity increased sharply with the development of the disease in the susceptible clone. The enzyme had a pH optimum of 8.0–8.5.

Protein Agglutinins

A phase of phytopathological research currently in its embryonic stages concerns microbial cell surface interactions with plant cell surfaces (203). The interest in small proteins, phytoagglutinins, and lectins has significant tenure in studies of *Rhizobium*-legume associations. However, relatively few studies have revealed a significant role for them vis-à-vis bacterial pathogens of plants. Nevertheless, proteins that have an affinity for

specific carbohydrate receptors are the subject of both active and contemplated research. Basically, the reactants are the surface-located proteins or glycoproteins that recognize specific carbohydrate moieties on compatible bacterial surfaces. Proteins with specific binding affinities have been detected on the roots of tree legume species for their compatible nodulating *Rhizobium* species (419).

Although lectin-phytobacterium affinities were postulated (162, 401), none has been established as influencing recognition. However, early literature contains implicating evidence of a class of low-molecular-weight proteins known as thionins that are antimicrobial by virtue of their agglutinating activity (54). The thionins, so named because of their high level of cysteine content (3–6/mol), also contain approximately 30% lysine and arginine and, thus, are highly positively charged. Thionins have been reported in wheat, corn, soybeans, and rice (217). Romeiro et al. (370, 371) found one in apple seed, stem, and leaf tissue (named malin), and Addy, Goodman and Karr (unpublished data) have detected thioninlike proteins in tomato and melon seed. Malin agglutinates the avirulent, unencapsulated mutant strain of *E. amylovora* at much higher dilutions than the virulent encapsulated parent strain (1024 vs. 36). The nature of the agglutinin action has been established as a charge-charge phenomenon with the highly positively charged protein being attracted to the negatively charged bacterial surface. It was further shown (371) that the negatively charged extracellular polysaccharide (EPS) of the virulent strain of *E. amylovora* effectively neutralized the agglutinating activity of the protein. This suggests that EPS may act as a virulence determinant by being able to form a complex with the protein. It is also probable that the granules agglutinating *E. amylovora* in Jonathan petiole xylem vessels detected by Huang et al. (203) are one and the same. It has been possible to raise antisera to malin which, when conjugated with colloidal gold, clearly stained malin-agglutinated *E. amylovora* in vitro.

Plants Infected by Fungi

Pathological Nitrogen Metabolism of Compatible Host-Pathogen Pairs

Proteins. In fungus-infected plants the total nitrogen and protein content of the host-pathogen complex generally increases during the early stage of disease. This increase may well be due to the presence of the pathogen or to newly synthesized proteins in the *host* (452). As shown in Table 5–12, in rust (*Uromyces phaseoli*)-infected beans the incorporation of labeled glycine into the protein fraction of the leaf is considerably higher at the infection centers than in the uninfected parts of the leaf, an indication of intensified protein synthesis in the diseased tissues.

This protein increase is due principally to protein synthesis of the pathogen rather than of the host, since it has been shown that tritium-labeled glycine fed to infected leaves is incorporated mainly into the fungal mycelium and uredospores (422). On the other hand, stimulated protein synthesis in the healthy tissues of the host around the infection center has also been observed in the infection of potato with *Phytophthora*, as well as in the infection of sweet potato with *Ceratocystis* and *Helicobasidium* fungi (452). Actually, the synthesis of some proteins and the degradation of others is stimulated. The enzymatic reactions may be influenced by changes in concentrations of activators, inhibitors, and effectors as well as coenzymes and substrates. The numbers of mitochondria are increased in both infected cells and noninfected cells around the infection sites. Enzyme complexes in the mitochondria may be directly involved in protein changes of the infected plant.

It is important to point out that protein changes in infected plants may be similar to those induced by *mechanical* injury or *senescence* or sometimes by *ripening* (116, 452, 453). Mechanical injury caused by infection is responsible for: (1) increase in respiration rate and in activity of cytochrome oxidase, glucose-6-P-dehydrogenase, 6-phosphogluconate dehydrogenase, invertase; (2) enhancement of peroxidase, polyphenoloxidase, ascorbic oxidase; (3) increased ribonuclease activity; (4) activated phenylalanine ammonia-lyase (PAL), a key

Table 5–12. Incorporation of glycine-C_{14} into the protein fraction of leaf tissue rings infected with *Uromyces phaseoli* (after Király et al., 237).

	μCi/g fr. wt.
Healthy leaf	79.0×10^{-3}
Infected leaf	
Green island, pustule removed	129.7×10^{-3}
Senescent part near the green islands	32.1×10^3

enzyme in aromatic biosynthesis; (5) increase in number and activity of mitochondria. All these processes also characterize wounded tissues (e.g. cut or injured plant cells) and are connected to the activation of RNA synthesis. The biological significance of this type of protein change is perhaps in the phenomenon of repair of wounded (injured) cells.

It is important to consider whether new types of pathogen-induced proteins are synthesized in the diseased host. Several lines of evidence indicate that these "new" proteins are not totally the same as those in uninfected plants. For example, Heitefuss et al. (192), using immuno-chemical analysis with rabbit antisera, demonstrated three new antigenic proteins, produced by the host, in susceptible cabbage infected by *Fusarium oxysporum* f.sp. *conglutinans*. The same was shown in sweet potatoes infected by *Ceratocystis fimbriata* (454). These examples indicate a general perturbation in the protein metabolism of the fungus-diseased plants.

However, de novo synthesis of new or accumulated proteins cannot be assumed, even if the contribution of the pathogen can be ruled out. Probably the normal protein turnover is disturbed when, for example, the degradation of proteins is prevented. This prevention would increase tissue levels of protein (enzymes). Studies by Ryan and co-workers (378, 379) revealed that in infected or wounded plants the activity of proteinase enzyme is inhibited. The inhibitor-inducing activity resides in cell wall fragments that can be released after infection as a result of enzymatic (polygalacturonase) degradation of host cell wall. Proteinase inhibition can be detected in noninfected parts of diseased plants at a distance from the site of infection or wounding (see also Chap. 4).

In the late stages of disease development the total nitrogen content of the fungus-infected plant organs usually decreases (253). As a rule, a decreased nitrogen content is associated with the degradation of proteins, due mainly to the breakdown of the cell structure of infected tissues, and correlates with an increased activity of proteolytic enzymes. At the same time, deaminases and deamidases also may be activated in tissues, as a consequence of which the free ammonia content of the tissue increases. Table 5–13 illustrates the results of experiments in which various nitrogenous compounds were infiltrated into healthy and rust-infected wheat leaves. The amount of ammonia liberated was much greater in the diseased tissue.

In an analogous study by Farkas and Király (117) the free ammonia content of the leaves increased only in wheat leaves infected by a race of stem rust that resulted in a susceptible reaction. Host-parasite combinations that showed a resistant (hypersensitive) reaction did not exhibit an increased ammonia content, while partially resistant combinations yielded intermediate amounts of ammonia. With a concomitant activation of deaminase and deamidase enzymes the free ammonia content always increases in rust disease at the time of sporulation of the fungus.

On the other hand, enzymes involved in amino acid

Table 5–13. Deamidase and Deaminase activity of wheat leaves infected with stem rust (after Farkas and Király, 117).

Infiltrated compound 10⁻² M	NH₃ liberated, expressed in µg/g fresh weight	
	healthy	infected
Glutamine	30	270
Asparagine	7	230
Glutamic acid	−7[a]	0
Aspartic acid	0	32
Tryptophan	0	28
Serine	0	0

a. Infiltration of glutamic acid resulted in a slight decrease of the endogenous ammonia concentration.

and amide *synthesis* are also activated in rust-infected plants; as a consequence, the levels of various amino acids and amides increase. Thus, parallel proteolysis and protein synthesis can occur in rust-infected tissues. This process was convincingly demonstrated by Rudolph (376), who compared the nitrogen metabolism of wheat infected with rust (*Puccinia graminis*) in intact (attached) and detached leaves. With infected leaves attached to the plant, a partial proteolysis was observed, which resulted in an accumulation of free amino acids and amides, while a relative increase in protein synthesis was suggested after infection in detached leaves accompanied a decrease in free amino acids and amides. Detached healthy leaves are normally characterized by a high rate of proteolysis with an accumulation of amino acids and amides, but in this study infection by the rust pathogen appears to have obscured this phenomenon. A parallel juvenility and senescence effect was also reported in attached leaves of bean infected with rust (237). Infection with *Uromyces phaseoli* resulted in increased protein synthesis in tissues of the green islands surrounding the infection center; on the other hand, the tissue between the green islands became senescent and was characterized by a low capacity for protein synthesis (see also Chap. 7).

Barley leaves infected by *Erysiphe graminis* (powdery mildew) accumulate ammonium ions and also evolve ammonia gas in an early stage of disease development. Enzymes of ammonium assimilation are activated (Fig. 5–54) and the content of amides and their amino acids is increased, showing that diseased leaves have a normal capacity to assimilate ammonium ions (380). Despite an increase in ammonium ions, mildewed leaves show an increased content of glutamic acid, glutamine, and asparagine (Fig. 5–55). The most likely source of the ammonium ions is from hydrolysis of asparagine. One can conclude that a parallel proteolysis and protein synthesis is characteristic also for the powdery mildew disease.

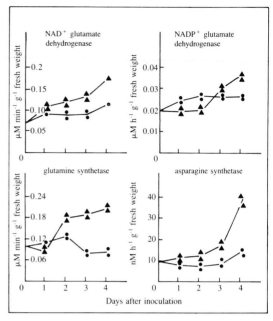

Fig. 5–54. Activities of some enzymes of ammonium assimilation in (●) noninfected and (▲) powdery mildew-infected barley leaves (after Sadler and Scott, 380).

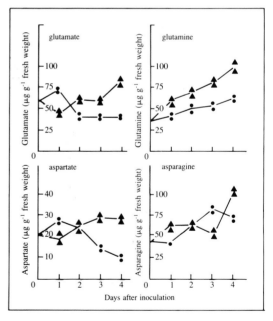

Fig. 5–55. Effect of powdery mildew infection on the levels of free amino acids in barley leaves: (●) noninfected; (▲) infected (after Sadler and Scott, 380).

Table 5–14. The effect of stem rust infection on the size of the nuclei and nucleoli in the mesophyll cells of Little Club wheat (after Whitney et al., 474).

	Nuclei		Nucleoli	
Stage of infection	uninfected radius^3	infected radius^3	uninfected radius^3	infected radius^3
6 d after inoculation	35	53	0.2	0.8
6 d after inoculation; nuclei located beneath center of pustule	—	128	—	0.8
15 d after inoculation	94	80	0.2	0.4

Note: Nuclear and nucleolar radii are in microns and are cubed.

Nucleic acids. Infection of hosts by rusts and mildews, by the agent of cabbage clubroot (*Plasmodiophora brassicae*), and probably by many other microorganisms generally increases the amount of RNA in the susceptible but not in the resistant plant. This increase is accompanied, at least in the case of diseases caused by obligate parasites (biotrophs), by enlargement of the host nucleolus, the site of RNA and protein accumulation (28, 29, 190, 193, 344, 367, 474, 477). It is surprising that practically no data are available concerning diseases caused by facultative parasites.

In an early stage of the rust disease of wheat, both nuclei and nucleoli in the rusted mesophyll cells were much larger than those in uninfected tissue, and they attained an especially large size immediately beneath the pustules. Some 15 d after inoculation with the rust fungus, many of the nuclei in diseased cells had collapsed, and their average size had decreased (Table 5–14).

In cabbage infected by *Plasmodiophora brassicae*, by 2 h after penetration the nucleolus of the host cell had begun to enlarge. Later, by ca. 20 h after infection, it had increased sixfold in volume compared to nucleoli in healthy cells (476, 477). The increase in the nucleolar volume was accompanied by increases in nuclear RNA and nonhistone protein content. Host DNA content was not changed with infection (Table 5–15). Alterations in DNA in rust- and mildew-infected plants are also negligible.

The higher RNA content of mildewed barley and wheat leaves is also an established fact (273, 288, 338). Four days after inoculation of barley with mildew there was 20% more RNA/μg DNA in diseased leaves than in noninfected ones. It is not known with certainty, however, how much the fungal component of the infected leaves contributes to the high RNA content of diseased leaves. Probably the high RNA content and the enhanced ribosome level of mildewed and rusted leaves represent largely ribosomes of the pathogens, which are in a stage of intense haustorial development.

The reported changes in ribosomal RNA (rRNA) are, however, contradictory. Wolf (480) observed a rise in the synthetic activity of rust-infected wheat leaves. Similarly, increased synthesis of rRNA was observed with bean and oat rusts (191, 437) and with barley infected by powdery mildew (27). On the other hand, rRNA and total ribosome content were reported to decrease in cucumber leaves infected by powdery mildew (55, 375).

There is less controversy concerning changes in the molecular species of rRNA in infected plants. The loss of chloroplast rRNA is a generally observed feature of infected plants when symptoms are expressed. However, both the synthesis and breakdown of chloroplast rRNA might be accelerated after infection (311). It is important to differentiate between the action of infection at early and later stages of disease. Tani et al. (437) have shown that rust infection (by *Puccinia coronata*) of oats exerts a juvenile effect on matured leaves at an early stage of infection: in this instance *both* the cytoplasmic and the chloroplast rRNA contents may rise slightly. But at the end of the vegetative stage of fungal growth, in a late stage of the disease, there is a senescence action: the *chloroplast* rRNA content is dramatically re-

Table 5–15. The effect of *P. brassicae* infection on cabbage root hair cell, nuclear DNA, RNA, and protein-bound lysine and arginine (after Bhattacharya and Williams, 29).

Cell condition	DNA+RNA	DNA	RNA	Lysine	Arginine	$\dfrac{DNA}{lysine}$	$\dfrac{DNA}{arginine}$	$\dfrac{Lysine}{arginine}$
noninfected[a]	28.3±0.65	16.3±0.32	12.0	7.2±0.20	8.1±0.27	2.26	2.0	0.88
infected[b]	38.4±1.04	17.0±0.42	21.4	5.4±0.09	8.0±0.52	3.14	2.1	0.67

a. Figures represent arbitrary units of fluorescence from berberine sulfate (DNA+RNA) and (DNA after RNA extraction), dansylchloride (lysine), and ninhydrin (arginine) corrected for background fluorescence. The values of RNA were obtained by subtracting DNA values from the combined DNA plus RNA values. Averages and standard errors are for 20 to 30 nuclear measurements.
b. Infected cells were 42 h old and contained one to five plasmodia with two to four nuclei/plasmodium.

duced. The increase in the cytoplasmic rRNA at this stage is attributed to the presence and intensive growth of the fungus (Fig. 5–56).

On the basis of the investigations of Tani et al. (437), one can summarize the changes in the ratios of different rRNAs between rust-infected and healthy oat leaves as follows:

	Ratio between inoculated/healthy leaves days after inoculation			
	2	3	4	6
Cytopl. rRNA	1.10	1.17	1.62	2.14
Chloropl. rRNA	1.02	1.06 ↑	0.70	0.53
	(Chlorotic stage)			

The loss of chloroplast rRNA and the decrease in chloroplast polysomes in powdery mildewed plants (55, 105) could result from a diffusable "toxic" material(s) of the fungus, since powdery mildew is an ectoparasite (Figs. 5–57 and 5–58). These fungi infect only the epidermal cells, which do not contain chloroplasts. Thus, the loss of ribosomes and nucleic acids in the chloroplast is the result of a conspicuous "toxic" effect. Subsequent experiments led to the conclusion that reduction in leaf polysome content is due to polysomal mRNA hydrolysis because there was an increase in the population of monosomes that are bound to fragments of mRNA. The enzyme that hydrolyzes polysomal mRNA is perhaps an insoluble ribonuclease. The changes occur only in the susceptible host; in an isogenic resistant line of barley there was no polysomal mRNA hydrolysis (411). This phenomenon occurs only in the very early stage of disease and is one of the earliest biochemical symptoms of powdery mildew infection.

Regarding the role of increased mRNA synthesis in infected plants, information is very limited. The changes in mRNA template activity seem insignificant, and the question of whether an exchange of mRNA takes place between the host and pathogen cannot be answered.

The increase in ribonuclease (RNase) activities in fungus-infected plants in late stages of disease parallels increases in RNase after mechanical stimuli and may be regarded as a nonspecific reaction to injury (116). Late symptoms at the biochemical level detected after sporulation may be nonspecific expressions of senescence and stress.

The few data available regarding changes in soluble RNA (sRNA) in diseased plants refer to nonspecific and seemingly unimportant alterations of sRNA fraction (193).

Pathological Nitrogen Metabolism of Incompatible Host-Pathogen Pairs

Nitrogen forms. Many investigators have endeavored to find a relation between the resistance of host tissues and their nitrogen metabolism. Thus, Lakshminarayanan (253) stressed that the ratio of nonprotein to protein nitrogen remains unchanged under the

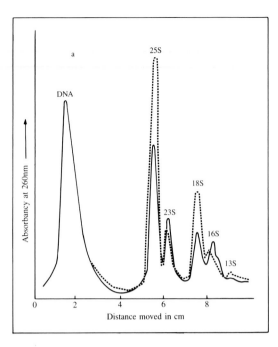

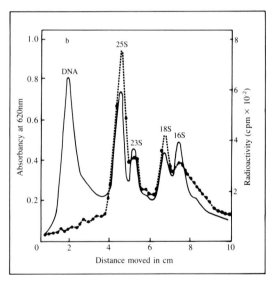

Fig. 5–56. (a) Disc electrophoretic profiles typical of RNA extracted from Victoria leaves 4 d after inoculation with *Puccinia coronata*. Noninoculated (———); inoculated (- - -). (b) Disc electrophoretic profile illustrating the coincidence of the area of rRNA and radioactivity. The absorbancy was determined by a Toyo Densitorol-2. RNA was extracted from the primary leaves of 9-d-old Victoria seedlings. The gel was frozen at –20 C and cut into 1mm slices using a gel slicer. Absorbancy (———); radioactivity (- - -) (after Tani et al., 437).

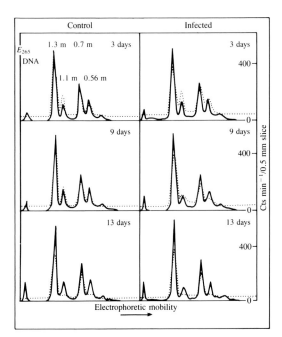

Fig. 5–57. Gel electrophoresis scans at 265 nm of RNA from uninfected leaves, and leaves infected when 3 d old (———), and incorporation of ^{32}P (- - -). 25 μg nucleic acid was loaded into each gel (2.4%), and separation was for 2.5 h at 4 mA/tube (after Callow, 55).

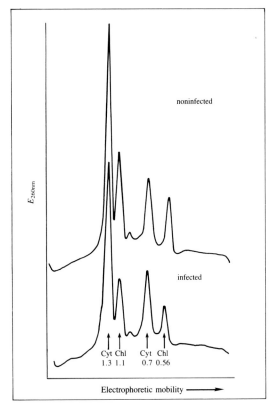

Fig. 5–58. RNA of polysomes of noninfected and infected M1622 (susceptible barley leaves). Samples were harvested 24 h after inoculation with powdery mildew fungus and ribosome fraction prepared from them (after Dyer and Scott, 105).

effect of infection in cotton cultivars resistant to the *Verticillium* wilt, whereas a substantial decrease in this ratio may be noted in the susceptible cultivars. When the disease is induced by an obligate parasite (rust), the resistant cultivar, which exhibits an incompatible (hypersensitive) reaction, is characterized by a nitrogen metabolism different from that of a susceptible one. In contrast to the susceptible cultivars, the resistant ones display, upon infection, a decrease in total nitrogen, soluble nitrogen, and protein nitrogen (403).

Enzymes. Sometimes "new" proteins are produced in the incompatible host in response to infection. These may be isozymes of respiratory enzymes, peroxidases, polyphenoloxidases, etc. (420). The activity of the key enzyme in aromatic biosynthesis, PAL, is also associated with tissue necrosis as a symptom of resistant host reaction. However, all these protein changes are consequences of damage to plant tissue (necrosis). (See also Chap. 6.) In comparative studies, tissue necroses are experienced much earlier in the resistant host/pathogen combination (if necrosis is produced at all) than in the compatible (susceptible) one. This is why enzymatic changes were related to host resistance in the past. It is important to stress here, too, that changes in peroxidase, polyphenoloxidase, and PAL activities are similar to those induced by mechanical injury and by many chemical treatments of plant tissues. The whole phenomenon may be better related to wound healing than to the resistance mechanisms of plants.

Changes in *ribonuclease* activity of wheat leaves infected by rust (*Puccinia graminis*) were causally associated also with susceptibility and resistance (57, 58). The ability of the pathogen to grow in hosts was correlated with the appearance of a so-called late RNase in infected leaf tissues. This enzyme has a substrate preference different from the RNases of uninfected host tissues and rust mycelium grown in axenic culture. In resistant host/pathogen combinations late RNase activity does not develop. Although the appearance of late RNase in susceptible tissue seems to correlate with pathogen growth in the host, it does not correlate with the number of successful infections or with the size of the pustules. This indicates that the appearance of the late RNase is a consequence, not the cause, of host susceptibility.

Nucleic acids. An increase in the size and subsequent collapse of the nuclei in the infected resistant host cells also occur, but these are more rapid than in compatible host plants. Therefore, an increase in *RNA content* is not experienced, and the incorporation of P^{32} into the RNA is unaltered by infection (Table 5–16). There is no conspicuous change in rRNA and in the ratio of cytoplasmic and chloroplast rRNA.

Table 5–16. P^{32} taken up by noninoculated and 4.5 d rusted wheat leaves and specific activity of RNA preparations isolated from this material (after Rohringer and Heitefuss, 367).

Sample	Rust reaction	Specific activity of isolated RNA[a] (c/m/mg RNA P)	Healthy control =100
Little Club, healthy		10.7 × 10^4	100
Little Club, rusted	susceptible	29.3 × 10^4	274
Khapli, healthy		9.3 × 10^4	100
Khapli, rusted	resistant	9.5 × 10^4	102

a. Corrected for activity taken up per mg total P of leaf material.

Gene-specific RNAs from resistant rust-infected wheat were related to the expression of host resistance (in fact to necrosis) against stem rust by Rohringer et al. (368). RNA prepared from a given host/pathogen complex exhibiting the hypersensitive reaction was claimed to stimulate tissue necrosis at a higher rate in an incompatible host/pathogen pair than in a compatible one. The stimulation of necrosis is a gene-specific effect. The RNA, upon RNA injection, stimulated cell necrosis only in the same host/pathogen pair from which it was prepared—not in another incompatible pair. No necrosis was induced in the uninfected leaf upon injection. However, the RNA also exerted a nonspecific but weaker necrotizing effect on compatible combinations. It was surmised that, because host-cell necrosis paralleled the resistance phenomenon and the two are associated in time, the RNA that stimulates host cell necrosis would be involved in the host resistance mechanism. Later the hypothesis was considered untenable because experimental data were too variable (366).

Amino acids. Experiments on the effect of amino acids, amino acid analogs, and kinetinlike compounds on leaves indicate the importance of protein metabolism in the reaction of the host to rust disease. Certain amino acid analogs such as p-fluorophenylalanine, canavanine, and ethionine, when applied to the host, inhibit rust infection, and these effects may be reversed by the corresponding amino acids. In addition, an excess of a particular amino acid, such as *DL*-serine, *DL*-histidine, methionine, and isoleucine, can induce rust resistance (13, 14, 97, 343, 385). The same is true for cucumber infected with *Cladosporium cucumerinum* (455, 456), apple infected with *Venturia inaequalis* (199, 248), and peas infected with *Aphanomyces euteiches* (323). The evidence suggests that an excess of a single amino acid may induce resistance, probably by disturbing normal host protein metabolism.

The resistance induced by an excess of some *L*- and *DL*-amino acids must be a special case of induced resistance, because this mechanism seems to be effective against both compatible and incompatible races of rust pathogen. Inhibitors of protein synthesis are able to reverse the amino acid-induced resistance in both compatible and incompatible cultivar-rust race combinations, thereby permitting the plant to express its original infection type. Here we have an unusual case of induced resistance that is always expressed whether the symptoms are susceptible or hypersensitive (13, 14).

Specific proteins. The possible role of certain specific *proteins* in disease resistance was deduced from experiments with kinetin and benzimidazole. Both compounds prevent protein breakdown and delay senescence in detached leaves. In addition, kinetin retards the degradation of RNA in detached leaves. As is known from several experiments (386, 466), Khapli wheat, which exhibits a resistant reaction to most races of stem rust, loses its resistance in detached-leaf cultures. However, both kinetin and benzimidazole prevent the breakdown of resistance in detached leaves of Khapli wheat. Barna et al. (11) have also shown that loss of resistance parallels the decomposition of certain proteins in detached Khapli wheat leaves. However, abscisic acid treatment, which induced protein decomposition (senescence) of attached leaves similar to that induced by detachment, did not affect rust resistance. Furthermore, treatment with maleic hydrazide did not influence protein composition, but rust resistance was still lost in Khapli seedlings (Fig. 5–59). Proteins were determined by serological as well as electrophoretic methods. One can conclude that specific protein changes in Khapli wheat leaves do not correlate with changes in resistance to stem rust (11).

Tani et al. (438) postulated that proteins synthesized in incompatible oat host very early after infection with a cultivar-nonpathogenic (avirulent) race of oat crown rust are associated with resistance. The synthesis of these proteins precedes the hypersensitive (resistant) response. Blasticidin S, a protein inhibitor, inhibited both the expression of resistance and the incorporation of a labeled amino acid into the protein fraction of leaves. Thus, de novo synthesis of host proteins seems to be involved in "induced" resistance to rust.

Using inhibitors of protein synthesis, Ouchi et al. (317) demonstrated that cereal powdery mildews require an active process for susceptibility that is associated with protein synthesis in a very early stage of infection. Hence susceptibility seems to be an active (induced) process with the mildew diseases of cereals. This does not mean, however, that protein synthesis cannot have a role in the process of expression of host resistance to powdery mildews. Ouchi and his coworkers think that the difference between rust and mildew in sensitivity to protein inhibitors may indicate that powdery mildews and rust fungi might be in different stages in the evolution of parasitic specificity.

It was hypothesized by Albersheim et al. (218) that the response of a plant to a pathogen depends upon the ability of the plant to inhibit the activity of the polygalacturonase enzyme secreted by the pathogen. Indeed, it was shown that *proteins* isolated from plant cell walls inhibited fungal polygalacturonases and in doing

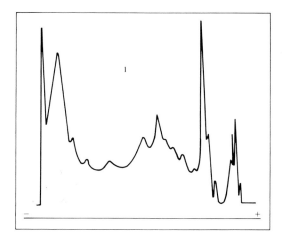
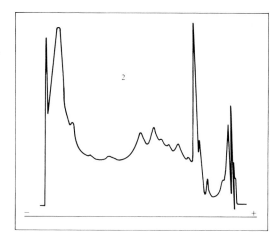
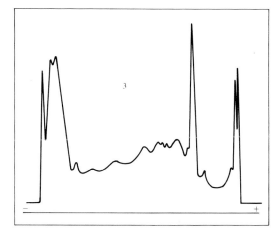
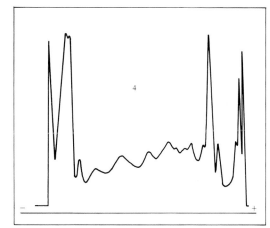

Fig. 5–59. Densitometer scans of polyacrylamide gels after electrophoresis of proteins (200 μg) of Khapli wheat leaf samples. (1) Nontreated leaf proteins (control). (2) Proteins from attached leaves treated with 50 mg maleic hydrazide/d for 8 d. (3) Proteins from attached leaves treated with 100 ppm abscisic acid for 8 d. (4) Proteins from attached leaves floated on water for 5 d (after Barna et al., 11).

so prevented cell wall degradation. However, proteins from walls of three bean cultivars that differ in their response to three pathogenic races of *Colletotrichum lindemuthianum* possessed equal abilities to inhibit each of the polygalacturonases secreted by the three races of the pathogen. It was concluded that the inhibitor proteins from different cultivars are identical, as are the endopolygalacturonases from three different races (3). Furthermore, a single inhibitor protein effectively inhibited the endopolygalacturonases secreted by both *C. lindemuthianum* (a pathogenic fungus) and *Aspergillus niger* (a saprophytic fungus) (120). One can conclude from the work of Albersheim et al. that the endopolygalacturonase/endopolygalacturonase-inhibitor system probably does not play a direct role in the interaction between host plants and pathogenic races leading to compatibility or incompatibility. The Albersheim school also claimed that a particular glycoprotein from incompatible races of *Phytophthora megasperma* f. sp.

glycinea has a role in inducing resistance in soybean seedlings from infection with compatible races (463). The hypothesis that "race-specific molecules" would have a role in resistance to *Phytophthora* was practically withdrawn because of the variability in the applied bioassay (87).

Common Antigens

According to this theory serologically similar proteins, and other antigens, occur in compatible host-pathogen relationships. In other words, the host is susceptible when it shares common proteins, or other antigenic materials, with the infecting agent. However, common antigens, or shared proteins, are lacking in incompatible host-parasite relations. Consequently the plant is resistant because it recognizes the pathogen as "nonself." Resistance would depend upon unsuccessful molecular mimicry by the infecting agent. When the mimicry is successful—this is the rare case—the plant is

Table 5–17. Association of rust reaction (R, resistant; S, susceptible) with rust antiserum titers against flax antigens (after Doubly et al., 100).

Flax variety	M. lini Race 1	M. lini Race 19	M. lini Race 22	M. lini Race 210
(Cass × Bison 7) cultivar	R 1:40	S 1:320	S 1:160	R 1:20
(Koto × Bison 7) cultivar	R 1:40	S 1:320	S 1:320	S 1:320
(Ottawa 770B × Bison 7) cultivar	R 1:40	R 1:40	S 1:320	R 1:20
(Bison) cultivar	S 1:320	S 1:320	S 1:160	S 1:320

susceptible. In the common case under natural conditions the infecting microorganism does not share serologically similar proteins with the higher plant and thus the plant is incompatible and resistant (100, 478).

Doubly et al. (100) prepared globular protein fractions (antigens) from the uredospores of each of four races of flax rust and from flax plants. Antisera to these antigens were prepared in rabbits. Antigens of virulent races of flax rust (*Melamspora lini*) seem to be present in susceptible hosts and absent in flax cultivars resistant to these particular races. As shown in Table 5–17, host-pathogen interaction was susceptible when titers of rust antiserum against flax antigens were relatively high (1:160 or 1:320) and resistant when the titers were low (1:20 or 1:40).

A similar relation was pointed out in an early work on wheat leaf rust (119). Furthermore, compatible *Xanthomonas* strains had more antigens in common with susceptible cotton cultivars than they had with the resistant cultivars (394). DeVay et al. (88) also observed that an antigenic relationship exists between isolates of *Ceratocystis fimbriata* pathogenic to *Ipomoea batatas*. Isolates of the fungus nonpathogenic to sweet potato but pathogenic to other hosts did not show a common antigenic relationship with sweet potato. Wimalajeewa and DeVay (478) showed that corn and corn smut (*Ustilago maydis*) lines had certain antigenic proteins in common. However, no resistant corn cultivars were included in this investigation to compare the situation of an incompatible corn/corn smut pair. Instead, barley and oat were used as incompatible hosts. Barley did not have any antigens in common with the fungus. Oats, when three days old, have a strong antigenic relationship with some fungus lines. At this stage of development, oat seedlings were susceptible to corn smut infection. Later, the 6-week-old seedlings showed a decrease in their common antigen relationship and, correspondingly, were unaffected by the pathogen.

Investigations with wheat rust fungi (12, 216) indicate that the common antigen relationship may explain compatibility/incompatibility between host plant and pathogen on a species level, if there is any, but not on a plant cultivar/pathogenic race level.

The latest papers from the DeVay school dealing with *Fusarium oxysporum* f. sp. *vasinfectum* conclude that the physiological function of cross-reactive antigens in specificity of host-parasite interaction is no longer tenable. They hypothesized, "Cross-reactive proteins may form a continuum between cells of host and parasite that favors the progressive growth and establishment of the parasite" (59, 89).

Toxin-binding proteins. Strobel and his associates (234, 425, 426) worked with the eye spot disease of sugarcane caused by *Helminthosporium sacchari*. The fungus produces a host-specific toxin that is responsible for the symptoms in susceptible host cultivars. This toxin, helminthosporoside, binds specifically to a protein present in susceptible but not in resistant strains of sugarcane. Actually, a similar protein exists in resistant sugarcane clones but in situ does not bind the toxin. The binding activity seems to be regulated by membrane lipids. According to Strobel (427), the binding protein from the susceptible clones occupies a specific site on the host plasma membrane. When the toxin binds to the protein, the ionic balance of the host cell is disrupted and symptoms develop. So the hypothesis contends that susceptible but not resistant host plants have the toxin-binding protein, and specificity depends on a specific "receptor" protein on the surface of susceptible cells of the host.

Lectins (phytoagglutinins). Plant lectins are proteins or glycoproteins that can bind specifically to carbohydrates or to molecules containing carbohydrates on cell surfaces. This may cause dramatic processes, for example agglutination of microorganisms or the stimulation of cell division of lymphocytes (mitogenesis). The possible role of lectins in recognition, including host-pathogen interactions, would be in binding the carbohydrates on the surface of the pathogen to host surface protein, preventing multiplication or inhibition of growth of the pathogen (113, 400). However, it has not been shown convincingly whether the membrane-bound lectins of the host bind, or recognize, the compatible or the incompatible pathogens. Thus, it is not clear whether the lectins recognize nonself or self. The most extensive research in this regard has been done with legume roots recognizing *Rhizobium* cells (35, 418). (See Chap. 1.)

Experimental evidence on lectins is meager at present, but research is active in this field, and possible roles for lectins have been claimed in the following phenomena: gamete-gamete interactions in algae, pollen-style and pollen-stigma incompatibility, recognition of nodule bacteria by legumes, recognition of plant pathogenic fungi and bacteria, recognition of toxins of plant pathogens.

A glycoprotein (lectin?) was claimed to be associated with general (cultivar non-specific) resistance. The enrichment of melon cell walls in hydroxyproline-rich glycoproteins after infection of the seedlings with *Col-*

letotrichum lagenarium was purported to parallel increased resistance to the pathogen. On the other hand, inhibiting the synthesis of this glycoprotein was correlated with increased susceptibility. The accumulation of hydroxyproline-rich glycoproteins upon infection usually happens too late to be efficient in defense. However, pretreatment of plants with ethylene resulted in more cell wall glycoproteins and in increased resistance, because of early triggering of enrichment of cell walls in hydroxyproline-rich glycoproteins. The possible role of hydroxyproline response in disease resistance to bacteria, viruses, and nematodes was also investigated, but the exact role and the mechanism (by a lectinlike action) have not been clarified (112, 443, 444).

Comparative Analysis of Disease Physiology— Nitrogen and Nucleic Acid Metabolism

The infection of a plant by any pathogen inevitably results in the appearance in infected tissue of new nucleic acids, pathogen-specific DNA and/or RNA. In the case of fungi and bacteria, these nucleic acids are synthesized and contained within the cell wall of the pathogen, which may or may not be itself contained within a host cell. Exceptions include T-DNA, a segment of the Ti plasmid of *Agrobacterium tumefaciens*, which becomes integrated into the host DNA. Viral nucleic acid is, of necessity, synthesized within the host cell.

The three types of pathogen can also cause alterations in the rate of biosynthesis and accumulation of host nucleic acids, to a degree that often correlates with disease severity. There may be an early rise in rate of RNA biosynthesis, particularly of cyt rRNA (in oat rust and in TMV-infected tobacco, e.g.), while later stages of infections are often characterized by a decrease in ch rRNA and an accompanying decrease in concentration of chloroplast ribosomes. In many of the earlier reports on nucleic acid variations, however, there was no distinction made between nucleic acids of the host and those of the pathogen. Infection, in common with any treatment that causes cell damage, usually causes an increase in hydrolytic enzymes, particularly ribonucleases.

Infection by all three types of pathogens is also accompanied by the biosynthesis of new pathogen proteins, which may be few in number in the case of the simpler viruses, or several hundreds in the case of bacteria and fungi. Additionally, there may be both qualitative and quantitative changes in the pattern of host protein biosynthesis or degradation, which may change as infection progresses. The early stages of infection, particularly by fungi, may be characterized by a stimulation in host protein synthesis. In some cases, there may be a redirection of host protein synthesis to permit extensive synthesis of pathogen proteins. *A. tumefaciens*, however, stimulates host cell division, causing a substantial increase in host protein synthesis. In addition, the pathogen's Ti plasmid codes specifically for an array of unique amino acids, the opines.

The activity of several host enzymes, including PAL

and peroxidases, may increase markedly, particularly when tissue responds hypersensitively. In the later stages of infection, a decrease in overall rate of protein synthesis and enhancement of proteolytic enzyme activity may be observed as host cells senesce. Some bacteria, including *Pseudomonas tabaci*, also produce toxins that appear to accelerate protein degradation.

Infection may also be characterized by a stimulation of ammonia production and, particularly in infection by viruses, an enhancement in concentration of some amino acids and amides.

Additional Comments and Citations

1. Recent evidence suggests that the two major subunits of the replicase enzyme involved in the replication of TYMV RNA in Chinese cabbage are polypeptides with MW 115,000 and 45,000, coded for by virus and host respectively (T. Candresse, C. Mouchès, and J. M. Bové. 1984. VIth Int. Congr. Virol., Sendai. Abs. p. 221).

2. Sequence data on RNAs 1 and 2 of CMV indicate that the primary translation products have MW ca. 110,000 and 94,300, respectively (M. A. Rezaian, R. H. V. Williams, K. H. J. Gordon, and R. H. Symons. 1984. VIth Int. Congr. Virol., Sendai. Abs. p. 227).

References

1. Agranovsky, A. A., V. V. Dolja, V. M. Kavsan, and J. G. Atabekov. 1978. Detection of polyadenylate sequences in RNA components of barley stripe mosaic virus. Virology 91:95–105.

2. Al Ani, R., P. Pfeiffer, O. Whitechurch, A. Lesot, G. Lebeurier, and L. Hirth. 1980. A virus specified protein produced upon infection by cauliflower mosaic virus (CaMV). Ann. Virol. 131E:33–53.

3. Anderson, A. J., and P. Albersheim. 1972. Host pathogen interactions. V. Comparison of the abilities of proteins isolated from three varieties of *Phaseolus vulgaris* to inhibit the endo-polygalacturonases secreted by three races of *Colletotrichum lindemuthianum*. Physiol. Plant Path. 2:339–46.

4. Ansa, O. A., J. W. Bowyer, and R. J. Shepherd. 1982. Evidence for replication of cauliflower mosaic virus DNA in plant nuclei. Virology 121:147–56.

5. Aoki, A., and I. Takebe. 1975. Replication of tobacco mosaic virus RNA in tobacco mesophyll protoplasts inoculated in vitro. Virology 65:343–54.

6. Astier-Manifacier, S., and P. Cornuet. 1971. RNA-dependent RNA polymerase in Chinese cabbage. Biochim. Biophys. Acta 232:484–93.

7. Atabekov, J. G., and S. Y. Morozov. 1979. Translation of plant virus messenger RNAs. Adv. Virus Res. 25:1–76.

8. Atchison, B. A. 1973. Division, expansion and DNA synthesis in meristematic cells of French bean (*Phaseolus vulgaris* L.) root-tips invaded by tobacco ringspot virus. Physiol. Plant Path. 3:1–8.

9. Bagi, G., and G. L. Farkas. 1967. On the nature of increase in ribonuclease activity in mechanically damaged tobacco leaf tissues. Phytochemistry 6:161–69.

10. Bancroft, J. B. 1958. Temperature and temperature-light

effects on the concentration of squash mosaic virus in leaves of growing cucurbits. Phytopathology 48:98–102.

11. Barna, B., E. Balázs, and Z. Király. 1975. Proteins and resistance of wheat to stem rust: Non-involvement of some serologically and electrophoretically determined proteins. Physiol. Plant Path. 6:137–43.

12. Barna, B., E. Balázs, and Z. Király. 1978. Common antigens and rust diseases. Phytopath. Z. 93:336–41.

13. Barna, B., and É. Janos. 1981. The role of protein metabolism of wheat in amino acid induced resistance to rust. Acta Phytopath. Hung. 16:49–57.

14. Barna, B., S. F. Mashaal, and Z. Király. 1977. Problems of involvement of protein metabolism in resistance of wheat to rust. In: Current topics in plant pathology. Z. Király (ed.). Akadémiai Kiadó, Budapest, pp. 35–40.

15. Barnett, A., and K. R. Wood. 1978. Influence of actinomycin D, ethephon, cycloheximide and chloramphenicol on the infection of a resistant and a susceptible cucumber cultivar with cucumber mosaic virus (Price's No. 6 strain). Physiol. Plant Path. 12:257–77.

16. Barta, A., I. Sum, and F. J. Foglein. 1981. 2-Thiouracil does not inhibit TMV replication in tobacco protoplasts. J. Gen. Virol. 56:219–22.

17. Bashan, Y., Y. Okon, and Y. Henis. 1980. Ammonia causes necrosis in tobacco leaves *Pseudomonas* tomato (Obabe) Alstatt. Physiol. Plant Path. 17:111–19.

18. Basler, E., Jr., and B. Commoner. 1956. The effect of tobacco mosaic virus biosynthesis on the nucleic acid content of tobacco leaf. Virology 2:13–28.

19. Bastin, M., and P. Kaesberg. 1976. A possible replicative form of brome mosaic virus RNA 4. Virology 72:536–39.

20. Bawden, F. C., and B. Kassanis. 1954. Some effects of thiouracil on virus-infected plants. J. Gen. Microbiol. 10:160–73.

21. Beachy, R. N., and M. Zaitlin. 1975. Replication of tobacco mosaic virus. VI. Replicative intermediate and TMV-RNA-related RNAs associated with polyribosomes. Virology 63:84–97.

22. Beachy, R. N., and M. Zaitlin. 1977. Characterization and in vitro translation of the RNAs from less-than-full-length virus-related nucleoprotein rods present in tobacco mosaic virus preparations. Virology 81:160–69.

23. Beachy, R. N., M. Zaitlin, G. Bruening, and H. W. Israel. 1976. A genetic map for the cowpea strain of TMV. Virology 73:498–507.

24. Becker, W. M. 1979. RNA polymerases in plants. In: Nucleic acids in plants. T. C. Hall and J. W. Davies (eds.). CRC Press, Hollywood, FL, 1:111–41.

25. Bedbrook, J. R., J. Douglas, and R. E. F. Matthews. 1974. Evidence for TYMV-induced RNA and DNA synthesis in the nuclear fraction from infected Chinese cabbage leaves. Virology 58:334–44.

26. Bénicourt, C., J.-P. Péré, and A.-L. Haenni. 1978. Translation of TYMV RNA into high molecular weight proteins. FEBS Lett. 86:268–72.

27. Bennett, J., and K. J. Scott. 1971. Ribosome metabolism in mildew-infected barley leaves. FEBS Lett. 16:93–95.

28. Bhattacharya, P. K., J. M. Naylor, and M. Shaw. 1965. Nucleic acid and protein changes in wheat leaf nuclei during rust infection. Science 150:1605–6.

29. Bhattacharya, P. K., and P. H. Williams. 1971. Microfluorometric quantitation of nuclear proteins and nucleic acids in cabbage root hair cells infected by *Plasmodiophora brassicae*. Physiol. Plant Path. 1:167–75.

30. Bisaro, D., and A. Siegel. 1980. Sequence homology between chloroplast DNAs from several higher plants. Plant Physiol. 65:234–37.

31. Bisaro, D. M., and A. Siegel. 1982. Subgenomic components of tobacco rattle virus infected tissue. Virology 118:411–18.

32. Black, D. R., and C. A. Knight. 1970. Ribonucleic acid transcriptase activity in purified wound tumor virus. J. Virol. 6:194–98.

33. Bockstahler, L. E. 1967. Biophysical studies on double stranded RNA from turnip yellow mosaic virus-infected plants. Mol. Gen. Genet. 100:337–48.

34. Bogorad, L., and J. H. Weil (eds.). 1977. Nucleic acids and protein synthesis in plants. NATO Advanced Study Institute Series, Vol. 12. Plenum Press, New York.

35. Bohlool, B. B., and E. L. Schmidt. 1974. Lectins: A possible basis for specificity in the *Rhizobium*-legume root nodule symbiosis. Science 185:269–71.

36. Boulter, D. 1970. Protein synthesis in plants. Ann. Rev. Plant Physiol. 21:91–114.

37. Bozarth, R. F., and R. E. Browning. 1970. Effect of infection with southern bean mosaic virus and the diurnal cycle on the free nucleotide pool of bean leaves. Phytopathology 60:852–55.

38. Bozarth, R. F., and T. O. Diener. 1963. Changes in concentration of free amino acids and amides induced in tobacco plants by potato virus X and potato virus Y. Virology 21:188–93.

39. Bourque, D. P., A. Hagiladi, and S. G. Wildman. 1975. Experimental test of a TMV replication model. Attempts to identify molecular precursors of TMV-RNA and determine the time required to synthesize a TMV-RNA molecule. Virology 63:135–46.

40. Bradbeer, J. W., Y. E. Atkinson, T. Börner, and R. Hagemann. 1979. Cytoplasmic synthesis of plastid polypeptides may be controlled by plastid-synthesized RNA. Nature 279:816–17.

41. Bradley, D. W., and M. Zaitlin. 1971. Replication of tobacco mosaic virus. II. The in vitro synthesis of high molecular weight virus-specific RNAs. Virology 45:192–99.

42. Braun, A. C. 1962. Tumor inception and development in the crown gall disease. Ann. Rev. Plant Physiol. 13:533–58.

43. Briand, J.-P., G. Keith, and H. Guilley. 1978. Nucleotide sequence at the 5′ extremity of the turnip yellow mosaic virus genome RNA. Proc. Nat. Acad. Sci. 75:3168–72.

44. Brishammar, S., and N. Juntti. 1974. Partial purification and characterization of soluble TMV replicase. Virology 59:245–53.

45. Brock, T. D. 1970. Biology of microorganisms. Prentice-Hall, Englewood Cliffs, NJ.

46. Broker, J., and A. Marcus. 1977. Ribosome binding of capped satellite tobacco necrosis virus RNA. FEBS Lett. 82:118–24.

47. Brown, E. G., and R. P. Newton. 1981. Cyclic AMP and higher plants. Phytochemistry 20:2453–63.

48. Bruening, G. 1977. Plant covirus systems: Two-component systems.In: Comprehensive virology. H. Fraenkel-Conrat and R. R. Wagner (eds.). Plenum Press, New York, 11:55–141.

49. Bruening, G., R. N. Beachy, R. Scalla, and M. Zaitlin. 1976. In vitro and in vivo translation of the ribonucleic acids of a cowpea strain of TMV. Virology 71:498–517.

50. Bujarski, J. J., S. F. Hardy, W. A. Miller, and T. C. Hall. 1982. Use of dodecyl- - D-maltoside in the purification and stabilization of RNA polymerase from brome mosaic virus-infected barley. Virology 119:465–73.

51. Burdon, R. H., M. A. Billeter, C. Weissmann, R. C. Warner, S. Ochoa, and C. A. Knight. 1964. Replication of viral RNA. V. Presence of a virus-specific double-stranded RNA in leaves infected with tobacco mosaic virus. Proc. Nat. Acad. Sci. 52:768–75.

52. Butler, P. J. G., J. T. Finch, and D. Zimmern. 1977. Config-

uration of tobacco mosaic virus RNA during virus assembly. Nature 265:217–19.

53. Butler, P. J. G., and A. Klug. 1978. The assembly of a virus. Sci. Amer. 239:52–59.

54. Caleya, R., B. Gonzalez-Pascual, F. Garcia-Olmedo, and P. Carbonero. 1972. Susceptibility of phytopathogenic bacteria to wheat purothionins in vitro. Appl. Microbiol. 23:998–1000.

55. Callow, J. A. 1973. Ribosomal RNA metabolism in cucumber leaves infected by Erysiphe cichoracearum. Physiol. Plant Path. 3:249–57.

56. Chadha, C. C., and B. I. S. Srivastava. 1971. Evidence for the presence of bacteria-specific proteins in sterile crown gall tumor tissue. Plant Physiol. 48:125–29.

57. Chakravorty, A. K. 1974. Ribonuclease activity of wheat leaves and rust infection. Nature 247:577–80.

58. Chakravorty, A. K., and M. Shaw. 1977. The role of RNA in host-parasite specificity. Ann. Rev. Phytopath. 15:135–51.

59. Charudattan, R., and J. E. DeVay. 1981. Purification and partial characterization of an antigen from Fusarium oxysporum f. sp. vasinfectum that cross-reacts with antiserum to cotton (Gossypium hirsutum) root antigens. Physiol. Plant Path. 18:289–95.

60. Chifflot, S., P. Sommer, D. Hartmann, C. Stussi-Garaud, and L. Hirth. 1980. Replication of alfalfa mosaic virus RNA: Evidence for a soluble replicase in healthy and infected tobacco leaves. Virology 100:91–100.

61. Chilton, M. D., T. C. Currier, S. K. Farrand, A. J. Bendick, M. P. Gordon, and E. W. Nester. 1974. Agrobacterium tumefaciens DNA and PS8 bacteriophage DNA not detected in crown gall tumors. Proc. Nat. Acad. Sci. 71:3672–76.

62. Chroboczek, J., L. Puchkova, and W. Zagorski. 1980. Regulation of brome mosaic virus gene expression by restriction of initiation of protein synthesis. J. Virol. 34:330–35.

63. Chu, P. W. G., G. Boccardo, and R. I. B. Francki. 1981. Requirement of a genome-associated protein of tobacco ringspot virus for infectivity but not for in vitro translation. Virology 109:428–30.

64. Ciferri, O., O. Tiboni, M. L. Munoz-Calvo, and G. Camerino. 1977. Protein synthesis in plants: Specificity and role of cytoplasmic and organellar systems. In: Nucleic acids and protein synthesis in plants. L. Bogorad and J. H. Weil (eds.). NATO Advanced Study Institutes Series, Plenum Press, New York, 12:155–66.

65. Clerx, C. M., and J. F. Bol. 1978. Properties of solubilized RNA-dependent RNA polymerase from alfalfa mosaic virus-infected and healthy tobacco plants. Virology 91:453–63.

66. Collmer, C. W., V. M. Vogt, and M. Zaitlin. 1983. H. Protein, a minor protein of TMV Virions, contains sequences of the viral coat protein. Virology 126:429–47.

67. Comai, L., and T. Kosuge. l980. Involvement of plasmid deoxyribonucleic acid in indoleacetic acid synthesis in Pseudomonas savastanoi. J. Bact. 143:950–57.

68. Comai, L., and T. Kosuge. 1983. Transposable element that causes mutations in a plant pathogenic Pseudomonas sp. J. Bact. 154:1162–67.

69. Condit, C., and H. Fraenkel-Conrat. 1979. Isolation of replicative forms of 3' terminal subgenomic RNAs of tobacco necrosis virus. Virology 97:122–30.

70. Cooper, P., and I. W. Selman. 1974. An analysis of the effects of tobacco mosaic virus on growth and the changes in the free amino compounds in young tomato plants. Ann. Bot. 38:625–38.

71. Coutts, R. H. A., and K. R. Wood. 1976. The infection of cucumber mesophyll protoplasts with tobacco mosaic virus. Arch. Virol. 52:59–69.

72. Covey, S. N., and R. Hull. 1981. Transcription of cauliflower mosaic virus DNA. Detection of transcripts, properties, and location of the gene encoding the virus inclusion body protein. Virology 111:463–74.

73. Covey, S. N., R. Hull, and G. P. Lommonossoff. 1981. Characterization of cauliflower mosaic virus transcripts. Vth Int.. Virol. Cong., Strasbourg, p. 207. (Abstr.)

74. Crosbie, E. S., and R. E. F. Matthews. 1974. Effects of TYMV infection on growth of Brassica pekinensis Rupr. Physiol. Plant Path. 4:389–400.

75. Cullimore, J. V., and A. P. Simms. 1980. An association between photorespiration and protein catabolism: Studies with Chlamydomonas. Planta 150:392–96.

76. Daubert, S., R. Richins, R. J. Shepherd, and R. C. Gardner. 1982. Mapping of the coat protein gene of cauliflower mosaic virus by its expression a prokaryotic system. Virology 122:444–49.

77. Davies, J. W. 1979. Translation of plant virus ribonucleic acids in extracts from eukaryotic cells. In: Nucleic acids in plants. T. C. Hall and J. W. Davies (eds.). CRC Press, Hollywood, FL, 2:111–49.

78. Davies, J. W., and R. Hull. 1982. Genome expression of plant positive-strand RNA viruses. J. Gen. Virol. 61:1–14.

79. Dawson, W. O. 1976. Synthesis of TMV RNA at restrictive high temperatures. Virology 73:319–26.

80. Dawson, W. O. 1978 Time-course of actinomycin D inhibition of tobacco mosaic virus multiplication relative to the rate of spread of the infection. Intervirology 9:304–9.

81. Dawson, W. O. 1983. Tobacco mosaic virus protein synthesis is correlated with double-stranded RNA synthesis and not single-stranded RNA synthesis. Virology 125:314–23.

82. Dawson, W. O., and A. L. Grantham. 1983. Effect of 2-thiouracil on RNA and protein synthesis in synchronous and asynchronous infections of tobacco mosaic virus. Intervirology 19:155–61.

83. Dawson, W. O., and C. W. Kuhn. 1972. Enhancement of cowpea chlorotic mottle virus biosynthesis and in vivo infectivity by 2-thiouracil. Virology 47:21–29.

84. Dawson, W. O., and D. E. Schlegel. 1976. Enhancement of tobacco mosaic virus spread in mechanically inoculated leaves by pre-incubation at a non-permissive temperature. Phytopathology 66:625–28.

85. Dawson, W. O., and D. E. Schlegel. 1976. The sequence of inhibition of tobacco mosaic virus synthesis by actinomycin D, 2-thiouracil and cycloheximide in a synchronous infection. Phytopathology 66:177–81.

86. Dawson, W. O., D. E. Schlegel, and M. C. Y. Lung. 1975. Synthesis of tobacco mosaic virus in intact tobacco leaves systemically inoculated by differential temperature treatment. Virology 65:565–73.

87. Desjardins, A. E., L. M. Ross, M. W. Spellman, A. G. Darvill, and P. Albersheim. 1982. Host-pathogen interactions. XX. Biological variation in the protection of soybeans from infection by Phytophthora megasperma f. sp. glycinea. Plant Physiol. 69:1046–50.

88. DeVay, J. E., W. C. Schnathorst, and M. S. Foda. 1967. Common antigens and host-parasite interactions. In: The dynamic role of molecular constituents in plant-parasite interaction. C. J. Mirocha and I. Uritani (eds.). Amer. Phytopath. Soc. St. Paul, MN, pp. 313–28.

89. DeVay, J. E., R. J. Wakeman, J. A. Kavanagh, and R. Charudattan. 1981. The tissue and cellular location of a major cross-reactive antigen shared by cotton and soil-borne fungal parasites. Physiol. Plant Path. 18:59–66.

90. Diaz-Ruiz, J. R., and J. M. Kaper. 1978. Isolation of viral

double-stranded RNAs using a LiCl fractionation procedure. Prep. Biochem. 8:1–17.

91. Dickson, E. 1979. Viroids: Infectious RNA in plants. In: Nucleic acids in plants. T. C. Hall and J. W. Davies (eds.). CRC Press, Hollywood, FL, 2:153–93.

92. Diener, T. O. 1961. Virus infection and other factors affecting ribonuclease activity of plant leaves. Virology 14:177–89.

93. Diener, T. O. 1979. Viroids: Structure and function. Science 205:859–66.

94. Diener, T. O. 1981. Viroids: Abnormal products of plant metabolism. Ann. Rev. Plant Physiol. 32:31–25.

95. Diener, T. O. 1982. Viroids and their interactions with host cells. Ann. Rev. Microbiol. 36:239–58.

96. Diener, T. O., and D. R. Smith. 1975. Potato spindle tuber viroid. XIII. Inhibition of replication by actinomycin D. Virology 63:421–27.

97. Donchev, N. 1969. The influence of some amino acids on the reaction type of wheat varieties to various races of leaf rust. Acta Phytopath. Hung. 4:103–10.

98. Dorssers, L., J. van der Meer, A. van Kammen, and P. Zabel. 1983. The cowpea mosaic virus RNA replication complex and the host-encoded RNA-dependent RNA polymerase-template complex are functionally different. Virology 125:155–74.

99. Dorssers, L., P. Zabel, J. van der Meer, and A. van Kammen. 1982. Purification of a host-encoded RNA-dependent RNA polymerase from cowpea mosaic virus-infected cowpea leaves. Virology 116:236–49.

100. Doubly, J. A., H. H. Flor, and C. O. Clagett. 1960. Relation of antigens of *Melampsora lini* and *Linum usitatissimum* to resistance and susceptibility. Science 131:229.

101. Duda, C. T., M. Zaitlin, and A. Siegel. 1973. In vitro synthesis of double-stranded RNA by an enzyme system isolated from tobacco leaves. Biochim. Biophys. Acta 319:62–71.

102. Dudley, R. K., J. T. Odell, and S. H. Howell. 1982. Structure and 5′-termini of the large and 19s RNA transcripts encoded by the cauliflower mosaic virus genome. Virology 117:19–28.

103. Durbin, R. D. 1971. Chlorosis-inducing pseudomonad toxins: Their mechanism of action and struture. In: Morphological and biochemical events in plant-parasite interactions. S. Akai and S. Ouchi (eds.). Phytopath. Soc. Jap., Tokyo, pp. 369–85.

104. Dyer, T. A., and C. J. Leaver. 1981. RNA: Structure and metabolism. In: The biochemistry of plants. Proteins and nucleic acids. A. Marcus (ed.). Academic Press, New York, 6:111–68.

105. Dyer, T. A., and K. J. Scott. 1972. Decrease in chloroplast content of barley leaves infected with powdery mildew. Nature 236:237–38.

106. Edelman, M. 1981. Nucleic acids of chloroplasts and mitochondria.In: The biochemistry of plants. Proteins and nucleic acids. A. Marcus (ed.). Academic Press, New York, 6:249–301.

107. Edwards, M. D., J. Gonsalves, and R. Provvidenti. 1983. Genetic analysis of cucumber mosaic virus in relation to host resistance: Location of determinants for pathogenicity to certain legumes and *Lactuca saligna*. Phytopathology 73:269–73.

108. Ellis, R. J. 1969. Chloroplast ribosomes. Stereospecificity of inhibition by chloramphenicol. Science 163:477–78.

109. Ellis, R. J. 1977. Protein synthesis by isolated chloroplasts. Biochim. Biophys. Acta 463:185–215.

110. Ellis, R. J. 1981. Chloroplast proteins: Synthesis, transport and assembly. Ann. Rev. Plant Physiol. 32:111–37.

111. Ellis, R. J., and R. Barraclough. 1978. Synthesis and transport of chloroplast proteins inside and outside the cell. In: Chloroplast development. G. Akoyunoglu and J. H. Argyroudi-Akoyunoglu (eds.). Elsevier/North Holland, Amsterdam, pp. 185–94.

112. Esquerré-Tugayé, M.-T., C. Lafitte, D. Mazau, A. Top-

pan, and A. Touzé. 1979. Cell surfaces in plant-microorganism interactions. II. Evidence for the accumulation of hydroxyproline-rich glycoproteins in the cell wall of diseased plants as a defense mechanism. Plant Physiol. 64:320–26.

113. Etzler, M. E. 1981. Are lectins involved in plant-fungus interactions? Phytopathology 71:744–46.

114. Falaschi, A., F. Cobianchi, and S. Riva. 1980. DNA-binding proteins and DNA-unwinding enzymes in eukaryotes. Trends Biochem. Sci 5:154–57.

115. Farkas, G. L. 1972. Nucleic acids and proteins in higher plants. Proc. of Tihany Symp., 1971. Akadémiai Kiadó, Budapest.

116. Farkas, G. L., L. Dézsi, M. Horváth, K. Kisbán, and J. Udvardy. 1964. Common pattern of enzymatic changes in detached leaves and tissues attacked by parasites. Phytopath. Z. 49:343–54.

117. Farkas, G. L., and Z. Király. 1961. Amide metabolism in wheat leaves infected with stem rust. Physiol. Planta. 14:344–53.

118. Farkas, G. L., and L. Lovrekovich. 1963. Counteraction by kinetin of the toxic effect of *Pseudomonas tabaci*. Phytopath. Z. 47:391–98.

119. Fedotova, T. I. 1940. Immunological character of various species and varieties of wheat in relation to wheat leaf rust (*Puccinia triticina*). Vestnik Zashch. Rast. 4:123–30 (in Russian).

120. Fisher, M. L., A. J. Anderson, and P. Albersheim. 1973. Host-pathogen interactions. VI. A single plant protein effectively inhibits endopolygalacturonases secreted by *Colletotrichum lindemuthianum* and *Aspergillus niger*. Plant Physiol. 51:489–91.

121. Föglein, F. J., C. Kalpagam, D. C. Bates, G. Premecz, A. Nyitrai, and G. L. Farkas. 1975. Viral RNA synthesis is renewed in protoplasts isolated from TMV- infected Xanthi tobacco leaves in an advanced stage of virus infection. Virology 67:74–79.

122. Fraenkel-Conrat, H. 1976. RNA polymerase from tobacco necrosis virus infected and uninfected tobacco. Purification of the membrane-associated enzyme. Virology 72:23–32.

123. Fraenkel-Conrat, H. 1983. RNA-dependent RNA polymerases of plants. Proc. Nat. Acad. Sci. 80:422–24.

124. Fraenkel-Conrat, H., M. Salvato, and L. Hirth. 1977. The translation of large plant viral RNAs. In: Comprehensive virology. H. Fraenkel-Conrat and R. R. Wagner (eds.). Plenum Press, New York, 11:201–35.

125. Fraenkel-Conrat, H., B. Singer, and R. C. Williams. 1957. Infectivity of virus nucleic acid. Biochim. Biophys. Acta 25:87–96.

126. Fraenkel-Conrat, H., Y. Takanami, and S. Stein. 1981. Comparative studies on plant RNA-dependent RNA polymerases. Vth Int. Virol. Cong., Strasbourg, p. 248. (Abstr.)

127. Fraenkel-Conrat, H., and R. C. Williams. 1955. Reconstitution of active tobacco mosaic virus from its inactive protein and nucleic acid components. Proc. Nat. Acad. Sci. 41:690–98.

128. Franck, A., H. Guilley, G. Jonard, K. Richards, and L. Hirth. 1980. Nucleotide sequence of cauliflower mosaic virus DNA. Cell 21:285–94.

129. Francki, R. I. B. 1962. Infectivity of tobacco mosaic virus from tobacco leaves treated with 2-thiouracil. Virology 17:9–21.

130. Francki, R. I. B., E. W. Kitajima, and D. Peters. 1981. Rhabdoviruses. In: Handbook of plant virus infections: Comparative diagnosis. E. Kurstak (ed.). Elsevier/North Holland, Amsterdam, pp. 455–89.

131. Francki, R. I. B., and J. W. Randles. 1972. RNA-dependent RNA polymerase associated with particles of lettuce necrotic yellows virus. Virology 47:270–75.

132. Francki, R. I. B., and J. W. Randles. 1973. Some properties of lettuce necrotic yellows virus RNA and its in vitro transcription by virion-associated transcriptase. Virology 54:359–68.

133. Francki, R. I. B., and J. W. Randles. 1975. Composition of the plant rhabdovirus lettuce necrotic yellows virus in relation to

its biological properties.In: Negative strand viruses and the host cell. B. W. J. Mahy and R. D. Barry (eds.). Academic Press, New York, pp. 223–42.

134. Francki, R. I. B., and J. W. Randles. 1980. Rhabdoviruses infecting plants. In: Rhabdoviruses. D. H. L. Bishop (ed.). CRC Press, Hollywood, FL, pp. 135–65.

135. Franssen, H., R. Goldbach, M. Broekhuijsen, M. Moerman, and A. van Kammen. 1982. Expression of middle component RNA of cowpea mosaic virus: In vitro generation of a precursor to both capsid proteins by a bottom component RNA- encoded protease from infected cells. J. Virol. 41:8–17.

136. Fraser, R. S. S. 1969. Effects of two TMV strains on the synthesis and stability of chloroplast ribosomal RNA in tobacco leaves. Mol. Gen. Genet. 106:73–79.

137. Fraser, R. S. S. 1972. Effects of two strains of tobacco mosaic virus on growth and RNA content of tobacco leaves. Virology 47:261–69.

138. Fraser, R. S. S. 1973. The synthesis of tobacco mosaic virus RNA and ribosomal RNA in tobacco leaves. J. Gen. Virol. 18:267–79.

139. Fraser, R. S. S., and A. Gerwitz. 1980. Tobacco mosaic virus infection does not alter the polyadenylated messenger RNA content of tobacco leaves. J. Gen. Virol. 46:139–48.

140. Friedman, B. A. 1962. Physiological differences between a virulent and weakly virulent radiation-induced strain of *Erwinia carotovora*. Phytopathology 52:328–32.

141. Fritsch, C., M. A. Mayo, and L. Hirth. 1977. Further studies on the translation products of tobacco rattle virus RNA in vitro. Virology 77:722–32.

142. Frost, R. R., and B. D. Harrison. 1967. Comparative effects of temperature on the multiplication in tobacco leaves of two tobacco rattle viruses. J. Gen. Virol. 1:455–64.

143. Fry, P. R., and R. E. F. Matthews. 1963. Timing of some early events following inoculation with tobacco mosaic virus. Virology 19:461–69.

144. Fukuda, M., T. Meshi, Y. Okada, Y. Otsuki, and I. Takebe. 1981. Correlation between particle multiplicity and location on virion RNA of the assembly initiation site for viruses of the tobacco mosaic virus group. Proc. Nat. Acad. Sci. 78:4231–35.

145. Furusawa, I., N. Yamaoka, T. Okuno, M. Yamamoto, M. Kohno, and H. Kunoh. 1980. Infection of turnip protoplasts with cauliflower mosaic virus. J. Gen. Virol. 48:431–35.

146. Gianinazzi, S., and J.-C. Vallée. 1969. Température et synthèse de materiel proteique viral chez le *Nicotiana Xanthi* nc infecté par le virus de la mosaïque du tabac. Compte. Rend. Acad. Sci. D 269:593–95.

147. Giegé, R., J.-P. Briand, R. Mengual, J.-P. Ebel, and L. Hirth. 1978. Valylation of the two RNA components of turnip-yellow mosaic virus and specificity of the tRNA aminoacylation reaction. Eur. J. Biochem. 84:251–56.

148. Gierer, A., and G. Schramm. 1956. Infectivity of ribonucleic acid fom tobacco mosaic virus. Nature 177:702–3.

149. Gill, D. S., R. Kumarasamy, and R. H. Symons. 1981. Cucumber mosaic virus-induced RNA replicase: Solubilization and partial purification of the particulate enzyme. Virology 113:1–8.

150. Gilliland, J. M., and R. H. Symons. 1968. Properties of a plant virus-induced RNA polymerase in cucumbers infected with cucumber mosaic virus. Virology 36:232–40.

151. Glover, J. F., and T. M. A. Wilson. 1982. Efficient translation of the coat protein cistron of tobacco mosaic virus in a cell-free system from *Escherichia coli*. Eur. J. Biochem. 122:485–92.

152. Goelet, P., and J Karn. 1982. Tobacco mosaic virus induces the synthesis of a family of 3' coterminal messenger RNAs and their complements. J. Mol. Biol. 154: 541–50.

153. Goelet, P., G. P. Lomonossoff, P. J. G. Butler, M. E. Akam, M. J. Gait, and J. Karn. 1982. Nucleotide sequence of tobacco mosaic virus RNA. Proc. Nat. Acad. Sci. 79:5818–22.

154. Goldbach, R., and G. Rezelman. 1983. Orientation of the cleavage map of the 100-kilodalton polypeptide encoded by the bottom-component RNA of cowpea mosaic virus. J. Virol. 46:614–19.

155. Goldbach, R., G. Rezelman, and A. van Kammen. 1980. Independent replication and expression of B-component RNA of cowpea mosaic virus. Nature 285:297–300.

156. Goldbach, R., G. Rezelman, P. Zabel, and A. van Kammen. 1982. Expression of the bottom component RNA of cowpea mosaic virus: Evidence that the 60- kilodalton Vpg precursor is cleaved into single Vpg and a 28-kilodalton polypeptide. J. Virol. 42:630–35.

157. Goldbach, R. W., J. G. Schitthuis, and G. Rezelman. 1981. Comparison of in vivo and in vitro translation of cowpea mosaic virus RNAs. Biochem. Biophys. Res. Comm. 99:89–94.

158. Gonda, T. J., and R. H. Symons. 1979. Cucumber mosaic virus replication in cowpea protoplasts: Time course of virus, coat protein and RNA synthesis. J. Gen. Virol. 45:723–36.

159. Goodman, R. M. 1981. Geminiviruses. J. Gen. Virol. 54:9–21.

160. Goodman, R. N. 1972. Electrolyte leakage and membrane damage in relation to bacterial population, pH, and ammonia production in tobacco leaf tissue inoculated with *Pseudomonas pisi*. Phytopathology 62:1327–31.

161. Goodman, R. N. 1972. Phytotoxin-induced ultrastuctural modifications of plant cells. In: Phytotoxins in plant diseases. R. K. S. Wood, A. Ballio, and A. Graniti (eds.). Academic Press, New York, pp. 311–29.

162. Goodman, R. N., P.-Y. Huang, J. S. Huang, and V. Thaipanich. 1976. Induced resistance to bacterial infection. In: Biochemistry and cytology of plant-parasite interaction. K. Tomiyama, J. M. Daly, I. Uritani, H. Oku, and S. Ouchi (eds.). Kodancha, Tokyo, pp. 35–42.

163. Gordon, K. H. J., D. A. Gill, and R. H. Symons. 1982. Highly purified cucumber mosaic virus induced RNA-dependent RNA polymerase does not contain any of the full-length translation products of the genomic RNAs. Virology 123:284–95.

164. Gordon, K. H. J., and R. H. Symons. 1983. Satellite RNA of cucumber mosaic virus forms a secondary structure with partial 3'-terminal homology to genomal RNAs. Nucl. Acid Res. 11:947–60.

165. Gould, A. R., P. Palukaitis, R. H. Symons, and D. W. Mossop. 1978. Characterization of a satellite RNA associated with cucumber mosaic virus. Virology 84:443–55.

166. Gould, A. R., and R. H. Symons. 1982. Cucumber mosaic virus RNA 3. Determination of the nucleotide sequence provides the amino acid sequences of protein 3A and viral coat protein. Eur. J. Biochem. 126:217–26.

167. Grill, L. K., and J. S. Semancik. 1980. Viroid synthesis: The question of inhibition by actinomycin D. Nature 283:399–400.

168. Gronenborn, B., R. C. Gardner, S. Schaefer, and R. J. Shepherd. 1981. Propagation of foreign DNA in plants using cauliflower mosaic virus as vector. Nature 294: 773–76.

169. Gross, H. J., G. Krupp, H. Domdey, G. Steger, D. Riesner, and H. L. Sänger. 1981. The structure of three plant viroids. Nucl. Acid Res. 10:91–98.

170. Guilfoyle, T. J. 1981. DNA and RNA polymerases. In: The biochemistry of plants. Proteins and nucleic acids. A. Marcus (ed.). Academic Press, New York, 6:208–47.

171. Guilfoyle, T. J., and J. L. Key. 1977. Purification and characterization of soybean DNA-dependent RNA polymerases and the modulation of their activities during development. In: Nu-

cleic acids and protein synthesis in plants. L. Bogorad and J. H. Weil (eds.). NATO Advanced Study Institutes Series, 12:37–63.

172. Guilley, H., G. Jonard, B. Kukla, and K. E. Richards. 1979. Sequence of 1000 nucleotides at the 3'-end of tobacco mosaic virus RNA. Nucl. Acid Res. 6:1287–1308.

173. Guilley, H., K. E. Richards, and G. Jonard. 1983. Observations concerning the discontinuous DNAs of cauliflower mosaic virus. EMBO J. 2:277–82.

174. Habili, N., and R. I. B. Francki. 1974. Comparative studies on tomato aspermy and cucumber mosaic viruses. III. Further studies on relationship and construction of a virus from parts of the two viral genomes. Virology 61:43–49.

175. Hadidi, A., and H. Fraenkel-Conrat. 1973. Characterization and specificity of soluble RNA polymerase of brome mosaic virus. Virology 52:363–72.

176. Haenni, A.-L., and F. Chapeville. 1982. tRNA-like structures in the genomes of RNA viruses. Prog. Nuc. Acid Res. and Mol. Biol. 27:85–104.

177. Hahn, P., and R. J. Shepherd. 1982. Evidence for a 58-kilodalton polypeptide as precursor of the coat protein of cauliflower mosaic virus. Virology 116:480–88.

178. Hall, T. C. 1979. Plant messenger RNA. In: Nucleic acids in plants. T. C. Hall and J. W. Davies (eds.). CRC Press, Hollywood, FL, 1:217–51.

179. Hall, T. C., and J. W. Davies (eds.). 1979. Nucleic acids in plants. Vol. 1. CRC Press, Hollywood, FL.

180. Hall, T. C., and J. W. Davies (eds.). 1979. Nucleic acids in plants. Vol. II. CRC Press, Hollywood, FL.

181. Hall, T. C., W. A. Miller, and J. J. Bujarski. 1982. Enzymes involved in the replication of plant viral RNAs. Adv. Plant Path. 1:179–211.

182. Hardy, S. F., T. L. German, S. Loesch-Fries, and T. C. Hall. 1979. Highly active template-specific RNA-dependent RNA polymerase from barley leaves infected with brome mosaic virus. Proc. Nat. Acad. Sci. 76:4956–60.

183. Hari, V., A. Siegel, C. Rozek, and W. E. Timberlake. 1979. The RNA of tobacco etch virus contains poly (A). Virology 92:568–71.

184. Harpaz, I., and S. W. Applebaum. 1961. Accumulation of asparagine in maize plants infected by maize rough dwarf virus and its significance in plant virology. Nature 192:780–81.

185. Harrison, B. D., and A. F. Murant. CMI/AAB descriptions of plant viruses. Commonw. Mycol. Inst. and Assn. Appl. Biol., Kew.

186. Harrison, B. D., and D. J. Robinson. 1978. The tobraviruses. Adv. Virus Res. 23:25–77.

187. Hatta, T., S. Bullivant, and R. E. F. Matthews. 1973. Fine structure of vesicles induced in chloroplasts of Chinese cabbage leaves by infection with turnip yellow mosaic virus. J. Gen. Virol. 20:37–50.

188. Hatta, T., and R. I. B. Francki. 1979. Enzyme cytochemical method for identification of cucumber mosaic virus particles in infected cells. Virology 93: 265–68.

189. Hatta, T., and R. I. B. Francki. 1981. Cytopathic structures associated with tonoplasts of plant cells infected with cucumber mosaic and tomato aspermy viruses. J. Gen. Virol. 53:343–46.

190. Heitefuss, R. 1966. Nucleic acid metabolism in obligate parasitism. Ann. Rev. Phytopath. 4:221–44.

191. Heitefuss, R., and M. Bauer. 1969. Der Einfluss von 5-Fluorouracil auf den Nukleinsäurestoffwechsel von *Phaseolus vulgaris* nach Infektion mit *Uromyces phaseoli*. Phytopath. Z. 66:25–37.

192. Heitefuss, R., D. J. Buchanan-Davidson, M. A. Stahmann, and J. C. Walker. 1959. Electrophoretic and immunochemical studies of proteins in cabbage infected with *Fusarium oxysporum* f. *conglutinans*. Phytopathology 50:198–205.

193. Heitefuss, R., and G. Wolf. 1976. Nucleic acids in host-parasite interactions. In: Physiological plant pathology. R. Heitefuss and P. H. Williams (eds.). Springer-Verlag, Berlin, pp. 480–508.

194. Higgins, T. J. V., P. B. Goodwin, and P. R. Whitfield. 1976. Occurrence of short particles in beans infected with the cowpea strain of TMV. II. Evidence that short particles contain the cistron for coat-protein. Virology 71:486–97.

195. Higgins, T. J. V., P. R. Whitfield, and R. E. F. Matthews. 1978. Size distribution and in vitro translation of the RNAs isolated from turnip yellow mosaic virus nucleoproteins. Virology 84:153–61.

196. Hirai, A., and S. G. Wildman. 1969. Effect of TMV multiplication on RNA and protein synthesis in tobacco chloroplasts. Virology 38:73–82.

197. Hirai, T., and A. Hirai. 1964. Tobacco mosaic virus: Cytological evidence of the synthesis in the nucleus. Science 145:589–91.

198. Hirth, L., and K. E. Richards. 1981. Tobacco mosaic virus: Model for structure and function of a simple virus. Adv. Virus Res. 26:145–99.

199. Holowczak, J., J. Kuć, and E. B. Williams. 1962. Metabolism of DL- and L- phenylalanine in *Malus* related to susceptibility and resistance to *Venturia inaequalis*. Phytopathology 52:699–703.

200. Honda, Y., and C. Matsui. 1974. Electron microscopy of cucumber mosaic virus-infected tobacco leaves showing mosaic symptoms. Phytopathology 64:534–39.

201. Honda, Y., C. Matsui, Y. Otsuki, and I. Takebe. 1974. Ultrastructure of tobacco mesophyll protoplasts inoculated with cucumber mosaic virus. Phytopathology 64: 30–34.

202. Howell, S. H. 1982. Plant molecular vehicles: Potential vectors for introducing foreign DNA into plants. Ann. Rev. Plant Physiol. 33:609–50.

203. Huang, P.-Y., J. S. Huang, and R. N. Goodman. 1975. Resistance mechanisms of apple shoots to an avirulent strain of *Erwinia amylovora*. Physiol. Plant Path. 6: 283–87.

204. Huber, R. 1979. Proteins synthesized in tobacco mosaic virus infected protoplasts. Meded. Landbouw., Wageningen, pp. 1–103.

205. Hull, R. 1979. The DNA of plant DNA viruses. In: Nucleic acids in plants. T. C. Hall and J. W. Davies (eds.). CRC Press, Hollywood, FL, 2:3–29.

206. Hull, R., and S. N. Covey. 1981. Timecourse of appearance of cauliflower mosaic virus transcripts. Vth Int. Virol. Cong., Strasbourg, p. 209. (Abstr.)

207. Hull, R., and S. N. Covey. 1983. Does cauliflower mosaic virus replicate by severe transcription? TIBS 8:119–21.

208. Hunter, T. R., T. Hunt, J. Knowland, and D. Zimmern. 1976. Messenger RNA for the coat protein of tobacco mosaic virus. Nature 260:759–64.

209. Ikegami, M., and H. Fraenkel-Conrat. 1978. RNA-dependent RNA polymerase of tobacco plants. Proc. Nat. Acad. Sci. 75:2122–24.

210. Ikegami, M., and H. Fraenkel-Conrat. 1978. The RNA-dependent RNA polymerase of cowpea. FEBS Lett. 96:197–200.

211. Ikegami, M., and H. Fraenkel-Conrat. 1979. Characterization of double-stranded ribonucleic acid in tobacco leaves. Proc. Nat. Acad. Sci. 76:3637–40.

212. Ikegami, M., and H. Fraenkel-Conrat. 1979. Characterization of the RNA-dependent RNA polymerase of tobacco leaves. J. Biol. Chem. 254:149–54.

213. Ikegami, M., and H. Fraenkel-Conrat. 1980. Lack of specificity of virus-stimulated plant RNA-dependent RNA polymerases. Virology 100:185–88.

214. Jackson, A. O., D. M. Mitchell, and A. Siegel. 1971. Replication of tobacco mosaic virus. I. Isolation and characteriza-

tion of double-stranded forms of ribonucleic acid. Virology 45:182–91.

215. Jackson, A. O., M. Zaitlin, A. Siegel, and R. I. B. Francki. 1972. Replication of tobacco mosaic virus. III. Viral RNA metabolism in separated leaf cells. Virology 48:655–65.

216. Johnson, R. 1962. Serological studies on wheat and wheat stem rust. Master's thesis, Univ. of Saskatchewan, Saskatoon.

217. Jones, B. L., and D. B. Cooper. 1980. Purification and characterization of a corn (*Zea mays*) protein similar to purothionins. Agr. Food Chem. 28:904–8.

218. Jones, T. M., A. F. Anderson, and P. Albersheim. 1972. Host-pathogen interactions. IV. Studies on the polysaccharide-degrading enzymes secreted by *Fusarium oxysporum* f. sp. *lycopersici*. Physiol. Plant Path. 2:153–66.

219. Joshi, S., C. W. A. Pleij, A. L. Haemmi, F. Chapeville, and L. Boch. 1983. Properties of the tobacco mosaic virus intermediate length RNA-2 and its translation. Virology 127:100–111.

220. Kado, C. I. 1975. Studies on *Agrobacterium*. III. A concept on the role of *Agrobacterium tumefaciens* DNA in plant tumorigenesis. In: Proc. First Intersect. Cong. Microbiol. Soc. Tokyo. T. Hasegawa (ed.). 1:100–130.

221. Kado, C. I., and P. F. Lurquin. 1975. Studies on *Agrobacterium tumefaciens*. IV. Nonreplication of bacterial DNA in mung bean (*Phaseolus aureus*). Biochem. Biophys. Res. Commu. 64:175–83.

222. Kado, C. I., and P. F. Lurquin. 1976. Studies on *Agrobacterium tumefaciens*. V. Fate of exogenously added bacterial DNA in *Nicotiana tabacum*. Physiol. Plant Path. 8:73–82.

223. Kadouri, A., D. Atsmon, and M. Edelman. 1975. Satellite-rich DNA in cucumber: Hormonal enhancement of synthesis and subcellular identification. Proc. Nat. Acad. Sci. 72:2260–64.

224. Kamalay, J. C., and R. B. Goldberg. 1980. Regulation of structural gene expression in tobacco. Cell 19:935–46.

225. Kaper, J. M. 1982. Rapid synthesis of double-stranded cucumber mosaic virus- associated RNA 5: Mechanism controlling viral pathogenesis? Biochem. Biophys. Res. Comm. 105:1014–22.

226. Kaper, J. M., and M. E. Tousignant. 1977. Cucumber mosaic virus-associated RNA 5. 1. Role of host plant and helper strain in determining amount of associated RNA 5 with virions. Virology 80:186–95.

227. Kaper, J. M., and M. E. Tousignant. 1978. Cucumber mosaic virus-associated RNA 5. V. Extensive nucleotide sequence homology among CARNA 5 preparations of different CMV strains. Virology 85:323–27.

228. Kaper, J. M., M. E. Tousignant, and H. Lot. 1976. A low molecular weight replicating RNA associated with a divided genome plant virus. Biochem. Biophys. Res. Comm. 72:1237–43.

229. Kaper, J. M., M. E. Tousignant, and S. M. Thompson. 1981. Cucumber mosaic virus-associated RNA 5. VIII. Identification and partial characterization of a CARNA 5 incapable of inducing tomato necrosis. Virology 114:526–33.

230. Kaper, J. M., and H. E. Waterworth. 1977. Cucumber mosaic virus associated RNA 5: Causal agent for tomato necrosis. Science 196:429–31.

231. Kaper, J. M., and H. E. Waterworth. 1981. Cucumoviruses. In: Handbook of plant virus infections: Comparative diagnosis. E. Kurstak (ed.). Elsevier/North Holland, Amsterdam, pp. 258–332.

232. Kassanis, B. 1954. Heat therapy of virus-infected plants. Ann. Appl. Biol. 41: 470–74.

233. Kassanis, B., and C. Bastow. 1971. The relative concentration of infective intact virus and RNA of four strains of tobacco mosaic virus as influenced by temperature. J. Gen. Virol. 11:157–70.

234. Kenfield, D. S., and G. A. Strobel. 1981. α-Galactoside

binding proteins from plant membranes: Isolation, characterization, and relation to helminthosporoside binding proteins of sugarcane. Plant Physiol. 67:1174–80.

235. Keys, A. J., I. F. Bird, M. J. Carnelius, P. J. Lea, R. M. Wallsgrove, and B. J. Miflin. 1978. Photorespiratory nitrogen cycle. Nature 275:741–43.

236. Kim, K. S., T. L. Shock, and R. M. Goodman. 1978. Infection of *Phaseolus vulgaris* by bean golden mosaic virus: Ultrastructural aspects. Virology 89:22–33.

237. Király, Z., B. I. Pozsár, and M. El Hammady. 1966. Cytokinin activity in rust-infected plants: Juvenility and senescence in diseased leaf tissues. Acta Phytopath. Hung. 1:29–37.

238. Klein, R. M. 1952. Nitrogen and phosphorus fractions, respiration, and structure of normal and crown gall tissues of tomato. Plant Physiol. 27:335–54.

239. Koike, M., T. Hibi, and K. Yora. 1977. Infection of cowpea mesophyll protoplasts with cucumber mosaic virus. Virology 83:413–16.

240. Kolodner, R., and K. K. Tewari. 1972. Molecular size and conformation of chloroplast deoxyribonucleic acid from pea leaves. J. Biol. Chem. 247:6355–64.

241. Kolodner, R., and K. K. Tewari. 1972. Physiochemical characterization of mitochondrial DNA from pea leaves. Proc. Nat. Acad. Sci. 69:1830–34.

242. Kolodner, R. D., and K. K. Tewari. 1975. Denaturation mapping studies on the circular chloroplast deoxyribonucleic acid from pea leaves. J. Biol. Chem. 250: 4888–95.

243. Konarska, M., W. Filipowicz, H. Domdey, and H. J. Gross. 1981. Binding of ribosomes to linear and circular forms of the 5′-terminal leader fragment of tobacco-mosaic-virus RNA. Eur. J. Biochem. 114:221–27.

244. Koper-Zwarthoff, E. C., and J. F. Bol. 1980. Nucleotide sequence of the putative recognition site for coat protein in the RNAs of alfalfa mosaic virus and tobacco streak virus. Nucl. Acid Res. 8:3307–18.

245. Kosuge, T., M. S. Heskett, and E. E. Wilson. 1966. Microbial synthesis and degradation of indole-3-acetic acid. I. The conversion of L-tryptophan to indole- 3-acetamide by an enzyme system from *Pseudomonas savastanoi*. J. Biol. Chem. 241:3738–44.

246. Kozak, M. 1980. Evaluation of the 'scanning model' for initiation of protein synthesis in eucaryotes. Cell 22:7–8.

247. Kozak, M. 1983. Comparison of initiation of protein synthesis in procaryotes, eucaryotes and organelles. Microbiol. Rev. 47:1–45.

248. Kuć, J., E. Barnes, A. Daftsios, and E. B. Williams. 1959. The effect of amino acids on susceptibility of apple varieties to scab. Phytopathology 49:313–15.

249. Kumarasamy, R., and R. H. Symons. 1979. Extensive purification of the cucumber mosaic virus-induced RNA replicase. Virology 96:622–32.

250. Kummert, J., and J. Semal. 1974. Polyacrylamide gel electrophoresis of the RNA products labeled in vitro by extracts of leaves infected with bromegrass mosaic virus. Virology 60:390–97.

251. Kupila, S., and H. Stern. 1961. DNA content of broad bean (*Vicia faba*) internodes in connection with tumor induction by *Agrobacterium tumefaciens*. Plant Physiol. 36:216–19.

252. Laflèche, D., and J. M. Bové. 1968. Sites d'incorporation de l'uridine tritiée dans les cellules du parenchyme foliare de *Brassica chinensis*, saines ou infectées par le virus de la mosaïque jaune du navet. Compte. Rend. Acad. Sci. 266:1839–41.

253. Lakshminarayanan, K. 1955. Some biochemical aspects of vascular wilts. Proc. Indian Acad. Sci. Sect. B 41:132–44.

254. Lamar, C., Jr., S. L. Sinden, and R. D. Durbin. 1969. The inhibition in vitro of rat cerebral glutamine synthetase by an ex-

otoxin from *Pseudomonas tabaci*. Biochem. Pharmacol. 18:521–29.

255. Lane, L. C. 1979. The nucleic acids of multipartite, defective and satellite plant viruses. In: Nucleic acids in plants. T. C. Hall and J. W. Davies (eds.). CRC Press, Hollywood, FL, 2:65–110.

256. Langenberg, W. G., and D. E. Schlegel. 1969. Localization of tobacco mosaic virus protein in tobacco leaf cells during the early stages of infection. Virology 37:86–93.

257. Leaver, C. J. 1979. Ribosomal RNA of plants. In: Nucleic acids in plants.T. C. Hall and J. W. Davies (eds.). CRC Press, Hollywood, FL, 1:193–215.

258. Leaver, C. J. (ed.). 1980. Genome organization and expression in plants. Plenum Press, New York.

259. Lebeurier, G., and L. Hirth. 1966. Effect of elevated temperatures on the development of two strains of tobacco mosaic virus. Virology 29:385–95.

260. Lebeurier, G., L Hirth, T. Hohn, and B. Hohn. 1980. Infectivities of native and cloned DNA of cauliflower mosaic virus. Gene 12:139–46.

261. Le Roy, C., C. Stussi-Garaud, and L. Hirth. 1977. RNA dependent RNA polymerases in uninfected and in alfalfa mosaic virus infected tobacco plants. Virology 82:48–62.

262. Linthorst, H. J. M., J. F. Bol, and E. M. J. Jaspars. 1980. Lack of temperature sensitivity of RNA-dependent RNA polymerases isolated from tobacco plants infected with ts mutants of alfalfa mosaic virus. J. Gen. Virol. 46:511–15.

263. Lis, H., and S. Nathan. 1981. Lectins in higher plants. In: The biochemistry of plants. Proteins and nucleic acids. A. Marcus (ed.). Academic Press, New York, 6:371–433.

264. Loesch-Fries, L. S., and T. C. Hall. 1980. Synthesis, accumulation and encapsidation of individual brome mosaic virus RNA components in barley protoplasts. J. Gen. Virol. 47:323–32.

265. Loesch-Fries, L. S., and T. C. Hall. 1982. In vivo aminoacylation of brome mosaic and barley stripe mosaic virus RNAs. Nature 248:771–73.

266. Lomonossoff, G. P., and P. J. G. Butler. 1979. Location and encapsidation of the coat protein cistron of tobacco mosaic virus. A bidirectional elongation of the nucleoprotein rod. Eur. J. Biochem. 93:157–64.

267. Lot, H., and J. M. Kaper. 1976. Further studies on the RNA component distribution among the nucleoproteins of cucumber mosaic virus. Virology 74:223–26.

268. Lot, H., G. Marchoux, J. Marrou, J. M. Kaper, C. K. West, L. van Vloten-Doting, and R. Hull. 1974. Evidence for three functional RNA species in several strains of cucumber mosaic virus. J. Gen. Virol. 22:81–93.

269. Lovrekovich, L., and G. L. Farkas. 1963. Kinetin as an antagonist of the toxic effect of *Pseudomonas tabaci*. Nature 198:710.

270. Lovrekovich, L., Z. Klement, and G. Farkas. 1964. Toxic effect of *Pseudomonas tabaci* on RNA metabolism in tobacco and its counteraction by kinetin. Science 145:165.

271. Lovrekovich, L., H. Lovrekovich, and R. N. Goodman. 1970. The relationship of ammonia to symptom expression in apple shoots inoculated with *Erwinia amylovora*. Can. J. Bot. 48:999–1000.

272. Mainwaring, W. I. P., J. H. Parish, J. D. Pickering, and N. H. Mann. 1982. Nucleic acid biochemistry and molecular biology. Blackwell Scientific, Oxford.

273. Malca, I., F. P. Zscheile, and R. Gulli. 1964. Nucleotide composition of RNA in healthy and mildew-infected leaves of barley. Phytopathology 54:1112–16.

274. Marchoux, G., J.-C. Devergne, J. Marrou, L. Douine, and H. Lot. 1974. Complémentation entre ARN de différentes souches du virus de la Mosaïque du Concombre. Localisation sur l' ARN-3,

de plusieurs propriétés, dont certaines liées à la nature de la capside. Compte. Rend. Acad. Sci. D 279:2165–68.

275. Marcus, A. (ed.).1981. The biochemistry of plants. Vol. 6. Proteins and nucleic acids. Academic Press, New York.

276. Marrou, J., G. Marchoux, and A. Migliori. 1971. Étude du rythme des variations périodiques de la concentration en virus de la mosaïque du concombre: Influence des différents facteurs sur l'évolution de l'infection. Ann. Phytopath. 3:37–52.

277. Matthews, R. E. F. 1953. Chemotherapy and plant viruses. J. Gen. Microbiol. 8: 277–88.

278. Matthews, R. E. F. 1973. Induction of disease by viruses, with special reference to turnip yellow mosaic virus. Ann. Rev. Phytopath. 11:147–70.

279. Matthews, R. E. F. 1978. Are viroids negative-strand viruses? Nature 276:850.

280. Matthews, R. E. F. 1981. Plant virology. 2d ed. Academic Press, New York.

281. Maule, A. J., M. I. Boulton, and K. R. Wood. 1980. Resistance of cucumber protoplasts to cucumber mosaic virus: A comparative study. J. Gen. Virol. 51: 271–79.

282. Mayo, M. A., H. Barker, and B. D. Harrison. 1982. Specificity and properties of the genome-linked proteins of nepoviruses. J. Gen. Virol. 59:149–62.

283. McIntyre, J. L., J. Kuć, and E. B. Williams. 1975. Protection of Bartlett pear against fireblight with deoxyribonucleic acid from virulent and avirulent *Erwinia amylovora*. Physiol. Plant Path. 7:153–70.

284. McMahon, D. 1975. Cycloheximide is not a specific inhibitor of protein synthesis in vivo. Plant Physiol. 55:815–21.

285. Mellemer, J.-R., C. Bénicourt, A.-L. Haenni, A. Noort, C. W. A. Pleij, and L. Bosch. 1979. Translational studies with turnip yellow mosaic virus RNAs isolated from major and minor virus particles. Virology 96:38–46.

286. Ménissier, J., G. Lebeurier, and L. Hirth. 1982. Free cauliflower mosaic virus supercoiled DNA in infected plants. Virology 117:322–28.

287. Ménissier, J., A. de Murcia, A. Lebeurier, and L. Hirth. 1983. Electron microscopic studies of the different topological forms of the cauliflower mosaic virus DNA: Knotted encapsidated DNA and nuclear minichromosome. EMBO J. 2:1067–71.

288. Millerd, A., and K. S. Scott. 1963. Host-pathogen relations in powdery mildew of barley. III. Utilization of respiratory energy. Aust. J. Biol. Sci. 16:775–83.

289. Misawa, T., and S. Kato. 1967. Changes in nuclear contents of host cells infected by virus. In: The dynamic role of molecular constituents in plant-parasite interaction. C. J. Mirocha and I. Uritani (eds.). Amer. Phytopath. Soc., St. Paul, MN, pp. 256–69.

290. Misawa, T., S. Kato, and T. Suzuki. 1966. Studies on infection and the multiplication of plant viruses. III. Relationship between the cell nucleus and cucumber mosaic virus replication. Tohoku J. Agr. Res. 16:265–74.

291. Mohamed, N. A., and J. W. Randles. 1972. Effect of tomato spotted wilt virus on ribosomes, ribonucleic acids and Fraction 1 protein in *Nicotiana tabacum* leaves. Physiol. Plant Path. 2:235–45.

292. Mohier, E., L. Pinck, and L. Hirth. 1974. Replication of alfalfa mosaic virus RNAs. Virology 58:9–15.

293. Morch, M.-D., and C. Bénicourt. 1980. Post-translation proteolytic cleavage of in vitro synthesized turnip yellow mosaic virus RNA-coded high molecular weight proteins. J. Virol. 34:85–94.

294. Morch, M.-D., G. Drugeon, and C. Bénicourt. 1982. Analysis of the in vitro coding properties of the 3' region of turnip yellow mosaic virus genomic RNA. Virology 119:193–98.

295. Morch, M.-D., W. Zagorski, and A.-L. Haenni. 1982.

Proteolytic maturation of the turnip yellow mosaic virus polyprotein coded in vitro occurs by internal catalysis. Eur. J. Biochem. 127:259–65.

296. Morris-Krsinich, B. A. M., and R. Hull. 1981. Translation of turnip rosette virus RNA in rabbit reticulocyte lysates. Virology 114:98–112.

297. Mossop, D. W., and R. I. B. Francki. 1977. Association of RNA 3 with aphid transmission of cucumber mosaic virus. Virology 81:177–81.

298. Mossop, D. W., and R. I. B. Francki. 1979. Comparative studies on two satellite RNAs of cucumber mosaic virus. Virology 95:395–404.

299. Mouchès, C., C. Bové, C. Barreau, and J. M. Bové. 1975. TYMV-RNA replicase. Colloq. Instit. Nat. Sante Rech. Med. (INSERM) 47:109–20.

300. Mouchès, C., C. Bové, and J. M. Bové. 1974. Turnip yellow mosaic virus-RNA replicase: Partial purification of the enzyme from the solubilized enzyme-template complex. Virology 58:409–23.

301. Mühlbach, H. P., and H. L. Sänger. 1979. Viroid replication is inhibited by α- amanitin. Nature 278:185–88.

302. Murant, A. F., and M. A. Mayo. 1982. Satellites of plant viruses. Ann. Rev. Phytopath. 20:49–70.

303. Nassuth, A., F. Alblas, A. J. M. van der Geest, and J. F. Bol. 1983. Inhibition of alfalfa mosaic virus RNA and protein synthesis by actinomycin D and cycloheximide. Virology 126:517–24.

304. Neish, A. C., and H. Hibbert. 1943. Studies on plant tumors: III. Nitrogen metabolism of normal and tumor tissues of the beet root. Arch. Biochem. 3:159–66.

305. Nilsson-Tillgren, T. 1970. Studies on the biosynthesis of TMV. III. Isolation and characterization of the replicative form and replicative intermediate RNA. Mol. Gen. Genet. 109:246–56.

306. Nilsson-Tillgren, T., M. C. Kielland-Brandt, and B. Bekke, 1974. Studies on the biosynthesis of tobacco mosaic virus. VI. On the subcellular localization of double-stranded viral RNA. Mol. Gen. Genet. 128:157–69.

307. Nilsson-Tillgren, T., L. Kolehmainen-Sevéus, and D. von Wettstein. 1969. Studies on the biosynthesis of TMV. 1. A system approaching a synchronized virus synthesis in a tobacco leaf. Mol. Gen. Genet. 104:124–41.

308. Novacky, A., and C. I. Ullrich-Eberius. 1982. Relationship between membrane potential and ATP level in *Xanthomonas campestris* pv. *malvacearum* infected cotton cotyledons. Physiol. Plant Path. 21:237–49.

309. Oberg, B., and L. Philipson. 1972. Binding of histidine to tobacco mosaic virus RNA. Biochem. Biophys. Res. Comm. 48:927–32.

310. Odell, J. T., and S. H. Howell. 1980. The identification, mapping, and characterization of m RNA for P66, a cauliflower mosaic virus-coded protein. Virology 102:349–59.

311. Oku, H., S. Ouchi, and T. Shiraishi. 1972. Response to obligate parasite infection of nucleic acid metabolism in host chloroplast. Symp. Biol. Hung. 13:277–82.

312. Okuno, T., and I. Furusawa. 1979. RNA polymerase activity and protein synthesis in brome mosaic virus-infected protoplasts. Virology 99:218–25.

313. Osborne, D. J. 1962. Effect of kinetin on protein and nucleic acid metabolism in *Xanthium* leaves during senescence. Plant Physiol. 37:595–602.

314. Otsuki, Y., and I. Takebe. 1973. Infection of tobacco mesophyll protoplasts by cucumber mosaic virus. Virology 51:433–38.

315. Otsuki, Y., I. Takebe, Y. Honda, S. Kajita, and C. Matsui. 1974. Infection of tobacco mesophyll protoplasts by potato virus X. J. Gen. Virol. 22:375–85.

316. Otsuki, Y., I. Takebe, Y. Honda, and C. Matsui. 1972. Ultrastructure of infection of tobacco mesophyll protoplasts by tobacco mosaic virus. Virology 49:188–94.

317. Ouchi, S., K. Inada, M. Fujiwara, H. Oku, and I. Matsubara. 1977. Effect of inhibitors of protein synthesis of the infection establishment in powdery mildew of barley. In: Current topics in plant pathology. Z. Király (ed.). Akadémiai Kiadó, Budapest, pp. 41–48.

318. Owens, R. A., and G. Bruening. 1975. The pattern of amino acid incorporation into two cowpea mosaic virus proteins in the presence of ribosome-specific protein synthesis inhibitors. Virology 64:520–30.

319. Owens, R. A., and T. O. Diener. 1982. RNA intermediates in potato spindle tuber viroid replication (complementary RNA/replication model). Proc. Nat. Acad. Sci. 79:113–17.

320. Owens, R. A., and J. M. Kaper. 1977. Cucumber mosaic virus-associated RNA 5. II. In vitro translation in a wheat germ protein-synthesis system. Virology 80: 196–203.

321. Oxelfelt, P. 1971. Development of systemic tobacco mosaic virus infection. II. RNA metabolism in systemically infected leaves. Phytopath. Z. 71:247–56.

322. Oxelfelt, P. 1972. Development of systemic tobacco mosaic virus infection. III. Metabolism of soluble proteins in systemically infected leaves. Phytopath. Z. 74: 131–40.

323. Papavizas, G. C., and C. B. Davey. 1963. Effect of sulfur-containing amino compounds and related substances on *Aphanomyces* root rot of peas. Phytopathology 53:109–15.

324. Patel, P. N., and J. C. Walker. 1963. Changes in free amino acids and amide content of resistant and susceptible beans after infection with the halo blight organism. Phytopathology 53:522–28.

325. Paterson, R., and C. A. Knight. 1975. Protein synthesis in tobacco protoplasts infected with tobacco mosaic virus. Virology 64:10–22.

326. Patil, S. S. 1974. Toxins produced by phytopathogenic bacteria. Ann. Rev. Phytopath. 12:259–79.

327. Peden, K. W. C., J. T. May, and R. H. Symons. 1972. A comparison of two plant virus-induced RNA polymerases. Virology 47:498–501.

328. Peden, K. W. C., and R. H. Symons. 1973. Cucumber mosaic virus contains a functionally divided genome. Virology 53:487–92.

329. Pegg, G. F., and L. Sequeira. 1968. Stimulation of aromatic biosynthesis in tobacco plants infected with *Pseudomonas solanacearum*. Phytopathology 58:476–83.

330. Pelham, H. R. B. 1978. Leaky UAG termination codon in tobacco msaic virus RNA. Nature 272:469–71.

331. Pelham, H. R. B. 1979. Synthesis and proteolytic processing of cowpea mosaic virus proteins in reticulocyte lysates. Virology 96:463–77.

332. Pelham, H. R. B. 1979. Translation of tobacco rattle RNAs in vitro: Four proteins from three RNAs. Virology 97:256–65.

333. Person, C. 1960. A preliminary note on the histochemical localization of DNA and RNA in rust-infected wheat leaves. Can. J. Genet. Cytol. 2:103–4.

334. Piazzolla, P., M. E. Tousignant, and H. M. Kaper. 1982. Cucumber mosaic virus- associated RNA 5. IX. The overtaking of viral RNA synthesis by CARNA 5 and ds CARNA 5 in tobacco. Virology 122:147–57.

335. Pinck, L., M. Genevaux, U. P. Bouley, and M. Pinck. 1975. Amino acid accepter activity of replicative form from some tymovirus RNAs. Virology 63:589–90.

336. Pirone, T. P., and G. S. Pound. 1962. Molybdenum nutrition of *Nicotiana tabacum* in relation to multiplication of tobacco mosaic virus. Phytopathology 52:822–26.

337. Pleij, C. W. A., J. R. Mellema, A. Noort, and L. Bosch. 1977. The occurrence of the coat protein messenger RNA in the minor components of turnip yellow mosaic virus. FEBS Lett. 80:19–22.

338. Plumb, R. T., J. G. Manners, and A. Myers. 1968. Behavior of nucleic acids in mildew-infected wheat. Trans. Brit. Mycol. Soc. 51:563–73.

339. Porter, C. A., and L. H. Weinstein. 1960. Altered biochemical patterns induced in tobacco by cucumber mosaic virus infection, by thiouracil, and by their interaction. Contrib. Boyce Thompson Inst. 20:307–16.

340. Pound, G. S., and C. Garces-Orejuela. 1959. Effect of photoperiod on the multiplication of turnip mosaic virus in rape. Phytopatology 49:16–17.

341. Pound, G. S., and K. Helms. 1955. Effects of temperature on multiplication of potato virus X in *Nicotiana* species. Phytopathology 45:493–99.

342. Pound, G. S., and L. G. Weathers. 1953. The relation of host nutrition to multiplication of turnip virus 1 in *Nicotiana glutinosa* and *N. multivalvis*. Phytopathology 43:669–74.

343. Pozsár, B. I., K. Kristev, and Z. Király. 1966. Rust resistance induced by amino acids: A decrease of the enhanced protein synthesis in rust-infected bean leaves. Acta Phytopath. Hung. 1:203–8.

344. Quick, W. A., and M. Shaw. 1964. The physiology of host-parasite relations: XIV. The effect of rust infection on the nucleic acid content of wheat leaves. Can J. Bot. 42:1531–40.

345. Ralph, R. K., S. Bullivant, and S. J. Wojcik. 1971. Evidence for the intracellular site of double-stranded turnip yellow mosaic virus RNA. Virology 44:473–79.

346. Ralph, R. K., and S. J. Wojcik. 1966. Synthesis of double-stranded viral RNA by cell-free extracts from turnip yellow mosaic virus-infected leaves. Biochim. Biophys. Acta 119:347–61.

347. Ralph, R. K., and S. J. Wojcik. 1969. Double-stranded tobacco mosaic virus RNA. Virology 37:276–82.

348. Randles, J. W. 1968. Ribonuclease isozymes in Chinese cabbage systemically infected with turnip yellow mosaic virus. Virology 36:556–63.

349. Randles, J. W., and D. F. Coleman. 1970. Loss of ribosomes in *Nicotiana glutinosa* L. leaves infected with lettuce necrotic yellows virus. Virology 41:459–64.

350. Randles, J. W., and D. F. Coleman. 1972. Changes in polysomes in *Nicotiana glutinosa* L. infected with lettuce necrotic yellows virus. Physiol. Plant Path. 2: 247–58.

351. Rao, A. L. N., and R. I. B. Francki. 1982. Distribution of determinants for symptom production and host range on the three RNA components of cucumber mosaic virus. J. Gen. Virol. 61:197–205.

352. Rasch, E., H. Swift, and R. M. Klein. 1959. Nucleoprotein changes in plant tumor growth. J. Biophys. Biochem. Cytol. 6:11–34.

353. Reddi, K. K. 1963. Studies on the formation of tobacco mosaic virus. II. Degradation of host ribonucleic acid following infection. Proc. Nat. Acad. Sci. 50:75–81.

354. Reddy, D. V. R., and L. M. Black. 1977. Isolation and replication of mutant populations of wound tumor virions lacking certain genome segments. Virology 80:336–46.

355. Reddy, D. V. R., D. P. Rhodes, J. A. Lesnaw, R. MacLeod, A. K. Banerjee, and L. M. Black. 1977. In vitro transcription of wound tumor virus RNA by virion-associated RNA transcriptase. Virology 80:356–61.

356. Reddy, M. N., D. L. Stuteville, and E. L. Sorensen. 1971. Protease production during pathogenesis of bacterial leaf spot of alfalfa and by *Xanthomonas alfalfae* in vitro. Phytopathology 61:361–65.

357. Reid, M. S., and R. E. F. Matthews. 1966. On the origin of the mosaic induced by turnip yellow mosaic virus. Virology 28:563–70.

358. Renaudin, J., and J.-M. Bové. 1977. Effet de l'actinomycine D sur la production de virions par les protoplastes de Chou de Chine infectés in vitro par le virus de la mosaïque jaune du navet. Compte Rend. Acad. Sci. D 284:783–86.

359. Renaudin, J., J.-M. Bové, Y. Otsuki, and I. Takebe. 1975. Infection of *Brassica* leaf protoplasts by turnip yellow mosaic virus. Mol. Gen. Genet. 141:59–68.

360. Ricard, B., C. Barreau, H. Renaudin, C. Mouchés, and J.-M. Bové. 1977. Messenger properties of TYMV RNA. Virology 79:231–35.

361. Richards, K. E., G. Jonard, M. Jacquemond, and H. Lot. 1978. Nucleotide sequence of cucumber mosaic virus-associated RNA 5. Virology 89:395–408.

362. Richmond, A. E., and A. Lang. 1957. Effect of kinetin on protein content and survival of detached *Xanthium* leaves. Science 125:650–51.

363. Roberts, P. L., and K. R. Wood. 1981. Decrease in ribosome levels in tobacco infected with a chlorotic strain of cucumber mosaic virus. Physiol. Plant Path. 19:99–111.

364. Robinson, D. J. 1977. A variant of tobacco rattle virus: Evidence for a second gene in RNA-2. J. Gen. Virol. 35:37–43.

365. Robinson, D. J., M. A. Mayo, C. Fritsch, A. T. Jones, and J. H. Raschke. 1983. Origin and messenger activity of two small RNA species found in particles of tobacco rattle virus strain SYM. J. Gen. Virol. 64:1591–99.

366. Rohringer, R. 1976. Evidence for direct involvement of nucleic acids in host-parasite specificity. In: Specificity in plant diseases. R. K. S. Wood and A. Graniti (eds.). Plenum Press, New York and London, pp. 185–98.

367. Rohringer, R., and R. Heitefuss. 1961. Incorporation of P[32] into ribonucleic acid of rusted wheat leaves. Can. J. Bot. 39:263–67.

368. Rohringer, R., N. K. Howes, W. K. Kim, and D. J. Samborski. 1974. Evidence for a gene-specific RNA determining resistance in wheat to stem rust. Nature 249: 585–88.

369. Romaine, C. P., and M. Zaitlin. 1978. RNA-dependent RNA polymerases in uninfected and tobacco mosaic virus-infected tobacco leaves: Viral-induced stimulation of a host polymerase activity. Virology 86:241–53.

370. Romeiro, R., A. Karr, and R. N. Goodman. 1981. *Erwinia amylovora* cell wall receptor for apple agglutinin. Physiol. Plant Path. 19:383–90.

371. Romeiro, R., A. Karr, and R. N. Goodman. 1981. Isolation of a factor from apple that agglutinates *Erwinia amylovora*. Plant Physiol. 68:772–77.

372. Röttger, B. 1965. Ribonucleic acids of healthy and tobacco mosaic virus infected tobacco leaves. Biochim. Biophys. Acta 95:525–31.

373. Rottier, P. J. M., G. Rezelman, and A. van Kammen. 1979. The inhibition of cowpea mosaic virus replication by actinomycin D. Virology 92:299–309.

374. Rubenstein, I., R. L. Phillips, C. E. Green, and B. G. Gengenbach. 1979. Molecular biology of plants. Academic Press, New York.

375. Rubin, B. A., V. A. Axenova, and N. D. Huyen. 1971. Some peculiarities of protein synthesis in infected plant tissues. Acta Phytopath. Hung. 6:61–64.21.

376. Rudolph, K. 1963. Weitere biochemische Untersuchungen zum Wirt-Parasit- Verhältnis am Beispiel von *Puccinia graminis tritici*: I. Der Einfluss der Infektion auf den Säuerestoffwechsel. Phytopath. Z. 46:276–90.

377. Rudolph, K., and M. A. Stahmann. 1966. The accumula-

tion of L-ornithine in halo-blight infected bean plants (*Phaseolus vulgaris* L.) induced by the toxin of the pathogen *Pseudomonas phaseolicola* (Burkh.) Dowson. Phytopath. Z. 57:29–46.

378. Ryan, C. A. 1973. Proteolytic enzymes and their inhibitors in plants. Ann. Rev. Plant Physiol. 24:173–96.

379. Ryan, C. A., P. Bishop, G. Pearce, A. G. Darvill, M. McNeil, and P. Albersheim. 1981. A sycamore cell wall polysaccharide and a chemically related tomato leaf polysaccharide possess similar proteinase inhibitor-inducing activities. Plant Physiol. 68:616–18.

380. Sadler, R., and K. J. Scott. 1974. Nitrogen assimilation and metabolism in barley leaves infected with the powdery mildew fungus. Physiol. Plant Path. 4:235–47.

381. Sakai, F., and I. Takebe. 1974. Protein synthesis in tobacco mesophyll protoplasts induced by tobacco mosaic virus infection. Virology 62:426–33.

382. Salomon, R., M. Bar-Joseph, H. Soreq, I. Gozes, and U. Z. Littauer. 1978. Translation in vitro of carnation mottle virus RNA. Regulatory function of the 3′ region. Virology 90:288–98.

383. Salomon, R., I. Sela, H. Soreq, D. Giveon, and U. Z. Littauer. 1976. Enzymatic acylation of histidine to tobacco mosaic virus RNA. Virology 71:74–84.

384. Salvato, M. S., and H. Fraenkel-Conrat. 1977. Translation of tobacco necrosis virus and its satellite in a cell-free wheat germ system. Proc. Nat. Acad. Sci. 74:2288–92.

385. Samborski, D. J., and F. R. Forsyth. 1960. Inhibition of rust development on detached wheat leaves by metabolites, antimetabolites, and enzyme poisons. Can. J. Bot. 38:467–76.

386. Samborski, D. J., F. R. Forsyth, and C. Person. 1958. Metabolic changes in detached wheat leaves floated on benzimidazole and the effect of these changes on rust reactions. Can. J. Bot. 36:591–601.

387. Sanger, F. 1981. Determination of nucleotide sequences in DNA. Biosci. Repts. 1:3–18.

388. Sänger, H. L., and C. A. Knight. 1963. Action of actinomycin D on RNA synthesis in healthy and virus infected tobacco leaves. Biochem. Biophys. Res. Comm. 13:455–61.

389. Scalla, R., E. Boudon, and J. Rigaud. 1976. Sodium dodecyl sulphate-polyacrylamide-gel electrophoretic detection of two high molecular weight proteins associated with tobacco mosaic virus infection in tobacco. Virology 69:339–45.

390. Schilperoort, R. A. 1972. Integration of *Agrobacterium tumefaciens* DNA in the genome of crown gall tumor cells and its expression. In:Proc. Third Int. Conf. Plant Pathogenic Bact. H. P. Maas Geesteranus (ed.). Pudoc, Wageningen, Neth., pp. 223–38.

391. Schilperoort, R. A., S. J. Van Stittert, and J. Schell. 1973. The presence of both phage PS 8 and *Agrobacterium tumefaciens* DNA base sequence in AG-induced sterile crown gall tissue cultured in vitro. Eur. J. Biochem. 33:1–7.

392. Schilperoort, R. A., H. Veldstra, H. Warnaar, G. Mulder, and J. A. Cohen. 1967. Formation of complexes between DNA isolated from tobacco crown gall tumors and RNA complimentary to *Agrobacterium tumefaciens* DNA. Biochim. Biophys. Acta 145:523.

393. Schlegel, D. E., and S. H. Smith. 1966. Sites of virus synthesis in plant cells. In: Viruses of plants. A. B. R. Beemster and J. Dijkstra (eds.). Elsevier/ North Holland, Amsterdam, pp. 54–65.

394. Schnathorst,W. C., and J. E. DeVay. 1963. Common antigens in *Xanthomonas malvacearum* and *Gossypium hirsutum* and their possible relationship to host specificity and disease resistance. Phytopathology 53:1142. (Abstr.)

395. Schwinghamer, M. W., and R. H. Symons. 1975. Fractionation of cucumber mosaic virus RNA and its translation in a wheat embryo cell-free system. Virology 63:252–62.

396. Schwinghamer, M. W., and R. H. Symons. 1977. Translation of the four major RNA species of cucumber mosaic virus in plant and animal cell-free systems and in toad oocytes. Virology 79:88–108.

397. Scovassi, A. I., P. Plevani, and U. Bertazzoni. 1980. Eukaryotic DNA polymerases. Trends Biochem. Sci. 5:335–37.

398. Selman, I. W., M. R. Brierley, G. F. Pegg, and T. A. Hill. 1961. Changes in the free amino acids and amides in tomato plants inoculated with tomato spotted wilt virus. Ann. Appl. Biol. 49:601–15.

399. Semal, J., and J. Kummert. 1969. Effects of actinomycin D on the incorporation of uridine into virus infected leaf fragments. Phytopath. Z. 65:364–72.

400. Sequeira, L. 1978. Lectins and their role in host-pathogen specificity. Ann. Rev. Phytopath. 16:453–81.

401. Sequeira, L., and T. L. Graham. 1977. Agglutination of avirulent strains of *Pseudomonas solanacearum* by potato lectin. Physiol. Plant Path. 11:43–54.

402. Shalla, T. A., and A. Amici. 1964. The distribution of viral antigen in cells infected with tobacco mosaic virus as revealed by electron microscopy. Virology 31:78–91.

403. Shaw, M., and N. Colotelo. 1961. The physiology of host-parasite relations. VII. The effect of stem rust on the nitrogen and amino acids in wheat leaves. Can. J. Bot. 39:1351–72.

404. Sheimin, R., J. Humbert, and R. E. Pearlman. 1978. Some aspects of eukaryotic DNA replication. Ann. Rev. Biochem. 47:277–316.

405. Shepherd, R. J. 1979. DNA plant viruses. Ann. Rev. Plant Physiol. 30:405–23.

406. Shih, D.-S., R. Dasgupta, and P. Kaesberg. 1976. 7-Methyl-guanosine and efficiency of RNA translation. J. Virol. 19:637–42.

407. Shipp, W., and R. Haselkorn. 1964. Double-stranded RNA from tobacco leaves infected with TMV. Proc. Nat. Acad. Sci. 52:401–8.

408. Siegel, A., V. Hari, and K. Kolacz. 1978. The effect of tobacco mosaic virus infection on host and virus-specific protein synthesis in protoplasts. Virology 85:494–503.

409. Siegel, A., V. Hari, I. Montgomery, and K. Kolacz. 1976. A messenger RNA for capsid protein isolated from tobacco mosaic virus-infected tissue. Virology 73: 363–71.

410. Siegel, A., D. Lightfoot, O. G. Ward, and S. Keener. 1973. DNA complementary to ribosomal RNA: Relation between genomic portion and ploidy. Science 179: 682–83.

411. Simpson, R. S., A. K. Chakravorty, and K. J. Scott. 1979. Selective hydrolysis of barley leaf polysomal messenger RNA during the early stages of powdery mildew infection. Physiol. Plant Path. 14:245–58.

412. Sinden, S. L., and R. D. Durbin. 1968. Glutamine synthetase inhibition: Possible mode of action of wildfire toxin from *Pseudomonas tabaci*. Nature 219:379–80.

413. Singer, B., and C. Condit. 1974. Protein synthesis in virus-infected plants. III. Effects of tobacco mosaic virus mutants on protein synthesis in *Nicotiana tabacum*. Virology 57:42–48.

414. Skoog, F. 1954. Substances involved in normal growth and differentiation of plants. Brookhaven Symp. Biol. 6:1–21.

415. Skotnicki, A., A. Gibbs, and D. C. Shaw. 1976. In vitro translation of polyribosome-associated RNAs from tobamovirus-infected plants. Intervirology 7: 256–71.

416. Smidt, M. L., and T. Kosuge. 1978. The role of indole–3-acetic acid accumulation by alpha-methyl tryptophan-resistant mutants of *Pseudomonas savastanoi* in gall formation on oleanders. Physiol. Plant Path. 13:203–14.

417. Smith, R. E., and J. M. Clark, Jr. 1979. Effect of capping

upon the mRNA properties of satellite tobacco necrosis virus ribonucleic acid. Biochemstry 18:1366–71.

418. Solheim, B., and J. Raa. 1973. Characterization of the substances causing the formation of root hairs of *Trifolium repens* when inoculated with *Rhizobium trifolii*. J. Gen. Microbiol. 77:241–47.

419. Stacey, G., and W. J. Brill. 1982. Nitrogen-fixing bacteria: Colonization of the rhizosphere and roots. In:Phytopathogenic pro-karyotes. M. S. Mount and G. H. Lacy (eds.). Academic Press, New York, 1:225–48.

420. Stahmann, M. A., and D. M. Demorest. 1973. Changes in enzymes of host and pathogen with special reference to peroxidase interaction. In: Fungal pathogenicity and the plant's response. R. J. W. Byrde and C. V. Cutting (eds.). Academic Press, London and New York, pp. 405–22.

421. Stanley, J., P. Rottier, J. W. Davies, P. Zabel, and A. van Kammen. 1978. A protein linked to the 5′ termini of both RNA components of the cowpea mosaic virus genome. Nucl. Acid Res. 5:4505–22.

422. Staples, R. C., and M. C. Ledbetter. 1958. A study of microautoradiography of the distribution of tritium-labeled glycine in rusted pinto bean leaves. Contrib. Boyce Thompson Inst. 19:349–54.

423. Stephens, G. J., and R. K. S. Wood. 1975. Killing of protoplasts by soft-rot bacteria. Physiol. Plant Path. 5:165–81.

424. Stern, D. B., and D. M. Lonsdale. 1982. Mitochondrial and chloroplast genomes of maize have a 12-kilobase DNA sequence in common. Nature 299:698–702.

425. Strobel, G. A. 1973. The helminthosporoside-binding protein of sugarcane. J. Biol. Chem. 284:1321–28.

426. Strobel, G. A. 1974. The toxin-binding protein of sugarcane, its role in the plant and in disease development. Proc. Nat. Acad. Sci. 71:4232–36.

427. Strobel, G. A. 1982. Phytotoxins. Ann. Rev. Biochem. 51:309–33.

428. Stroun, M., P. Anker, and L. Ledoux. 1967. Translocation of DNA of bacterial origin in *Lycopersicon esculentum* by ultra-centrifugation in cesium chloride gradient. Nature 215:975–76.

429. Sugimura, Y., and R. E. F. Matthews. 1981. Timing of the synthesis of empty shells and minor nucleoproteins in relation to turnip yellow mosaic virus synthesis in *Brassica* protoplasts. Virology 112:70–80.

430. Sulzinski, M. A., and M. Zaitlin. 1982. Tobacco mosaic virus replication in resistant and susceptible plants: In some resistant species, virus is confined to a small number of initially infected cells. Virology 121:12–19.

431. Symons, R. H. 1975. Cucumber mosaic virus RNA contains 7-methylguanosine at the 5′-terminus of all four RNA species. Mol. Biol. Repts. 2:277–85.

432. Symons, R. H. 1979. Extensive sequence homology at the 3′-termini of the four RNAs of cucumber mosaic virus. Nucl. Acid Res. 7:825–38.

433. Symons, R. H., D. S. Gill, and K. H. J. Gordon. 1981. Characterization of the soluble and particulate forms of the RNA replicase of cucumber mosaic virus. Vth Int. Virol. Cong., Strasbourg. p. 250. (Abstr.)

434. Takanami, Y. 1981. A striking change in symptoms on cucumber mosaic virus-infected tobacco plants induced by a satellite RNA. Virology 109:120–26.

435. Takanami, Y., S. Kubo, and S. Imaizumi. 1977. Synthesis of single- and double-stranded cucumber mosaic virus RNAs in tobacco mesophyll protoplasts. Virology 80:376–89.

436. Takebe, I. 1977. Protoplasts in the study of plant virus replication.In: Comprehensive virology. H. H. Fraenkel-Conrat and R. R. Wagner (eds.). Academic Press, New York, 11:237–83.

437. Tani, T., M. Yoshikawa, and N. Naito. 1973. Effect of rust infection of oat leaves on cytoplasmic and chloroplast ribosomal ribonucleic acids. Phytopathology 63:491–94.

438. Tani, T., H. Yamamoto, G. Kadota, and N. Naito. 1977. Induction of susceptibility in oat leaves to avirulent rust fungi by blasticidin S, a protein synthesis inhibitor. In: Current topics in plant pathology. Z. Király (ed.). Akadémiai Kiadó, Budapest, pp. 25–34.

439. Tewari, K. K. 1979. Chloroplast DNA: Structure, transcription, and replication. In: Nucleic acids in plants. T. C. Hall and J. W. Davies (eds.). CRC Press, Hollywood, FL, 1:41–108.

440. Tewari, K. K., and S. G. Wildman. 1970. Information content in the chloroplast DNA. Symp. Soc. Exp. Biol. 24:147–79.

441. Thomas, A. A. M., R. Benne, and H. O. Voorma. 1981. Initiation of eukaryotic protein synthesis. FEBS Lett. 128:177–85.

442. Thompson, W. F., and M. G. Murray. 1981. The nuclear genome: Structure and function. In: The Biochemistry of plants. Proteins and nucleic acids. A. Marcus (ed.). Academic Press, New York, 6:1–81.

443. Toppan, A., D. Roby, and M.-T. Esquerré-Tugayé. 1982. Cell surfaces in plant-microoorganism interactions. III. In vivo effect of ethylene on hydroxyproline-rich glycoprotein accumulation in the cell wall of diseased plants. Plant Physiol. 70:82–86.

444. Touzé, A., and M.-T. Esquerré-Tugayé. 1982. Defense mechanisms of plants against varietal non-specific pathogens. In: Active defense mechanisms in plants. NATO Advanced Study Institute Series, Vol. 37. R. K. S. Wood (ed.). Plenum Press, New York and London, pp. l03–17.

445. Tseng, T. C., and M. S. Mount. 1974. Toxicity of endopolyglacturonate transeliminase, phosphatidase and protease to potato and cucumber tissue. Phytopathology 64:229–36.

446. Tu, J. C. 1977. Cyclic AMP in clover and its possible role in clover yellow mosaic virus-infected tissue. Physiol. Plant Path. 10:117–23.

447. Tu, J. C., and R. E. Ford. 1970. Free amino acids in soybeans infected with soybean mosaic virus, bean pod mottle virus, or both. Phytopathology 60:660–64.

448. Tu, J. C., and R. E. Ford. 1970. Maize dwarf mosaic virus infection in susceptible and resistant corn: Virus multiplication, free amino acid concentrations and symptom severity. Phytopathology 60:1605–8.

449. Turner, J. G. 1981. Tabtoxin, produced by *Pseudomonas tabaci*, decreases *Nicotiana tabacum* glutamine synthetase in vivo and causes accumulation of ammonia. Physiol. Plant Path. 19:57–67.

450. Turner, J. G., and J. M. Debbage. 1982. Tabtoxin-induced symptoms are associated with the accumulation of ammonia formed during photorespiration. Physiol. Plant Path. 20:223–33.

451. Ullrich, W. R., and A. Novacky. 1982. Ammonia accumulation and its effect on the membrane potential in bacteria-inoculated cotton and tobacco. Phytopathology 72:981. (Abstr.)

452. Uritani, I. 1971. Protein changes in diseased plants. Ann. Rev. Phytopath. 9: 211–34.

453. Uritani, I. 1976. Protein metabolism. In: Physiological plant pathology. R. Heitefuss and P. H. Williams (eds.). Springer-Verlag, Berlin, pp. 509–25.

454. Uritani, I., and M. A. Stahmann. 1961. Changes in nitrogen metabolism in sweet potato with black rot. Plant Physiol. 36:770–82.

455. van Andel, O. M. 1958. Investigations on plant chemotherapy. II. Influence of amino acids on the relation plant-pathogen. Tijdschr. Plantenziekten 64:307–27.

456. van Andel, O. M. 1966. Amino acids and plant diseases. Ann. Rev. Phytopath. 3: 349–69.

457. Van Larebeke, N., G. Engler, M. Holsters, S. Van den Elsacker, I. Zaenen, R. A. Schilperoort, and J. Schell. 1974. Large plasmid in *Agrobacterium tumefaciens* essential for crown gall inducing ability. Nature: 252:169–70.

458. van Loon, L. C., and A. van Kammen. 1970. Poly-acrylamide disc electrophoresis of the soluble leaf proteins from *Nicotiana tabacum* var. 'Samsun' and 'Samsun NN'. II. Changes in protein constitution after infection with tobacco mosaic virus. Virology 40:199–211.

459. van Vloten-Doting, L., and E. M. J. Jaspars. 1977. Plant covirus systems: Three-component systems. In: Comprehensive virology. H. H. Fraenkel-Conrat and R. R. Wagner (eds.). Plenum Press, New York, 11:1–53.

460. von Wettstein, D., and H. Zech. 1962. The structure of the nucleus and cytoplasm in hair cells during tobacco mosaic virus reproduction. Z. Naturforsch. 176:376–79.

461. van Wezenbeek, P., J. Verver, J. Harmsen, P. Vos, and A. van Kammen. 1983. Primary structure and gene organization of the middle-component RNA of cowpea mosaic virus. EMBO J. 2:941–46.

462. Vedel, F., F. Quetier, M. Bayen, A. Rode, and J. Dalmon. 1972. Intramolecular heterogeneity of mitochondrial and chloroplastic DNA. Biochem. Biophys. Res. Comm. 46:972–78.

463. Wade, M., and P. Albersheim. 1979. Race-specific molecules that protect soybeans from *Phytophthora megasperma* var. *sojae*. Proc. Nat. Acad. Sci. 76: 4433–37.

464. Waitz, L., and W. Schwartz. 1956. Untersuchungen über die von *Pseudomonas phaseolicola* (Burkh.) hervorgerufene Fett-fleckenkrankheit der Bohne. Phytopath. Z. 26:297–312.

465. Walbot, V., and R. Goldberg. 1979. Plant genome organization and its relationship to classical plant genetics. In:Nucleic acids in plants. T. C. Hall and J. W. Davies (eds.). CRC Press, Hollywood, FL, 1:3–40.

466. Wang, D., M. S. H. Hao, and E. R. Waygood. 1961. Effect of benzimidazole analogues on stem rust and chlorophyll metabolism. Can. J. Bot. 39:1029–36.

467. Wasuwat, S. L., and J. C. Walker. 1961. Relative concentration of cucumber mosaic virus in a resistant and a susceptible cucumber variety. Phytopathology 51:614–16.

468. Weathers, L. G., and G. S. Pound. 1954. Host nutrition in relation to multiplication of tobacco mosaic virus in tobacco. Phytopathology 44:74–80.

469. Weeks, D. P. 1981. Protein biosynthesis: Mechanisms and regulation. In: The biochemistry of plants. Proteins and nucleic acids. A. Marcus (ed.). Academic Press, New York, 6:491–529.

470. Weil, J. H., G. Burkard, P. Guillemaut, G. Jeannin, R. Martin, and A. Steinmetz. 1976. tRNAs and aminoacyl-tRNA synthetases in plant cytoplasm, chloroplasts and mitochondria. In: Nucleic acids and protein synthesis in plants. L. Bogorad and J. H. Weil (eds.). NATO Advanced Study Institutes Series, Vol. 12. New York, pp. 97–120.

471. Welkie, G. W., S. F. Yang, and G. W. Miller. 1967. Metabolite changes induced by cucumber mosaic virus in resistant and susceptible strains of cowpea. Phytopathology 57:472–75.

472. White, J. L., and W. O. Dawson. 1978. Characterization of RNA-dependent RNA polymerases in uninfected and cowpea chlorotic mottle virus-infected cowpea leaves: Selective removal of host RNA polymerase from membranes containing CCMV RNA replicase. Virology 88:33–43.

473. Whitfield, P. R., and T. J. V. Higgins. 1976. Occurrence of short particles in beans infected with the cowpea strain of TMV. I. Purification and characterization of short particles. Virology 71:471–85.

474. Whitney, H. S., M. Shaw, and J. M. Naylor. 1962. The physiology of host-parasite relations: XII. A cytophotometric study of the distribution of DNA and RNA in rust-infected leaves. Can. J. Bot. 40:1533–44.

475. Wildman, S. G., C. C. Cheo, and J. Bonner. 1949. The proteins of green leaves. III. Evidence of the formation of tobacco mosaic virus protein at the expense of a main protein component in tobacco leaf cytoplasm. J. Biol. Chem. 180:985–1001.

476. Williams, P. H. 1966. A cytochemical study of hypertrophy in clubroot of cabbage. Phytopathology 56:521–24.

477. Williams, P. H., J. R. Aist, and P. K. Bhattacharya. 1973. Host-parasite relations in cabbage clubroot. In: Fungal pathogenicity and the plant's response. R. J. W. Byrde and C. V. Cutting (eds.). Academic Press, London and New York, pp. 141–58.

478. Wimalajeewa, D. L. S., and J. E. DeVay. 1971. The occurrence and characterization of a common antigen relationship between *Ustilago maydis* and *Zea mays*. Physiol. Plant Path. 1:523–35.

479. Wolanski, B. S., and T. C. Chambers. 1971. The multiplication of lettuce necrotic yellows virus. Virology 44:582–91.

480. Wolf, G. 1968. On the incorporation of ^{32}P into various nucleic acid fractions of rust-infected primary leaves of wheat. Neth. J. Plant Path. (Suppl. 1):19–23.

481. Wollum, J. C., G. B. Shearer, and B. Commoner. 1967. The biosynthesis of tobacco mosaic virus RNA: Relationships to the biosynthesis of virus-specific ribonuclease resistant RNA. Proc. Nat. Acad. Sci. 58:1197–1204.

482. Wood, H. A. 1973. Viruses with double-stranded RNA genomes. J. Gen. Virol. 20:61–85.

483. Wood, K. R., and D. J. Barbara. 1971. Virus multiplication and peroxidase activity in leaves of cucumber (*Cucumis sativus* L.) cultivars systemically infected with the W strain of cucumber mosaic virus. Physiol. Plant Path. 1:73–82.

484. Wyen, H. V., J. Udvardy, S. Erdei, and G. L. Farkas. 1972. The level of a relatively purine-specific ribonuclease increases in virus-infected hypersensitive or mechanically injured tobacco leaves. Virology 48:337–41.

485. Yamaoka, N., I. Furusawa, and M. Yamamoto. 1982. Infection of turnip protoplasts with cauliflower mosaic virus DNA. Virology 122:503–5.

486. Yamaoka, N., T. Morita, I. Furusawa, and M. Yamamoto. 1982. Effect of temperature on the multiplication of cauliflower mosaic virus. J. Gen. Virol. 61:283–87.

487. Yarwood, C. E. 1952. Latent period and generation time for two plant viruses. Amer. J. Bot. 39:613–18.

488. Zabel, P., I. Jongen-Neven, and A. van Kammen. 1979. In vitro replication of cowpea mosaic virus RNA. III. Template recognition by cowpea mosaic virus RNA replicase. J. Virol. 29:21–33.

489. Zabel, P., H. Weenen-Swaans, and A. van Kammen. 1974. In vitro replication of cowpea mosaic virus RNA. I. Isolation and properties of the membrane-bound replicase. J. Virol. 14:1049–55.

490. Zagorski, W., M.-D. Morch, and A.-L. Haenni. 1983. Comparison of three different cell-free systems for turnip yellow mosaic virus RNA translation. Biochimie 65:127–33.

491. Zaitlin, M., C. T. Duda, and M. A. Petti. 1973. Replication of tobacco mosaic virus. V. Properties of the bound and solubilized replicase. Virology 53:300–311.

492. Zaitlin, M., and V. Hariharasubramanian. 1970. Proteins in tobacco mosaic virus-infected tobacco plants. Biochem. Biophys. Res. Comm. 39:1031–36.

493. Zaitlin, M., C. L. Niblett, E. Dickson, and R. B. Goldberg. 1980. Tomato DNA contains no detectable regions complementary to potato spindle tuber viroid as assayed by solution and filter hybridization. Virology 104:1–9.

494. Zelcer, A., K. F. Weaber, E. Balázs, and M. Zaitlin. 1981. The detection and characterization of viral-related double-stranded

RNAs in tobacco mosaic virus-infected plants. Virology 113:417–27.

495. Zelcer, A., M. Zaitlin, H. D. Robertson, and E. Dickson. 1982. Potato spindle tuber viroid-infected tissues contain RNA complementary to the entire viroid. J. Gen. Virol. 59:139–48.

496. Ziemiecki, A., and K. R. Wood. 1976. Proteins synthesized by cucumber cotyledons infected with two strains of cucumber mosaic virus. J. Gen. Virol. 31:373–81.

497. Zimmern, D. 1975. 5′-end group of tobacco mosaic virus RNA is $m^7G^{5'}$ ppp$^{5'}$ Gp. Nucl. Acid Res. 2:1189–1201.

6

Secondary Metabolites

The Healthy Plant

Many infected plant tissues, and particularly locally infected and resistant ("hypersensitive") tissues, show a common shift in the metabolic pattern that includes accumulation of an array of secondary substances (phenolics, flavonoids, coumarins, terpenoids, steroids, etc.), auxins, ethylene production, respiratory increase, and activation of peroxidase and other phenoloxidizing enzymes.

Plants can synthesize and accumulate a comprehensive spectrum of these secondary substances in response to infectious agents as well as physiologic stimuli and stresses. Low molecular weight (MW) antimicrobial compounds that are both synthesized by and accumulated in plants after exposure to microorganisms and to stresses are called *phytoalexins*. Since some resistance phenomena of plants are associated with cell and tissue necrotization, the corresponding activation of secondary metabolism and certain enzymes has been considered responsible for the resistance mechanism itself (11, 46, 48, 68, 80, 140, 164). The analysis of secondary substances is also important, from a pathophysiological point of view at least, in understanding some mechanisms of symptom expression (132).

Historically, secondary plant substances have been considered as terminal compounds not further utilized by the plant. However, it has been shown that some of these substances can be metabolized in plant tissues. These substances are not universally found in higher plants but are restricted to certain plant species. Emphasis here will be on phenolic compounds (hydroxybenzoic acids, coumarins, flavonoids, lignins, and lignans) and on isoprenes (terpenoids, steroids), as well as on some phenoloxidizing enzymes (secondary metabolites with hormonal activity are discussed in Chap. 7). Fig. 6–1 presents an abbreviated scheme of metabolic pathways that is intended to reveal the route of origin and synthesis of secondary substances, highlighting the shikimic acid, mevalonate, and malonate pathways.

Biosynthesis of Phenols

The initial steps of biogenesis of aromatic compounds are the same in microorganisms as in higher plants. Their ontogeny is from aromatic amino acids, which are produced from carbohydrates. These aromatic amino acids, in turn, serve as precursors for the synthesis of phenolics (18, 80, 101, 166, 186).

Three pathways have been demonstrated for the synthesis of phenolics in higher plants from carbohydrates, the shikimate and the two acetate pathways (malonate and mevalonate). The most important of these, leading to the biosynthesis of aromatic amino acids, is the shikimic acid pathway. By combining a three-carbon compound (phosphoenol pyruvic acid) with a four-carbon compound (erythrose phosphate), dehydroshikimic acid is formed. The latter compound forms shikimic acid. Of the two components, phosphoenol pyruvic acid is formed by the glycolytic pathway and erythrose phosphate via the pentose phosphate respiratory cycle (Fig. 6–2).

Shikimate. Shikimic acid produced by this pathway forms, through several intermediates, prephenic acid, which serves as a branching point, being converted to either phenylaline or tyrosine (Fig. 6–3). The intermediates of the phenylaline and tyrosine pools serve as precursors of various aromatic compounds (Fig. 6–4).

Both phenylalanine and tyrosine, the latter found in the *Gramineae*, are enzymatically deaminated by the deaminase enzyme phenylalanine ammonia lyase (PAL). Through deamination of phenylalanine, cinnamic acid is formed, and its hydroxylation by cinnamic

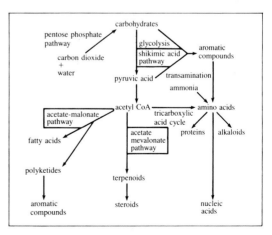

Fig. 6–1. The major pathways of secondary metabolite synthesis.

211

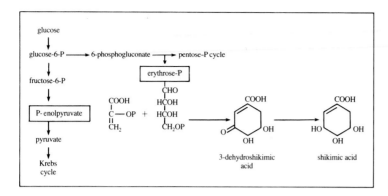

Fig. 6–2. The shikimic acid pathway.

acid 4-hydroxylase leads to the formation of cinnamoyl phenolics, initially *p*-coumaric acid. From the latter compound other cinnamoyl phenols are formed, such as caffeic acid, ferulic acid, and sinapic acid. By the reduction of ferulic and sinapic acids to the corresponding cinnamyl alcohols and by their oxidative polymerization, lignins are produced (see Fig. 6–4). The synthesis of coumarin takes place via the cinnamic acid—*o*-coumaric acid—coumarin route (Fig. 6–5).

Malonate

The biosynthesis of additional complex phenols may also be accomplished via one of the *acetate pathways*. It has been suggested by tracer methods that some phenol compounds may be formed from polyacetic acids. These compounds are derived from head-to-tail condensation of acetate units, for example, acetyl-CoA units are condensed to form malonyl-CoA. There is good evidence that the acetate pathway also operates in higher plants and has a role in the biosynthesis of *flavonoids*, for example. The enzymes involved in different steps of the pathway from phenylalanine to flavonoids and isoflavonoids are indicated in Fig. 6–6. In Fig. 6–7 we see the addition of malonyl CoA to a cinnamic acid derivative (probably *p*-coumaric acid) catalyzed by the en-

zyme flavonone synthase. Studies in this field have shown that the A-rings of, for example, quercetin, cyanidin, and phloretin are derived from acetate molecules. On the other hand, the B-rings of these compounds are synthesized by the shikimic acid pathway (100). Examples of chalcone derivatives are shown in Fig. 6–7c.

Phenoloxidizing Enzymes

The two most important phenoloxidases are laccase and tyrosinase. Both are widely distributed in the plant kingdom and occur in fungi, as well as in higher plants, and, according to Kubowitz (138), both contain copper

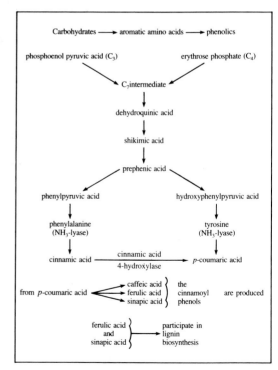

Fig. 6–4. Formation of various aromatic compounds from the phenylalanine and tyrosine pools.

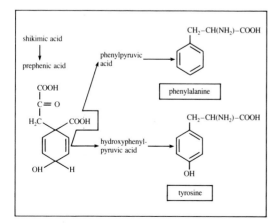

Fig. 6–3. Formation of aromatic amino acids.

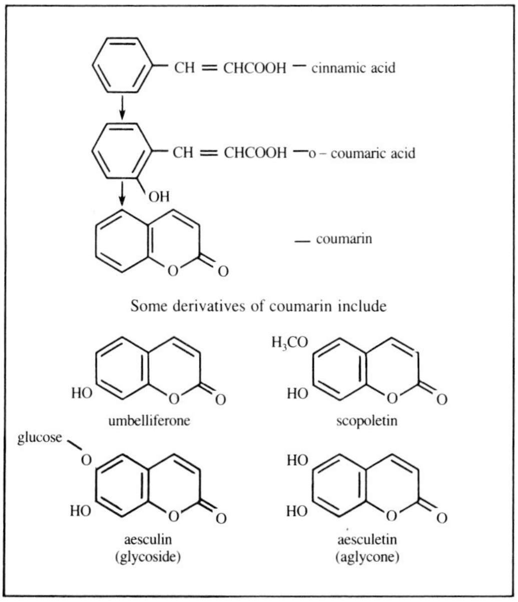

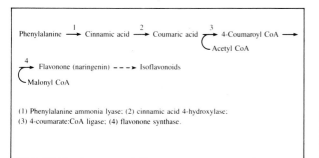

Fig. 6–5. The synthesis and derivatives of coumarin.

(Cu). Laccase is able to oxidize *ortho-* and *para*-dihydroxyphenols in the presence of oxygen. Tyrosinase is also an aerobic phenol oxidase; however, neither *para-* nor *meta*-dihydroxyphenols are suitable substrates for this enzyme, since it is able to oxidize only *ortho*-dihydroxyphenols. A *meta*-polyphenoloxidase has been detected in both *Pyricularia* and *Polyporus* fungi (168). The conventional substrates of the phenoloxidase enzymes are presented in Table 6–1.

Tyrosinase, variously referred to as phenolase, phenoloxidase, polyphenoloxidase, DOPA-oxidase, and catecholase (*o*-diphenol:O_2 oxidoreductase, EC 1.10.3.1), is a complex enzyme. The enzymic activity that makes *o*-diphenol from monophenol oxidatively (monophenol + $\frac{1}{2}O_2 \rightarrow$ *o*-diphenol) is frequently referred to as cresolase activity. On the other hand, catecholase activity oxidizes the produced diphenol to quinone (*o*-diphenol + $\frac{1}{2}O_2 \rightarrow$ *o*-quinone + H_2O). Hence, the complex phenolase enzyme has two kinds of activity (Fig. 6–8).

These two catalytic activities occur together, they are inhibited to the same degree by metal-binding reagents, and each activity has been found proportional to the available Cu^{++}. Both activities are lost when Cu^{++} is removed from the preparations but reestablished when Cu^{++} is added to the inactive enzyme. *o*-Quinone, derived, for example, from DOPA and similar related amino acids by the activity of catecholase, produces "melanine." This is a dark-colored pigment produced in the course of further enzymatic and nonenzymatic reactions. Melanine is considered an end product, but the details of the formation of this complex compound are not yet known.

Importance of Phenoloxidation in Plant Physiology

An earlier view of polyphenoloxidase enzymes held that they play an important role in plant respiration as terminal oxidases (8). However, this hypothesis has not proved valid. More convincing is the modern view that the phenoloxidase system plays a role in respiration by transferring electrons from respiratory substrates to other hydrogen or electron acceptors. The nature of some of these hydrogen and electron donors and acceptors has not been definitively characterized.

Oxidation products of phenolics (quinones) may be reduced to their original phenol form by respiratory carriers. $NADH_2$ and $NADPH_2$ generated by the oxidation of respiratory substrates may reduce quinones in the presence of quinone reductase enzyme (257). According to Mason (157), the most important role of polyphenolases in the physiology of plants is their ability to oxidize monophenols to polyphenols. This function was shown above, in the discussion of the mechanism of phenol-*o*-hydroxylase (cresolase) activity. In this reaction, hydroxylation of monophenols leads to the formation of complicated polyphenols—flavonoids, tannins, lignins, and others. Furthermore, oxidation of polyphenols by the catecholase activity (dehydrogenation of diphenol) has a role in the browning reaction of plant tissues. As we have seen before, both reactions involve the consumption of molecular O_2. If biosynthesis of *o*-diphenol and related compounds is very intensive, the share of the total consumed O_2 catalyzed by the phenolase complex may be relatively high. This serves to explain the role of phenolase enzyme in the control of O_2 consumption without implicating terminal oxidation.

There seems also to be a connection between protein

Phenylalanine $\xrightarrow{1}$ Cinnamic acid $\xrightarrow{2}$ Coumaric acid $\xrightarrow{3}$ 4-Coumaroyl CoA $\longrightarrow$
                                                                      Acetyl CoA

$\xrightarrow{4}$ Flavonone (naringenin) $----\blacktriangleright$ Isoflavonoids
   Malonyl CoA

(1) Phenylalanine ammonia lyase; (2) cinnamic acid 4-hydroxylase;
(3) 4-coumarate:CoA ligase; (4) flavonone synthase.

Fig. 6–6. The biosynthetic pathway from phenylalanine to the flavonone naringenin (and to isoflavonoids) and the enzymes involved in different steps (after Bailey and Mansfield, 11).

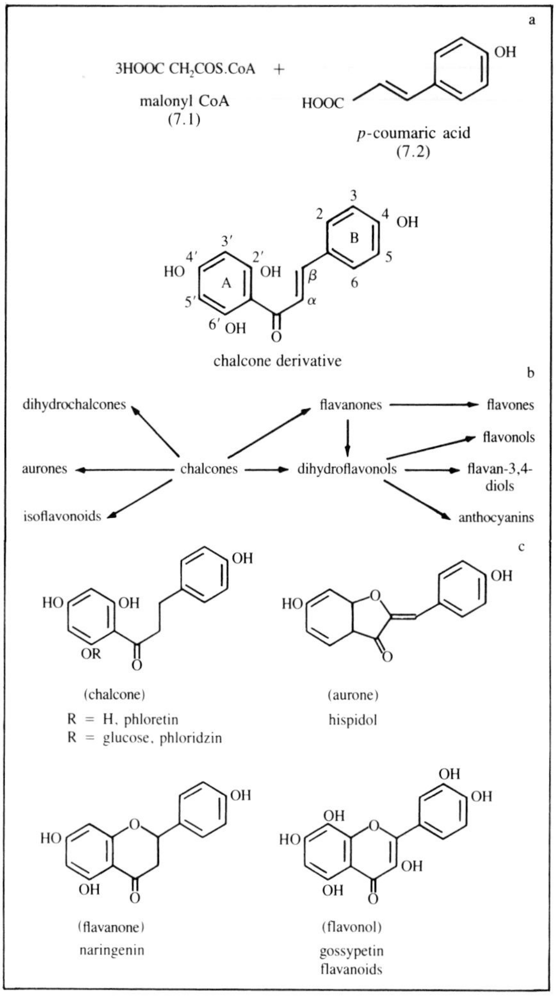

Fig. 6–7. The biosynthesis of chalcones (a), the biogenetic relationship of the flavonoids (b), and some related compounds (c) (after Vickery and Vickery, 245).

metabolism and the phenol-phenolase system. Hess (103) pointed out the role of some oxidized phenol compounds with amino acids, peptides, and simple proteins. These "adducts" (complexes) are catalysts for the oxidative deamination of amino acids:

$$\text{Polyphenol} + \text{amino acid} + \tfrac{1}{2}O_2 \xrightarrow{\text{polyphenolase}} \text{adduct}$$

$$\text{Amino acid} + \tfrac{1}{2}O_2 \xrightarrow{\text{adduct}} \alpha\text{-keto acid} + \text{ammonia}$$

By these reactions, the phenolase enzymes may influence amino acid metabolism of plants.

Phenolics, and their oxidation products, seem also to play a role in the auxin metabolism of plants. For example, the activity of indoleacetic acid oxidase may be stimulated or inhibited by the phenol-phenolase system, thereby controlling the level of auxin in tissues (90, 189). Regarding the enzymatic oxidation of phenolics, it should be borne in mind that peroxidase is also able to

Table 6–1. Characteristics of phenoloxidases.

Phenoloxidase enzyme	Substrate			
	Monophenols	o-dihydroxyphenols	m-dihydroxyphenols	p-dihydroxyphenols
Laccase	—	Oxidized	—	Oxidized
Tyrosinase	Oxidized	Oxidized	—	—
meta-Polyphenoloxidase	—	—	Oxidized	—

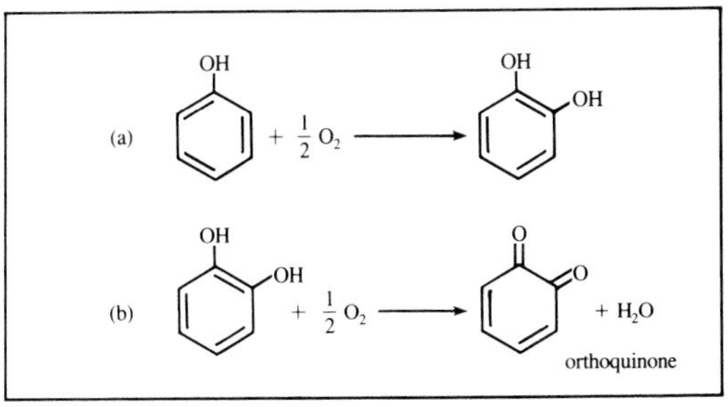

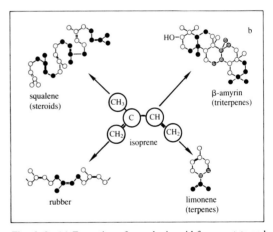

Fig. 6–8. (a) Cresolase activity of the phenolase enzyme. (b) Catecholase activity of the phenolase enzyme.

oxidize several phenol substrates. Hydrogen atoms derived from phenol substrates are combined in these cases with the oxygen derived from peroxide.

Complex phenolics of high MW and their oxidation products are known as effective inhibitors of pectic enzymes and, hence, influence cell wall metabolism (254).

Mevalonic Acid Pathway—Formation of Isoprenes (Terpenoids and Steroids)

This very diverse group of plant substances, among which are some plant growth regulators and phytoalexins, is made up of isoprene or isopentane units.

$$-CH_2 = C - CH = CH-$$
$$\underset{CH_3}{|}$$

These units link together in various ways and with different types of ring closures. All carbon atoms of the isoprene residue are derived from acetate (23). Three acetate residues are condensed, with the elimination of one carbon atom, to form the five-carbon isoprene molecule. The manner in which acetyl-CoA molecules are linked together is shown in Fig. 6–9a. The basic building block of the isoprenoids is isopentenyl pyrophosphate to form monoterpenes. The further biosynthetic steps usually are as follows: C_5, C_{10}, C_{20}, C_{30}. The structural relationships of the isoprenoids are seen in Fig. 6–9b. According to the number of isoprene units, one can differentiate between monoterpenoids (C_{10} compounds), diterpenoids (C_{20}), and triterpenoids (C_{30}). The isoterpenoids include steroids, the carotene pigments, vitamin A, terpenes, the phytol chain in chlorophyll, essential oils, rubber, gibberellic acid, and abscisic acid. Included are some phytoalexins that are associated with tissue necrosis. Alves et al. (6) have suggested that cyanide-resistant respiration is involved in the biosynthesis of the terpenoid phytoalexins (26). (See also Chaps. 3, 10.)

Changes Induced by Viruses

For the purposes of this section, it is again convenient to distinguish between the virus infection that leads to necrosis of cells at and adjacent to the site of infection,

Fig. 6–9. (a) Formation of mevalonic acid from acetate and formation of isopentenyl pyrophosphate from mevalonic acid (after Bonner and Varner, 24). (b) Structural relationships of the isoprenoids (after Bonner and Varner, 24).

where virus movement is often (though not always) restricted, and that which leads to a systemic infection, usually in the absence of necrosis. The reason for this distinction is that virus infection can lead to significant changes in the concentration of polyphenols, of their oxidation products, and of the enzymes involved in their biosynthesis and oxidation. As with so many changes in virus-infected plants (e.g., respiration, photo-

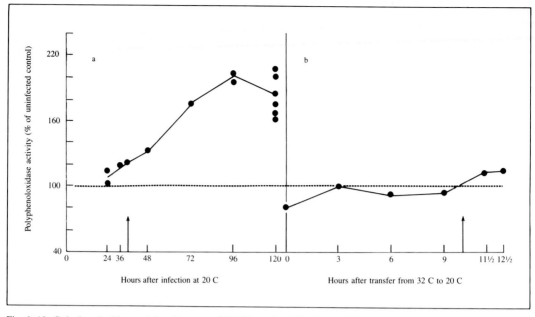

Fig. 6–10. Polyphenoloxidase activity of extracts of TMV inoculated Xanthi nc half-leaves expressed as a percentage of that of corresponding healthy half-leaves. (a) During infection at 20 C. (b) After transfer from 32 to 20 C 72 h after infection at 32 C. Substrate: chlorogenic acid. The arrow shows time of appearance of local lesions (after Cabanne et al. 34).

synthesis), their magnitude is often related to the severity of the symptoms produced and is greatest when the tissue becomes necrotic. Because of the putative role of changes in phenol metabolism in the necrogenic process and its consequences (see Chap. 10), and also because of the simplicity with which virus-infected tissue can be identified, facilitating observation of the comparatively early events in symptom development, attention has inevitably focused on those virus/host combinations that exhibit hypersensitivity. It is more meaningful, therefore, to consider localized and systemic reactions under separate sections since, in the former case, we need to address ourselves to questions that do not present themselves when virus is not localized. Changes in concentration of secondary metabolites and in the levels of relevant enzymes are, therefore, evaluated here in relation to necrogenesis, and in Chap. 10 we examine the consequences of this reaction in relation to the limitation of symptom spread and virus movement.

Changes Accompanying the Expression of Hypersensitivity

In considering changes in the metabolism of phenolics, other secondary metabolites, and the associated enzymes that accompany lesion formation, it is important to bear in mind some reservations that apply, as we have already seen, to the assessment of many other metabolic changes. First, the magnitude of these changes in the early stages of infection has not been measured, since it has not proved possible to identify those initially infected cells destined to become necrotic. By the time necrosis is visible, it is too late to measure early changes

in those cells that may be responsible for these necroses. Instead, changes have usually been measured in whole leaf tissue. The best approach of this kind has been to infect leaves with a high virus inoculum, so that as many as possible of the cells become infected. Alternatively, changes may be measured in cells at the periphery of expanding lesions, cells destined to become necrotic as a lesion develops (145). It may be, therefore, that in the majority of cases important early changes taking place in a small proportion of the cells in the leaf have not been observed. Additionally, since the moisture content of tissue usually decreases on lesion formation, changes relative to uninoculated control tissue will be artificially high if calculated on a wet-weight basis (244).

Polyphenoloxidases

The hypersensitive reaction is very often, though not universally, accompanied by a stimulation of activity of polyphenoloxidases (PPO) and a concomitant increase in the concentration of quinones, which by combination with proteins are potentially cytotoxic (69, 185, 252). Observations on several virus/host combinations in which PPO was stimulated (69, 71, 211, 212, 241) led Farkas and his colleagues to the view that stimulation of PPO activity and, hence, quinone concentration was a primary cause of necrosis (for cytological details, see Chap. 10). The evidence, however, though not contradictory to this view, did not convincingly support such a causal relationship. If accumulation of quinones was a primary cause of lesion formation, then it was argued that keeping them in the reduced, less toxic state with a reducing agent should reduce, delay, or even abolish lesion formation. Indeed, when ascorbic acid or glutathione, for example, were infiltrated into *N. glutinosa* prior to inoculation with TMV, many fewer lesions were produced, while virus replication was reduced only slightly. It is quite likely, however, that these reducing agents were interfering with biochemical processes in addition to quinone reduction (211). Additionally, there was little increase in PPO activity in TMV- inoculated *N. glutinosa* maintained a 36 C, conditions under which a systemic rather than a hypersensitive response resulted. This was, however, no more than a correlation. Also, administration of the substrate of the stimulated enzymes, phenols, to *N. sylvestris* inoculated with TMV-U2 resulted in darkening of existing lesions (176).

Other, more contradictory evidence, led Solymosy to a modified view (211). It was observed (115) that wounding Java tobacco plants by cutting above the 11th leaf caused an increase in PPO activity. Enzyme stimulation was, therefore, occurring as a result of tissue damage, not vice-versa. More importantly, however, stimulation of enzyme activity resulting from virus infection occurred around the time of lesion development, rather than prior to it, so that unless there were rises in the initially infected cells, difficult to detect (see Chap. 3), the timing would suggest that PPO stimulation and the associated rise in quinones were a consequence, rather than a cause, of necrosis. This was the view reached by Solymosy and supported by subsequent observations. PPO enzymes are stimulated in Xanthi nc tobacco (34, 35) and pinto bean (136) by TMV inoculation, in Ky 16 tobacco inoculated with aucuba and yellow strains of TMV (219), in Samsun NN tobacco infected with the U1 strain of TMV (243), and in Chinese cabbage infected with CaMV (44) (Table 6–2). Both polyacrylamide gel electrophoresis (243) and estimation of Michaelis constant (241) of enzymes from healthy and TMV-infected tobacco suggest that, in this case at least, the nature of the enzymes, though changed quantitatively, remains qualitatively unchanged after infection. Apart from transient increases soon after inoculation, which are also apparent in situations where necrosis does not develop (243), an increase in PPO activity in advance of lesion formation has never been observed, and the general pattern is similar to that in Fig. 6–10.

Similarly, when inoculated plants were maintained at 32 C to allow development of systemic infection and then returned to 20 C, the rise in PPO did not precede lesion formation. AMV, though inducing lesions in pinto bean, did not stimulate PPO (244), adding further weight to the view that PPO is unlikely to play a causal role in lesion formation.

Peroxidases

Peroxidases also oxidize phenols to quinones and have, consequently, been assigned a role analogous to that originally proposed for polyphenoloxidases in lesion formation. A stimulation of activity of these enzymes in the soluble fraction of extracts from virus-inoculated leaves has been observed in all hypersensitive virus/host combinations studied (Table 6–2). Catalase, too, may be stimulated (243). Again, however, peroxidase stimulation either parallelled or followed lesion development (35, 244; Fig. 6–11) and, generally, remained at a high level after lesion maturity, arguing against a primary role for these enzymes in necrogenesis.

Procedures such as polyacrylamide gel electrophoresis and isoelectricfocusing, which permit the separation of complex mixtures of proteins or polypeptides, have revealed the presence in crude enzyme preparations of several enzymes (isoenzymes) with peroxidase activity. In general, the stimulation of peroxidase activity observed in extracts of infected leaves is accounted for by increases in activity or concentration of enzymes already present in uninfected tissue (12, 14, 170), suggesting that the increased levels result from de novo synthesis rather than activation of preexisting enzymes (72). There have, however, been reports of the appearance of new isoenzymes consequent upon virus infection, both in cytoplasmic (72, 85, 213) and in cell-wall-bound (85) fractions. It is not clear, however, whether or not the "new" isoenzymes are present in healthy tissue at concentrations too low to be detected under the conditions of the experiment. Nor is it clear whether these putative "new" isoenzymes fulfill a role not already fulfilled by preexisting enzymes. As observed by Vegetti et al. (244), a stimulation of peroxidase activity may not be simply an accidental consequence of cell death; it may be that peroxidases play a role in degradation of H_2O_2 produced as a result of enhanced respiration, in ethylene production (85) and, as we shall see in Chap. 10, in virus limitation.

Phenylalanine Ammonia Lyase

Since phenols and polyphenols are substrates for PPO and peroxidases, a stimulation of enzymes of the phenylpropanoid pathway might also be expected to accompany necrotic lesion formation, and, indeed, for those investigated, such is the case. A stimulation of activity of phenylalanine ammonia lyase (PAL) has been observed on several occasions (Table 6–2; 73). Typically, there is no detectable enhancement during the first few hours after infection, but then activity begins to rise marginally before the appearance of visible lesions. Stimulation extends for a brief period, reaching a maximum during lesion maturation but before complete development, and then decreases as lesions mature (244; Fig. 6–12).

The degree of stimulation is directly related to the number of lesions that appear (81, 244) and is a function of the degree of tissue necrosis, rather than of virus multiplication. This observation has been supported by maintaining TMV-inoculated Xanthi nc tobacco plants at 32 C and then returning them to 20 C. When necrosis develops following transfer to the lower temperature, leaves contain more virus, but extracts have the same PAL activity as extracts from plants maintained at the

218 · **Biochemistry and Physiology of Plant Disease**

Table 6–2. Enzyme changes following virus infection of hypersensitive hosts.

Host	Virus	Qualitative or quantitative changes in Peroxidase isoenzymes		Increase in activity		
				Peroxidases	Polyphenol-oxidases	Phenyl-alanine ammonia lyase
Phaseolus	SBMV	(+)(a)	72	72		
vulgaris cv.		(−)	170			
Pinto	TMV;TRSV	(−)	170			
	AMV	(+)	213	244	244[b]	244
	TNV	(+)	213			
cv. Prince	TNV	(−)	14	14		
cv. Fürj	TNV					73
Nicotiana	TMV	(−)	37	37,70	70,129	43
glutinosa	TMV-U5					
	TNV	(+)	213		70	
	PVX	(−)	14			
Nicotiana	TMV				176	
silvestris						
Nicotiana						
tabacum cv.						
Xanthi nc	TMV	(+)	85	34	35	180
Samsun NN	TMV	(+)	243	22	243	53,81,156
		(−)	22			
Ky 12	TMV	(−)	170			
Ky Iso 1 Ky16	TMV			219	219	
Xanthi(c)	TMV			12		
Vigna	CMV	(−)	170		70	
sinensis	TRSV	(−)	170			
	TNV	(−)	14	14,248	248	
Datura	TMV				70	
stramonium						
Brassica	CaMV				44	44
chinensis						

a. (+) Report of a "new" isoenzyme; (−) no "new" isoenzyme detected.
b. No increase observed in this investigation.
c. heat-induced lesions

lower temperature, which have developed lesions at the usual time.

In a more comprehensive extension of their earlier observations, Fritig and his colleagues have investigated the timing of the stimulation of PAL and other enzymes of the phenylpropanoid pathway following TMV inoculation of Samsun NN tobacco (40, 137, 145, 146, 156). Activities of PAL and cinnamic acid-4-hydroxylase (CAH) begin to rise before lesion appearance, PAL reaching its maximum slightly before CAH. Caffeic acid- 0-methyltransferase (CMT) activity begins to rise later, along with peroxidase, and is still increasing when activities of PAL and CAH are declining (Fig. 6–13).

Fritig et al. (156) have also made a more realistic estimate of the extent of PAL stimulation in individual cells. PAL activity was enhanced six- to tenfold in leaves with ca. 300 lesions, 48 h after inoculation. However, since only 5% of the cells showed a stimulation of the phenylpropanoid pathway, as revealed by an increase in UV fluorescence, with 0.5% destroyed by necrosis, it follows that ca. 95% were unaffected. Therefore, in the affected cells, PAL stimulation by de novo synthesis (63) was as high as 100- to 200-fold.

Estimation of activities in small discs centered on the lesions and on a ring of tissue surrounding these discs, revealed a stimulation in activity of PAL, CMT, and CAH spreading radially from the center. Moreover,

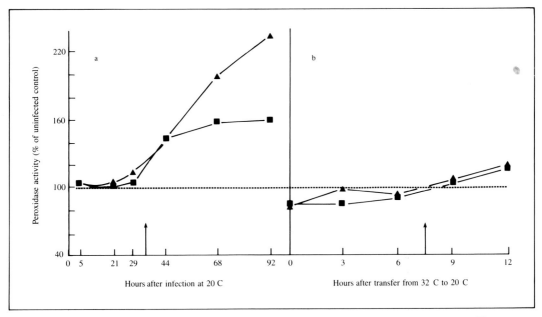

Fig. 6–11. Percentage of peroxidase activity determined under conditions of Fig. 6–10. Substrate: ▲, guaiacol, ■, pyrogallol. The arrow shows time of appearance of local lesions (after Cabanne et al., 34).

there was a stimulation in activity in cells at the lesion periphery that were beyond the area of visible necrosis and contained little virus. Almost all cells with stimulated enzyme activity were destined to become necrotic.

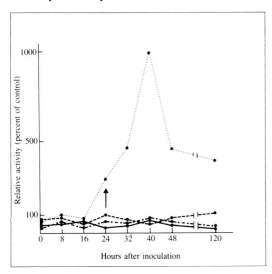

Fig 6–12. Changes in phenylalanine ammonia-lyase (PAL) activity in *Phaseolus vulgari* var. Columbia, during the first hours of AMV-MI infection and after simulated inoculation. The arrow on horizontal axis indicates approximate average time of lesion appearance. The number of local lesions averaged about 1000 per leaf. Each point is the mean of two replicates. ······inoculated; ----·opposite previously inoculated; ·——— rubbed; ·-·-··opposite previously rubbed (after Vegetti et al., 244).

It appears, therefore, that there is a close association between the enzymes and necrosis and that, although a causal relationship is not established with certainty, the enzymes are good biochemical markers of the necrotic reaction. Experiments in which PAL activity is inhibited have suggested, however, that this enzyme may play a role in virus limitation (see Chap. 10).

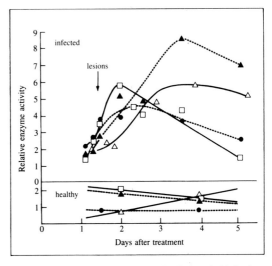

Fig. 6–13. Changes in enzyme activity during viral infection. Leaves have been inoculated with TMV at a concentration of 0.2 μg/ml in order to develop 250 local lesions. The vertical arrow indicates the time of appearance of the lesions. □—— —□ PAL; ●----● CAH; ▲---▲ OMT; △---△ peroxidase (after Legrand et al., 145).

Phenolics and Other Metabolites.

Accumulation of polyphenolic compounds in and around local lesions is a well-established feature of the hypersensitive response. Their presence in tissue that is either necrotic or surrounds necrotic cells can be revealed by their fluorescence in UV light (e.g. 145). A stimulation in concentration of the polyphenol scopoletin, for example, was observed by Best (19–21) as long ago as 1936.

Two unidentified compounds appeared in a narrow band around TMV lesions in Ky12 tobacco (99); more recently, several compounds, including ferulylputrescine and diferulylputrescine, have been identified in tobacco reacting hypersensitively to TMV (155).

Compounds of this type could be produced as a consequence of necrosis, of metabolic processes leading to necrosis, or of metabolic disturbances related to the infection process. They may play a role in restriction of virus to the lesion area or may be involved at an earlier stage, in lesion formation. Tanguy and Martin (222) have recorded an enhancement of the concentration of several compounds, including flavonol glycosides, caffeoylquinic and feruloylquinic acids, and glucosides of cinnamic and benzoic acids in TMV-inoculated leaves of Xanthi nc tobacco. Their concentration did not begin to rise until lesion development was well advanced; they were, therefore, regarded as a secondary consequence of necrosis.

However, since activities of enzymes involved in both anabolism and catabolism are stimulated during the hypersensitive response, measurement of concentration only is clearly insufficient, and the observations of Legrand and his colleagues (82, 145, 146) on TMV-inoculated Samsun NN tobacco are at variance with this view. They measured both concentration and rate of synthesis of chlorogenic acid (a good substrate for PPO) and of scopoletin (a good substrate for peroxidases) before, during, and after lesion formation (Fig. 6–14).

A stimulation in the rate of synthesis of both compounds was observed as lesions appeared, at about 33–37 h, though this increased rate was short-lived and decreased 15–20 h after lesions became visible. However, the concentration of chlorogenic acid declined as its rate of synthesis decreased, suggesting an active catabolism, while scopoletin concentration continued to increase, reaching a high concentration in relation to healthy tissue. The glycoside of scopoletin, scopolin, behaved similarly, while the behavior of the isomers of chlorogenic acid and 4-caffeoylquinic and 5-caffeoylquinic acids paralleled that of chlorogenic acid, suggesting these observations could be extended to other orthodiphenols and monophenols.

It is of interest that when the normally observed stimulation of PAL activity was reduced by treatment of TMV-inoculated Samsun NN tobacco leaves with the PAL inhibitor, α-aminooxyacetate (156), then the accumulation of PAL-derived phenolics, including chlorogenic acids and scopoletin, was considerably reduced. There was also a significant increase in lesion size, establishing a negative correlation between accumulation of phenolics and extent of necrosis (Fig. 6–15). The larger the lesions, the lower the production of PAL-derived phenolics.

There is evidence (110) that Xanthi nc tobacco inoculated with TMV produces a compound, not yet characterized, that is toxic to tobacco protoplasts and may well be directly involved in the hypersensitive response. It appears before the onset of necrosis, and its concentration is correlated with rate of lesion expansion and lesion number.

In efforts to establish a mechanism for virus localization, Redolfi et al. (190) have investigated the formation of new phenolic compounds in *G. globosa* infected with TBSV. There was an accumulation of three compounds, one of which (III) was present in healthy plants; all appeared following infection by other viruses. Compound I, suggested to be a cinnamic acid derivative, appeared before necrosis, was present in tissue that did not become necrotic, and was considered to be the result of a nonspecific response to injury. Accumulation of compounds II, suggested also to be a cinnamic acid derivative, and III, possibly a flavanoid, was a specific response to virus infection. Compound III was detectable shortly before the appearance of necrosis and was found in both chlorotic and necrotic tissue and in tissue surrounding lesions. Along with II, it has been assigned a tentative role in virus localization, though the evidence is, at best, circumstantial.

It seems, therefore, that the activation of PAL and CAH closely parallels the rise in concentration of phenolic compounds, and as the activities of these enzymes decline, so the rate of synthesis of phenolic compounds falls. Their concentration may also decline, or remain at a relatively high level. Although the concentration of some methoxylated phenols is stimulated in advance of necrosis, the general stimulation in phenol biosynthesis closely parallels lesion formation. Much more needs to be learned, particularly about the very early stages of lesion formation, before a causal role for these changes in either necrosis or virus limitation can be established.

Changes Accompanying Systemic Infection

Virus infections leading to systemic invasion also promote changes in activity or concentration of peroxidases (14, 150) and PPO (70) (Table 6–3), although stimulation, when observed, is much less pronounced than in hosts reacting hypersensitively, and reflects symptom severity. There was, for example, a significant stimulation of these enzymes in tobacco inoculated with the aucuba and yellow strains of TMV, but not with the green strain (219), which elicited a barely distinguishable green mottle.

The degree of stimulation is not related to virus replication per se (14), though in cucumber, for example, where there is a parallel between symptom severity and virus multiplication, peroxidase stimulation does also reflect virus accumulation, if indirectly. In tobacco (85),

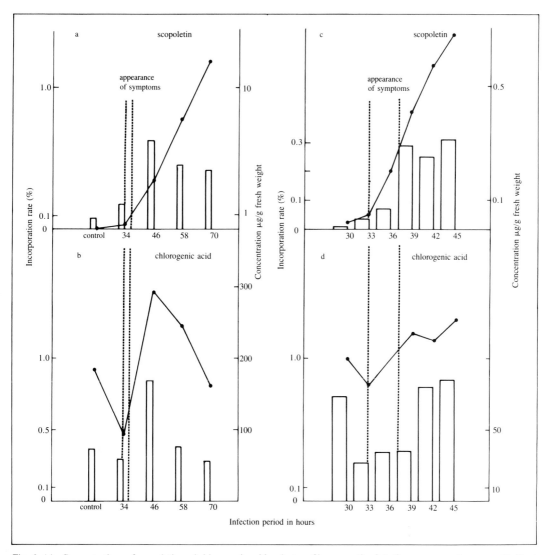

Fig. 6–14. Concentrations of scopoletin and chlorogenic acid and rates of incorporation into these compounds measured in 2 sets of parallel experiments. Times of infection were between 34 and 70 h after inoculation for the first and between 30 and 45 h for the second. Experiments carried out with either uninoculated plants or plants taken anytime later than 12 h after water inoculation or plants taken between 12 and 28 h after TMV inoculation gave the same results and may serve as controls. Appearance of local lesions occurred in the zone of time of infection between the vertical dotted lines: the first local lesions were detectable after 33 h of infection, and there were lesions on all inoculated leaves 37 h after TMV inoculation. Concentration curves are represented by ●━━━━━●. Rates of incorporation are represented by □. These were measured during the last 2 h of infection during which leaves were detached from the plants and labeled by uptake of [^{14}C]phenylalanine through the cut petioles (after Legrand et al., 146).

cucumber (255, 256), bean, and cowpea (14), rise in peroxidase activity is attributable to quantitative changes in preexisting isoenzymes. In tobacco cultivars Xanthi nc and Samsun NN, there is little change in either peroxidase or PPO above 30 C, when plants become systemically infected (e.g. 35 C; Figs. 6–11 and 6–12). PAL activity may, however, be slightly below that of healthy plants (179, 181).

Systemic infection of Xanthi nc tobacco (above 29 C) was accompanied by an early decrease in content of phenols, including chlorogenic, rutin, and caffeoylquinic acids, with a rise to levels in excess of those in healthy plants as infection progressed beyond 7 d (222); in Samsun, systemic infection was accompanied by increases in levels of chlorogenic acid, scopolin, and rutin. However, many compounds found in TMV-inoculated Xanthi nc were undetectable in plants with systemic infection.

In the stock plant, *Mathiola incana*, the red and purple colors in the flower petals are due either to pel-

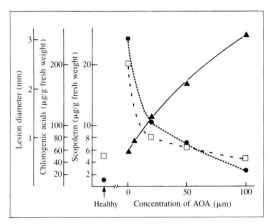

Fig. 6–15. Competitive inhibition in vivo of the flux of phenolic compounds after treatment with α-aminooxyacetate (AOA). *N. tabacum* Samsun NN leaves were inoculated with the common strain of TMV in order to develop 500 local lesions per leaf. Healthy control leaves were water-inoculated. Twenty-four hours after inoculation, control leaves and TMV-infected leaves were detached from the plants and fed through their cut petioles either with mineral medium alone or with a mineral medium containing AOA. Fifty-four h after inoculation, lesion diameter (▲) and amount of scopoletin (●) and chlorogenic acid (□) were measured (after Massala et al., 156).

argonidin or cyanidin glycosides, which may also be acylated with any of several hydroxylated cinnamic acids, including p-coumaric, caffeic, ferulic, or sinapic acid. Virus infection causes a loss of color on red or purple flower types but may cause either loss of color or intensification of color on pink and lilac types. Feenstra et al. (76), monitoring changes in these compounds, and in flavonols, in colored and noncolored areas of infected flowers, found a tendency for the content of these compounds in the white areas on genetically pink or red flowers to resemble that of genetically white flowers. This was particularly apparent in the case of anthocyanin content, which was markedly reduced in white compared to colored areas. In white areas, cinnamic acids were also demonstrated to be free, while in colored flowers they were bound to anthocyanins. In red areas of genetically pink flowers there was a pronounced increase in concentration in both anthocyanins and cinnamic acids. However, as these experiments were performed on plants naturally infected with unidentified viruses, it is not clear whether increase and decrease in color intensity were caused by the same virus.

Uninoculated Parts of Hypersensitive Hosts

Although the activities of both PPO (67, 208) and PAL (43, 81, 209) and the concentration of phenols (209) generally remain unchanged, there is a significant stimulation in activity of peroxidase (207, 208, 242, 244) and, to a lesser extent, of catalase (208, 242). The

stimulation of peroxidase in upper leaves of Samsun NN tobacco inoculated with TMV on the lower leaves led Simons and Ross (207) to implicate peroxidase in the expression of induced resistance, which developed simultaneously with enhancement of enzyme activity. More recent evidence, however, considered in Chap. 10, suggests this to be unlikely.

We may conclude, therefore, that, although virus infection leads to a general stimulation of enzymes involved in the biosynthesis and oxidation of phenols, a stimulation that is much greater when tissue becomes necrotic, it remains to be resolved whether this stimulation is instrumental in causing necrosis or the result of it. Current evidence would favor the latter view. The involvement of phenols in virus limitation and resistance is discussed in Chap. 10. See Table 6–3, p. 223.

Changes Induced by Bacteria

Whereas plant secondary metabolites were once considered metabolic waste products, we now know that they include such exceedingly functional compounds as auxins, cytokinins, gibberellic acid, abscisic acid, and ethylene. These compounds are crucial to the healthy plant and, as we describe at length in Chap. 7, their imbalance in diseased tissue may be either the cause or the result of infection. Kuć (142) has observed that specificity, order, and integration of structure and metabolic activity are characteristic of healthy plants. Hence, it is not surprising that the disruption of their integrated and precisely controlled development by bacterial pathogens leads to sharp changes in metabolism.

In this discussion, too, we shall focus on nonhormonal compounds derived from the acetate-malonate, acetate-mevalonate, and shikimate pathways (245). We shall examine, initially, those aromatic compounds indigenous to the plant species in question but whose levels may be increased by bacterial pathogenesis. Secondly, we will discuss those compounds that may be minimally detectable in healthy plants but accumulate as a consequence of bacterial infection or some stress phenomenon. This discussion of aromatic-associated compounds is a restricted one, as it deals with them in a rather clinical way, i.e. detecting their presence and describing the manner in which they are synthesized and/or accumulate. The possible function of these compounds is mentioned cursorily in Chap. 1 and in greater detail in Chap. 10.

Potentiated Aromatic Synthesis

Aromatic amino acid synthesis. One of the clearest signs of increased levels of aromatization in bacterially infected plant tissue was recorded by Pegg and Sequeira (182). Tobacco leaves and stems showed increases in L-phenylalanine of 240% at 24 h after inoculation with *Pseudomonas solanacearum*. The level of this amino acid increased 605% over controls after 36 h in plants that were, until that time, symptomless. Commensurate with increased levels of phenylalanine, significant increases in L-tyrosine, 3,4 dihydroxy-phenylalanine

Secondary Metabolites · 223

Table 6–3. Enzymes in systemic hosts and in uninoculated parts of hypersensitive hosts.

Host	Virus	Peroxidase isoenzymes	Peroxidases	Polyphenol-oxidases	Phenyl-alanine ammonia lyase
A. Systemic Infection					
N. tabacum cv.					
Samsun	TMV	85	85,150	70	181
	PVX		150		
Samsun NN[a]	TMV				181
	TMV-HR		243	243	
Xanthi nc	CMV	85	85		
(a)	TMV		35	35	145,180,181
White Burley	CMV	255		70	
Bright Yellow	CMV			225	
Ky Isol Ky16	TMV		219	219	
Ky 12[a]	TMV	170			
Turkish					
Wisconsin 38	TMV	22	22		
N. glutinosa	CMV			70	
	PVX	37	37,150		
D. stramonium	CMV			70	
	PVX		150		
Cucumis sativus	CMV	255	13,161,256	13	
P. vulgaris cv. Prince	TMV	14	14		
Vigna sinensis	TMV	14	14		
Capsicum annuum	TEV	84	84		
B. Uninoculated Tissue of Hypersensitive Hosts					
N. tabacum cv. Samsun NN	TMV		22,207,208, 209,242,243	208	81,209
N. glutinosa	TMV-U5				43
P. vulgaris cv. Pinto	AMV		244	244	244
Glycine max cv. Bragg	CCMV		15	15	

a. Plants maintained at 30–32 C.

(DOPA), and L-tryptophan were also recorded (Fig. 6–16). The data suggested that these increases reflected increased synthesis by the host rather than the pathogen. Earlier, Sequeira (203) revealed that pathogenesis by *P. solanacearum* results in a vast accumulation of IAA (see Chap. 7). This is now at least partially explained by the increased levels of tryptophan that were synthesized. In addition, the accumulation of phenylalanine may also account for the report that *P. solanacearum*–infected plants also contain greater-than-normal amounts of scopoletin, which is a potent inhibitor of IAA oxidase (205). As described in Chap. 7, the inhibition of this enzyme probably accentuates the accumulation of IAA in tobacco infected by *P. solanacearum*.

Shikimate-malonate-derived chalcones, aurones, and isoflavones. Addy and Goodman (1) monitored the increase in extractable phenols from Jonathan apple foliage infiltrated with either a virulent (V) or an avirulent (AV) isolate of *Erwinia amylovora*. The conditions of

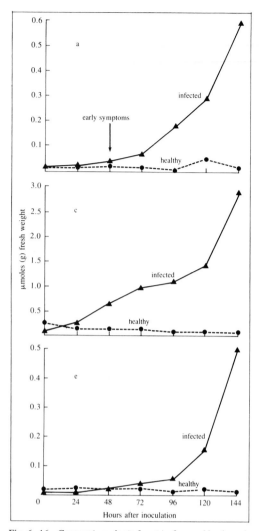
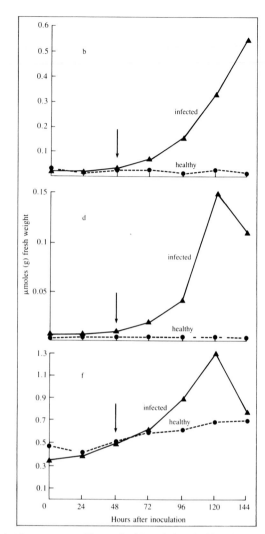

Fig. 6–16. Content (μmoles/g fr. wt.) of several basic amino acids in tobacco stems and leaves, healthy or infected with a virulent strain (K60) of *Pseudomonas solanacearum*. Each represents the mean of two determinations. Plants were collected at 24-h intervals after inoculation. (a) L-Tryptophan, (b) L-tyrosine, (c) L-phenylalanine, (d) 3,4- dihydroxyphenylalanine (DOPA),(e) L-arginine, and (f) γ-aminobutryric acid (after Pegg and Sequeira, 182).

the experiment permitted nearly parallel growth by both strains as populations increased from 5×10^6 to 10^9 cells/cm² leaf tissue. Both strains caused fourfold increases in extractable phenols to develop in inoculated leaves as compared with controls. Characterization of these phenols established the glycoside phloridzin as the compound present in highest concentrations in both inoculated and uninoculated leaves. The extracts of inoculated leaves contained a number of compounds partially characterized as chalcones, aurones, or isoflavones. Positive identification of three aglycones—phloretin, quercitin, and podospicatin—was also made from the acid-hydrolized extracts of *E. amylovora*-inoculated leaves. The rate of appearance of the phenols and population trends of the two bacterial isolates responsible for the increases in detectable phenols are presented in

Figs. 6–17 and 6–18. Although both inoculated and control tissues had large endogenous levels of the glycoside phloridzin, only the inoculated leaves revealed an array of phenols and a fourfold increase in water-extractable phenolics. That modifications of endogenous phenolics may occur is suggested by the increases in enzyme levels registered in additional experiments. Figs. 6–17 and 6–18 reveal that both the V and AV strains of *E. amylovora* caused marked increases in the levels of both polyphenoloxidase and peroxidase. Differences in the extent of the enzyme increase were recorded, although rates of increase induced by V and AV strains paralleled one another. Whereas the AV strain induced an approximate twofold increase in the level of these two enzymes, the V isolate fostered a threefold increase. It might be conjectured that the differences

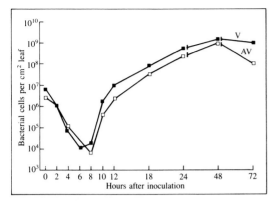

Fig. 6–17. Rate of growth of avirulent (AV) and virulent (V) strains of *E. amylovora* in 4-week-old Jonathan apple leaf tissue inoculated with 10^8 cells/ml (after Addy and Goodman, 1).

between V and AV strains reflected differences in bacterial population attained in susceptible Jonathan leaf tissue. However, during the initial 48 h of the 72-h experiment, the bacterial populations in apple leaves that developed from the two inocula were identical.

Induction of phytoalexin synthesis. It is becoming ever more apparent that a number of bacterial species in either nonhost or resistant tissue induce the production of phytoalexins. For example, soybean leaves infected by an incompatible race of *Pseudomonas glycinea* caused the rapid accumulation of glyceollin and the constitutive isoflavonoids coumestrol, daidzien, and sojagol. When soybean plants were inoculated with the nonpathogen *P. lachrymans*, the same compounds were recovered (120, 124). Similarly, Stholasuta et al. (217)

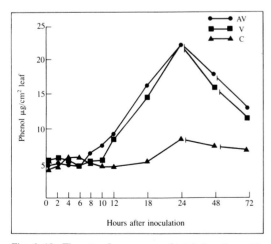

Fig. 6–18. The rate of appearance of total phenols as chlorogenic acid equivalents in 4-week-old Jonathan apple leaf leachate. Apple leaves had been inoculated with 10^8 cells/ml of avirulent (AV) and virulent (V) strains of *E. amylovora* and controls (C) with 0.05 M phosphate buffer (after Addy and Goodman, 1).

were apparently the first to detect phytoalexin synthesis in plants exposed to pathogenic bacteria. They reported the production of the pterocarpanoid phaseollin in bean and pea leaves in response to inoculation with incompatible bacteria.

Studies concerning the induction of phytoalexins by various pathogenic fungi, as well as bacteria, were earlier hampered by the comparatively small amounts of these compounds that could be obtained from the favored test tissue, hypocotyls (121). This technical problem was remedied when Keen (120) inoculated germinating seeds from several species and recovered 30–50 mg of phytoalexin from 100 g of seed. Seeds that were killed by freezing failed to produce detectable amounts of phytoalexin, although they became heavily colonized by microorganisms, supporting the contention of induced active synthesis. This technique was still further refined by Keen in order to obtain larger amounts of glyceollin (120). Soybean seeds permitted to imbibe water for a few hours were quartered and then exposed to *P. glycinea* with the resultant recovery of large amounts of glyceollin. Song, Karr, and Goodman (unpublished data) were able to increase glyceollin yields when water-imbibed soybean seeds were sliced thinly prior to inoculation with an incompatible race of *P. glycinea*. In subsequent experiments, *P. tabaci* was also able to induce production of significant quantities of glyceollin with little evidence of pathogenesis. On the other hand, sliced water-swollen soybean seeds inoculated with *P. syringae* revealed during the 144 h of the experiment intense tissue destruction and unexpectedly an even greater amount of glyceollin. This apparent paradox may be explained by the insensitivity of *P. syringae* to glyceollin as compared to *P. tabaci*. Apparently, the exposure of greater numbers of soybean cells to the pathogen, achieved by slicing the seed thinly, amplified phytoalexin production. Of importance here is not the increase in yield of glyceollin but evidence supporting the report of Yoshikawa et al. (258) that in soybean hypocotyls glyceollin is produced rapidly and in high concentration in those host cells nearest the advancing fungal hyphae. This supports an essential feature of phytoalexin synthesis, i.e. these compounds tend to accumulate at or near the site of induction (trauma) (121, 123).

An interesting experimental tool was exploited by Holliday and Keen (108) in their study of glyceollin synthesis. Synthesis was induced in soybean leaves inoculated with a hypersensitive reaction (HR)–inducing strain of *P. glycinea*. They observed that the herbicide glyphosate (N-phsophonomethylglycine), which inhibits the shikimic acid pathway and consequently aromatic amino acid synthesis, also reduced glyceollin production in soybean leaves inoculated with an incompatible strain of *P. glycinea*. Additional experiments indicated that aminooxyacetic acid (AOA), aminophenylproprionic acid (AOPP), and benzyloxycarbonyl aminooxyphenylproprionic acid (N-BOC-AOPP), which inhibit PAL, were unable to reduce glyceollin production. In companion experiments, it was shown that glyphosate suppression of glyceollin accumulation in planta could be reversed by pretreatment of plants with nutrient solutions containing phenylalanine and tyrosine. Although Holliday and Keen (108) recognized that glyphosate had an endogenous antibacterial activity in vitro, its activity in planta was not due to a direct effect upon the bacteria. The antibacterial effect of glyphosate in vitro was believed due to its capacity to block the shikimic acid pathway.

Lyon and Wood (151) reported differential accumulation of phytoalexin compounds in French bean foliage as a consequence of inoculations with *Pseudomonas mors-prunorum*, *P. phaseolicola*, and *P. fluorescens*. Phaseollin, coumestrol, and other unidentified compounds accumulated more rapidly following inoculation with the incompatible pathogens and were not detected following inoculation with the saprophyte *P. fluorescens*.

Suberin synthesis. Suberin is a secondary metabolite that consists of both aromatic and aliphatic domains. It is the normal sheathing mantle of the underground surfaces of plants and a component of the Casparian strip of the root endodermis that modulates the flow of water to the stele. In addition, it is an integral factor in the wound-healing response. Some cells and tissues, when injured by microorganisms, insects, or abiotic agents, are capable of rapidly synthesizing suberin to limit water loss or provide a barrier against microbial penetration. The alleged antimicrobial agent in suberin is the aromatic component of the polymer, which is derived from phenylalanine. This amino acid is, subsequently, altered by PAL and hydrolases (see Chap. 4). The precise aromatic monomer integrated into suberin is not known. However, in experiments by Cottle and Kolattukudy (45), where specific labeled phenols were fed as precursors, chlorogenic acid appeared *not* to be the monomer of choice. Instead, the sustained accumulation of this extractable compound after wounding may occur for defense purposes rather than becoming integrated into the suberin polymer. There is some evidence that ferulic acid is the monomer regularly covalently linked by ester linkages to the aliphatic acid domains.

Zucker and Hankin (260) performed a series of classic experiments that showed how suberization diminished degradation of potato tuber slices by *E. carotovora*- and *P. fluorescens*-produced pectate lyases. Their experiments indicated that cycloheximide almost totally suppressed suberization and permitted bacterial pectate lyase to macerate potato tissue. This suppressive effect was due to the inhibition of PAL synthesis by the antibiotic. Hence, the integration of phenylalanine into a suberin component and, ostensibly, the synthesis of suberin was precluded by cycloheximide. Similarly, lignin, which is essentially an aromatic polymer, may be considered a complex secondary metabolite. Like suberin, lignin synthesis may also be accelerated by infection (191).

Compatible and Incompatible (Hypersensitive) Necrotization

Necrotization. Visual evidence for phenoloxidation, the development of brown or black pigmentation in plant tissue, is characteristic of large numbers of local lesion, leaf- spotting bacterial pathogens (*Xanthomonas vesicatoria*, *X. pruni*, *X. malvacearum*, *Pseudomonas tomato*, etc.) upon infecting susceptible (compatible) host tissue. Although many of the oxidation products have not been identified, some are predictable from the aromatic compounds (in glycoside and aglycone form) that are detected in healthy tissue, i.e. chlorogenic acid and tomatin in tomato, amygdalin and prunetin in peach, and gossypetin and gossypol in cotton (16, 80, 245).

When pathogenesis proceeds so that symptom development is typical, necrogenesis is comparatively slow. Necrotic spots 1–5 mm in diameter develop in 72–96 h or longer, depending upon virulence of the pathogen, susceptibility of the host, and temperature. In effect, plasmalemma, lysosomes, and tonoplast become leaky, and phenols concentrated in the vacuole react with cytoplasmic and membrane-bound oxidases and glycosidases.

Hypersensitivity. The HR is induced by some bacterial plant pathogens in nonhost (incompatible) plant tissue. The effect of HR-inducing bacteria in nonhost tissue is a swift and irreversible loss of cell membrane permeability and structural integrity. Significant electrolyte leakage is detected in tobacco inoculated with *Pseudomonas pisi* in 3 h, whereas the compatible pathogen, *P. tabaci*, causes comparable leakage after 48–72 h (89).

Neither the bacterial inducer nor the plant cell receptor responsible for the rapid and total disorientation of all subcellular membranes is known. However, unpublished experiments from Novacky's laboratory in Columbia, Missouri, suggest the possibility of singlet oxygen (free radical) attack on the lipid components of membranes. Sequeira (206) has suggested that an autolytic process is set into motion that activates a series of oxidative and hydrolytic enzymes. However, Németh et al. (167) were unable to detect significant increases in protease, polyphenoloxidase, and phosphatidase. However, they did detect increases in glucose-6-phophate, 6-phosphogluconate, and shikimate dehydrogenases, suggesting a shift in respiration to greater pentose phosphate shunt activity and increased shikimic acid pathway syntheses. It is of interest that in HR (in tobacco) the collapsed desiccated tissue becomes a light tan, rather than the brown-black necrosis caused by compatible pathogens. It would seem either that the reaction develops too rapidly for some oxidation reactions to occur, or another type of aromatization develops. Support for the contention of modification of indigenous aromatic compounds was derived by Essenberg et al. (66), who detected and characterized two sesquiterpenoid phytoalexins that are produced during the course of hypersensitive necrosis. These phytoalexins have the same carbon skeleton as other sesquiterpenoids detected in healthy cotton foliage, such as hemigossypol and deoxyhemigossypol. The two compounds characterized by Essenberg et al., 2,7 dehydroxycadeline and laciniline C, are shown below.

2, 7, Dehydroxycadeline   Lacinilene C

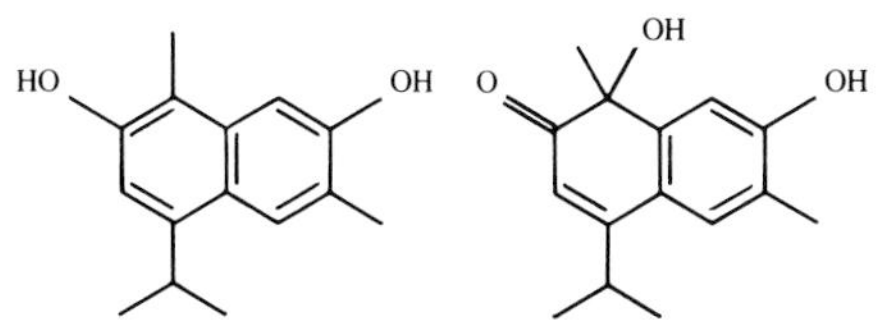

Finally, attention should be called to the almost axiomatic condition that accumulation of phytoalexins in plants is greater in resistant (incompatible) reactions than in susceptible ones. A rationale for this and a more detailed account of where, how, and why phytoalexins accumulate are presented in Chap. 10.

Changes Induced by Fungi

Secondary Substances in Healthy Plants as Preformed Resistance Factors

Phenol content in protective layers of plant organs. It has been shown, in many experiments, that a correlation may exist between the degree of resistance and the phenol level in healthy plants (197). In this field, the classic work of Walker and Link (251) produced interesting results. According to their investigations, onion varieties resistant to *Colletotrichum circinans* accumulate flavones, anthocyanins, and simple phenolics, such as protocatechuic acid and catechol, in the dyed bulb scales that are yellow- or red-pigmented. Phenol compounds in the scales are water-soluble. They can diffuse into the inoculum-containing droplet from the scales and inhibit germination and penetration of infective fungi.

The penetration of infection hyphae is also inhibited by phenolic acids found in waxy coverings of apple and other leaves (154). Waxy material, derived from leaves resistant to mildew caused by *Podosphaera leucotricha*, was deposited onto the leaves of susceptible varieties and imparted resistance to the mildew fungus.

Other instances are known in which phenolic compounds in the protective layers of the healthy plant may be associated with resistance to certain microorganisms (240). It has been found that the seed coat of pea varieties resistant to the causal organisms of root diseases contains larger amounts of phenolic compounds than the seed coat of susceptible varieties. One of the main components was identified as leucoanthocyanin (38). A nonphenol compound, the glucoside of γ-hydroxy car-

boxylic acid ("tuliposide") was shown to be responsible for resistance of tulip to *Botrytis cinerea* (198).

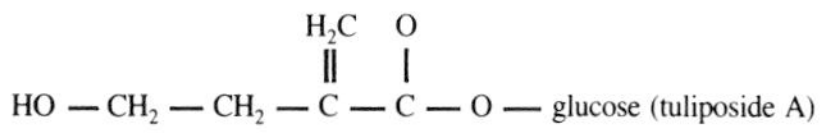

$HO - CH_2 - CH_2 - \overset{\overset{\displaystyle H_2C}{\displaystyle \|}}{C} - \overset{\overset{\displaystyle O}{\displaystyle \|}}{C} - O - glucose$ (tuliposide A)

Phenol content in actively metabolizing organs and the degree of resistance. Interesting results reported by Lee and Le Tourneau (144) in *Verticillium* wilt disease of potatoes revealed that cultivars resistant to *Verticillium* infection contain higher amounts of chlorogenic acid in the roots than susceptible varieties (Tables 6–4 and 6–5).

According to McLean et al. (160), *Verticillium* wilt developed in the field more rapidly and more severely in the susceptible cultivars, coincident with or following the decrease in phenol compounds in the vascular systems. Patil et al. (177) also obtained very convincing results, showing that young potato roots that are practically resistant to infection by *Verticillium* have a relatively high level of phenolics until five weeks after sprouting. From the time of sprouting, chlorogenic acid content decreased continuously; this decrease was closely correlated with an increase in susceptibility to infection. Total phenols decreased significantly during the fourth and fifth weeks, which coincided precisely with the time that the plants became susceptible to *Verticillium*.

Subsequent studies by Patil et al. (178) revealed that the higher level of phenolic compounds in the resistant cultivar reflects, in fact, a greater *synthetic capacity* for these substances. The relative amounts of phenolics

Table 6–4. The chlorogenic acid content of potato leaves and roots (after Lee and Le Tourneau, 144).

		Percentage of chlorogenic acid[a]		
		Leaves		Roots
Cultivar	Resistance to *Verticillium* wilt	Wet basis	Dry basis	Wet basis
Populair	extremely resistant	0.17	2.1	0.08
41956	very resistant	0.25	2.2	0.07
Great Scott	resistant	0.12	1.3	0.11
Early Gem	susceptible	0.17	2.0	0.01
Kennebec	susceptible	0.14	1.6	0.05
Russet Burbank	susceptible	0.18	1.3	0.01
Bliss Triumph	very susceptible	0.08	0.9	0.03

a. Determined by paper chromatographic separation followed by ultraviolet spectrophotometric assay.

Table 6–5. Total visual ferric chloride units of tissues at 5 locations[a] of 6 cultivars of potatoes tested histochemically with 10% ferric chloride (after McLean et al., 160).

	Days from planting				
Cultivar	108	122	155	180	200
Susceptible					
Early Gem	8	8	5	4	3
Russet Burbank	8	4	5	4	3
Kennebec	7	7	4	2	
Average susceptible	7.7	6.3	4.7	3.3	2.0
Resistant					
Populair	11	12	13	11	10
41956	17	15	16	12	9
Great Scott	23	23	18	11	6
Average resistant	17.0	16.7	15.7	11.3	8.3

a. Test points were root stele, stem node, stem vascular, stolon vascular, and tuber vascular sections.

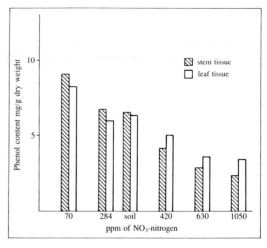

Fig. 6–20. Phenol content in tissues of tomato plants grown with different concentrations of NO_3-nitrogen (after Sarhan et al., 196).

synthesized by resistant and susceptible potato cultivars are presented in Fig. 6–19. In some instances, environmental factors may influence the phenol level of the plant tissues, leading to changes in resistance or susceptibility of host plants.

For example, application of nitrogen fertilizers in large amounts tends to increase susceptibility of cereals to rust disease and, at the same time, to decrease the total phenol level in leaf tissues (130). The same correlation is apparent in the rice blast disease caused by *Pyricularia oryzae*. Nitrogen fertilization increases disease susceptibility of the host; at the same time, the phenol level of leaf tissues significantly decreases

(250). On the other hand, phenol content and the activity of phenoloxidizing enzymes decrease in tomato plants after high nitrogen fertilization (Figs. 6–20 and 6–21) but increase resistance to infection caused by *Fusarium oxysporum* f. sp. *lycopersici*, as well as insensitivity to its toxin (196). The correlation between high phenol level and resistance in this instance is not a positive one.

It is also possible to increase the phenol level in plant tissues experimentally. By infusing young apple plants with *D*- and *DL*-phenylalanine, Kuć and his group (109) were successful in inducing resistance in two cultivars susceptible to *Venturia inaequalis*. Increase in resistance was correlated with increased amount of phenol

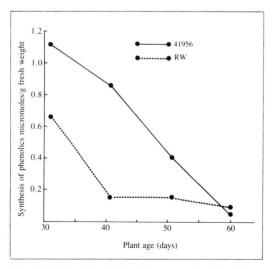

Fig. 6–19. Synthesis of phenolic compounds in 41956 (resistant) and Red Warba (susceptible) potato varieties (after Patil et al., 178).

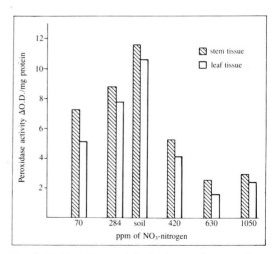

Fig. 6–21. Peroxidase activity in tissues of tomato plants grown with different concentrations of NO_3-nitrogen (after Sarhan et al., 196).

compounds. Two of these were identified as phloretin and phloretic acid. Both are components of phloridzin, a phenolic glucoside found in large amounts in certain apple tissues (109). It is well known that aromatic amino acids are produced through the shikimic acid pathway in the first phase of aromatic biosynthesis. As we have noted earlier, phenylalanine and tyrosine are the two most important aromatic amino acids in this pathway (see Fig. 6–3). A series of phenolic compounds are synthesized from the phenylalanine pool (Fig. 6–4). Several species belonging to the genus *Malus* are able to synthesize phloridzin, chlorogenic acid, quercetin glucosides, and other phenolic compounds from the phenylalanine pool (113). Hence, the resistance in apple induced by phenylalanine may be correlated with increased amounts of phloretin and phloretic acid. It seems that this is true for the Geneva and Russian apple cultivars. On the other hand, the Jonared cultivar is not able to increase phenol content after infusion with phenylalanine and, therefore, it is not possible to induce resistance in this variety with phenylalanine.

Genetic experiments by Defago and Kern (57) and Defago et al. (58) examined the role of the steroid alkaloid tomatine, a saponin, in the resistance of green tomato fruits to *Fusarium solani*. Wild-type (WT) isolates of *F. solani*, though able to rot ripe (red) fruit, stem and root tissue, are unable to rot green fruit tissue. Rot is ostensibly prevented by tomatine as WT isolates are inhibited in vitro by 100 µg/ml of tomatine. Spores of WT *F. solani* treated with mutagen nitrosoguanadine (NTS) provided mutants that were insensitive to 800 µg/ml of this preformed alkaloid. An analysis of F_1 strains showed tomatine insensitivity in the pathogenic mutants was due to a low content of a sterol, presumably ergosterol, in their mycelia. The site of action of tomatine is alleged to be the component of the fungal plasmamembrane.

Support for the hypothesis concerning the positive role of phenolics as resistance factors is, however, not unanimous (112, 114, 196, 254).

Postinfectional Accumulation of Secondary Substances

Parasitically enhanced aromatic biosynthesis. What kind of changes in aromatic metabolism occur after infection? Many experimental data, in pathophysiological research, support the theory that the phenolic level, and the level of other secondary substances, is higher in diseased plants than in healthy ones. A number of particular examples are available (11, 60, 68, 80, 193, 236, 238). In association with phenol biosynthesis, the activity of PAL and tyrosine ammonia-lyase (TAL) is enhanced. Both enzymes are involved in the deamination of aromatic amino acids and, therefore, in the synthesis of phenolics. It must be stressed, however, that the enhanced aromatic biosynthesis and the stimulated enzyme activities are characteristic not only of infections but also of injury and accelerated aging. These responses are, therefore, not specific for infection-based injury. Some important phenolic compounds synthe-

Phenol	Structure	Host
caffeic acid		sweet potato
chlorogenic acid		sweet potato / white potato / carrot
phaseollin		French bean
pisatin		garden pea
orchinol		orchid
3-methyl-6-methoxy-8-hydroxy-3,4-dihydroisocoumarin		carrot
phloretin		apple
hydroquinone		pear
umbelliferone		sweet potato
scopoletin		sweet pototo

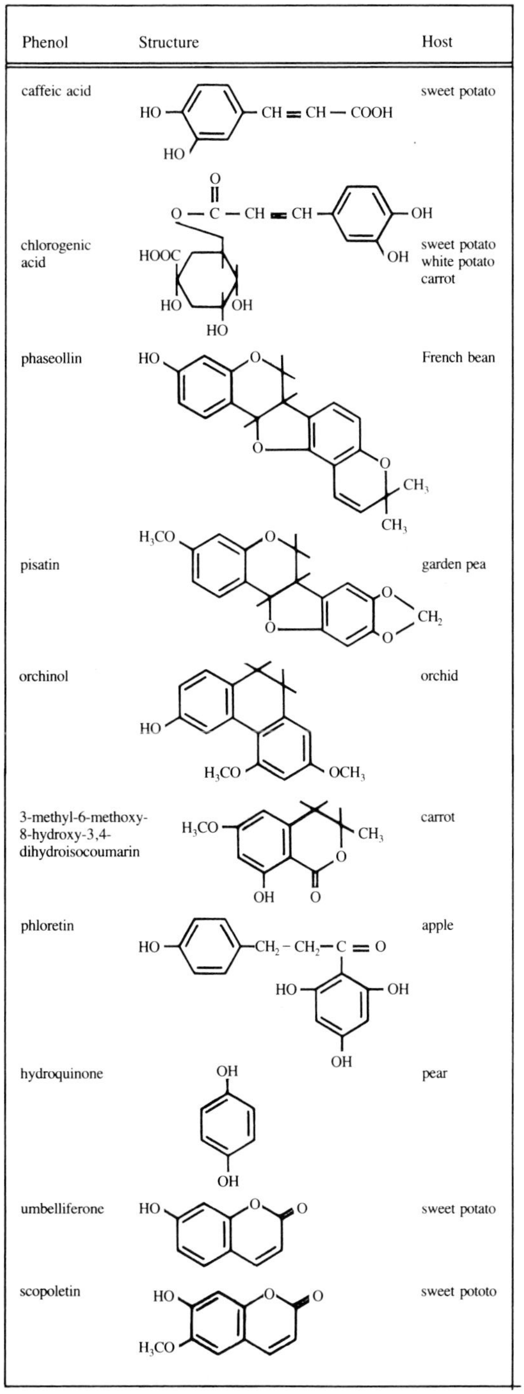

Fig. 6–22. Phenolic compounds synthesized by plants in response to injury or inoculation (after Kuć, 139).

sized by plants in response to injury as well as infection are presented in Fig. 6–22.

When comparing resistant and susceptible host/pathogen combinations, in many cases a more rapid accumulation of phenolics takes place in the resistant complexes than in the compatible ones (133, 230, 231). Comparison of susceptible and resistant infected cultivars has not, however, always revealed a positive correlation between phenol content and resistance (65). For example, it was shown by careful investigation, using near-isogenic lines instead of different cultivars, that phenolics appear not to have a role in wheat rust resistance—either as preformed resistance factors or as postinfectionally induced agents (54, 200).

Tissue concentrations of phenolics are usually measured after extraction with organic solvents, rather than in vivo, because the aromatic compounds are sparingly soluble in water. It is important also to note that high concentrations of phenolics, or other secondary plant constituents in tissue, do not necessarily constitute effective concentrations for microbial toxicity (53). Thus, the above-mentioned experimental evidence is only correlative. The enhanced aromatic biosynthesis probably results from the accompanying tissue necrosis. The process, therefore, may be a consequence of the resistance "symptom," rather than a cause of resistance. According to Daly (53), "there is no single instance where proof is complete for the precise role in disease resistance of any chemically-defined component produced during invasion."

Phytoalexins. These compounds are postinfectionally produced secondary plant substances exhibiting antimicrobial activity (2, 11, 28, 46, 48, 80, 92, 126). Some of the phytoalexins, however, could also be among the constituents of the plant before infection. Chemically, the phytoalexins may belong to the following groups:

(1) Phenolics
    (a) simple phenols, e.g. chlorogenic acid
    (b) flavonoids, e.g. phloretin, pisatin, phaseollin, pterocarpans
    (c) coumarins, e.g. methoxymellein
(2) Polyacetylenes, e.g. wyerone acid, safinol
(3) Isoprenes
    (a) terpenoids, e.g. ipomeamarone, rishitin, phytuberin, lubimin, gossypol
    (b) steroids, e.g. α-solanine, α-chaconine.

Approximately 70 phytoalexins have been isolated from 100 plants representing 21 families. The term *phytoalexin* was first used by Müller (162), who wanted to point out the nonspecific nature of the antimicrobial substances (alexins) produced by plants (phyton) upon infection. The first experiments were conducted earlier with *Phytophthora infestans* (164). It was shown that resistant potato tuber tissues infected by an incompatible race of *P. infestans* produced antifungal compound(s) in association with tissue necrosis. This was experimentally demonstrated by applying droplets containing zoospores of *P. infestans* on the exposed inner (aseptic) surface of young bean pods (which is a nonhost for this fungus). After incubation, the droplets were collected and tested for the presence of antifungal substances (phytoalexins) on the basis of their action on spore germination of different fungi.

The synthesis or accumulation of phytoalexins is not specifically induced only by pathogens. Nonbiological stimulation of these compounds by different stresses (chemical, physical, and mechanical) has also been reported (42, 80, 93, 98, 184, 237). In addition to certain fungal metabolites, synthetic compounds or UV light can trigger the de novo production of the phytoalexin pisatin. The synthesis of this phytoalexin depends on the synthesis of RNA and protein and is accompanied by dramatic increases in PAL activity (95). There are also characteristic changes in the rate of RNA and protein synthesis. Clearly, the inducer does not specifically act on PAL; rather, a broad range of proteins is influenced. It is suggested by Hadwiger et al. (98) that diverse groups of phytoalexin inducers affect multiple segments of nuclear DNA, directly affecting the normal expression of the plant's genome. As yet, no complete biosynthetic pathway for the production of a phytoalexin has been described. The activation of the cyanide-resistant respiration pathway may be applicable in the synthesis of terpenoid phytoalexins (6, 26). (See also Chaps. 3 and 10.) Apparently, ethylene activates the cyanide-resistant respiration pathway and, at the same time, promotes phytoalexin accumulation.

The role of PAL in several phytoalexin biosyntheses has been substantiated. Mayama et al. (158) applied a PAL inhibitor (α-amino-oxiacetate) to incompatible oat leaves inoculated with rust, *Puccinia coronata* f. sp. *avenae*. The inhibitor greatly suppressed the rapid accumulation of the phytoalexin avenalumin (Fig. 6–23).

It would seem that plants have the potential to synthesize phytoalexins and that the infecting fungi or other stresses to the cells determine their rate of synthesis by the plant. For example, the concentration of an isocoumarin varies from 5 to 342 μg/g carrot, depending upon the organism used for inoculation (42). Inoculation experiments using different races of *Colletotrichum lindemuthianum* have shown that the accumulation of the phytoalexin phaseollin was clearly associated with tissue necrosis (9, 187). A close association of tissue necrosis with the accumulation of phytoalexin (phaseollin) was experienced also in virus and rust infections (10). It is likely that phytoalexin production accompanied the process of tissue necrosis (browning) and that it is a consequence of the necrosis (resistant reaction) or other nonvisible adverse action, rather than the cause of it. Even in soybean infected by *Peronospora manshurica*, when resistance expressed itself by nonvisible necrotic lesions, Érsek et al. (64) have shown that phytoalexin accumulated around the infection site. Similarly, in rust infection the phytoalexin was produced apparently as a result of tissue necrosis and was not detected in a compatible host/pathogen combination, where chlorotic flecks were produced without necrosis (Table 6–6).

It has been reported (153) that wyerone acid, a nonphenol compound, was produced by living cells in

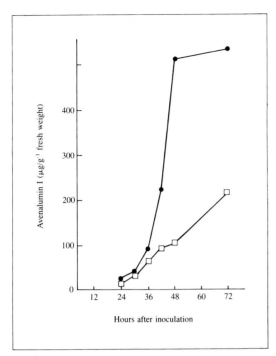

Fig. 6–23. Effect of α-aminooxyacetate on the accumulation of avenalumin I in Shokan 1 leaves infected with *P. coronata avenae* race 226. (□), AOA (200 μM); ●, control (after Mayama et al., 158).

Table 6–6. Phaseollin content of *P. vulgaris* leaves 6 d after inoculation with *Uromyces appendiculatus* (after Bailey and Ingham, 10).

	Fr. wt. of leaves (g)	Phaseollin content	
		(μg)	(μ/g fr. wt.)
Imuna, uninoculated	16.0	—[a]	—
Imuna, inoculated with *U. appendiculatus*[b]	17.5	—[a]	—
No. 765, uninoculated	24.0	—[a]	—
No. 765, inoculated with *U. appendiculatus*[c]	25.0	42	1.7

a. No phaseollin was detected by GLC.
b. Leaves bore chlorotic flecks, from which large uredosori erupted after 2 d.
c. Leaves bore small superficial brown flecks, from which small uredosori erupted after 8 d.

a low MW peptide (monilicolin A). However, in this case, no tissue necrosis was observed.

The accumulation of phytoalexins in potatoes seems to be connected to wound-healing response of host tissues. Around the site of injury in the tuber, phenols and steroid glycoalkaloids accumulate because of the activation of the acetate-mevalonate pathway. However, when *Phytophthora infestans* invades, wounds, and infects tuber tissues, elicitors from the cell wall of the fungus shift the wound-healing response from steroid glycoalkaloids (α-solanine, α-chaconine) to terpenoids (rishitin, phytuberin, lubimin). The latter compounds are highly fungitoxic and accumulate rapidly, particularly in the resistant host.

Acetate-mevalonate pathway
→ *wounding* → steroid glycoalkaloids
→ *infection (elicitor)* → terpenoid phytoalexins

In *susceptible* hosts, the situation seems to be more complicated, according to the hypothesis of Doke and his associates (61, 62, 86). Susceptibility would be a result of suppression of phytoalexin elicitation by low MW water soluble glucans, *suppressors*, released by the compatible fungus. In this case, nonfungitoxic terpenoids will be produced, instead of fungitoxic terpenoids (phytoalexins).

Acetate-mevalonate pathway
→ *wounding* → steroid glycoalkaloids
→ *infection (elicitor* ← → *suppressor)* → nonfungitoxic terpenoids

broad bean leaves infected with *Botrytis cinerea*. It is not known whether metabolites released from dead plant cells induced the biosynthesis of wyerone acid in adjacent living cells or whether the phytoalexin in live cells reached phytotoxic concentrations and, thus, caused plant cell death.

The Elicitor-Suppressor Hypothesis for the Synthesis of Phytoalexins

Treating tissues with high MW wall components of pathogenic fungi can induce accumulation of phytoalexins, phenolics, and the development of necrotic symptoms (hypersensitive reaction). These wall products have been termed *elicitors* (3, 119) and are derived from glucan cellulose or from glucan-chitin fungal cell walls. They have been characterized from only three fungal genera: *Phytophthora*, *Colletotrichum*, and *Fusarium* (7). However, some of these glucans also exert elicitor activity on plants that are not hosts for these fungal species. Thus, hyphal cell wall glucan (elicitor) recognition seems to be neither specific nor universal. In addition to glucans, high MW glycoproteins, pectic enzymes, and unsaturated fatty acids were also found to have phytoalexin elicitor activity (25, 125, 216). Analogously, Cruickshank and Perrin (52) reported earlier that phytoalexin accumulation is stimulated in beans by

The chemical nature of the nontoxic terpenoids has not yet been characterized. According to that hypothesis, the interaction of *elicitors*, high MW glucans from cell wall of the fungus, and *suppressors*, low MW water-soluble glucans of the fungus, would control either the accumulation or suppression of phytoalexins. Specificity would lie in the binding of the suppressor to a hypothetical receptor on the host cell membrane. The elicitors from *Phytophthora infestans* seem to be nonspecific. It is also noteworthy that no effect of the suppressor on the growth of infection hyphae was observed, at least in the penetrated cells.

The elicitor-suppressor hypothesis for phytoalexin synthesis was also proposed in the case of soybean-*Phytophthora* complex (259). According to the proposed mechanism, the fungus produces glucan elicitors and glycoprotein suppressors of phytoalexin. The suppressor is, in fact, an extracellular invertase, and the carbohydrate portion of the molecule is responsible for the inhibitory effect. The inhibition is race-specific. Therefore, not the accumulation of the phytoalexin but, rather, its suppression is specifically induced in the susceptible plant by the pathogen.

Phytoalexin elicitors can easily be derived from fungal cultures near the end of or after their logarithmic growth phase, when degradation products are usually present in culture filtrates. When pea endocarp tissues were infected by an incompatible f. sp. of *Fusarium solani*, hydrolytic enzymes produced by the host released fungal wall carbohydrates, including glucosamine polymers. The latter compounds are chitosans and have an antifungal character. In addition, it was shown by Hadwiger et al. (96) that glucosamine polymers released from the fungus accumulate in the host cells and may have phytoalexin elicitor activity. To initiate these and other metabolic events in the host and the pathogen, an early intimate contact seems to be essential between the germinating conidia and the host cells (169). A diagrammatic summary of the release, migration, and accumulation of chitosan into host cell, and an increase in active mRNA specific for the synthesis of PAL and other metabolic events, is presented in Fig. 10–24.

Sensitivity of the infecting fungus to the phytoalexins produced after inoculation was claimed to be an important factor in disease resistance. Cruickshank presented evidence that fungi pathogenic to the pea are relatively insensitive to pisatin, whereas nonpathogens are highly sensitive (46, 47). Cruickshank and Perrin (51) compared various host-pathogen combinations of several pea cultivars and several strains of the fungus; the concentration of pisatin formed by the plant varied with the specific strain of *Ascochyta pisi* used. On the other hand, when one strain of the fungus was used to inoculate several cultivars of the host, the concentration of the phytoalexin varied with the variety. Hence, both the specific cultivars and the strain of the pathogen determine the extent to which phytoalexin production is stimulated.

Fungi and bacteria can degrade phytoalexins, apparently mediating the toxicity of those compounds. It is known that host plants, too, can metabolize phytoalexins (218, 223, 239, 253).

The most important phytoalexins are grouped as follows on the basis of the plants in which they are produced (80, 122, 140):

Orchid (*Orchis militaris*) infected by *Rhizoctonia repens* (Fig. 6–22). *Orchinol*'s structure has been established as 9,10-dihydro-2,4-dimethoxy-6-hydroxyphenanthrene (88).

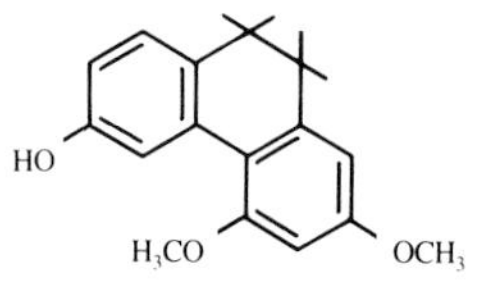

Orchinol

Potato (*Solanum tuberosum*) infected with *Phytophthora infestans* (4,5,36,111,224,227,229) (Fig. 6–22).

Caffeic acid ⎫
Chlorogenic acid ⎬ Phenol derivatives
Scopolin ⎭
Rishitin ⎫
Rishitinol ⎪
Phytuberin ⎬ Terpenoids ⎫
Lubimin ⎭ ⎬ Isoprenoids
α-Solanine ⎫ ⎭
α-Chaconine ⎬ Steroids

Chlorogenic acid

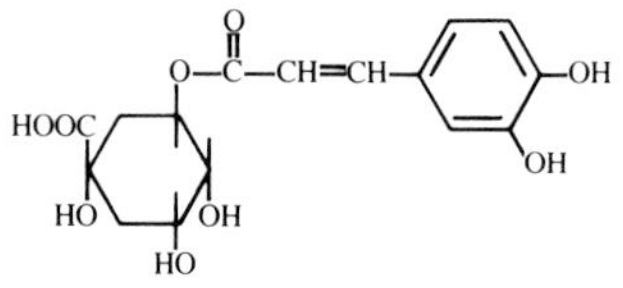

Caffeic acid

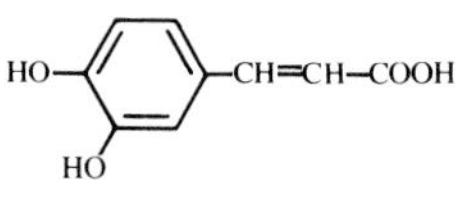

Pea (*Pisum sativum*) infected with *Monilia fructicola*.

Pisatin is a flavonoid synthesized by the acetate and shikimic acid pathways (49). PAL has an important role in the synthesis of this phytoalexin. The activity of this enzyme was found to increase tenfold when pea pods were induced to produce pisatin (97). It was proposed that the induction of the synthesis of this phytoalexin occurs via the derepression of certain genes in the plant. This would be the induction hypothesis for disease resistance (199), which needs further clarification.

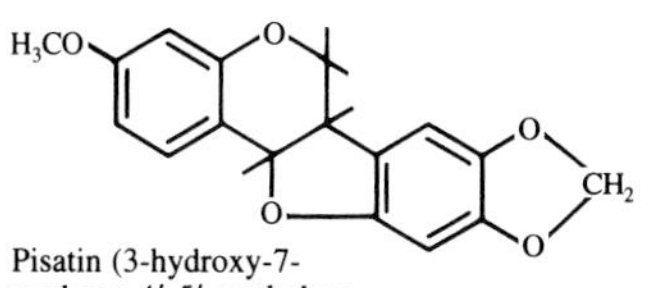

Pisatin (3-hydroxy-7-
methoxy-4',5'-methylene-
dioxychromanocoumaran)

Bean (*Phaseolus* vulgaris) infected with *Monilia fructicola*, *Collectotrichum lindemuthianum*, and *Rhizoctonia solani*.

Phaseollin (Fig. 6–22) is a flavonoid (7-hydroxy-3',4'-dimethylchromanocoumaran) (50, 183). It is synthesized via the joint participation of the acetate and shikimic acid pathways. There is evidence that an increase in the amount of PAL takes place 24 h earlier than either lesion formation by *Colletotrichum lindemuthianum* or changes in the concentrations of phenolic phytoalexins (188). *Kievitone*, an isoflavanone, occurs in hypocotyls of beans infected by *Rhizoctonia solani* (30, 210).

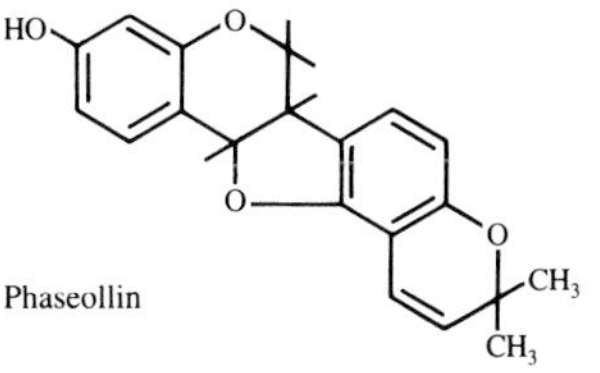

Phaseollin

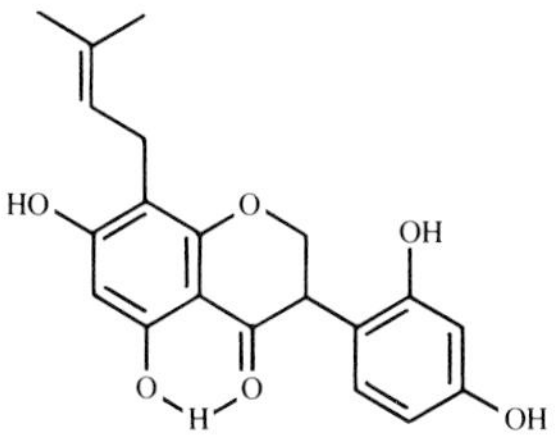

Kievitone (2',4',5,7-
tetrahydroxy-8-isopen-
tenylisoflavanone)

Broad bean (*Vicia faba*) infected with *Botrytis fabae* and *B. cinerea.Wyerone acid* (and wyerone, which is an acetylenic furanoid keto-ester) is a preformed fungitoxic compound (74, 249). As a result of infection with *B.*

$CH_3OOC — CH = CH$ $CO — C = O — CH = CHC_2H_3$

Wyerone

fabae a 400- to 500-fold increase in this phytoalexin occurs. Wyerone acid inhibits *B. cinerea* but much less so *B. fabae*, which is not as sensitive to wyerone acid and actively degrades this phytoalexin (75, 147).

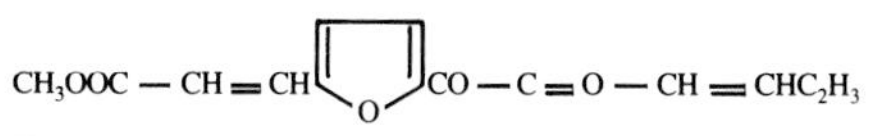

Soybean (*Glycine max*) infected with *Phytophthora megasperma* var. *sojae* and with a series of nonpathogenic fungi induces glyceollin (29, 143) synthesis. The compound is closely related to phaseollin and pisatin (126) (Fig. 6–22).

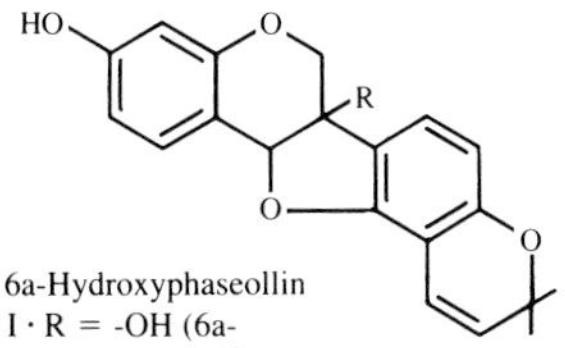

6a-Hydroxyphaseollin
I · R = -OH (6a-
hydroxyphaseollin)
II · R = -H (phaseollin)

Carrot (*Daucus carota*) infected by *Ceratocystis fimbriata* and other nonpathogens (41, 42) produces 6-*methoxymellein* (Fig. 6–22). Its structure has been established as 3-methyl-6-methoxy-8-hydroxy-3,4-dihydroxy-isocoumarin. This phytoalexin is synthesized via the acetate pathway. The accumulation of 6-methoxymellein can also be induced by chemicals, cold treatment, and ethylene; thus it seems to be a response to stress, not specifically to infection (140).

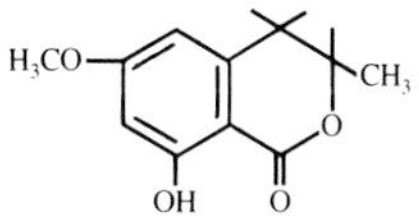

6-Methoxymellein

Sweet potato (*Ipomoea batatas*) infected by *Ceratocystis fimbriata* (235, 237) synthesizes the terpenoid *ipomeamarone* via the acetate pathway. The carbon sources utilized in the infected sweet potato for respiration and for the biosynthesis of phenolics may be diverted to the synthesis of ipomeamarone by inhibition of the tricarboxylic acid cycle and phenol biosynthesis (234, 235). This phytoalexin is also induced by injury and many chemical agents. In addition to ipomeamarome, coumarins such as scopoletin, esculetin, and umbelliferone and simple phenols such as

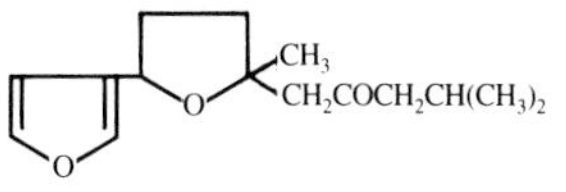

Ipomeamarone—Furano-terpenoid

chlorogenic acid, isochlorogenic acid, and caffeic acid are produced in sweet potato (Fig. 6–22).

Cotton (*Gossipium* spp.) infected with *Verticillium albo-atrum* (16,17). *Gossypol* is a diterpenoid compound normally found in glands of most cotton cultivars.

Gossypol

Alfalfa (*Medicago sativa*) infected with non-pathogens to alfalfa like *Helminthosporium turcicum*, *Colletotrichum* spp., *Stemphylium* spp., *Ascochyta* spp., etc. (104–6) produces a pterocarpane (flavonoid) and derivative of coumesterol.

Red clover (*Trifolium pratense*) infected by *Sclerotinia trifoliorum* and *Helminthosporium turcicum* (27, 246, 247) produces several pterocarpans. The pterocarpan structure: (a) trifolirhizin R_1 = -$OC_6H_{12}O_5$,R_2-R_3 = -O-CH_2O-; (b) maackiain R_1 = -OH, R_2-R_3 = -O-CH_2-O-; (c) medicarpin R_1 = -OH, R_2 = -H, R_3 = -OCH_3. The general formula is modified to a series of different compounds by changes at R_1, R_2, and R_3.

Grapevine (*Vitis vinifera*) leaves infected with *Botrytis cinerea* produce a stilbene compound, viniferin.

Gramineae. A number of phytoalexins have been isolated from barley, corn, rice, and oats. For example, avenalumins are nitrogen-containing phenolics produced in oats infected by rust. Powdery mildew and *Helminthosporium* induce the synthesis of several compounds in rice, corn, and barley (149, 158, 159, 175, 232).

Release of bound phenolics from glucosidic linkages by fungal glucosidases. Several authors have demonstrated in vitro release of bound phenolics by the enzymatic activities of pathogens. Phenols liberated from glucosidic compounds may become very toxic to the pathogen. For example, Oku (172) has shown that *Cochliobolus miyabeanus*, the causal agent of *Helminthosporium* disease of rice, is able to produce β-glucosidase and to liberate phenol compounds from glucosides in concentrations inhibitory to the fungus. *Venturia pirina*, the causal agent of pear scab, also pos-

sesses an enzymatic activity by which hydroquinone is liberated from the glucoside arbutin in vitro (78). β-Glucosidase activity has also been demonstrated in the sap of *Fusarium*-infected tomato. This activity was absent in healthy plants (55).

Most of these experiments represent in vitro conditions, so the in vivo role of these processes remains unknown. It is also not clear whether these "released" phenols are really more toxic in vivo to the fungal pathogens than their glucosidic forms.

Activation of Enzymes Oxidizing Phenol Compounds in Infected Plants (Polyphenoloxidases, Peroxidases)

In many instances, latent polyphenoloxidases are activated as a consequence of fungus infections. *Botrytis cinerea*, the causal agent of chocolate spot disease of beans, produces pectic enzymes that liberate galacturonic acid derivatives and polygalacturonic acid derivatives from cell walls. These compounds were shown by Deverall (59, 220) to activate the latent polyphenoloxidase of the host plant. Ophiobolin, a toxin produced by a fungus, *Cochliobolus miyabeanus*, also activates polyphenoloxidase of rice leaves (165).

The mechanism of activation of latent phenoloxidizing enzymes remains unresolved. According to Kenten (127, 128), inhibitors in plants are proteinaceous in nature and may combine with latent phenolases and thereby mask these enzymes. A number of substances and treatments are able to influence the effectiveness of inhibitors, thereby evoking an "activating" effect. According to Robb et al. (192), the prosthetic group of the "latent" tyrosinase of broad bean is masked by virtue of its tertiary structure. Alteration of this structure by denaturing agents results in an active configuration of similar molecular weight. Whether these mechanisms have some role in the activation of latent polyphenoloxidases in diseased plants remains to be clarified.

The increased activity of phenoloxidizing enzymes in diseased plants cannot be explained exclusively by the activation of latent enzymes. Net synthesis of enzyme proteins in infected tissues may play an equally important role. This activity has been observed in the case of the potato-*Phytophthora* and sweet potato-*Ceratocystis* complexes. One of the proteins synthesized during pathogenesis was identified as a peroxidase (214, 215).

The role of the activated peroxidase in infected tissue was reinvestigated by Seevers et al. (201, 202). Rust-infected wheat seedlings containing the Sr6 gene were held at 20 C for 6 d after inoculation to allow the development of necrosis and the activation of peroxidase (resistant reaction type). The seedlings were then transferred to 26 C and the disease reaction changed from resistant infection type 0 to susceptible infection type 3. However, there was no correlated decline in total peroxidase activity (Table 6–7).

Increased activity of polyphenoloxidase and peroxidase might also be due to the production of these enzymes by the infecting fungi, *Cochliobolus miyabeanus* and *Puccinia graminis* (70, 173).

Table 6–7. Disease-induced peroxidase activity in wheat leaves showing resistant or susceptible reactions to *Puccinia graminis*[a] (after Seevers and Daly, 201).

Tissue[b]	No. of visible sites[c]	OD/min per leaf on day								
		1	2	3	4	5	6	7	9	11
		Held continuously at 20 C								
SI	226	2	4	6	6	7		8		1
RI	136	1	5	12	19	20	30	22	31	32
		Transferred to 26 C on day 4								
RI	135					22	20	19	15	18

a. Peroxidase activity of infected leaves minus peroxidase activity of control leaves.
b. R = resistant leaves; S = susceptible leaves; I = inoculated leaves.
c. SI pustules were of type 3; RI showed type 0 at 20 C; RI at 26 C showed type 3 with some flecking.

A number of authors have noted a correlation between the activity of peroxidase of healthy host plants and their resistance to disease. Resistance in the field of cultivars of *Solanum tuberosum* to late blight, caused by *P. infestans*, may be indicated, with some restrictions, by the peroxidase test (118, 233). The correlation between peroxidase activity and resistance to late blight does not apply for *Solanum demissum*. The whole concept was criticized by Henniger and Bartel (102), who found only a loose relationship between resistance and peroxidase activities of more than 50 potato varieties. However, this relationship has been observed by Gretchushnikov and Popkova (91), who have shown a decrease of peroxidase activity in potato leaves, upon ringing the stem, and a decrease in resistance to *Phytophthora* disease. Therefore, it was concluded that in special host-pathogen relations, the activity of phenoloxidizing enzymes in healthy plants may be correlated with the degree of resistance. This question, important from a practical (plant breeding) point of view, has been reinvestigated from time to time. Sakai and Tomiyama (195) came to the conclusion that there exists a high degree of correlation between host resistance to *Phytophthora* and high peroxidase activity, but only in the adult stage of the plant. In the young plant, this correlation cannot be detected. They claimed that resistance of potato cultivars can be evaluated on the basis of peroxidase test only in the adult stage. In further investigations, Fehrmann and Dimond (77) showed that the peroxidase activity in different plant organs is correlated with resistance to *Phytophthora*. The higher enzyme activity was found in the root tissues of potato, which is the most resistant organ to infection. Peroxidase activity gradually decreased in young leaves, matured leaves, tuber peel, and flesh, respectively. The resistance of the particular organs to infection decreased in correlation with the decreased enzyme activity. This is a correlative phenomenon that does not necessarily imply a causal relationship between re-

sistance and enzyme activity, but it could possibly be used in screening resistant plants. Resistance, in these cases, may be more closely associated with hypersensitivity—i.e. a greater capacity for a fast necrotic reaction of the host plant. Production of necrotic reactions is, in turn, closely associated with polyphenol-oxidation.

The Hypersensitive Reaction

The HR is a significant and common reaction in the plant kingdom (87, 132). Usually a small amount of tissue becomes necrotic after infection, and the pathogen is unable to spread into the healthy parts of the plant. It has been shown, in a number of instances, that phenolic compounds, and other secondary metabolites (phytoalexins), accumulate in the tissues of hypersensitive plants. Appearance of necrosis induced by HR is correlated with oxidative reactions.

HR was first investigated in *Phytophthora* disease of potato. In potato tissues treated with narcotics, HR and related enzymatic processes were inhibited (163, 228). Poisons of heavy metal enzymes and inhibitors of both oxidation and phosphorylation and protein synthesis also reduced the speed of HR (83, 226). Probably phenoloxidizing enzymes have an important role in the process leading to tissue necrosis. According to Rubin et al. (194), ionizing radiations decrease the activity of polyphenoloxidase and peroxidase in potato tissues and, simultaneously, decrease the necrotic HR to *P. infestans*. Furthermore, poisons of Cu enzymes also decrease the necrotic symptom (83) in potato, and a high Cu level in the tissues is correlated with greater resistance and with more pronounced HR to *P. infestans*. It is significant that polyphenoloxidase is a typical Cu enzyme.

It also has been shown that increase of enzymatic- and nonenzymatic-reducing activity in host tissue results in a lowered probability for the appearance of HR by reducing oxidation products of phenols. A necrotic reaction is usually correlated with the disappearance, or lowering, of the level of reducing compounds such as ascorbic acid. This situation has been observed in stem rust disease of wheat and in bacterial scab of potatoes (116, 133). Reducing compounds such as glutathione and ascorbic acid applied directly to the tissues reduced the degree and intensity of necrosis of rice infected with *Cochliobolus* and apples infected by *Sclerotinia* (31, 174). When circumstances did not allow the appearance of HR, the activity of dehydrogenases usually increased. This was the case with *Phytophthora*-infected potatoes and in stem rust disease of wheat.

One can conclude that the appearance of HR is, in part at least, a consequence of the disturbance of the balance between oxidative and reductive processes. The result is an excessive oxidation of polyphenol compounds and a breakdown of cellular structure (necrosis), a sequence that has been called the *redox theory* of hypersensitivity. In stem rust of wheat, for example, a rapid, irreversible oxidation of redox systems takes place in the incompatible ("resistant") host-pathogen combination; on the other hand, susceptible ("compatible") combinations are able to maintain a balanced redox potential in the diseased tissue (117).

There are many experimental results to show that quinones are extremely fungitoxic compounds (148, 171). Furthermore, it is also possible that natural quinones, in some cases, are closely linked with a resistant reaction of the host plant (152, 240).

Aromatic (Phenol) Metabolism and Physiological Processes in the Diseased Plant

Respiration. It is generally accepted that polyphenoloxidizing enzymes are not involved in the terminal oxidation process of plants. The polyphenolpolyphenolase system can, nevertheless, transfer electrons to the oxygen of air in vitro, and the quinone reductases can catalyze the transfer of electrons from NADPH to quinones. Furthermore, reduced quinone arising in this way presumably reduces cytochrome c, which may be oxidized by cytochrome oxidase; however, these processes have not been demonstrated in vivo.

Nitrogen metabolism. In agricultural practice, it is well known that the application of nitrogenous fertilizers in large amounts tends to increase the plants' susceptibility to disease caused by biotrophic pathogens (131). Accumulation of soluble nitrogen compounds in resistant wheat and potato results in a shift toward susceptibility (79, 91). A significant consequence of nitrogen fertilization of wheat is the decrease of total phenol level of wheat leaf tissues; this may increase susceptibility to stem rust, *Puccinia graminis* (130). Moreover, Kirkham (134) has shown that an increase of the ratio of soluble nitrogen to phenol compounds in a medium lowers the toxicity of polyphenols toward *Venturia* fungi, which cause leaf and fruit lesions of apples and pears. According to Szweykowska (221), aromatic biosynthesis and nitrogen metabolism may be regarded as competing pathways. (See also Chap. 5.)

Pectic enzymes. Oxidized phenols are strong inhibitors of extracellular pectolytic enzymes of fungi that cause wilt diseases and soft rots (*Fusarium, Verticillium, Sclerotinia*). Deese and Stahmann (56) have shown (Tables 6–8, 6–9, 6–10) that *Verticillium*, the causal agent of wilt in tomatoes and potatoes, is unable to produce the pectolytic enzymes in resistant plant tissues that are necessary for its pathogenicity. Hence, the suppression of pectin-degrading enzymes seems to be closely linked with high activity of the polyphenoloxidase enzyme in resistant cultivars (177).

An analogous condition has been reported for *Sclerotinia fructigena*, the causal organism of the brown rot disease of apple (31–33). Tannins of high MW, as well as oxidation products of catechins and particularly those of leucoanthocyanins, were found to be powerful inhibitors of pectolytic enzymes (32, 39, 254).

Auxin metabolism. It has been demonstrated in diseases caused by biotrophic parasites that auxin content markedly increases in infected leaves (see Chap. 7). The increased indoleacetic acid level of diseased tissues may be correlated with an inhibition of IAA oxidase in the

Table 6–8. Pectic enzyme activity of filtrates from *Verticillium albo-atrum* cultured on stem tissues from resistant and susceptible tomato plants (after Deese and Stahmann, 56).

Reaction time hours	Pectic polygalacturonase[a]	
	Bonny Best susceptible	Loran Blood resistant
2	8.8	0.0
4	10.4	0.0
6	12.0	1.0
Pectin methyl esterase[b]	1.0	0.0

a. Microequivalents of reducing groups liberated from sodium polypectate at 30 C by 1 ml of culture filtrate.
b. Microequivalents of methoxyl groups liberated from pectin, pH 4.5 in 3 h at 30 C by 1 ml of culture filtrate.

Table 6–9. Pectic polygalacturonase activity in filtrates from 2 isolates of *Verticillium* cultured on stem tissues of susceptible and resistant tomato plants (after Deese and Stahmann, 56).

Tomato cultivar	Enzyme activity[a]	
	isolate 4 (T-16)	isolate 50
Bonny Best, susceptible	6.5	6.0
Loran Blood, resistant	2.0	0.6
VR Moscow, highly resistant	0.0	0.0

a. Microequivalents of reducing groups liberated from sodium polypectate by 1 ml culture filtrate in 6 h at 30 C.

Table 6–10. Modified "quinone" test given by aqueous extracts of stems and *Verticillium* culture filtrates from stem tissues of susceptible and resistant tomato plants (after Deese and Stahmann, 56).

Tomato cultivar	Quinone test[a]		
		culture filtrate[c]	
	extract[b]	isolate 4 (T-16)	isolate 50
Bonny Best, susceptible	0	0	0
Loran Blood, resistant	2	3	2
VR Moscow, highly resistant	3	5	5

a. Color change produced in acidified potassium iodide-starch solution: 0 = none, 1 = weak, 2 = medium, 3 = strong, 5 = very strong, rapid.
b. Aqueous solution from uninoculated stem sections.
c. Aqueous extract from stem sections inoculated with isolates of *Verticillium albo-atrum*, preadapted to glucose plus 0.1% pectin.

plants. IAA oxidase is a typical peroxidase-mediated system, and phenolic inhibitors may play an important role in regulating its activity.

Although there is no direct evidence for fungal pathogens, the working hypothesis that suggests itself is that alteration of the auxin level and activity of IAA oxidase in the diseased tissues are closely associated with the altered phenol metabolism of the diseased plant (see also Chap. 7).

Comparative Analysis of Disease Physiology—Secondary Metabolites

It seems to be a general phenomenon that an enhanced metabolism of secondary substances parallels symptom development. As a rule, more rapid accumulation of secondary substances takes place in incompatible host-pathogen pairs than in compatible ones. In the last 50 years, a series of experiments have called attention to a possible cause-and-effect relationship between accumulation or synthesis of various secondary substances, including phenolics, and tissue necrosis associated with hypersensitive resistance. This is characteristic of plants infected with viruses, bacteria, or fungi. Infection leads to a general stimulation of enzymes involved in the biosynthesis (PAL) and oxidation (polyphenoloxidase, peroxidase) of phenols. However, there is as yet not enough evidence to establish a causal relationship between potentiated aromatic synthesis and either tissue necrosis or resistance.

Considerable research has called attention to the synthesis and accumulation of *phytoalexins* after infection. These compounds are generally postinfectionally produced secondary plant substances exhibiting antimicrobial activity. Chemically, they may be phenolics, isoprenes (terpenoids and steroids), polyacetylenes, etc. They are synthesized and accumulate not only after infection but also upon exposure to different stresses. The cyanide-resistant respiratory pathway and the activation of PAL and other enzymes may be involved in the biosynthesis of some phytoalexins. It is possible that phytoalexin production accompanies the process of tissue necrosis and that it is a consequence of necrosis, or some other nonvisible adverse action rather than the cause of it. However, this question is unresolved.

Treating plant tissues with high MW cell wall components of pathogenic fungi and bacteria can induce accumulation of phytoalexins, phenolics, and the development of necrotic symptoms. These have been termed *elicitors*. It has been shown, in the case of *Phytophthora infestans*, that many races of the pathogen can release elicitors if the fungus is damaged or adversely influenced. This occurs, under natural conditions, in the resistant host. However, no elicitation occurs in compatible host-pathogen combinations ostensibly because the fungus is not damaged in the susceptible host. This hypothesis presupposes prior "contact" by the pathogen with a host recognition feature. Damage to the incompatible pathogen and release of the elicitor would be a secondary response phenomenon. In some instances,

necrosis and the accumulation of phytoalexins are inhibited by low MW *suppressors* released by compatible fungi. This interaction of phytoalexin elicitor and suppressor results in either the accumulation or suppression of phytoalexins. According to this hypothesis, specificity would lie in the binding of the suppressor to a hypothetical receptor on the host cell membrane, and the phytoalexin would play a primary role in inducing necrosis and hypersensitivity. Evidence has also been presented that some fungi are successful pathogens because they produce enzymes that specifically degrade phytoalexins.

It has been shown that it is possible to decrease tissue levels of secondary substances, including phytoalexins, by treatment with N-fertilizers and inhibitors of the biosynthesis of some key compounds, thereby increasing host susceptibility. It is not known whether these treatments affect primarily or only the accumulation or synthesis of secondary substances. A significant body of data clearly establishes a shift in the redox potential to one favoring a reducing milieu accompanied by a suppression of both HR and necrosis.

Additional Comments and Citations

1. J. Kuć and J. S. Rush 1985. Phytoalexins. Arch. Biochem. Biophys. 236 (in press). An incisive review of phytoalexins, their structural and biosynthetic features, their elicitation at very low concentrations (10^{-13}M), enzymes and the primary pathways involved in their synthesis. Included are comments on plant immunization, "definitions" of HR, suppressors and enhancers of elicitation, and the mechanism and site of phytoalexin synthesis. However the point is made that the crucial rationale for the efficacy of phytoalexin synthesis is the speed with which the gene products are produced, the specific activity of the gene products, and the magnitude of the resistance response.

2. Evidence has been presented dealing with the effects of stresses, including infections and treatments with fungal elicitors, on the synthesis of secondary metabolites (phytoalexins) in parsley and bean cells. UV treatment or infection with *Colletotrichum lindemuthianum* or treatment with fungal elicitors of phytoalexins cause marked coordinated increases in messenger RNA activities encoding enzymes of phytoalexin biosynthesis. Activation of enzymes, such as phenylalanine ammonia-lyase (PAL), 4-coumarate: CoA ligase (4CL), and chalcone synthase (CHS), is based on rapid changes in amounts and activities of the corresponding messenger RNAs. These changes are due to transient increases in the transcription rates of the genes that are involved in the biosynthesis of phytoalexin-type stress compounds. Some recent papers that report on these phenomena are:

J. Chappell and K. Hahlbrock. 1984. Transcription of plant defence genes in response to UV light or fungal elicitor. Nature 311:76–78.

J. N. Bell, R. A. Dixon, J. A. Bailey, P. M. Rowell, and C. J. Lamb. 1984. Differential induction of chalcone synthase mRNA activity at the onset of phytoalexin accumulation in compatible and incompatible plant-pathogen interactions. Proc. Natl. Acad. Sci. USA 81:3384–88.

T. B. Ryder, C. L. Cramer, J. N. Bell, M. P. Robbins, R. A. Dixon, and C. J. Lamb. 1984. Elicitor rapidly induces chalcone synthase mRNA in *Phaseolus vulgaris* cells at the onset of the phytoalexin defense response. Proc. Nat. Acad. Sci. USA 81:5724–28.

References

1. Addy, S. K., and R. N. Goodman. 1974. Phenols in relation to pathogenesis induced by avirulent and virulent strains of *Erwinia amylovora*. Acta Phytopath. Hung. 9:277–86.

2. Afzal, M., and G. Alorqua. 1982. Biosynthesis of isoflavonoid and related phytoalexins. Heterocycle 19:1295–1318.

3. Albersheim, P., and B. Valent. 1978. Host-pathogen interactions in plants. Plants when exposed to oligosaccharides of fungal origin, defend themselves by accumulating antibiotics. J. Cell Biol. 78:627–43.

4. Allen, E., and J. Kuć. 1964. Steroid alkaloids in the disease resistance of white potato tubers. Phytopathology 54:886.

5. Allen, E., and J. Kuć. 1968. α-Solanine and α-chaconine as fungitoxic compounds in extracts of Irish potato tubers. Phytopathology 58:776–81.

6. Alves, L. M., E. G. Heisler, J. C. Kissinger, J. M. Petterson, and E. B. Kalan. 1979. Effects of controlled atomospheres on production of sesquiterpenoid stress metabolites by white potato tuber. Plant Physiol. 63:359–62.

7. Anderson, A. J. 1980. Studies on the structure and elicitor activity of fungal glucans. Can. J. Bot. 58:2343–48.

8. Arnon, D. J. 1950. Functional aspects of copper in plants. In: Copper metabolism: A symposium. W. D. McElroy and B. Glass (eds.). Johns Hopkins Univ. Press, Baltimore, pp. 89–110.

9. Bailey, J. A., and B. J. Deverall. 1971. Formation and activity of phaseollin in the interaction between bean hypocotyls (*Phaseolus vulgaris*) and physiological races of *Colletotrichum lindemuthianum*. Physiol. Plant Path. 1:435–49.

10. Bailey, J. A., and J. L. Ingham. 1971. Phaseollin accumulation in bean (*Phaseolus vulgaris*) in response to infection by tobacco necrosis virus and the rust *Uromyces appendiculatus*. Physiol. Plant Path. 1:451–56.

11. Bailey, J. A., and J. W. Mansfield (eds.). 1982. Phytoalexins. Blackie, London.

12. Balázs, E., B. Barna, and Z. Király. 1977. Heat-induced local lesions with high peroxidase activity in a systemic host of TMV. Acta Phytopath. Hung. 12:151–56.

13. Barbara, D. J., and K. R. Wood. 1972. Virus multiplication, peroxidase and polyphenoloxidase activity in leaves of two cucumber (*Cucumis sativus* L.) cultivars inoculated with cucumber mosaic virus. Physiol. Plant Path. 2:167–73.

14. Bates, D. C., and S. R. Chant. 1970. Alterations in peroxidase activity and peroxidase isozymes in virus-infected plants. Ann. Appl. Biol. 65:105–10.

15. Batra, G. K., and C. W. Kuhn. 1975. Polyphenoloxidase and peroxidase activities associated with acquired resistance and its inhibition by 2-thiouracil in virus-infected soybean. Physiol. Plant Path. 5:239–48.

16. Bell, A. 1967. Formation of gossypol in infected or chemically irritated tissues of *Gossypium* species. Phytopathology 57:759–64.

17. Bell, A., and J. Presley. 1969. Heat-inhibited or heat-killed conidia of *Verticillium albo-atrum* induce disease resistance and phytoalexin synthesis in cotton. Phytopathology 59:1147–51.

18. Bell, E. A., and B. V. Charlwood (eds.). 1980. Secondary plant products. Springer-Verlag, New York.

19. Best, R. J. 1936. Studies on a fluorescent substance present in plants: 1. Production of the substances as a result of virus infection and some applications of the phenomenon. Aust. J. Exp. Biol. Med. Sci. 14:199–213.

20. Best, R. J. 1944. Studies on a fluorescent substance present in plants. 2. Isolation of the substance in a pure state and its identification as 6-methoxy-7-hydroxy 1:2 benzo-pyrone. Aust. J. Exp. Biol. Med. Sci. 22:251–55.

21. Best, R. J. 1948. Studies on a fluorescent substance present in plants. 3. The distribution of scopoletin in tobacco plants and some hypotheses on its part in metabolism Aust. J. Exp. Biol. Med. Sci. 26:225–29.

22. Birecka, H., J. L. Catalfamo, and P. Urban. 1975. Cell wall and protoplast isoperoxidases in tobacco plants in relation to mechanical injury and infection with tobacco mosaic virus. Plant Physiol. 55:611–19.

23. Bonner, J., and B. Arreguin. 1949. The biochemistry of rubber formation in the guayule. I. Rubber formation in seedlings. Arch. Biochem. 21:109–24.

24. Bonner, J., and J. E. Varner. 1965. Plant biochemistry. Academic Press, New York and London.

25. Bostock, R. M., R. A. Laine, and J. Kuć. 1982. Factors affecting the elicitation of sesquiterpenoid phytoalexin accumulation by eicosapentaenoic and arachidonic acids in potato. Plant Physiol. 70:1417–24.

26. Bostock, R. M., E. Nuckles, J. W. Henfling, and J. Kuć. 1983. Effects of potato tuber age and storage on sesquiterpenoid stress metabolic accumulation, steroid glycoalkaloid accumulation and response to abscisic and arachidonic acids. Phytopathology 73:435–38.

27. Bredenberg, J., and P. Hietala. 1961. Investigation of the structure of trifolirhizin, an antifungal compound from *Trifolium pratense* L. Acta Chem. Scand. 15:696–99.

28. Breugger, B. B., and N. T. Keen. 1979. Specific elicitors of glyceollin accumulation in the *Pseudomonas glycinea*-soybean host-parasite system. Physiol. Plant Path. 15:43–51.

29. Burden, R. S., and J. A. Bailey. 1975. Structure of phytoalexin from soybean. Phytochemistry 14:1389–90.

30. Burden, R. S., J. A. Bailey, and G. W. Dawson. 1972. Structures of three new isoflavonoids from *Phaseolus vulgaris* infected with tobacco mosaic virus. Tetrahedron Lett. 4175–78.

31. Byrde, R. J. W. 1957. The varietal resistance of fruits to brown rot. II. The nature of resistance in some varieties of cider apple. J. Hort. Sci. 32:227–38.

32. Byrde, R. J. W. 1963. Natural inhibitors of fungal enzymes and toxins in disease resistance. In: Perspectives of biochemical plant pathology. S. Rich (ed.). Conn. Agr. Exp. Sta. Bull. 663:31–47.

33. Byrde, R. J. W., A. H. Fielding, A. H. Williams. 1960. The role of oxidized polyphenols in the varietal resistance of apples to brown rot. In: Phenolics in plants in health and disease. J. B. Pridham (ed.). Pergamon Press, New York, pp. 95–99.

34. Cabanne, F., R. Scalla, and C. Martin. 1968. Activité de la polyphenoloxidase au cours de la réaction d'hypersensibilité chez *Nicotiana* Xanthi n.c. infectée par le virus de la mosaïque du tabac. Ann. Physiol. Veg. 10:199–208.

35. Cabanne, F., R. Scalla, and C. Martin. 1971. Oxidase activities during the hypersensitive reaction of *Nicotiana xanthi* to tobacco mosaic virus. J. Gen. Virol. 11:119–22.

36. Chalova, L., N. Vasyukova, O. Ozeretskovskaya, and L. Metlitskii. 1971. Chemical identification of one of the potato phytoalexins. Prikl. Biokhim. Microbiol. 7:55–58.

37. Chant, S. R., and D. C. Bates. 1970. The effect of tobacco mosaic virus and potato virus X on peroxidase activity and peroxidase isozymes in *Nicotiana glutinosa*. Phytochemistry 9:2323–26.

38. Clauss, E. 1961. Die phenolischen Inhaltsstoffe der Samenschalen von *Pisum sativum* L. und ihre Bedeutung für Resistenz gegen die Erreger der Fusskrankheit. Naturwissenschaften 48:106.

39. Cole, M. 1958. Oxidation products of leuco-anthocyanins as inhibitors of fungal polygalacturonase in rotting apple fruit. Nature 181:1596–97.

40. Collendavelloo, J., M. Legrand, and B. Fritig. 1982. Plant disease and regulation of enzymes involved in lignification. De novo synthesis controls 0-methyltransferase activity in hypersensitive tobacco leaves infected by tobacco mosaic virus. Physiol. Plant Path. 21:271–81.

41. Condon, P., and J. Kuć. 1960. Isolation of a fungitoxic compound from carrot root tissue inoculated with *Ceratocystis fimbriata*. Phytopathology 50:267–70.

42. Condon, P., J. Kuć, H. N. Draudt. 1963. Production of 3-methoxy-8-hydroxy-3, 4-dihydroisocoumarin by carrot tissue. Phytopathology 53:1244–50.

43. Conti, G. G., E. Bellini, G. Vegetti, and M. A. Favalli. 1974. Changes in phenylalanine ammonia-lyase and development of induced resistance in *Nicotiana glutinosa* leaves infected with the U5 strain of tobacco mosaic virus. Rivista di Patologia Vegetale 10:177–93.

44. Conti, G. G., G. Vegetti, M. A. Favali, and M. Bassi. 1974. Phenylalanine ammonia-lyase and polyphenoloxidase activities correlated with necrogenesis in cauliflower mosaic virus infection. Acta Phytopath. Hung. 9:185–93.

45. Cottle, W., and P. E. Kolattukudy. 1982. Biosynthesis, deposition and partial characterization of potato suberin phenolics. Plant Physiol. 69:794–97.

46. Cruickshank, I. A. M. 1963. Phytoalexins. Ann. Rev. Phytopath. 1:351–74.

47. Cruickshank, I. A. M. 1965. Phytoalexins in the *Leguminosae* with special reference to their selective toxicity. Proc. Conf. "Biochemische Probleme der kranken Pflanze." Aschersleben, Aug. 1964. Tagungsber. Deutsch. Akad. Landw. Berlin, pp. 331–32.

48. Cruickshank, I. A. M. 1980. Defenses triggered by the invader: Chemical defenses. In: Plant disease. An advanced treatise. J. G. Horsfall and E. B. Cowling (eds.). Academic Press, New York, 5:247–67.

49. Cruickshank, I. A. M., and Perrin, D. R. 1960. Isolation of a phytoalexin from *Pisum sativum* L. Nature 187:799–800.

50. Cruickshank, I. A. M., and Perrin, D. R. 1963. Phytoalexins of the leguminosae. Phaseollin from *Phaseolus vulgaris* L. Life Sci. 2:680–82.

51. Cruickshank, I. A. M., and Perrin, D. R. 1965. Studies on Phytoalexins. IX. Pisatin formation by cultivars of *Pisum sativum* L. and several other *Pisum* species. Aut. J. Biol. Sci. 18:829–35.

52. Cruickshank, I. A. M., and D. R. Perrin. 1968. The isolation and partial characterization of monilicolin A, a polypeptide with phaseollin-inducing activity from *Monilinia fructicola*. Life Sci. 7 (pt. 2): 449–58.

53. Daly, J. M. 1972. The use of near-isogenic lines in biochemical studies of the resistance of wheat to stem rust. Phytopathology 62:392–400.

54. Daly, J. M., P. Ludden, and P. Seevers. 1971. Biochemical comparisons of stem rust resistance controlled by the Sr6 and Sr11 alleles. Physiol. Plant Path. 1: 397–407.

55. Davis D., P. E. Waggoner, A. E. Dimond. 1953. Conjugated phenols in the *Fusarium* wilt syndrome. Nature 172:959.

56. Deese, D. C., and M. A. Stahmann. 1962–63. Formation of pectic enzymes by *Verticillium albo-atrum* on susceptible and resistant tomato stem tissues and on wheat bran. Phyopath. Z. 46:53–70.

57. Defago, G., and H. Kern. 1983. Induction of *Fusarium solani* mutants insensitive to tomatine, their pathogenicity and aggressiveness to tomato fruits. Physiol. Plant Path. 22:29–37.

58. Defago, G., H. Kern, and L. Sedlar. 1983. Genetic analysis of tomatine insensitivity, sterol content, and pathogenicity for green tomato fruits in mutants of *Fusarium solani*. Physiol. Plant Path. 22:39–43.

59. Deverall, B. J. 1961. Phenolase and pectic enzyme activity in the chocolate spot disease of bean. Nature 184:311.

60. Deverall, B. J. 1976. Current perspectives in research on phytoalexins. In: Biochemical aspects of plant-parasite relationships. J. Friend and D. R. Threlfall (eds.). Academic Press, London, New York, and San Francisco, pp. 207–23.

61. Doke, N. 1975. Prevention of the hypersensitive reaction of potato cells to infection with an incompatible race of *Phytophthora infestans* by constituents of the zoospores. Physiol. Plant Path. 7:1–7.

62. Doke, N., N. A. Garas, and J. Kuć. 1980. Effect on host hypersensitivity of suppressors released during the germination of *Phytophthora infestans* cystospores. Phytopathology 70:35-39.

63. Duchesne, M., B. Fritig, and L. Hirth. 1977. Phenylalanine ammonia-lyase in tobacco mosaic virus-infected hypersensitive tobacco. Density-labelling evidence of de novo synthesis. Biochim. Biophys. Acta 485:465–81.

64. Érsek, T., M. Holliday, and N. T. Keen. 1982. Association of hypersensitive host cell death and autofluorescence with a gene for resistance to *Peronospora manshurica* in soybean. Phytopathology 72:628–31.

65. Érsek, T., Z. Király, and A. Dobrovolszky. 1977. Lack of correlation between tissue necrosis and phytoalexin accumulation in tubers of potato cultivars. J. Food Saf. 1:77–85.

66. Essenberg, M., M. d'Arcy Doherty, B. Hamilton, V. Henning, E. Cover, S. McFaul, and W. Johnson. 1982. Identification and effects on *Xanthomonas campestris* pv. *malvacearum* of two phytoalexins from leaves and cotyledons of resistant cotton. Phytopathology 72:1349–56.

67. Faccioli, G. 1979. Relation of peroxidase, catalase and polyphenol-oxidase to acquired resistance in plants of *Chenopodium amaranticolor* locally infected by tobacco necrosis virus. Phytopath. Z. 95:237–49.

68. Farkas, G. L., and Z. Király. 1962. Role of phenolic compounds in the physiology of plant diseases and disease resistance. Phytopath. Z. 44:105–50.

69. Farkas, G. L., Z. Király, and F. Solymosy. 1960. Role of oxidative metabolism in the localization of plant viruses. Virology 12:408–21.

70. Farkas, G. L., and G. A. Ledingham. 1959. Studies on the polyphenol-polyphenol-oxidase system of wheat stem rust uredospores. Can. J. Microbiol. 5:37–46.

71. Farkas, G. L., and F. Solymosy. 1965. Host metabolism and symptom production in virus-infected plants. Phytopath. Z. 53:85–93.

72. Farkas, G. L., and M. A. Stahmann. 1966. On the nature of changes in peroxidase isoenzymes in bean leaves infected by southern bean mosaic virus. Phytopathology 56:669–77.

73. Farkas, G. L., and J. Szirmai. 1969. Increase in phenylalanine ammonia-lyase activity in bean leaves infected with tobacco necrosis virus. Neth. J. Plant Path. 75:82–85.

74. Fawcett, C. 1968. Natural acetylenes Part XXVII. An antifungal acetylenic furanoid keto-ester (wyerone) from shoots of the

broad bean (*Vicia faba* L.; Fam. Papillonaceae). J. Chem. Soc. (c) 2455–62.

75. Fawcett, C., D. Spencer, and R. Wain. 1969. The isolation and properties of a fungicidal compound present in the seedling of *Vicia faba*. Neth. J. Plant Path. 75:72–81.

76. Feenstra, W. J., B. L. Johnson, P. Ribereau-Gayon, and T. A. Geissman. 1963. The effect of virus infection on phenolic compounds in flowers of *Matthiola incana* R. BR. Phytochemistry 2:273–79.

77. Fehrmann, H., and A. E. Dimond. 1967. Peroxidase activity and *Phytophthora* resistance in different organs of the potato plant. Phytopathology 57:69–72.

78. Flood, A. E., and D. S. Kirkham. 1960. The effect of some phenolic compounds on the growth and sporulation of two *Venturia* species. In: Phenolics in plants in health and disease. J. B. Pridham (ed.). Pergamon Press, New York, pp. 81–85.

79. Forsyth, F. R., and D. J. Samborski. 1958. The effect of various methods of breaking resistance on stem rust reaction and content of soluble carbohydrate and nitrogen in wheat leaves. Can. J. Bot. 36:717–23.

80. Friend, J. 1981. Plant phenolics, lignification and plant disease. Prog. Phytochem. 7:197–261.

81. Fritig, B., J. Gosse, M. Legrand, and L. Hirth. 1973. Changes in phenylalanine ammonia-lyase during the hypersensitive reaction of tobacco to TMV. Virology 55:371–79.

82. Fritig, B., M. Legrand, and L. Hirth. 1972. Changes in the metabolism of phenolic compounds during the hypersensitive reaction of tobacco to TMV. Virology 47:845–48.

83. Fuchs, W. H., and E. Kotte. 1954. Zur Kenntnis der Resistenz von *Solanum tuberosum* gegen *Phytophthora infestans* de By. Naturwissenschaften 41:169–70.

84. Gáborjányi, R., and T. F. Fernandez. 1976. Induced alteration of peroxidase activities and the growth of peppers inoculated by tobacco etch virus. Acta Phytopath. Hung. ll:277–81.

85. Gáborjányi, R., F. Sagi, and E. Balázs. 1973. Growth inhibition of virus-infected plants: Alterations of peroxidase enzymes in compatible and incompatible host-parasite relations. Acta Phytopath. Hung. 8:81–90.

86. Garas, N. A., N. Doke, and J. Kuć. 1979. Suppression of the hypersensitive reaction in potato tubers by mycelial components from *Phytophthora infestans*. Physiol. Plant Path. 15:117–26.

87. Gäumann, E. 1956. Über Abwehrreaktionen bei Pflanzenkrankheiten. Experientia 12:411–18.

88. Gäumann, E., R. Braun, and B. Bazzigher. 1950. Über induzierte Abwehrreaktionen bei *Orchideen*. Phytopath. Z. 17:36–62.

89. Goodman, R. N. 1968. The hypersensitive reaction in tobacco: A reflection of changes in host cell permeability. Phytopathology 58:872–73.

90. Gortner, W. A., M. J. Kent, G. K. Sutherland. 1958. Ferulic and p-coumaric acids in pine-apple tissue as modifiers of pineapple indoleacetic acid oxidase. Nature 181:630–31.

91. Gretchushnikov, A. I., and K. V. Popkova. 1958. Reuced resistance to *Phytophthora infestans* of ringed leaves of *Phytophthora*-resistant potato varieties. Izv. Akad. Nauk SSR. Ser. Biol. 4:456–62. (In Russian.)

92. Gross, D. C. 1977. Phytoalexine und verwandte Pflanzenstoffe. In: Progress in the chemistry of organic natural products, no. 34. Springer-Verlag, Vienna, and New York, pp. 187–247.

93. Hadwiger, L. A. 1972. Increased levels of pisatin and phenylalanine ammonia lyase activity in *Pisum sativum* treated with antihistaminic, antiviral, antimalarial, tranquilizing, or other drugs. Biochem. Biophys. Res. Comm. 46:71–79.

94. Hadwiger, L. A., and D. C. Loschke. 1981. Molecular communication in host-parasite interactions: Hexosamine polymers (chitosan) as regulator compounds in race-specific and other interactions. Phytopathology 71:756–62.

95. Hadwiger, L. A., and M. Schwochau. 1971. Ultraviolet light-induced formation of pisatin and phenylalanine ammonia lyase. Plant Physiol. 47:588–90.

96. Hadwiger, L. A., J. M. Beckman, and M. J. Adams. 1981. Localization of fungal components in the pea-*Fusarium* interaction detected immunochemically with antichitosan and antifungal cell wall antisera. Plant Physiol. 67:170–75.

97. Hadwiger, L. A., S. L. Hess, and S. von Broembsen. 1970. Stimulation of phenylalanine ammonia-lyase activity and phytoalexin production. Phytopathology 60:332–36.

98. Hadwiger, L. A., A. Jafri, S. von Broembsen, and R. Eddy Jr. 1974. Mode of pisatin induction. Increased template activity and dye-binding capacity of chromatin isolated from polypeptide-treated pea pods. Plant Physiol. 53:52–63.

99. Hampton, R. E., R. Suseno, and D. M. Brumagen. 1964. Accumulation of fluorescent compounds in tobacco associated with tobacco mosaic virus hypersensitivity. Phytopathology 54:1062–64.

100. Harborne, J. B., T. J. Mabry, and H. Mabry. 1975. The flavonoids. Chapman and Hall, London.

101. Haslam, E. 1974. The shikimate pathway. Wiley, New York and Toronto.

102. Henniger, H., and W. Bartel. 1963. Die Eignung des Peroxydaseaktivität-Testes zur Bestimmung der "relativen *Phytophthora* Resistenz" (Feldresistenz) bei Kartoffeln. Züchter 33:86–91.

103. Hess, E. G. 1958. The polyphenolase of tobacco and its participation in amino acid metabolism. I. Manometric studies. Arch. Biochem. Biophys. 74:198–208.

104. Higgins, V. J., and R. Millar. 1968. Phytoalexin production by alfalfa in response to infection by *Colletotrichum phomoides, Helminthosporium turcicum, Stemphylium loti* and *S. botryosum*. Phytopathology 58:1377–1383.

105. Higgins, V. J., and R. Millar. 1970. Degradation of alfalfa phytoalexin by *Stemphylium loti* and *Colletotrichum phomoides*. Phytopathology 60:269–71.

106. Higgins, V. J., and D. Smith. 1972. Separation and identification of two pterocarpanoid phytoalexins produced by red clover leaves. Phytopathology 62:235–38.

107. Hildebrand, D. C., and M. N. Schroth. 1964. Arbutin-hydroquinone complex in pear as a factor in fireblight development. Phytopathology 54:640–45.

108. Holliday, M. J., and N. T. Keen. 1982. The role of phytoalexins in the resistance of soybean leaves to bacteria: Effects of glyphosate on glyceollin accumulation. Phytopathology 72:1470–74.

109. Holowczak, J., J. Kuć, and E. B. Williams. 1962. Metabolism of DL- and L- phenylalanine in *Malus* related to susceptibility and resistance to *Venturia inaequalis*. Phytopathology 52:699–703.

110. Hooley, R., and D. McCarthy. 1980. Extracts from virus infected hypersensitive tobacco leaves are detrimental to protoplast survival. Physiol. Plant Path. 16: 25–38.

111. Hughes, J., and T. Swain. 1960. Scopolin production in potato tubers infected with *Phytophthora* infestans. Phytopathology 50:398–400.

112. Hulme, A. C., and K. L. Edney. 1960. Phenolic substances in the peel of Cox's orange pippin apples with reference to infection by *G. perennans*. In: Phenolics in plants in health and disease. J. B. Pridham (ed.). Pergamon Press, New York, pp. 87–94.

113. Hutchinson, H., C. D. Taper, and G. H. N. Towers. 1959. Studies on pfloridzin in *Malus*. Can. J. Biochem. Physiol. 37:901–10.

114. Jarvis, W. R. 1961. Growth of isolates of *Phytophthora fragariae* Hickman in the presence of various polyphenols. Trans. Brit. Mycol. Soc. 44:357–64.

115. Jockusch, R. N. 1966. The role of host genes, temperature and polyphenoloxidase in the necrotization of TMV-infected tobacco tissue. Phytopath. Z. 55:185–92.

116. Johnson, G., and L. A. Schaal. 1957. Accumulation of phenolic substances and ascorbic acid in potato tuber tissue upon injury and their possible role in disease resistance. Amer. Potato J. 34:200–9.

117. Kaul, R., and M. Shaw. 1960. The physiology of host-parasite relations. VI. Oxidation-reduction changes in wheat leaf sap caused by rust infection. Can. J. Bot. 38:399–407.

118. Kedar (Kammermann), N. 1959. The peroxidase test as a tool in the selection of potato varieties resistant to late blight. Amer. Potato J. 36:215–24.

119. Keen, N. T. 1975. Specific elicitors of plant phytoalexin production: Determination of race specificity in pathogens. Science 187:74–75.

120. Keen, N. T. 1975. The isolation of phytoalexins from germinating seeds of *Cicer arietinum*, *Vigna sinensis*, *Arachis hypogaea* and other plants. Phytopathology 65:91–92.

121. Keen, N. T. 1978. Phytoalexins: Efficient extraction from leaves by a facilitated diffusion technique. Phytopathology 68:1237–39.

122. Keen, N. T. 1981. Evaluation of the role of phytoalexins. In: Plant Disease Control, Resistance and Susceptibility. R. C. Staples and G. Toenniessen (eds.). Wiley-Interscience, New York, pp. 155–77.

123. Keen, N. T., T. Érsek, M. Long, B. Bruegger, and M. Holliday. 1981. Inhibition of the hypersensitive reaction of soybean leaves to incompatible *Pseudomonas* sp. by blasticidin S, streptomycin or elevated temperature. Physiol. Plant Path. 18:325–37.

124. Keen, N. T., and B. W. Kennedy. 1974. Hydroxyphaseollin and related isoflavanoids in the hypersensitive reaction of soybeans to *Pseudomonas glycinea*. Physiol. Plant Path. 4:173–85.

125. Keen, N. T., and M. Legrand. 1980. Surface glycoproteins: Evidence that they may function as the race-specific phytoalexin elicitors of *Phytophthora megasperma* f. sp. *glycinea*. Physiol. Plant Path. 17:175–92.

126. Keen, N. T., J. Sims, D. Erwin, E. Rice, and J. Partridge. 1971. 6a-Hydroxyphaseollin: An antifungal chemical induced in soybean hypocotyls by *Phytophthora megasperma* var. *sojae*. Phytopathology 61:1084–89.

127. Kenten, R. H. 1957. Latent phenolase in extracts of broadbean (*Vicia faba* L.) leaves. I. Activation by acid and alkali. Biochem. J. 300–307.

128. Kenten, R. H. 1958. Latent phenolase in extracts of broad bean (*Vicia faba* L.) leaves. II. Activation by anionic wetting agents. Biochem. J. 68:244–51.

129. Kikuchi, M., and A. Yamaguchi. 1960. Polyphenol oxidase activity of *Nicotiana glutiosa* leaves infected with tobacco mosaic virus. Nature 187:1048–49.

130. Király, Z. 1964. Effect of nitrogen fertilization on phenol metabolism and stem rust susceptibility of wheat. Phytopath. Z. 51:252–61.

131. Király, Z. 1976. Plant disease resistance as influenced by biochemical effects of nutrients in fertilizers. In: Fertilizer use and plant health. Proc. 12th Coll. Int. Potash Inst. Izmir., pp. 33–46.

132. Király, Z. 1980. Defenses triggered by the invader: Hypersensitivity. In: Plant disease. An advanced treatise. J. G. Horsfall and E. B. Cowling (eds.). Academic Press, New York, 5:201–24.

133. Király, Z., and G. L. Farkas. 1962. Relation between phenol metabolism and stem rust resistance in wheat. Phytopathology 52:657–64.

134. Kirkham, D. S. 1954. Significance of the ratio between the water soluble aromatic and nitrogen constituents of apple and pear in the host-parasite relationships of *Venturia* species. Nature 173:690–91.

135. Klement, Z. 1982. Hypersensitivity. In: Phytopathogenic prokaryotes. M. S. Mount and G. H. Lacy (eds.). Academic Press, New York, pp. 149–77.

136. König, D., and F. Nienhaus. 1970. Stoffwechselphysiologische Veränderungen in der Pflanze nach Virusinfektion unter Einfluss von Wundreiz. 1. Polyphenoloxidase-, Peroxidase- und Cytochromoxidase-aktivität. Phytopath. Z. 68:193–205.

137. Kopp, M., P. Geoffroy, and B. Fritig. 1983. Plant disease and the regulation of enzymes involved in lignification. Effects of osmotic pressure on phenylpropanoid enzymes and on the hypersensitive response of tobacco to tobacco mosaic virus. Planta 157:180–89.

138. Kubowitz, F. 1938. Spaltung und Resynthese der Polyphenoloxydase und des Hämocyanins. Biochem. Z. 299:32–57.

139. Kuć, J. 1964. Phenolic compounds and disease resistance in plants. In: Phenolics in normal and diseased fruits and vegetables. V. C. Runeckles (ed.). Imperial Tobacco Co., Montreal, pp. 63–81.

140. Kuć, J. 1972. Phytoalexins. Ann. Rev. Phytopath. 10:207–32.

141. Kuć, J. 1978. Changes in intermediary metabolism. In: Plant disease: An advanced treatise. J. G. Horsfall and E. B. Cowling (eds.). Academic Press, New York, 3:349–74.

142. Kuć, J. 1981. Multiple mechanisms, reaction rates and induced resistance in plants. In: Plant disease control. R. C. Staples and G. H. Toenniessen (eds.). Wiley, New York, pp. 259–72.

143. Lazarovits, G., P. Stossel, and E. W. B. Ward. 1981. Age-related changes in specificity and glyceollin production in the hypocotyl reaction of soybeans to *Phytophthora megasperma* var. *sojae*. Phytopathology 71:94–97.

144. Lee, S., and D. J. Le Tourneau. 1958. Chlorogenic acid content and *Verticillium* wilt resistance of potatoes. Phytopathology 48:268–74.

145. Legrand, M., B. Fritig, and L. Hirth. 1976. Enzymes of the phenyl-propanoid pathway and the necrotic reaction of hypersensitive tobacco to tobacco mosaic virus. Phytochemistry 15:1353–59.

146. Legrand, M., B. Fritig, and L. Hirth. 1978. O-Diphenol-O-methyl-transferases of healthy and tobacco-mosaic-virus-infected hypersensitive tobacco. Planta 144:101–8.

147. Letcher, R., B. Widdowson, B. Deverall, and J. Mansfield. 1970. Identification and activity of wyerone acid as a phytoalexin in broad bean (*Vicia faba*) after infection by *Botrytis*. Phytochemistry 9:249–52.

148. Le Tourneau, D. J., J. G. McLean, and J. W. Guthrie. 1957. Effect of some phenols and quinones on growth in vitro of *Verticillium albo-atrum*. Phytopathology 47: 602–6.

149. Lim, S. M., A. L. Hooker, and J. D. Paxton. 1970. Isolation of phytoalexins from corn with monogenic resistance to *Helminthosporium turcicum*. Phytopathology 60:1071–75.

150. Loebenstein, G., and N. Linsey. 1966. Alteration of peroxidase activity, associated with disease symptoms, in virus-infected plants. Israel J. Bot. 15:163–67.

151. Lyon, F. M., and R. K. S. Wood. 1975. Production of

phaseollin, coumestrol and related compounds in bean leaves inoculated with *Pseudomonas* spp. Physiol. Plant Path. 6:117–24.

152. Mace, M. E., and T. T. Hebert. 1963. Naturally occurring quinones in wheat and barley and their toxicity to loose smut fungi. Phytopathology 53:692–700.

153. Mansfield, J. W., J. A. Hargreaves, and F. C. Boyle. 1974. Phytoalexin production by live cells in broad bean leaves infected with *Botrytis cinerea*. Nature 252:316–17.

154. Martin, J. T., R. F. Batt, and R. T. Burchill. 1957. Defense mechanism of plants against fungi: Fungistatic properties of apple leaf wax. Nature 180:796–97.

155. Martin-Tanguy, J., C. Martin, and M. Gallet. 1973. Présence de composés aromatiques liés à la putrescine dans divers *Nicotiana* virosés. Compte. Rend. Acad. Sci. 276D:1433–35.

156. Massala, R., M. Legrand, and B. Fritig. 1980. Effect of α-aminooxyacetate, a competitive inhibitor of phenylalanine ammonia-lyase, on the hypersensitive resistance of tobacco to tobacco mosaic virus. Physiol. Plant Path. 16:213–26.

157. Mason, H. S. 1955. Comparative biochemistry of the phenolase complex. Adv. Enzymol. 16:105–84.

158. Mayama, S., S. Hayashi, R. Yamamoto, and T. Tani. 1982. Effects of elevated temperature and α-aminooxyacetate on the accumulation of avenalumins in oat leaves infected with *Puccinia coronata* f. sp. *avenae*. Physiol. Plant Path. 20: 305–12.

159. Mayama, S., T. Tani, Y. Matsuura, T. Ueno and H. Fukami. 1981. The production of phytoalexins by oat in response to crown rust, *Puccinia coronata* f. sp. *avenae*. Physiol. Plant Path. 19:217–26.

160. McLean, J. G., D. J. Le Tourneau, and J. W. Guthrie. 1961. Relation of hitochemical tests for phenols to *Verticillum* wilt resistance of potatoes. Phytopathology 51:84–89.

161. Menke, G. H., and J. C. Walker. 1963. Metabolism of resistant and susceptible cucumber varieties infected with cucumber mosaic virus. Phytopathology 53: 1349–55.

162. Müller, K. O. 1956. Einige einfache Versuche zum Nachweis von Phytoalexinen. Phytopath. Z. 27:237–54.

163. Müller, K. O., and L. Behr. 1949. Mechanism of *Phytophthora* resistance of potatoes. Nature 163:498–99.

164. Müller, K. O., and H. Börger. 1940. Experimentelle Untersuchungen über die *Phytophthora*-Resistenz der Kartoffel. Arb. Biol. Reichsanst. Land-Forstw. 23:189–31.

165. Nakamura, M., and H. Oku. 1960. Biochemical studies on *Cochliobolus miyabeanus*. IX. Detection of ophiobolin in the diseased rice leaves and its toxicity against higher plants. Ann. Rept. Takamine Lab. 12:266–71.

166. Neish, A. C. 1960. Biosynthetic pathways of aromatic compounds. Ann. Rev. Plant Physiol. 11:55–80.

167. Németh, J., Z. Klement, and G. L. Farkas. 1969. An enzymological study of the hypersensitive reaction induced by *Pseudomonas syringae* in tobacco leaf tissues. Phytopath. Z. 65:267–78.

168. Neufeld, H. A., F. M. Latterell, L. F. Green, and R. L. Weintraub. 1958. Oxidation of *meta*-polyhydroxyphenols by enzymes from *Pyricularia oryzae* and *Polyporus versicolor*. Arch. Biochem. Biophys. 76:317–27.

169. Nichols, J., J. M. Beckman, and M. A. Hadwiger. 1980. Glycosidic enzyme activity in peas tissue and pea-*Fusarium solani* interactions. Plant Physiol. 66: 199–204.

170. Novacky, A., and R. E. Hampton. 1968. Peroxidase isozymes in virus-infected plants. Phytopathology 58:301–5.

171. Oku, H. 1958. Biochemical studies on *Cochliobolus miyabeanus*. III. Some oxidizing enzymes of the rice plant and its parasites and their contribution to the formation of the lesions. Ann. Phytopath. Soc. Jap. 23:169–75.

172. Oku, H. 1959. Biochemical studies on *Cochliobolus miyabeanus*: V. β- Glucosidase activity of the fungus and the effect of phenolglucoside on the mycelial growth. Ann. Rept. Takamine Lab. 11:190–92.

173. Oku, H. 1960. Biochemical studies on *Cochliobolus miyabeanus*. VIII. Properties of the polyphenoloxidase produced by the fungus. Ann. Rept. Takamine Lab. 12:261–65.

174. Oku, H. 1960. Biochemical studies on *Cochliobolus miyabeanus*. VI. Breakdown of disease resistance of rice plant by treatment with reducing agents. Ann. Phytopath. Soc. Jap. 25:92–98.

175. Oku, H., S. Ouchi, T. Shiraishi, Y. Komoto, and K. Oki. 1975. Phytoalexin activity in barley powdery mildew. Ann. Phytopath. Soc. Jap. 41:185–91.

176. Parish, C. L., M. Zaitlin, and A. Siegel. 1965. A study of necrotic lesion formation by tobacco mosaic virus. Virology 26:413–18.

177. Patil, S. S., R. L. Powelson, and R. A. Young. 1962. The relation of chlorogenic acid and total phenols in potato to infection by *Verticillium albo-atrum*. Phytopathology 52:364.

178. Patil, S. S., M. Zucker, and A. E. Dimond. 1966. Biosynthesis of chlorogenic acid in potato roots resistant and susceptible to *Verticillium albo-atrum*. Phytopathology 56:971–74.

179. Paynot, M., C. Martin, and M. Giraud. 1973. Activité phenylalanine ammonia lyase du *Nicotiana tabacum* var. Xanthi n.c. et infection systémique par le virus de la mosaique du Tabac. Compte. Rend. Acad. Sci. 277D: 1713–15.

180. Paynot, M., and C. Martin. 1974. Effet d'un transfert de 20 à 32°C sur l'activité phenylalaline-ammoniac lyase de *Nicotiana tabacum* var. Xanthi n.c. sains et inoculés par le virus de la mosaïque du tabac. Compte. Rend. Acad. Sci. 278D:533–36.

181. Paynot, M., C. Martin, and F. Javelle. 1975. Activité phenylalanine ammoniac lyase de divers *Nicotiana tabacum* et réaction nécrotique d'hypersensibilité au virus de la mosaïque du Tabac. Compte. Rend. Acad. Sci. 280D:1841–44.

182. Pegg, G. F., and L. Sequeira. 1968. Stimulation of aromatic biosynthesis in tobacco plants infected by *Pseudomonas solanacearum*. Phytopathology 58:476–83.

183. Perrin, D. R. 1964. The structure of phaseollin. Tetrahedron Lett. 1:29–35.

184. Perrin, D. R., and I. A. M. Cruickshank. 1965. Studies on phytoalexins. VII. Chemical stimulation of pisatin formation in *Pisum sativum* L. Aust. J. Biol. Sci. 18:803–16.

185. Pierpoint W. S. 1970. Formation and behavior of O-quinones in some processes of agricultural importance. Rept. Rothamsted Exp. Sta., 199.

186. Pridham, J. B., and T. S. Swain. 1965. Biosynthetic pathways in higher plants. Academic Press, New York.

187. Rahe, J., J. Kuć, C. Chuang, and E. Williams. 1969. Correlation of phenolic metabolism with histological changes in *Phaseolus vulgaris* inoculated with fungi. Neth. J. Plant Path. 75:58–71.

188. Rathmell, W. G. 1973. Phenolic compounds and phenylalanine ammonia lyase activity in relation to phytoalexin biosynthesis in infected hypocotyls of *Phaseolus vulgaris*. Physiol. Plant Path. 3:259–67.

189. Ray, P. M. 1958. Destruction of auxin. Ann. Rev. Plant Physiol. 9:81–118.

190. Redolfi, P., A. Cantisani, and S. Pennazio. 1978. Unusual phenolic compounds in the hypersensitive reaction of *Gomphrena globosa* to tomato bushy stunt virus. Phytopath. Z. 93:325–35.

191. Ride, J. P. 1983. Cell walls and other structural barriers in defence. In: Biochemical plant pathology. J. A. Callow (ed.). Wiley, London, Chichester, pp. 215–36.

192. Robb, D. A., L. W. Mapson, and T. Swain. 1964. Activation of the latent tyrosinase of broad bean. Nature 201:503–4.

193. Rohringer, R., and D. J. Samborski. 1967. Aromatic compounds in the host-parasite interaction. Ann. Rev. Phytopath. 5:77–86.

194. Rubin, B. A., L. V. Metlitsky, E. G. Salkova, E. N. Muhin, H. P. Korableva, and N. P. Morozova. 1959. Application of ionizing radiation for the regulation of dormancy of potato tubers during storage. Biokhim. Plodov i Ovoshch. 5:5–101. (In Russian.)

195. Sakai, R., and K. Tomiyama. 1964. Relation between physiological factors related to phenols and varietal resistance of potato plant to late blight. I. Field observations. Ann. Phytopath. Soc. Jap. 29:33–38.

196. Sarhan, A. R. T., B. Barna, and Z. Király. 1982. Effect of nitrogen nutrition on *Fusarium* wilt of tomato plants. Ann. Appl. Biol. 101:245–50.

197. Schlösser, E. W. 1980. Preformed internal chemical defenses. In: Plant disease: An advanced treatise. J. G. Horsfall and E. B. Cowling (eds.). Academic Press, New York, 5:161–77.

198. Schönbeck, F., and C. Schroeder. 1972. Role of antimicrobial substances (tuliposides) in tulips attacked by *Botrytis* spp. Physiol. Plant Path. 2:91–99.

199. Schwochau, M., and L. Hadwiger. 1970. Induced host resistance—a hypothesis derived from studies of phytoalexin production. Rec. Adv. Phytochem. 3: 181–89.

200. Seevers, P. M., and J. M. Daly. 1970. Studies on wheat stem rust resistance controlled at the Sr6 locus. I. The role of phenolic compounds. Phytopathology 60:1322–28.

201. Seevers, P. M., and J. M. Daly. 1970. Studies on wheat stem rust resistance controlled at the Sr6 locus. II. Peroxidase activities. Phytopathology 60:1642–47.

202. Seevers, P. M., J. M. Daly, and F. F. Catedral. 1971. The role of peroxidase isozymes in resistance to wheat stem rust disease. Plant Physiol. 48:353–60.

203. Sequeira, L. 1964. Inhibition of indoleacetic acid oxidase in tobacco plants infected by *Pseudomonas solanacearum*. Phytopathology 54:1078–83.

204. Sequeira, L. 1965. Origin of indoleacetic acid in tobacco plants infected by *Pseudomonas solanacearum*. Phytopathology 55:1232–36.

205. Sequeira, L. 1969. Synthesis of scopolin and scopoletin in tobacco plants infected by *Pseudomonas solanacearum*. Phytopathology 59:473–78.

206. Sequeira, L. 1976. Induction and suppression of the hypersensitive reaction induced by phytopathogenic bacteria: Specific and nonspecific components. In: Specificity in plant diseases. R. K. S. Wood and A. Graniti (eds.). Plenum Press, New York, pp. 289–306.

207. Simons, T. J., and A. F. Ross. 1970. Enhanced peroxidase activity associated with induction of resistance to tobacco mosaic virus in hypersensitive tobacco. Phytopathology 60:383–84.

208. Simons, T. J., and A. F. Ross. 1971. Metabolic changes associated with induced resistance to tobacco mosaic virus in Samsun NN tobacco. Phytopathology 61:293–300.

209. Simons, T. J., and A. F. Ross. 1971. Changes in phenol metabolism associated with induced systemic resistance to tobacco mosaic virus in Samsun NN tobacco. Phytopathology 61:1261–65.

210. Smith, D. A., H. D. Van Etten, and D. F. Bateman. 1973. Kievitone: The principal antifungal component of "substance II" isolated from *Rhizoctonia*-infected bean tissues. Physiol. Plant Path. 3:179–86.

211. Solymosy, F. 1970. Biochemical aspcts of hypersensitivity to virus infection in plants. Acta Phytopath. Hung. 5:55–63.

212. Solymosy, F., G. L. Farkas, and Z. Király. 1969. Biochemical mechanism of lesion formation in virus-infected plant tissues. Nature 184:706–7.

213. Solymosy, F., J. Szirmai, L. Beczner, and G. L. Farkas. 1967. Changes in peroxidase-isozyme patterns induced by virus infection. Virology 32:117–21.

214. Stahmann, M. A., and D. M. Demorest. 1973. Changes in enzymes of host and pathogen with special reference to peroxidase interaction. In: Fungal pathogenicity and the plant's response. R. J. W. Byrde and C. V. Cutting (eds.). Academic Press, London and New York, pp. 405–22.

215. Stahmann, M. A., I. Uritani, and K. Tomiyama. 1960. Changes in potato proteins induced by fungus infections. Phytopathology 50:655. (Abstr.)

216. Stekoll, M., and C. A. West. 1978. Purification and properties of an elicitor of castor bean phytoalexin from culture filtrates of the fungus *Rhizopus stolonifer*. Plant Physiol. 61:38–45.

217. Stholasuta, P., J. Bailey, V. Severin, and B. Deverall. 1971. Effect of bacterial inoculation of bean and pea leaves on the accumulation of phaseollin and pisatin. Physiol. Plant Path. 1:177–83.

218. Stoessl, A., C. H. Unwin, and E. W. B. Ward. 1973. Postinfectional fungus inhibitors from plants: Fungal oxidation of capsidiol in pepper fruit. Phytopathology 63:1225–31.

219. Suseno, H., and R. E. Hampton. 1966. The effect of three strains of tobacco mosaic virus on peroxidase and polyphenoloxidase activity in *Nicotiana tabacum*. Phytochemistry 5:819–22.

220. Sutton, D. C., and B. J. Deverall. 1982. Liberation of antifungal activity during the extration of soybean leaves. Physiol. Plant Path. 20:365–67.

221. Szweykowska, A. 1959. The effect of nitrogen feeding on anthocyanin synthesis in isolated red cabbage embryos. Acta Soc. Bot. Polon. 28:539–49.

222. Tanguy, J., and C. Martin. 1972. Phenolic compounds and the hypersensitivity reaction in *Nicotiana tabacum* infected with tobacco mosaic virus. Phytochemistry 11: 19–28.

223. Tegtemeier, K. J., and H. D. Van Etten. 1982. The role of pisatin tolerance and degradation in the virulence of *Nectria haematococca* on peas: A genetic analysis. Phytopathology 72:608–12.

224. Tjamos, E. C., and I. M. Smith. 1974. The role of phytoalexins in the resistance of tomato to *Verticillium* wilt. Physiol. Plant Path. 4:249–59.

225. Tomaru, K., T. Shiroya, and Y. Takanami. 1969. Changes in polyphenol-oxidase activity in the cucumber mosaic virus infected tobacco leaves and its non-responsibility for the changes in virus concentration in vivo. Ann. Phytopath. Soc. Jap. 35:41–46.

226. Tomiyama, K. 1957. Cell-physiological studies on the resistance of potato plants to *Phytophthora infestans*. V. Effect of 2, 4-dinitrophenol upon the hyper-sensitive reaction of potato plant cell to infection by *P. infestans*. Ann. Phytopath. Soc. Jap. 22:75–78.

227. Tomiyama, K., N. Ishizaka, N. Sato, T. Masamune, and N. Katsui. 1968. Rishitin, a phytoalexin-like substance, its role in the defense reaction of potato tubers to infection. In: Biochemical regulation in diseased plants or injury. T. Hirai (ed.). Phytopath. Soc. Jap., Tokyo, pp. 287–92.

228. Tomiyama, K., R. Sakai, N. Takase, and M. Takakuwa. 1957. Physiological studies on the defence reaction of potato plant to the infection by *Phytophthora infestans*. IV. The influence of preinfectional ethanol narcosis upon the physiological reaction of potato plant to the infection by *P. infestans*. Ann. Phytopath. Soc. Jap. 21:153–58.

229. Tomiyama, K., T. Sakuma, N. Ishizaka, N. Sato, N. Katsui, M. Takasugi, and T. Masamune. 1968. A new antifungal substance isolated from resistant potato tuber tissue infected by pathogens. Phytopathology 58:115–16.

230. Tomiyama, K., N. Takase, R. Sakai, and M. Takakuwa. 1955. Physiological studies on the defence reaction of potato plant to infection by *Phytophthora infestans*. II. Changes in the physiology of potato tuber induced by the infection of different strains of *P. infestans*. Ann. Phytopath. Soc. Jap. 29:59–64.

231. Tomiyama, K., N. Takase, R. Sakai, and M. Takakuwa. 1956. Physiological studies on the defense reaction of potato plant to the infection by *Phytophthora infestans*. I. Changes in the physiology of potato tuber induced by the infection of *P. infestans* and their varietal differences. Res. Bull. Hokkaido Agr. Exp. Sta. 71:32–50.

232. Trivedi, N., and A. K. Shinha. 1978. Production of a fungitoxic substance in rice in response to *Drechslera* infection. Trans. Brit. Mycol. Soc. 70:57–60.

233. Umaerus, V. 1959. The relationship between peroxidase activity in potato leaves and resistance to *Phytophthora infestans*. Amer. Potato J. 36:124–31.

234. Uritani, I. 1962. Quantitative relationship between respiration and ipomeamarone in sweet potato infected by *Ceratostomella fimbriata*. Ann. Phytopath. Soc. Jap. 27:73. (In Japanese.)

235. Uritani, I. 1963. The biochemical basis of disease resistance induced by infection. In: Perspectives of biochemical plant pathology. S. Rich (ed.). Conn. Agr. Exp. Sta. Bull. 663:4–19.

236. Uritani, I. 1965. Molecular pathology in the plant field with special regard to defense action of the host. Tagungsber. DAL. zu Berlin 74:201–18.

237. Uritani, I., M. Uritani, and H. Yamada. 1960. Similar metabolic alterations induced in sweet potato by poisonous chemicals and by *Ceratostomella fimbriata*. Phytopathology 50:30–34.

238. Vance, C. P., T. K. Kirk, and R. T. Sherwood. 1980. Lignification as a mechanism of disease resistance. Ann. Rev. Phytopath. 18:259–88.

239. Van Den Heuvel, J. 1976. Sensitivity to, and metabolism of, phaseollin in relation to the pathogenicity of different isolates of *Botrytis cinerea* to bean (*Phaseolus vulgaris*). Neth. J. Plant Path. 82:153–60.

240. Van Fleet, D. S. 1972. Histochemistry of plants in health and disease. In: Structural and functional aspects of phytochemistry. V. C. Runeckles and T. C. Tso (eds.). Academic Press, New York and London, pp. 165–95.

241. van Kammen, A., and D. Brouwer. 1964. Increase of polyphenol oxidase activity by local virus infection in uninoculated parts of leaves. Virology 22:9–14.

242. van Loon, L. C. 1976. Systemic acquired resistance, peroxidase activity and lesion size in tobacco reacting hypersensitively to tobacco mosaic virus. Physiol. Plant Path. 8:231–42.

243. van Loon, L. C. and J. L. M. C. Geelen. 1971. The relation of polyphenol-oxidase and peroxidase to symptom expression in tobacco var. 'Samsun NN' after infection with tobacco mosaic virus. Acta Phytopath. Hung. 6:9–20.

244. Vegetti, G., G. G. Conti, and P. Pesci. 1975. Changes in phenylalanine ammonia- lyase, peroxidase and polyphenoloxidase during the development of local necrotic lesions in pinto bean leaves infected with alfalfa mosaic virus. Phytopath. Z. 84:153–71.

245. Vickery, M. L., and B. Vickery. 1981. Secondary plant metabolism. Univ. Park Press, Baltimore.

246. Virtanen, A., and P. Hietala. 1957. Additional information on the antifungal factor in red clover. Acta Chem. Fenn. 30B:99.

247. Virtanen, A., and P. Hietala. 1958. Isolation of an anti-*Sclerotinia* factor, 7- hydroxy-4'methoxy-isoflavone from red clover. Acta Chem. Scand. 12:597.

248. Waigh, E. E., and R. H. A. Coutts. 1982. Peroxidase, polyphenol oxidase and ribonuclease in tobacco necrosis virus infected or mannitol osmotically-stressed cowpea and cucumber tissue. I. Quantitative alterations. Phytopath. Z. 104:1–12.

249. Wain, R., D. Spencer, C. Fawcett. 1961. Atifungal compounds in seedling of *Vicia faba*. Fungicides in agriculture and horticulture. Soc. Chem. Ind. Monograph 15:109–31.

250. Wakimoto, S., and H. Yoshii. 1958. Relation between polyphenols contained in plants and phytopathogenic fungi. I. Polyphenols contained in rice plants. Ann. Phytopath. Soc. Jap. 23:79–84.

251. Walker, J. C., and K. P. Link. 1935. Toxicity of phenolic compounds to certain onion bulb parasites. Bot. Gaz. 96:468–84.

252. Walker, J. R. L. 1975. The biology of plant phenolics: Studies in Biology. Edward Arnold, London.

253. Ward, E. W. B., and A. Stoessl. 1977. Phytoalexins from potatoes: Evidence for the conversion of lubimin to 1,5-dihydrolubimin by fungi. Phytopathology 67: 468–71.

254. Williams, A. H. 1963. Enzyme inhibition by phenolic compounds. In: Enzyme chemistry of phenolic compounds. J. B. Pridham (ed.). Pergamon Press, New York, pp. 87–93.

255. Wood, K. R. 1971. Peroxidase isoenzymes in leaves of cucumber (*Cucumis sativus* L.) cultivars systemically infected with the W strain of cucumber mosaic virus. Physiol. Plant Path. 1:133–40.

256. Wood, K. R., and D. J. Barbara. 1971. Virus multiplication and peroxidase activity in leaves of cucumber (*Cucumis sativus* L.) cultivars systemically infected with the W strain of cucumber mosaic virus. Physiol. Plant Path. 1:73–82.

257. Wosilait, W. D., and A. Nason. 1954. Pyridine nucleotide quinone reductase. I. Purification and properties of the enzyme from pea seeds. J. Biol. Chem. 206: 255–70.

258. Yoshikawa, M., K. Yamauchi, and H. Masago. 1978. Glyceollin: Its role in restricting fungal growth in resistant soybean hypocotyls infected with *Phytophthora megasperma* var. *sojae*. Physiol. Plant Path. 12:73–82.

259. Ziegler, E., and R. Pontzen. 1982. Specific inhibition of glucan-elicited glyceollin accumulation in soybeans by an extracellular mannan-glycoprotein of *Phytophthora megasperma* f. sp. *glycinea*. Physiol. Plant Path. 20:321–31.

260. Zucker, M., and L. Hankin. 1970. Physiological basis for a cycloheximide-induced soft rot of potatoes by *Pseudomonas fluorescens*. Ann. Bot. 34:1047–62.

7

Growth Regulator Metabolism

The Healthy Plant

Growth and development in plants is under hormonal control by such biologically active endogenous compounds as 3-indole-acetic acid (IAA, auxin), indolic-related compounds, gibberellins, cytokinins, ethylene, abscisic acid, and others. A departure from normal levels of these compounds in the plant, such as might be caused by pathogenic attack, could alter the growth habit of the host. The subject of hormonal control over plant metabolism is complicated by the fact that the natural plant growth regulators generally do not act singly, rather two or more may act in a cooperative effort in which, for example, auxin is supported by gibberellins, cytokinins, ethylene, or abscisic acid and some phenols that may act as hormone modifiers. Sachs (224) postulated group hormonal action as early as 1882.

More specifically, the interplay between auxin and gibberellin controls elongation; auxin, cytokinin, and sometimes gibberellin control cell division and subsequent cell differentiation; the combination of auxin and ethylene influences growth "habit"; and auxin, abscisic acid, and ethylene control leaf abscission and senescence. Thus, pathologically altered hormonal levels in the above combinations reflect, respectively, such symptoms as stunting, tumorigenesis, epinasty, and premature leaf drop.

Hence it is of more than academic interest to study the relationships of altered growth regulator metabolism to pathogenesis. First, however, let us review some salient features of these plant hormones. As we have already noted, IAA influences both the promotion and inhibition of cell elongation, organ formation, and cell division; as a consequence, it has been referred to as a pleotropic "gene." It is the most ubiquitous of the endogenous plant growth regulators (48).

IAA Synthesis, Site of Formation, and Movement

The synthesis of auxin is centered in young leaves and shoot tips. Concentrations of auxin in green plants have been variously estimated to be between 10 and 0.1 μg/ml, with a falling gradient from young to older tissues. The movement of auxin in the plant is rapid and usually basipetally polar in the phloem (193). Movement may also occur at a much slower rate by cellular diffusion, and it too is basipetal (92). Auxin is also produced by a wide spectrum of microorganisms, particularly if a precursor such as tryptophan is available.

Some routes for IAA synthesis are presented in Fig. 7–1.

At least two pathways have been established in healthy plants for the biosynthesis of IAA. One, most frequently referred to and fully established by Wightman and Cohen (301) in mungbean, is tryptophan (TTP) conversion to indole pyruvic acid (IPyA) by TPP transaminase wherein glutamic acid is formed from α-ketoglutaric acid. A decarboxylase converts IPyA to indole acetaldehyde (IAAld). The aldehyde in the presence of $NADH_2$ and an alcohol dehydrogenase forms tryptophol (Tol) or by the participation of an oxidase or an aldehyde dehydrogenase and $NADH_2$ forms IAA. The second pathway, studied by Muir and Lantican (179) in dwarf pea apical tissue, is the conversion of TTP to tryptamine (TNH_2) and subsequently by a "presumed" amine oxidase directly to IAA. It is of interest that gibberellic acid (GA) promotes both steps in this pathway. In Avena coleoptiles, these workers proposed also the TTP → TNH_2 → IAAld → IAA pathway.

It has been difficult to define the "primary" mechanism by which IAA exerts its effect. One may consider that there is but a single fundamental site of action at which IAA is able to elicit or initiate a number of different responses (31). However, it would seem that wall-loosening is the primary response of plant cells to auxin (281). A review of our understanding of auxin receptors or auxin-binding proteins or sites, which when occupied initiate the biochemical or biophysical events ascribed to IAA, was prepared by Rubery (221).

Growth (cell wall effects). Auxin affects growth at concentrations of 10^{-7} to 10^{-5}M, with an optimum of 10^{-6} to 10^{-5}M. The plant cell may be regarded as enclosed in a semirigid box, the cell wall. The restraints placed on the cell by its wall appear to be at least in part under the control of IAA. The capability of the wall to stretch has been studied by Tagawa and Bonner (268) and Cleland (49) and has been separated into elastic and plastic components. The recovery from applied tension is designated the *elastic*, and the difference, or permanent deformation, the *plastic* component. The relationship of these two components to growth and IAA is apparent from the experiments of Bonner (31), who observed that by increasing IAA from 10^{-8} to 10^{-5}M, the plastic component of the wall stretched in nearly linear fashion. Hence auxin seems to play a role in the wall-loosening phenomenon so essential for cell growth

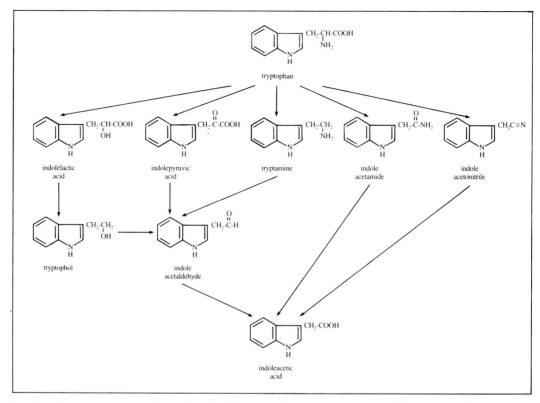

Fig. 7–1. Possible pathways in the biosynthesis of 3-indoleacetic acid from tryptophan (Phelps and Sequeira, 203).

and expansion. The loosening is metabolically controlled because it is reversed by KCN and 2,4-DNP. The subsequent step of elongation does not appear to require metabolic energy as the loosening-induced "slack" is taken up by the diffusion of water into the cell per se.

How auxin does this is not fully understood. There is some uncertainty over the possibility that IAA exerts its effect upon a genetic target, through gene derepression and subsequent enzyme synthesis. Clevage of cross-linkages between cellulose microfibrils or between cellulose and other wall components may be enzymatic. That m-RNA and enzyme protein synthesis may be involved is inferred from the results of Key et al. (121). However, the enzymes affecting wall polymers (e.g. cellulase, β-glucanase, hemicellulase) may not be generated rapidly enough to account for the phenomenon. The first increase in RNA synthesis is detectable in 10 min, and subsequent detectable enzyme protein synthesis requires at least 1 h. Since auxin-induced growth is almost immediate, the gene target theory is suspect.

Rayle (213) has presented evidence that auxin initiates a flow of H^+ ions into the cell wall region and that the subsequent wall acidification triggers cell-wall loosening (Fig. 7–2). The relationship of H^+ ions to growth was suggested long ago by Bonner (30). That the effect may be on cellulose microfibril hydrogen bonding was suggested by Keegstra et al. (120). Valent and Albersheim (279) attempted to link the wall-loosening

phenomenon to an acid-catalyzed cleavage of H-bonds between xyloglucan chains that are connected to cellulose via hydrogen bonds. However, their measurements on rate of hydrogen bond formation between cellulose and 7–9 sugar xyloglucan fragments failed to detect changes between pH 2 and 7. Although their data

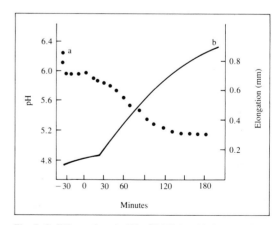

Fig. 7–2. Effects of auxin (10^{-5} M IAA, added at zero time) on acidification of the incubation medium, a, and on elongation, b, of *Avena* coleoptile segments (redrawn with modifications, by permission, from Rayle, 213).

support the hypothesis that xyloglucan chains are connected to cellulose by hydrogen bonds, the connection between xyloglucan and cellulose may not be the precise site in the wall that regulates cell elongation. Penny (200) has presented data that place in doubt the hypothesis that acidification of the cell wall is the main response to auxin. He does consider the possiblity that enzymes in the cell wall are important to cell expansion and have a definite dependence on a pH optimum. In this regard Darvill et al. (60) have measured compositional changes in cell walls of maize coleoptiles after treatment with auxin or at pH values of 4.5 and 7.0. Most striking was the release of free arabinose and galactose, 5-arabinose and free glucuronic, and galacturonic acid from tissue treated with auxin or at pH 4.5. They contend that the linkages that are broken (releasing the sugars) are not acid labile, and hence breakage must be enzyme mediated.

The basis for the auxin-induced acidification of the cell wall is the operation of a membrane-bound H^+ or proton pump. The actual process by which acidification may proceed may be twofold: (1) acid (H^+)-induced growth that is probably of short duration and (2) a second wave of elongation that is dependent on protein synthesis (281). Data presented by Ray (212) suggest that auxin binding sites are primarily on the endoplasmic reticulum (ER). It is possible that these sites are not "physiological auxin receptor sites" but rather are concerned with auxin transport in the ER. Ray suggests that auxin with its receptor sites on ER could induce H^+ transport from cytoplasm into ER cisternal space. In this manner, the acid could be subsequently transported via Golgi vesicles and excreted into the cell wall. Libbenga (143) reported that tobacco pith cell cultures have both soluble and membrane-bound auxin receptors.

Other Proposed Modes of Action of IAA

Effect on cell permeability. It has been suggested that IAA influences the permeability of the cell membrane. As a consequence of the high lipophilic surface activity of auxins, Veldstra (288) has suggested that IAA is adsorbed to the surface of the protoplasmic cell membrane, which would change boundary potentials (reducing them) and thus implement membrane transport. In testing this theory it was found that IAA increases the exodiffusion of anthocyanin pigments from the cells of red beet root tissue. Similarly, concentrations of 10^{-5} to 10^{-8} M IAA have potentiated the absorption of urea and glycerine by onion-scale protoplasts.

Complex formation. Bandurski and Piskornik (22) have reported the occurrence of IAA as an ester of a cellulosic glucan. In fact they found that approximately one-half the total IAA of mature sweet corn kernels is in ester linkage with the glucan. The physiological role of this ester is unknown. They suggest that it could be a storage, detoxification, transport, or even physiologically active form of the hormone. Actually auxin forms complexes rather readily with sugars such as glucose, arabinose, and galactose as glycosides. In addition,

IAA also forms peptides, linking with aspartic and glutamic acids. These bound forms of auxin are considered storage forms of the hormone (140, 176).

Effect of auxin on RNA and protein synthesis. Although auxin has been linked with cell-wall metabolism, phenol metabolism, respiration, and a number of other metabolic processes, the evidence suggests that IAA exerts an effect, albeit in an unspecific way, on both RNA and protein synthesis. This is accomplished by increasing RNA polymerase I activity (rRNA synthesis) (121, 122, 176, 187). It is also clear, however, that some auxin responses occur too rapidly to be dependent upon nucleic acid and subsequent protein synthesis. For example, protoplasmic streaming is accelerated within 20 sec, and membrane potentials are altered by auxin in 10–15 min. Nevertheless, the twofold process of wall acidification provides for a protein synthesis step (122, 187) that mediates cell elongation (95).

Implications of Auxin Synthesis and Degradation

The levels of auxin in the plant are not only a consequence of its synthesis but also reflect the rate at which it is degraded or inactivated. It is apparent from Fig. 7–1 that IAA is synthesized from tryptophan, which is formed via the shikimic acid pathway (see Chap. 6). Hence shifts in aromatization patterns clearly impinge upon precursor levels of IAA. Auxin may also be affected by direct enzymatic oxidative catalysis, as well as complex formation.

The pathway of direct oxidative destruction of IAA, by IAA oxidase, has been elucidated by Hinman and Lang (102), and the end product is 3-methylene oxindole (MOX). Their study revealed that MOX is formed in the absence of added H_2O_2 when IAA is present at levels of 2×10^{-4}M or lower. The oxidation route appears also to be concentration dependent, since at higher concentrations neutral indole is the principal end product (Fig. 7–3).

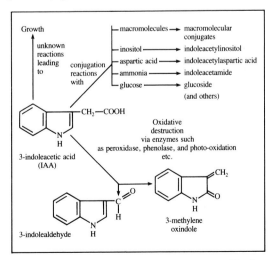

Fig. 7–3. The metabolism of IAA (after Galston and Davies, 89).

The very nature of auxin oxidase was for a long time a perplexing and confusing matter. However, an elegant study by Galston et al. (90) revealed the nature of the dual function of the enzyme, i.e. auxin oxidation and peroxidative capacity, and the control that IAA has over the induction and repression of specific isoperoxidases (144). It was shown that removal of the heme group destroyed the enzymes' peroxidative activity. However, the apoprotein retained its auxin oxidative activity when supplied with the two usual cofactors Mn^{++} and 2,4-dichlorophenol. Hence "IAA oxidase" is a single enzyme with two active sites.

The study also presented evidence that IAA could repress the synthesis of two isoperoxidases and subsequently induce the synthesis of a third. Since both of these effects are largely inhibited by actinomycin D, the appearance of these enzymes reflects de novo protein synthesis. Thus it appears that IAA does indeed control enzyme synthesis at a gene locus; however, whether the enzymes controlled bear on hormonally induced growth-regulating events remains unresolved. The appearance of the new enzymes could signal new biochemical capacities such as the synthesis of lignin from phenyl propanoid precursors (242).

Finally, it is known that the peroxidase-IAA oxidase activity of tissue may indicate not only enzyme levels per se but also that the relative relationship between aromatization and auxin levels is of particular significance.

Gibberellins

Gibberellic acid (GA_3) and a number of other gibberellins were first isolated and purified from culture filtrates of the fungus *Gibberella fujikuroi*, which attacks the rice plant. In fact, the growth modification induced by the pathogen in the host contributed to the discovery of these compounds and their eventual characterization as plant growth regulators. According to Moore (176), more than 60 gibberellins (GAs) have been isolated from higher plants and fungi.

Structure and nomenclature. The GAs are diterpenoid acids derived from the tetracyclic diterpene hydrocarbon *ent*-kaur-16-ene [(-)-kaurene] depicted in Fig. 7–4a. About half of the GAs retain the full 20 carbon atoms of kaurene. They are known as the C_{20}-GAs and have the carbon skeleton shown in Fig. 7–4b. The remainder have lost the number 20 carbon during biosynthesis, and they are referred to as C_{19}-GAs and have the carbon skeleton shown as Fig. 7–4c. The system of nomenclature for the classification of GAs of Rowe (220) has generally been accepted and replaces an earlier one by Paleg (195). In this system, the GAs fall into two groups, the C_{20} gibberellins (ent-gibberellanes) and the C_{19} gibberellins (ent-20-norgibberellanes). The numerical order of the GAs is not systematic but, in general, represents the order of their discovery (139). The various GAs are characterized, in part, by hydroxylation patterns at carbons 3, 12, and 13, the

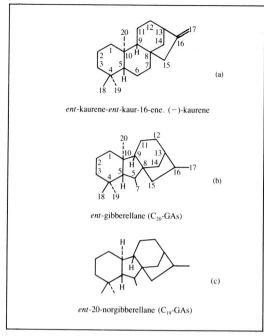

ent-kaurene-*ent*-kaur-16-ene. (−)-kaurene (a)

ent-gibberellane (C_{20}-GAs) (b)

ent-20-norgibberellane (C_{19}-GAs) (c)

Fig. 7–4. The basic structures of the gibberellins.

state of oxidation of carbon 20, or the presence of a carboxyl group at carbon 6.

The biosynthetic pathway of the GAs from acetyl-CoA and mevalonic acid has been worked out by West and his colleagues (133, 293). Their scheme (Fig. 7–5) carries acetyl CoA and mevalonic acid through kaurene and kaurenoic acid. Thus, the GAs are placed in biosynthetic sequence, and the derivation of other terpenes and sterols is also made apparent.

The route from mevalonate to geranylgeranyl pyrophosphate is common to all terpenes. Branching at geraniol gives rise to the monoterpenes, whereas farnesol is the precursor and branch point for sesquiterpenes (including abscisic acid) and sterols. Geranylgeranyl pyrophosphate serves as the common precursor of the noncyclic and cyclic terpenes as well as side chains of quinones and carotenoids and the phytyl side chain of chlorophyll and higher terpenoids.

Although there is significant diversity in chemical configuration of GAs, little is known of the relationship of specific structure to biological activity. In addition the highly active compounds have an intact gibberellane ring system (rings A, B, C, D), a carboxyl group at C-7, and a lactone ring in ring A. There is a good deal of evidence that the more biologically active 19-C GAs are formed from 20-GAs by contraction of the B- ring and elimination of a carbon atom at position 7 to form the C-4—C-10 lactone. The GAs are interconverted mainly from nonhydroxylated to mono- and dihydroxylated forms.

Sites of synthesis. Studies on the site of GA synthesis suggest that some if not all of the hormone is produced

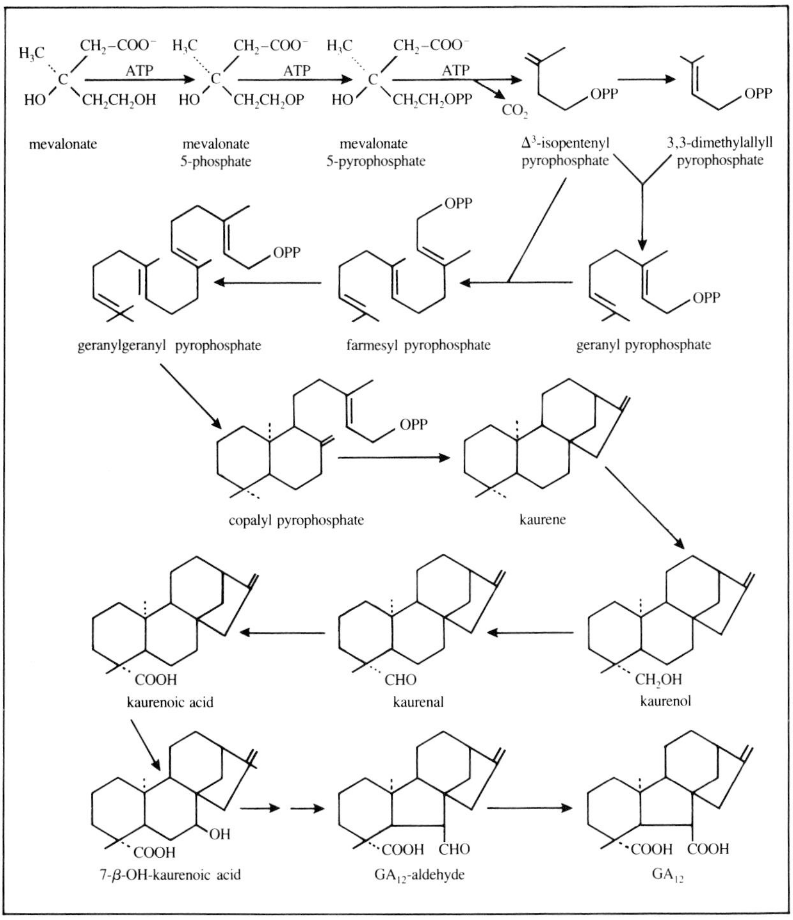

Fig. 7–5. Pathway of GA biosynthesis from mevalonate as far as elucidated for seed plants (after Moore, 176).

in plastids, primarily the active chloroplasts, with the key intermediate being kaurene. Stoddart (260) reported that kaurenoic acid is partially converted to a GA_3-like compound by sonicated chloroplasts and other plastids. In fact, Cooke and Kendrick (54) have reported that GA_4 and GA_9 were closely associated with chloroplast membranes. They also revealed a phytochrome-dependent increase in GA-like activity associated with wheat leaf etioplasts. There is indeed a direct relation between red light and the increased extraction of GA from leaves of several species. The transport of GA across the plastid envelope appears to be regulated by phytochrome.

The concentration necessary to evoke a response in plant tissue is low. For example, the concentration of GA_3 required to induce amylolytic activity in barley endosperm is slightly in excess of $0.001 \mu g/ml$, with maximum induction occurring at $0.1 \mu g/ml$ (51).

Site and Mode of Action

Two sites of action have been studied in detail, the aleurone layer of the endosperm of cereal seeds and the apical meristem.

Aleurone layer. Gibberellin found in the embryo of seeds is secreted into the aleurone layer where it induces the synthesis of enzymes that move into the endosperm liberating the metabolites required for the growth and development of the embryo. Gibberellin diffuses across the endosperm and induces the formation of α-amylase, proteinase, and certain cell-wall hydrolyzing enzymes (endo-β-glucanase and endoxylanase) in the aleurone layer.

The apparent capacity of gibberellin to induce the synthesis of some enzymes implies that this hormone exerts an influence upon both protein and RNA synthesis (not unlike the conclusion reached for one of the functions of IAA). That both protein and RNA synthesis are to some extent controlled by gibberellin has been established with certainty. Experiments utilizing intact endosperm reveal that the gibberellin-induced release of reducing sugars by amylase is blocked by protein-synthesis inhibitors. Varner and Ram Chandra (286, 287) have postulated that a gene is derepressed by gibberellin, thus permitting DNA-controlled m-RNA synthesis to proceed, which culminates in the production of α-amylase and other hydrolases. The precise manner in which this hormone derepresses genes or "restarts" interrupted enzyme synthesis is not known. A study by Evans (77) revealed that GA potentiates ribosome synthesis and hence may be responsible for selective m-RNA synthesis. Interestingly, abscisic acid, which suppresses the hormonal effect of GA, inhibits RNA synthesis and necessarily inhibits polyribosome formation.

Meristems. Gibberellins also exert potent hormonal stimuli on terminal meristematic tissues, which are mitotically active, and it is here that the growth-promoting effects of gibberellin are readily discerned.

The effects of gibberellins on meristematic tissues seem on cursory examination to duplicate those previously ascribed to auxin. For example, the stimulation of cell elongation and cell division of cambium, the maintenance of apical dominance caused by a restrictive effect upon the growth of lateral buds, and the induction of geotropic responses (stem bending) are all recognized effects of auxin. Yet each of these may be induced by exogenous application of gibberellin to appropriate tissues. Paleg (195) has interpreted these apparently similar effects as an enhancement of auxin action by gibberellin. This interpretation is substantiated by a number of reports that tissues treated with gibberellin contain higher levels of auxin than untreated ones (91). However, it has been well established that normalized growth of genetic dwarfs is a genuine GA effect. Genetic dwarfs of some plant species are deficient in GAs due to their inability to efficiently perform one reaction in GA biosynthesis. Specifically, Hedden and Phinney (99) found the dwarf-5 mutant of *Zea mays* unable to produce sufficient amounts of *ent-* kaurene, the GA precursor, instead producing isokaurene, which is not converted to GA.

The possibility that gibberellins may increase auxin levels by stimulating auxin synthesis, first proposed by Brian and Hemming (39), also deserves consideration. This concept seems even more plausible in the light of a report by Muir (178) in which dwarf pea apices treated with gibberellins yielded enzyme preparations capable of converting tryptophan to auxin at rates greater than comparable control preparations.

In summary it would appear that gibberellin exerts its influence upon meristematic tissues through an enhancement of auxin action. This enhancement may in turn be due to increased levels of auxin, which are controlled by gibberellin through an effect upon either auxin degradation or synthesis (186). Gibberellins are also a class of hormones in their own right (99, 156).

The effect of gibberellin on the aleurone layer, which is incapable of growth, is according to Paleg (195) "upon modes of physiological expression, none of which auxin could be expected to duplicate." This analysis seems to have been borne out by a report of Cleland and McCombs (51) indicating that the action of gibberellin on increasing the amylolytic activity of barley endosperm is independent of either the *presence* or *absence* of auxin.

It should be recalled, however, that there is evidence that both auxin and gibberellins exert at least a part of their effects by altering DNA-controlled RNA synthesis. Hence, the enzymes that each influence may differ, but the ways in which they derepress genes or induce enzyme activity may be similar.

Cytokinins

Cell division is not a continuous process but rather one that starts, stops, and may start again. The hormones most clearly in control of the cell division cycle are the cytokinins. They influence additional features of plant cell metabolisms as well, often in concert with other plant growth regulators. Together with auxin and gibberellins they influence, in addition to cell division, cell enlargement, differentiation, and organogenesis.

Chemistry and source. According to Skoog and Armstrong (246), cytokinins must be N^6 substituted adenyl compounds. Substitution of the N^6 position by either oxygen or sulfur as detected in some compounds greatly diminishes biological activity. Activity varies greatly with the length and other properties of the side chain of the N^6 substituent. Thus, adenine is much less active than the other cytokinins shown in Fig. 7–6. Skoog and Leonard (247) have shown that 6-γ,γ-dimethylallylamino purine (6-γ,γ DMAAP) is an exceedingly active cytokinin, however the 3-DMAAP is inactive. If, however, the 3-substituted compound is autoclaved, the active 6-DMAAP is formed. They concluded that wherever nucleic acids are present, trace quantities of N^6-substituted adenines are apt to be formed, thus providing a ubiquitous source of cytokinins.

Free cytokinins can arise through substitution of the characteristic side chains onto carbon 6 of adenine, adenine riboside, or adenine ribotide. They may also be formed by hydrolysis of certain tRNA species, i.e. those that contain adenine and to which an isopentenyl side chain is attached at carbon 6. The latter possibility appears to be more widely held at the present time (176) (Fig. 7–6).

Although no systematic search has been made, cytokinins have been detected in germinating seed, roots, and "bleeding" sap, as well as in some microorganisms (141). It is of interest that the sap of water-deficient and *wilted* plants has lower amounts of cytokinin than plants not under moisture stress (107). Wilting has been shown to cause significant qualitative and quantitative changes in plant protein and RNA levels. As might be expected, cytokinins are found in highest concentration in those tissues undergoing intense cell division. Biosynthesis may occur in tissues that are meristematic or possess mitotic potential. Cytokinins are generally synthesized in the roots and translocate acropetally in xylem fluid. As such they have been detected in the xylem sap of grape, sunflower, tobacco, and other species. Moore (176) suggests that cytokinins moving upward in xylem counter the effect of auxin, which migrates in the opposite direction. Whereas auxin inhibits differentiation of the vascular tissues in bud

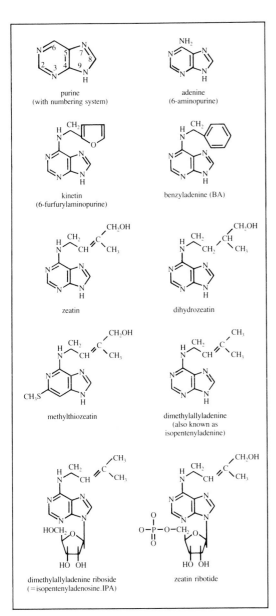

Fig. 7–6. Basic structure of some cytokinins (after Moore, 176).

traces connecting bud and stem, cytokinin promotes differentiation. Hence, where auxin tends to suppress lateral end development and growth, cytokinins release this suppression.

Regulatory effects. The most obvious affect of cytokinins is their capacity to promote cell division. Jablonski and Skoog (108) clearly demonstrated that tobacco pith cells did not divide readily without cytokinin. Early studies with cytokinin (kinetin) revealed that it induced the development of specific organelles such as chloroplasts and the differentiation of

groups of cells into tissues and organs such as vascular elements, embryos, or roots.

Some morphogenetic phenomena such as bud development, although influenced by cytokinin, are not under the singular control of this hormone. Differentiation in this instance is affected by the concentrations of both cytokinin and auxin. *High* molar ratios of cytokinin to auxin lead to bud formation; at *equal* concentrations undifferentiated callus develops; *high* auxin to *low* cytokinin ratios foster the initiation of root growth (248).

Another striking effect of cytokinins is their ability to direct transport, accumulation, and retention of metabolites in tissues and organs. These properties are doubtless related to their capacity to delay senescence (172, 177). The evidence suggests that cytokinins in some way help the plant to sustain an effective level of protein synthesis. This may reflect the presence of cytokinin-ribosides in tRNA. However, free cytokinins are believed to be synthesized in plants by mechanisms independent of the degradation of tRNA (176). The precise mode of action in this respect, whether upon nucleic acid or protein synthesis per se, is not known.

When Osborne (191) supplied leaf discs with the antibiotic puromycin, an analog of tRNA, incorporation of amino acids into the RNA template on the ribosomes was reduced 50%. Under these conditions kinetin was no longer able to stimulate protein synthesis or delay senescence. This suggests that kinetin does not regulate protein synthesis at the stage of peptide bond formation. If the leaf discs were exposed to actinomycin D, kinetin was still capable of stimulating protein synthesis. Furthermore, the degree of stimulation is approximately proportional to the amount of DNA-dependent RNA synthesis occurring in the tissues. The interpretation given to these results by Osborne (189) is that the action of kinetin in delaying senescence is, perhaps, at the stage of mRNA synthesis.

Osborne has likened senescence in the leaf to a "progressive turning off of genetic information in the nucleus" of cells normally engaged in anabolic processes. As fewer and fewer genes remain functional as templates for DNA-dependent RNA synthesis, the cells stop making protein and pass into a state of dormancy. During dormancy the genes may only be temporarily repressed; if reactivated, for example by kinetin, they are derepressed, and the cells resume synthesis of both RNA and protein. Thus Osborne proposed that the role of kinetin in delaying senescence is to keep the genes derepressed, preventing "the genetic material from being irreversibly turned off."

Hence it would appear that cytokinins retard senescence by sustaining chloroplast nucleic acid synthesis. However, Shibaoka and Thimann (241) contend that cytokinins maintain membrane integrity by inhibiting proteolysis rather than promoting protein synthesis. Rijiven (218) recently presented evidence that cytokinins increase the activity of the ribosome fraction in in vitro protein synthesis. Ribosomes from fenugreek cotyledons (*Trigonella foenum-graecyn* L.) treated with a number of cytokinins are more active in the initiation of poly (U)-directed polyphenylalanine synthesis than untreated controls. Improved ribosome activity was interpreted as an enchancement of peptide chain *initiation*.

Ethylene

Perhaps the first known endogenous plant growth regulator is ethylene (C_2H_4). It exhibits biological activity at concentrations of 1.0 ng/l. Some of the more conspicuous effects of C_2H_4 are epinasty, tissue proliferation or swelling, marked increase in respiration, the stimulation of adventitious root formation, abscission, and chlorosis (all of these have been recorded as symptoms of pathogenesis).

Biochemical origin. It appears that ethylene is more restricted in its occurrence than the other plant hormones. It is, however, ubiquitous among seed plants and is produced by a number of plant pathogenic fungi and bacteria. The proposed pathway as reported by Yang (307) is presented in Fig. 7–7.

In the biosynthetic pathway described by Yang (307), methionine is converted to an intermediate S-adensosylmethionine (SAM). Both methylthioadenosine (MTA) and metholthioribose (MTR), derived from the CH_3S group of methionine as it is converted to ethylene, derive from an apparent sulfur recycling reaction supporting the resynthesis of methionine. Under the

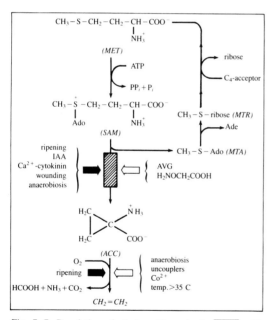

Fig. 7–7. Regulation of ethylene biosynthesis. ▨ : this reaction is normally suppressed and is the rate-limiting step in the pathway; (♦): induction of synthesis of the enzyme; (◊): inhibition of the reaction. Met, Ado, and Ade stand for methionine, adenosine, and adenine, respectively; SAM: S-adenosylmethionine; MTA: methylthioadenosine; and MTR: methylthioribose (after Yang, 307).

stresses of wounding and anaerobiosis and the presence of IAA and cytokinin, SAM is converted to a second intermediate 1-aminocyclopropane-1-carboxylic acid (ACC), which in the presence of air is converted to ethylene. The conversion of SAM to ACC is blocked by aminoethoxyvinylglycine (AVG) with pyridoxal phosphate, and the final step of oxidizing ACC to ethylene may be blocked by uncouplers, heat, and anaerobiosis. However, it appears now that ACC is the immediate precursor to the synthesis of ethylene.

A study by Kao and Yang (113) links ethylene production to the presence or absence of light. Light inhibits ethylene production at the conversion step of ACC to ethylene. Light inhibition is mediated by the internal level of CO_2, which directly modulates the enzyme converting ACC to ethylene. According to Kao and Yang (113), ethylene regulates leaf senescence as maximum ethylene production precedes chlorophyll degradation, ACC promotes chlorophyll degradation, and inhibitors of ethylene retard chlorophyll degradation.

Influence of ethylene on abscission. Jackson and Osborne (110) have defined the role that ethylene plays in *natural* leaf abscission. From their study it may be inferred that pathological leaf abscission reflects a similar sequence of events.

Abeles (1, 3) concluded that the timing of abscission is determined not by the rate of C_2H_4 production but by an increase in sensitivity of the abscission zone (2–5 rows of cells) shown in Fig. 1–7 to this hormone. Jackson and Osborne (110) detected a rise of C_2H_4 that precedes normal leaf abscission. In *Parthenocissus* the concentration of C_2H_4 produced by petiolar tissue adjacent to abscission is 13 times greater at the time of abscission than "some days" before. Hence C_2H_4 production is a feature of senescence that is potentiated at a specific stage of aging and is responsible for the biochemical processes that actually cause leaf separation.

The actual separation of cells in the abscission zone is caused by cellulase and pectinase, which are activated by C_2H_4. The sensitivity of the cells in the abscission zone to cellulase is to some extent influenced by levels of pectin-methyl esterase (PME). Up to the time of abscission the PME level falls continuously on the pulvinar side of the zone but not on the petiolar side. Osborne has suggested that the methylated pectins in the pulvinar side of the zone render the cellulose walls of the cells there particularly susceptible to cellulase degradation (190). Abeles and Leather (3) have reported that ethylene not only regulates the amount of cellulase synthesized but also controls its movement from the cytoplasm to the cell wall and its ultimate secretion outside the cell.

Ethylene-induced hypertrophy. Hypertrophy, or intumescences, of lenticels, cortex, and other tissues may be induced by C_2H_4 (1). Wallace (291) recorded some swelling resulting from concentrations as low as 0.01 $\mu g/ml$. The most characteristic change in tissue undergoing intumescence formation is the loss of coherence of affected cells. The middle lamella disappears, cell walls are digested, and the cells appear as swollen rounded protoplasts. These developments suggest intense pectinase and cellulase activity and seem similar to the events that occur during normal leaf abscission.

Influence of ethylene on cell wall. Ridge and Osborne (217) have suggested that C_2H_4 regulates the level of hydroxyproline-rich proteins (hyp) in cell walls. Furthermore, the role of hyp in the wall is perhaps a dual one: structural, linked to arabinose through glycosidic bonds, and functional as a constituent of peroxidase enzyme (239). Ridge and Osborne (217) speculate that C_2H_4 increases the level of hyp-peroxidase and that this enzyme is one of a group of hyp-containing proteins that determine cell wall extensibility and growth.

Influence of ethylene on the enzyme cellulase. Studies by Geballe and Galston (91) appear to have discovered an important relationship between ethylene and the enzyme cellulase. In studies concerned with the release of leaf cells with the enzyme preparation "cellulysin," they noted that oat leaves wounded in a number of ways became resistant to the enzyme. More specifically, when leaf epidermis was peeled, the oat leaves became resistant to digestion by cellulysin. Peeling induced a fourfold increase in ethylene synthesis. The release of ethylene began between 1 and 2 h after wounding and continued for 5 h. Maximum resistance to the enzyme cellulase is recorded 11 h after wounding. The nature of the changes in wall architecture and chemistry due to the wounding would seem to be a reasonable research endeavor. The release of ethylene appears to require the synthesis of protein, as it can be suppressed by actinomycin D. The antibiotic cycloheximide suppresses both ethylene release and cellulase resistance; hence the latter appears to require the synthesis of RNA.

Additional effects of ethylene are: (1) C_2H_4 synthesis speeds fruit ripening; however, this action requires the presence of adequate O_2, and ripening reflects activation of hydrolytic and oxidative enzyme activity. The C_2H_4-forming mechanism may operate continuously at low levels in some plants but is not activated until the tissues reach a critical physiological age. (2) C_2H_4 also changes membrane permeability, which leads to altered compartmentation and the release of respiratory and hydrolytic enzymes. Permeability changes are also signaled by leakage of cell solute into free space. (3) Whether the de novo synthesis of enzymes associated with increases in C_2H_4 accumulation is evoked by the hormone per se remains uncertain. (4) IAA stimulates ethylene production in roots, stems, flowers, fruits, and leaves. As a consequence it has been possible to ascribe the "inhibitory" effects of IAA to the influence of C_2H_4.

Abscisic Acid (ABA)

In 1964 Eagles and Wareing (73) discovered an inhibitory substance in *Betula* that increased in concentration with onset of dormancy. Subsequently El-Antably et al. (74) and Aspinall et al. (9) simultaneously purified growth regulators with inhibitory capabilities and

named them respectively *dormin* and *abscisin* II. They turned out to be identical, and by mutual consent the compound was named abscisic acid, or ABA.

Essentially, ABA prepares the plant for dormancy by shutting down its synthetic machinery. Specifically, ABA applied to actively growing stems induces the typical symptoms of dormancy, i.e. shortened internodes, formation of bud scales, reduced meristematic activity, and some abscission. This hormone also inhibits bud break and elongation of genetically tall corn. Hence it is apparently an antagonist of GA and may be reasonably interpreted to be the active component in the plant's response to a shortening photoperiod. In addition ABA sustains dormancy in seeds that might normally germinate in response to GA. ABA may therefore be considered a "metabolic brake" brought on by a reduction of the photoperiod.

It is more than coincidental that ABA and GA, which exert at least a part of their effects in opposition to each other, are both isoprenoids and are derived via the mevalonic acid pathway. Galston and Davies (89) have suggested that they are perhaps governed by a common switch, *phytochrome*, which is activated by day length. The scheme is presented diagrammatically in Fig. 7–8.

Mode of action. Perhaps the most direct action of ABA is on chlorophyll and carotenoid synthesis as they are inhibited 25–58 and 30%, respectively, by this hormone (171). In fact, El Antably et al. (74) suggest that "dormin" accelerates chlorophyll breakdown. Hence the chloroplast may be at least one of the organellar sites of action. However, the mode at this site seems to be on the inhibition of specific RNA synthesis. For example, since ABA inhibits the synthesis of amylase it seems possible that it acts by inhibiting the synthesis of enzyme-specific RNA or by preventing its organization into an enzyme-synthesizing unit.

Other ultrastructural changes fostered by ABA, according to Mittelheuser and Van Stevenick (172), are those that reflect an accelerated senescence-associated degeneration of chloroplast ribosomes, formation of large lipid bodies in cytoplasm, and vacuole and osmiophilic fibrils in microbodies. It is also possible that chloroplast aging provides larger amounts of precursor to be

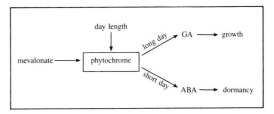

Fig. 7–8. Both abscisic acid and gibberellin are isoprenoid in structure, deriving from mevalonate. Abscisic acid is frequently found under short-day conditions and gibberellin under long days. Phytochrome may be the means by which the synthetic pathway is switched between these two compounds, depending upon the photoperiod.

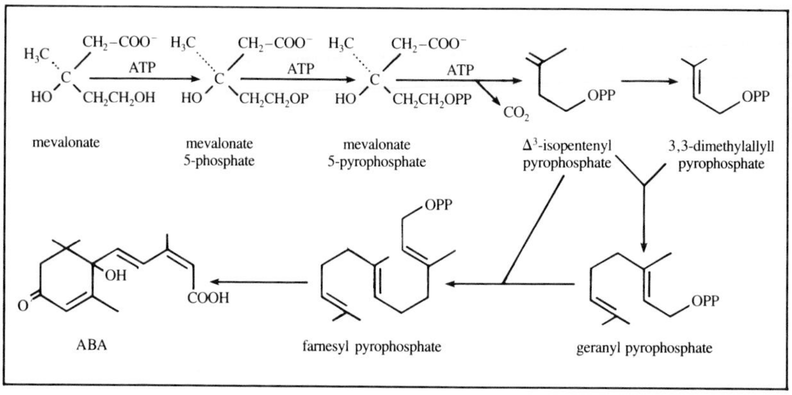

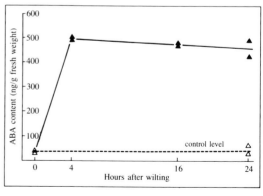

Fig. 7–9. Proposed pathway for the biosynthesis of (S)-abscisic acid from mevalonate (after Moore, 176).

converted to ABA. Since plastids and chloroplasts are a major site of GA synthesis, it seems biologically sound that the counter hormone, ABA, should participate in chloroplast destruction.

Udvardy and Farkas (278) have ascribed the aging or senescence induction of plant tissues to the influence ABA has on aging specific purine-specific RNAases. According to them ABA monitors the level of RNA by controlling those nucleases responsible for voiding or destroying m-RNA.

Biosynthesis and site of ABA synthesis. There is good evidence that ABA is synthesized from mevalonic acid via farnesyl pyrophosphate (70, 173, 176), an intermediate in the isoprenoid biosynthetic pathway. It is biologically "reasonable" since both ABA and GA, "counter" hormones, are synthesized from farnesyl phyrophosphate. Both are to some extent synthesized in chloroplasts. A possible alternate pathway for ABA biosynthesis is from the carotenoid violaxanthin, with xanthin as an intermediate (Fig. 7–9).

Attention is called to the large and rapid increase of ABA in leaves when plants are placed under stresses such as drought, flooding, mineral starvation, and injury. Moore (176) contends that ABA increases plants' resistance to stress, a factor meriting futher study (Fig. 7–10).

Some physiological effects of ABA. Cottle and Kolattukudy (55) and Soliday et al. (253) established ABA as a primary mediator of plant cell wall suberization. ABA stimulated aliphatic and phenol deposition three- and fourfold and that of related wax components, fatty alcohols, and hydrocarbons five- and ninefold respectively. The implication of these findings in plant resistance to disease development remains largely unexplored.

ABA plays a major role in regulating stomatal closure. The response to low-level exogenous applications (l) is rapid, occurring within 3–9 min. The basis for this is in the guard cell turgor-regulating system, which depends upon an H^+/K^+ exchange process. ABA inhibits K^+ uptake by the guard cells, which in turn keeps them flaccid and hence closed (292).

Bud dormancy is controlled by day length with long days promoting vegetative growth and short ones effect-ing the onset of dormancy. ABA levels increase in leaves in late summer and fall, and ABA transported to the axillary and terminal buds causes dormancy to develop. Concomitantly, GA levels decrease. The balance between the two hormones is reversed in spring as the photoperiod lengthens and the chilling requirement is met. This balance between the two hormones seems, nevertheless, not to be the complete control system of bud dormancy and growth. The interaction between ABA and GA becomes apparent from the levels maintained by the plant during the course of the growing season. Their manipulation might provide some interesting advances in our understanding of induced resistance.

Additional Multiple Hormonal Interactions

Other examples of multiple hormone control of physiological phenomena include leaf senescence and the plant's response to water stress; both reveal an ABA-cytokinin interaction.

Leaf senescence. There is extensive literature (271) indicating that cytokinins are general senescence-retarding growth regulators. An almost equal array of

Fig. 7–10. Increase in ABA content in wilting bean (*Phaseolus vularis*) leaves. Dashed line indicates level of ABA in control plants (redrawn from Harrison and Walton by Moore, 176).

references reveal ABA to be a senescence promoter and that the two hormones interact in a competitive manner (292). Jackson et al. (109) advanced the notion that ethylene produced by senescing organs triggers processes in the 3–5 cell layer abscission zone that foster the synthesis of pectinases and cellulases. Abeles and Holm (2) have reported enhanced RNA and protein synthesis by ethylene. Additionally, they noted that the synthesis of these enzymes could be blocked by actinomycin D and by IAA and cytokinin. Hence, it is possible that four growth regulators actively influence leaf senescence.

Water stress. Plants subjected to drying conditions reveal a marked increase in ABA, which results from increased de novo synthesis of mevalonic acid (176). As ABA levels increase, cytokinin levels decrease. It is believed that since the root system is the primary site of cytokinin synthesis, reduced xylem transport of water would also reduce the upward delivery of cytokinin. However, there is evidence that a reduction in cytokinin synthesis in the roots also results when leaf water potential drops. The foregoing sequence of events suggests that a signal to reduce cytokinin synthesis is received from the leaves. Although the precise nature of the signal is not known, it has been suggested that the signal is *increased* ABA coming from the leaves. The terminal phase of water stress is often leaf abscission, which implies an increased synthesis of ethylene. However, the mechanism by which increased ethylene synthesis occurs in situations of intense water stress is not known. One might speculate that there is a relationship between vascular occlusion, reduced water transport, and ethylene accumulation.

Leaf abscission. There has been evidence for some time of a leaf abscission accelerator (190), and for a while ABA was believed to be the factor. Subsequently, another abscission accelerator, termed senescence factor (SF), was described by Osborne et al. (192).

It has also been apparent that elevated levels of C_2H_4 are crucial to the abscission process. Although both SF and ABA promote abscission, the former stimulates a sustained production of C_2H_4, and ABA does not. Paradoxically IAA, which retards abscission, also promotes C_2H_4 synthesis. However, IAA-fostered C_2H_4 synthesis may be a wound or stress phenomenon that temporarily increases permeability of membranes that enclose the SF factor, releasing it to sites of C_2H_4 synthesis. Subsequent membrane repair accompanying wound healing makes this a transitory phenomenon.

Hence SF may be regarded as distinct from ABA, and it is described by Osborne et al. (192) as *the* in vivo regulator of C_2H_4 production. In this respect Jackson et al. (109) advance the notion that C_2H_4 produced by senescing organs triggers processes in the 3–5 cell abscission zone that foster the synthesis of both pectinases and cellulases. Abeles and Holm (2) have reported enhanced RNA and protein synthesis by C_2H_4. Furthermore, these syntheses may be blocked by actinomycin D and by IAA and cytokinin.

Plants Infected by Viruses

Many of the symptoms of virus diseases suggest that alterations of endogenous levels or in the metabolism of plant growth substances may be involved. Their application to healthy plants may lead to altered growth habit, hyperplastic and hypoplastic changes in leaves, clearing of veins, tumor formation, mottling, and epinasty. These changes are often characteristic of virus infection, so that there has, not unnaturally, been a certain interest either in examining changes in endogenous hormone levels following virus infection or in attempting to modify the effects of virus infection by applying growth substances to infected plants. Partly because of the difficulties in measuring endogenous levels accurately, the latter has, in many cases, proved the popular course. However, growth hormones may interact with each other, either antagonistically or synergistically; their concentrations may vary in different parts of the plant and with the age of the plant or extent of infection, and virus infection may affect their synthesis, degradation, or translocation. The situation is, therefore, a complex one.

Auxins

Experiments designed to determine auxin levels in virus-infected plants have yielded conflicting results; in some instances, virus infection has been reported to decrease auxin level (94, 196, 210, 251), while in others enhancement has been reported (112, 188). However, the observed concentration depends on a number of factors, including plant growth conditions, method of extraction, and type, sensitivity, and specificity of the assay used. In addition, determinations have usually been made on crude extracts of plant tissue, which may contain both growth promoters and inhibitors. Since auxin concentrations have, until relatively recently, invariably been assessed by bioassay on the basis of growth-promoting activity, the presence of growth inhibitors can clearly prejudice the results. Also, auxins may be free or bound to tissue components, and results may be influenced by the relative proportions of these forms present at the time of assay. One must be cautious, therefore, in drawing conclusions from these observations.

In attempting to relate auxin levels to the expression of overt symptoms, an additional complication is that the effect of auxin is dependent on its concentration. At high concentration, growth is inhibited, while it is enhanced at low concentration. The optimum concentration for growth also varies considerably between tissues in the same plant; it is ca. 10^5 times greater in stems than in roots, so that auxin concentrations that stimulate stem growth may inhibit the growth of roots and buds. It is important, therefore, to determine auxin distribution in different host-tissues and indications of the optimum concentration required for growth of the tissue in question. Pavillard, for example (196), concludes that infection of potato plants either with PVX, PVX + PVY, or PLRV leads to a reduction in endogenous free auxin concentration, though he finds that PLRV, particularly in the advanced stages of infection, causes an increase in auxin concentration in leaf tissue.

Effects of virus infection on auxin transport have been recorded, CTV infection of tomato apparently causing a decrease in basipetal transport, an effect correlated with the deleterious effect of virus infection on the phloem tissue (252). Ethylene can also affect auxin transport (42), and since ethylene production in etiolated cowpea hypocotyls infected by CPMV was greater than in healthy controls, Lockhart and Semancik (149) proposed interference with auxin transport as one of several possible explanations for the virus-induced stunting.

In those instances where direct comparisons have been made, auxin concentration does not seem to be related to symptom severity. While there was a reduced auxin concentration in leaf tissue of *Beta vulgaris*, *Phaseolus vulgaris*, and *Lycopersicon esculentum* systemically infected with CTV, the decrease was not correlated with symptom expression but was observed both in susceptible plants, in which there were severe symptoms, and in resistant plants, in which there was limited virus replication but in which symptoms were minimal (251). When auxin was supplied to infected plants, it served only to enhance symptom expression (251), lending support to earlier conclusions that it was hormone imbalance that was important (25). Though auxin is required for growth, and virus infection may lead to reduced auxin levels, often accompanied by stunting, application of exogenous auxin does not normally overcome the virus-induced reduction in growth; application of several growth substances to SBMV-infected bean enhanced leaf deformation and stunting even further (96).

Growth regulators can also enhance and modify the symptoms of wound tumor virus. Pricking the stems of infected plants with a pin produces tumors; application of several growth regulators, including NAA, leads to the formation of larger and more numerous tumors and eliminates the requirement for wounding (28).

In certain circumstances, however, virus-induced symptoms can be alleviated. In bean plants infected with SBMV (96), axillary buds develop, suggesting that apical dominance, controlled by auxin, is inhibited. Hartman and Price (96) observed that application of growth substances 2,4-D, β-naphthoxyacetic or p-chlorophenoxyacetic acids, while enhancing other symptoms, inhibited the development of axillary shoots. They point out, "It is not clear whether the virus interferes with activity of the growth substance that normally inhibits development of axillary buds, whether it interferes with production or movement of this substance, or whether reduced growth of terminal portions of the plant, due to the presence of the virus, inhibits production of the growth substance." Similarly, the growth of axillary buds was inhibited by IAA treatment of peanut plants infected with "witches broom" disease (273). The severity of symptoms on tomatoes infected with TAMV (290) could be reduced by treatment with IAA and on TMV-infected tobacco with NAA or IBA (185). Virus-induced premature leaf abscission could also be

delayed by spraying *Physalis floridana* with IAA or NAA (72).

By comparison with these investigations on the systemic effects of virus infection, there have been relatively few studies on plants responding hypersensitively. A notable exception is the work of van Loon on Samsun NN tobacco infected with TMV (283, 285), in which he has investigated endogenous IAA levels in various leaves prior to virus inoculation, changes following virus inoculation, and the size of lesions that develop on various leaves with or without exogenous auxin treatment. He observed a substantial increase in IAA concentration in inoculated leaves during the period of lesion enlargement, compared to levels in uninoculated controls. In marked contrast were the decreases observed in cultivars that react systemically to TMV infection (210). In noninoculated leaves (in which systemic resistance was induced), there was a slight but insignificant decrease in IAA level. Endogenous IAA levels were greater in younger than in older, senescing leaves or in leaves with systemic resistance. On TMV inoculation, young leaves developed larger lesions than older or systemically resistant ones. However, since inoculated leaves with high IAA levels also had acquired resistance, it was concluded that there was no direct relationship between endogenous IAA level and final lesion size. The small lesion size on old leaves or on young, systemically resistant leaves was not considered a consequence of low auxin level at the time of inoculation.

Treatment with exogenous auxins affected lesion development, application of IAA or 2-naphthylacetic acid decreasing final lesion size when applied to young, expanding leaves before or after infection (Fig. 7–11), though having no effect on older or detached leaves.

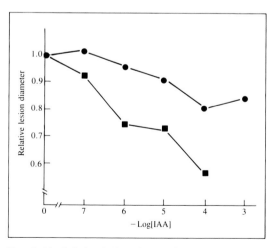

Fig. 7–11. Relative lesion size on Samsun NN tobacco leaves, 14–18 cm long, sprayed with different concentrations of IAA 6–24 h after TMV inoculation (●), and on leaves 10–13 cm long, sprayed 12 h after inoculation (■) (after van Loon, 284).

Experiments with 2,4-D were somewhat anomalous; it reduced lesion size when applied at low concentration but caused an increase at high concentration, in confirmation of a previous report by Simons et al. (244, 245). Though IAA treatment decreased peroxidase activity of leaves prior to virus inoculation and at the same time led to the formation of smaller lesions, the two effects were considered unrelated. The role of peroxidase in virus limitation will be considered more fully in Chap. 10. The observation that the effects of auxin could be simulated with ethephon (Ethrel, 2-chloroethylphosphonic acid), which releases ethylene on contact with the leaf surface, led van Loon to suggest (283) that the action of auxin was mediated by production of ethylene.

Application of exogenous auxins may also affect virus replication, though again the effect depends on several factors including the age of the leaf, the virus/host combination, and the tissue in question (whether inoculated or systemically infected) (228). IAA, 2-naphthylacetic acid, and indolebutyric acid reduced virus accumulation both in tobacco systemically infected with TMV (135) and in tissue culture cells (137, 138). The increase in accumulation of TMV in inoculated tobacco leaves observed by Kluge (132) was at variance with the decrease reported by van Loon (282) following IAA treatment of Samsun tobacco. However, the auxin-induced reduction observed in these experiments led van Loon to speculate that the influence of IAA on lesion size in young, hypersensitive tobacco leaves could be mediated both by a restriction of virus multiplication and by a stimulation of factors involved in virus replication. He did not, however, measure virus accumulation in his hypersensitive tobacco, Samsun NN, and the assumption that virus content is linearly related to lesion diameter may not always be warranted (17, 21, 80). 2,4-D also increased TMV accumulation in tobacco (132, 135) and in *Physalis floridana* (45), while in cotton, which normally allows only a minimal degree of virus replication, considerably increased replication was induced by treatment with IAA or 2,4-D (46). 2,4- D was also able to influence TMV multiplication in tobacco protoplasts (150), stimulating replication in protoplasts derived from the hypersensitive hosts Samsun NN and Xanthi nc; in protoplasts from Samsun, which reacts systemically to TMV, multiplication was reduced.

Several enzymes, the IAA-oxidases, and peroxidases are able to oxidize auxins in vivo. Also, virus infection generally leads to an increase in peroxidase activity, correlating with symptom severity, when IAA level might be expected to decrease. Gaborjanyi et al. (88) found, for example, that the virus-induced stunting of Xanthi nc tobacco systemically infected with CMV or Samsun infected with TMV was correlated with an increased peroxidase activity in leaf and stem tissues; the stunting of CPMV-infected cowpea hypocotyls was also accompanied by increases in IAA oxidase and peroxidase activity. Increases in peroxidase activity are more pronounced in hypersensitive hosts where necrosis occurs and, according to van Loon's experiments on Samsun NN tobacco, correlate with reduced auxin levels. While auxin treatment led to a reduction in lesion size, infiltration of horseradish peroxidase (which has IAA-oxidase activity) increased lesion size. However, though correlations of this sort exist, they are not universal (eg. 204), and the role of enzymes with IAA-oxidase activity in symptom expression remains unclear. The coumarin, scopoletin, is an inhibitor of IAA-oxidase in vitro, though in vivo it may either stimulate or inhibit the oxidation of IAA depending on the conditions (7, 8). Its concentration has been observed to increase in TMV-infected tobacco (197), although this was also accompanied by a decrease in auxin content. The role of scopoletin in the auxin balance of infected plants, therefore, also remains unresolved (see Chap. 6, 10).

Gibberellins

In those instances where virus infection is accompanied by stunting, there is often a decrease in the gibberellin content of infected plants. Both Bailiss (12, 13) and Aharoni et al. (4) have, for example, observed a decrease in gibberellins or gibberellinlike activity in CMV-infected cucumber, a decrease that correlated with a retardation of stem elongation. A similar decrease in gibberellic acid (GA_3) was observed in BYDV-infected barley (222), but not in TAV-infected tomato plants (11). There are also suggestions that the gibberellins in CMV-infected cucumber plants differ qualitatively as well as quantitatively from those in healthy plants (26) and may also be metabolized differently. In infected tissue, active forms may be converted to inactive forms or, alternatively, inactive forms may fail to be converted to active forms.

In addition, there are several reports of the partial reversal of virus-induced stunting by exogenous application of gibberellins (usually GA_3). Maramorosch (163) observed the stunting associated with, inter alia, WTV infection of crimson clover to be partially overcome by spraying with gibberellic acid, while stunting caused by infection of tobacco with severe etch virus (47, 259), of tomato with TAV (11), and of tobacco or pepper with TEV (79) could also be reduced by gibberellic acid treatment. Application of gibberellic acid also increased the number of lesions produced by TMV in leaves of a hypersensitive tobacco (227).

It is doubtful, however, that application of gibberellins counteracts the effect of virus infection, a conclusion reached by Yerkes (308) from studies on bean infected with stunting bean virus. He found that gibberellic acid treatment applied at the time plants were infected was effective in reducing stunting, while application later in the development of the disease was less effective. In addition, exogenous gibberellins stimulate stem elongation whether plants are infected or not (e.g. 259). It also appears that infected plants treated with gibberellic acid do not attain the same height as treated healthy plants.

More importantly, it appears that, at least in TAV infection of tomato (11) or BYDV infection of barley

Table 7–1. Effect of spraying 10 ppm GA_3 on mean cell number in third leaf blade of healthy and BYDV-infected barley (after Russell and Kimmins, 222).

	C	C/GA$_3$	IA	IA/GA$_3$
Length of third leaf blade (mm)	199.2	240.6*	162.1	214.8*
Fresh weight of leaf blade (mm)	287.5	321.5*	226.6	231.7*
Number of cells/blade	37.39×10^5	37.15×10^5	26.32×10^5	26.38×10^5

C, healthy plants; C/GA$_3$, healthy plants sprayed with 10 ppm GA$_3$; IA, infected plants; IA/GA$_3$, infected plants sprayed with 10 ppm GA$_3$; *, different at 1% level.

(222), stunting is the result of a reduction in cell number compared to control plants, rather than a difference in cell size, suggesting that virus infection affects mitosis. Application of gibberellic acid, on the other hand, rather than counteracting the effects of virus infection by stimulating cell division, stimulates cell expansion (Table 7–1). This is not to say, however, that endogenous gibberellins may not influence cell division. Although exogenous GA$_3$ did not influence cell division in either healthy or BYDV-infected barley (222), or in TAV-infected tomato (11), cell division was affected in healthy tomato plants by GA$_3$ treatment. Since it has also not proved possible to detect changes in endogenous gibberellins prior to the onset of symptoms (4,12,41,222), it is difficult to assign a causal role to gibberellins in virus-induced stunting. It cannot, however, be ruled out that the depression of gibberellin levels either directly or indirectly, possibly through interaction with other hormones, contributes to the retardation of plant growth.

Cytokinins

In spite of the potential importance of cytokinins in the mediation of symptom expression or virus replication, there have been few investigations on the changes in cytokinin concentration resulting from virus infection. The available information is to some degree conflicting. The cytokinin content of cowpea root tissue, of cowpea root exudates (136), and of inoculated leaves of *N. glutinosa* (269) was observed to decrease following inoculation of cowpea or *N. glutinosa* leaves with TRSV. By contrast, Sziráki and his colleagues (21, 265) found an enhancement of the total cytokinin level in half-leaves of the hypersensitive host Xanthi nc tobacco inoculated with TMV and in uninoculated leaves of Xanthi nc plants previously inoculated on the lower leaves. In leaves of Samsun tobacco systemically infected with TMV, cytokinin content was approximately twice that in leaves of comparable healthy plants one month after infection and was greater in dark green than in light green tissue (263). Chromatography of crude extracts suggested that at least part of the activity in these two types of tissue was due to different cytokinins, though they were not identified.

Although the presence of residues with cytokinin activity in TMV RNA has been reported (263, 264), more recent gas chromatographic studies indicate this possibility to be unlikely.

The effect of *treatment* of plants with cytokinins is critically dependent on the timing of the treatment in relation to infection; the type of cytokinin used and its concentration; whether attached leaves, detached leaves, or leaf discs are used; and the age of the plant. The effect, which may be on virus multiplication, symptom expression, or both, is generally one of reduction, when plants are treated before or very soon after infection, and of stimulation when treatment is begun after infection. There are, of course, exceptions.

Treatment of *Datura stramonium* (215), Xanthi (127) or White Burley (necrotic) (180) tobacco, or pinto bean (127, 182) with kinetin prior to inoculation with TMV led to an apparent reduction in the number of local lesions. Similarly, application of N^6-benzyladenine to petunia leaf strips from 24 h before to 72 h after TSWV inoculation, or to detached *N. rustica* leaves from 48 h before to 72 h after inoculation, also reduced lesion number (6). Milo and Srivastava (175) also found treatment of TMV-inoculated pinto bean leaves with several cytokinins, including kinetin, kinetin riboside, and N^6-benzyladenine to decrease lesion number when applied for 3 d after inoculation, although Nakagaki (182) found the reduction in local lesion number produced by kinetin treatment of TMV-infected pinto bean leaves to be effective only up to 3 h after infection.

Application of cytokinins may also influence the size of lesions. Application to *D. stramonium* plants prior to TMV inoculation has been reported to reduce lesion size (215), although Mukherjee (180) found pretreatment of tobacco with kinetin to increase TMV lesion size (while at the same time decreasing the number). The size of lesions produced by TSWV on petunia or *N. rustica* was also reduced by N^6-benzyladenine (6).

Accumulation of infective virus was reduced by treatment of *D. stramonium* (215) or tobacco (128) with kinetin before inoculation, or by treatment of tomato with N^6-benzyladenine before TSWV infection (6). Pretreatment of White Burley tobacco plants with methyl benzimidazol-2-yl-carbonate (carbendazim), the fungitoxic principle of the systemic fungicide benomyl, not only reduced the severity of symptoms produced by TMV infection but also significantly reduced virus accumulation (82, 83, 274), an effect possibly related to its cytokininlike activity.

The results of postinoculation treatments, however, seem to be less predictable. Virus production was stimulated in inoculated leaves of the systemic hosts *N. rust-*

ica and Wisconsin 38 tobacco, excised and floated on solutions of several cytokinins for 3 d, starting 10–20 min after infection (175). These results confirmed earlier reports by Daft (56), who treated excised tobacco leaves with kinetin 24 h after infection and observed a stimulation in TAMV production. Treatment of attached leaves of tobacco with N[6]-benzyladenine also increased virus accumulation (5), though a decrease in accumulation was reported by Reunov et al. (216) following kinetin treatment of TMV-inoculated Samsun tobacco leaves, and by Király and Poszár (128) using higher kinetin concentrations. These discrepancies may, however, be partly explained by differences in the time after inoculation at which infectivity was assayed; since this can significantly affect the results (115), it must also be borne in mind that leaf detachment per se can influence events.

The importance of timing has also been emphasized by the studies of Aldwinckle (5), in which attached leaves of tobacco were inoculated with TMV and half-leaves were subsequently treated with N[6]-benzyladenine for 6 d, beginning at various times after inoculation. When treatment was begun as soon as 1 min after inoculation, virus yield compared to that in untreated half-leaves was reduced, while treatment begun 5 min after inoculation led to increased accumulation of infective virus. Maximum stimulation was achieved when treatment was begun 60 min after inoculation.

Postinoculation treatment of pinto bean with various cytokinins decreased accumulation of TMV (175), at the same time decreasing lesion number. Similarly, Király and Szirmai (129) observed a decrease in both virus accumulation and lesion number following kinetin treatment of TMV-inoculated *N. glutinosa* leaf discs; Milo and Srivastava, however, found a stimulation of virus production in *N. glutinosa*, in spite of the complete suppression of local lesion formation (175).

The importance of cytokinin concentration has been demonstrated by treatment of *N. glutinosa* for 9 d prior to inoculation with TRSV (269). Solutions containing 0.1 μg/ml kinetin stimulated virus production, while at 1–10 μg/ml virus accumulation was reduced. Floating tomato discs inoculated with TMV on solutions of kinetin ranging in concentration from 20 to 0.002 μg/ml led to an increased virus accumulation at the lower concentration but decreased accumulation at the higher (15). High concentrations of kinetin, of the order of those used by Király et al. (127), have been found by others to be cytotoxic (175).

Clearly, cytokinin treatment could be affecting one or more of several events, from the initial establishment of infection to subsequent virus multiplication and, in the case of hypersensitive hosts, the development of necrosis; in many experiments, it is not possible to distinguish between them. It may be that a reduced virus accumulation resulting from cytokinin treatment of plants before inoculation arises from a stimulation of host RNA and protein synthesis, circumstances in which virus RNA may have difficulty in competing. Conversely, treat-

ment after virus inoculation, when virus RNA is established intracellularly, may stimulate virus RNA and protein synthesis to an extent at least comparable with the stimulation in host RNA and protein synthesis.

There are indications, however, that the primary influence of cytokinins, whether applied before or after virus inoculation, is on events subsequent to the establishment of infection. Balázs et al. (17), for example, found the number of visible lesions produced in *Xanthi nc* tobacco leaves by TMV infection to be reduced by spraying with kinetin solution prior to inoculation. However, staining of the necrotic cells with Evans blue and microscopic examination of the stained tissue revealed that the total number of lesions was not reduced, but rather that some consisted only of single necrotic cells, invisible on visual inspection. Virus infection would appear to be established to similar extents in treated and untreated tissue, but in tissue receiving kinetin some lesions are not permitted to develop. Although there was no detectable infectivity associated with the microlesions, infectivity recovered from lesions that did develop was, in fact, greater than that from controls. Extending the hypothesis of others, Balázs et al. suggested (17) that there are cells on the leaf surface that are more susceptible to infection and, in these, kinetin, by stimulating RNA and protein synthesis, stimulates virus synthesis. However, in cells that, for unspecified reasons, may be less susceptible to virus infection, host RNA and protein synthesis are stimulated, but virus RNA is unable to compete successfully. Virus multiplication is negligible, and a necrotic lesion does not develop.

Recent studies by Kasamo and Shimomura (115), in which the epidermis is removed from TMV-inoculated Samsun NN tobacco leaves at various times after inoculation, also suggest that the role of kinetin applied after infection may differ in different cells. Virus accumulation was depressed by kinetin treatment of isolated lower epidermis but was stimulated in areas of the leaf where the epidermis had been removed. They suggest that the effect of kinetin on the multiplication of TMV in intact discs is the result of a balance between stimulation in mesophyll cells and inhibition in epidermal cells.

Balázs and his colleagues (21, 266) have also proposed a role for cytokinins in the expression of systemic induced resistance (see Chap. 10). *Xanthi nc* tobacco plants, inoculated on their lower leaves with TMV, CMV, or *Pseudomonas pisi* or treated with HgCl$_2$, become systemically resistant. Upper leaves contain a slightly greater cytokinin concentration than controls and produce fewer visible lesions when subsequently challenged. However, infectivity recovered from "protected" leaves was little different from that from controls and, in the case of plants previously inoculated with TMV, several microlesions were also observed on challenged leaves. This led them to suggest that increased cytokinin concentration was inhibiting the development of necrosis rather than virus multiplication. Their view was strengthened by the observation that exogenous application of kinetin reduced necrosis induced by $HgCl_2$. However, their data were not unequiv-ocal, since there was no evidence to suggest whether or not virus was multiplying in and around the microle-sions. It is not clear whether, in leaves showing induced resistance, necrosis is inhibited at some sites while rep-lication proceeds normally, although this would appear to be the case in TMV-inoculated White Burley tobacco (Chap. 10). Nor is it apparent whether at some sites both replication and necrosis are inhibited, with an increase in virus multiplication at other sites that develop normal lesions. Also, while the evidence is consistent with the view that cytokinins are directly involved in the ex-pression of systemic induced resistance, it is as yet no more than a correlation and further studies are required.

Ethylene

Although experiments on ethylene in this context are limited, evidence suggests that a stimulation in ethylene production is associated with virus infection and that the degree of stimulation depends on the severity of the symptoms. In tobacco reacting hypersensitively to TMV infection with the formation of necrotic lesions, there is a significant increase in ethylene production (19, 88, 183, 209). Pritchard and Ross, for example, found ethylene to be stimulated "at or about" the time of lesion appearance, 36–42 h after infection, and to reach a maximum stimulation 3–5 d after infection. Fifteen days after infection, ethylene production in virus-infected plants was still 5–10 times that of con-trols. (By contrast, ethylene stimulation as a result of injury was observed only during necrotization, and for-mation quickly declined to control levels.) The rate of ethylene synthesis was, in the majority of cases, directly related to the rate at which lesions developed, and the amount produced was linearly related to the extent of necrosis. Inoculation of "resistant" upper leaves on plants previously inoculated on their lower leaves led to a similar pattern of ethylene stimulation, though in this case the extent of the enhancement was rather less than in nonresistant healthy tissue.

In attempting to assign a role for ethylene in necrosis, the timing of its stimulation is critical. However, since necrosis of initially infected cells may occur some time before the lesion is visible, and ethylene produced by a few cells is difficult to detect, the relative timing of the two events has not been determined with confidence. Stimulation of ethylene production has usually been ob-served to occur during lesion formation (19, 67, 68, 183, 209), although van Loon (284) found an increase prior to lesion appearance in TMV-inoculated Samsun NN tobacco (Fig. 7–12), and Pritchand and Ross ob-served stimulation slightly before lesions appeared in resistant leaves (209). Kato (116) found that, in CMV-inoculated cowpea leaves, both precursors of ethylene and a pathway of ethylene biosynthesis were affected before and after lesion appearance. It seems likely that, in virus-infected tissue, ethylene is a product of reac-tions leading to necrosis, though not a cause of it. It is apparent, however, that in some circumstances ethylene can be involved in necrosis, since administration of 2-

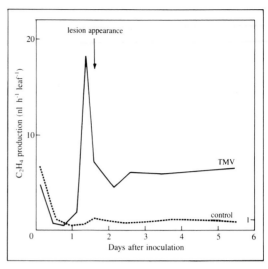

Fig. 7–12. Ethylene production of water-inoculated and TMV-infected Samsun NN tobacco leaves (after van Loon, 284).

chloroethylphosphonic acid (Ethrel, ethephon) to to-bacco leaves by pricking with a pin (284) led to the formation of necrotic lesions similar to those induced by virus infection.

It is possible that ethylene may influence lesion en-largement; although an inhibitor of ethylene synthesis did not affect the size of TMV lesions in tobacco (209), preinfection treatment of Xanthi nc or Samsun NN to-bacco with ethylene or 2-chloroethyl phosphonic acid led to an increase in the number of visible TMV lesions (14, 19), a decrease in lesion size (209), and a decreased accumulation of infectivity in those lesions that did ap-pear. Treatment after infection increased lesion number (19) and size (19, 209). However, as indicated pre-viously, it is difficult, particularly with preinfection treatments, to be sure which stage of the infection pro-cess is being influenced. The possible role of ethylene in lesion limitation will be considered, along with other factors, in Chap. 10.

While ethylene does not seem to be stimulated in tobacco systemically infected with TMV (19, 86, 183), there is evidence for the involvement of ethylene in the expression of other types of symptoms that do not in-volve necrosis. Enhancement of ethylene production accompanied the formation of chlorotic lesions in *Tetragonia expansa* inoculated with BYMV (14); ex-posure of plants to ethylene promoted the development of necrotic spots in the lesions, while treatment of plants with CO_2, an inhibitor of ethylene action, restricted lesion development. Cucumber seedlings develop chlo-rotic lesions on cotyledons inoculated with CMV, the plants become epinastic, and hypocotyl elongation is inhibited. Symptom development was associated with a stimulation of ethylene production (142, 164, 165). An increase in endogenous ethylene occurred about 24 h before epinasty was apparent; exogenous application of ethylene promoted epinasty in uninoculated seedlings, while removal of ethylene from infected seedlings prevented epinasty (142). Similarly, enhanced ethylene production was detected in advance of visible chlorosis, and, again, when ethylene was removed development of chlorosis was suppressed (165). Although the difficulties involved in assessing the timing of the start of ethylene stimulation and symptom expression were recognized, the authors were of the opinion that ethylene was probably causally involved in the expression of these symptoms.

There are also indications that ethylene can influence virus replication in systemic hosts, though its effect is likely to be indirect. For example, an increase in virus accumulation was observed following postinoculation Ethrel treatment of Xanthi tobacco (18), but not after treatment of TMV-inoculated *Physalis floridana* (46). TMV accumulation was also stimulated by postinfection treatment of cotton seedlings, in which this virus normally accumulates only to a very low level. Multiplication of CMV was not stimulated by postinoculation Ethrel treatment of resistant cucumber plants (24).

The mechanism by which ethylene is stimulated is unknown, although enhancement may result from stimulated abscisic acid concentrations and in turn may lead to alterations in auxin metabolism or transport (149).

Abscisic Acid

In contrast to the observations on gibberellins, the relatively few investigations of changes in the concentration of the growth inhibitor ABA have, with few exceptions, been concerned with tobacco and have revealed a stimulation by virus infection. The two exceptions are the studies by Rajagopal (210), who found a decrease in ABA concentration in the early stages of systemic infection of tobacco by TMV, and by Bailiss (13), who found there was no difference in ABA levels between CMV-infected and control cucumber plants, in spite of considerable virus-induced stunting. Marco et al., however, found ABA to be stimulated in CMV-infected cucumber (4).

Bailiss did, however, find a progressive increase in endogenous ABA concentration in TMV-inoculated leaves of Xanthi nc tobacco as lesions developed (14). Whenham and Fraser (297) also found ABA to be stimulated in White Burley tobacco, both by the *vulgare* strain of TMV multiplying systemically (Fig. 7–13) and also by the *flavum* strain, which produced local lesions (Fig. 7–14). In systemically infected tissue, free ABA was stimulated up to sixfold compared to controls, while the bound form (which they presumed to be the glucosyl ester) was unaffected. In leaves inoculated with the *flavum* strain, stimulation of both forms was observed, with free ABA concentrations up to 18 times those of uninfected leaves. It was suggested that in these tissues the rise in ABA could be a consequence of water loss, caused by tissue necrosis.

Leaf growth was curtailed in both systemically infected plants and uninfected leaves of plants inoculated with the *flavum* strain; these leaves also had stimulated

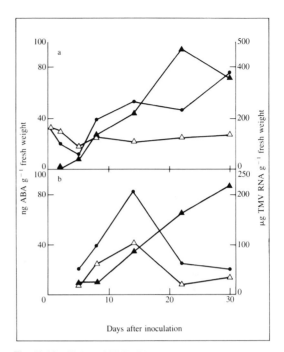

Fig. 7–13. ABA and TMV RNA concentration in *vulgare*-infected tobacco leaves. Plants were inoculated with TMV strain *vulgare* diluted 1:2x10³ (●) or with buffer (△). Free ABA concentration was compared in (a) inoculated leaves and in (b) the 3 leaves above the upper inoculated leaf. (▲) TMV RNA concentration (after Whenham and Fraser, 297).

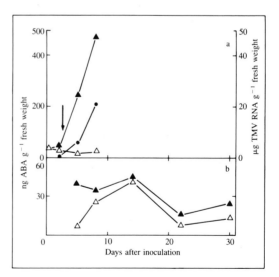

Fig. 7–14. ABA and TMV RNA concentration in *flavum*-infected tobacco plants. Plants were inoculated with TMV strain *flavum* diluted 1:2 × 10² (▲) or with buffer (△). Free ABA concentration was compared in (b) inoculated leaves and in (a) the 3 leaves above the upper inoculated leaf. (●) TMV RNA concentration in *flavum*-inoculated leaves. The arrow indicates the approximate time of lesion appearance (after Whenham and Fraser, 297).

262 · **Biochemistry and Physiology of Plant Disease**

ABA levels. Since both ABA and virus infection inhibit cell division, it was suggested that the stimulation in ABA was a major cause of the observed reduction in growth of virus-infected leaves. This view was strengthened by the observation that application of ABA to plants in a quantity sufficient to bring the endogenous concentration to a level comparable to that of virus-infected plants caused also a comparable inhibition of growth.

In addition to its effect on plant growth, ABA can influence other aspects of the interaction between virus and host. For example, treatment of tobacco (Xanthi nc) discs or attached leaves prior to infection led to a stimulation in the number of lesions that appeared (14, 20). However, when physiological rather than phytotoxic levels were applied, Fraser found both lesion number and size to be reduced (81). Clearly, pretreatment could be influencing the establishment of infection, although the reduction in virus accumulation in lesions in treated leaf tissue suggests that subsequent events are also influenced. This is supported by work on systemic hosts. In contrast to the reduction in infectivity associated with lesions in the hypersensitive host, a stimulation of viral RNA synthesis (296) and virus accumulation (20) was observed on treatment of TMV-inoculated discs of Samsun or White Burley tobacco. Although ABA can influence growth and the development of senescence and, more specifically, can affect RNA synthesis (82), its role in virus multiplication and virus-induced symptom expression is unresolved. Again, it may act in concert with other hormones, such as ethylene.

Although some progress has been made, it will be apparent that the role of hormones in virus infection largely remains to be elucidated (84).

Plants Infected by Bacteria

Just as an augmented respiration rate is an almost to be expected symptom of pathogenesis, so is hyperauxiny. Four general questions are usually raised in this regard: (1) Does the increased level of growth regulator reflect production of auxin by the pathogen? (2) Does it reflect increased synthesis by the host? (3) Does it indicate a suppression of the normal degradative process of the growth regulator (e.g. IAA oxidase)? (4) Is there a causal relationship between growth regulator accumulation and the establishment of the pathogen in host tissue pathogenesis?

Synthesis of IAA by the Bacterial Pathogen

In an exhaustive study of the biosynthesis of IAA by *Agrobacterium tumefaciens*, Kaper and Veldstra (114) concluded that tryptophan was the principal precursor for the production of IAA. The pathways most commonly considered for the formation of IAA from tryptophan were presented above in Fig. 7–1 (184).

Kaper and Veldstra (114) have shown that IAA synthesis by *A. tumefaciens* proceeds via indole-3-pyruvic acid and indole-3-acetaldehyde. Their study also indicates that under anaerobic conditions indolepyruvic acid might also serve as a precursor for indolelactic acid

and tryptophol. However, they were unable to detect tryptamine, another possible precursor, as an intermediate in the synthesis of IAA by *A. tumefaciens*.

It has also been conclusively demonstrated by Magie (157) that *Pseudomonas savastanoi* converts tryptophan to indoleacetic acid, with indoleacetamide as an intermediate. His experiments also revealed that when 10^{-3} M D-tryptophan was used as substrate, neither indoleacetamide nor indoleacetic acid was produced by the enzyme preparations obtained from sonicated *P. savastanoi*. The first reaction is catalyzed by tryptophan 2-monooxygenase and the second by indole acetamide hydrolase (53, 230, 234, 301). Both indoleacetamide and IAA were apparent, however, in reaction mixtures when either DL or L-tryptophan was used as substrate. These results suggest that this pseudomonad utilizes the L isomer of tryptophan preferentially.

Additional understanding of the possible role of IAA, and in particular its accumulation in pathogenesis, is derived from studies of the Granville wilt organism, *Pseudomonas solanacearum*. One of the symptoms of this disease is a twisting and bending of stem and leaves (epinasty), which led Grieve (93) in early studies to suspect an auxin effect. Sequeira and Kelman (235) attempted to define more clearly whether IAA actually plays a part in pathogenesis by *P. solanacearum*. They observed that artificially infected tobacco plants contain significantly greater amounts of a growth-promoting substance not found in healthy plants. The growth-regulating substance proved to be IAA, and the extent of the increase is from 0.03 μg/100 g in healthy tissue to 3.3 μg/100 g in diseased tissue. Similar increases were recorded in banana tissue inoculated with a strain of the pathogen capable of attacking that species.

Origin of increased auxin levels in the host. An obvious question at this point concerns the nature of the source of IAA detected in plants infected with *P. solanacearum* (233, 237). Was it host-synthesized or was it produced by the pathogen? An analysis of the IAA content of leaf, stem, and root tissue of healthy and diseased plants revealed that auxin could be detected in healthy plants with certainty only in the leaves whereas in the diseased plants the greatest amounts of the hormone were located in the stems and roots, which were precisely the sites in which the pathogen had proliferated most actively.

An accumulation of auxin at the site of bacterial proliferation fails to implicate either the pathogen or the host as the synthesizer of this "greater-than-normal amount of IAA." As we shall see later on, the fact that synthesis is increased is also not fully established, since the increase may reflect, for example, a suppression of auxin degradation by IAA oxidase.

Studying the basis for the increased auxin levels in tobacco infected by *P. solanacearum*, Sequeira and Kelman (235) recorded 33% increases 48 h after inoculation. Later Pegg and Sequeira (199) reported that during a similar period the level of tryptophan increased 500%. There is, therefore, no shortage of precursor.

Table 7–2. Specific activity of IAA from liquid culture of *Pseudomonas solanacearum* grown 48 h in the presence of C^{14} chain- and ring-labeled DL-tryptophan (after Sequeira and Williams, 237).

Label	Strain	Count/min - μ M $\times$ 1,000
Chain	K-60 (wild-virulent)	0.9
	B-1 (mutant-avirulent)	7.6
Ring	K-60	6.0
	B-1	10.4

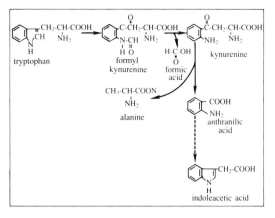

Fig. 7–15. Aromatic pathway of tryptophan metabolism as employed by the K-60 strain of *P. solanacearum*. The broken line to IAA indicates that the intermediates are not known (after Phelps and Sequeira, 203).

To determine the origin of the auxin increase, Sequeira and Williams (237) sought to exploit the known differences in the pathways of IAA synthesis that had been reported for higher plants and microorganisms. Thus, with radioactive precursors that only one member of the host-pathogen complex could convert to IAA, the contribution of each to the observed increase could be estimated.

Preliminary experiments showed that the tobacco plant could convert tryptamine to IAA, whereas the pathogen was unable to do so. Of additional interest was their discovery that a wild type virulent strain of *P. solanacearum* (K-60) and an avirulent mutant strain (B-1) converted trytophan to IAA by different pathways. The data substantiating this conclusion appear in Table 7–2 and are interpreted as indicating that the K-60 strain synthesizes IAA from tryptophan by a completely different pathway, not involving incorporation of the side chain of tryptophan.

Chromatography of the reaction products from K-60-fed chain-labeled tryptophan revealed low specific activity of labeled IAA and high concentrations of unlabeled anthranilic and 3-OH anthranilic acids. These results suggest that the K-60 strain incorporates the benzene ring or a portion of it into IAA and that this breakdown is accomplished via the kynurenine pathway which has been found operative in a number of microorganisms (see Fig. 7–15).

Increased auxin synthesis by the host. It is difficult to state unequivocally the degree to which host and parasite each contributes to the increase in IAA detected in the diseased plant. It is clear that a number of bacterial pathogens are capable of producing large quantities of IAA in vitro. To what extent the pathogenic attack stimulates the host plant to elevate its production of IAA remains partially obscured. The fact that auxin accumulation was detected in tomato plants within 5 d after inoculation and continued to increase even 20 d after inoculation, when the plants had completely wilted, strongly suggests that hyperauxiny is due to some extent to synthesis by the pathogen (203).

Sequeira (232) reinvestigated the inability of *P. solanacearum* to utilize tryptamine in synthesizing of IAA in order to determine the origin of auxin in *P. solanacearum*-infected tobacco. Feeding tryptamine-2-C^{14} to inoculated and healthy plants, he found that IAA

increased more than 100% over the healthy controls, by the fifth and sixth day after inoculation. Parallel increases in labeled IAA were also recorded through the fourth and fifth days; since the pathogen is unable to metabolize tryptamine, these increases must be due to host synthesis.

In addition, the amount of radioactive IAA was sufficiently higher during the early phases of infection. This suggests that IAA synthesis by the host is correlated with host increases in respiratory metabolism and dry weight. Such increases, it will be recalled, were reported by Maine (159, 160; see also Chap. 3).

During the later stages of infection (sixth to ninth day) the amount of detectable radioactive IAA lessened. Since auxin was still found in large amounts in diseased tissue during this period, it was suggested that a part of the auxin present was of bacterial origin.

Suppression of normal degradative processes of auxin. The phenomenon of hyperauxiny in plants infected by bacteria that are capable of synthesizing large amounts of IAA has two possible causes: (1) the increase in IAA reflects its synthesis by either the pathogen and/or the host, or (2) the normal degradation of IAA, which occurs in healthy tissue, is impaired in the infected host by some inhibitor or inhibitors. Although the data we have at present allowing us to discriminate between the contribution of host and pathogen are meager, there is indirect evidence that hyperauxiny reflects, at least in part, an inhibition of the normal degradative process of this hormone (206).

The evidence that hyprauxiny is due to a suppression of IAA oxidase has been supplied by Sequeira (231). He reported that the activity of IAA oxidase in tobacco, 48 h after inoculation with *P. solanacearum*, was considerably depressed. This reduction in activity continued over a 14-d period, by which time IAA destruction in roots of diseased plants was one-third that in healthy plants. Degradation of IAA by the oxidase in tobacco

Table 7–3. IAA destroyed after half-hour incubation in dialyzed and undialyzed extracts from tobacco roots (after Sequeira, 231).

Extract	Source	μg/ml IAA destroyed after ½ h incubation
Undialyzed	Healthy	18.5
	Diseased[b]	6.0
Dialyzed[a]	Healthy	28.5
	Diseased[b]	28.0

a. Extract dialyzed against distilled water for 24 h at 4 C.
b. Plants inoculated with *P. solanacearum* and maintained at 28 C for 10 d in the greenhouse.

plants seemed dependent on the presence of two cofactors, Mn^{+2} and 2,4-dichlorophenol, and this dependence was made evident in dialyzed enzyme-containing extracts. Dialysis evidently removed certain factors responsible for suppressing IAA oxidase activity, as shown in Table 7–3. It was noted that the effect of dialysis on IAA oxidation was greater on extracts of diseased than healthy roots, suggesting a prevalence of IAA oxidase inhibitors in the former. The data in Table 7–3 also indicate that, in fact, there is only about one-third the IAA oxidase activity in extracts of diseased tobacco tissue.

Extracts from infected plants that showed marked activation of IAA oxidase following dialysis were examined for the presence of water-soluble inhibitors. The dialyzates revealed a number of phenols, and scopoletin (7-hydroxy-6-methoxycoumarin) was present in highest concentration. Furthermore, this phenol increased progressively in concentration in diseased root tissue from the second to the tenth day after inoculation, at which time the scopoletin level was ten times that found in comparable healthy tissue.

Sequeira also attempted to determine whether the levels of scopoletin detected in diseased tissue were actually capable of suppressing IAA oxidase activity. It was calculated that extracts from roots of diseased tobacco plants, which contained 3–12 μg/ml scopoletin, significantly delayed the destruction of auxin. In parallel in vitro experiments, it was observed that 9 μg/ml almost totally suppressed the enzymatic destruction of auxin.

The enzyme preparations used by Sequeira (231) contained considerable catalase, polyphenoloxidase, and ascorbic acid oxidase activity in addition to IAA oxidase. Attention is called to this fact since Maine (159) and Maine and Kelman (160) found that oxidation rates of both polyphenols and ascorbic acid increased rapidly in tobacco stem tissue infected with *P. solanacearum* (see Chaps. 3 and 6). Thus it would seem that IAA oxidase is selectively suppressed early in pathogenesis and, unlike other oxidative enzymes, suffers increasing inhibition of its activity during the course of disease development.

It may also be postulated, therefore, that at least a part of the observed increase in IAA is due to the suppression of IAA oxidase, probably by scopoletin.

The relationship of IAA accumulation to pathogenesis in the Granville wilt disease. A relationship of hyperauxiny to pathogenesis is discernible from experiments on artificial inoculation of susceptible and resistant varieties of tobacco with *P. solanacearum* (235). Although proliferation of the pathogen is vastly greater in susceptible tissues, auxin also accumulated in tissues of resistant plants in the limited area of infection. That hyperauxiny seems to be a consequence of bacterial proliferation in host tissue was demonstrated by Sequeira and Kelman (235). They observed IAA accumulation to occur in diseased stem tissues of both resistant and susceptible varieties where suspension of the pathogen had been uniformly taken up through cut stem bases. With this technique, the influence of superior rate of multiplication of the pathogen in susceptible tissue on IAA accumulation was minimized. It seems clear, therefore, that hyperauxiny is in some way linked to pathogenesis. Sequeira and Kelman (235) have suggested four ways in which hyperauxiny might influence pathogenesis.

(1) An abundance of IAA may increase or maintain cell-wall plasticity. This could conceivably be accomplished by the binding of pectin methylesterase to the cell wall by auxin. The normal increase in cell-wall rigidity, imparted by Ca^{++} ion cross-linkages (replacing methyl ester linkages), would be delayed. Succulence of the issues involved would thereby be maintained, and these would in turn be more susceptible to degradation by pectin-, cellulose-, and perhaps protein-degrading enzymes.

(2) IAA inhibits lignification (243). The concomitant accumulation of intermediates and precursors of lignin synthesis (chlorogenic acid, scopoletin, and other phenols) and IAA supports this concept. Thus, in still another way, the maintenance of tissue in the nonlignified succulent condition favors more rapid tissue disintegration.

(3) An increase in transpiration rate has been well documented for plants infected by *P. solanacearum* by Grieve (93) (Chap. 8). It was, therefore, suggested by Sequeira and Kelman (235) that abundant IAA in diseased stem tissue could be carried to the leaves where cell permeability is increased, perhaps by binding pectin methylesterase (62), producing a high rate of transpiration.

(4) Finally, the observation that the amino acid content increases in tobacco cuttings treated with auxin suggests that the release of IAA by the pathogen fosters accumulation of tryptophan, thus priming a mechanism for an augmented production of auxin. This seems to be the case in the later stages of infection (235). It is also possible that the increased amino acid level simply provides an environment more conducive to the proliferation of the pathogen.

The foregoing hypotheses and observations provide the rationale for linking hyperauxiny to pathogenesis in the case of Granville wilt.

Growth-Regulator Imbalance and Pathogenesis in the Crown Gall Disease

The essentiality of IAA to tumor formation. It has been amply demonstrated that IAA plays an essential role in tumor formation in plants inoculated with *Agrobacterium tumefaciens*. Braun and Laskaris (36), as well as Klein and Link (131), demonstrated that decapitated plants inoculated with *attenuated* strains of *A. tumefaciens* would develop galls only if IAA was supplied exogenously to the cut surface. Where IAA was not provided, the attenuated inoculum failed to induce typical tumors.

It is appropriate to offer here a more detailed appraisal of this unusual disease, incited by *A. tumefaciens*, which is perhaps the most widely known and intensively studied bacterial plant pathogen. The capacity of this pathogen to induce development tumor in plants has made it a variety of plant species a favorite model system among scientists studying cancer and carcinogenesis. Its influence upon normal cell growth and division has made the bacterium an object of intense study by plant physiologists and biochemists. Pathogenesis is reflected in sustained uncontrolled growth by the host. The relationship of auxin to the etiology of crown gall and the finding that this phytobacterial pathogen evokes at least a portion of its symptoms by altering the hormonal balance of the host give the crown gall disease special interest and importance.

The relationship of IAA to tumorigenesis. That the regulatory processes of growth are altered as a consequence of infection is readily apparent. The knowledge that some facets of plant cell growth are controlled by IAA has implicated this plant growth regulator in pathogenesis. As mentioned in the section on auxin-cytokinin metabolism of the healthy plant, cell enlargement of healthy tobacco pith tissue is fostered by supplying auxin to the tissue, and an even greater growth stimulus is noted when both IAA and cytokinin are provided (111). Braun (33) corroborated these findings with tobacco pith parenchyma cells and in addition revealed that, following the transformation of these cells to tumor cells, both substances were produced in greater than regulatory amounts.

That a hyperauxinic metabolism was at least partly responsible for tumor formation was established in a number of ways. For example, active tumor grafted into healthy tomato plants elicited symptoms such as epinasty, adventive root formation, suppression of lateral-bud growth, and the inhibition of abscission of senescent leaves (148). The failure of additional auxin to stimulate rapidly growing tumors suggested ample synthesis of the hormone by tumor cells. Conversely, as previously mentioned, tumors initiated by an attenuated culture continue to develop actively only when growth

substances are supplied exogenously or by a neighboring tumor incited by a virulent bacterial culture. Finally, as direct evidence, much greater amounts of free auxin were detected in cultured tumor tissue than in normal tissue (69).

What is the source of the apparent increase in auxin level? It was shown that crown gall tumor cells have acquired the capacity to synthesize larger amounts of IAA. Certainly the influence of bacterially synthesized IAA is not always a factor in this respect, since hyperauxiny exists in tumors that are *freed* of the bacterial pathogen. A number of early workers in this field (35, 60) concluded that crown gall tumor cells had acquired as a consequence of their (genetic) transformation the capacity to synthesize auxin in quantities greater than needed for normal regulatory purposes. However, proof for this contention remained inconclusive until recently when the nature of the long-sought TIP (tumor-inducing, or transforming, principle), was satisfactorily provided (169, 184). The apparent source of the increased amounts of IAA in tumor tissue is neatly demonstrated by Lui and colleagues (146, 147). The oncogenic and attenuated strains of *A. tumefaciens* (C-58 and NT1, respectively) are both able to produce large amounts of IAA when tryptophan is applied to the culture medium. Under these circumstances C-58 produces approximately twice as much auxin as NT1. However, when tyrosine is the substrate (see Table 7–4), C-58 synthesizes 7–10 times more IAA than the attenuated strain.

The basis for this is found in the recently gained understanding (169, 184) that tumorigenicity is conferred upon a host plant cell by the "insertion" of a plasmid (Ti) from an onconogenic strain of the pathogen (see Chap. 1). The results of Liu and colleagues (146, 147) indicate that IAA production is regulated by the Ti plasmid of strains like C-58. When this plasmid was inserted into the plasmidless NT1 strain by mating it with C-58, the resulting transconjugant (strain ID 1274) was fully onconogenic and produced IAA levels comparable to the wild-type onconogenic strain C-58. Furthermore when the transconjugant was "cured" of this plasmid (the plasmid was inactivated), the IAA levels it produced were equivalent to those of NT1. It would appear, therefore, that IAA production is regulated by the Ti plasmid of *A. tumefaciens*. The lower levels of IAA produced by the plasmidless strain suggest that there may be two operons for IAA synthesis, one chromosomal and the other on the plasmid. It is clear, however, that, to a significant measure, onconogenesis is due to inordinate amounts of IAA produced by the transformed plant cell.

The Mechanics of Tumorigenesis

The almost legendary TIP studied so arduously by Braun (33, 34) and Braun and White (38) was shown to be Ti plasmid DNA that could infect plant protoplasts (97) and induce the phenotypic responses characteristic of the crown gall disease (61, 130, 131, 169, 184). The tissue derived from the transformed protoplasts con-

Table 7–4. Indoleacetic acid production by *A. tumefaciens* in relation to plasmid content and its phenotype (after Liu and Kado, 146).

Strain	pTi plasmid present	MW of pTi plasmid ($\times 10^6$d)	Nopaline Synthesis	Nopaline Utilization	Indoleacetic acid in culture medium and cell extract (total μg[a]) with cells grown in specified aromatic amino acids Culture medium Tryptophan	Tyrosine	None	Cell extract Tryptophan	Tyrosine	None
C-58	yes	117	yes	yes	352.4	21.0	0.0	1,494.0	108.4	1.0
NT1	no	0	no	no	167.6	3.0	0.0	680.0	9.6	3.0
1D1274	yes	117	yes	yes	343.5	29.0	0.0	1,490.0	109.0	1.0

a. Total average amount extractable from 250 ml of culture medium freed of cells that had grown for 48 h at 25 C with shaking. Averages based on three experiments. Oncogenicity: C-58(+), NT1 (−), ID1274 (+).

tained the precise segment of the plasmid T-DNA that is integrated into plant DNA (184).

The hypertrophy derived from this transformation is a reflection of autonomous cell proliferation that ensues as a consequence of continuous and potentiated auxin and cytokinin biosynthesis by the transformed plant cell. These metabolic excesses are a part of the new information coded for on the transferred plasmid fragments.

It is generally conceded that wounding of host tissue is necessary for tumor induction. It seems that wound-related metabolites are required for tumor formation (34) and wounding may expose or induce the formation of binding sites. Wounding is associated with the "conditioning" requirement for transformation. According to Braun (33, 34), there is a variable time period, with an optimum of 24 h after wounding, when inoculated stem tissue of *Kalanchoe daigremontiana* becomes fully autonomous. Bacteria applied 24 h after wounding develop tumors that achieve maximum size.

Following wounding the bacteria attach to plant cell walls, through the lipopolysaccharide (LPS) component of their outer cell wall to a polysaccharide component of the plant cell wall (294, 295). Since pectinesterase potentiates binding it is plausible that binding is to the polygalacturonic acid facet of the plant cell wall (145).

The insertion of plasmid DNA into the plant cell, perhaps through the wound, is an area that remains very poorly defined. It may perhaps be advantageously studied by means of ultrahistochemistry that could reveal precise binding sites and the apparent conjugational event (184). That plasmid DNA (T-DNA), actually the T-region of the Ti plasmid, is actually inserted into a plant cell chromosome rather than remaining as a free replicon was fully established by Hernalsteens et al. (101). The apparent "conjugation" phenomenon is believed to be thermosensitive (35), yet much concerning the plasmid insertion process, through the plant cell wall and into the plant's genome, remains to be clarified. The development of sterile secondary tumors at

distances from initial sites of infection supports the possibility of movement of a transmissable agent (T-DNA or a fragment thereof) from tumorous to normal tissue (33, 37). However, evidence has been developed suggesting that secondary tumors developing in root-inoculated grape plants are initiated by *A. tumefaciens* per se. The tumors develop in aseptically wounded grape stem tissue that was initially root-infected with a strain of *A. tumefaciens* (biovar 3) resistant to streptomycin and rifampicin. The "labeled" strain of the pathogen was readily recovered from the secondary tumors (unpublished data, Tarbah and Goodman).

Hyperauxiny and the olive knot pathogen. The olive knot pathogen caused by *Pseudomonas savastonoi*, which causes tumor formation on its hosts, oleander and olive, reflects a hyperauxinic condition in infected tissue. Magie (157) presented evidence that *P. savastonoi* produced large amounts of IAA in culture. Subsequently, Smidt and Kosuge (250) established that IAA synthesis by the pathogen is necessary for gall formation and virulence on oleander. IAA$^-$ mutants failed to cause tumors in oleander, and isolates with heightened capacity to synthesize IAA caused the severest symptoms. Comai and Kosuge (52, 53) presented evidence that the enzymes concerned in IAA synthesis are coded for on a 34 md plasmid (PIAA-l) (Fig. 7–16). The enzymes involved in the regulation of these enzymes have been described by Nester and Kosuge (184).

Although IAA is clearly linked to pathogenesis in oleander, there is no evidence that infected plant cells are transformed as in the case of *A. tumefaciens*-infected plants.

The Influence of Cytokinins on Pathogenesis

Although kinetin per se has not been found in plants, compounds analogous to 6-furfurylaminopurine in physiological action, referred to as *cytokinins* (249), have been found in numerous higher plants. It will be recalled that healthy tobacco pith parenchyma needs

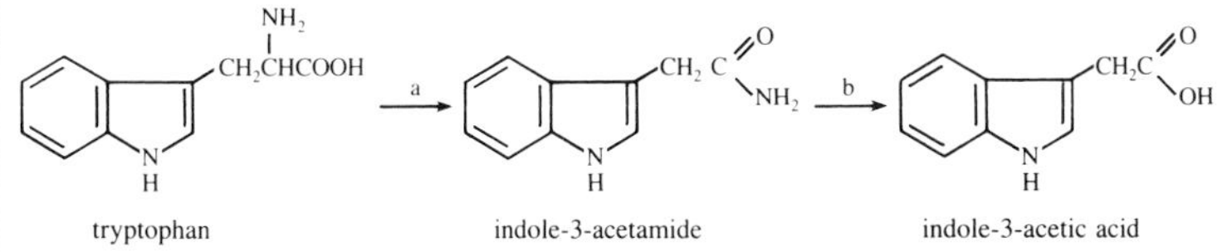

Fig. 7–16. Pathway of IAA synthesis in *P. savastanoi*. In reaction a, catalyzed by Trp monooxygenase, molecular oxygen and Trp react to yield indoleacetamide, CO_2, and H_2O. In reaction b, catalyzed by indoleacetamide hydrolase, indoleacetamide is hydrolyzed to yield indole-3-acetic acid and ammonia (after Comai and Kosuge, 52).

both auxin and cytokinin to promote both cell enlargement and division (108, 248). However, although minimally transformed crown gall tumor cells (34-h conditioning) required a number of growth factors and nucleic acid precursors, they had no absolute exogeneous requirement for cytokinin (32). It would seem that even weakly active tumor cells can synthesize enough kinetinlike material to accomplish cell division. Braun contended that this requirement by the normal cell for cytokinins constitutes a basic difference between normal and tumor cells (31).

An interesting effect of kinetin has been described by Lovrekovich and Farkas (151). Noting reports that kinetin could reverse protein degradation and chlorosis, they attempted to reverse these conditions as they are induced by a toxin elaborated by *Pseudomonas tabaci*. This toxin is described fully in Chap. 9. In their experiments, intact tobacco leaves were injected with culture filtrates containing the toxin. Halves of those leaves that were sprayed with 10^{-5} M kinetin failed to develop typical toxin-induced chlorosis. Protein levels were approximately doubled in leaf tissue that was treated with kinetin. The resultant restoration of near normal protein synthesis is presumed to be due to kinetin-fostered primary changes in RNA metabolism.

Thimann and Sachs (270) have observed that infection of sweet peas by *Corynebacterium fascians* induced outgrowths of a large number of buds from a node whose subsequent elongation was severely restricted. The similarity of this effect of a bacterial pathogen to an effect induced by the application of kinetin to lateral buds caused them to seek a cytokinin in the culture medium and in the bacterial cells per se. Using the accepted biological assay for cytokinin activity of retention of chlorophyll in yellowing barley leaves and the release of lateral buds from apical dominance, they found that these cytokinin effects could be induced by bacterial cells applied en masse and by partially purified chloroform extracts of the bacteria.

Matsubara et al. (170) reported that the cytokinin associated with the sweet pea pathogen is invariably detected as a component of its tRNA fraction. Helgeson and Leonard (100) established the chemical identity of this cytokinin as γ-(γdimethylallylamino) purine. However, Rathbone and Hall (211) subsequently found this purine to be only one of several cytokinins synthesized

by *C. fascians*. A positive correlation between cytokinin accumulation in culture by *C. fascians* and virulence was reported by Murai et al. (181). Their results also suggest that a large plasmid coded for cytokinin, that isolates bearing the plasmid were more virulent than one containing a small plasmid, and that one that was plasmidless was avirulent.

These experiments suggest that a cytokininlike effect is responsible for symptoms of the witches' broom type that are characteristic of a number of diseases. Wilson (303) has suggested that cytokinin may be involved in the cell proliferation stage of tumor development in oleander infected by *P. savastanoi*.

Ethylene and Its Influence on Pathogenesis

The growth-modifying effects of ethylene and its occasional detection in parasitized and wounded tissue suggest that this compound may be more prevalent at the site of host-parasite interactions than has been suspected. The probable production of this gas in trace (but physiologically active) amounts made detection difficult. However, with the advent of gas chromatography this problem has been resolved.

The view that ethylene is directly related to bacterial pathogenesis was initially reported by Freebairn and Buddenhagen (85). They conclusively linked premature ripening of banana fruit infected with *P. solanacearum* to high internal levels of ethylene. The differences in ethylene content between infected and healthy fruit were large. For example, severely infected fruit that had not yet yellowed contained 3 μg/ml ethylene, whereas infected half-yellow fruits contained 20 μg/ml. The gas was not detectable in healthy fruits.

This study also provided the first evidence of ethylene production in vitro by a bacterium, with rates as high as 0.1 μl/h per flask containing 20 ml of a 10^8 cells/ml bacterial suspension. Ethylene production was recorded for strains of *P. solanacearum* that infect both tobacco and banana, with the tobacco strain being most productive in this respect. Freebairn and Buddenhagen (85) also reported that a number of pseudomonads, xanthomonads, and *Erwinia* species produced significant amounts of ethylene in vitro.

Bonn et al. (29) also studied the synthesis of C_2H_4 by *P. solanacearum*. They observed that the virulent K-60 strain produced twice as much C_2H_4 as the avirulent

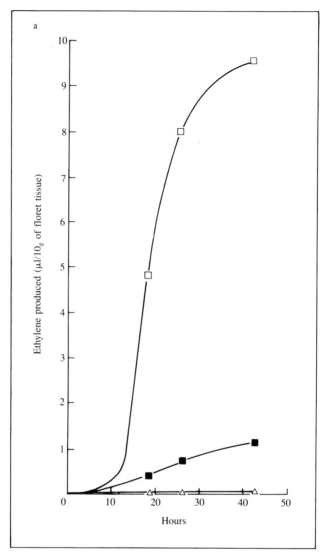

Fig. 7–17. (a) Stimulation of ethylene production dependent on both plant and bacterial enzymes. Floret tissue (10 g) was incubated in 8 ml of phosphate buffer (0.1 M), pH 6.8, to which the following additions were made: □, 2 ml of bacterial enzyme preparation; ■, 2 ml of autoclaved bacterial enzyme preparation; △, 2 ml of bacterial enzyme preparation, but enzymes of floret tissue had been inactivated by heat before addition (after Lund and Mapson, 154).

(b) Stimulation by bacterial enzyme preparation from *E. carotovora* of the synthesis of ethylene from 4-methylmercapto-2-oxobutyric acid by homogenate of cauliflower tissue. Homogenate was prepared from floret tissue (20 g) with 15 ml of a solution containing 0.4 M sucrose, 2 mM EDTA, and 0.1 M phosphate buffer, pH 6.5. Homogenate (5g) was placed in each flask with the following additions: ●, oxo acid (3 mM) and bacterial enzyme preparation (0.5 ml); □, oxo acid (3 mM); ■, no addition; x, bacterial enzyme preparation (after Lund and Mapson, 154).

mutant B1 strain. The relationship of the difference to virulence per se or pathogenesis was not determined.

The relative rapidity with which exogenously applied ethylene is lost from healthy tissue suggests that under normal conditions endogenously synthesized ethylene does not accumulate (1). However, it is possible that when the main pathway for the movement of this gas is blocked, for example by polysaccharidal slime or wilt-

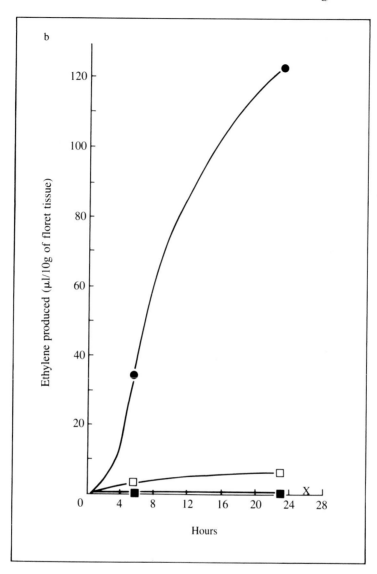

b

Ethylene produced (μl/10g of floret tissue)

Hours

induced stomatal closure, accumulation occurs and epinasty becomes apparent.

Lund (153) and Lund and Mapson (154) studied C_2H_4 production in cauliflower florets infected by bacteria. Although cauliflower florets produce C_2H_4 endogenously in small quantities, only tissues infected with pectolytic species like *Erwinia carotovora* or *Xanthomonas campestris* induced the production of large quantities of this growth regulator (Fig. 7–17A). Their experiments with bacteria-free culture filtrates revealed that bacterial enzymes stimulated heat labile components of the cauliflower tissue to produce the observed large quantities of C_2H_4 (Fig. 7–17B). These studies suggested that stimulation of ethylene production was dependent on the action of both bacterial and plant enzymes and actually reflected an increased rate of C_2H_4 synthesis via the pathway normally occurring in cauliflower florets. This pathway, worked out initially

by Mapson et al. (162) as conceived by Mapson and Hulme (161), suggested that carbon atoms 3 and 4 of methionine are converted to C_2H_4.

The *E. carotovora* isolate used by Lund and coworkers produced C_2H_4 in culture filtrate that, on average, contained pectate lyase 1.5 units/ml, polygalacturonase 0.25 units/ml, pectin-methyl-esterase 0.05 units/ml, and had some protease activity. The enzyme specificity data vis-à-vis C_2H_4 stimulation suggest pectate lyase was particularly effective in this respect and that polygalacturonase was less so. Protease apparently played no direct role in C_2H_4 stimulation.

The biochemical events pertaining to in vivo C_2H_4 synthesis are postulated as follows. The influence of bacterial pectate lyase seems to be on the solubilization of wall-adsorbed glucose oxidase (161), which together with a transaminase converts methionine to 4-methyl-mercapto-2-oxobutyric acid. The former enzyme becomes detached from the wall and accelerates H_2O_2 production from glucose. In the presence of peroxidase and certain cofactors such as phenolic and sulphinic acids, H_2O_2 is oxidized with the subsequent reduction of carbon atoms 3 and 4 of the oxoacid to C_2H_4 + other products. Although the sequence of events for the synthesis of ethylene postulated by Lund and coworkers (153, 154) predicts methionine to be the substrate, their contention that the intermediate precursor is 4-methylmercepto-2-oxobutyric acid is at variance with more recent data provided by Yang (307), which present l-aminocyclopropane-l-carboxylic acid (ACC) as the key intermediate to ethylene biosynthesis.

The relative importance of polygalacturanase in C_2H_4 synthesis is not known, and Lund has not explored the role of cellulase in this matter. From the studies of Lund and associates it would appear that the rise in C_2H_4 detected by Jackson and Osborne (110) that precedes leaf abscission, for example, is a reflection of pectin lyase activity and perhaps other wall-degrading enzymes. The pathological effects of these large amounts of C_2H_4 are not known.

Jackson et al. (109) reported that C_2H_4 applied to cucumber fruits causes their abscission and that the C_2H_4 also caused an increase in cellulase levels. The question is raised here as to whether in some pathological situations pectin lyase triggers C_2H_4 synthesis, which in turn stimulates host as well as pathogen-produced cellulase activity. This is an area that bears further investigation, particularly since there are data that suggest ethylene induces, in wounded tissues, a resistance to cellulase (91).

The Role of ABA in Pathogenesis

It is reasonably safe to generalize that with the development, maturity, and senescence of tissues there is a concomitant increase in resistance to bacterial pathogens. It is presumed that under proper conditions one could turn on and off growth and anabolic metabolism by the application of ethylene, ABA, or cytokinins to a plant model system. Similarly, one should be able to switch the system from a susceptible to a resistant condition by applying these and other growth-regulating substances. It is also apparent that research has not progressed to the stage where manipulations of this type are practicable. They should, however, be studied experimentally.

Steadman and Sequeira (257, 258) have presented the first evidence of increased levels of ABA as a consequence of pathogenesis. Apparently ABA is present in greater concentrations in *P. solanacearum*-infected tobacco than in healthy tissue. Since the pathogen did not synthesize ABA in vitro, it was presumed that the observed twofold increase reflected a host response to the pathogen. The authors suggested that increases in concentration of the growth inhibitor may be involved in the stunting of tobacco caused by *P. solanacearum*. The authors also postulated a relationship between the increased levels of ABA and the wilt symptom caused by the pathogen. It is known the ABA levels increase as

much as 40-fold in healthy plants placed under water stress.

The relationship between stress-elevated levels of both ethylene and ABA suggests that monitoring these two hormones in infected tissue might improve our understanding of inducible resistance mechanisms. Studies of this kind might also look at levels of GA, the counter hormone of abscisic acid.

Plants Infected by Fungi

Growth substances may, in a restricted sense, be regarded as a special group of toxic substances produced in the host by some pathogens. The greater than normal production and/or accumulation of otherwise naturally occurring hormonal substances may result in exaggerated growth responses, such as those observed in safflower rust, bakanae effect on rice, and malformations such as hypertrophy and hyperplasia induced by *Albugo candida* or *Taphrina deformans*. On the other hand, rapid leaf drop of fungus-infected plants is correlated with a lowered content of growth regulators or with the stimulated action of ethylene and abscisic acid.

Hyperauxiny in the Infected Plant

In numerous instances plant infections caused by fungi result in hyperauxiny. In some special diseases, such as white rust (*A. candida*) and *Peronospora* on *Capsella bursapastoris* (123), clubroot disease of *Cruciferae* (43), and *Phytophthora* on potato (78), in addition to IAA the level of indole acetonitrile is also increased.

Daly and Inman (58) have recorded conspicuous changes in auxin levels in safflower (*Carthamus*) hypocotyls infected with rust. Hypocotyls infected by *Puccinia carthami* show marked elongation (8–13 d after inoculation) during the period of intense mycelial development. Auxin content also increased during the period of growth of the hypocotyl, and a good correlation was obtained between growth and the auxin content of the rust-infected hypocotyls (Fig. 7–18 and Table 7–5). When sporulation of the fungus took place, growth inhibitors accumulated and elongation of the hypocotyls ceased.

In a compatible combination of the potato variety Irish Cobbler with *Phytophthora infestans* race 4, IAA increased five- to tenfold in the layers of tuber tissues invaded by the fungus, as reported by Fehrmann (78). He also noted that in an incompatible variety (Kennebec), both IAA and indole acetonitrile were inactivated in tissues invaded by the pathogen. Significantly, the level of IAA in this resistant combination increased 15–20 fold in tissues under the infected layers. A lower level of auxin in incompatible combinations of potato and *P. infestans* was reported earlier by Suchorukov and Strogonov (262). It would appear, therefore, that the stage of infection and the specific plant tissue involved must be carefully considered when alterations in auxin levels are evaluated.

As might have been expected, tumorous tissue of

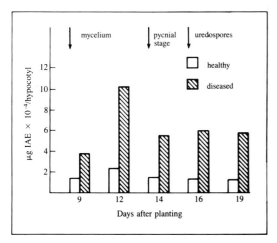

Fig. 7–18. Auxin levels in healthy and rust-affected hypo-cotyls of safflower during development of the pathogen *Puccinia carthami* (after Daly and Inman, 58).

corn infected by corn smut (*Ustilago zeae*) was found to contain a higher level of IAA than normal stalk tissue (275–77). Hence, overgrowths caused by smut as well as rust fungi seem somehow linked to hyperauxiny resulting from an impairment of normal auxin metabolism. According to Reingard and Pashkar (214), the same is true for the increased auxin content in tumors caused by *Synchytrium endobioticum*.

The clubroot disease of *Cruciferae* caused by *Plasmodiophora brassicae* represents an interesting case. The damage associated with infection alters the integrity of host cells in such a way that auxins are released from indole complexes of the host (43). Species of *Cruciferae* accumulate large amounts of indole glucosinolates in their tissues. The host also contains enzymes that hydrolize indole glucosinolates to 3-indoleacetonitrile, glucose, sulphate, hydrogen sulphide, and sulphur. Another enzyme, nitrilase, converts 3-in-

Table 7–5. Growth patterns of healthy and of rust-affected safflower by hypocotyls (after Daly and Inman, 58).

Days after plantinga	Kind of tissue	Length (cm)	Fresh weight (mg)	Dry weight (mg)	Infection
8	Healthy	1.43	45	3.0	
	Diseased	1.42	44	2.8	Not visible
10	Healthy	1.60	47	2.7	
	Diseased	1.99	62	3.2	Mycelial
13	Healthy	1.92	58	2.8	Initial
	Diseased	2.41	71	3.7	Pycniospores
16	Healthy	1.89	54	2.6	Initial
	Diseased	2.46	81	3.9	Uredospores

a. Inoculation was effected at the time of planting.

doleacetonitrile to 3-indoleacetic acid. After infection, indole glucosinolates are degraded and relatively large amounts of 3-indoleacetonitrile and 3- indoleacetic acid are released (Fig. 7–19).

The Origin of Increased Levels of IAA

Increased levels of auxin in infected tissues may, as we have already noted, result from either increased synthesis of IAA by either host or pathogen or a decreased rate of auxin degradation and release of auxin from other indole compounds.

There is no direct evidence to reveal whether the host plant or the pathogen is the actual source of increased IAA level in fungus-host plant complexes. However, a number of in vitro studies indicate a possibly preeminent role of the pathogen in IAA synthesis and consequent hyperauxiny in vivo. Fungus pathogens such as *Ustilago zeae* (corn smut), *Melampsora lini* (flax rust), a special strain of *Gymnosporangium juniperi-virginianae* (the cedar apple rust fungus), and *Taphrina deformans* are all able to produce auxin in axenic culture (254, 255, 304, 305). Mace (155) has demonstrated that *Fusarium oxysporum* f. *cubense*, causal agent of wilt of banana, also produces an auxin in vitro; the growth-regulating substance was purified and characterized as indoleacetic acid. Both *F. vasinfectum* (158) and *P. infestans* were also shown to produce IAA in vitro (178).

It became evident that at least two phytotoxins have auxinlike actions. *Victorin*, the host-specific toxin of *Helminthosporium victoriae*, can induce a very rapid stimulation of cell elongation of *Avena* coleoptile segments (see Chap. 9). The timing and magnitude of growth promotion by indoleacetic acid and victorin are similar (77, 225, 226). It was suggested that the IAA-like effect was mediated by altering membrane properties in host cells.

Fusicoccin, a non-specific toxin, is more effective in stimulating cell growth than several auxins (IAA, NAA, 2, 4-D) and has been regarded as a "superauxin." The growth-promoting activity of both fusicoccin and IAA is mediated by a decrease of pH in the cell wall and thus by an increase in wall plastic extensibility (167). The pH decrease effect is caused by a change in H^+ concentration in the cell wall per se (166–68). This type of proton extrusion frequently accompanies growth stimulation caused by auxin and cytokinins, as well as fusicoccin (168). The effect may be causally related to a negative membrane potential (E_m) in plant cells. Both proton extrusion and E_m changes are discernible within minutes after plant tissues are exposed to fusicoccin and other hormones. However, the action of fusicoccin differs from that of IAA in two ways: (1) fusicoccin, but not IAA, can stimulate both growth and H^+ excretion in tissues whose protein synthesis is blocked by cycloheximide; (2) fusicoccin but not IAA stimulates H^+ excretion in tissues at incipient plasmolysis. It would appear, therefore, that these two hormones exert their effects at different sites within the plant cell (50, 306) (see also Chap. 9).

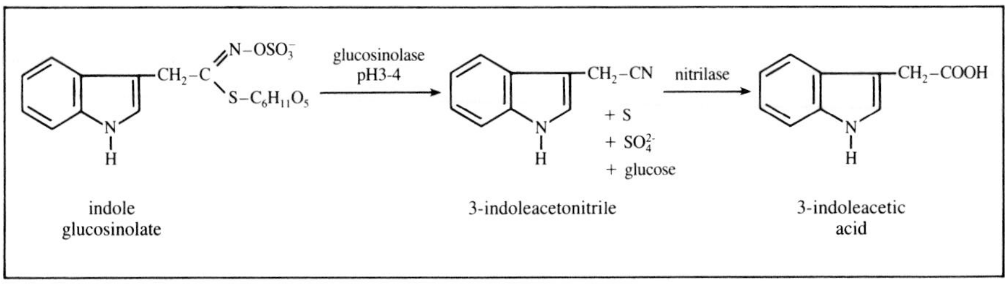

Fig. 7–19. The enzymatic release of auxins from indole glucosinolate (after Butcher et al., 43).

Biosynthesis of IAA by Pathogenic Fungi

The biosynthesis of IAA by pathogenic fungi proceeds along pathways similar to those of the host plant. In general, the pathway for IAA production is:

Tryptophan → indole-3-acetaldehyde → IAA

There are, however, two important divergent pathways leading to the production of indoleacetaldehyde (see Fig. 7–1). For example, auxin synthesis in *Melampsora* was suggested by Srivastava and Shaw (255) to proceed via the indole pyruvic acid → indoleacetaldehyde route, whereas in *Gymnosporangium* the tryptamine → indoleacetaldehyde route is followed (305). From a study by Wightman (300) it would appear that in cabbage, indoleacetonitrile and not indoleacetaldehyde may be the precursor of IAA. In general, however, tryptophan is the common precursor, and indoleacetaldehyde is the immediate precursor of IAA. The unusual pathways of IAA synthesis in fungal-diseased plants have not been demonstrated satisfactorily.

Inhibition of Auxin Degradation

We have mentioned that increased levels of auxin in diseased plants could also result from the decreased degradation of auxin—namely, by the inhibition of IAA oxidase. That inhibition of auxin oxidase is a mechanism for auxin accumulation has been clearly shown by Pilet (205) in the rust disease of *Euphorbia cyparissias* infected with *Uromyces pisi*. Enzyme preparations from healthy stems of the host exhibited higher IAA oxidation activity than those from rust-infected plants. In the early stages of disease, the increase in IAA corresponded closely with the calculated decrease in enzyme activity (see Table 7–6). A similar mechanism was suggested later by Wiese and DeVay (299) for cotton wilt caused by *Verticillium albo-atrum*.

According to Shaw and Hawkins (240), IAA oxidase activity is inversely correlated with IAA content in the rust-infected wheat. In addition it was inferred that the destruction of IAA in infected leaves of wheat is accomplished by the pathogen per se (see Fig. 7–20). In this experiment the rate of decarboxylation of IAA-[14]COOH was taken as a measure of the activity of IAA oxidase. However, the decarboxylation of IAA was correlated with the uptake of IAA from the ambient solution. In later experiments by the same authors, the rates of both uptake and decarboxylation decreased at the sporulation

stage. Therefore, it is impossible at present to decide whether the rate of decarboxylation reflects the activity of IAA oxidase or the rate of uptake from the solution.

The results of Shaw and Hawkins (240), represented by the idealized curve of Fig. 7–20, were critized by Daly and Deverall (57), who point out that the claim that IAA accumulation may arise from inhibition of IAA oxidase is valid only at the "sporulation stage" of wheat infection by stem rust. In earlier stages of the disease, mechanisms other than inhibition of IAA oxidase must be considered in order to understand the disease-induced increase in IAA. In vitro, peroxidase with monophenols as cofactors can decarboxylate IAA. Diphenolic compounds usually inhibit IAA oxidase. IAA decarboxylation in rust-infected wheat can be regarded as an expression of increased peroxidase activity (229).

It would appear axiomatic that, as the rate of degradation of IAA increases in a host-pathogen complex, the level of auxin will be lowered. Sequeira (230), Sequeira and Steeves (236), and Kazmaier (119) have explained the accelerated premature leaf drop caused by the fungi *Omphalia flavida* and *Diplocarpon rosae* in coffee and rose plants as due to enzymes produced by these fungi, which are similar to IAA oxidase. These enzymes rapidly degrade IAA, thus interfering with the normal flow of auxin from leaf blade to petiole, causing pre-

Table 7–6. Comparison between auxin content and auxin destruction in stems of *Euphorbia* from healthy and diseased plants at the mycelium and aecidiospores stages. Auxin content: in μg of IAA equivalent per 100 g fresh weight. Auxin destruction: in μg of IAA destroyed during 60 min per 200 g fresh weight (after Pilet, 205).

		Stages	
		Mycelium	Aecidiospores
IAA content	Diseased	22.8	39.7
	Healthy	12.4	7.8
	Diseased/healthy	1.9	5.1
IAA	Diseased	20.2	21.1
destruction	Healthy	38.1	41.6
	Healthy/diseased	1.8	1.9

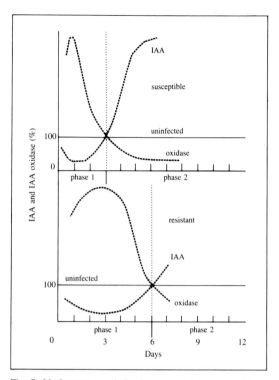

Fig. 7–20. Inverse correlation between indoleacetic acid oxidase activity and IAA content in wheat leaves susceptible and resistant to stem rust (after Shaw and Hawkins, 240).

mature development of an abscission layer and the characteristic leaf drop symptom.

The *Omphalia* enzyme is similar to IAA oxidase of higher plants and has been used as a model system to study the steps involved in the oxidative degradation of auxin. The enzyme has a pH optimum of 3.5 and is potentiated by Mn^{++} but does not require it, a condition noted by Sacher (223) in beans. However, unlike some higher plant auxin oxidases, the *Omphalia* enzyme does not seem to require the presence of phenolic cofactors such as 2,4-dichlorophenol, chlorogenic, or ferulic acids for IAA oxidation. Oxidation of IAA by the enzyme seems to be stoichiometric, 1 mole of O_2 consumed and 1 mole of CO_2 released per mole of IAA degraded. The enzyme exhibits peroxidase activity with pyrogallol as substrate, but no polyphenolase activity. The presence of an active IAA oxidase in coffee leaf tissues infected by *Omphalia flavida* was demonstrated by Rodrigues and Arny (219). Another instance is the witches' broom disease of cacao induced by *Marasmius perniciosus* (134). Culture filtrates of the fungus destroy IAA with two enzymes: a peroxidase and a phenoloxidase (laccase).

As we have already intimated, IAA oxidases are peroxidase-mediated enzymes; i.e. they may require H_2O_2, which removes the lag period for the activation of the enzyme. Young corn smut tumors and rust-infected *Euphorbia cyparissias* tissues have high catalase

activity, which destroys H_2O_2 and, according to Turian (275, 276), inhibits IAA oxidase, thereby increasing the level of IAA in the plant. This is an interesting phenomenon, as it clearly reveals the dependence of the auxin balance in infected plants on another oxidative enzyme, peroxidase.

Cytokinins in Fungus-Infected Plants

The diverse action of cytokinins on plant cell division, nutrient accumulation and nutrient transport, the inhibition of senescence, etc., is expressed in different disease symptoms because this hormone is produced in excess in many infected tissues (126, 64, 65, 265, 289). Symptoms like green island production, induced senescence in uninfected tissues of diseased plants, abnormal phloem transport, and accumulation of nutrients at the infection sites are characteristic of many diseases. All of these symptoms can be simulated by cytokinins (kinetin, benzyladenine) applied to the leaves (208). Because of the effect of cytokinins on cell division, this hormone could be causally related to fungus diseases characterized by disorganized growth, such as tumors or galls caused by *Plasmodiophora brassicae* or leaf curl induced by *Taphrina deformans* (63, 230, 234, 267, 280).

Disorganized growth. A great amount of attention has been paid to the clubroot disease of cabbage, caused by *P. brassicae*. Katsura et al. (117) and Kavanagh et al. (118) obtained the initial evidence of an active role for cytokinin in the tumorous growth in clubroot-infected cabbage. It has been shown that the callus from infected roots is completely independent of an exogenous source of cytokinins for growth in culture. However, healthy plant callus tissues always require an external source of cytokinin to support growth (106, 302). The cytokinin produced in infected turnip was identified as zeatin (64). The levels of cytokinins in clubroot tissues were estimated to be 10 to 100 times higher than in healthy turnip tissues (Fig. 7-21). However, healthy and clubroot-in-

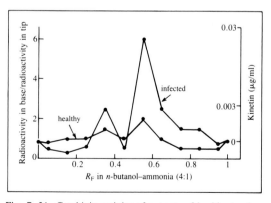

Fig. 7–21. Cytokinin activity of extracts of healthy turnips and clubroots. Activity on paper chromatograms determined by mobilization bioassay (after Dekhuijzen and Overeem, 64).

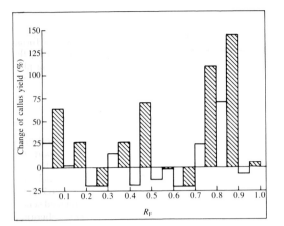

Fig. 7–22. Paper chromatographic separation of cytokinin activity from purified extracts of healthy (☐) and infected (▨) peach leaves by soybean callus bioassay (% change of callus yield over water control). Purified extracts were chromatographed on Whatman No. 1 paper in *t*-butanol-conc. NH_4OH-water (3:1:1 v/v/v). Cultures were grown on basal medium supplemented with portions of chromatograms corresponding to R_F regions. Cultures maintained at 27 C for 28 d (after Sziráki et al., 265).

fected turnips contain similar cytokinins, on the basis of R_F values.

Neoplastic tissues of curled peach leaves infected by *Taphrina deformans* also contain increased cytokinin activity. Three chromatographically similar cytokinins are present in both healthy and infected leaves, but those in infected leaves are more active (in the soybean callus test). Diseased leaf tissues contain an additional cytokinin not present in healthy leaves (265). Figure 7-22 shows the analysis of chromatograms of purified cytokinin extracts of peach leaves by the soybean callus test. It is clear that the cytokinin synthesized only in the *Taphrina*-infected leaves is at R_F region 0.4–0.5.

It is significant that high IAA levels in both clubroot and curled peach leaf tissues contribute to and perhaps accentuate the cell response to high levels of cytokinin. It is probable that the striking morphological changes in *Taphrina*-infected peach leaves reflect the detected hormonal imbalance. Increased cell growth and water uptake may be related to the increased auxin content, and the stimulated cell division and abnormal growth pattern of infected leaves could be a consequence of increased cytokinin activity.

It is noteworthy that mycorrhizal fungus, *Rhizopogon roseolus,* also produces cytokinins (zeatin and zeatin riboside) but without causing tumorous overgrowths (174).

The juvenility effect of cytokinins and their action on nutrient mobilization. It has long been known that infections caused by rusts and powdery mildews result in the development of green islands around the infection sites. In these islands of juvenile tissues chlorophyll is

retained, and RNA as well as protein synthesis is stimulated (see Chap. 5). Indeed, as a result of infections nutrients are mobilized, and they accumulate around the infection sites (in green islands). Tissues near and between the infection sites become senescent. This senescence effect is induced by the export of metabolites from the uninfected tissues toward the green islands (65, 71, 124, 207, 208, 272) (see also Fig. 7-23). Additional signs of juvenility are the increased leaf growth of rust-infected plants (75) and the inhibition of leaf abscission (Fig. 7-24).

All these symptoms resemble a cytokinin effect and can be simulated by cytokinin treatments. That purified extracts of rust-infected bean and broad bean leaves show increased cytokinin activity was first demonstrated by Király et al. (126). Bean rust spores, too, were shown to contain relatively high levels of cytokinin. Germinating spores of wheat stem rust release cytokinin into the germination fluid. Dekhuijzen and Staples (65) confirmed these investigations and demonstrated that the increased cytokinin activity in rusted bean leaves was due primarily to one fraction that is active in mobilization and in stimulating growth of callus tissue. Cytokinin compounds from spores and rust mycelium had different R_F values from those from infected leaves. Thus it was concluded that cytokinin increase in infected bean leaves is of host origin.

Király et al. (unpublished) have shown that the cytokinin effect of rust infection is a very early action on the host. When the rust fungus was selectively killed in leaves 24 h after infection, diseased bean leaves still exhibited higher cytokinin activity in the soybean callus test than uninfected ones. Green islands also appear, in spite of the absence of the fungus in infected leaves. In incompatible host-pathogen combinations (with necrotic flecks), the cytokinin effect is also conspicuous (267). One can conclude that the infection process itself is enough to induce high cytokinin production in the host. However, this factor is obscured by a simultaneous detrimental effect of infection on cell organelles. Green islands are seen on leaves only in later stages of disease, when the infection site is green relative to the general senescence of the whole leaf.

Gibberellins in Fungus-Infected Plants

In the bakanae disease of rice caused by *Gibberella fujikuroi*, the "toxic" product of the fungus is a naturally occurring plant growth hormone, gibberellic acid (261). The affected rice plants grow rapidly and are conspicuous for their unusual height, hence the Japanese term *foolish seedling disease* for this disorder. However, no specific determinations have been made on the actual content of gibberellic acid (GA_3) or other gibberellins in diseased rice tissues.

Gibberellin production is probably common among pathogenic fungi, but little analytical work has been done in this field. Aubé and Sachston (10) have shown that a series of *Verticillium* species, including pathogenic fungi like *V. albo-atrum* and *V. dahliae*, can produce gibberellinlike substances. The rust infection

Fig. 7–23. Senescence of the uninfected half-leaves of pinto bean plants near the site of rust infection or treatment with a cytokinin. Top row: control leaves. Middle row: half-leaves infected by *Uromyces phaseoli*; the opposite half-leaves are senescent (bleached) at 14 d after infection. Note the relative juvenility of the infected half-leaves because of the "green islands" around the infection centers. Bottom row: half-leaves treated with a 30 μg/ml aqueous solution of benzyladenine daily for 14 d; the opposite half-leaves are senescent (bleached). Note the relative juvenility and green color of the benzyladenine- treated half-leaves (after Király et al., 126).

(*Puccinia punctiformis*) of *Cirsium arvense* resulted in consistently higher levels of gibberellins, which were correlated with the period when the infected shoots grew faster than the healthy ones (16). It was possible here, too, to simulate a series of the symptoms with exogenously applied gibberellic acid. It was supposed that some of the morphological changes in diseased plants and the lack of inflorescence may be the result of a combined effect of increased IAA and gibberellin con-

tent. Systemic infection increased cell division in the stem subapex and also the length of pith and cortical cells. Similar effects could be induced by treating healthy plants with gibberellic acid.

The dwarfing syndrome on wheat caused by *Tilletia controversa* has been investigated by Vanicková-Zemlová et al. (282). Inhibited growth of diseased plants could not be overcome by the application of GA_3. Differences in concentrations of endogenous gib-

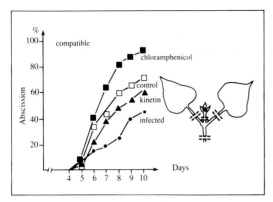

Fig. 7–24. Inhibition of abscission of leaf petioles of pinto bean leaf explants infected by *Uromyces phaseoli* (compatible host pathogen combination) as compared to various treatments. Preparation of explants is shown on right (after Gáborjányi et al., 87).

Table 7–7. Abscisic acid in healthy and in *Verticillium albo-atrum*-infected cotton tissues (after Wiese and DeVay, 299).

Plant part	Method of assay	Healthy	Diseased[a] Nondefoliating Isolate	Defoliating Isolate
			$\mu g/g$ dry weight	
Leaves	Gas-liquid chromato-graphy	4.0	3.8	8.6
	Thin-layer chromato-graphy	1.2	1.6	2.8
Stems	Thin-layer chromato-graphy	0.6	0.3	0.4

a. Tissue harvested 12 d after inoculation.

berellins or inhibitors cannot explain the retarded growth. Probably, a relative insensitivity of smut-infected plants to gibberellin is responsible for the dwarfing syndrome.

The phytotoxin helminthosporal also exerts gibberellinlike effects on plant growth and on amylase synthesis of embryoless seeds (see also Chap. 9). This toxin with hormonal action was isolated from *Helminthosporium sativum,* the causative agent of root rot and leaf spot in wheat and barley. Helminthosporol, the reduced analog of helminthosporal, promotes the elongation of stems of rice seedlings but not those of wheat (98). This would explain why the infection of wheat by this fungus does not result in conspicuous stem elongation by the host. Helminthosporal, the other phytotoxin, can stimulate the synthesis of amylase by embryoless seeds of barley. On the other hand, it inhibits the gibberellic acid–induced synthesis of amylase (298). It is to be recalled that gibberellic acid has two primary regulatory effects, cell elongation and induction of enzymes concerned with starch synthesis and cell wall modification.

Role of Senescence Hormones: Abscisic Acid and Ethylene

Although synthesis of ABA by plant pathogenic fungi has not been reported, it would seem that this hormone produced in the diseased plant has an important role in expression of such symptoms as leaf abscission, inhibition of extension growth, and inhibition of flowering.

In fact, the only fungus disease in which an increased tissue level of abscisic acid has been demonstrated is *Verticillium albo-atrum,* which showed a fivefold increase of this hormone (198). The level of ABA doubled in cotton leaves infected with the defoliating strain of *V. albo-atrum.* On the other hand, the nondefoliating strain did not affect ABA levels (299). It was concluded

that this hormone is responsible for the defoliation of infected plants (see Table 7-7 and Fig. 7-25).

Ethylene production also increased with both isolates, but the influence of the defoliating strain was more pronounced than that of the nondefoliating isolate (Fig. 7-26). It was proposed that ethylene is involved in inducing epinasty and, in part, the defoliation symptom. It was mentioned earlier that IAA increases vascular wilt caused by *Verticillium,* most probably as a consequence of reduced rate of IAA-decarboxylation. It is, of course, not easy to determine how these hormones are related in symptom expression—in other words, how they control increased leaf abscission and other symptoms associated with vascular wilt.

Increased ethylene production is characteristic of numerous fungus diseases. Recent research has thrown light on the role of ethylene in premature abscission, epinastic responses, adventitious root formation, and several abnormal growth phenomena. The increased production of this hormone is a consequence of injury of host cells by infecting fungi, although ethylene itself is able to damage or induce damage or senescence in plant cells. In fact, it is not possible to determine with surety whether ethylene increases follow or precede the injury associated with disease development. It seems likely that ethylene has no role in host-pathogen compatibility or incompatibility. The role of ethylene as a key trigger in resistance mechanisms (105, 256) has not been confirmed (44, 59, 103, 104).

Because ethylene stimulates the synthesis of phytoalexins and other aromatic compounds by increasing the activity of PAL, it seems logical that this hormone plays an important role in disease resistance (95). Furthermore, ethylene treatment of plant tissues results in increased peroxidase activity, which has been fre-

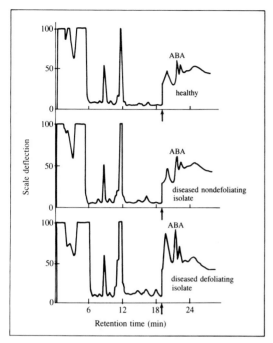

Fig. 7–25. Gas-liquid chromatograms obtained from injection of 3 μl (3% of the total sample) of silylated crude acidic fractions equivalent to 10 g dry weight of cotton leaves into a Varian aerograph model 204 equipped with a flame-ionization detector and linear temperature programmer. A 5-ft × 1/8-in column of 5% QF-1 coated on 60 to 80 mesh acid-washed dimethylchlorosilane-treated Chromosorb W was used. Column temperature was set at 30 C for 6 min and increased to 195 C at a rate of 9.1 C/min. Attenuation was set at × 128 and changed to × 32 at 19 min (after Wiese and DeVay, 299).

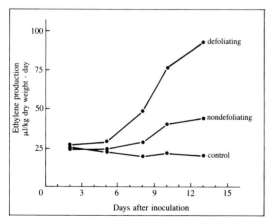

Fig. 7–26. Ethylene production by healthy and infected plants at intervals after inoculation (after Wiese and DeVay, 299).

quently related to plant resistance to fungus, virus, and bacterial pathogens.

The original claim that ethylene treatment of sweet potato increases resistance to *Ceratocystis fimbriata* was not confirmed (44). Furthermore, since phytoalexin production can be stimulated with a wide variety of chemical agents and mechanical injury, ethylene does not seem to be a specific inducer of phytoalexin production. The proposed causal relationship between ethylene, peroxidase, and host resistance has been questioned by Daly et al. (59). It was demonstrated that, although ethylene treatment of wheat leaves induced high peroxidase activity, the treated resistant leaves became completely susceptible to rust infection. Ethylene production by infected susceptible leaves was higher than that by resistant leaves. The same is true for barley infected with *Erysiphe graminis* (104). In both instances, ethylene evolution and increased peroxidase activity probably result from damage to tissues. Injury is usually more intensive in the compatible combinations than in the incompatible ones, so ethylene evolution is also more intensive in the compatible case.

A series of abnormal growth phenomena previously attributed to other hormones are probably due, in part at least, to increased ethylene production. Future investigations will provide deeper insight into the mechanism and role of ethylene in symptom expression. Hislop et al. (103) give an excellent summary of this problem.

The Influence of Exogenously Applied Growth Regulators on Disease Control

The reports that artificial applications of high levels of natural and synthetic auxins induced disease resistance in tomato plants to *Fusarium* wilt (62) and potato to *Phytophthora infestans* (238) have far-reaching implications. The evidence presented suggests that changes in the nature of pectic substances wrought by these plant-growth regulators were responsible for the observed increase of resistance to *Fusarium oxysporum* f. *lycopersici*. Specifically, tomato plants treated with naphthaleneacetic acid contained less water-soluble pectin than the untreated plants. Noted also was an increased calcium bonding as reflected by higher levels of insoluble pectates. It was hypothesized that these insoluble pectates might have imparted to the plant tissue greater resistance to fungal pectolytic enzymes. There is support for this contention in the fact that resistance induced by naphthaleneacetic acid is dependent upon the presence of Ca^{++}. It will be recalled that the Ca^{++} ions form rather rigid linkages between pectin chains.

There are only a few reports concerning the influence of gibberellic acid and kinetin on the control of fungus diseases. Király et al. (125) found that gibberellic acid decreased the susceptibility of wheat to *Tilletia foetida*, whereas Petersen et al. (201) reported that GA increased the susceptibility of red kidney beans to infection by *Rhizoctonia solani*. It was shown that the charcoal rot disease of soybean caused by *Macrophomina phaseolina* can be successfully controlled by treatments with gibberellic acid even under field conditions (194). A study by Dekker (66) has revealed that kinetin fully protected leaf discs of cucumber from infection by *Erysiphe cichoracearum*. At present it is not possible to give good explanations in biochemical terms for this type of protection.

Ethylene treatment may increase or decrease disease susceptibility in different host-parasite interactions. In the case of pretreated melon seedlings, the increased resistance to *Colletotrichum lagenarium* was attributed to an early triggering of enrichment of host cell walls in hydroxyproline-rich glycoproteins (76). Whether the accumulated glycoproteins have a lectinlike activity or strengthen the host cell walls is not known at present (see also Chaps. 4 and 10).

Comparative Analysis of Disease Physiology— Growth Regulators

Stunting, abscission, yellowing, epinasty, hyper- and hypoplasia are caused by viral, bacterial, and fungal pathogens and may be traced to altered hormone levels in infected plants. In some instances these symptoms will reflect an interaction between two or more growth regulators. For example, the increased levels of ethylene generated at the site of virus-induced necrotic lesion development may be due to elevated abscisic acid concentrations. The ABA increase may reflect water loss, which in turn leads to alterations in either auxin metabolism (oxidation or synthesis) or transport.

The effect of CTV on phloem tissue is to impede the basipetal transport of IAA and may explain the stunting caused by this and other viruses. Similarly, the blockage of IAA movement to the abscission zone caused by *Omphalia flavida* and perhaps *Xanthomonas pruni* appears to be responsible for premature abscission of *Coleus* and peach leaves. Some virus infections result in a reduced GA level and a concomitant increase in the "counter" hormone ABA. These two events may be causally related, or they may develop as a consequence of water stress.

The influence of cytokinins on the unique situation of host cell–directed virus replication reveals this hormone favoring either host cell or virus protein and nucleic acid synthesis. Which of these alternatives is stimulated appears to depend on whether exogenous cytokinin is applied prior to or after inoculation.

Whereas IAA levels are frequently much higher in plant tissue infected by bacterial or fungal pathogens, a distinct pattern has not evolved from the many studies of auxin levels in virus-infected plants. Certain bacterial pathogens, namely *Agrobacterium tumefaciens* and *Pseudomonas savastanoi*, derive their virulence from their capacity to produce inordinate amounts of plasmid-directed IAA synthesis in host tissues. Similarly, plasmid coding for cytokinin synthesis appears to be the basis for the "witches broom" symptom development in pea plants infected by *Corynebacterium fascians*. The grotesque foliar shapes of peach foliage infected by *Taphrina deformans* and gnarled ears of corn infected

by *Ustilago zea* or crucifers parasitized by *Plasmodiophora brassica* or *Albugo candida* exemplify the hypo- and hypertrophy resulting from abnormal tissue levels of IAA.

Ethylene concentrations seem generally to be elevated in virus-infected plants, and stimulation appears to coincide with, if not being causally related to, lesion development. Of particular interest is the divergence between virus infection and mechanical injury- induced ethylene production. Whereas infection and mechanical injury are often considered analagous as plant stresses, the ethylene synthesis they provoke suggests otherwise. TMV-induced ethylene production proceeds at levels 5–10 times that in controls, 15 d after necrosis becomes apparent. Mechanical injury, on the other hand, stimulates ethylene only during necrotization, and the ethylene then quickly declines to control levels.

Although some bacterial and fungal plant pathogens cause symptoms because they are able to synthesize large amounts of IAA, others foster above-normal levels in tissues by suppressing IAA oxidation. Infection-stimulated synthesis of aromatic compounds like scopoletin block auxin oxidase activity and permit IAA to accumulate in *P. solanacearum*-infected tissues.

A symptom often evoked by rust and powdery mildew fungi is the development of "green islands at infection sites." These zones reveal a stimulated RNA and protein synthesis, whereas the neighboring tissues become yellow and senescent. It has been shown that cytokinin activity increases at the infection site and mobilizes nutrients from distant healthy tissues to the infection site. Thus, juvenility is prolonged at the site of obligate parasitism.

The literature concerning altered hormone levels in virus-infected plants suggests that virologists have considered disease suppression or symptoms thereof by manipulating growth regulator levels in vivo. This avenue of disease control has not been as assiduously studied in diseases caused by either bacteria or fungi. It might prove useful to devise procedures that promote juvenility, degrade excess auxin, or expedite ethylene evolution. It may also be of interest to attempt to increase the permeability of plant tissues to systemic fungicides and bactericides with either natural or synthetic auxins.

Additional Comments and Citations

Research findings concerning crown gall, caused by *Agrobacterium tumefaciens*, continue to enter the literature in increasing rapidity and numbers. It had previously been held that monocots were not infectible by *A. tumefaciens* because of an absence of receptor sites for attachment. This concept has been made untenable by C. Douglas, W. Halperin, M. Gordon, and E. Nester. 1985. Specific attachment of *Agrobacterium tumefaciens* to bamboo cells in suspension cultures. J. Bact. 161:764–66. Apparently failure to transform monocot plant cells is due to a block in subsequent steps in pathogenesis. The question of whether tumor cells are homologous with respect to tranaformation has been answered by G. van Slogteren, J. Hoge, P. Hooykaas, and R. Schilperoort. 1983. Clonal analysis of heterogenous gall tumor tissues induced by wild-type and shooter mutant strains of *Agrobacterium tumefaciens*—expression of T-DNA genes. Plant Mol. Biol. 2:321–33. Cloning of isolated axenic tumor tissues revealed that in all cases they consisted of 10–26% tumor cells neighboring a majority of normal cells. Hence great phenotypic variability of a particular tumor may be explained by the presence of varying ratios of normal and genetically different types of tumor cells. Variation in the kind of tumor appears to be in part controlled by growth regulator genes 1 and 2 (auxin) and 4 (cytokinin). Host range of limited host range isolates (like biotype 3 that infects grape) can be extended by incorporating the *cyt* 4 gene into an isolate lacking it. A. Hoekema, A. B. de Pater, A. Fillinger, P. Hooykaas and R. Schilperoort. 1984. Limited host range of an *Agrobacterium tumefaciens* strain extended by a cytokinin gene from a wide host range T-region. EMBO Journal 3:3043–47. W. Buchholz and M. Thomashow. 1984. Host range encoded by the *Agrobacterium tumefaciens* tumor-inducing plasmid PTiAg63 can be expanded by modification of its T-DNA oncogene complement. J. Bact. 160:327–32.

References

1. Abeles, F. B. 1973. Ethylene in Biology. Academic Press, New York.

2. Abeles, F. B., and R. E. Holm. 1966. Enhancement of RNA synthesis; protein synthesis and abscission by ethylene. Plant Physiol. 41:1337–43.

3. Abeles, F. B., and G. R. Leather. 1971. Abscission: Control of cellulase secretion by ethylene. Planta 97:87–91.

4. Aharoni, N., S. Marco, and D. Levi. 1977. Involvement of gibberellins and abscisic acid in the suppression of hypocotyl elongation in CMV-infected cucumbers. Physiol. Plant Path. 11:189–94.

5. Aldwinckle, H. S. 1975. Stimulation and inhibition of plant virus infectivity in vivo by 6-benzylaminopurine. Virology 66:341–43.

6. Aldwinckle, H. S., and I. W. Selman. 1967. Some effects of supplying benzyladenine to leaves and plants inoculated with viruses. Ann. Appl. Biol. 60:49–58.

7. Andreae, W. A. 1952. Effect of scopoletin on indoleacetic acid metabolism. Nature 170:83–84.

8. Andreae, W. A., and S. R. Andreae 1953. Studies on indoleacetic acid metabolism. I. The effect of methyl umbelliferone, maleic hydroxide and 2,4 D on indoleacetic acid oxidation. Can. J. Bot. 31:426–37.

9. Aspinall, D. L., G. Paleg, and F. T. Addicott. 1967. Abscisin II and some hormone- regulated plant responses. Aust. J. Biol. Sci. 20:869–82.

10. Aubé, C., and W. E. Sachston. 1965. Distribution and prevalence of *Verticillium* species producing substances with gibberellin-like biological properties. Can. J. Bot. 43:1335–42.

11. Bailiss, K. W. 1968. Gibberellins and the early disease syndrome of asperny virus in tomato (*Lycopersicon esculentum* [Mill.]). Ann. Bot. 32:543–51.

12. Bailiss, K. W. 1974. The relationship of gibberellin content to cucumber mosaic virus infection of cucumber. Physiol. Plant Path. 4:73–79.

13. Bailiss, K. W. 1977. Gibberellins, abscisic acid and virus-induced stunting. In: Current topics in plant pathology. Z. Király (ed.).Akademiai Kiado, Budapest, pp. 361–73.

14. Bailiss, K. W., E. Balázs, and Z. Király. 1977. The role of ethylene and abscisic acid in TMV-induced symptoms in tobacco. Acta Phytopath.Hung. 12:133–40.

15. Bailiss, K. W., F. M. Cocker, and A. C. Cassells. 1977. The effect of Benlate and cytokinins on the content of tobacco mosaic virus in tomato leaf discs and cucumber mosaic virus in cucumber cotyledon discs and seedlings. Ann. Appl. Biol. 87:383–92.

16. Bailiss, K. W., and I. M. Wilson. 1967. Growth hormones and the creeping thistle rust. Ann. Bot. 31:195–211.

17. Balázs, E., B. Barna, and Z. Király. 1976. Effect of kinetin on lesion development and infection sites in Xanthi-nc tobacco infected by TMV: Single-cell local lesions. Acta Phytopath. Hung. 11:1–9.

18. Balázs, E., and R. Gáborjányi. 1974. Ethrel-induced leaf senescence and increased TMV susceptibility in tobacco. Z. Pflanzenkr. Pflanzensch. 81:389–92.

19. Balázs, E., R. Gáborjányi, A. Tóth, and Z. Király. 1969. Ethylene production in Xanthi tobacco after systemic and local virus infections. Acta Phytopath. Hung. 4:355–58.

20. Balázs, E., R. Gáborjányi, and Z. Király. 1973. Leaf senescence and increased virus susceptibility in tobacco: The effect of abscisic acid. Physiol. Plant Path. 3:341–46.

21. Balázs, E., I. Sziraki, and Z. Király. 1977. The role of cytokinins in the systemic acquired resistance of tobacco hypersensitive to tobacco mosaic virus. Physiol. Plant Path. 11:29–37.

22. Bandurski, R. S., and Z. Piskornik. 1973. An indole-3-acetic acid ester of a cellulosic glucan. In: Biogenesis of plant cell wall polysaccharides. F. Loewus (ed.). Academic Press, New York, pp. 297–314.

23. Banerjee, D., M. Basu, I. Choudhury, and G. C. Chatterjee. 1981. Cell surface carbohydrates of *Agrobacterium tumefaciens* involved in adherence during crown gall tumor initiation. Biochem. Biophys. Res. Comm. 100:1384–88.

24. Barnett, A., and K. R. Wood. 1978. Influence of actinomycin D., ethephon, cycloheximide and choramphenicol on the infection of a resistant and a susceptible cucumber cultivar with cucumber mosaic virus (Price's No. 6 strain). Physiol. Plant Path. 12:257–77.

25. Bawden, F. C. 1959. Physiology of virus diseases. Ann. Rev. Plant Physiol. 10:239–56.

26. Ben-Tal, Y., and S. Marco. 1980. Qualitative changes in cucumber gibberellins following cucumber mosaic virus infection. Physiol. Plant Pathol. 16:327–36.

27. Bitancourt, A. A., A. Nogueira, and K. Schwarz. 1961. Pathways of decomposition of indole derivatives. In: Plant growth regulation. R. M. Klein (ed.).Iowa State Univ. Press, Ames, pp. 181–98.

28. Black, L. M., and C. L. Lee, 1957. Interaction of growth-regulating chemicals and tumefacient virus on plant cells. Virology 3:146–59.

29. Bonn, W. G., L. Sequeira and C. D. Upper. 1972. Determination of the rate of ethylene production by *Pseudomonas solanacearum*. Proc. Can. Phytopath. Soc. 39:26.

30. Bonner, J. 1934. The relation of hydrogen ions to the growth rate of the *Avena* coleoptile. Protoplasma 21:406–23.

31. Bonner, J. 1961. On the mechanics of auxin-induced growth. In: Plant growth regulation. R. M. Klein (ed.). Iowa State Univ. Press, Ames. pp. 307–28.

32. Braun, A. C. 1952. Conditioning of the host cell as a factor in the transformation process in crown gall. Growth 16:65–74.

33. Braun, A. C. 1956. The activation of two growth-substance systems accompanying the conversion of normal to tumor cells in crown gall. Cancer Res. 16:53–56.

34. Braun, A. C. 1962. Tumor inception and development in the crown gall disease. Ann. Rev. Plant Physiol. 13:533–58.

35. Braun, A. C. 1978. Plant tumors. Biochim. Biophys. Acta 516:167–91.

36. Braun, A. C., and T. Laskaris. 1942. Tumor formation by attenuated crown gall bacteria in the presence of growth promoting substances. Proc. Nat. Acad. Sci., 28:468–77.

37. Braun, A. C., and T. Stonier. 1958. Morphology and physiology of plant tumors. Protoplasmatalogia 10:1–93.

38. Braun, A. C., and P. R. White. 1943. Bacteriological sterility of tissues derived from secondary crown-gall tumors. Phytopathology 33:85–100.

39. Brian, P. W., and H. B. Hemming. 1958. Complimentary action of gibberellic acid and auxin on pea internode extension. Ann. Bot. 22:1–17.

40. Bryan, W. H., and E. H. Newcomb. 1954. Stimulation of pectin methylesterase activity of cultured tobacco pith by indoleacetic acid. Physiol. Plant 7:290–97.

41. Budagyan, E. G., V. N. Lozhnikova, M. I. Golldin, and M. K. Chailakhyan. 1964. On the effect of gibberellin-like substances on tobacco mosaic virus. Rev. Appl. Mycol. 43:377.

42. Burg, S. P., and E. A. Burg. 1967. Inhibition of polar auxin transport by ethylene. Plant Physiol. 42:1224–28.

43. Butcher, D. N., S. El-Tigani and D. S. Ingram. 1974. The role of indole glucosinolates in the club root disease of the Cruciferae. Physiol. Plant Path. 4:127–40.

44. Chalutz, E., and J. E. DeVay. 1969. Production of ethylene in vitro and in vivo by Ceratocystis fimbriata in relation to disease development. Phytopathology 59:750–55.

45. Cheo, P. C. 1969. Effect of 2,4-dichlorophenoxyacetic acid on tobacco mosaic virus infection. Phytopathology 59:243–44.

46. Cheo, P. C. 1971. Effect of plant hormones on virus-replicating capacity of cotton infected with tobacco mosaic virus. Phytopathology 61:869–72.

47. Chessin, M. 1957. Growth substances and stunting in virus infected plants. Proc. Third Conf. Potato Virus Diseases, Lisse-Wageningen, 24–28 June 1957, pp. 80–84.

48. Cleland, R. E. 1961. The relation between auxin and metabolism. In: Handbuch du pflangen physiologie. W. Rhuland (ed.). Springer-Verlag, Berlin, 14:754–83.

49. Cleland, R. E. 1967. Auxin and the mechanical properties of the cell wall. Annals N.Y. Acad. Sci. 144, Art 1:3–18.

50. Cleland, R. E. 1979. Fusicoccin as a tool for studying the mechanism of auxin action. Plant Physiol. Ann. Suppl. Abs. 242.

51. Cleland, R. E., and N. McCombs. 1965. Gibberellic acid: Action in barley endosperm does not require endogenous auxin. Science 150:497–498.

52. Comai, L., and T. Kosuge. 1980. Involvement of plasmid deoxyribonucleic acid in indoleacetic acid synthesis in Pseudomonas savastanoi. J. Bact. 143:950–57.

53. Comai, L., and T. Kosuge. 1983. Transposable element that causes mutations in plant pathogenic Pseudomonas sp. J. Bact. 154:1162–67.

54. Cooke, R. J., and R. E. Kendrick. 1976. Phytochrome controlled gibberellin metabolism in etioplast envelopes. Planta 131:303–7.

55. Cottle, W., and P. E. Kolattukudy. 1982. Abscisic acid stimulation of suberization. Plant Physiol. 70:775–80.

56. Daft, M. J. 1963. Some effects of pancreatic ribonuclease and kinetin on tomato aucuba mosaic virus. Ann. Appl. Biol. 52:393–403.

57. Daly, J. M., and B. J. Deverall. 1963. Metabolism of indo-

58. Daly, J. M., and R. E. Inman. 1958. Changes in auxin levels in safflower hypocotyls infected with Puccinia carthami. Phytopathology 48:91–97.

59. Daly, J. M., P. M. Seevers and P. Ludden. 1970. Studies on wheat stem rust resistance controlled at the Sr6 locus. III. Ethylene and disease reaction. Phytopathology 60:1648–52.

60. Darvill, A. G., C. Smith, and M. Hall. 1977. Auxin induced proton release, cell wall structure and elongation growth. A hypothesis. In: Regulation of cell membrane activities in plants. E. Marrè and O. Cifferi (eds.). Elsevier North Holland, Biomedical Press, Amsterdam, pp. 275–82.

61. Davey, M. R., E. C. Cocking, J. Freeman, N. Pearce, and I. Tudor. 1980. Transformation of petunia protoplasts by isolated Agrobacterium plasmids. Plant Sci. Lett. 18:307–13.

62. Davis, D., and A. E. Dimond. 1953. Inducing disease resistance with plant growth-regulators. Phytopathology 43:137–40.

63. Dekhuijzen, H. M. 1976. Endogenous cytokinins in healthy and diseased plants. In: Physiological plant pathology. R. Heitefuss and P. H. Williams (eds.). Springer-Verlag, Berlin, pp. 527–59.

64. Dekhuijzen, H. M., and J. C. Overeem. 1971. The role of cytokinins in clubroot formation. Physiol. Plant Path. 1:151–61.

65. Dekhuijzen, H. M., and R. C. Staples. 1968. Mobilization factors in uredospores and bean leaves infected with bean rust fungus. Contr. Boyce Thompson Inst. Pl. 24:39–52.

66. Dekker, J. 1963. Effect of kinetin on powdery mildew. Nature 197:1027–28.

67. de Laat, Ad. M. M., and L. C. van Loon. 1982. Regulation of ethylene biosynthesis in virus-infected tobacco leaves. II. The time course of levels of intermediates and in vivo conversion rates. Plant Physiol. 69:240–45.

68. de Laat, Ad. M. M., L. C. van Loon and C. R. Vonk. 1981. Regulation of ethylene biosynthesis in virus-infected tobacco leaves. I. Determination of the role of methionine as a precursor of ethylene. Plant Physiol. 68:256–60.

69. DeRopp, R. S. 1947. The growth promoting and tumefacient factors of bacteria-free crown-gall tumor tissue. Am. J. Bot. 34:248–61.

70. Dorfling, K. 1972. Recent advances in abscisic acid research. In: Hormonal regulation in plant growth and development. H. Kaldeway and Y. Vardar (eds.). Verlag Chemie, Weinheim, pp. 281–95.

71. Durbin, R. D. 1967. Obligate parasites: Effect on the movement of solutes and water. In: The dynamic role of molecular constituents in plant-parasite interaction. C. J. Mirocha and I. Uritani (eds.). Amer. Phytopath. Soc., St. Paul, MN, pp. 80–99.

72. Dyson, J. G., and M. Chessin. 1961. Effect of auxins on virus-induced leaf abscission. Phytopathology 51:195.

73. Eagles, C. F., and P. F. Wareing. 1964. The role of growth substances in the regulation of bud dormancy. Physiol. Plant. 17:697–709.

74. El-Antably, H. M. M., P. F. Wareing, and J. Hillman. 1967. Some physiological responses to d,1-abscisin (dormin). Planta 73:74–90.

75. El Hammady, M., B. I. Pozsár, and Z. Király. 1968. Increased leaf growth regulated by rust infections, cytokinins and removal of the terminal bud. Acta Phytopath. Hung. 3:157–64.

76. Esquerré-Tugayé, M.-T., C. Lafitte, D. Mazau, A. Toppan, and A. Touzé. 197o. Cell surfaces in plant-microorganism interactions. II. Evidence for the accumulation of hydroxyproline-rich glycoproteins in the cell wall of diseased plants as a defense mechanism. Plant Physiol. 64:320–26.

77. Evans, M. L. 1973. Rapid stimulation of plant cell elonga-

tion by hormonal and non-hormonal factors. BioScience 23:711–18.

78. Fehrmann, H. 1965. Zum Auxin-Stoffwechsel phytophthorakranker Kartoffelknollen. Naturwissenschaften 52:481.

79. Fernandez, T. F., and R. Gáborjányi. 1976. Reversion of dwarfing induced by virus infection; effect of polyacrylic and gibberellic acids. Acta Phytopath. Hung. 11:271–75.

80. Fraser, R. S. S. 1979. Systemic consequences of the local lesion reaction to tobacco mosaic virus in a tobacco variety lacking the N gene for hypersensitivity. Physiol. Plant Path. 14:383–94.

81. Fraser, R. S. S. 1982. Are 'pathogenesis-related' proteins involved in acquired systemic resistance of tobacco plants to tobacco mosaic virus? J. Gen. Virol. 58:305–13.

82. Fraser, R. S. S., and R. J. Whenham. 1978. Inhibition of the multiplication of tobacco mosaic virus by methyl benzimidazol-2-yl carbamate. J. Gen. Virol. 39:191–94.

83. Fraser, R. S. S., and R. J. Whenham. 1978. Chemotherapy of plant virus disease with methyl benzimidazol-2-yl-carbamate: Effects on plant growth and multiplication of tobacco mosaic virus. Physiol. Plant Path. 13:51–64.

84. Fraser, R. S. S., and R. J. Whenham. 1982. Plant growth regulators and virus infection: A critical review. Plant Growth Regulation 1:37–59.

85. Freebairn, H. T., and I. W. Buddenhagen. 1964. Ethylene production by Pseudomonas solanacearum. Nature 202:313–14.

86. Gáborjányi, R., E. Balázs, and Z. Király. 1971. Ethylene production, tissue senescence and local virus infections. Acta Phytopath. Hung. 6:51–56.

87. Gáborjányi, R., E. Balázs and Z. Király. 1972. Abscission of rust infected bean leaves. (In Hungarian, with English summary.) Bot. Kozlem. 59:79–83.

88. Gáborjányi, R., F. Sági, and E. Balázs. 1973. Growth inhibition of virus-infected plants: Alterations of peroxidase enzymes in compatible and incompatible host-parasite relations. Acta Phytopath. Hung. 8:81–90.

89. Galston, A. W., and P. J. Davies. 1970. Control mechanisms in plant development. Prentice-Hall, Englewood Cliffs, NJ.

90. Galston, A. W., S. Lavee, and B. Z. Siegel. 1968. The induction and repression of peroxidase isozymes by indole-3-acetic acid. In: Biochemistry and physiology of plant growth substances. F. Wightman and G. Setterfield (eds.). Runge Press, Ottawa, pp. 455–72.

91. Geballe, G. T., and A. W. Galston. 1982. Ethylene as an effector of wound-induced resistance to cellulase in oat leaves. Plant Physiol. 70:788–90.

92. Goldsmith, M. H. M. 1977. The polar transport of auxin. Ann. Rev. Plant. Physiol. 28:439–78.

93. Grieve, B. J. 1941. Studies in the physiology of host-parasite relations. I. The affect of Bacterium solanacearum on the water relations of plants. Proc. Royal Soc., Victoria 53:268–99.

94. Grieve, B. J. 1943. Studies in the physiology of host-parasite relations. 4. Some effects of tomato spotted wilt on growth. Aust. J. Exp. Biol. Med. Sci. 21:89–101.

95. Hadwiger, L. A., S. L. Hess, and S. von Broembsen. 1970. Stimulation of phenylalanine ammonia-lyase activity and phytoalexin production. Phytopathology 60:332–36.

96. Hartman, R. T., and W. C. Price. 1950. Synergistic effect of plant growth substances and southern bean mosaic virus. Amer. J. Bot. 37:820–28.

97. Hasezawa, S., T. Nagata, and K. Soyono. 1981. Transformation of Vinca protoplasts by Agrobacterium spheroplasts. Mol. Gen. Genet. 182:206–10.

98. Hashimoto, T., A. Sakurai, and S. Tamura. 1967. Physiological activities of helminthosporal and helminthosporic acid. I. Effects on growth of intact plants. Plant Cell Physiol. 8:23–34.

99. Hedden, P., and B. O. Phinney. 1976. The dwarf-5 mutant

of Zea mays: A genetic lesion controlling the cyclization step (B activity) in kaurene biosynthesis. In: Collected abstracts of the paper-demonstrations 9th Int. Conf. Plant Growth Substances, p. 136. (Abst.)

100. Helgeson, J. P., and N. J. Leonard. 1966. Indentification of compounds isolated from Corynebacterium faciens. Proc. Nat. Acad. Sci. 56:60–63.

101. Hernalsteens, J. P., F. Van Vliet, M. De Beuckeleer, A. Depicker, G. Engler, M. Lemmers, M. Holsters, M. Van Montagu, and J. Schell. 1980. The Agrobacterium tumefaciens Ti plasmid as a host vector system for introducing foreign DNA in plants. Nature 287:654–56.

102. Hinman, R. L., and J. Lang. 1965. Peroxidase-catalyzed oxidation of indole-3-acetic acid. Biochem. 4:144–58.

103. Hislop, E. C., G. V. Hoad, and S. A. Archer. 1973. The involvement of ethylene in plant diseases. In: Fungal pathogenicity and the plant's response. R. J. W. Byrde and C. V. Cutting (eds.). Academic Press, London and New York, pp. 87–117.

104. Hislop, E. C., and M. A. Stahmann. 1971. Peroxidase and ethylene production by barley leaf infected with Erysiphe graminis f. sp. hordei. Physiol. Plant Path. 1:297–312.

105. Imaseki, H., T. Teranishi, and I. Uritani. 1968. Production of ethylene by sweet potato roots infected by black rot fungus. Plant Cell Physiol. 9:769–81.

106. Ingram, D. S. 1969. Growth of Plasmodiophora brassicae in host callus. J. Gen. Microbiol. 55:9–18.

107. Itai, C., and Y. Vaadia. 1965. Kinetin-like activity in root exudates of water-stressed sunflower plants. Physiol. Plant. 18:941–44.

108. Jablonski, J. R., and F. Skoog. 1954. Cell enlargement and cell division excised tobacco pith tissue. Physiol. Plant. 7:16–24.

109. Jackson, M. B., I. B. Morrow, and D. J. Osborne. 1972. Abscission and dehiscence in the squirting cucumber, Ecaballium elaterium: Regulation by ethylene. Can. J. Bot. 50:1465–71.

110. Jackson, M. B., and D. J. Osborne. 1970. Ethylene, a natural regulator of leaf abscission. Nature 225:1019–21.

111. Jansen, E. F., R. Jang, P. Albersheim, and J. Bonner. 1960. Pectin metabolism of growing cell walls. Plant Physiol. 35:87–97.

112. Jones, J. P. 1956. Studies on the auxin levels of healthy and virus infected plants. Diss. Abstr. 16:1567–68.

113. Kao, C. H., and S. F. Yang. 1983. Role of ethylene in the senescence of detached rice leaves. Plant Physiol. 73:881–85.

114. Kaper, J. M., and H. Veldstra. 1958. On the metabolism of tryptophan by Agrobacterium tumefaciens. Biochim. Biophys. Acta 30:401–20.

115. Kasamo, K., and T. Shimomura. 1977. The role of the epidermis in local lesion formation and the multiplication of tobacco mosaic virus and its relation to kinetin. Virology 76:12–18.

116. Kato, S. 1976. Ethylene production during lipid peroxidation in cowpea leaves infected with cucumber mosaic virus. Ann. Phytopath. Soc. Jap. 43:587–89.

117. Katsura, K., H. Egawa, M. Masuko, and A. Ueyama. 1969. On the plant hormones in the healthy and Plasmodiophora infected roots of Brassica rapa var. Neosuguki Kitam. II. Proc. Kansai Plant Prot. Soc. 11:23–27.

118. Kavanagh, J. A., M. N. Reddy, and P. H. Williams. 1969. Growth regulators in clubroot of cabbage. Phytopathology 59:1035.

119. Kazmaier, H. E. 1960. Some pathophysiological aspects of premature defoliation associated with rose blackspot. Diss. Abstr. 21:21.

120. Keegstra, K., K. W. Talmadge, W. D. Bauer, and P. Albersheim. 1973. The structure of plant cell walls. III. A model of walls of suspension-cultured sycamore cells based on the intercon-

nections of the macromolecular components. Plant Physiol. 51:188–96.

121. Key, J. L., M. N. Barnett, and C. Y. Len. 1967. RNA and protein biosynthesis and the regulation of cell elongation by auxin. Ann. N.Y. Acad. Sci. 144:49–62.

122. Key, J. L., and J. C. Shannon. 1964. Enhancement of auxin of ribonucleic acid synthesis in excised soybean hypocotyl tissue. Plant Physiol. 39:360–64.

123. Kiermeyer, O. 1958. Papierchromatographische Untersuchungen über den Wuchsstoffgehalt von *Capsella bursa-pastoris* nach Infektion mit *Albugo candida* und *Peronospora parasitica*. Österr. Botan. Z. 105:518–28.

124. Király, Z. 1968. Role of cytokinins in biochemical and morphogenetic changes induced by rust infections. First Int. Congr. Plant Path., London, p. 105. (Abstr.)

125. Király, Z., É. Bócsa, and J. Vörös. 1962. Effect of treatment of wheat seed with gibberellic acid on bunt infection. Phytopathology 52:171–72.

126. Király, Z., M. El Hammady, and B. I. Pozsár. 1967. Increased cytokinin activity of rust-infected bean and broad bean leaves. Phytopathology 57:93–94.

127. Király, Z., M. El Hammady, and B. I. Pozsár. 1968. Susceptibility to tobacco mosaic virus in relation to RNA and protein synthesis in tobacco and bean plants. Phytopath. Z. 63:47–63.

128. Király, Z., and B. I. Pozsár. 1964. On the inhibition of TMV production by kinetin and adenine in intact tobacco leaves. In:Host-parasite relations in plant pathology. Z. Király and G. Ubrizsy (eds.).Res. Inst. Plant Prot., Budapest, pp. 61–64.

129. Király, Z., and J. Szirmai. 1964. The influence of kinetin on tobacco mosaic virus production in *Nicotiana glutinosa* leaf discs. Virology 23:286–88.

130. Klein, R. M. 1955. Resistance and susceptibility of carrot roots to crown-gall tumor formation. Proc. Nat. Acad. Sci. 41:271–74.

131. Klein, R. M., and G. K. K. Link. 1952. Studies on the metabolism of plant neoplasms. V. Auxin as a promoting agent in the transformation of normal to crown gall tumor cells. Proc. Nat. Acad. Sci. 38:1066–72.

132. Kluge, S., S. Paunow, and G. Schuster. 1977. On the action of some metabolically active substances on the protein content and the multiplication of viruses in leaves of *Nicotiana tabacum* L. Phytopath. Z. 88:11–17.

133. Knotz, J., R. C. Coolbaugh, and C. A. West. 1977. Regulation of biosynthesis of ent-kaurene from mevalonate in the endosperm of immature *Marah macrocarpus* seeds by adenylate energy charge. Plant Physiol. 60:81–85.

134. Krupasagar, V., and Sequeira, L. 1969. Auxin destruction by *Marasmius perniciosus*. Amer. J. Bot. 56:390–97.

135. Krylov, A. V., V. A. Smirnova, and G. A. Tarakanova. 1960. The effect of physiologically active substances on the reproduction of tobacco mosaic virus. Soviet Plant Physiol. 7: 253–58.

136. Kuriger, W. E., and G. N. Agrios. 1977. Cytokinin levels and kinetin-virus interactions in tobacco ringspot virus-infected cowpea. Phytopathology 67:604–9.

137. Kutsky, R. J. 1952. Effects of indolebutyric acid and other compounds on virus concentration in plant tissue cultures. Science 115:19–20.

138. Kutsky, R. J., and T. E. Rawlins. 1950. Inhibition of virus multiplication by naphthalene acetic acid in tobacco tissue cultures, as revealed by a spectrophotometric method. J. Bact. 60:763–66.

139. Lang, A. 1970. Gibberellins: Structure and metabolism. Ann. Rev. Plant Physiol. 21:537–70.

140. Leopold, A. C., and T. H. Plummer. 1961. Auxin phenol complexes. Plant Physiol. 36:589–92.

141. Letham, D. S. 1967. Regulators of cell division in plant tissue. III. The identity of zeatin. Tetrahedron 23:479–86.

142. Levy, D., and S. Marco. 1976. Involvement of ethylene in epinasty of CMV-infected cucumber cotyledons which exhibit increased resistance to gaseous diffusion. Physiol. Plant Pathol. 9:121–26.

143. Libbenga, K. R. 1978. Hormone receptors in plants. In: Frontiers of plants. T. A. Thorpe (ed.). Int. Assoc. Plant Tissue Cult., Calgary, pp. 325–33.

144. Lipetz, J., and A. W. Galston. 1959. Indole acetic acid oxidase and peroxidase in normal and crown gall tissue of *Parthenocissus tricuspedata*. Amer. J. Bot. 46:193–96.

145. Lippincott, J. A., and B. B. Lippincott. 1977. Nature and specificity of the bacterium-host attachment in *Agrobacterium* infection. In: Cell wall biochemistry related to specificity in host-plant pathogen interactions. B. Solheim and J. Raa (eds.). Universitesforlaget, Oslo, pp. 439–51.

146. Liu, S.-T., and C. I. Kado. 1979. Indoleacetic acid production, a plasmid function of *Agrobacterium tumefaciens* c 58. Biochem. Biophys. Res.Comm. 90:171–78.

147. Liu, S.-T., K. L. Perry, L. Schardl, and C. Kado. 1982. *Agrobacterium* Ti plasmid indoleacetic acid gene is required for crown gall oncogenesis. Proc. Nat. Acad. Sci 79:2812–16.

148. Locke, S. B., A. J. Riker, and B. M. Duggar. 1938. Growth substance and the development of crown gall. J. Agr. Res. 57:21–39.

149. Lockhart, B. E. L., and J. S. Semancik. 1970. Growth inhibition, peroxidase and 3-indole acetic acid oxidase activity, and ethylene production in cowpea mosaic virus-infected cowpea seedlings. Phytopathology 60:553–54.

150. Loebenstein, G., A. Gera, A. Barnett, S. Shabtai, and J. Cohen. 1980. Effect of 2,4- dichlorophenoxyacetic acid on multiplication of tobacco mosaic virus in protoplasts from local-lesion and systemic-responding tobaccos. Virology 100:110–15.

151. Lovrekovich, L., and G. L. Farkas. 1963. Kinetin as an antagonist of toxic effect of *Pseudomonas tabaci*. Nature 196:170.

152. Luberman, M. L., W. Mapson, A. T. Kunishi, and D. A. Wardale. 1965. Ethylene production from methionine. Biochem. J. 97:449–59.

153. Lund, B. M.1973. The effect of certain bacteria on ethylene production by plant tissue. In: Fungal pathogenicity and the plant's response. R. J. W. Byrde and C. V. Cutting (eds.). Academic Press, New York, pp. 69–86.

154. Lund, B. M., and L. W. Mapson. 1970. Stimulation by *Erwinia carotovora* of the synthesis of ethylene in cauliflower tissue. Biochem. J. 119:251–63.

155. Mace, M. E. 1965. Isolation and identification of 3-indoleacetic acid from *Fusarium oxysporum* f. *cubenese*. Phytopathology 55:240–41.

156. MacMillan, J. 1972. Current aspects of gibberellins and plant growth regulation. In: Hormonal regulation in plant growth and development. H. Kaldeevey and Y. Vardar (eds.). Proc. Adv. Study Inst., Izmir 1971 Verlag Chemie, Weinheim, pp. 175–87.

157. Magie, A. R. 1963. Physiological factors involved in tumor production by the oleandor knot pathogen *Pseudomonas savastanoi*. Ph.D. Diss., Univ. of California, Davis.

158. Mahadevan, A. 1965. *In vitro* production of indole acetic acid by *Fusarium vasinfectum* Atk. Experientia 21:433.

159. Maine, E. C. 1960. Physiological responses in the tobacco plant to pathogenesis by *Pseudomonas solanacearum*, causal agent of bacterial wilt. Ph.D. Diss., North Carolina State Univ., Raleigh.

160. Maine, E. C., and A. Kelman. 1960. Changes in oxidation rates of polyphenols and ascorbic acid in tobacco stem tissue invaded by *Pseudomonas solanacearum*. Phytopathology 50:645. (Abstr.)

161. Mapson, L. W., and A. C. Hulme. 1970. The bio-

synthesis, physiological effects and mode of action of ethylene. In: Progress in phytochemistry. L. Rheinhold and Y. Liwschitz (eds.).Interscience, New York, 2:343–384.

162. Mapson, L. W., J. F. March, J. J. C. Rhodes, and L. C. Wooltorton. 1970. A comparative study of the ability of methionine or linolenic acid to act as precursors of ethylene in plant tissues. Biochem. J. 117:473–79.

163. Maramorosch, K. 1957. Reversal of virus-caused stunting in plants by gibberellic acid. Science 126:651–52.

164. Marco, A., and D. Levy. 1979. Involvement of ethylene in the development of cucumber mosaic virus-induced chlorotic lesions in cucumber cotyledons. Physiol. Plant Path 14:235–44.

165. Marco, S., D. Levy, and N. Aharoni. 1976. Involvement of ethylene in the suppression of hypocotyl elongation in CMV-infected cucumbers. Physiol. Plant. Path. 8:1–7.

166. Marrè, E. 1980. Mechanism of action of phytotoxins affecting plasmalemma function. Prog. Phytochem. 6:253–84.

167. Marrè, E., P. Lado, F. R. Caldogno, and R. Colombo. 1973. Correlation between cell enlargement in pea internode segments and decrease in the pH of the medium of incubation. I. Effects of fusicoccin, natural and synthetic auxins and mannitol. Plant Sci. Lett. 1:179–84.

168. Marrè, E., P. Lado, A. Ferroni, and A. B. Denti. 1974. Transmembrane potential increase induced by auxin, benzyladenine and fusicoccin. Correlation with proton extrusion and cell enlargement. Plant Sci. Lett. 2:257–65.

169. Marton, L., G. W. Wullems, L. Molendijk, and R. A. Schilperoort. 1979. In vitro transformation of cultured cells from Nicotiana tabacum by Agrobacterium tumefaciens. Nature 277:129–31.

170. Matsubara, S., D. J. Armstrong, and F. Skoog. 1968. Cytokinins in tRNA of Corynebacterium fascians. Plant Physiol. 43:451-53.

171. Mercer, E. I., and J. E. Pughe. 1969. The effects of abscisic acid on the synthesis of isoprenoid compounds in Maize. Phytochemistry 8:115–22.

172. Mittelheuser, C. J., and R. F. M. Van Stevenick. 1971. The ultrastructure of wheat leaves. I. Changes due to senescence and the effects of kinetin and ABA on detached leaves incubated in the dark. Protoplasma 73:239–52.

173. Milborrow, B. V., and R. C. Noddle. 1970. Conversion of 5-(1,2-epoxy-2,6,6,-trimethyl-cyclohexyl)-3-methylpenta-cis-2-trans-4-dienoic acid into abscisic acid in plants. Biochem. J. 119:727–34.

174. Miller, C. D. 1967. Zeatin and zeatin riboside from a mycorrhizal fungus. Science 157:1055–56.

175. Milo, G. E., and B. I. S. Srivastava. 1969. Effect of cytokinins on tobacco mosaic virus production in local-lesion and systemic hosts. Virology 38:26–31.

176. Moore, T. C. 1979. Biochemistry and physiology of plant hormones. Springer-Verlag, New York.

177. Mothes, K., L. Engelbrecht, and H. R. Schutte. 1961. Über die akkumulation von α amino-isobittersäure—im Blattgewebe unter dem Einfluss von kinetin. Physiol. Plant. 14:72–75.

178. Muir, R. M. 1964. Conversion of tryptophan to auxin by plant tissue preparations. Plant Physiol. Suppl. XVII.

179. Muir, R. M., and B. P. Lantican. 1968. Purification and properties of the enzyme system forming indoleacetic acid. In: Biochemistry and physiology of plant growth substances. F. Wightman and G. Setterfield (eds.). Runge Press, Ottawa. pp. 259–72.

180. Mukherjee, A. K., L. C. Soans, and M. Chessin. 1967. Effects of kinetin and actinomycin D on the susceptibility of Nicotiana glutinosa L. to infection by tobacco mosaic virus. Nature 216:1344–45.

181. Murai, N., F. Skooge, M. Doyle, and R. Hanson. l979.

Relationship between cytokinin production, presence of plasmids and fasciation caused by strains of Corynebacterium fascians. Proc. Nat. Acad. Sci. 77:6l9–23.

182. Nakagaki, Y. 1971. Effect of kinetin on local lesion formation on detached bean leaves inoculated with tobacco mosaic virus or its nucleic acid. Ann. Phytopath. Soc. Jap. 37:307–9.

183. Nakagaki, Y., T. Hirai, and M. A. Stahmann. 1970. Ethylene production by detached leaves infected with tobacco mosaic virus. Virology 40:1–9.

184. Nester, E. W., and T. Kosuge. 1981. Plasmids specifying plant hyperplasias. Ann. Rev. Microbiol. 35:331–65.

185. Nichols, C. W. 1952. The retarding effect of certain plant hormones on tobacco mosaic symptoms. Phytopathology 42:579–80.

186. Nitsch, C., and J. P. Nitsch. 1963. Etude du mode d'action de l'acide gibbérellique A3 marquée. I. Action sur substances de croissance du haricot nain. Bull. Soc. Botan. France, 110:7–17.

187. Nooden, L. D., and K. V. Thimann. 1963. Evidence for a requirement for protein synthesis for auxin-induced cell enlargement. Proc. Nat. Acad. Sci. 49:194–200.

188. Ochs, G. 1958. Untersuchungen über den Einfluss eines phytopathogenen Virus auf den Wuchsstoff-haushalt der Rebe. Naturwiss. 14:343.

189. Osborne, D. J. 1962. Effect of kinetin on protein and nucleic acid metabolism in Xanthium leaves during senescence. Plant Physiol. 37:595–602.

190. Osborne, D. J. 1968. Ethylene as a plant hormone. Soc. Chem. Ind. (London) Monograph 31:249–63.

191. Osborne, D. J. 1968. Hormonal regulation of leaf senescence. Symp. Soc. Exp. Biol. 21:305–22.

192. Osborne, D. J., M. B. Jackson, and B. V. Milborrow. 1972. Physiological properties of abscission accelerator from senescent leaves. Nature 240:98–101.

193. Osborne, D. J., and M. G. Mullins. 1969. Auxin, ethylene and kinetin in a carrier protein model system for polar transport of auxins in petiole segments of Phaseolus vulgaris. New Phytol. 68:977–91.

194. Oswald, T. H., and T. D. Wyllie. 1973. Effects of growth regulator treatments on severity of charcoal rot disease of soybean. Plant Dis. Rept. 57:789–92.

195. Paleg, L. S. 1965. Physiological effects of gibberellins. Ann. Rev. Plant Physiol. 16:291–322.

196. Pavillard, J. 1955. La teneur en auxine des pommes de terre virosées. Proc. 2d Conf. Potato Virus Diseases, Wageningen, pp. 178–84.

197. Pavillard, J., and C. Beauchamp. 1957. La constitution auxinique de tabacs sains ou atteints de maladies à virus: Présence et rôle de la scopolétine. Compte. Rend. Acad. Sci. 244:1240–43.

198. Pegg, G. F., and I. W. Selman. 1959. An analysis of the growth response of young tomato plants to infection by Verticillium albo-atrum. The production of growth substances. Ann. Appl. Biol. 47:222–31.

199. Pegg, G. F., and L. Sequeira. 1968. Stimulation of aromatic biosynthesis in tobacco plants infected by Pseudomonas solanacearum. Phytopathology 58:476–83.

200. Penny, W. 1977. On the possible role of hydrogen ions in auxin induced growth. In: Regulation of cell membrane activities in plants. E. Marré and O. Cifferi (eds.). Elsevier North Holland, Biomedical Press, Amsterdam, pp. 283–90.

201. Petersen, L. J., J. E. DeVay, and B. R. Houston. 1963. Effect of gibberellic acid on development of hypocotyl lesions caused by Rhizoctonia solani on red kidney bean. Phytopathology 53:630–33.

202. Phelps, R. H., and L. Sequeira. 1967. Synthesis of indoleacetic acid by cell-free systems from virulent and avirulent strains of Pseudomonas solanacearum. Phytopathology 57:1182–90.

203. Phelps, R. H., and L. Sequeira. 1968. Auxin biosynthesis in a host parasite complex. In: Biochemistry and physiology of growth substances. F. Wightman and G. Setterfield (eds.).Runge Press, Ottawa, pp. 197–212.

204. Phipps, J., and C. Martin. 1968. Incidence de l'inoculation d'un virus à des hôtes sensibles sur l'activité des auxine-oxydases de leurs feuilles. Compte. Rend. Acad. Sci. 266:1796–99.

205. Pilet, P. E. 1960–61. Auxin content and auxin catabolism of stems of *Euphorbia cyparissias* L. infected by *Uromyces* (Pers). Phytopath. Z. 40:75–90.

206. Platt, R. S., Jr. 1954. The inactivation of auxin in normal and tumorous tissue. Année Biol. 3d ser. 30:349–59.

207. Pozsár, B. I., and Z. Király. 1964. Cytokinin-like effect of rust infections in the regulation on phloem-transport and senescence. Proc. Symp. Host-parasite relations in plant pathology, Budapest Res. Inst. for Plant Protection, pp. 199–210.

208. Pozsár, B. I., and Z. Király. 1966. Phloem-transport in rust-infected plants and the cytokinin-directed long-distance movement of nutrients. Phytopath. Z. 56:297–309.

209. Pritchard, D. W., and A. F. Ross. 1975. The relationship of ethylene to formation of tobacco mosaic virus lesions in hypersensitive responding tobacco leaves with and without induced resistance. Virology 64:295–307.

210. Rajagopal, R. 1977. Effect of tobacco mosaic virus infection on the endogenous levels of indoleacetic, phenylacetic and abscisic acids of tobacco leaves in various stages of development. Z. Pflanzenphysiol. 83:403–9.

211. Rathbone, M. P., and R. H. Hall. 1972. Concerning the presence of the cytokinin N^6-(δ^2-isopentenyl) adenine, in cultures of *Corynebacterium fascians*. Planta 108:93–102.

212. Ray, P. M. 1977. Auxin-binding sites of maize coleoptiles are localized on membranes of endoplasmic reticulum. Plant Physiol. 59:594–99.

213. Rayle, D. L. 1973. Auxin-induced hydrogen-ion secretion in *Avena* coleoptiles and its implications. Planta 114:63–73.

214. Reingard, T. A., and S. J. Pashkar. 1958. Participation of growth substances of the auxin type in formation of tumors on potato plants. Fiziologiya Rast. 5:512–18.

215. Reunov, A. V., G. D. Reunova, and G. V. Shmaidenko. 1976. Kinetin effect on tobacco mosaic virus (TMV) replication in *Datura stramonium* leaves. Virus diseases of plants in the Far East 25:174.

216. Reunov, A. V., G. D. Reunova, L. A. Vasilyeva, and V. G. Reifman. 1977. Effect of kinetin on tobacco mosaic virus and potato virus X replication in leaves of systemic hosts. Phytopath. Z. 90:342–49.

217. Ridge, I., and D. Osborne. 1971. Role of peroxidase when hydroxyproline-rich protein in plant cell walls is increased by ethylene. Nature 229:205–8.

218. Rijiven, A. H. G. C. 1974. Initiation of polyphenylalanine synthesis and the action of cytokinins in fenugreek cotyledons. Nature 252:257–59.

219. Rodrigues, C. J., and D. C. Arny. 1966. Role of oxidative enzymes in the physiology of the leaf spot of coffee caused by *Mycena citricolor*. Phytopath. Z. 57:375–84.

220. Rowe, J. R. (ed.). 1968. The common and systematic nomenclature of cyclic diterpenes, 3d rev. Proposal I.U.P.A.C. Commission Org. Nomen.

221. Rubery, P. H. 1981. Auxin receptors. Ann. Rev. Plant Physiol. 32:569–96.

222. Russell, S. L., and W. C. Kimmins. 1971. Growth regulators and the effect of barley yellow dwarf virus on barley (*Hordeum vulgare* L.). Ann. Bot. 35:1037–43.

223. Sacher, J. 1962. In IAA-oxidase inhibitor system in bean pods. II. Kinetic studies of oxidase and natural inhibitor. Plant Physiol. 37:74–82.

224. Sachs, J. 1887. Lectures on the physiology of plants. Clarendon Press, Oxford.

225. Saftner, R. A. 1972. Promotion of cell elongation by host-specific fungal toxins. Master's thesis, Ohio State Univ., Columbus.

226. Saftner, R. A., and M. L. Evans. 1974. Selective effects of victorin on growth and auxin response in *Avena*. Plant Physiol. 53:382–87.

227. Schuster, G. 1974. The effect of gibberellic acid, 2-chloroethyl phosphonic acid, and some growth retardants on virus multiplication in systemically infected plants. Biochem. Physiol. Pflanzen 166:393–400.

228. Schuster, G. 1975. The effect of several auxins and auxin herbicides on the development of local lesions of tobacco mosaic virus. Arch. Phytopathol. u. Pflanzensch., Berlin, 11:9–17.

229. Seevers, P. M., and J. M. Daly. 1970. Studies on wheat stem rust resistance controlled at the Sr6 locus. II. Peroxidase activities. Phytopathology 60:1642–47.

230. Sequeira, L. 1963. Growth regulators in plant disease. Ann. Rev. Phytopath. 1:5–30.

231. Sequeira, L. 1964. Inhibition of indoleacetic acid oxidase in tobacco plants infected by *Pseudomonas solanacearum*. Phytopathology 54:1078–83.

232. Sequeira, L. 1965. Origin of indoleacetic acid in tobacco plants infected by *Pseudomonas solanacearum*. Phytopathology 55:1232–36.

233. Sequeira, L. 1969. Synthesis of scoplin and scopoletin in tobacco plants infected with *Pseudomonas solanacearum*. Phytopathology 59:473–78.

234. Sequeira, L. 1973. Hormone metabolism in diseased plants. Ann. Rev. Plant Physiol. 24:353–80.

235. Sequeira, L., and A. Kelman. 1962. The accumulation of growth substances in plants infected by *Pseudomonas solanacearum*. Phytopathology 52:439–48.

236. Sequeira, L., and T. A. Steeves. 1954. Auxin inactivation and its relation to leaf drop caused by the fungus *Omphalia flavida*. Plant Physiol. 29:11–16.

237. Sequeira, L., and P. H. Williams. 1964. Synthesis of indoleacetic acid by *Pseudomonas solanacearum*. Phytopathology 54:1240–46.

238. Serrano, L. S. 1955. Posible inducción de resistencia del tomato al *Phytophthora infestans* (Mont.) De Barry mediante auxonas. Acta Agronomica (Palmira, Colombia) 5:117–37.

239. Shannon, L. M., E. Kay, and J. Y. Lew. 1966. Peroxidase isozymes from horseradish roots. I. Isolation and physical properties. J. Biol. Chem. 241:2166–73.

240. Shaw, M., and A. R. Hawkins. 1958. The physiology of host-parasite relations. V. A preliminary examination of the level of free endogenous indoleacetic acid in rusted and mildewed cereal leaves and their ability to decarboxylate exogenously supplied radioactive indoleacetic acid. Can. J. Bot. 36: 1–16.

241. Shibaoka, H., and K. V. Thimann. 1970. Antagonisms between kinetin and amino acids. Plant Physiol. 46:212–20.

242. Siegel, S. M. 1956. The chemistry and physiology of lignin formation. Quart. Rev. Biol. 31:1–18.

243. Siegel, S. M. 1962. The plant cell wall. MacMillan, New York.

244. Simons, T. J., H. W. Israel, and A. F. Ross. 1972. Effect of 2,4- dichlorophenoxyacetic acid on tobacco mosaic virus lesions in tobacco and on the fine structure of adjacent cells. Virology 48:502–15.

245. Simons, T. J., and A. F. Ross. 1965. Effect of 2,4-dichlorophenoxyacetic acid on size of tobacco mosaic virus lesions in hypersensitive tobacco. Phytopathology 55:1076–77.

246. Skoog, F., and D. J. Armstrong. 1970. Cytokinins. Ann. Rev. Plant Physiol. 21:359–84.

247. Skoog, F., and N. J. Leonard. 1968. Sources and structure: Activity relationship of cytokinins. In: Biochemistry and physiology of plant growth substances. F. Wightman and G. Setterfield (eds.).Runge Press, Ottawa pp. 1–18.

248. Skoog, F., and C. O. Miller. 1957. Chemical regulation of growth and organ formation in plant tissues cultured in vitro. Symp. Soc. Exp. Biol. 11:118–31.

249. Skoog, F., F. M. Strong, and C. O. Miller. 1965. Cytokinins. Science 148:532–33.

250. Smidt, M. L., and T. Kosuge. 1978. The role of indole-3-acetic acid accumulation by alpha-methyl tryptophan-resistant mutants of *Pseudomonas savastanoi* in gall formation on oleanders. Physiol. Plant Path. 13:203–14.

251. Smith, S. H., S. R. McCall, and J. H. Harris. 1968. Alterations in the auxin levels of resistant and susceptible hosts induced by the curly-top virus. Phytopathology 58:575–77.

252. Smith, S. H., S. R. McCall, and J. H. Harris. 1968. Auxin transport in curly-top virus-infected tomato. Phytopathology 58:1669–70.

253. Soliday, C. L., B. B. Dean, and P. E. Kolattukudy. 1978. Suberization: Inhibition by washing and stimulation by abscisic acid in potato disks and tissue culture. Plant Physiol. 61:170–74.

254. Sommer, N. F. 1961. Production of *Taphrina deformans* of substance stimulating cell elongation and division. Physiol. Plant. 14:460–69.

255. Srivastava, B. I. S., and M. Shaw. 1962. The biosynthesis of indoleacetic acid in *Melampsora lini (Pers.) Lev.* Can. J. Bot. 40:309–15.

256. Stahmann, M. A., B. J. Clare, and W. Woodbury. 1966. Increased disease resistance and enzyme activity induced by ethylene and ethylene production by black rot infected sweet potato tissue. Plant Physiol. 41:1505–12.

257. Steadman, J. R., and L. Sequeira. 1969. A growth inhibitor from tobacco and its possible involvement in pathogenesis. Phytopathology 59:499–503.

258. Steadman, J. R., and L. Sequeira. 1970. Abscisic acid in tobacco plants. Tentative identification and its relation to stunting induced by *Pseudomonas solanacearum*. Plant Physiol. 41:691–97.

259. Stein, D. B. 1962. The developmental morphology of *Nicotiana tabacum* "White Burley" as influenced by virus infection and gibberellic acid. Am. J. Bot. 49:437–43.

260. Stoddart, J. L. 1969. Incorporation of kaurenoic acid into gibberellins by chloroplast preparations of *Brassica oleraceae*. Phytochemistry 8:831–37.

261. Stowe, B. B., and T. Yamaki. 1957. The history and physiological action of the gibberellins. Ann. Rev. Plant Physiol. 8:181–216.

262. Suchorukov, K. T., and B. P. Strogonov. 1945. On the growth hormones of the diseased plant. Doklady Akad. Nauk USSR. 47:593–96. (In Russian.)

263. Sziráki, I., and E. Balázs. 1977. The effect of infection by TMV on cytokinin level of tobacco plants, and cytokinins in TMV-RNA. In: Current topics in plant pathology. Z. Király (ed.). Akadémiai Kiadó, Budapest, pp. 345–52.

264. Sziráki, I., and E. Balázs. 1979. Cytokinin activity in the RNA of tobacco mosaic virus. Virology 92:578–82.

265. Sziráki, I., E. Balázs, and Z. Király. 1975. Increased levels of cytokinin and indoleacetic acid peach leaves infected with *Taphrina deformans*. Physiol. Plant Path. 5:45–50.

266. Sziráki, I., E. Balázs, and Z. Király. 1980. Role of different stresses in inducing systemic acquired resistance to TMV and increasing cytokinin level in tobacco. Physiol. Plant Path. 16:277–4.

267. Sziráki, I., B. Barna, S. El Waziri, and Z. Király. 1976. Effect of rust infection on the cytokinin level of wheat cultivars susceptible and resistant to *Puccinia graminis* f. sp. *tritici*. Acta Phytopath. Hung. 11:155–60.

268. Tagawa, T., and J. Bonner. 1957. Mechanical properties of *Avena* coleoptile as related to auxin and ionic interactions. Plant Physiol. 32:207–12.

269. Tavantzis, S. M., S. H. Smith, and F. H. Witham. 1979. The influence of kinetin on tobacco ringspot virus infectivity and the effect of virus infection on the cytokinin activity in intact leaves of *Nicotiana glutinosa* L. Physiol. Plant Path. 14:227–33.

270. Thimann, K. V., and T. Sachs. 1966. The role of cytokinins in the "fasciation" disease caused by *Corynebacterium fascians*. Amer. J. Bot. 53:731–39.

271. Thomas, H., and J. L. Stoddart. 1980. Leaf senescence. Ann. Rev. Plant Physiol. 31:83–111.

272. Thrower, L. B. 1965. On the host-parasite relationships of *Trifolium subterraneum* and *Uromyces trifolii*. Phytopath. Z. 52:269–94.

273. Thung, T. H., and T. Hadiwidjaja. 1951. Growth substances in relation to virus diseases: Experiments with *Arachis hypogea*. Tijdschr. Plantenziekten 57:95–99.

274. Tomlinson, J. A., E. M. Faithfull, and C. M. Ward. 1976. Chemical suppression of the symptoms of two virus diseases. Ann. Appl. Biol. 84:31–41.

275. Turian, G. 1958. Recherches anatomo-physiologiques sur les cécides. Rev. Gén. Bot. 65:279–94.

276. Turian, G. 1961. Accumulation of 3-indole acetic acid and activation of enzymes in the *Ustilago* induced tumor of maize. Conf. on Scientific Problems of Plant Protection, Budapest, 1960. Phytopathology 1:35–39.

277. Turian, G., and R. H. Hamilton. 1960. Chemical detection of 3-indolylacetic acid in *Ustilago zeae* tumors. Biochim. Biophys. Acta 41:148–50.

278. Udvardy, J., and G. L. Farkas. 1972. Abscisic acid stimulation of the formation of an aging-specific nuclease in *Avena* leaves. J. Exp. Bot. 22:914–20.

279. Valent, B. S., and P. Albersheim. 1974. The structure of plant cell walls. V. On the binding of xyloglucan to cellulose fibers. Plant Physiol. 54:105–8.

280. van Andel, O. M., and A. Fuchs. 1972. Interference with plant growth regulation by microbial metabolites. In: Phytotoxins in plant diseases. R. K. S. Wood, A. Ballio and A. Graniti (eds.). Academic Press, New York, pp. 227–49.

281. Vanderhorf, L. N., T. S. Lu, and C. A. Williams. 1977. Comparison of auxin-induced and acid-induced elongation in soybean by hypocotyl. Plant Physiol. 59:1004–7.

282. Vanicková-Zemlová, D., H. W. Liebisch, and G. Sembdner. 1981. Role of gibberellins and endogenous inhibitors in the pathogenesis of dwarf smut infected winter wheat (*Triticum aestivum* L./*Tilletia controversa* Kuhn). Biochem. Physiol. Pflanzen. 176:291–304.

283. van Loon, L. C. 1979. Effects of auxin on the localization of tobacco mosaic virus in hypersensitively reacting tobacco. Physiol. Plant Path. 14:213–26.

284. van Loon, L. C. 1982. Regulation of changes in proteins and enzymes associated with active defence against virus infection. In: Active defense mechanisms in plants. NATO Advanced Study Institute Series, Vol. 37. R. K. S. Wood (ed.). Plenum Press, New York, pp. 247–73.

285. van Loon, L. C., and A. T. Berbée. 1978. Endogenous levels of indoleacetic acid in leaves of tobacco reacting hypersensitively to tobacco mosaic virus. Z. Pflanzenphysiol. 89:373–75.

286. Varner, J. E. 1964. Gibberellic acid controlled synthesis of the amylase in barley endosperm. Plant Physiol. 39:413–15.

287. Varner, J. E., and S. R. Chandra. 1964. Hormonal control

of enzyme synthesis in barley endosperm. Proc. Nat. Acad. Sci. 52:100–106.

288. Veldstra, H. 1947. Considerations on the interactions of ergons and their "substrates." Biochim. Biophys. Acta 1:364–78.

289. Vizárová, G. 1979. Changes in the level of endogenous cytokinins of barley during the development of powdery mildew. Phytopath. Z. 95:314–29.

290. Volken, P. 1972. Quelques aspects des relations hôte-parasite en fonction de traitments à l'acide indol-acetique et à l'acide gibberellique. Phytopath. Z. 75:163–74.

291. Wallace, R. H. 1927. The production of intumescences in transparent apple by ethylene gas as affected by external and internal conditions. Torrey Bot. Club Bull. 54:499–542.

292. Walton, D. C. 1980. Biochemistry and physiology of abscisic acid. Ann. Rev. Plant Physiol. 31:453–89.

293. West, C. A., M. Oster, D. Robinson, F. Leev, and P. Murphy. 1968. Biosynthesis of gibberellin precursors and related diterpenes. In: Biochemistry and physiology of plant growth substances. F. Wightman and G. Setterfield (eds.). Runge Press, Ottawa, pp. 313–32.

294. Whatley, M. H., J. S. Bodwin, B. B. Lippincott, and J. Lippincott. 1976. Role for *Agrobacterium* cell envelope polysaccharide infection site attachment. Infect. Immu. 13:1080–83.

295. Whatley, M. H., J. Margot, J. Schnell, B. B. Lippincott, and J. Lippincott. 1978. Plasmid and chromosomal determination of *Agrobacterium* adherence specificity. J. Gen. Microbiol. 107:393–96.

296. Whenham, R. J., and R. S. S. Fraser. 1981. Stimulation by abscisic acid of RNA synthesis in discs from healthy and tobacco mosaic virus-infected tobacco leaves. Planta 150:349–53.

297. Whenham, R. J., and R. S. S. Fraser. 1981. Effect of systemic and local-lesion- forming strains of tobacco mosaic virus on abscisic acid concentration in tobacco leaves: Consequences for the control of leaf growth. Physiol. Plant Path. 18:267–78.

298. White, G. A., and E. Taniguchi. 1969. The mode of action of helminthosporal. I. Effect on the formation of amylase by embryoless barley seeds. Can. J. Bot. 47:873–84.

299. Wiese, M. V., and J. E. DeVay. 1970. Growth regulator changes in cotton associated with defoliation caused by *Verticillium albo-atrum*. Plant Physiol. 45:304–9.

300. Wightman, F. 1962. Metabolism and biosynthesis of 3-indoleacetic acid and related indole compounds in plants. Can. J. Bot. 40:689–718.

301. Wightman, F., and D. Cohen. 1968. Intermediary steps in the enzymatic conversion of tryptophan to IAA in cell free systems from mung bean seedlings. In: Biochemistry and physiology of plant growth substances. F. Wightman and G. Setterfield (eds.).Runge Press, Ottawa. pp. 273–88.

302. Williams, P. H., M. N. Reddy, and J. C. Strandberg. 1969. Growth of non-infected and *Plasmodiophora brassicae* infected cabbage callus in culture. Can. J. Bot. 47:1217–20.

303. Wilson, E. E. 1965. Pathological histogenesis in oleander tumors induced by *Pseudomonas savastanoi*. Phytopathology 55:1244–49.

304. Wolf, F. T. 1952. The production of indoleacetic acid by *Ustilago zeae* and its possible significance in tumor formation. Proc. Nat. Acad. Sci. 38:106–11.

305. Wolf, F. T. 1955. The production of indoleacetic acid by the cedar apple rust fungus and its identification by paper chromatography. Phytopathology 26:219–23.

306. Yamagata, Y., and Y. Masuda. 1975. Comparative studies on auxin and fusicoccin actions on plant growth. Plant and Cell Physiol. 16:41–52.

307. Yang, S. F. 1980. Regulation of ethylene biosynthesis. Hortscience 15:238–43.

308. Yerkes, W. D., Jr. 1960. Interaction of potassium gibberellate and a stunting bean virus on beans, *Phaseolus vulgaris*. Phytopathology 50:525–27.

8

Transcellular and Vascular Transport

The Healthy Plant

Introduction

Transport in the plant may conveniently be characterized as proceeding over long and short distances, with the former occurring in the xylem and phloem and the latter occurring by cyclosis within the cell and active transport across membranes (230). Solutes and metabolites of the healthy plant traverse these pathways, and it is evident that, within the constraints of their dimensions, pathogens and their metabolites also move along these routes. The rate at which ions are transferred by roots of intact plants to the leaves may be controlled by the transpiration rate and their active uptake by the roots.

In considering the movement of large quantities of water in the xylem, the data suggest that the rate of movement can reach velocities in the transpiration stream of up to 75 cm/min (106). It is also clear that larger vessels in healthy condition carry water more efficiently than smaller ones. Water moves along a negative pressure gradient that becomes apparent once transpiration begins (sunup). The gradient required to overcome gravity is 0.1 atm/m. In addition, 0.1–0.2 atm/m is necessary to overcome internal resistance to flow at maximum velocities. The volume of water moved through a vessel at a given pressure gradient is proportional to the fourth power of the vessel radius. Hence, if one compares the volume of water passing through two vessels side by side, 50 μm and 200 μm wide, the latter will transport 256 times as much as the former, or 99.6% of the total.

It is apparent that pathological blockage of these pathways must, sooner or later, be translated into plant cell damage and eventually appear as a symptom of disease. Hence, vascular and transcellular transport of both solvent (H_2O) and solute (metabolites of plant and pathogen) are foci of study by plant pathologists.

Symplast-Apoplast Concept

The mass flow concept of cyclic flow within the plant was originally advanced by Munch (143) and later expanded by Arisz (2) to the symplast-apoplast concept. The concept visualizes transport over two distinctly different but interconnected pathways. The symplast is composed of all of the protoplasm (cytoplasm and organelles between the plasmalemma and tonoplast), which are connected to one another through adjoining cell walls by plasmodesmata (cytoplasmic strands). The apoplast consists of cell walls and the cavities of dead cells such as vessels and tracheids and corresponds roughly to "free space" (110).

Route of Water Movement

In the movement of water from soil to the atmosphere via the plant's transpiration stream, two distinct linear files of plant cells must be traversed; the first lies between the soil and the xylem elements of the root (epidermis, cortex, pericycle, and endodermis) and the second between leaf xylem and transpiring leaf cell (mesophyll, epidermis). Intervening these two files of cells are the tracheary elements (tracheids and vessel members) (65) of the stem.

Transpiration of water from leaf cells produces a water potential gradient through the leaf tissue in response to which water moves from the vein. This results in a reduction of hydrostatic pressure and therefore of water potential (WP) in the xylem, in response to which water moves across the root cortex from the soil-root interface. The question posed by Weatherley (217) is whether water moves in the plant from cell vacuole to vacuole crossing membranes and walls of each cell or moves around the cells staying within the cell walls. These are, in a sense, two parallel pathways. Water will move through both, but if there is a significant difference between them in resistance to water flow, the water will move through the one offering the least resistance.

Early Evidence for Water Movement within Cell Walls

Strugger (193), using fluorescent dyes, first suggested the possibility that water could move within plant cell walls. Levitt (128) concluded, on purely theoretical considerations, that movement through mesophyll cells cannot be diffusional and consists presumably of mass flow in the cell walls. Scott and Priestley (181) also mentioned the possibility of free water movement in the walls, and they postulated the endodermis as a barrier. Arisz (2) also called attention to the barrier to mass flow through cell walls that is posed specifically by the Casparian strip of the endodermis. Tyree (203) suggested that bridging the endodermis may be accomplished through plasmodesmata that exist between cortical cells, the endodermis and the pericycle.

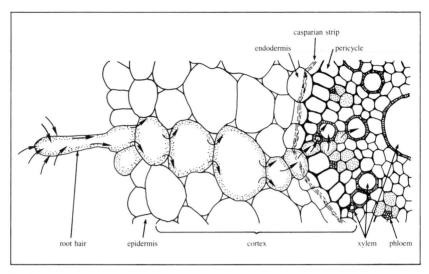

Fig. 8–1. The pathway of water and salts transport from soil to the tracheary elements (indicated by the stippled area), which in the process pass from root hair through cortex, endodermis and its casparian strip, the pericycle, and thence into the xylem. Note that the route is essentially apoplastic, through the (stippled) cell wall zone, excepting the endodermis where symplastic passage is required to circumvent the casparian strip (courtesy of Prof. Katherine Esau, although slightly modified).

In some species, the endodermis with its suberized Casparian strip may not pose a significant barrier as a consequence of the presence of "passage cells." These, according to Fahn (65), provide passage for substances between cortex and vascular cylinder and may be less heavily impregnated with suberin. However, Robards and Robb (168) emphasize that this is a transient phenomenon. With maturity, the endodermis becomes ever more restrictive to the passage of water and ions as a consequence of the intensification of suberized lamellar deposits. Apparently, in young cells, the passage of water and many ions across membranes and the endodermis is possible (45). However, as the Casparian strip barrier becomes more comprehensive even movement from apoplast to symplast is restricted. Such movement may occur via endocytotic processes (movement of vesicles containing water and ions from wall to cytoplasm). In general, however, movement from cortex or pericycle to endodermal cell occurs between respective symplasts via plasmodesmata, thereby circumventing the barrier. Water finally leaves the endodermal cell, entering the wall and lumen of the xylem (see Fig. 8–1).

It would seem, therefore, that there is strong and long-standing support for the concept of the cell wall per se as a conduit of water movement in plants. Combining some points made in the aforementioned literature, it would appear that water movement in roots is a three-step process: (1) mass flow in the cell walls up to the endodermis; (2) active uptake of ions across the cytoplasmic membrane, accompanied by osmotic movement of water; and (3) mass flow in the xylem.

Weatherley (217) constructed leaf and root models in the laboratory to test whether water moved from cell to cell or passed around the outside of cells (in the walls).

His *leaf* model revealed that water moved through the leaf approximately 60 times faster than could be accounted for by movement via "inner space" (symplastic movement). The Weatherley model was comprised of but a single cell, hence this difference could be even greater if one considers movement along a series of cells. The model could mimic both "outer space" (free or apoplastic) and "inner space" (symplastic) pathways. Water movement through the former was insensitive to temperature changes over the range of 3–23 C. In addition, it was insensitive to respiratory inhibitors. Hence, water movement into and out of the leaf was essentially totally via mass flow (apoplastic).

The *root* model indicated that water flow through it was sensitive to both temperature change and the metabolic inhibitor cyanide. In addition, water flux (flow) was calculated to be three-fourths osmotic and one-fourth mass flow. Weatherley suggested that metabolic control probably lies in the endodermis which functions as the "resistance link" (or valve preventing backflow) in the pathway of water in the root, where water must pass through the cell's symplastic component.

Summarizing the foregoing, it would appear that movement into and out of leaves is mass flow primarily through the walls. On the other hand, the passage of water through the roots from the epidermis to the xylem is via mass flow up to the endodermis. Movement through the endodermis is symplastic and to some extent metabolically controlled. Finally, movement out of the endodermis, once past its plasmalemma, and into xylem is again mass flow. The flow of water continues to foliar xylem vessel ends where it enters the apoplast of the leaf spongy and palisade mesophyll cells. Water exits from the surface of the walls of these cells into the substomatal chamber in the form of water vapor by diffusion and passes through the stomatal apertures into the atmosphere (180).

Radial and Tangential Transport in Xylem and Phloem

In addition to the horizontal (in root) and vertical (in xylem and phloem) movement of solutes, there is a radial and tangential movement in the vascular tissues. Ample evidence exists that radial and tangential transport systems are active ones that permit the movement of nutrients between phloem and xylem. For example, solutes can ascend the plant in the transpiration stream and then move from the xylem into the phloem and be transported downward. This is known as centrifugal movement and may occur as movement from xylem (lumen) to symplast of xylem (xylem parenchyma) to the ray cells and thence to the phloem per se. Gunning and Pate (87, 160) have characterized *transfer cells* as being associated with vascular tissues of certain plant species and possibly being involved in the symplastic transport of solutes between xylem and phloem. Specifically, the transfer cell by virtue of its "wall ingrowths" (irregular"fingers" of secondary wall material) provides a greater wall surface in contact with plasmalemma and cytoplasm. Its function is believed to be concerned with loading the conducting elements.

Gunning and Pate (87) characterized the fungal haustorium as a transfer cell and detected transfer cells adjacent to *Rhizobium* nodule xylem and phloem. The position and function of nodule-transfer cells is diagrammed in Fig. 8–2. They consider these transfer cells to link the phloem sugar flow to the bacteroid (its energy source) and to aid in the export of fixed amino nitrogen out of the nodule.

As early as 1928, Mason and Maskell (136) were able to demonstrate mass transfer of sugars between ray cells

and xylem parenchyma. The evidence supports the contention that radial movement can and does occur in both directions and is probably mediated via a symplastic system (161).

Sauter et al. (175) and Czaninski (31) have presented evidence for the activity of specialized "transfer" cells that they have designated *contact cells* and *vessel associated cells* (v.a.c.), respectively. Sauter et al. (175) credit the "contact cells" of rays and paratracheal parenchyma as the most likely site of solute exchange between parenchyma and vessels. They are, for example, the cells that deliver sucrose in the sugar maple (*Acer saccharum*) to the xylem vessels. Entry apparently takes place through unusually large and numerous pits and is energy controlled. Similarly, Czaninski (31) has indicated that v.a.c.s, in response to pathogenic attack, assume a secretory activity that leads to the accumulation of polysaccharides that are, subsequently, discharged into the vessels. The transfer of polysaccharides and some phenolic substances in this way implies, according to Czaninski (31), that the walls of v.a.c.s have considerable permeability to large molecules. This capability may explain, at least partially, polymer accumulation in vessels infected by a number of different plant pathogens.

The Xylem as a Site for Growth and Movement of a Pathogen

Water conduction takes place primarily in the mature annular and helical xylem vessels of primary growth (protoxylem) and through the large early wood vessels (produced by cambium before leaf expansion) of the current year's cambial activity (secondary xylem) (65; Fig. 8–3). Bacteria and fungi must enter the water-conducting xylem elements through a wound, and a hypothesis for this process has been reported by Goodman and White (79) for bacteria. Since there is a negative pressure engendered by transpiration, wounding permits the fungal spore or bacterium to be sucked into the vessel with the receding water. This could also occur where leaf scars are exposed by premature leaf abscission (e.g. in a driving rain and wind storm) (see Chap. 1). According to Zimmermann and Brown (231), a 1 μm break in a vessel wall would engender a pressure differential of 2 atm; however, much less tension would develop for breaks that are larger, e.g. breaks that expose a leaf midrib. At the moment the air enters the injured vessel, a substantial transient negative pressure is placed on the tensile water columns in all contiguous intact trachea until the injured cell is completely air filled. Once bacteria are sucked into a tracheid, probably to a pit membrane near one of the cell's ends, their multiplication could commence. Peel (161) has indicated that the osmotic concentration (pressure) of xylem fluid is ca. 0.8 atm, ideal for the replication of bacteria and fungi. Movement into the next cell by the bacteria or fungal spore, in the absence of perforations, must be effected by the pathogens themselves. However, movement of greater distances, i.e. through a file of vessels, is possi-

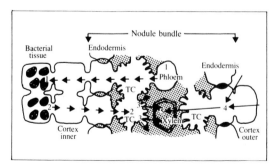

Fig. 8–2. Solute transport in the root nodule of a legume. 1, Symplast gradient for sugars utilized in nitrogen fixation; 2, symplast gradient for amino compounds produced in bacterial tissue and destined for export; 3, selective secretion by transfer cells (TC) of amino compounds to the bundle apoplast (stippled); 4, entry of water into bundle across the endodermis (Casparian strip shown in black); 5, discharge of concentrated solution of amino compounds in xylem (after Gunning and Pate, 87).

ble, and the propagule is arrested only when it reaches a vessel whose terminating end is not perforated.

In the event bacteria or fungi do get into one or several wounded tracheids, it is clear that emboli described by Zimmermann and McDonough (232) do not seriously interrupt the otherwise healthy water-conducting system. The neighboring tracheids continue to perform, as the system is "usually overefficient." The

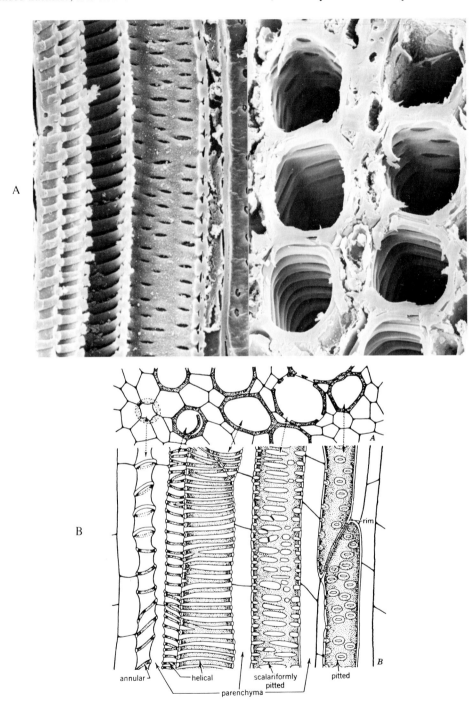

Fig. 8–3. Longitudinal and cross-sectional scanning electron micrographs (a) of apple stem tissue showing annular, helical, and pitted xylem vessels from apple (after Suhayda and Goodman, 195) and line drawing (b) (courtesy of Prof. Katherine Esau).

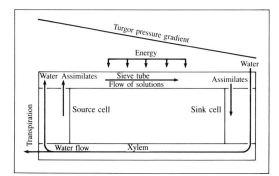

Fig. 8–4. A model illustrating the pressure flow mechanism in the vascular system of higher plants (after Peel, 161).

tracheids and vessels comprise a complex, multi-branched network that can isolate an area of injured cells by diverting water flow around the affected cells, thus localizing and minimizing their influence upon the water economy of the plant.

Phloem Transport

Peel (161) has stated that any transport system must consist of at least three parts: a source of solutes, the transporting conduits that are traversed, and the sink at which the solutes are removed from the conduits (Fig. 8–4). The source is defined as that tissue or organ in which export exceeds import. The opposite is true of a sink. In other words, solutes cannot be transported at a greater rate through the conduits than they can be loaded at the source. Obviously, the rate of translocation will depend on the rate at which the solutes are removed at the sink. In exudation experiments, Weatherley et al. (218) determined that a single aphid stylet could empty 100 sieve elements/min, and Dixon and Ball (43) have calculated velocity flow of sucrose in phloem at 40 cm/h.

It is also apparent that there are at least three environmental factors that govern transport: water potential, light, and temperature. Light affects transport as a consequence of its influence on photosynthesis and through a possible photocontrol of translocation that is dependent upon the quality of light received by the plant. Hartt (95) noted that basipetal transport was stimulated in red or blue light more than in green or cool white fluorescent illumination. Furthermore, far red light was not at all stimulatory of transport but rather was analogous to total darkness.

The influence of water stress on translocation may reflect either the loading effect of solute at the source or the unloading influence at the sink. There is some evidence, though not incontrovertible, that the lowering of water potential can directly affect transport.

Considering the phloem sieve tubes as a part of the symplast and the walls of these as apoplast, the role of water in phloem transport becomes clearer. For example, as a phloem sieve cell accumulates sucrose from its source its osmotic potential must decrease. Water from the wall enters the symplast, thus raising the turgor pressure within that cell. Tyree (203) has referred to this phenomenon as "active" water transport through the symplasm, which is coupled to solute flow. According to him, this increased turgor pressure causes mass flow through the plasmodesmata, forcing water-laden solute out of the cells further downstream. It is clear, therefore, that lowering the total water potential of the system can reduce the rate of flow of solute by reducing turgor pressure within.

In measuring temperature coefficients (Q 10s) of acropetal and basipetal transport in phloem in sugar cane, Hartt (94) noted the coefficient for basipetal transport over the temperature range of 24–34 C to be approximately 1, whereas acropetal movement had a value of 3.9. At 20–30 C, basipetal translocation was 1.7 and acropetal 16.2. These two sets of data appear to be the result of a reduction of the rate of enzymic processes at the growing point sink.

Role of Plasmodesmata in Transport

Plasmodesmata have been reported to occur between companion cells and parenchyma cells by Esau and Cheadle (55). Most of these intercellular connections occur between sieve elements and companion cells. These may have a branched structure as noted in Fig. 4–18. In addition to the movement of solute through plasmodesmata, there is evidence for the movement of virus particles (54) in them. The dimension of plasmodesmal pores or *desmotubules* has been calculated by Tyree (203) to be 20–200 nm in diameter. Their frequency has been estimated at the amazingly high value of 1.5×10^8 to 6×10^9 pores/cm². The mechanism of solute transport across plasmodesmata is uncertain. However, Tyree (203) seems to have simplified the problem by ruling out an active pumping system. Whereas active pumping seems credible across a membrane eight nm in diameter (plasmalemma), it is less so when the distance across the membrane and wall is 500 nm. Tyree concludes that the predominant transport mechanism across the plasmodesmatal pores is diffusion for small nonelectrolytes such as sucrose and amino acids. The same holds true for electrolytes, provided that no substantial electrostatic charges occur within the pore. It is apparent, therefore, that cell-to-cell movement requires a concentration gradient to maintain a net solute flux. Transport along a file of cells via plasmadesmata demands a source-sink relationship at the extremes of the file (Fig. 8–5).

Solute and Particulate Flow in Sieve Elements

The contents of the sieve elements are also apparently under great hydrostatic pressure. Any damage to the plant that results in the opening of a sieve tube to the atmosphere may result in a surging flow in the cells and the displacement of cellular content. The positive pressure here is opposite to what occurs in the exposure of the xylem vessel. In the latter, the release of tension results in a suction. A wide range of organic substances are transported in the phloem, and an exhaustive list has

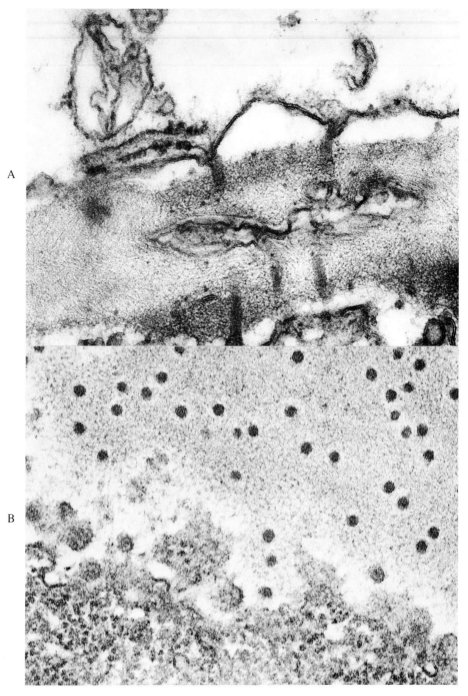

Fig. 8–5. (a) Cross section through pear fruit cell wall revealing plasmodesmata detail in the wall and connections to plasmalemma and rough endoplasmic reticulum (Goodman, umpublished). (b) Cross section through apple stem phloem parenchyma cell wall showing comparatively high frequency of plasmodesmata (P.-Y. Huang, unpublished).

been compiled by Crafts and Crisp (27). The most prevalent solute in the phloem is sucrose. In some species it may account for over 90% of the sugars present in sieve tube exudate. According to Peel and Weatherley (162) sucrose may be present in concentrations up to 20% w/v.

A number of workers have detected a slimelike substance, or P-protein (phloem protein), often described as filaments or strands in the sieve elements. This fibrillar system consists of tubules having a diameter of approximately 23 nm. According to Cronshaw and Esau (28), the role that this fibrillar material plays in the movement of solvent or particles in the phloem is not clearly understood. It has been asked whether or not the sieve pores with their apparently obstructive fibrillar material are really open, giving complete continuity from one element to the next, or whether they are to varying degrees occluded. Crafts and Crisp (27) favor the "open pore" concept and in their calculations on solute flow rates use 2.41 μm as the radius of sieve pores of *Cucirbita melopepo*. Hence, a diameter of nearly 5 μm would accommodate the passage of most phytopathogenic bacterial cells that usually measure $1.0–2.0 \times 0.5$ μm. A number of fungal spores could also pass through these openings.

If we return, for a moment, to the positive pressure developed by the exposure of a sieve element of the atmosphere, it may be noted that exudation from sieve elements very often ceases rapidly (138). The injured sieve elements may become plugged by some material and the plugging material is, at least in part, the P-protein, the exact structure of which is not known. Plugging also occurs as a result of deposition of callose, a $β-(1 \rightarrow 3)$ glucan, on the sieve plate.

Another thought concerning phloem and the two-way flow of solutes along the same pathway has, as an ideal model, the nitrogen-fixing nodule of legumes (88). In this case, carbohydrates could be moving, essentially in the symplast, to the nodules, and asparagine and amino acids could be moving away from the nodules in the same pathway.

With respect to the actual transport of substances in the phloem, it is apparent that energy will be necessary at some stage in the transport process. At the very least, energy will be required to maintain the integrity of the plasmamembrane to preclude excessive leakage of concentrated solutes. Energy will also be required at either end of the system to pump solutes into the conduit from the source and also out of the conduit into the sink (88). In this respect Gardner and Peel (72) have found sieve tube exudate to contain ATP at concentrations as high as 1 μg/μl. There is little doubt that the pressure flow hypothesis first defined by Munch (143) in 1930 and since rather significantly modernized is the most widely accepted theory (see Fig. 8-4; 161).

Wilt, the Sign of a Water Deficit

Impairments in the water economy of a plant as a result of either pathogenic attack or insufficient soil water are heralded by the same classic symptom, *wilt*.

Wilting indicates that the plant is losing water more rapidly through stomatal and cuticular transpiration than it can be supplied by the xylem. Bonner and Galston (20) assess transpiration through stomata as 97% of the total, with only 3% via the cuticle. As the cells of the leaf and shoot tip lose water, turgor pressure of the cells is also lost and the cells become flaccid and may collapse.

Incipient wilting, followed by collapse and recovery, generally results in no permanent injury to the plant. In fact, it occurs regularly in a number of species during the heat of the day. Full turgor is again achieved during the night. The loss of turgor by the vast numbers of photosynthesizing mesophyll cells occurs also in the guard cells of the stomates and, hence, these close. As a result of the closing of the stomates, much less CO_2 is absorbed from the atmosphere and less of this raw material of the photosynthetic process is available. Although stomatal closure is strongly influenced by light, it is also related to the CO_2 content of the substomatal cavity. When the CO_2 content of the atmosphere of the cavity falls below that of normal air (0.3%), presumably because of photosynthetic activity in the neighboring mesophyll cells, the guard cells characteristically respond to light by regaining their turgor and, as a result, the stomates open. However, this response to light is possible only when the guard cells can obtain sufficient water to regain their full turgor pressure. A number of researchers link moisture stress (low water potential) to growth, specifically to leaf elongation (207) and cell-wall metabolism (150). In the former study, decreasing water potential reduced the rate of elongation; in the latter, it decreased the incorporation of glucose into the various components of cell wall. Thus, a pathogen that somehow curtails the plant's water supply may have profound effects upon its growth and metabolism.

Plants Infected by Viruses

Water Content; Transpiration and Translocation

The water content of leaf tissue may be affected by several interacting factors, including efficiency of water absorption by the roots and its subsequent translocation, and by the rate of transpiration, all of which may be influenced either directly or indirectly by virus infection. Though not a symptom commonly associated with virus diseases, a severe wilt accompanies infection of tabasco pepper (*Capsicum frutescens* L.) by TEV. If newly wilted shoots are cut and maintained in water, wilting symptoms disappear (84), suggesting an important impairment of root function. There is necrosis of both phloem and cambial cells (222), accompanied by degeneration of cortical cells and plastids in roots of infected plants. Perhaps more important, there is a marked change in the permeability of root cells, accompanied by electrolyte loss, which is apparent before the onset of wilting and phloem degeneration (74). It may be concluded that wilting results from an impairment of root function, possibly mediated by changes in membrane permeability, rather than a reduced translocation.

Alternatively, changes in water content may be unrelated to alterations in root growth or function. The reduced water content of TMV-infected tobacco leaves has been suggested to be associated with rapid changes in intracellular water potential (153). TRSV-infected cowpea plants had a lower water content than uninfected ones, although it was not clear whether this was due to a reduced ability of roots to absorb water, to virus-induced damage to stem or leaf tissue, or indeed to other effects (125). Conversely, tomato leaves systemically infected with TAV (200) or TMV (25) had a higher water content than healthy leaves, effects that could be related to changes in cell permeability (200, 225).

Although there are exceptions, virus infection generally leads to a reduction in transpiration rate, a reduction often correlated with a reduced leaf stomatal aperture. It was suggested as early as 1932 (142) that this could account for the reduced transpiration rate of PLRV-infected plants and has since been observed in relation to BCTV-infected sugar beet (91) and MDMV-infected *Zea mays* (130), where the restriction in stomatal aperture was suggested to be caused by a reduced K^+ uptake by stomatal guard cells. There was, however, no difference in stomatal opening between healthy and TAV-infected tomato plants (200). In barley infected with BYDV, transpiration rate became lower in systemically infected leaf tissue than in uninfected leaves, a reduction that was more pronounced in susceptible plants showing severe symptoms than in resistant ones (53, 152). In susceptible plants, reduction was greatest in those areas of the leaf showing the most severe yellowing symptoms; it was suggested that reduced transpiration could result from reduced leaf water content. This could, in turn, be caused by either a reduced water absorption by the roots or a reduced rate of translocation, possibly occasioned by blockage of xylem vessels. Unfortunately, effect on stomatal opening was not investigated and, from the data presented, it is not possible to draw definitive conclusions. TAV infection did not apparently affect transpiration of tomato leaves (107).

Virus infection that does not cause wilting may, however, affect both absorption and release of solutes by the roots by altering membrane permeability and/or rate of translocation. There was, for example, a stimulation in solute leakage from roots of peas infected with BYMV or BCMV (19) and from squash hypocotyls infected with SMV (92). In tomato infected with BCTV, uptake of radiolabeled phosphate and SO_4^- was inhibited (63, 159). Ca^{++} absorption was stimulated (157), although transport was reduced, suggesting diversion into a new or enlarged root pool. There is also evidence (151) that TRSV infection of soybean can interfere with leghemoglobin synthesis and nitrogen fixation, while infection with SMV or BPMV can reduce nodulation (201).

We have already seen that the turgidity of a leaf prior to and immediately after inoculation can influence the creation or development of infectible sites and, hence, the apparent susceptibility of a plant to virus infection.

Similarly, it may be that changes in the water relations of upper leaves following virus inoculation of lower leaves contribute to the induced resistance, or susceptibility, of upper leaves to challenge inoculation when assessed by changes in lesion number. These considerations are discussed more fully in Chap. 10.

Carbohydrate Translocation

Accumulation of carbohydrates in leaf tissue is a characteristic of several virus diseases, including PLRV-infected potato (144), BCTV-infected tomato (156, 158), BYV-infected sugar beet (91, 216), and BYDV-infected barley (152) (see also Chap. 3). It is usually accompanied by phloem necrosis and/or gummosis, particularly in the later stages of infection, so that it is not unreasonable to suppose that an impairment of translocation is at least a contributory factor. However, a number of observations suggest that phloem degeneration is not the only factor involved. For example, carbohydrate accumulation preceded detectable changes in phloem tissue in potato infected with PLRV (144). Secondly, rate of loss of starch and carbohydrate during the period of darkness was the same in both healthy and BYV-infected sugar beet plants (216). Finally, sucrose sprayed onto beet leaves apparently transported to the roots equally effectively in healthy and yellows-infected plants (215). It has been proposed, therefore, that carbohydrate accumulation is due principally to an impaired activity of enzymes involved in carbohydrate interconversion rather than a reduction in translocation (215).

Photosynthates were retained in leaves of curly top-infected tomato (156) and PLRV-infected *Physalis floridana* (62), a retention that could not be explained by conversion of recent assimilates to insoluble products. In addition, rate of sucrose turnover was reduced in infected leaves, while the hexoses glucose and fructose accumulated. Since sucrose is normally the principal translocatable carbohydrate in these plants, it was suggested that its conversion to the less readily translocated hexoses could be the cause of the reduced carbohydrate transport. There was, however, no information on possible changes in concentration or activity of the enzymes involved. However, neither carbohydrate translocation nor activity of enzymes utilizing carbohydrates was affected in swollen shoot virus-infected cocoa leaves (1), and it was suggested that infected plants had a greater photosynthetic capacity and a higher diurnal rate of carbohydrate turnover than healthy ones. Comprehensive studies of these changes, including potentially important alterations in hormone balance, have not been done, and it is as yet impossible to define the primary cause(s) of the events referred to in the preceding paragraphs.

Virus Transport

The dissemination of a viral infection through plant tissue takes place by often related but distinct routes: a slow, cell-to-cell spread via plasmodesmata and a much more rapid, long-distance movement through the vascular tissue. We felt it logical to include both in this chapter on vascular transport; both have been the subject of extensive reviews (9, 17, 33, 52, 119, 179).

Cell-to-cell movement. Mechanical inoculation of leaf tissue introduces virus particles, which may be at least partially uncoated, into epidermal cells. In some hosts, replication may take place here before movement to the underlying mesophyll; in others, it may not (see also Chap. 1). It is possible, therefore, that the entity that first reaches the mesophyll cells is not a complete virus particle but one that is at least partially uncoated and, in extreme cases, may be the viral genome. Indeed, if we accept that at least partial uncoating of the less complex viruses takes place in association with membranes prior to or during penetration of the epidermal cells, it is not immediately obvious how intact virus particles could pass to adjacent mesophyll cells, and between mesophyll cells, without being uncoated. It would seem possible, therefore, that infection can progress through the intercellular transport of the naked viral genome (167, 229), which is usually ssRNA, but can be dsRNA, ssDNA or dsDNA (see Chap. 5). The genome of the majority of plant viruses is ss(+)RNA, which is infective but is also susceptible to ribonuclease digestion. If this is the form in which infection by these viruses progresses, then either the genome must not encounter ribonucleases or it must be protected from their action, possibly by association with membranes (33). The double-stranded forms found in infected cells, RF and RI, would be less susceptible to degradation and, perhaps, constitute a more practical form for genome translocation. However, RFs have not been demonstrated to be infective, though there is no reason to suppose that they would not be in the appropriate environment.

Additionally, there are several instances in which the spread of infection seems to occur via RNA. The *flavum* strain of TMV, for example, is temperature sensitive in capsid production and produces few virions at temperatures above 30 C. Nevertheless, it spreads between cells in leaves of Samsun tobacco maintained at the nonpermissive temperature of 35 C as rapidly as the *vulgare* strain, which forms normal virions at this temperature (154). Similarly, the TMV strains PM1 and PM2 are defective in capsid protein formation yet move intercellularly (3, 187). The genome of TRV consists of two RNA species, both of which are normally encapsidated in protein coded for by the smaller of the two, RNA-2. However, RNA-1 can infect plants, replicate, and move between cells in the absence of RNA-2, although under these circumstances it clearly cannot be encapsidated (93). There are also several viroids, naked ssRNAs (e.g. 36), that cause systemic infections and that are not encapsidated. There is, therefore, substantial evidence that unencapsidated RNAs can be translocated, although it is perhaps pertinent that the spread of infection is generally slower than might be expected with viruses that form capsids.

There is, however, contrary evidence that, at least for viruses able to form capsid, it is the complete virion that is transported. This evidence comes almost exclusively from electron microscope observations on plasmodesmata. There is little doubt that solute transport between cells is via plasmodesmata (89). They are of a size (20–200 nm; 89, 131, 182, 189), that can accommodate some virions (e.g. 75), although it may be that in order to allow passage of virus particles, the desmotubule needs to be displaced, or completely absent (61, 219), or the plasmodesmata otherwise enlarged (119). It is inferred, therefore, that infection also spreads via plasmodesmata in both leaf and callus tissue (189). Particles of many icosahedral viruses have been observed within them (219), including BWYV in beet (Fig. 8–6; 60, 61), SLRV in *C. amaranticolor* (169), and ToRSV in *Datura stramonium* (Fig. 8–7; 34), often in association with paramural bodies. Observations on elongated viruses, though less numerous, include PVY, TEV (Fig. 8–8; 219, 220), and TMV (Fig. 8–9; 220) in tobacco, TEV in *D. stramonium* (219), and BYV in beet (Fig. 8–10; 57–59). In many of these, virus particles appear to be in transit between cells, and it would seem that particles can remain intact within plasmodesmata. The evidence, therefore, indicates that when viruses can produce capsid protein, then virions are involved in cell-to-cell spread, although it may be that some of the more complex viruses are transported as cores (206). However, anomalies remain, and the situation is far from resolved. For the sake of simplicity, in the following paragraphs we refer to movement of "virus" with the implication that the entity translocated is not necessarily intact virions.

Although other methods have been employed (179), the time taken for virus to pass from the epidermal cells to the underlying mesophyll has usually been assessed by inoculation of local lesion hosts on either the upper or lower surface. Epidermis has, then, been removed from the inoculated surface at various times after inoculation and the subsequent development of lesions on the exposed mesophyll taken as evidence that virus had reached these cells before the epidermis was removed. Conversely, failure of lesions to appear was taken to indicate that virus had not left the epidermis. Using this procedure, the minimum time for CMV to reach mesophyll cells of cowpea maintained at 28 C was estimated at 2 h (221), or 3–4 h (50) after virus inoculation, or less than 3 h after inoculation with viral RNA (50). Timing is clearly temperature dependent, CMV taking 5 h to reach the mesophyll at 20 C. Similarly, TMV was present in mesophyll cells of *N. glutinosa* maintained at 20–22 C in approximately 8 h, but not until 10 h had elapsed at 17–19 C (37).

However, many of the observed times are overestimates of the minimum time. First, stripping the epidermal layer may damage the mesophyll cells immediately adjacent so that virus may have to penetrate further before its effect can be observed. Second, the presence of the epidermis may be required for lesion formation; virus may have passed to the mesophyll, but remain

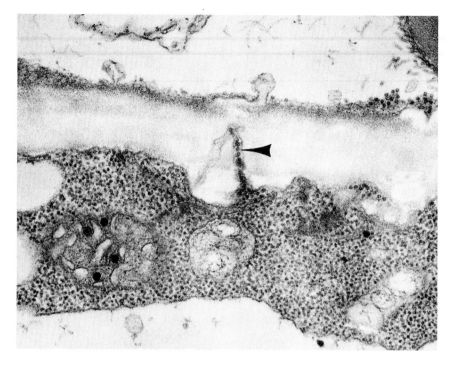

Fig. 8–6. BWYV particles (arrowed) in plasmodesma between two beet parenchyma cells (Mag. ×52,000) (after Esau and Hoefert, 60).

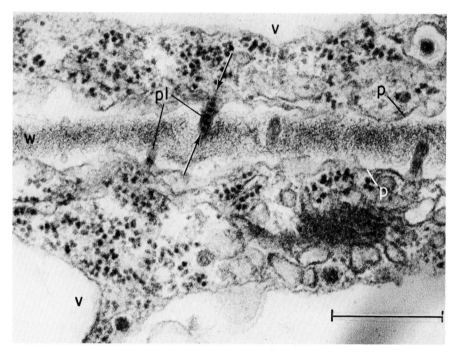

Fig. 8–7. ToRSV particles (arrows) in plasmodesma of *D. stramonium*. Mag. ×60,000; p, plasmalemma; pl, plasmodesmata; v, vacuole; w, cell wall (after De Zoeten and Gaard, 34).

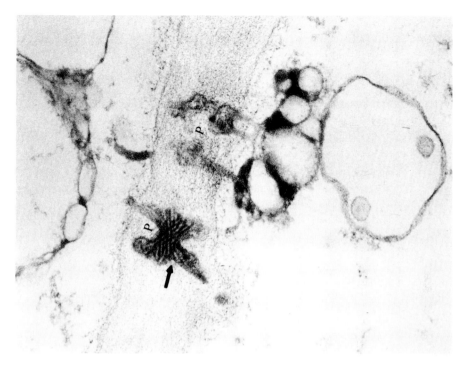

Fig. 8–8. PVY particles in *N. tabacum* plasmodesmata. Mag. ×62,000; P, plasmodesmata; arrow indicates virus particles (after Weintraub et al., 219).

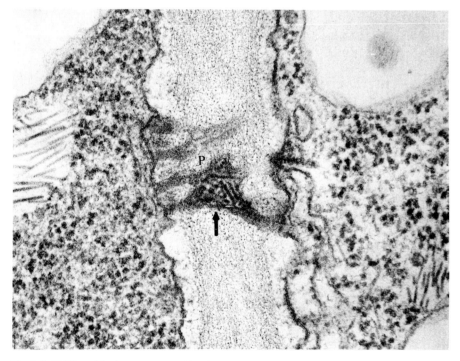

Fig. 8–9. TMV particles in transverse and longitudinal section within plasmodesmata and cytoplasm of *N. tabacum*.; P, plasmodesmata. Mag. ×73,000; arrow indicates virus particles (after Weintraub et al., 219).

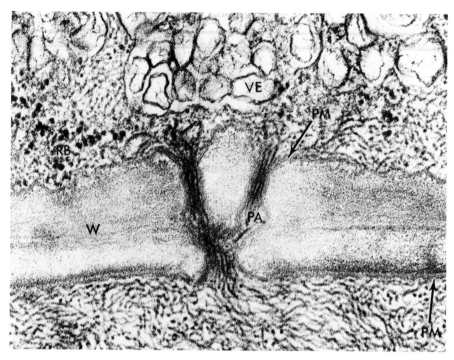

Fig. 8–10. Parenchyma cell (above) and sieve elements (below) of small vein from *Beta vulgaris* leaf infected with BYV, with virus particles (PA) in branched plasmodesmata and sieve element. Mag. ×78,000; PM, plasmalemma; RB, ribosomes; VE, vesicles; W, cell wall (after Esau et al., 57).

undetected, in areas where epidermis has been removed soon after inoculation. Usually, although lesions appear when epidermis is stripped at the times indicated above, the number may be increased considerably if epidermis is left intact for longer periods. Kasamo and Shimomura (114) found very few TMV lesions on tobacco leaves from which the lower epidermis had been stripped 6 h after infection, while in one extreme case Kontaxis (122) failed to get any lesions at all when epidermis was stripped from *N. glutinosa* as long as 21 h after inoculation, although virus was clearly present in the mesophyll after 3 h. It is also of interest that maintaining plants in darkness prior to inoculation has been found not only to increase susceptibility, as assessed by increase in lesion number (see also Chap. 1), but also to facilitate transport to the mesophyll. For example, when cowpea or Xanthi nc tobacco leaves were darkened for 24 h prior to TNV inoculation of the lower epidermis, virus entered mesophyll within 1 h, while in plants maintained in the light, it took 6 and 8 h, respectively (26, 223). The reduction in time caused by darkening was correlated with plasmodesmata number (223). Additionally, the fact that virus reached the mesophyll in this short time would seem to indicate that infection was initiated by the inoculum virus, or an uncoated form of it, rather than by products of replication in epidermal cells.

In comparable experiments in systemic hosts, movement to mesophyll tissue has been assessed by assay of virus accumulation in tissue from which the epidermis has again been removed at various times after inoculation. TMV was observed to reach mesophyll of *N. sylvestris* maintained at 24–30 C (204) or tobacco at 27 C (70) within 4 h.

Having reached the mesophyll cells, virus may spread, unless restricted, throughout the mesophyll tissue, and systemic infection is the likely result. In some cases, however, progress of translocation between cells may be curtailed and virus localized to cells at and adjacent to the site of infection. Possible explanations for this restriction are discussed in Chap. 10. Restriction may be accompanied by necrosis of infected cells (hypersensitivity), and the lesion diameter may continue to enlarge over a period of several days. If it is assumed that cells become visibly necrotic soon after infection and that the rate of radial movement of virus from the infection site is not appreciably faster than the rate of lesion expansion, then measurement of lesion diameters at several times after their appearance may give an approximation of the rate of virus spread. Using this procedure, Rappaport and Wildman (167) found three strains of TMV to move at different rates, from 6–13 μm/h. This was calculated to represent one cell/4 h for the U1 strain and one cell/6 h for U2. Estimation of movement by observation of progressive cytological changes indicated a rate of 4 μm/h, or one cell/3 h, for TYMV movement through Chinese cabbage leaves (96).

Clearly, these figures represent the result of both transport and replication in each cell infected. The time required for translocation, presumably mediated by a combination of cytoplasmic streaming and diffusion, is therefore much less, and differences among strains may represent differences in times required for replication, rather than differences in translocation rate.

There is little information on the determinants of rate of spread, although recent studies by Zaitlin and his colleagues (127, 228) reveal the potential importance of a virus-coded polypeptide in the translocation of TMV in tomato. The temperature-sensitive strain of TMV, LS-1 (148), is not translocated at the nonpermissive temperature. This deficiency correlates with a decreased number of plasmodesmata in leaves of infected plants transferred to the nonpermissive temperature compared to the number in healthy plants or plants maintained at the permissive temperature. The only observed difference between LS-1 and strains that spread normally was in a virus-coded nonstructural polypeptide of MW 30,000. If the presence of this defective polypeptide correlates with a reduced plasmodesmata number, then the inference is that a normal polypeptide is somehow involved in the prevention of loss of plasmodesmata. It is difficult to see how this could be so, and the function of this polypeptide in translocation is far from clear.

The rate of translocation may also be affected by host factors such as age (196), in addition to environmental factors such as temperature (155), and may be greater in a direction perpendicular to the leaf surface than parallel to it.

Long distance transport. Following relatively slow cell-to-cell movement from the site of infection through the parenchyma tissue, viruses capable of systemic invasion enter the vascular tissue and are, thence, transported much more rapidly to distant parts of the plant. Although the number of viruses studied has been limited, it appears that the majority are transported in the phloem (97, 179), entering sieve tubes via plasmodesmata from the vascular parenchyma.

Evidence for phloem transport comes from several sources (e.g. 137), original observations correlating movement of virus with transport of food (16, 17). In three-crowned beet plants, for example (15), BCTV was transported from inoculated leaves into crowns that were in darkness or defoliated, but did not reach crowns that were actively photosynthesizing until much later. The correlation between movement of virus and assimilates is not, however, particularly close (98). Also, destruction of the xylem tissue does not prevent movement of TMV in tomato, for example, suggesting movement in phloem rather than xylem, but not necessarily ruling out translocation in either (75, 97). When transmitted by vectors, some viruses, such as PLRV and BYDV, may be introduced directly into phloem tissue, though only rarely into sieve elements directly (14, 179). They may thus be transported from the site of infection at a rate much greater than when mechanically inoculated. Oth-

ers, such as BCTV, are unable to initiate infection unless introduced directly into phloem tissue.

It is not absolutely certain that infection spreads as intact virions, although the evidence suggests that this is the case. First, particles of viruses such as TMV, BYV, and BWYV have been observed within sieve elements (Figs. 8–10 and 8–11; 32, 56, 57, 61). Second, Oxelfelt (155) has observed a correlation between the ability of TMV strains to form virions and their ability to be translocated through the vascular system. When inoculated Samsun tobaco plants were maintained at 35 C, rate of translocation of the *vulgare* strain was greater than at 24 C, while movement of *flavum*, which is deficient in capsid production at this temperature, was curtailed. When plants were returned to the lower temperature, *flavum* produced virions and was translocated normally. Thus, although the RNA of the *flavum* strain appears to be able to move between cells, it does not appear to be translocated in the phloem. The same seems to apply to RNA-1 of TRV.

The rate of virus movement and the development of its distribution in plants have been estimated by essentially similar procedures. Since relatively few particles may be necessary to initiate infection at distant sites, transported particles have usually been identified by allowing further replication. This has usually involved cutting inoculated plants into several sections at various times after inoculation, incubation of the excised sections, and assessment of virus accumulation by homogenization and inoculation onto a local lesion host.

The apparent time required to move from inoculated leaves to stem tissue varies considerably and clearly depends on several factors, including rate and level of replication, ease of entry into the vascular system, and rate of transport within the system. Estimates have been made for several virus/host combinations (179), although many of the times may be overestimates due to the difficulties in detecting low levels of translocated virus. PVX, for example, was found to be translocated from inoculated *Physalis floridana* leaves in 40–48 h; PLRV took about the same time as PVX; TuMV and PVY took several hours longer.

Estimated times for movement of TMV from inoculated tomato leaves vary from 44 h (124) to 4–5 d (22, 174). Samuel's comprehensive study of the distribution of TMV following inoculation of a single leaflet of tomato plants (9, 137, 174) indicated virus to have moved to the roots within 4 d and to the apical leaves within 5 d. Not until 25 d had elapsed did all the leaves in intermediate positions along the stem become infected.

Helms and Wardlaw (98) have investigated the rate of movement of TMV from inoculated leaves of *N. glutinosa* maintained at 36 C and found that virus moved from inoculated upper leaves of otherwise defoliated plants within 18 h, while Oxelfelt (154) found TMV to move from inoculated tobacco leaves within 32–48 h. Others, however, have used a slightly different technique, allowing callus formation from the excised tissue sections by culture in vitro before virus assessment. This may be a more sensitive procedure. Kondo (121)

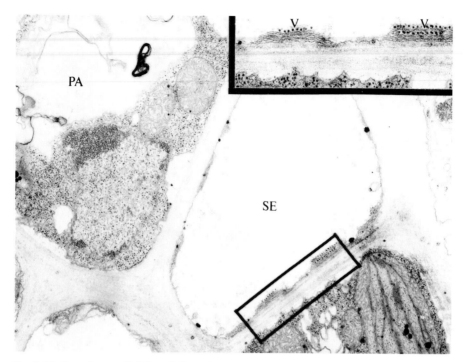

Fig. 8–11. Parenchyma cell (PA) and sieve element (SE) in phloem tissue of young leaf from BWYV-infected *Thlaspi anvense* (pennycress), showing virus particles (v). Mag. ×18,000; insert ×34,000 (after D'Arcy and de Zoeten, 32).

found infective TMV in the stem immediately below a midrib-inoculated tobacco leaf within 6 h. Virus had reached upper leaves within 15 h.

Translocation out of leaves may also depend on factors such as temperature (112, 155) and age of the plant; in extreme cases when very old leaves are inoculated, virus may remain in the inoculated leaf (18, 179). However, virus may be restricted by several factors other than inability to be transported in phloem tissue (Chap. 10).

The speed of movement of TMV through *N. glutinosa* phloem tissue has been estimated by Helms and Wardlaw to be 0.11–2.33 cm/h, lower than the 43 cm/h observed for ^{14}C-labeled photosynthetic assimilate and the 92 cm/h of ^{32}P-labeled phosphate, but approximating to previously obtained values for TMV (16, 22). It is suggested, however, that the virus moves in sieve tubes at the same rate as assimilate and the apparent difference is a reflection of the additional problems encountered by virus particles. First, because of their size, they are likely to have difficulty entering sieve tubes. Second, the irregular distribution of virus in stems of infected plants, and the presence of areas of apparently uninfected tissue separating infected areas, would seem to indicate that virus does not replicate during its passage through sieve tubes, although virus replication may take place in immature phloem elements. BCTV moves quickly after introduction into phloem tissue and does not have time to replicate. Before virus can be detected

at sites distant from inoculated leaves, therefore, it must move from mature sieve elements and into adjacent nucleate cells. It may be that a relatively small portion of translocated particles is detected. Indeed, some viruses, including the luteoviruses, of which BYDV is an example (76, 77), and BCTV are phloem limited. They have to be introduced directly into phloem tissue and apparently do not replicate outside it.

Although we have concentrated on phloem transport, there are some viruses that do appear to be translocated through the xylem vessels (75). PMTV, for example, moves from inoculated Xanthi nc tobacco leaves faster in an upward direction than downward but only rarely reaches apical leaves, a pattern of movement consistent with transport through xylem rather than phloem (113). Since the movement of the TMV strains PM1 and PM2, which fail to form functional capsid proteins, follows a similar pattern, it is possible that these, too, move through the xylem. LNYV also appears to be associated with the xylem (23), while PRV has been found to be associated with peanut xylem cells but not phloem cells (66). SBMV moves readily from one inoculated bean leaf to the opposite one (178, 224) and may also be xylem transmitted (75). Viruses normally transported through the phloem may be translocated through the xylem if introduced by artificial means (179).

Following transport to parts of the plant distant from the inoculation site, the virus may normally leave the phloem tissue via the vascular parenchyma cells and is then able to resume cell-to-cell spread through the mesophyll. Its pattern of distribution, however, may be very irregular, and many areas may remain uninfected (see also Chap. 10). There is evidence that in leaves maintained at low temperatures (5 C), virus may remain in the vascular system and not reach mesophyll cells (44). Additionally, although some viruses penetrate meristematic tissue (118, 210), others may fail to do so. Excision of meristematic tissue from virus-infected plants, followed by in vitro culture, often in combination with other treatments, can provide a valuable means of producing virus-free plants (166, 209).

Plants Infected by Bacteria

Introduction

There is ample evidence that bacteria can enter the vascular system, both xylem and phloem, through wounds (78, 79). Further, it has become increasingly apparent that some bacteria that enter the xylem move great distances and eventually cause significant damage there. This section examines how bacteria enter the vascular system, where they go, what they do, and how they do it.

Sites of Xylem Penetration and Rate of Migration

Entry of bacteria into xylem vessels through leaf scars (29), leaf wound (30), and leaf petioles (79) has been reported. Wind-driven rain may wash bacteria from surfaces of the aerial parts of plants into leaf scars exposed prematurely (68) and into vessels on torn leaves (129, 195). A vascular suction force of 10 to 20 atm (161, 232) engendered by the "pull" of foliar transpiration is probably the mechanism responsible for the penetration of pathogens through leaf scars in the olive knot disease caused by *Pseudomonas savastanoi* (99), *P. mors-prunorum* (29), and *Xanthomonas pruni* (68). Crosse et al. (30) confirmed that *Erwinia amylovora* entered the midrib of "clipped" apple leaves in the same way. Inoculation of the exposed midrib following excision of the apical cm of apple leaf tissue revealed that systemic infection was established in 50% of Jonathan shoots given an inoculum dose (ID 50) of 38 *E. amylovora* cells (Figs. 8–12 and 1–18). Most of the bacteria placed on the exposed midrib were probably impeded by a "filtering process," i.e. blockage by the exposed ground parenchyma cells, positive pressure of phloem cells, and retention at end walls of shorter vessels (83, 161, 194, 230). However, some bacteria moved great distances, as Crosse et al. (30) observed bacteria up to 3 mm below the cut edge of the leaf 5 min after inoculation and systemic infection of stem tissue 48 h after inoculation. In a subsequent study (194), an inoculum of 5×10^4 *E. amylovora* cells placed at the tip of an exposed leaf petiole resulted in systemic infection of the subtending shoot, even though the petiole was removed 2 min after inoculation. Some of the inoculum had to travel the 6–8 mm of the petiole into the stem in 2 min. Apple petioles, similarly inoculated by Goodman and White (79) and examined electron microscopically,

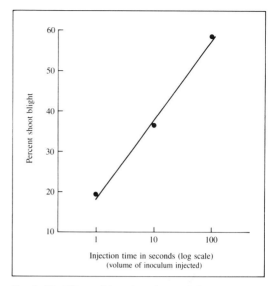

Fig. 8–12. Effects of inoculum dose and time of injection (inoculum volume) on the incidence of shoot and leaf infections of apple caused by *E. amylovora*. Inoculum injected into leaves by submerging cut ends into bacterial suspensions (after Crosse et al., 30).

revealed that, excepting the apical mm upon which the inoculum was placed, *E. amylovora* was confined to and multiplied only in the xylem vessels during the 48-h period of observation.

The movement of *Corynebacterium michiganense* in tomato plants through a 0.5- cm-long leaf petiole, inoculated with a scalpel wet with a suspension of the pathogen, revealed initial transport to take place only in the xylem. Pine et al. (164) reported that, although the phloem sieve tubes were similarly exposed, the pathogen could not be detected there. Fifteen min after inoculation, *C. michiganense* had moved 2.3 mm in two of three leaf traces (Fig. 8–13A) and was confined to the spiral vessel elements of the protoxylem. Twenty-four h later (8–13B), the bacteria were evident in all three traces, had moved down the main axis of the stem a distance of 25 mm, and completely filled the lumina of some vessels with bacteria. After 5 d, the bacteria were observed 13 mm above and 32 mm below the inoculated petiole. At this time, however, the bacteria had moved through cavities in the thin primary xylem cell walls into bordering parenchyma. An ultrastructural study of the same host-pathogen interaction by Wallis (212) recorded migratory patterns vis-à-vis time and distance that were almost identical to those described by Pine et al. (164). Further, Wallis noted that *C. michiganense* moved past open perforation plates and remained confined to the xylem vessels for at least 96 h. Entry of pathogen into the phloem region via ground parenchyma tissue was not recorded until 8 d after inoculation.

Suhayda and Goodman (195) monitored the movement of *E. amylovora* from a needle-prick inoculation

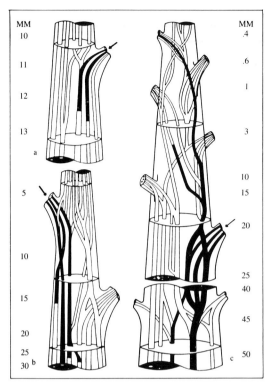

Fig. 8–13. (a) The movement of *C. michiganense* 15 min after inoculation is indicated by the blackened leaf traces; (b) the distance the bacterium had moved in 24 h; and (c) after 5 d (after Pine et al., 164).

of apical stem tissue. Exclusive of the 1 mm region above and below the point of inoculation, the bacteria were confined to the mature annular and helical vessels of the protoxylem (Fig. 8–14). This pathogen labeled with ^{32}P could also move upward through the xylem vessels of cut bases of 5–7 cm long Jonathan apple shoots at the rate of 34 mm/h (194).

A number of other studies have detected long-distance movement of bacteria in xylem vessels. Stapp et al. (190, 191) isolated *Agrobacterium tumefaciens* in vessels of *Datura tatulo* 110 cm from the site of a primary tumor. Similarly, Lehoczky (126) reported that post-pruning "bleeding" (xylem exudation) in *A. tumefaciens*-infected grape canes yielded as many as 1.5×10^3 cells/ml, 80 cm away from the nearest gall. The force moving these bacteria was, apparently, root pressure. These data support the observation by El Khalifa (51) that secondary tumors developing in plants infected with the crown gall pathogen may be incited by cells of the bacterium migrating in xylem vessels. Experiments by Tarbah and Goodman (unpublished results), monitoring the movement of double-labeled (streptomycin- and rifampicin-resistant) biotype 3 *A. tumefaciens* in cv. Seyval grape, recorded bacterial cells moving at a rate of 1.5 cm/h. In addition, inoculating roots of the highly susceptible cv. Chancellor and, subsequently, wound-

ing stem tissue aseptically 30 cm from the roots resulted in tumor development at the wound site. It was possible to isolate the labeled pathogen from the tumors. These observations, and others made by Lehoczky (126), tend to diminish the likelihood of naked bacterial plasmids being the basis for the development of secondary tumors. Some rickettsialike bacteria, e.g. those reported by Mollenhauer and Hopkins (140) to cause Pierce's Disease in grape, move through and appear to be totally confined to the lumina of the xylem vessels during the entire course of pathogenesis.

Pathologic Activity of Bacteria in Xylem

Informative ultrastructural studies by Wallis and co-workers (212–14) have provided detailed evidence of significant bacterial growth and the extent of damage caused in xylem vessels and surrounding cells. Cabbage seedlings, inoculated in a tertiary leaf vein with *Xanthomonas campestris*, spread in both directions from the point of inoculation. This was, perhaps, due to entrance of air into the vessels at the time of inoculation as suggested by Zimmermann (230–32). The bacteria were, at the outset, unevenly distributed among and within vessels but tended to aggregate in the interspiral regions where they multiplied. Electron micrographs (213) gave the impression that some bacteria were sheltered by the spiral thickenings, whereas the bacteria in the lumina were being carried by the transpiration flow (Fig. 8–15). The bacteria successively caused plasmolysis of adjacent xylem parenchyma cells (which they had not entered), initiated primary wall dissolution between some contiguous vessels, and degraded the primary wall where it connected with secondary thickening, which caused detachment of secondary thickenings. The shredding of the primary wall released masses of fibrillar material that surrounded the bacteria and appeared to associate with bacterial capsule. The density of this material was positively correlated with the concentration of bacteria in the vessels. Protruberances formed on the weakened vessel walls, permitting bacteria to enter intercellular space between adjacent parenchyma cells. These wall-degrading activities are probably due to the pectic lyase and cellulase synthesized by the pathogen (120, 145). Release of bacteria from xylem vessels into intercellular space has also been reported by Nelson and Dickey (146) in carnations infected with *Pseudomonas caryophylli*, which also produces wall-degrading enzymes. However, the cause of the apparent rupture of apple xylem vessels containing *E. amylovora* cells is obscured by the fact that this pathogen is alleged not to be able to synthesize wall-macerating enzymes. Goodman and White (79) have postulated that pathogenesis may engender the induction of host elaborated wall-degrading enzyme activity. Recent evidence by Wallis and Goodman (unpublished) has found *E. amylovora* capable of using highly purified pectin as a sole carbon source.

In tomato roots inoculated with *P. solanacearum*, Wallis and Truter (213) observed that tracheid fiber cells (xylem parenchyma) filled with bacteria while neigh-

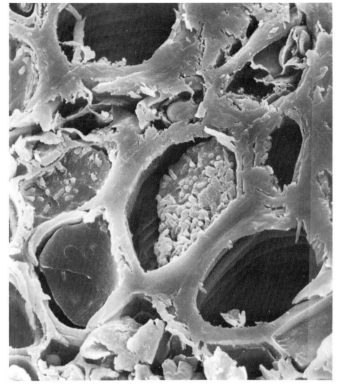

Fig. 8–14. Three completely occluded xylem vessels. The vessel lumens contain bacteria in an extremely dense matrix. The site shown is 5–10 mm from the inoculation point (×2,645) (after Suhayda and Goodman, 195).

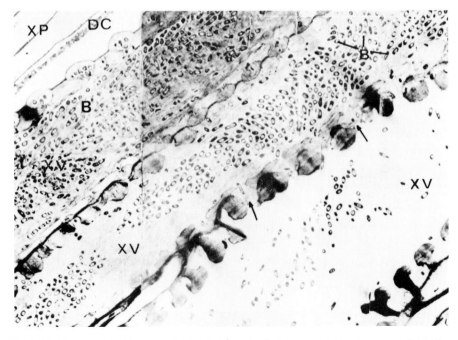

Fig. 8–15. Montage of a cabbage vascular bundle infected with *X. campestris* showing uneven distribution of bacteria (B) within and between xylem vessels (XV). Note xylem parenchyma (XP) are free of bacteria and between secondary thickening (see arrows) (Mag. ×6,000) (after Wallis et al., 213).

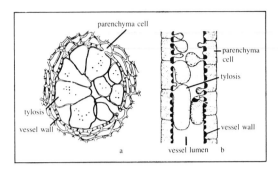

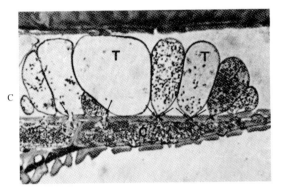

Fig. 8–16. Tylose formation in (a) *Robinia pseudoacacia*, (b) a vessel of *Vitis vinifera*, and (c) a *Nicotiana tabacum* stem xylem vessel infected with *P. solanacearum* showing Tyloses (T) in the vessel lumen. The tyloses have pushed through pits (arrows) and bacteria are apparent in the tyloses as well as the parenchyma cells whose plasmalemmas have embolized into the vessel lumen (after Fahn, 65; Wallis and Truter, 213).

boring xylem vessels remained free of the organism for ca. 24 h after inoculation. At that time, both invaded and some uninvaded cells (3–4 cm from the inoculated root tip) formed massive tyloses that bulged into xylem vessels through pit membranes, a process that continued for ca. 72 h (Fig. 8–16). No bacteria were seen in the xylem vessels prior to the collapse of tyloses 48–72 h after inoculation. Up to that time, movement of bacteria had been slow, only 7 cm in 72 h. However, during the next 24 h bacteria were detected an average distance of 22 cm from the inoculated root tip. Initial signs of wilting coincided with the collapse of tyloses and the release of bacteria, slime, and plant cell-wall debris into the vessels. This study, like some of its predecessors (69, 85, 108, 109), implicated vessel plugging by bacterial cells, their extracellular materials, and wall-macerating enzymes as the basis for the observed wilt. It also added

to the disease syndrome the impact of intense tylosis development as a contributor to water flow impairment.

Main and Walker (135) found that *Erwinia tracheiphila* reduced the transpiration rate significantly three days after petiole inoculation of cucumber plants. The rate fell from approximately 100 ml/plant to near zero in 8 d. Bacterial masses were observed in the lower stem xylem, 2 d after inoculation. These tended to plug the petiole xylem, and wilt was coincident with this accumulation of bacterial mass. The bacteria were confined to the xylem for ca. 4–5 d, after which they broke through the xylem walls and spread into the parenchyma. Vascular deterioration was not deemed integral to the wilt symptoms that developed. Rather, blockage, or plugging of the transpirational stream, was judged to be the cause of this symptom. This contention gained support from the inability of Main and Walker to detect wall-macerating enzymes. It might be noted that their experiments, in this regard, were not performed under optimum conditions to detect pectin lyase activity. The situation described for *E. tracheiphila* is not dissimilar from that we have noted for *E. amylovora*, as the former pathogen, like the latter, is encapsulated and neither is reputed to have an effective pectin-degrading capability. To date, no studies have indicated whether avirulent strains of *E. tracheiphila* are acapsular.

A precise or unequivocal perception of vessel occlusion by a single bacterial or plant product has not yet been described. This is probably because vessel occlusion is a reflection of bacterial mass action. As a consequence, it has been difficult to determine the *real* or *primary* cause of wilt in the milieu of bacterial cells, enzymes, substrates, and metabolites (67, 101, 102).

Until recently, it seemed that *E. amylovora* caused vessel occlusion and wilt by its synthesis of copious amounts of a high MW polysaccharide, amylovorin (78, 105), which had a reported MW of 40×10^6 (4). A subsequent study has more accurately determined the MW of amylovorin to be 100×10^6 (188). Suhayda and Goodman detected helical and annular vessel occlusion caused by amylovorin per se that seemed identical to that caused by the pathogen and its enveloping polysaccharidal mass (195) (Fig. 8–17). Nevertheless, it has recently become possible to induce wilt in Jonathan apple shoots by bacteriophage enzyme hydrolysate of amylovorin that has a MW of ca. 20,000. It is not known whether this comparatively low MW polysaccharide fragment of amylovorin can physically occlude xylem vessels (188). Van Alfen and Turner (205) have found the xylem of alfalfa (*Medicgo sativa* L.) to be susceptible to obstruction by picomoles of dextrans. Wilt was due to the plugging of pit membranes, and it was con-

Fig. 8–17. Xylem vessels of Jonathan apple stem tissue. (a) Vessel occluded with bacteria (*E. amylovora*) and polysaccharide matrix. Shrinkage of bacterial(EPS) mass from vessel wall is caused by dehydration during sample preparation (×2,645). (b) Two xylem vessels occluded by EPS produced by *E. amylovora*. The EPS, at 100 μg/ml, was absorbed through the bases of the cut shoots. Samples were fixed 4 h after treatment (×4,232) (after Suhayda and Goodman, 195).

A

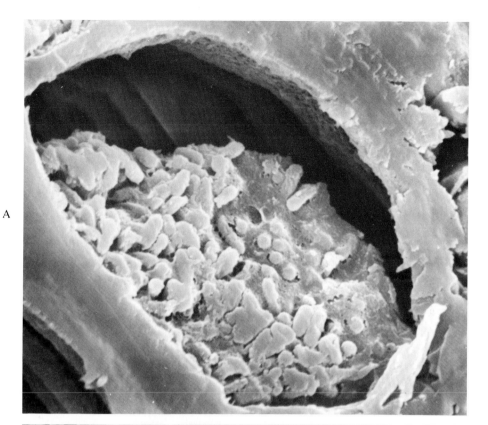

B

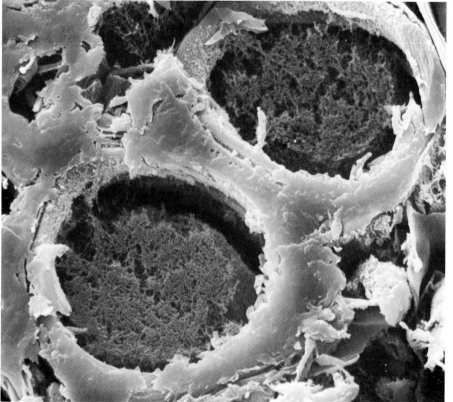

tended that dextrans with MW of 250,000 or less had little ability to stop vascular flow. It is possible that the 20,000 amylovorin fragment causes wilt in a fashion other than by vessel occlusion. This subject is discussed further in Chap. 9.

From the foregoing, it is clear that plant pathogenic bacteria can enter xylem vessels through wounds, move great distances therein, and often in relatively short periods. Peel (161) contends that water can move in the transpiration stream at velocities up to 83 cm/min. Once in the xylem, the bacteria seem to multiply there preferentially, probably because of the water environment and the xylems' conducively low but nutritive osmotic concentration of 0.8 atm. It is also apparent that, ultrastructural studies notwithstanding, the basis for bacteria-caused wilt has yet to be fully resolved.

Entry and Transport of Bacteria in Phloem

The study of Gowda and Goodman (83), as well as an earlier one by Lewis and Goodman (129), noted that downward migration of a streptomycin-resistant mutant of *E. amylovora* from apple shoot apex inoculation could be precluded by girdling. These reports suggest that some bacteria are able to migrate in the phloem following penetration via a wound. With the exception of rickettsialike and mycoplasmalike organisms, plant pathogenic bacteria have rarely been detected in the phloem (105). This is probably due to the fact that the high osmotic concentration, 10–12 atm, of the phloem prevents extensive multiplication (161, 183).

Rickettsialike organisms (RLOs) and mycoplasmalike organisms (MLOs) have specialized walls and membranes that permit them to survive and grow in environments with high osmotic concentrations. The RLOs have unusually strong cell walls that appear uniformly wrinkled and ostensibly permit stretching of their protoplasts. On the other hand, the MLOs have no cell wall and, hence, their plasmamembranes can stretch and contract in conformity with the solute concentration of their environment.

Since 1967 (111), it has become evident that more than 200 diseases of plants held to be of viral etiology, for want of more definitive information, are in fact of mycoplasma origin (147). Perhaps one of the first of these to be properly recognized was *Spiroplasma citri*, the causal agent of "citrus stubborn" disease (173). In general, the MLO pathogens are confined to the sieve elements of the phloem. They disrupt normal flow of growth regulators, which probably accounts for such symptoms as phyllody, witches broom, and adventive bud formation. In addition, wilting, yellowing, and general decline are symptoms ascribed to these pathogens. It is difficult to associate all these symptoms with aberrant phloem transport, hence the suspicion and developing evidence that toxigenesis may be involved (103, 104).

Penetration of the MLOs into phloem is accomplished by their vectors. These are, in the main, leaf hoppers; however, several other insect species are also effective in this regard.

The RLOs appear to fall into two groups, those that are generally confined to the xylem and those that are found in the phloem. Perhaps the best understood of the former is the RLO that causes Pierce's Disease (PD) in grapes (103). It causes a wilting and drying of leaves and fruit, necrosis or scorching of leaves, and general decline in vigor. Although bacterial aggregates, gums, and tyloses occlude vessels and limit water movement in the vessels, in this case, too few vessels were actually occluded to account for the observed symptoms (140). However, Hopkins monitored PD bacteria in wood of the previous seasons' growth and in current season's stem, petiole, and leaf vein tissue. Between 1 March and 21 April, most bacteria were detected in the petioles, and from 21 April to 7 June the major increase in percentage of vessel plugging was detected. It was most intense in leaf veins, and marginal necrosis was detected on 15 May. Of particular importance was Hopkins's comparison of the percentage of vessels occluded by bacteria observed through a series of sections in a 5 mm length of stem, petiole, or leaf vein tissue with the percentage occluded in a single 15 mm cross-section on the three observation dates. He observed the percentage of occluded vessels could be as much as 13 times greater than in a single 1.0 mm cross-section. When marginal necrosis was most intense ca. 80% of the leaf vein vessels were completely occluded somewhere in a 5 mm length of leaf vein tissue. Plugging of leaf veins was highly correlated with marginal necrosis and appeared to be sufficient for the observed water stress symptoms. A similar sequence of events probably occurs in the almond leaf scorch diseases (133, 139).

Phloem-inhabiting RLOs have been detected in several *Trifolium* species as well as trees affected by "citrus greening" disease (24). The phloem of potato and other plants have also yielded these organisms (100). Both xylem- and phloem-limited RLOs have been discussed in a general way by Hopkins (103). This area of research on MLOs and RLOs is young, and data are just beginning to accumulate. Apparently, there is reason to believe that factors other than vessel blockage, perhaps phytotoxins, cause some of the symptoms.

The force-moving solute and bacteria in the phloem are associated with the source-conduit-sink system. Solutes are loaded from synthesizing cells (source) into sieve tubes (conduit) with the expenditure of energy. Movement in the sieve tube follows a concentration gradient, and unloading the conduit into the sink (also energy requiring) provides the gradient. If the bacteria move passively in the phloem sieve tubes, they can be carried along at velocities of 30–70 cm/h (230, 231).

In summary, it is clear that several ultrastructural studies have revealed what appears to be bacterial or bacterially induced vessel occlusion. Random cross-sections have, on occasion, exposed too few occluded xylem vessels to account for the observed "pathogen-induced wilt." Zimmermann (230) has indicated that the penetration of vessels by fungi or bacteria permits air to enter and, as a consequence, the vessels embolize. The emboli render the vessels, according to Zimmermann, nonfunctional in water transport. Wilt, therefore, is a reflection of numerous emboli rather than plugs, which, according to Zimmermann, probably occur after the fact. However, the study of Hopkins (104) that monitored vessel plugging in grapes over a length of 5 mm tends to lend credence to comprehensive vessel occlusion-induced wilt. It would appear that precise experiments with bacteria and/or other pathogens must be performed in order to answer the embolism-pathogen occlusion question regarding impairment of the plant's water economy and the observed symptom of wilt.

Plants Infected by Fungi

Water Economy

The infectious diseases of plants caused by fungi that infect roots, specifically the vascular system, generally influence the *lifting force* by which the water moves through the xylem from root and soil (10). The water transpired from the leaf generates a negative water potential toward which the water flows. Fungi such as *Pythium*, *Rhizoctonia*, *Fusarium*, *Verticillium*, *Ophiobolus*, *Fusicoccum*, *Ceratocystis*, and *Cytospora* all have been shown to influence the water lifting force as a consequence of pathogenesis.

The water transported in healthy plants is influenced by resistance to water flow, e.g. frictional forces improved by microfibrils of the cell wall as well as the walls of the xylem lumen. This resistance in the system may be exacerbated by fungal pathogens. Reduction in the diameter of channels in the root, in the endodermis, and in xylem, as well as the increase of resistance to water flow in leaves, may alter water economy of diseased plants. These aspects of two important water relations in diseased plants have been summarized by Ayres (6) and Talboys (197).

How the pathogens and their toxins and enzymes are involved in the mechanism of the impaired water economy was studied by Dimond (40) and continues to receive intensive scrutiny (46, 230). Apart from wilting per se, tissue necrosis, accelerated respiration, ethylene and auxin production, and phenol and aromatic compound accumulation characterize wilts. These responses of the host tend to be more generally distributed in wilt than in local lesion diseases.

The Dysfunction of Stomates and the Cuticle (Transpiration): Influencing the Lifting Force

Dimond and Waggoner (42) have revealed that asymptomatic leaves from *Fusarium*-infected tomato transpired one-third as rapidly as leaves from healthy plants supplied with the same amount of water. In this regard, they behaved much like leaves of drought-hardened plants, which transpired one-sixth as rapidly as succulent leaves. In both instances, the stomates remained closed during the day, while stomates of healthy leaves were fully opened.

Powdery mildew (*Erysiphe graminis* f. sp. *hordei*) strongly inhibits stomatal opening, thereby reducing transpiration rate (7) in barley (Fig. 8–18). Similarly,

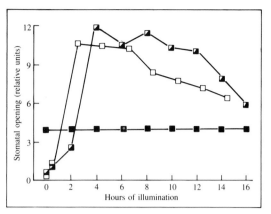

Fig. 8–18. Effect of powdery mildew on stomatal behavior in flag leaves of 55-d-old barley plants illuminated for 16 h/d. (□) Uninfected; (◩) uninfected region of an infected leaf; (■) infected leaf (after Ayres and Zadoks, 7).

the mildews of sugar beet (*E. polygoni*) and peas (*E. pisi*) also reduce stomatal aperture (5, 80).

In other infectious diseases (mildews, rusts), enhanced transpiration is a common feature. This increase is due to ruptured epidermis, as well as to the enhanced permeability to water of the injured cuticle. In these instances, according to Yarwood (226), the primary effect appears to be on transcuticular, rather than stomatal, transpiration. This was shown also by Duniway and Durbin (47), who reported that the primary cause of wilting in beans infected by *Uromyces phaseoli* was the continuous water loss through damaged portions of the cuticle. The sporulation process damages the cuticle even in slight infections, and the subsequent water vapor loss upsets the water economy of rust-infected plants under mild drought conditions. After sporulation, there is no longer a correlation between stomatal aperture and transpiration. The transpiration rate remains at a high level in rust-infected leaves, even in the dark when stomata are normally closed (48). It was also demonstrated that bean rust inhibited stomatal opening in the light, and both the transpiration rate and average stomatal aperture were significantly reduced during the "fleck stage" of rust development, i.e. before sporulation (Fig. 8–19). Subsequently, however, transpiration became excessive at the stage of sporulation because the epidermis was ruptured (Fig. 8–20).

Transpiration rate has been increased by toxic effects of pathogens. Fusicoccin, a fungal toxin, was shown to inhibit stomatal closure (202). Toxin-induced opening of the stomata caused increased transpiration and subsequent wilting in plants infected by *Fusicoccum amygdali*.

Permeability Changes in Leaf Cells: Influencing the Lifting Force

Injury to the host cell by toxins and enzymes of the pathogen may occur and are probably dependent on toxin concentration and the intensity of enzyme activity.

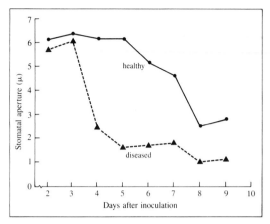

Fig. 8–19. Average stomatal apertures of healthy and rusted bean leaf discs in the light on different days after inoculation. A difference of 1.1μ between apertures indicates significance at the 1% level (after Duniway and Durbin, 48).

These perturbations influence permeability barriers such as the cell wall and plasma membrane.

The experiments of Gottlieb (81) clearly demonstrated that an enhanced permeability to water takes place in tomato infected with *Fusarium* wilt fungus. Parenchymatous cells were exposed to the tracheal sap from infected tomato, which decreased the time required to recover their turgor after plasmolysis. The sap from drought-wilted plants had only a slight effect. However, the effect of the tracheal sap from the diseased tomato was reversible, since it did not destroy cell function. Severely wilted tomato leaves also recovered their turgor when they were immersed in water, provided that the discolored and plugged portion of the petiole was removed (42). On the other hand, Sadasivan and associates (171, 172) found that an ionic imbalance in the leaf cells takes place in wilted cotton plants infected with *Fusarium*. Apparently, the fungus alters cell permeability as well as the turgor pressure. According to Gäumann et al. (73), increased permeability in leaves is caused by various nonspecific wilt toxins such as fusaric acid, lycomarasmin, and alternaric acid.

Pegg and Selman (163) have shown accumulation of auxins in tomato seedlings infected with *Verticillium*. Since auxin has been reported to increase the permeability of protoplasts to water (90), this effect may be implicated in the development of vascular diseases as well.

The Dysfunction of the Root: Increased Resistance to Water Flow

In the damping-off, foot-rot, and take-all diseases, rotting of the root and collar is caused by necrotrophic parasites. In these diseases, absorption of water by the diseased roots is more or less inhibited. However, in the vascular wilt diseases the symptoms of pathological water economy appear long before the root damage is extensive (123, 134), and the injury to roots did not

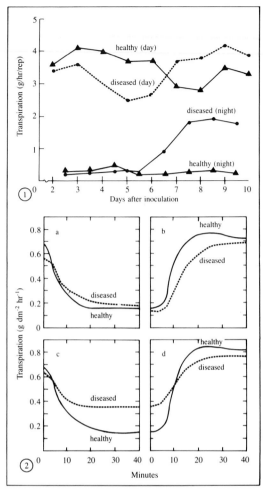

Fig. 8–20. (1) The transpiration rates of healthy and rust diseased bean leaves measured gravimetrically at different times after inoculation; (2) the transpiration rates of healthy and diseased leaves measured continuously during the first 40 min of the night or day period. The 0 time corresponds to the beginning of the 5th night (a); the 6th day (b); the 9th night (c); and the 10th day after inoculation (d) (after Duniway and Durbin, 48).

appear to effectively reduce the absorption of water. Still, in the vascular wilts a loss in selective permeability and absorption may occur in the root cells as a consequence of the partial dysfunction of the root system (171, 172).

The Dysfunction of Conductive Elements: Increased Resistance to Water Flow

There are many indications of reduced water flow through the vessels of diseased stems in vascular wilt diseases (38, 40). The resistance to flow of water through vascular tissues of plants infected by a wilt-inducing pathogen is always increased. Two factors responsible for increased resistance to flow are: (1) Increased hydrodynamic resistance to flow through the lumina of vessels caused by gels, mycelia, spores, collapsed vessels, and tyloses. In tomato plants infected by *Fusarium*, an increase in the xylem resistance to water flow was shown by Duniway (46) to be sufficient to account for leaf wilting. (2) Colloidal polysaccharides, hydrolytic products of pectin, and cellulose and melanoid pigments may be trapped in pit membranes. Interchange of water between vessels and redistribution of water within the plant, as a consequence, are reduced.

Several theories have been suggested to account for the impaired water transport. The viscous tracheal fluid and the gas embolism theories have been proposed in the study of vascular wilts. The physical presence of the pathogen in vessels is sometimes considered a cause of wilt; mycelia contribute to, but are not solely responsible for, the increased resistance to water flow through vessels (208).

Generally, in plants with vascular disease, transpiration is significantly less than in healthy plants (134, 176). This lowered transpiration is closely related to the plugging of vessels and the resultant shortage of water in the leaves. The latter has received extensive support from a series of studies by Zimmermann (230).

Currently, the occlusion theory seems to be satisfactory. Pectic and cellulolytic enzymes of pathogens, by an action in vivo, liberate partially hydrolyzed gels, gums, and phenolic compounds from the host-cell walls (see Table 8–1). The hydrolyzed compounds may contribute to the observed vascular occlusion, browning, and reduced rate of water movement (71, 86). However, most of these alterations occur at a later stage of disease development and, hence, there is less evidence for cause-and-effect relationships.

Obstruction of vessels can also result from nonspecific pathogen-induced wound reaction by the host.

Table 8–1. Production of wall-hydrolyzing enzymes by wilt pathogens (after Dimond, 40).

Pathogen	Found in	PME	Endo-PG	Endo-PTE	C_1/C_x
P. solanacearum	f/ip	+	+		+
V. albo-atrum	f/ip	tr	+	tr	$+/+_i$
V. dahliae	f	tr	$+_p$		
C. ulmi	f	—	+		+/+
C. fagacearum	f	—	+		
F. oxysporum f. sp.					
lycopersici	f/ip	+	$+_i$		+/+
cubense	f/ip	+	$+_i$		+
vasinfectum	f/ip	+	$+_i$		
niveum	f	tr	+		

PME: pectin methylesterase; Endo-PG: endopolygalacturonase; Endo-PTE: endopolygalacturonate transeliminase; C_1/C_x: cellulases attacking native cellulose/carboxymethylcellulose; f: culture filtrates; ip: infected plant tissues; $+_i$: inducible; $+_p$: protopectinase; tr: trace.

Ethylene may play a role in the development of the wound reaction. The observed host response includes the production of gums or the formation of tyloses. The gums are polysaccharides containing glucuronic acid associated with phenolic compounds.

Polysaccharides produced by *Fusarium* may also be involved in the obstruction of normal water flow in some plants, as has been suggested by Dimond and Waggoner (41). Tyloses in vessels may also reduce water flow (13, 134, 192), and this abnormality seems to be the cause of oak wilt (*Ceratocystis fagacearum*) and wilt of melon, tomato, etc., caused by *Fusarium* spp., as well as canker diseases. In some instances, the tyloses are very large and numerous and may completely occlude the vessels.

Dimond (38) noted that in the early phases of wilt disease the conductive dysfunction in the stem may not reduce the water supply to leaves to the point of producing "desiccation symptoms," since transpiration is automatically lowered. However, desiccation did occur in older infections, probably where water loss had become excessive. It has also been shown by Dimond (39) that the cross-sectional area available for conductivity decreases as the terminus of the conductive system is reached. Consequently, an occlusion of the same size has greater influence when it occurs in the tissues of a petiolar bundle or of a leaf vein than when it occurs in vessels in the stem.

Accumulation and the Phloem-Transport of Nutrients

A consequence of infection with obligate fungal parasites is the accumulation of certain metabolites at the infection site. This phenomenon was first clearly shown by Gottlieb and Garner (82), who fed radioactive phosphate to wheat infected with stem rust (*Puccinia graminis* f. sp. *tritici*). Investigation of distribution of radioactivity demonstrated that radioactive phosphorus accumulated in the rust-infected parts of leaves and tended to concentrate in those areas invaded by the parasite. It was not established whether the radiophosphorus was accumulating in host tissue per se or in mycelium. However, the concentration of radioactive phosphorus in uredospores was lower than in infected or healthy leaves. The question of solute movement and accumulation was reinvestigated by Durbin (49) (see Table 8–2). His study showed an immediate and sustained accumulation of ^{32}P in primary leaves of infected bean plants. However, the first trifoliate did not reveal a reduced total ^{32}P level until the eighth day after inoculation.

Selective heat-killing of the pathogen (*Uromyces phaseoli*) in bean leaves after infection did not fully decrease the mobilization of materials in the plant (227). The same conditions prevailed in wheat infected with stem rust (Király, unpublished). Incorporation of radioactive phosphate into the acid-insoluble fraction continued at the same rate for two days after heat-killing of the fungus, as in leaves with fully developed symptoms. A similar observation was reported when powdery mildew was removed from barley leaves (185).

It is of interest that plants infected with necrotrophic

Table 8–2. Percentage of total ^{32}P content of primary and first trifoliate leaves of healthy and rust-infected bean plants following a 1-h absorption period 4 d after inoculation (after Durbin, 49).

		^{32}P content	
Days after inoculation	Bean plant	Primary leaf %	First trifoliate leaf %
5	healthy	39	14
	infected	48	13
6	healthy	41	22
	infected	51	20
7	healthy	48	20
	infected	53	21
8	healthy	43	22
	infected	56	17
9	healthy	49	19
	infected	64	16
10	healthy	45	22
	infected	69	10
11	healthy	44	22
	infected	68	15
12	healthy	44	22
	infected	53	13
13	healthy	39	22
	infected	81	5

Table 8–3. Transport of labeled compounds in healthy bean plant (after Pozsár and Király, 165).

	Mean cpm/100 mg dry weight		
Compound	Upward[a]	Downward[a]	Downward upward
$Na_2HP^{32}O_4$	960	655	0.6
$Na_2S^{35}O_4$	2724	1940	0.7
Glycine-C^{14}	1708	852	0.5

a. Upward transport represents translocation of nutrients out of primary leaves to the young secondary ones. Downward transport represents translocation of materials out of secondary leaves to the primary ones.

parasites do not accumulate nutrients around their infection courts. This seems to occur only if they are infected by biotrophic parasites. Shaw and Samborski (186) have pointed out that accumulation depends upon the aerobic metabolism of the diseased plant. Inhibitors of respiration or oxidative phosphorylation interfere with the mobilization of nutrients and, apparently, their subsequent transport. Thus, accumulation of radioactivity was inhibited by anaerobiosis via the respiratory inhibitors azide and dinitrophenol. On the other hand, it has been suggested that the frequently observed increase of respiratory activity in infected plants may be a consequence of the utilization of accumulating nutrients as well as of potentiated synthetic activities. Active starch and chlorophyll syntheses in "green islands," increased synthesis of aromatic compounds and proteins, and renewed growth have been regularly observed consequences of infection.

The characteristic green islands around the infection sites reflect enhanced vigor of the rust- or mildew-infected host tissue. The mesophyll cells in these green islands of rust-infected wheat are rich in starch, and increased incorporation of RNA precursors occurred in uninvaded tissues surrounding the infection sites (170). In general, infections with biotrophic parasites (rusts and mildews) are characterized both by increased syn-

theses and dry weight of the host tissue and by enhanced respiration as well as an apparent decreased rate of photosynthesis. Thus, there is an increased demand for nutrients in infected tissue, which is satisfied by their long-distance movement to sites of utilization. Indeed, there is evidence that an intensive transport of nutrients occurs from distant leaves toward the infected ones (132, 165, 198, 199). Rust infection (*Uromyces phaseoli*) induced abnormal phloem transport of nutrients in bean plants. The translocation of labeled substances out of uninfected secondary leaves to infected primary leaves increased considerably, but mobilization of nutrients out of infected leaves was almost fully inhibited.

It would appear that transport occurs toward the locus of infection. In experiments in which only the mature primary bean leaves were infected and the younger leaves were not, the ratio of downward to upward transport was always greater in the diseased plants than in the healthy ones (Tables 8–3 and 8–4). It was also shown by Pozsár and Király (165) that kinetin exerts a similar effect on long-distance movement of nutrients, inhibit-

Table 8–4. Transport of labeled compounds in bean plant infected by *Uromyces phaseoli* (after Pozsár and Király, 165).

	Mean cpm/100 mg dry weight		
Compound	Upward[a]	Downward[a]	Downward upward
$Na_2HP^{32}O_4$	168	1872	11.1
$Na_2S^{35}O_4$	108	620	5.7
Glycine-C^{14}	88	592	6.7

a. Upward transport represents translocation of nutrients out of primary leaves to the young secondary ones. Downward transport represents translocation of materials out of secondary leaves to the primary ones.

ing mobilization of materials out of kinetin-treated leaves and increasing the phloem transport from other leaves toward the treated ones. Supposedly, rust infection is able to influence phloem transport by a kinetinlike mechanism (see Chap. 7).

In this regard, it is noteworthy that Bushnell and Allen (21) could induce the formation of green islands and accumulation of starch in barley leaves when aqueous extracts of powdery mildew and rust spores were applied locally to the leaves. Mobilization factors were shown in uredospores by Dekhuijzen and Staples (35). One may speculate that the mechanism for this action is kinetinlike and similar to the one described by Mothes and Englebrecht (141). The control upon accumulation and transport of nutrients in plants infected by biotrophic parasites seems to be twofold: (1) The parasite colony behaves as a nutrient "sink." The infection site may be considered as a "field of dominance" (184), where the infected tissue competes with the rest of the plant for nutrients (198, 199). (2) Cytokinin hormones produced at the infection sites (35, 116, 117) exert a juvenility effect on tissues (the life of the infected tissue is prolonged in green islands) and directly govern short- and long-distance movement of nutrients (165).

Vascular Transport of Fungi

Only a few fungi are transported passively in the sap of xylem, and these are primarily vascular pathogens. For example, Banfield (8) has shown that conidia of *Ceratocystis ulmi* (the agent of Dutch elm disease) are carried upward in the vessels of the spring wood. This rise may be as high as five meters in 24 h. The same is true for *Fusarium* and *Verticillium* conidia in the conductive elements of watermelon and tomato (8, 12, 115, 149, 177). When detached shoots absorbed conidia of *Fusarium* and *Verticillium* from an artificial suspension, after an incubation period, a general invasion by mycelium occurred throughout the water-conducting system, indicating the migration of conidia with the transpiration stream. The banana wilt pathogen *Fusarium oxysporum* f. sp. *cubense* distributes itself systemically by successive spore passages in the transpiration stream. Resistance of some banana varieties depends upon vascular occlusion, which immobilizes the spore distribution front (11, 12).

Comparative Analysis of Disease Physiology— Transcellular and Vascular Transport

With the exception of transport from endodermis to pericycle, solute and solvent movement in the xylem is apoplastic. Some energy is required to circumvent the barrier imposed by the Casparian strip. Thus, water and solute from the soil move in response to a gradient created when the loss of water from leaves exerts a transpiration-based "pull" on the water in the xylem.

Symplastic phloem transport is a three-part system: (1) source of solutes, (2) conduit, and (3) sink, the site of solute removal. Actual transport is a function of increased sieve tube water uptake in response to the accumulation of solute. The resultant buildup of cell turgor moves solute toward the sink. Radial movement of solute from xylem to ray cells to phloem unifies the transport system. With the aid of specialized transfer cells, the efficiency of conduit loading is heightened.

Viral, bacterial, and fungal pathogens are capable of the complete array of pathologic disturbances that influence the water economy of the plant. Root cell degeneration, permeability changes, and vessel blockage leading to wilt have been reported for each pathogen type.

Peculiar to virus infection, however, appears to be the accumulation of carbohydrate at the sites of infection. This could be due to the aforementioned vascular blockage, although there is evidence for a paucity of the enzyme needed to convert the carbohydrate to its translocative form, sucrose. Similarly, cytokinin-directed mobilization of metabolites seems to be a feature of biotrophic fungus infections. Radioisotope studies for the three major pathogen types in regard to their influence on long-distance metabolite movement appear to be warranted.

Whereas rate of movement of fungi in the vascular system has not been intensively studied, the efforts by virologists and bacteriologists have been more definitive. Once inside the phloem, virus particles have been monitored at migration rates of 2 to 3 cm/h. Of interest is the fact that photosynthate may move two orders of magnitude faster. Apparently, particle size, sieve tube obstructions such as P-protein, and sieve plates account for the reduced rate. Although some viruses are xylem limited, most appear to be more prevalent in the phloem, which is probably a reflection of vector feeding and deposition. Exclusive with some viruses at least is their capacity to traverse the plasmodesmata, which have diameters of 20–200 mm. In addition, their cell-to-cell movement is aided by cyclosis and diffusion. Virus movement is, perhaps, most precisely controlled by replication rate. Of interest is the capacity of virus particles, though usually protein coated, to migrate as naked viral genomes. Protection within vesicles or partial protein-coating may explain this capability.

Emergence from the vascular conduits into intercellular space seems to be prerequisite for pathogenic bacteria to cause intense symptoms. The necessary enzyme compliment, of either pathogen or host origin, seems available for the process. However, movement of viruses to nonliving cells or to an extracellular environment seems counterproductive for viral replication. Bacterially induced activation of host cell wall-altering enzymes appears to be an ideal area for fruitful research.

Perhaps the best understanding relative to the movement of pathogens in the vascular system has been recorded for bacteria. A break in the xylem sucks bacteria into vessels under a negative pressure of -2.0 atm. Their movement in water utilizing only flagella as their propellant is about 10^4 μm/h (based on a 25 μm/sec measurement). However, bacteria may move in the xylem under the pull of transpiration at the startling rate of 500 μm/sec (30, 195). It is understandable, therefore, why flagella are unimportant vis-à-vis pathogenesis once the bacteria have penetrated the host. The xylem vessel appears to be the primary site for initial rapid multiplication of bacterial pathogens. It is apparently preferred to phloem sieve tubes. The basis for this is the divergent osmotic concentration of the two vascular fluids, xylem being 0.8 atm and phloem 6.0–12.0 atm.

There is uncertainty regarding the effect of vessel penetration by bacterial cells. Zimmermann (230) contends, logically, that the entry of air simultaneously with the bacteria results in an embolism that precludes the entered vessel from further water transport. Thus, the bacteria must remain in that vessel unless they possess degrading enzymes. Experiments monitoring the movement of *Erwinia amylovora* and other pathogens without these enyzmes are at apparent variance. The matter requires further study.

Occlusion of the vascular conduits by pathogen or by plant debris and intrusion of vessels by tyloses are clearly impediments to long-distance movement of pathogen propagules. However, observations of vessel diameter reduction at nodes and apparent active localization of viruses, bacteria, and fungi warrant further ultrastructural studies of pathogen migration and barriers thereto.

Finally, the insufficiency of water, regardless of its cause, ultimately manifests itself in a reduction of transpiration. Stomatal closure and eventual foliar abscission are safeguards against or at least delay plant death by desiccation. Unique, however, is the fungal toxin fusicoccin that overrides this safety device preventing stomatal closure and exacerbating water loss. Rupture of the cuticular layer by some rust mildew and other ascomycetous fungi also fosters excessive water loss.

There is still much to be learned regarding the movement and fate of large macromolecules, such as extracellular polysaccharides in the vascular system. Similarly, a better understanding of the physiology and ontogeny of tylose formation is needed. Finally, the impact of reduced water potential on photosynthesis and on the transport of photosynthate and other metabolites emphasizes the need for a better understanding of vascular transport as it is altered by pathogenesis. For example, biotrophic fungi influence transport and accumulation of nutrients in infected tissues in two ways: (1) the parasite colony behaves as a nutrient sink, and (2) cytokinin hormones produced at the infection site directly govern the movements of nutrients.

Additional Comments and Citations

1. Recent studies indicate that the deficiency in transport of the LS-1 strain of TMV can be restored by other strains of TMV and by PVX, suggesting that, whatever modification of host cells is involved, it can be performed by viruses other than TMV (M. E. Taliansky, S. I. Malyshenko, E. S. Pshennikova, I. B. Kaplan, E. F. Ulanova, and J. G. Atabekov. 1982. Plant virus-specific transport functions. I. Virus genetic control required for systemic spread. *Virology,* 122:318–26. M. E. Taliansky, S. I. Malyshenko, E. S. Pshennikova, and J. G. Atabekov. 1982. Plant virus-specific transport function

II. A factor controlling virus host range. *Virology* 122:327–31.) It is also of interest that certain plants considered to be "nonhosts" of a virus, or immune, can be rendered susceptible by preinfection with a "helper" virus. Preinfection of tomato and bean plants with TMV or DEMV, respectively, permits BMV to be transported from vascular tissue and to be replicated in mesophyll cells, presumably by complementing a transport deficiency in BMV.

2. M. Mahjoub, D. LePicard, and M. Moreau. 1984. Origin of tyloses in melon (*Cucumis melo* L.) in response to a vascular *Fusarium.* IAWA Bull. 5:397–11. This study reveals, ultrastructurally, the growth response of "contact cells" (transfer cells between xylem vessels) to infection and the resultant forcing of the plasmalemma of these cells through pit-lined primary wall into the vessel lumen as tyloses.

3. G. E. Vander Molen, J. Labavitch, L. Strand, and J. DeVay. 1983. Pathogen-induced vascular gels: Ethylene as a host intermediate. Physiol. Plant. 59:573–80. Vascular gel formation leads to vascular occlusion and wilt. The authors suggest that these morphological responses are a direct consequence of ethylene action. Vascular gel formation did not occur when ethylene production in response to pectin-degrading enzymes is prevented.

References

1. Adomako, D., and W. V. Hutcheson. 1974. Carbohydrate metabolism and translocation in healthy and cocoa swollen shoot virus-infected cocoa plants. Physiol. Plant 30:90–96.

2. Arisz, W. H. 1959. Symplasm theory of salt uptake into transport in parenchymatic tissue. In: Recent advances in botany, from lectures and symp. presented to the IX Int. Bot. Cong., Univ. of Toronto Press, pp. 1125–28.

3. Atabekov, J. G. 1977. Defective and satellite plant viruses. In: Comprehensive virology. H. H. Fraenkel-Conrat and R. R. Wagner (eds.). Plenum Press, New York, 11:143–200.

4. Ayers, A. R., S. B. Ayers and R. N. Goodman. 1979. Extracellular polysaccharide of *Erwinia amylovora*: A correlation with virulence. Appl. Environ. Microbiol. 38:659–66.

5. Ayres, P. G. 1976. Patterns of stomatal behavior, transpiration and CO_2 exchange in pea following infection by powdery mildew (*Erysiphe pisi*). J. Exp. Bot. 27:1196–205.

6. Ayres, P. G. 1978. The water relations of diseased plants. In: Water deficits and plant growth. T. T. Kozlowski (ed.). Academic Press, New York, 5:1–60.

7. Ayres, P. G., and J. C. Zadoks. 1979. Combined effects of powdery mildew disease and soil water level on the water relations and growth of barley. Physiol. Plant Path. 14:347–67.

8. Banfield, W. M. 1941. Distribution by the sap stream of spores of three fungi that induce vascular wilt diseases of elm. J. Agr. Res. 62:637–81.

9. Bawden, F. C. 1964. Plant viruses and virus diseases. 4th ed. Ronald Press, New York.

10. Beckman, C. H. 1964. Host responses to vascular infection. Ann. Rev. Phytopath. 2:231–52.

11. Beckman, C. H., and S. Halmos. 1962. Relation of vas-

cular occluding reactions in banana roots to pathogenicity of root-invading fungi. Phytopathology 52:893–97.

12. Beckman, C. H., S. Halmos, and M. E. Mace. 1962. The interaction of host, pathogen, and soil temperature in relation to susceptibility to *Fusarium* wilt on bananas. Phytopathology 52:134–40.

13. Beckman, C. H., J. E. Kuntz, A. J. Riker, and J. G. Berbee. 1953. Host responses associated with the development of oak wilt. Phytopathology 43:448–54.

14. Bennett, C. W. 1934. Plant tissue relations of sugar beet curly top virus. J. Agr. Res. 48:665–701.

15. Bennett, C. W. 1937. Correlation between movement of the curly top virus and translocation of food in tobacco and sugar beet. J. Agr. Res. 54:479–502.

16. Bennett, C. W. 1940. Relation of food translocation to movement of virus of tobacco mosaic. J. Agr. Res. 60:361–90.

17. Bennett, C. W. 1956. Biological relations of plant viruses. Ann. Rev. Plant Physiol. 7:143.

18. Best, R. J. 1956. Resistance of higher plants to viruses. J. Aust. Inst. Agr. Sci. 23:174–79.

19. Beute, M. K., and J. H. Lockwood. 1968. Mechanism of increased root rot in virus-infected peas. Phytopathology 58:1643–51.

20. Bonner, J., and A. W. Galston. 1952. Principles of plant physiology. W. H. Freeman, San Francisco.

21. Bushnell, W. R., and P. J. Allen. 1962. Induction of disease symptoms in barley leaves produced by single colonies of powdery mildew. Plant Physiol. 37:50–59.

22. Capoor, S. P. 1949. The movement of tobacco mosaic viruses and potato virus X through tomato plants. Ann. Appl. Biol. 36:307–19.

23. Chambers, T. C., and R. I. B. Francki. 1966. Localization and recovery of lettuce necrotic yellows virus from xylem tissues of *Nicotiana glutinosa*. Virology 29:673–76.

24. Cole, R. M., J. G. Tully, T. J. Popkin, and J. M. Bové. 1973. Morphology, ultrastructure, and bacteriophage infection of the helical mycoplasma-like organism (*Spiroplasma citri* gen. nov., sp. nov) cultured from 'stubborn' disease of citrus. J. Bact. 115:367–86.

25. Cooper, P., and I. W. Selman. 1974. An analysis of the effects of tobacco mosaic virus on growth and the changes in the free amino compounds in young tomato plants. Ann. Bot. 38:625–38.

26. Coutts, R. H. A. 1980. Virus induced local lesions on cowpea leaves: The role of the epidermis and the effects of kinetin. Phytopath. Z. 97:307–16.

27. Crafts, A. S., and C. E. Crisp. 1971. Phloem transport in plants. W. H. Freeman, San Francisco.

28. Cronshaw, J., and K. Esau. 1968. P-protein in the phloem of *Cucurbita* II. The P- protein of mature sieve elements. J. Cell Biol. 38:292–303.

29. Crosse, J. E. 1965. Bacterial canker of stone fruit. VI. Inhibition of leaf scar infection of cherry by a saprophytic bacterium from leaf surfaces. Ann. Appl. Biol. 56:149–60.

30. Crosse, J. E., R. N. Goodman, and W. H. Shaffer, Jr. 1972. Leaf damage as a predisposing factor in the infection of apple shoots by *Erwinia amylovora*. Phytopathology 62:176–82.

31. Czaninski, Y. 1977. Vessel-associated cell. IAWA Bull. 3:51–55.

32. D'Arcy, C. J., and G. A. de Zoeten. 1979. Beet western yellows virus in phloem tissue of *Thlaspi arvense*. Phytopathology 69:1194–98.

33. de Zoeten, G. A. 1976. Cytology of virus infection and virus transport. In: Encyclopedia of plant physiology. New Series. R. Heitefuss and P. H. Williams (eds.). Springer-Verlag, Berlin, 4:129–49.

34. de Zoeten, G. A., and G. Gaard. 1969. Possibilities for inter and intracellular translocation of some icosahedral plant viruses. J. Cell Biol. 40:814–23.

35. Dekhuijzen, H. M., and R. C. Staples. 1968. Mobilization factors in uredospores and bean leaves infected with bean rust fungus. Contrib. Boyce Thompson Inst. 24:39–52.

36. Diener, T. O., and A. Hadidi. 1977. Viroids. In: Comprehensive virology. H. H. Fraenkel-Conrat and R. R. Wagner (eds.). Plenum Press, New York, 2:285–337.

37. Dijkstra, J. 1962. On the early stages of infection by tobacco mosaic virus in *Nicotiana glutinosa* L. Virology 18:142–43.

38. Dimond, A. E. 1955. Pathogenesis in the wilt diseases. Ann. Rev. Plant Physiol. 6:329–50.

39. Dimond, A. E. 1966. Pressure and flow relations in vascular bundles of the tomato plant. Plant Physiol. 41:119–31.

40. Dimond, A. E. 1970. Biophysics and biochemistry of the vascular wilt syndrome. Ann. Rev. Phytopath. 8:301–22.

41. Dimond, A. E., and P. E. Waggoner. 1953. On the nature and role of vivotoxins in plant disease. Phytopathology 43:229–35.

42. Dimond, A. E., and P. E. Waggoner. 1953. The water economy of *Fusarium* wilted tomato plants. Phytopathology 43:619–23.

43. Dixon, H. H., and N. G. Ball. 1922. Transport of organic substances in plants. Nature 109:236–37.

44. Dorokhov, Y. L., N. A. Miroshnichenko, N. M. Alexandrova, and J. G. Atabekov. 1981. Development of systemic TMV infection in upper noninoculated tobacco leaves after differential temperature treatment. Virology 108:507–9.

45. Dumbroff, E. B., and D. R. Peirson. 1971. Probable sites for passive movement of ions across the endodermis. Can. J. Bot. 49:35–38.

46. Duniway, J. M. 1971. Resistance to water movement in tomato plants infected with *Fusarium*. Nature 230:252–53.

47. Duniway, J. M., and R. D. Durbin. 1971. Detrimental effect of rust infection on the water relations of bean. Plant Physiol. 48:69–72.

48. Duniway, J. M., and R. D. Durbin. 1971. Some effects of *Uromyces phaseoli* on the transpiration rate and stomatal response of bean leaves. Phytopathology 61:114–19.

49. Durbin, R. D. 1967. Obligate parasites: Effect on the movement of solutes and water. In: The dynamic role of molecular constituents in plant-parasite interaction. C. J. Mirocha and I. Uritani (eds.). Amer. Phytopath. Soc., St. Paul, MN, pp. 80–99.

50. Ehara, Y., and T. Misawa. 1967. Studies on the infection of cucumber mosaic virus. III. Time of the infection after inoculation with whole virus or with viral nucleic acid. Tohoku J. Agr. Res. 17:193–200.

51. El Khalifa, M. D. 1974. Crown gall castor beans. III. Some quantitative aspects of secondary tumor initiation. Physiol. Plant Path. 4:435–42.

52. Esau, K. 1956. An anatomist's view of virus diseases. Amer. J. Bot. 43:739–48.

53. Esau, K. 1957. Phloem degeneration in *Gramineae* affected by the barley yellow-dwarf virus. Amer. J. Bot. 44:245–51.

54. Esau, K. 1968. Viruses in Plant Hosts. Univ. of Wisconsin Press, Madison.

55. Esau, K., and V. I. Cheadle. 1965. Cytologic studies on phloem. Univ. of Calif. Publs. Bot. 36:253–344.

56. Esau, K., and J. Cronshaw. 1967. Relation of tobacco mosaic virus to the host cells. J. Cell Biol. 33:665–78.

57. Esau, K., J. Cronshaw, and L. L. Hoefert. 1967. Relation of beet yellows virus to the phloem and to movement in the sieve tube. J. Cell Biol. 32:71–87.

58. Esau, K., and L. L. Hoefert. 1971. Cytology of beet yellows virus infection in *Tetragonia*. II. Vascular elements in infected leaf. Protoplasma 72:459–76.

59. Esau, K. and L. L. Hoefert. 1971. Cytology of beet yellows virus infection in *Tetragonia*. III. Conformations of virus in infected cells. Protoplasma 73:51–65.

60. Esau, K., and L. L. Hoefert. 1972. Development of infection with beet western yellows virus in the sugarbeet. Virology 48:724–38.

61. Esau, K., and L. L. Hoefert. 1972. Ultrastructure of sugarbeet leaves infected with beet western yellows virus. J. Ultrastruc. Res. 40:556–71.

62. Faccioli, G., N. J. Panopoulos, and A. H. Gold. 1971. Kinetics of the carbohydrate metabolism in potato leafroll infected *Physalis floridana*. Phytopath. Medit. 10:1–25.

63. Faccioli, G., N. J. Panopoulos, and A. H. Gold. 1972. Uptake and translocation of phosphorus in healthy and curly top-diseased tomatoes. Phytopathology 62:524–29.

64. Fahn, A. 1964. Some anatomical adadptations of desert plants. Phytomorphology 14:93–102.

65. Fahn, A. 1969. Plant anatomy. Pergamon Press, Oxford, London, New York.

66. Favali, M. A. 1977. Ultrastructural study of systemic lesions induced by peanut rosette virus in peanut leaves. Phytopath. Z. 89:68–75.

67. Feder, W. A., and P. A. Ark. 1951. Wilt-inducing polysaccharides derived from crown gall, bean blight and soft rot bacteria. Phytopathology 41:804–8.

68. Feliciano, A., and R. H. Daines. 1970. Factors influencing ingress of *Xanthomonas pruni* through peach leaf scars and subsequent development of spring cankers. Phytopathology 60:1720–26.

69. Foster, R. C. 1964. Fine structure of tyloses. Nature 204:494–95.

70. Fry, P. R., and R. E. F. Matthews. 1963. Timing of some early events following inoculation with tobacco mosaic virus. Virology 19:461–69.

71. Gagnon, C. 1967. Histochemical studies on the alteration of lignin and pectic substances in white elm infected by *Ceratocystis ulmi*. Can. J. Bot. 45:1619–23.

72. Gardner, D. C. J., and A. J. Peel. 1969. ATP in sieve tube sap from willow. Nature 222:774.

73. Gäumann, E., S. Naef-Roth, P. Reusser, and A. Amman. 1952. Über den Einfluss einiger Welketoxine und Antibiotica auf die osmotischen Eigenschaften pflanzlicher Zellen. Phytopath. Z. 91:160–220.

74. Ghabrial, S. A., and T. P. Pirone. 1964. Physiological changes which precede virus-induced wilt of tabasco pepper. Phytopathology 54:893.(Abstr.)

75. Gibbs, A. J. 1976. Viruses and plasmodesmata. In: Intercellular communication in plants: Studies on plasmodesmata. B. E. S. Gunning, and A. W. Robards (eds.). Springer-Verlag, Berlin, pp. 149–64.

76. Gill, C. C., and J. Chong. 1975. Development of infection in oat leaves inoculated with barley yellow dwarf virus. Virology 66:440–53.

77. Gill, C. C., and J. Chong. 1981. Vascular cell alterations and predisposed xylem infection in oats by inoculation with paired barley yellow dwarf viruses. Virology 114:405–13.

78. Goodman, R. N., J. S. Huang, and P.-Y. Huang. 1974. Host specific phytotoxic polysaccharide from apple tissue infected by *Erwinia amylovora*. Science 183:1081–82.

79. Goodman, R. N., and J. A. White. 1981. Xylem parenchyma plasmolysis and vessel wall disorientation are early signs of pathogenesis caused by *Erwinia amylovora*. Phytopathology 71:844–52.

80. Gordon, T. R., and J. M. Duniway. 1982. Effects of powdery mildew infection on the efficiency of CO_2 fixation and light utilization by sugar beet leaves. Plant Physiol. 69:139–42.

81. Gottlieb, D. 1944. The mechanism of wilting caused by *Fusarium bulbigenum* var. *lycopersici*. Phytopathology 34:41–59.

82. Gottlieb, D., and J. M. Garner. 1946. Rust and phosphorus distribution in wheat leaves. Phytopathology 36:557–64.

83. Gowda, S. S., and R. N. Goodman. 1970. Movement and persistence of *Erwinia amylovora* in shoot, stem and root of apple. Plant Dis. Reptr. 54: 576–80.

84. Greenleaf, W. H. 1953. Effects of tobacco etch virus on peppers (*Capsicum* spp.). Phytopathology 43:564–70.

85. Grieve, B. J. 1941. Studies in the physiology of host parasite relations: I. The affect of *Bacterium solanacearum* on the water relations of plants. Proc. Royal Soc., Victoria 53:268–99.

86. Grossmann, F. 1962. Untersuchungen über die Hemmung pektolytischer Enzyme von *Fusarium oxysporum* f. *lycopersici*. III. Wirkung einiger Hemmstoffe in vivo. Phytopath. Z. 45:139–59.

87. Gunning, B. E. S., and J. S. Pate. 1969. Transfer cells—plant cells with wall ingrowths, specialized in relation to short distance transport of solutes—their occurrence, structure, development. Protoplasma 68:107–33.

88. Gunning, B. E. S., J. S. Pate, R. R. Minchin, and I. Marks. 1974. Quantitative aspects of transfer cell structure in relation to vein loading in leaves and solute transport in legume nodules. In: Transport at the cellular level. Symp. Soc. Exp. Biol. 28:87–126.

89. Gunning, B. E. S., and A. W. Robards. 1976. In: Transport and transfer processes in plants. I. F. Wardlaw and J. B. Passioura (eds.). Academic Press, New York, pp. 15–41.

90. Guttenberg, H., and L. Kropelin. Über den Einfluss des Heteroauxins auf das Laminargelink von *Phaseolus coccineus*. Planta 35:257–80.

91. Hall, A. E., and R. S. Loomis. 1972. An explanation for the difference in photosynthetic capabilities of healthy and beet yellows virus-infected sugar beets (*Beta vulgaris* L.). Plant Physiol. 50:576–80.

92. Hancock, J. G., and A. Magyarosy. 1972. The influence of squash mosaic on squash hypocotyl permeability. Phytopathology 62:762. (Abstr.)

93. Harrison, B. D., and D. J. Robinson. 1978. The tobraviruses. Adv. Virus Res. 23:25–77.

94. Hartt, C. E. 1965. The effect of temperature upon translocation of ^{14}C in sugar cane. Plant Physiol. 40:718–24.

95. Hartt, C. E. 1966. Translocation in coloured light. Plant Physiol. 41:369–72.

96. Hatta, T., and R. E. F. Matthews. 1974. The sequence of early cytological changes in Chinese cabbage leaf cells following systemic infection with turnip yellow mosaic virus. Virology 59:383–96.

97. Helms, K., and I. F. Wardlaw. 1976. Movement of viruses in plants: Long distance movement of tobacco mosaic virus in *Nicotiana glutinosa*. In: Transport and transfer processes in plants. I. F. Wardlaw and J. B. Passioura (eds.). Academic Press, New York, pp. 283–93.

98. Helms, K., and I. F. Wardlaw. 1978. Translocation of tobacco mosaic virus and photosynthetic assimilate in *Nicotiana glutinosa*. Physiol. Plant Path. 13:23–36.

99. Hewitt, W. B. 1938. Leaf scar infection in relation to olive-knot disease. Hilgardia 12:41–71

100. Hirumi, H., M. Kimura, K. Maramorosch, J. Bird, and R. Woodbury. 1974. Rickettsia-like organisms in the phloem of little leaf-diseased *Sida cordifolia*. Phytopathology 64:518–82. (Abstr.)

101. Hodgson, R., A. J. Riker, and W. H. Peterson. 1947. A wilt-inducing toxic substance from crown-gall bacteria. Phytopathology 37:301–18.

102. Hodgson, R., A. J. Riker, and W. H. Peterson. 1949. The toxicity of polysaccharides and other large molecules to tomato cuttings. Phytopathology 39:47–62.

103. Hopkins, D. L. 1977. Diseases caused by leafhopper-borne, rickettsia-like bacteria. Ann. Rev. Phytopath. 15:277–94.

104. Hopkins, D. L. 1980. Seasonal concentration of the Pierce's disease bacterium in grapevine stems, petioles, and leaf veins. Phytopathology 71:415–18.

105. Huang, P.-Y., and R. N. Goodman. 1976. Ultrastructural modifications in apple stems induced by *Erwinia amylovora* and the fire blight toxin. Phytopathology 66:269–76.

106. Huber, B. 1932. Beobachtung und Messung pflanzlichen Saftstrome Ber. Deut. Bot. Ges. 50:89–109.

107. Hunter, C. S., and W. E. Peat. 1973. The effect of tomato aspermy virus on photosynthesis in the young tomato plant. Physiol. Plant Path. 3:517–24.

108. Husain, A., and A. Kelman. 1958. Relation of slime production to mechanism of wilting and pathogenicity of *Pseudomonas solancearum*. Phytopathology 48:155–65.

109. Husain, A., and A. Kelman. 1958. The role of pectic and cellulolytic enzymes in pathogenesis by *Pseudomonas solanacearum*. Phytopathology 48:377–86.

110. Ingleston, B., and B. Hylmo. 1961. Apparent free space and surface file determined by a centrifugation method. Physiol. Plant. 14:157–70.

111. Ishiie, T., Y. Doi, K. Yora, and H. Asuyama. 1967. Suppressive effects of antibiotics of tetracycline group on symptom development of mulberry dwarf disease. Ann. Phytopath. Soc. Jap. 33:267–75.

112. Jensen, S. G. 1973. Systemic movement of barley yellow dwarf virus in small grains. Phytopathology 63:854–56.

113. Jones, R. A. C. 1975. Systemic movement of potato moptop virus in tobacco may occur through the xylem. Phytopath. Z. 82:352–55.

114. Kasamo, K., and T. Shimomura. 1977. The role of the epidermis in local lesion formation and the multiplication of tobacco mosaic virus and its relation to kinetin. Virology 76:12–18.

115. Keyworth, W. G. 1950. Inoculation of plants with vascular pathogens. Nature 166:955.

116. Király, Z., M. El Hammady, and B. I. Pozsár. 1967. Increased cytokinin activity of rust-infected bean and broad bean leaves. Phytopathology 57:93–94.

117. Király, Z., B. I. Pozsár, and M. El Hammady. 1966. Cytokinin activity in rust-infected plants: Juvenility and senescence in diseased leaf tissues. Acta Phytopath. Hung. 1:29–37.

118. Kitajima, E. W., and A. S. Costa. 1969. Association of pepper ringspot virus (Brazilian tobacco rattle virus) and host cell mitochondria. J. Gen. Virol. 4:177–81.

119. Kitajima, E. W., and J. A. Lauritis. 1969. Plant virions in plasmodesmata. Virology 37:681–85.

120. Knosel, D., and E. D. Garber. 1967. Pektolytische und cellulolytische Enzyme bei *Xanthomonas campestris* (Pammel) Dowson. Phytopath. Z. 59:194–202.

121. Kondo, A. 1974. Long-distance movement of tobacco mosaic virus within the tobacco plant. Ann. Phytopath. Soc. Jap. 40: 299–303.

122. Kontaxis, D. G. 1961. Movement of tobacco mosaic virus through epidermis of *Nicotiana glutinosa* leaves. Nature 192:581–82.

123. Kramer, P. J. 1933. The intake of water through dead root systems and its relation to the problem of absorption by transpiring plants. Amer. J. Bot. 20:481–92.

124. Kunkel, L. O. 1939. Movement of tobacco mosaic virus in tomato plants. Phytopathology 29:684–700.

125. Kuriger, W. E., and G. N. Agrios. 1977. Cytokinin levels and kinetin-virus interactions in tobacco ringspot virus-infected cowpea. Phytopathology 67:604–9.

126. Lehoczky, L. 1968. Spread of *Agrobacterium tumefaciens*

in the vessels of the grapevine, after natural infection. Phytopath. Z. 63:239–46.

127. Leonard, D. A., and M. Zaitlin. 1982. A temperature sensitive strain of tobacco mosaic virus defective in cell-to-cell movement generates an altered viral coded protein. Virology 117:416–24.

128. Levitt, J. 1956. The physical nature of transpiration pull. Plant Physiol. 31:248–51.

129. Lewis, S., and R. N. Goodman. 1965. Mode of penetration and movement of fireblight bacteria in apple leaf and stem tissue. Phytopathology 55:719–23.

130. Lindsey, D. W., and R. T. Gudanskas. 1975. Effect of maize dwarf mosaic virus on water relations in corn. Phytopathology 65:434–40.

131. Livingston, L. G. 1964. The nature of plasmodesmata in normal (living) plant tissue. Amer. J. Bot. 51:950–57.

132. Livne, A., and J. M. Daly. 1966. Translocation in healthy and rust-affected beans. Phytopathology 56:170–75.

133. Lowe, S. K., G. Nyland, and S. M. Mircetich. 1976. The ultrastructure of the almond leaf scorch bacterium with special reference to topography of the cell wall. Phytopathology 66:147–51.

134. Ludwig, R. A. 1952. Studies on the physiology of hadromycotic wilting in tomato plant. MacDonald Coll. J. Ser. Tech. Bull. 20.

135. Main, C. E., and J. C. Walker. 1971. Physiological responses of susceptible and resistant cucumber to *Erwinia tracheiphila*. Phytopathology 61:518–22.

136. Mason, R. G., and E. J. Maskell. 1928. Studies on the transport of carbohydrates in the cotton plant. II. The factors determining the rate and direction of movement of sugars. Ann. Bot. 42:571–636.

137. Matthews, R. E. F. 1981. Plant virology. 2d ed. Academic Press, New York.

138. Milburn, J. A. 1972. Phloem physiology and protective sealing mechanisms. Nature 236:82.

139. Mircetich, S. M., S. K. Lowe, W. J. Moller, and G. Nyland. 1976. Etiology of almond leaf scorch disease and transmission of the causal agent. Phytopathology 66:17–24.

140. Mollenhauer, H. H., and D. L. Hopkins. 1974. Ultrastructural study of Pierce's disease bacterium in grape xylem tissue. J. Bact. 119:612–18.

141. Mothes, K., and L. Engelbrecht. 191. Kinetin-induced directed transport of substances in excised leaves in the dark. Phytochemistry 1:58–62.

142. Müller, D. 1932. Die Assimilation der blattrollkranken Karttoffelpflanzen. Planta 16:10–16.

143. Munch, E. 1930. Die Stoffbewegungen in der Pflanze. Gustav Fischer Verlag, Jena.

144. Murphy, P. A. 1923. On the cause of rolling in potato foliage and on some further insect carriers of the disease. Sci. Prog. Roy. Dublin Soc. 17:163–84.

145. Nasuno, S., and M. P. Starr. 1967. Polygalacturonic acid transeliminase of *Xanthomonas campestris*. Biochem. J. 104:178–85.

146. Nelson, P. E., and R. S. Dickey. 1966. *Pseudomonas caryophylli* in carnation. II. Histological studies of infected plants. Phytopathology 56:154–63.

147. Nienhaus, F., and R. A. Sikora. 1979. Mycoplasmas, spiroplasmas, and rickettsia-like organisms as plant pathogens. Ann. Rev. Phytopath. 17:37–58.

148. Nishiguchi, M., F. Motoyoshi, and N. Oshima. 1980. Further investigation of a temperature-sensitive strain of tobacco mosaic virus: Its behaviour in tomato leaf epidermis. J. Gen. Virol. 46:497–500.

149. Nishimura, S. 1960. Pathochemical studies on water-

melon wilt. 7. The physiology of *Fusarium* wilt of watermelon plant. Trans. Tottari Soc. Agr. Sci. 12:13–17.

150. Orden, L. 1958. The effects of water stress on the cell wall metabolism of plant tissue. In: Radioisotopes in scientific research. Pergamon Press, London, pp. 553–64.

151. Orellana, R. G., F. Fan, and C. Sloger. 1978. Tobacco ringspot virus and *Rhizobium* interactions in soybean: Impairment of leghemoglobin accumulation and nitrogen fixation. Phytopathology 68:577–82.

152. Orlob, G. B., and D. C. Arny. 1961. Some metabolic changes accompanying infection by barley yellow dwarf virus. Phytopathology 51:768–775.

153. Owen, P. C. 1958. Some effects of virus infection on leaf water contents of *Nicotiana* species. Ann. Appl. Biol. 46:205–9.

154. Oxelfet, P. 1970. Development of systemic tobacco mosaic virus infection. I. Initiation of infection and time course of virus multiplication. Phytopath. Z. 69:202–11.

155. Oxelfelt, P. 1975. Development of systemic tobacco mosaic virus infection. IV. Synthesis of viral RNA and intact virus and systemic movement of two strains as influenced by temperature. Phytopath. Z. 83:66–76.

156. Panopoulos, N. J., G. Faccioli, and A. H. Gold. 1972. Translocation of photosynthate in curly top virus-infected tomatoes. Plant Physiol. 50:266–70.

157. Panopoulos, N. J., G. Faccioli, and A. H. Gold. 1972. Effect of curly top disease on uptake, transport and compartmentation of calcium. Phytopathology 62:43–49.

158. Panopoulos, N. J., G. Faccioli, and A. H. Gold. 1972. Kinetics of carbohydrate metabolism in curly top virus-infected tomato plants. Phytopath. Medit. 11:48–58.

159. Panopoulos, N. J., G. Faccioli, and A. H. Gold. 1975. Sulfate uptake and translocation in curly top infected tomatoes. Phytopathology 65:77–80.

160. Pate, J. S., and B. E. Gunning. 1969. Vascular transfer cells in angiosperm leaves, a taxonomic and morphological survey. Protoplasma 68:135–56.

161. Peel, A. J. 1974. Transport of nutrients in plants. Wiley, New York.

162. Peel, A. J., and P. E. Weatherley. 1959. Composition of sieve tube sap. Nature 184:1955–56.

163. Pegg, G. F., and I. W. Selman. 1959. An analysis of the growth response of young tomato plants to infection by *Verticillium albo-atrum*. The production of growth substances. Ann. Appl. Biol. 47:222–31.

164. Pine, T. S., R. G. Grogan, and W. B. Hunt. 1955. Pathological anatomy of bacterial canker of young tomato plants. Phytopathology 45:267–71.

165. Pozsár, B. I., and Z. Király. 1966. Phloem-transport in rust infected plants and the cytokinin-directed long-distance movement of nutrients. Phytopath. Z. 56:297–309.

166. Quak, F. 1977. Meristem culture and virus-free plants. In: Plant cell tissue and organ culture. J. Reinert and Y. P. S. Bajaj (eds.). Springer-Verlag, Berlin, pp. 598–615.

167. Rappaport, I., and S. G. Wildman. 1957. A kinetic study of local lesion growth on *Nicotiana glutinosa* resulting from tobacco mosaic virus infection. Virology 4:265–74.

168. Robards, A. W., and M. E. Robb. 1974. The entry of ions and molecules into roots: An investigation using electron-opaque tracers. Planta 120:1–12.

169. Roberts, I. M., and B. D. Harrison. 1970. Inclusion bodies and tubular structures in *Chenopodium amaranticolor* plants infected with strawberry latent ringspot virus. J. Gen. Virol. 7:47–54.

170. Rohringer, R., and R. Heitefuss. 1961. Incorporation of P[32] into ribonucleic acid of rusted wheat leaves. Can. J. Bot. 39:263–67.

171. Sadasivan, T. S., and R. Kalyanasundaram. 1956. Spectrochemical studies on the uptake of ions by plants. I. The Lundgardh flame technique in ash analysis of toxin/antibiotic invaded cotton plants. Proc. Indian Acad. Sci. 343:271–75.

172. Sadasivan, T. S., and L. Saraswathi-Devi. 1957. Vivotoxins and uptake of ions by plants. Current Sci. (India) 26:74–75.

173. Saglio, P., D. Laflèche, C. Bonissol, and J. M. Bové. 1971. Isolement et culture in vitro des mycoplasmes associées au 'stubborn' des agrumes et leur observation au microscope electronique. C.R., Acad. Sci. Ser. D. 272:1387–90.

174. Samuel, G. 1934. The movement of tobacco mosaic within the plant. Ann. Appl. Biol. 21:90–111.

175. Sauter, J., W. Iten, and M. Zimmermann. 1973. Studies on the release of sugar into the vessels of sugar maple (*Acer saccharum*). Can. J. Bot. 51:1–8.

176. Scheffer, R. P., and J. C. Walke. 1953. The physiology of *Fusarium* wilt of tomato. Phytopathology 43:116–25.

177. Scheffer, R. P., and J. C. Walker. 1954. Distribution and nature of *Fusarium* resistance in the tomato plant. Phytopathology 44:94–101.

178. Schneider, I. R. 1964. Difference in the translocatability of tobacco ringspot and southern bean mosaic viruses in bean. Phytopathology 54:701–5.

179. Schneider, I. R. 1965. Introduction, translocation and distribution of viruses in plants. Adv. Virus Res. 11:163–221.

180. Schwarz, M., P. E. Sonnet, and N. Wakabayashi. 1970. Leaf hydraulic system: Rapid epidermal and stomatal responses to changes in water supply. Science 167:189–92.

181. Scott, L. I., and J. H. Priestley. 1928. The root as an absorbing organ. I. A reconsideration of the entry of water and salts in the absorbing region. New Phytol. 27:125–41.

182. Shalla, T. A. 1959. Relations of tobacco mosaic virus and barley stripe mosaic virus to their host cells as revealed by ultrathin tissue-sectioning for the electron microscope. Virology 7:193–219.

183. Shaw, L. 1935. Intercellular humidity in relation to fireblight susceptibility in apple and pear. Cornell Univ. Agr. Exp. Sta. Mem. 181:1–40.

184. Shaw, M. 1963. The physiology and host-parasite relations of the rusts. Ann. Rev. Phytopath. 1:259–94.

185. Shaw, M., S. A. Brown, and D. Rudd-Jones. 1954. Uptake of radioactive carbon and phosphorus by parasitized leaves. Nature 173:768.

186. Shaw, M., and D. J. Samborski. 1956. The physiology of host-parasite relations. I. The accumulation of radioactive substances at infections of facultative and obligate parasites including tobacco mosaic virus. Can. J. Bot. 34:389–405.

187. Siegel, A., M. Zaitlin, and O. P. Sehgal. 1962. The isolation of defective tobacco mosaic virus strains. Proc. Nat. Acad. Sci. 48:1845–51.

188. Sijam, K., R. N. Goodman, and A. L. Karr. 1985. The effect of salts on the viscosity and wilt inducing capacity of the capsular polysaccharide of *Erwinia amylovora*. Physiol. Plant Path. 26:231–39.

189. Spencer, D. F., and W. C. Kimmins. 1969. Presence of plasmodesmata in callus cultures of tobacco and carrot. Can. J. Bot. 47:2049–50.

190. Stapp, C. 1936. Der pflanzenkrebs und sein erriger *Pseudomonas tumefaciens*. IV. Eine neue Wirtspflanze (*Dahlia variabilies* Desf.) mit hochvirulentem Erreger. Zentralb. Bakt. 95:273–82.

191. Stapp, C., H. Muller, and F. Dame. 1938. Der pflanzenkrebs und sein Erreger *Pseudomonas tumefaciens*. VII Mitt. Untersuchen über die Möglichkeit einer wirksamen bekamfung on Kernobstgeholzen Zentralb. Bakt. II 99:210–76. (Abstr.)

192. Struckmeyer, B. E., C. H. Beckman, J. E. Kuntz, and A.

J. Riker. 1954. Plugging of vessels by tyloses and gums in wilting oaks. Phytopathology 44:148–53.

193. Strugger, S. 1938–39. Die lumineszenzmikroskopische Analyse des Transpirationsstromes in Parenchymen. Flora, Jena, 133:56–68.

194. Suhayda, C. H., and R. N. Goodman. 1981. Infection courts and systemic movement of ^{32}P-labeled *Erwinia amylovora* in apple petioles and stems. Phytopathology 71:656–60.

195. Suhayda, C. H., and R. N. Goodman. 1981. Early proliferation and migration and subsequent xylem occlusion by *Erwinia amylovora* and the fate of its extracellular polysaccharide (EPS) in apple shoots. Phytopathology 71:697–707.

196. Takahashi, T. 1974. Studies on viral pathogenesis in plant hosts. VI. The rate of primary lesion growth in the leaves of 'Samsun NN' tobacco to tobacco mosaic virus. Phytopath. Z. 79:53–66.

197. Talboys, T. W. 1978. Dysfunction of the water system. In: Plant disease. An advanced treatise. J. G. Horsfall and E. B. Cowling (eds).Academic Press, New York, 2:141–62.

198. Thrower, L. B. 1965. On the host-parasite relationships of *Trifolium subterraneum* and *Uromyces trifolii*. Phytopath. Z. 52:269–94.

199. Thrower, L. B. 1965. Host physiology and obligate fungal parasites. Phytopath. Z. 52:319–34.

200. Tinklin, R. 1970. Effects of aspermy virus infection on the water status of tomato leaves. New Phytol. 69:515–20.

201. Tu, J. C. 1977. Effects of soybean mosaic virus infection on ultrastructure of bacteroidal cells in soybean root nodules. Phytopathology 67:199–205.

202. Turner, N. C., and A. Graniti. 1969. Fusicoccin: A fungal toxin that opens stomata. Nature 223:1070–71.

203. Tyree, M. T. 1970. The symplast concept—a general theory of symplastic transport according to the thermodynamics of irreversible proscesses. J. Theor. Biol. 26:181–214.

204. Uppal, B. N. 1934. The movement of tobacco mosaic virus in leaves of *Nicotiana sylvestris*. Indian J. Agr. Sci. 4:865–73.

205. Van Alfen, N., and V. Allard-Turner. 1979. Susceptibility of plants to vascular disruption by macromolecules. Plant Physiol. 63:1072–75.

206. Vela, A., and P. E. Lee. 1975. Infection of leaf epidermis by wheat striate mosaic virus. J. Ultrastruc. Res. 52:227–34.

207. Wadliegh, C. H., and H. G. Gauch. 1948. Rate of leaf elongation as affected by the intensity of the total soil moisture stress. Plant Physiol. 23:485–95.

208. Waggoner, P. E., and A. E. Dimond. 1954. Reduction in water flow by mycelium in vessels. Amer. J. Bot. 41:637–40.

209. Walkey, D. G. A. 1980. Production of virus-free plants by tissue culture. In: Tissue culture methods for plant pathologists.D. A. Ingram and J. P. Helgeson (eds.). Blackwell, Oxford, pp. 109–17.

210. Walkey, D. G. A., and M. J. W. Webb. 1970. Tubular inclusion bodies in plants infected with viruses of the Nepo type. J. Gen. Virol. 7:159–66.

211. Wallis, F. M. 1975. A histopathological study on selected bacterial vascular diseases with emphasis on ultrastructure. Ph.D. diss. Univ. of Natal, Pietermaritzburg, S.A.

212. Wallis, F. M. 1977. Ultrastructural histopathology of tomato plants infected with *Corynebacterium michiganense*. Physiol. Plant Path. 11:333–42.

213. Wallis, F. M., and S. J. Truter. 1978. Histopathology of

tomato plants infected with *Pseudomonas solanacearum*, with emphasis on ultrastructure. Physiol. Plant Path. 13:307–17.

214. Wallis, F. M., F. H. J. Rijkenberg, J. J. Joubert, and M. M. Martin. 1973. Ultrastructural histopathology of cabbage leaves infected with *Xanthomonas campestris*. Physiol. Plant Path. 3:371–78.

215. Watson, M. A. 1955. The effect of sucrose spraying on symptoms caused by beet yellows virus in sugar beet. Ann. Appl. Biol. 43:672–85.

216. Watson, M. A., and D. J. Watson. 1951. The effect of infection with beet yellows and beet mosaic viruses on the carbohydrate content of sugar beet leaves, and on translocation. Ann. Appl. Biol. 38:276–88.

217. Weatherley, P. E. 1963. The pathways of water movement across the root cortex and the leaf mesophyll of transpiring plants. In: The water relations of plants. A. J. Rutter and R. H. Whitehead (eds.). Blackwell, London, pp. 85–100.

218. Weatherley, P. E., A. J. Peel, and G. P. Hill. 1959. The physiology of the sieve tube. J. Exp. Bot. 10:1–16.

219. Weintraub, M., H. W. J. Ragetli, and E. Leung. 1976. Elongated virus particles in plasmodesmata. J. Ultrastruc. Res. 56:351–64.

220. Weintraub, M., H. W. J. Ragetli, and E. Lo. 1974. Potato virus Y particles in plasmodesmata of tobacco leaf cells. J. Ultrastruc. Res. 46:131–48.

221. Welkie, G. W., and G. S. Pound. 1958. Temperature influence on the rate of passage of cucumber mosaic virus through the epidermis of cowpea leaves. Virology 5:362–71.

222. White, J. C., and N. L. Horn. 1965. The histology of tabasco peppers infected with tobacco etch virus. Phytopathology 55:267–69.

223. Wieringa-Brants, D. H. 1981. The role of the epidermis in virus-induced local lesions on cowpea and tobacco leaves. J. Gen. Virol. 54:209–12.

224. Worley, J. F., and I. R. Schneider. 1963. Progressive distribution of southern bean mosaic virus antigen in bean leaves determined with a fluorescent antibody stain. Phytopathology 53:1255–57.

225. Wynd, F. L. 1943. Metabolic phenomena associated with virus infection in plants. Bot. Rev. 9:395–465.

226. Yarwood, C. E. 1947. Water loss from fungus cultures. Amer. J. Bot. 34:514–20.

227. Yarwood, C. E., and L. Jacobson. 1955. Accumulation of chemicals in diseased areas of leaves. Phytopathology 45:43–48.

228. Zaitlin, M., D. A. Leonard, T. A. Shalla, and L. J. Petersen. 1981. A possible function for the 30,000 MW protein coded by TMV RNA. Vth Int. Virol. Cong., Strasbourg, p. 254. (Abstr.)

229. Zech, H., and L. Vogt-Kohne. 1956. Untersuchungen zur Reproduktion des Tabakmosaikvirus I. Elektronmikroskopische Beobachtungen. Exp. Cell Res. 10: 458–75.

230. Zimmermann, M. H. 1974. Long distance transport. Plant Physiol. 54:472–79.

231. Zimmermann, M. H., and C. L. Brown. 1971. Trees: Structure and function. Springer-Verlag, Berlin. (3d printing, 1977.)

232. Zimmermann, M. H., and J. McDonough. 1978. Dysfunction in the flow of food. In: Plant Disease: An advanced treatise. J. G. Horsfall and E. B. Cowling (eds.). Academic Press, New York, 3:117–40.

9

Toxins

General Characteristics and Classification

The success of the pathogen in becoming established in a host or causing symptoms may depend on the action of antimetabolites secreted by the pathogen. Some of these pathogenic products have been designated as toxins. In fact, the host-parasite relationship and the development of disease syndrome of the plant are the result of the interplay of several factors induced or produced by the pathogen. Thus, pathogens may release cell-wall-degrading enzymes, growth hormones or, in one instance, genetic determinants (e.g. *Agrobacterium tumefaciens*). These metabolites may also destroy growth hormones enzymatically, block the water-conducting vessels physically, or produce chemical substances (antimetabolites) that perturb specific plant function. The products of the pathogen that adversely affect plant function in "physiological" concentrations and do not have enzyme, hormone, or nucleic acid character are called *plant-pathogenic toxins* (or phytotoxins).

The early studies of Gottlieb (63), Dimond and Waggoner (34), Braun (16), and others clearly demonstrated that plant pathogens are capable of producing metabolites (exotoxins) in culture that cause symptoms similar to those found in natural infections. However, our knowledge is limited regarding the role of endotoxins (which are liberated from the damaged pathogen or only after the death of the pathogen) in the disease syndrome (cf. 40, 43, 94). In a few cases the mechanism of action and the chemical structure of toxins have been elucidated (cf. 37, 199). One cannot assume, however, that all plant-pathogenic microorganisms cause disease by means of toxins. For example, the production and role of toxins in diseases caused by biotrophs (rusts, mildews, powdery mildews, etc.) is questionable at present.

Classification

It is appropriate to differentiate between host-specific and nonhost-specific toxins, and this grouping will be followed in this book (232). There are disagreements in the usage of specific words for well-defined concepts. Thus, the terms *toxin*, *toxic substance*, *poison*, etc., may have different meanings in the literature. We submit some definitions that appear to have a degree of consensus.

Poison is a general term for any chemical substance that, in relatively small amounts, may cause damage to structure or disturbance of function of the living organism. Poisons are usually chemicals of nonliving origin and may affect both the plant and the pathogen.

Toxic substances are chemicals of plant or microbial origin that, in relatively small amounts, may cause damage to organisms other than the producer. For example, some phenolic glycosides of plant origin are "toxic" to some pathogens, and in some instances this may be the basis for the preformed resistance of plants to pathogens. Similarly, the pathogenically or chemically induced "phytoalexins" are also toxic to a series of microorganisms. When the toxic substance is produced by a higher plant and is effective against other species of higher plants, we are faced with the phenomenon of *allelopathy*. The chemicals produced by plant pathogens that are "toxic" to higher plants are called *toxins* (phytotoxins or plant-pathogenic toxins).

Phytotoxins (or plant-pathogenic toxins) are produced by plant-pathogenic *fungi* or *bacteria*; they adversely affect plant *function*; they do not attack structural integrity of cell-wall-related materials; they are usually but not always of low molecular weight and are *mobile*; and are active at low (physiological) concentrations; and they do not have *enzymatic*, *hormonal*, or *genetic* (nucleic acid) character.

Phytoaggressins are enzymes, hormones, or genetic determinants produced by plant pathogens that cause damage to structure or otherwise disturb plant function.

Phytoncide is a misleading expression for chemical substances produced by higher plants that damage or disturb the function of microorganisms. (See also the definition for *toxic substances*.)

Antibiotic is a substance produced by microorganisms that adversely affect other microorganisms.

Vivotoxin (34). The toxic principle is isolated from the diseased plants in reproducible experiments, purified or characterized chemically, and when applied to the healthy plant induces at least a portion of the disease syndrome. The above requirements would resemble Koch's postulates.

Pathotoxin (230). This term is widely used and is practically equivalent to *host-specific toxin* (see below). Pathotoxins are toxins that have a causal role in disease.

Primary determinant of pathogenicity (168). Equivalent to *host-specific toxin* (see below).

Secondary deterimant of pathogenicity. Equivalent to *non-host-specific toxin* (see below).

Host-specific toxins fulfill the following requirements:

1. The pathogen and the toxin exhibit similar *host specificity*. Resistance or susceptibility of the host plant to the pathogen parallels insensitivity or sensitivity, respectively, to the toxin.

2. The *virulence* of the pathogenic strains varies with their capacity to produce the toxin.

3. The toxin is able to produce, in susceptible plants, *symptoms* that are characteristic of the disease.

Thus, host-specific toxins may have a role in the establishment of the pathogen in the host plant.

Non-host-specific toxin. In this case susceptibility of the host plant to the pathogen does not parallel sensitivty to the toxin. The toxin does not have a role in the establishment of the pathogen in the host; however, it is able to induce some characteristics of the disease syndrome.

Exotoxin (extracellular toxin) refers to toxins that diffuse from the living pathogen. Extracellular toxins have been most thoroughly investigated in plant pathology (232).

Endotoxin (intracellular toxin) is not liberated from the cells of the pathogen until the fungus or bacterium is damaged or dies. The importance of endotoxins is more apparent in medical pathology. Few reports have recorded the role of endotoxins in plant disease.

Chemically, plant pathogenic toxins may be glycosides, amino acid derivatives, peptides, terpenoids, sterols, pyridines, quinones, etc. In fact, in only a few cases is the precise chemical nature of plant-pathogenic toxins known. Most of the toxins have, however, been partially characterized chemically (cf. 97, 108, 118, 190).

Virus-Coded Toxins

There is no unequivocal evidence for the involvement, either directly or indirectly, of virion components or of virus-coded polypeptides in the expression of symptoms in susceptible hosts. There are, however, important precedents; many cytopathic animal viruses inhibit host cell protein synthesis at an early stage of infection (7).

Vaccinia virus, for example, causes a shut-off of host protein synthesis in the first few minutes after virus infection at a high multiplicity. This inhibition occurs in Ehrlich ascites tumor cells in the presence of actinomycin D and cycloheximide, and also in the presence of cordycepin (3'-deoxyadenosine), which suppresses both viral transcription and viral poly-A synthesis (and therefore viral protein synthesis). The effect is also observed in vitro in cell-free extracts and in rabbit reticulocyte lysates, and it is suggested that it is mediated by a component of the virion, which may be phosphorylated in vivo and which irreversibly inhibits the initiation of protein synthesis (145, 146, 161, 169).

Similarly, poliovirus selectively inhibits translation of host messengers in tissue culture cells (157, 216). In uninfected cells, a cap-binding protein interacts with the 5'-end of host messenger RNAs, which possess a 5'-cap structure (see Chap. 5) and participates in the formation of the initiation complex with 40s ribosome subunits. It is suggested that poliovirus infection leads to an inactivation of the binding protein, an inactivation that at least contributes to the shut-off of host protein synthesis (216). It is presumably mediated by a virus protein. Since poliovirus does not possess a 5'-cap structure, translation of viral RNA is not affected by inactivation of cap-binding protein. One or more components of the virion of frog virus 3 also appears to mediate the early inhibition of translation of host mRNA (151).

There are indications that capsid protein of TMV (87) or CMV (156) may contribute to the expression of symptoms in tobacco, though results are inconclusive. It is possible, however, that either a plant virion component or a virus-coded polypeptide is able to interact with a plant cell and cause significant changes in its metabolism. We still know very little about virus-coded polypeptides and their function, and it would be premature to conclude, from the lack of evidence for such interactions, that they do not exist.

Toxins Produced by Bacteria

It is apparent that a number of bacterial plant pathogens cause disease by producing toxic substances. Whether the presumption by Gäumann (50) that all plant disease is the result of toxins is correct will depend on one's definition of a toxin. In general, their effects upon host tissue include blockage of water-conducting vessels, damage to the plasmamembrane (45, 56) and interference with specific facets of host metabolism wherein the toxin acts as an antimetabolite. Owens (133) has defined a bacterial toxin as a nonenzymic substance that injures plant cells or disrupts their metabolism. In recent years the specific chemical structure of some toxins has been established, as well as their modes of action. In one instance the activity of a macromolecule has been duplicated by a mere fragment of its parent polymer (173). Attention is given here to a broad range of toxins, some more intensively studied and understood than others, that affect parasitism by plant pathogenic bacteria.

Chlorosis-Inducing Toxins

The chlorosis-inducing toxins produced by plant pathogenic bacteria were among the first detected and have been the most intensively studied. The characteristic yellow halo that develops around the necrotic lesion produced by the pathogen early attracted the attention of pathophysiologists. Perhaps the apparent damage to the photosynthetic machinery in the affected tissues offered a more interesting feature of pathogenesis to study than necrosis. This situation remains, as we shall see in our summaries of the effects of the toxins produced by several pseudomonads (177, 211).

Wildfire toxin. Pseudomonas tabaci produces a toxin that causes foliar chlorosis in the form of a halo around a necrotic lesion. Woolley et al. (234) were first to isolate this toxin in relatively pure form. In initial studies of the so-called wildfire toxin, Braun (17) suggested that the toxin exerted its effect as an antimetabolite of methionine. Support for this theory came from the discovery that a structural analog of methionine, methionine sulfoximine (MSO), induced a chlorotic halo in tobacco foliage similar to that caused by the toxin (109). Further, MSO is toxic to *Chlorella vulgaris*, and, like the toxin, this toxicity could be reversed by methionine. Curiously, the effects on tobacco by either toxin or MSO were not reversible with methionine. As a consequence, the question of the mode of action of the wildfire toxin was reexamined by Sinden and Durbin (176). They found that MSO was an antimetabolite of glutamic acid, thus casting serious doubt on the original theory of Wooley et al. Sinden and Durbin's experiments indicated that the wildfire toxin and MSO both inhibited glutamine synthetase (GS), which converts glutamic acid to glutamine. The newer hypothesis is supported by the observation that tobacco leaves infiltrated with either toxin or MSO plus glutamine do not develop chlorosis. However, subsequent studies by Durbin (36, 37) with the purified toxin and purified pea GS failed to disclose inhibition of the enzyme by the toxin, whereas MSO was effective in this respect.

If inhibition of GS, a key enzyme in the principal pathway of nitrogen assimilation, is not the target of the toxin, what then is responsible for the inordinately large accumulation of ammonia in the chlorotic tissue? A causal relationship between ammonia and chlorosis has been suggested by Turner (218), who reported that the inoculation of tobacco with either purified tabtoxin or 2.8×10^7 cells/ml of *P. tabaci* quickly results in a decrease in GS and an accumulation of ammonia (Table 9–1). To support this contention Turner cites loss of GS activity as the earliest detected physiological change in

tabtoxin-treated tissue, followed by a measurable increase in ammonia. The accumulating ammonia could be accounted for quantitatively by the loss of GS activity.

Turner and Debbage (219) then showed that both chlorosis and necrosis caused by the pathogen were associated with the accumulation of ammonia. In addition, ammonia production by tabtoxin-treated tissue ceased immediately when the tissue was darkened or when tissue was treated with either DCMU or excess CO_2. The authors concluded that inhibition of GS by tabtoxin interrupts the photorespiratory-N cycle and the symptoms are associated with an intermediate of this cycle, which is ammonia (29). Hence Turner and Debbage concluded that tabtoxin caused both chlorosis and necrosis in tobacco leaves and that these symptoms reflected the accumulation of ammonia rather than the depletion of the product of GS, glutamine.

The currently accepted structure of the *P. tabaci* toxin was established by Stewart (189), who suggested that it was composed of tabtoxinine and threonine, and this structure was subsequently verified by Taylor et al. (213) and Lee and Rapoport (101). Stewart deduced the structure of the purified threonine analog by infrared, NMR, and mass spectrometry. Structure of the toxin is shown in Fig. 9–1. The bioactive component is tabtoxinine-μ-lactam[2-amino-4-(3-hydroxy-2-oxo-azacyclobutan-3-yl)-butanoic acid linked to either threonine or serine via a peptide bond. Hydrolyses of the amino acids by an enzyme produced by either the plant or some isolates liberates the bioactive lactam form. It is, according to Durbin and his associates, tabtoxinine-β-lactam that inhibits GS, not tabtoxin. This is accomplished by irreversible binding of the toxin to the enzyme.

A study by Sinden and Durbin (178) revealed that a chlorosis-inducing toxin produced by *Pseudomonas coronafaciens* was identical to that produced by *P. tabaci*. Subsequently, Taylor et al. (212, 213) reported that a third *Pseudomonas* species pathogenic to timothy also produces this same toxin (211). These studies confirmed the study of Stewart (189) and characterized all three toxins as a tabtoxinine lactam linked to either threonine or serine. In most instances, both analogs are present; however, the former predominates.

Since at least three pseudomonads produce the same toxin complex, the term *wildfire* is no longer applicable;

Table 9–1. Glutamine synthetase and ammonia levels in *Nicotiana tabacum* leaf tissue infiltrated with *P. tabaci* (after Turner, 218).

Time after infiltrating[a] leaves	Glutamine synthetase (μmol min^{-1} cm^{-2} leaf)	Ammonia (μmol cm^{-2} leaf)
0[b]	0.084	0.081
6	0.102	0.059
12	0.102	0.065
18	0.108	0.077
24	0.116	0.053
30	0.091	0.069
36	0.047	0.063
48	0.008	0.074

a. A suspension of 2.8×10^7 cells *P. tabaci* ml^{-1} was infiltrated into an area of leaf approximately 6 cm^2.
b. Samples taken at each interval comprised 6 disks, 1.5-cm diameter, cut from each of 6 leaves on 3 different plants.

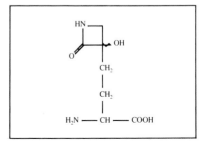

Fig. 9–1. Tabtoxinine-β-lactam ("tabtoxin").

hence the Durbin group has suggested the terms *tabtoxin* for the threonine-containing compound and *2-serine-tabtoxin* for the minor active analog. Patil (139) has referred to the biologically active mixture of the two for the sake of simplicity as *tabtoxin*. The question is raised here as to the nature of pathogenesis engendered by the nontabtoxin-producing isolates of *P. tabaci*, historically referred to as *P. angulata* (23). These certainly cause severe symptoms in the field, particularly when the inoculum is forced into leaf stomates during driving rains. Under these conditions large angular leaf spots or whole leaves become necrotic.

Phaseolotoxin. Pseudomonas phaseolicola, the causal organism in the halo blight disease of bean, also produces an extracellular toxin. Jensen and Livingston (86) described strains of this pathogen that are capable of producing the halo effect and others that are not. It will be recalled that *P. tabaci* and *P. angulata* are allegedly distinguishable from each other only by virtue of the fact that the former elaborates the halo-inducing tabtoxin. However, the halo-type isolate of *P. phaseolicola* also caused symptoms in the bean plant such as marked stunting, vein clearing, wilting, and premature death as a consequence of systemic infection. The "haloless" types, on the other hand, caused small leaf lesions and little or no stunting, vein clearing, or premature death. According to Jensen and Livingston (86), the more virulent strains are those able to produce the halo-inducing toxin.

Waitz and Schwartz (224) demonstrated the presence of a toxin in culture filtrates of *P. phaseolicola*, and Patel and Walker (137) presented evidence that the toxin influenced the amino acid metabolism of bean plants. As a consequence of pathogenesis, a number of amino acids accumulate, particularly ornithine.

In the experiments of Waitz and Schwartz (224), foliar inoculation of bean plants with *P. phaseolicola* resulted in the development of halo symptoms around the point of inoculation and an intense vein clearing (mosaic pattern) in the trifoliate leaves above those that were inoculated. The mobility of this toxin is indicated by the fact that the leaves that showed vein clearing were frequently free of the bacterium (238).

Hoitink et al. (78, 79) partially purified the toxin and found it to be thermo-stable, nonspecific, and, like tabtoxin, capable of inducing chlorosis in nonhost tissue.

A series of studies by Patil (138) and Patil et al. (140) reported the toxin to be a peptide containing two unknown amino acids as well as serine, glycine, and valine. They contended that mode of action was, as originally noted by Patel and Walker (137), due to the dysfunction of the ornithine cycle.

Rudolph and Stahmann (159) confirmed the observation that ornithine accumulates in halo-blight-infected plants. As ornithine is involved in the formation of citrulline and arginine, it is understandable why the studies of Patel and Walker, as well as Rudolph and Stahmann, revealed lowered levels of arginine in chlorotic tissues of infected plants. With these observations in mind,

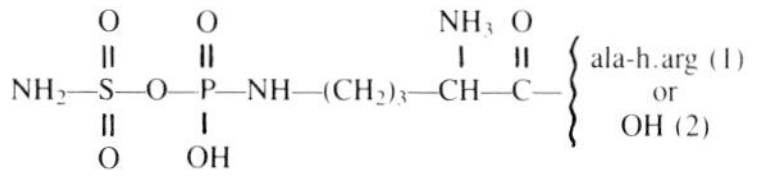

Fig. 9–2. Phaseolotoxin and Octicidin (see Additional Citations and Comments, p. 339).

Patil and coworkers examined the effect of the *P. phaseolicola* toxin on ornithine carbamoyltransferase (OCT), the enzyme that catalyzes the carbamylation of ornithine to citrulline. Substrate for the enzyme OCT is carbamoylphosphate (CAP).

Tam and Patil (205) studied the kinetics of this enzyme and observed that the toxin, which they named *phaseotoxin* (139), induced allosteric competitive inhibition vis-à-vis CAP. It is, however, noncompetitive with respect to ornithine, and the toxin influences the CAP site of the OCT molecule. Apparently, the induction of chlorosis does not involve irreversible structural changes in bean cell organelles (141). (See also Chap. 2.)

Subsequently, this chlorosis-inducing toxin was found to be a tripeptide by Mitchell (117, 118), its structure was fully elucidated as (N^8-phosphosulfamyl) ornithylalanylhomoarginine, and it was given the trivial name *phaseolotoxin*. This structure was subsequently confirmed by Staskawicz and Panopoulos (185). How the inhibition of OCT results in chlorosis remains to be clarified. However, both citrulline and arginine alleviate the chlorosis. Suggestions have been made that arginine deficiency is the important consequence of the toxic action. The net effect here could be reduced synthesis of proteins and chlorophyll (205). Several studies have linked toxicity to aberrant RNA synthesis based on a failure of affected host cells to synthesize orotic acid from carbamayl-L-aspartate, which uses citrulline as a carbamoylphosphate donor (102, 139). Rudolph and Stahmann (159) presented evidence that the effect of the toxin on *Euglena gracilis* could be negated by citrulline and partially by orotic acid.

Rhizobitoxine. Rhizobium japonicum, which fixes nitrogen in the nodules of roots of soybean, is generally nonpathogenic to its host. However, some strains of this bacterium are indeed toxigenic. They simultaneously fix nitrogen in normal fashion and synthesize a toxin that induces chlorosis in the new leaf growth of soybean (135). Rhizobitoxine, like the other halo-inducing toxins, has been shown to be nonhost specific.

According to Owens (133), low concentrations of the toxin cause all the disease symptoms, and toxin production in the nodule is essential to disease expression. Hence Owens et al. (135) classified rhizobitoxine as a primary disease determinant. The toxin has been isolated from chlorotic leaves and nodules of infected plants and has been recovered from culture filtrates. Of additional interest is the discovery that soybean vari-

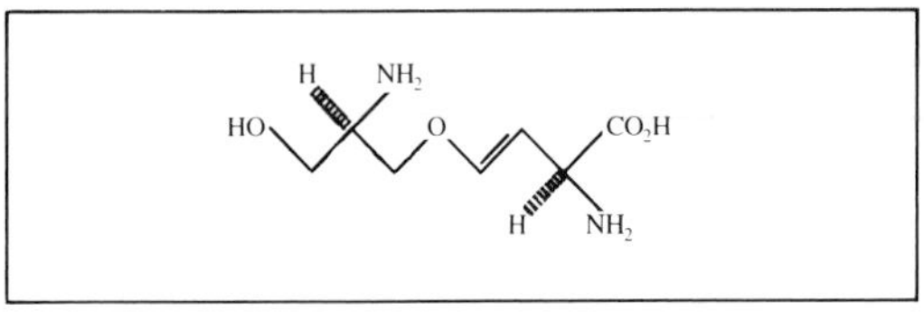

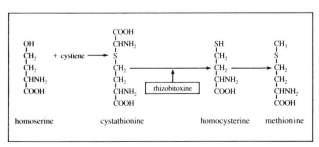

Fig. 9–3. Rhizobitoxine.

eties resistant to toxigenic strains of *Rhizobium* require four times as much rhizobitoxine to develop foliar chlorosis (135).

Owens and coworkers (134, 136) defined the structure of rhizobitoxine as 2-amino-4-(2-amino-3-hydroxypropoxy)-*trans*-but-3-enoic acid (Fig. 9–3). Owens and his associates studied the mode of action of rhizobitoxine and focused on its influence on methionine biosynthesis. Their studies revealed that the toxin blocked the methionine biosynthetic pathway. Specifically, the blockage occurs at the conversion of cystathionine to homocysteine (Fig. 9–4). Apparently, rhizobitoxine exerts its influence specifically upon β-cystathionase.

The manner in which the toxin causes chlorosis has been studied by Giovanelli et al. (53, 54) in spinach and corn seedlings. Their data, however, are inconclusive with respect to the presumed mode of action in soybean. The toxin, when applied to either of these two test plants, inhibits cystathionase only partially. Although the enzyme caused a 22- fold increase in cystathionine, the accumulation of radioative methionine was slight, and this suggests the participation of another pathway. For example, the direct sulfur hydration pathway, which bypasses the cystathionine-homosyteine-methionine pathway, may be affected. Hence, if the sulfur hydration pathway is operative, it may explain why toxin-treated corn can escape the influence of the toxin.

The toxin is equally active against the enzyme extracted from either the bacterium *Salmonella typhimurium* or spinach, *Spinacia olerae*. A spinach seedling develops chlorosis when 1 μg of toxin is added to its root-bathing solution (133). Since the toxin blocks the active site on the enzyme for cystathionine synthesis, it would appear that there is competition between cystathionine and rhizobitoxine for the active site on the enzyme.

Wilt-Inducing Bacterial Polysaccharides

The release of polysaccharides by phytopathogenic bacteria into their external environment, in vitro and in vivo, has been linked to the virulence and toxigenicity of a number of bacterial species (158). These organic polymers are commonly referred to as *capsules* if they are relatively compact and well defined. However, the term *slime* or *slime layer* has been used when the polysaccharide is loosely held and diffuse. The diffuse nature of the extracellular polysaccharidal slime (EPS) that surrounds *Erwinia amylovora* and the tightly bound capsule (147) is depicted in Fig. 9–5. The discussion here will be concerned primarily with the loosely organized EPS of a number of pathogens, which is released into the culture medium or into the host's tissues.

Relationship of EPS to virulence. The production of EPS is a hereditary characteristic (35) that a particular strain may lose and in some instances regain. Its production is also strongly influenced by the environment in which the bacterium is growing. For example, increasing the concentration of glucose or varying the carbon source may result in significant increases in polysaccharide production. According to Anderson and Rogers (1), the addition of electrolytes such as chlorides and sulfates of Na^+, K^+, NH_4^+ and Mg^{++}, at concentrations up to 0.4–0.5 M, potentiated EPS production markedly in a number of enterobacteria.

It appeared, from their study, that the osmotic concentration of the bacterium's environment plays an important role in activating "the slime-production pathway." Hence it seems that EPS production is a reflection of both the *genetic* constitution of the bacterium and its *environment*.

If EPS production can be linked to pathogenesis, and we shall see that it has been, then the factors that control its production may be considered to control virulence. Thus it is possible to relate virulence of the bacterium to both its inherent capacity to produce EPS and the nutritional qualities of the host as substrate for EPS production (57, 58).

A causal relationship between EPS production and virulence was established by Corey and Starr (27) in bean plants inoculated with *Xanthomonas phaseoli*. Four colony types of the pathogen—rough, smooth, semimucoid, and mucoid—produced, in the order listed, increasing amounts of polysaccharide in culture as well as greater size and numbers of lesions on artificially inoculated bean leaves (Table 9–2). The same order was maintained for colony size on agar. Since the numbers of bacteria detected in inoculated leaf tissue were the same for all variants, it was concluded that the dif-

Fig. 9–4. Pathway of methionine biosynthesis in *Salmonella typhimurium* and site inhibition by rhizobitoxine (after Owens, 133).

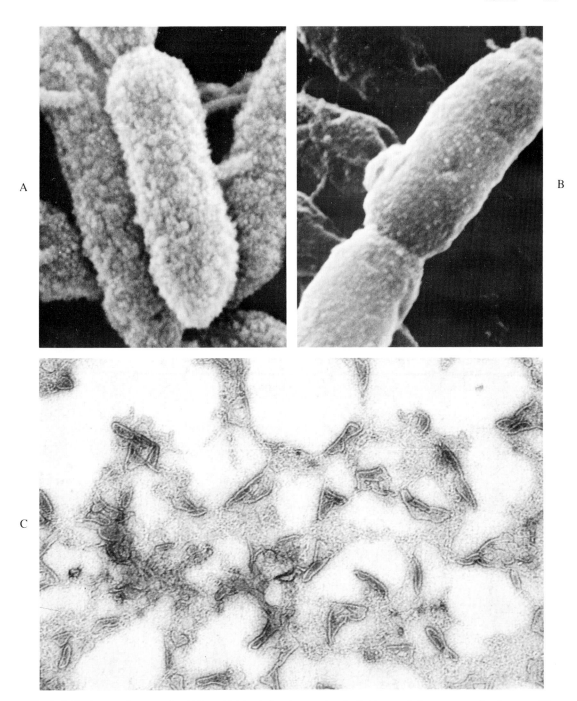

Fig. 9–5. *Erwinia amylovora* strains (a) E_9, which is virulent (encapsulated); (b) E_8, an avirulent unencapsulated mutant of E_9 (after Politis and Goodman, 147); and (c) extracellular polysaccharide (EPS) from E_9 negatively stained with phosphotungstic acid ($\times$ 228,000) (Goodman unpublished).

ferences in virulence, as judged by lesion size and number, were dependent upon the amount of polysaccharide produced. Corey and Starr also reported that heat-killed mucoid cells were also capable of inducing leaf lesions. However, studies by Roantree (155), concerned with the relationship of lipopolysaccharide (LPS) to virulence in *Salmonella typhimurium* and *Shigella dysenteriae*, suggest that qualitative dif-

Table 9–2. Comparison of four variants of *X. phaseoli* with respect to their ability to produce polysaccharide, multiply within the host, and produce symptoms in the host (after Corey and Starr, 27).

Type of colony	Polysaccharide in mg/ml of culture	Avg. no. of bact. per 44-mm disk of leaf surface	Avg. size of lesions as diameter in mm	Avg. no of lesions per cm^2 of leaf surface
Rough	4.40 ± 0.01	2×10^6	1.5 ± 0.1	6 ± 0.1
Smooth	4.53 ± 0.02	2×10^6	1.8 ± 0.1	7 ± 0.1
Semimucoid	4.70 ± 0.02	2×10^6	2.1 ± 0.1	19 ± 0.2
Mucoid	5.52 ± 0.06	2×10^6	2.5 ± 0.1	12 ± 0.1

ferences in the polysaccharide component of capsule are more closely linked to virulence than quantitative ones.

Variation in EPS production among three strains of *P. solanacearum* differing in virulence was also described by Husain and Kelman (83). Virulence measured by wilt-inducing potential of the three variants seemed also to reflect differences in polysaccharide production. The weakly virulent and avirulent strains produced little or no polysaccharide, formed smaller colonies in triphenyl tetrazolium chloride agar, and did not induce wilt (84). It was found that the relative viscosity of the culture filtrates was an excellent index of the polysaccharide production and wilt-inducing capacity (Table 9–3) of the three bacterial strains. Subsequent studies by Sequeira and Graham (171) and Husain and Kelman (84), have clearly linked EPS production by *P. solanacearum* and relative viscosities to virulence. However, as we shall see in studies concerning EPS from *E. amylovora*, the viscosity imparted to a solution by capsular polysaccharide is not necessarily an indicator of virulence (173).

Amylovorin

Wilting is often obscured as part of the disease syndrome associated with fireblight. The first clear symptom caused by *Erwinia amylovora* is the development of the "shepherds crook" at the apex of affected shoots followed by nontransient loss of turgor. This is followed by tissue browning, discharge of bacterial ooze from lenticels of infected stem tissue, and desiccation (73).

Goodman et al. (57) established a causal relationship

Table 9–3. Relative viscosity and wilt-inducing effect of culture filtrates of three strains of *P. solanacearum* (after Husain and Kelman, 84).

Culture filtrate	Relative viscosity[a]	Wilting index[b]
F-1 highly virulent	83.0	5
B-1 weakly pathogenic	50.4	0
B-2 avirulent	51.5	0

a. Relative viscosity expressed in terms of flow times in seconds in size 100 Fenske-Ostwald viscosimeter.
b. Average wilting index in 5 tomato cuttings after 12 h (0 = no wilting, 5 = complete wilting).

between a bacteria-free preparation of the polysaccharidal ooze, called amylovorin, and the wilt facet of pathogenesis. The MW of the "partially" purified polysaccharide, essentially a galactose-glucuronic acid-glucose polymer (Albersheim, personal communication), was initially estimated at 165,000 and reproduced the wilt symptom and ultrastructural modifications caused by the pathogen per se (57). The amylovorin molecule also contains ca. 15% pyruvate (Goodman, unpublished data). Further, the susceptibility of host species to the pathogen parallels their sensitivity to the toxin. Wilting is irreversible following a 2-h exposure of shoots to the toxin, and the effect of EPS is nonspecific.

In subsequent studies the MW of amylovorin was calculated to be 40×10^6 (5) and then 100×10^6 (172). In the latter study, data were accumulated that showed amylovorin, like *P. solanacearum* EPS, greatly increased the viscosity of solutions containing these polysaccharides. Sijam et al. (173) have used *E. amylovora* bacteriophage depolymerase enzyme preparations to depolymerize amylovorin and have reduced it from 10^8 to 2×10^4. The much maller polymer induced wilt of Jonathan apple shoots at concentrations comparable to the parent molecule but did not increase the viscosity of the solution (see Table 9–4).

Additional evidence of the relationship of virulence to toxigenicity is derived from studies with virulent (V) and avirulent (AV) strains of *E. amylovora* (5). Not only are acapsular (EPS deficient) strains of *E. amylovora* avirulent, but isolates that produce less amylovorin (EPS) are proportionally less virulent (Fig. 9–5).

Electron microscopic examination of tissue from Jonathan apple shoot apices revealed identical symptom development whether shoots were inoculated with the bacteria or exposed to the amylovorin. In both instances xylem parenchyma cells revealed intense plasmolysis. This symptom appeared 48 h following inoculation with bacteria and 4 h following exposure to 100 µg/ml amylovorin (204; see Fig. 8–17). In both bacteria-infected and toxin-treated shoots, plasmolysis was followed by disorganization of subcellular organelles. From these observations Huang and Goodman (81, 82) postulated that amylovorin exerts its primary effect upon the plasmalemma of xylem parenchyma. Following plasmolysis of the xylem parenchyma, the xylem

Table 9–4. Time in min. to wilt for three concentrations of EPS (amylovorin) and their respective φdp[c] produced by isolate E$_9$. Each number was derived from 6 shoots. A shoot was considered wilted when the first leaf from the cut end lost turgor (after Sijam et al., 173).

Conc. μ/ml	E$_9$-EPS[a] or E$_9$A[b]	E$_9$-EPS-φdp or E$_9$A-φdp
1,250	60	90
500	90	150
250	120	180

a. EPS derived from *E. amylovora* grown on solid medium.
b. EPS (amylovorin) derived from *E. amylovora* grown on green apple slices.
c. φdp = phage depolymerized EPS (MW ≃ 20,000).

vessels collapse and release their bacteria into intercellular space. The plasmolyzed xylem parenchyma shrink to form lysigenous cavities that become filled with bacteria (28, 59).

Several studies have attributed the wilt-inducing activity of amylovorin to its molecular size, which ostensibly causes vessel occlusion. Such vessel occlusion was shown by Suhayda and Goodman (204). Experiments by Van Alfen and associates on the absorption of dextrans by alfalfa shoots suggested that molecular size determines whether vessel occlusion is achieved and that the limiting MW approximated 2.5×10^5. It remains to be determined whether the 2×10^4 fragment of amylovorin causes wilt by vessel occlusion. It is possible that the smaller amylovorin polymer may be responsible for the oft-reported parenchyma plasmolysis-associated fireblight (173). Thus the wilt symptom may reflect plasmalemma damage rather than vessel occlusion per se (Fig. 9–6).

Other polysaccharide toxins. A polysaccharide was isolated from *Agrobacterium tumefaciens* culture filtrates by Hodgson et al. (75–77) and McIntire et al. (116) that induced wilt in tomato cuttings. Its causal relationship to pathogenesis is open to question, since both virulent and attenuated strains are able to produce the polysaccharide and wilting is not a part of the syndrome associated with crown gall. However, it may play a role in the recognition between the bacterium and host or perhaps in the ultimate plasmid transfer process. This seems increasingly plausible in light of the report by Matthysse (see Fig. 1–11) of the possible participation of bacteria-produced cellulose fibrils in the attachment process and the older finding of McIntire et al. (116) that the polysaccharide of *A. tumefaciens* was a glucose polymer.

Other phytobacterial pathogens have been shown to produce polysaccharides, e.g. *Xanthomonas hyacinthi,* *X. transluscens*, *X. maculifolium-gardeniae* (62, 179), *X. oryzae* (60, 61), and *X. phaseoli* (27). However, with the exception of *X. phaseoli* (27), their polysaccharides have not been associated with symptom induction.

It is apparent that the relationship between toxigenicity and virulence vis-à-vis plant pathogenic bacteria has been established in a number of instances. Nevertheless, the picture that emerges from the foregoing commentary suggests that these high MW toxic substances play more than an occasional role in pathogenesis. In some instances, e.g. wilt of chrysanthemum caused by *Erwinia chrysanthemi* (144) and wilt of solanaceous plant species by *P. solanacearum* (83), the capacity of the pathogen to also synthesize wall-degrading enzymes may obscure the relationship of the high MW compounds to the observed wilt.

The probability that *E. chrysanthemi* produces a toxin is supported by at least two pieces of experimental evidence. Garibaldi and Bateman (48) found that some isolates of the pathogen produce pectin lyase, and Pennypacker et al. (144) reported that only 30% of the bacterial population were competent in this respect. Furthermore, their histological observations failed to record good evidence for tissue maceration. The occurrence of collapsed and degraded xylem parenchyma cells and subsequent development of lysigenous cavities may be due to a plasmolytic influence of EPS (similar to amylovorin). These observed lysigenous cavities were devoid of bacteria that could be regarded as presumptive evidence of the participation of a toxin. An analagous situation has been reported by Main and Walker (107) for *E. tracheiphila*, although they discounted the possibility of a toxigenic effect. That interaction remains an ideal subject for additional study.

Glycopeptide Toxins

Several corynebacteria produce high MW wilt toxins that are predominantly carbohydrate in composition with a small peptide component (118, see Chap. 7). *Corynebacterium insidiosum* (alfalfa wilt) and *C. sepedonicum* (potato ring rot) have been reported to cause wilting via the production of toxic "polysaccharides" (183). However, subsequent studies by Strobel (192–94, 201) and Ries and Strobel (153, 154) characterized the toxic principles as glycopeptides. Rai and Strobel (152) isolated a family of three glycopeptides from culture filtrates of *C. michiganense*. All of these high MW compounds are nonspecific and at rather high concentrations impair the water economy of tomato cuttings.

Corynebacterium sepedonicum was reported by Strobel (199) to have an MW of 2,400 and was composed of 33.4% mannose, 18.8% glucose, 1.1% rhamnose, 0.7% arbinose, 3.8% ribose, 4.5% galactose, 9.9% 2-keto-3-deoxygluconic acid, and 4.6% peptide as well as several other new sugars. The oligosaccharide portion appeared to be linked glycosidically from mannose to the OH of threonine in the small peptide. The toxin of *C. insidiosum* is much larger, with an MW of 5

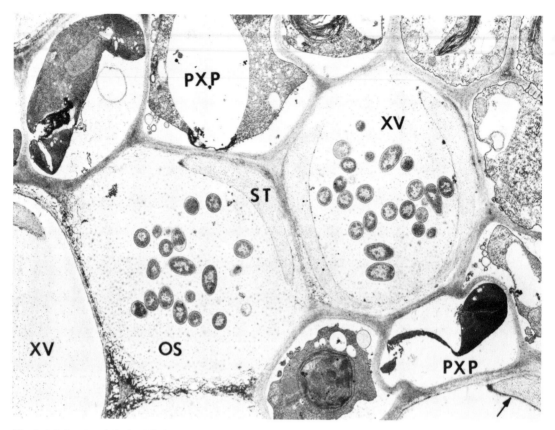

Fig. 9–6. Infected apple leaf petiole tissue 24 h after inoculation with 5×10^4 *E. amylovora* cells (E$_9$). Xylem vessels (XV) with secondary thickening (ST) are surrounded by 8 plasmolyzed xylem parenchyma cells (PXP). The vessels are partially occluded with bacteria and an occluding substance (OS) ($\times$ 2,600) (after Goodman and White, 59).

$\times$ 10^6 but essentially the same sugar residues and a proportionately smaller (2.5%) peptide component.

Peptide Toxins

Erikson and Montogomery (40) presented evidence that a proteinaceous *endotoxin* from culture filtrates of *Pseudomonas mors-prunorum* might be causally linked to the cankers in cherry and plum caused by this pathogen. Histological examination of diseased tissue by Erikson (38, 39) suggested that pathogenesis reflected the participation of bacterially produced substances that are toxic to host tissue. A degree of specificity was shown by culture filtrates of *P. mors-prunorum* in that they mimicked pathogen-induced symptoms in a susceptible plum variety whereas damage was negligible on a resistant variety. Furthermore, bacteria grown on media containing bark extracts from a resistant variety produced a more toxic filtrate than that grown on susceptible-bark media. The greater activity of older culture filtrates suggested the effect of metabolites resulting from bacterial autolysis. Unfortunately these experiments have never been pursued in greater detail.

Syringomycin. In 1968, DeVay et al. (9, 32) isolated an exotoxic polypeptide from *P. syringae* that had broad spectrum antibiotic activity (antimicrobial as well as phytotoxic) (174–75). Antimicrobial activity was exhibited against *Geotrichum candidum* at 24 µg/ml, and peach shoots exposed to 600 µg/ml developed phytotoxic symptoms. This *P. syringae*-produced toxin, designated syringomycin, may be similar to the substance described earlier by Erikson and Montogomery (40). The rationale for this stems from Erikson's data (38–40) that culture filtrates from *P. syringae* isolates induced a phytotoxic response in plum, but one less intense than that produced by filtrates from *P. mors-prunorum*.

Syringomycin causes water soaking, chlorosis, and necrosis in maize (67), which are symptoms the pathogen causes in this host. Linkage between symptom induction by the toxin taken up by plants and symptoms caused by the pathogen per se cannot as yet be categorically established, as syringomycin has not been re-isolated from the toxin-treated or inoculated plants. Gross et al. (68) have identified the toxin as a molecule with a strong positive charge from its constituent four amino acids, serine (2 moles), phenylalanine (1 mole), arginine (1 mole), and a lysinelike amino acid (2 moles). The toxin is believed to be a hexapeptide with an MW of ca. 1,253.

Strains of *P. syringae* pathogenic in *Citrus* spp. have been reported to produce syringotoxin rather than syringomycin. The biological activity of the two compounds is similar, both causing water-soaking chlorosis and necrosis on maize leaves and both inhibitory to *Geotrichum candidum*. The chemical composition of syringotoxin includes threonine, serine, glycine, ornithine, and a lysinelike amino acid similar to that detected in syringomycin.

Tagetitoxin. Mitchell et al. (119) have isolated, purified, and chemically characterized a monocyclic, heavily oxygenated phosphate ester toxin produced in marigold by *P. syringae* pv. *tagetis* and have given it the trivial name *tagetitoxin*. The pathogen causes necrotic leaf spots and chlorosis of the plant's apical leaves (217). Chlorosis is caused by the toxin, whose stereochemical structure has been elucidated (119) as $C_{11}H_{18}NO_{13}PS$ with an MW of 435.

It is apparent that bacterial pathogens have, through time, developed an array of toxic metabolites that participate in or are primary features of pathogenesis and symptom development. These toxins are biochemically diverse, and clearly their chemical composition and modes of action are far from understood. More attention must be paid to molecular configuration, relationship to receptor sites, and elicitor activities of these bacterial metabolites. Bacterial cell-host cell surface attractiveness and interactions must also be studied more intensively.

Toxins Produced by Fungi

Host-Specific Toxins

Helminthosporoside (HS-toxin). This host-specific toxin is the product of *Helminthosporium sacchari*, the causal organism of eye spot disease of sugarcane. It is a sesquiterpene glycoside, containing galactose. Its structure has been variously reported (13, 187, 188), its authenticity challenged (97), and it would seem to have been that at last purified by Macko et al. (106). The structure established by Macko et al. (106) appears below as three isomeric sesquiterpene aglycones, A, B, and C. The isomeric forms account for some of the earlier uncertainty of structure and the fact that nontoxic lower homologs also are produced by *H. sacchari* as tri-, di-, and monoglycosides. The MW of the toxin (isomers A, B, and C) is 844, and for tri-, di-, and monoglycoside the MWs established are 722, 560, and 398, respectively. It is noteworthy that galactose is unusually toxic to plants, and it is tempting to relate the effect of the toxin to that carbohydrate. It is interesting that helminthosporal, a toxin of *H. sativum*, is also a sesquiterpenoid.

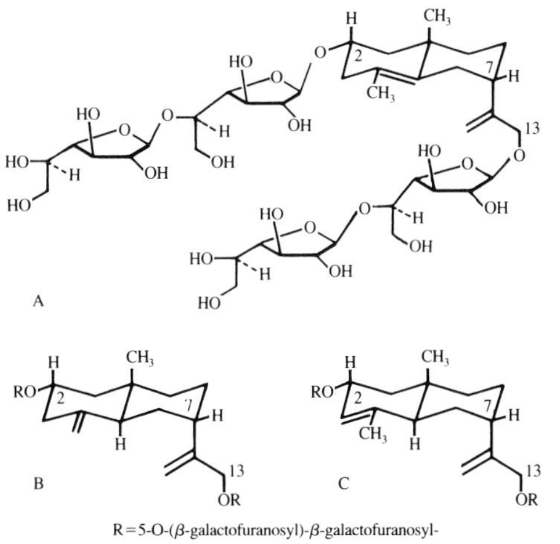

R = 5-O-(β-galactofuranosyl)-β-galactofuranosyl-

The HS-toxin produces reddish-brown streaks only on those clones of sugarcane that are susceptible to the fungus. From this it is clear that helminthosporoside plays a primary role in the production of that disease symptom. Infection by the fungus induces water-soaking symptoms and abnormalities in host cellular membranes, particularly of chloroplasts. The disintegration of the outer chloroplast membrane is the earliest cytological disturbance in the susceptible plant, and this is accompanied by a strong decrease in CO_2 fixation (19). However, as a result of treatment with the toxin, chloroplasts in resistant leaves remain generally unaltered (203). Only occasional chloroplasts are damaged, with ruptured membranes and separating lamellae. The cytological disturbances appear after the water-soaking symptoms and thus are probably secondary.

Regarding toxin sensitivity (susceptibility) and insensitivity (resistance) of different clones of sugarcane, it was purported that binding of helminthosporoside to membranes was the determining event. It was shown that membrane preparations (proteins) from clones insensitive to the toxin did not possess binding activity for helminthosporoside. On the other hand, sensitive clones do possess a membrane protein that recognizes and, therefore, binds the toxin (90, 91, 195, 196). Isotope labeling experiments have shown that the toxin also binds to host membranes in vivo. Different α-galactosides compete with helminthosporoside for binding sites and inhibit the binding of the toxin by host membrane preparations (Table 9–5). Similarly, they inhibit symptom expression in leaves after application of helminthosporoside. Thus, susceptibility to helminthosporoside is highly correlated with the ability to bind this toxin. Some of these data have been questioned by Wheeler (225) and Daly (30).

It was also reported that the binding protein and the susceptibility to toxin have been successfully transferred to protoplasts of an insensitive sugarcane and to

Table 9–5. Effects of various sugars and sugar derivatives on binding of helminthosporoside to membrane preparations of a susceptible sugarcane clone (after Strobel, 195).

Compound	Relative activity
None	100
Methyl-β-galactopyranoside	91
Methyl-α-galactopyranoside	52
Lactose	100
Glucose	100
Raffinose	40
Galactose	100
Melibiose	30
Sucrose	100

All tests were conducted at 10^3 M concentration of the treatment compound on both sides of the dialysis membrane. The standard assay for binding was conducted with 4.3 mg of membrane protein and helminthosporoside at 1.4×10^{-4} M.

tobacco protoplasts (200). The resistant clone of sugarcane contains a modified toxin receptor similar to the receptor from the susceptible clone. The "resistant receptor" does not bind the toxin in vivo because its binding activity is obscured by a glycolipid in the host plasmalemma (90, 91). It would seem that the binding activities of the two proteins are regulated by membrane lipids.

Some experimental results of Strobel (196) suggest that the binding protein is bound to the plasma membrane. The most important of these is that treatment of susceptible sugarcane leaves with an antiserum prepared against the pure binding protein afforded protection against the effects of helminthosporoside. Sugarcane cells and free protoplasts react with (^{3}H) antiserum prepared against the binding protein. Furthermore, the antiserum specific for the binding protein agglutinates sugarcane protoplast preparations. These data were believed to provide evidence for the presence of the toxin-binding protein on the host plasma membrane (202). The binding protein is involved in α-galactoside transport. Upon binding the toxin, this complex perturbs the membrane K^+ and Mg^{++} ATPase. Thus, there is an indirect effect on the ATPase via the toxin-binding protein complex (197, 198). Probably other membrane proteins are also influenced by the helminthosporoside binding protein, and one of the detrimental effects of the toxin to the host cell is an altered ion balance (increased permeability), which leads to necrosis.

Victorin (HV-toxin). This toxin, like the fungus *Helminthosporium victoriae* that produces it, is specific for certain oat cultivars. Thus, victorin affects only those oat cultivars susceptible to *H. victoriae*, which causes foot and root rot and leaf blight. High yields of toxin can be obtained from virulent isolates of the fungus, where-

as avirulent strains of *H. victoriae* fail to produce victorin.

Chemically, this toxin was suggested to be a polypeptide linked to a nitrogen-containing sesquiterpene (148). The latter moiety of the toxin is called *victoxinine*. It is non-host-specific and is produced by both virulent and avirulent strains of the fungus. The peptide moiety of victorin is nontoxic; it probably contributes to toxin specificity by its affinity for toxin receptor sites. However, the identity and cellular location of the toxin-sensitive site remain unresolved. It is not known whether the hypothetical receptor molecule occupies a site in the plasmalemma or within the cell (30, 71, 227, 228, 229). Further possibilities are that the toxin is inactivated in resistant (insensitive) tissues or that both sensitive and insensitive tissues are initially affected by victorin, but insensitive tissues possess a recovery or self-repair mechanism (229). This means host-selectivity instead of host-specifity.

Victorin causes some general changes in the physiology of the host that are common to infectious plant diseases (cf. 125). The earliest detectable effect of victorin on a susceptible plant is the increase in permeability of the plasma membrane to electrolytes (226). This occurs only a few minutes after exposure to the toxin (Fig. 9–7). However, the very rapid changes in permeability do not result in lethal effects in toxin-treat-

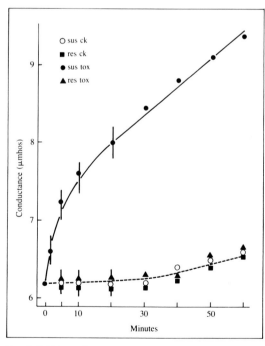

Fig. 9–7. Effect of *Helminthosporium victoriae* toxin on loss of electrolytes from cut tissue. sus: susceptible tissue; res: resistant tissue; ck: control; tox: toxin-treated; minutes refers to time after exposure of tissue to toxin (after Scheffer and Samaddar, 167).

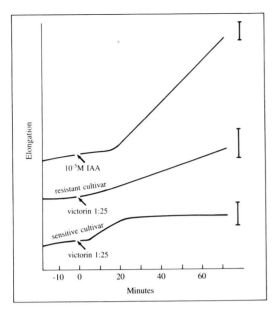

Fig. 9–8. Comparison of the timing and magnitude of growth promotion by IAA and victorin in oat coleoptile tissue. Upper curve, growth medium changed from buffer to 0.01 mM IAA at arrow (resistant cultivar of *Avena*). Middle and lower curves, growth medium changed from buffer to a 25-fold dilution of victorin. Middle curve, resistant cultivar of *Avena*. Lower curve, sensitive cultivar of *Avena*. The response to IAA is the same in resistant and sensitive cultivars (after Evans, 42).

ed cells (ll5). In addition, drastic changes in transmembrane potential, respiration, and ethylene production can occur without causing cell death.

Another early effect of victorin seems to be an auxinlike action: a rapid stimulation of cell elongation of *Avena* coleoptile segments (Fig. 9–8). The latent period in toxin-stimulated cell elongation is similar to that in toxin-induced electrolyte leakage (about four minutes). It would appear that the effects of victorin on cell growth and membrane permeability may both be mediated by some change in membrane properties (42, 160). Novacky and Hanchey (124) measured the transmembrane potential of individual cells in roots treated with victorin and found a depolarization of membrane potentials as a result of the toxic action of victorin (Fig. 9–9). However, this seems not to be in direct relation with the initial action of the toxin.

Susceptible protoplasts begin to burst almost immediately after exposure to victorin, whereas protoplasts from resistant cultivars are not affected (l62). Electron microscopic observations have provided evidence for selective toxicity of victorin to the cell's plasmalemma. The endoplasmic reticulum, nuclear envelope, and mitochondrial membranes were more resistant to damage caused by victorin (72, 105), as was the tonoplast (89).

One of the most striking secondary effects of victorin is the increased respiration of host tissues that is directly

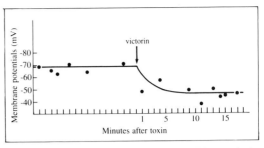

Fig. 9–9. Cell membrane potentials of susceptible roots in the absence and presence of victorin. Each point represents one measurement from one cell (after Novacky and Hanchey, 124).

proportional to the concentration of toxin applied (Fig. 9–10). However, the oxygen and phosphorus uptake by mitochondria from sensitive tissues is not affected by the toxin, indicating that the action on respiration is indirect. According to Wheeler and Hanchey (228), victorin causes a reduction in efficiency of the energy- generating system of mitochondria isolated from toxin-treated susceptible plants. A loss of respiratory control (the respiratory control ratio is calculated as the rate of substrate oxidation in the presence of suitable phosphate acceptor divided by the rate in the absence of such an acceptor) and a low ADP to oxygen ratio in mitochondria may result from host cell breakdown products. It is

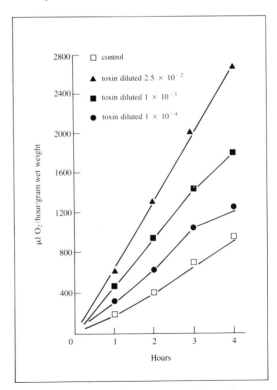

Fig. 9–10. Increase in respiration of susceptible oat tissue as a function of toxin concentration (after Krupka, 98).

not known whether mitochondria in intact diseased tissues are similarly affected by victorin.

The recorded increases in ascorbic acid oxidation as well as peroxidase activity and a general decrease in protein synthesis seem also to be secondary effects of the toxin.

Host sensitivity to victorin can be decreased by pretreatment of oat tissues with cycloheximide.It was proposed that a protein of the plasma membrane having a receptor character for the toxin is inhibited by cycloheximide (47). It is hypothesized that toxin-insensitive cells lack receptor sites, but there is no direct evidence. Furthermore, insensitivity to victorin seems to be a quantitative rather than a qualitative phenomenon. Higher concentrations of victorin were required to cause alterations in peroxidase isoenzymes in resistant oat leaves than in susceptible leaves (126). This contradicts the concept of a lack of receptors as a basis for insensitivity to the toxin. Similarly, there is no good evidence that insensitivity to victorin depends on a rapid inactivation of the toxin by the resistant tissue. Victorin acts on resting tissues as well as on metabolically active tissues. Oat seeds exposed to victorin and stored until the toxin was no longer detected did not germinate under conditions conducive to seed germination (163). The aleurone cells in half-seeds devoid of embryos were not activated by gibberellin for α-amylase synthesis when they were pretreated with victorin. Here, too, victorin acted on resting tissues, blocking synthesis of α-amylase, whereas aleurone exposed to water synthesized this enzyme upon being treated with gibberellin (Table 9–6).

Seeds of resistant plants were insensitive to both seed germination-inhibition influence and the aleurone-inhibiting action of the toxin. Consequently, it would seem that toxin insensitivity is a preformed character of the plant.

An additional argument for a passive, preformed characteristic of toxin insensitivity comes from the results of Nishimura and Scheffer (123), indicating that small, equal amounts of phytoalexins are produced in both susceptible and toxin-insensitive plants. In other words, phytoalexins (see Chap. 10) do not seem to play a role in the mechanism of resistance to H. victoriae or in the mechanism that makes the plant insensitive to the toxin.

Resistance or susceptibility to infection of H. victoriae is controlled by one gene pair, and insensitivity or sensitivity to the toxin seems to be controlled by the same gene pair. It was concluded (168) that resistance to the fungus and insensitivity to its toxin are genetically the same. However, incompatible oat leaves that are resistant to the fungus and insensitive to the toxin still produce distinct necrotic flecks upon infection. This suggests that there is a mechanism in addition to insensitivity of tissues to the toxin that explains host resistance to the fungus; in other words, there is a factor responsible for confinement of the fungus (localization of the pathogen). (See also *Unsolved Problems* below.)

It is well documented that victorin is required for growth of the fungus in tissues and for the development

Table 9–6. Effect of immersion in victorin and drying on subsequent production of α-amylase by oat aleurone cells (after Samaddar and Scheffer, 163).

Immersion time (h)	Treatment	Toxin-sensitive		Toxin-insensitive	
		In medium (μg)	In cells (μg)	In medium (μg)	In cells (μg)
0[a]	water control	25	62	24	58
	toxin control	0	3	29	59
2	water	29	59	—	—
	toxin	2	18	—	—
4	water	26	60	—	—
	toxin	2	10	—	—
8	water	22	63	—	—
	toxin	1	8	—	—
12	water	27	58	27	58
	toxin	0	9	29	57

Half-seeds were immersed in water or toxin solution (0.16 μg/ml) for times indicated, washed in running tap water for 1 h, dried over $CaCl_2$ and P_2O_2 under vacuum for 48 h, and stored for 14 d. Half-seeds were then reimbubated for 24 h and incubated in the reaction mixture containing acetate buffer (pH 4.8) and gibberellic acid (1 μM).
a. These controls were not pretreated and dried.

of symptoms. This toxin is released by H. victoriae very early in the infection process during spore germination. Thus, the toxin may change host plasma membrane characteristics before penetration of the pathogen. Physiological changes that are more or less injurious to the host are certainly advantageous for the pathogen, the latter being a very weak parasite. This would be a good explanation for the successful colonization and growth of H. carbonum (a pathogen of maize) on toxin-sensitive oat leaves treated with victorin. Similarly, the presence of victorin in sensitive tissues helps the nontoxin-producing mutants of H. victoriae to colonize and grow in oat leaves (168, 236, 237). Probably the natural resistance mechanism of oat leaves to H. carbonum is abolished by victorin, which damages the plant's plasma membrane and thereby exerts many secondary effects. The damaged oat leaf cells then serve as substrate (not "host") for the "pseudo"-saprophytic H. carbonum. It would be of interest to know whether additional Helminthosporium species or other weak parasites grow on victorin-damaged oat leaf tissues.

Helminthosporium maydis T-toxin (HMT-toxin). Race T of the fungus is the causal agent of Southern corn leaf blight, one of the most serious diseases in the recent history of plant pathology. H. maydis race T exhibits a high level of virulence on corn lines having Texas male-sterile cytoplasm. Hence, host susceptibility is linked to a cytoplasmically inherited gene (Texas or T-cytoplasm) that causes male sterility in corn (cf. 80). It was shown by Smedegaard-Petersen and Nelson (182) that the fungus produces a host-specific toxin that preferentially inhibits root growth of T-cytoplasm corn but has little effect on resistant, "normal" corn having N-cytoplasm (cf. 230). Studies of mechanism of action of the HMT-toxin revealed membrane permeability changes resulting in ion leakage to be one of the first symptoms, as it is with several other host-specific toxins (142).

In addition, T-toxin effects growth, stomatal closure, respiration, dark CO_2 fixation, and photosynthesis. Probably, there are multiple receptor sites for the toxin. An alteration of the structural integrity of cell membranes also occurs in T-cytoplasm corn naturally infected by race T of *H. maydis*. The first detectable ultrastructural change in infected host cells was the rupture of the tonoplast, which occurred within six hours (Fig. 9–11). The presence of this toxin in the T-cytoplasm corn tissues promotes the growth of the otherwise non-pathogenic race O and *H. carbonum* and also stimulates electrolyte leakage in leaf tissues. Neither *H. maydis* race O nor *H. carbonum* is capable of increasing electrolyte leakage in the N (normal)-cytoplasm corn simultaneously treated with the HMT-toxin.

Halloin et al. (70) stressed that, although the HMT-toxin is host-specific in increasing leakage of electrolytes, it is non-host-specific in increasing leakage of carbohydrates from root and leaf tissues of corn (Table 9–7). Thus, this toxin is not entirely host-specific in its effects at the physiological level, though it may be in producing macroscopic symptoms (64, 220).

Mitochondria isolated from T-cytoplasm plants, but not from resistant N-cytoplasm seedlings, were found to be sensitive to the HMT-toxin. Sensitive mitochondria swelled rapidly and had reduced respiratory rates and lower ADP to O_2 ratios after treatment with the toxin-containing culture filtrates. Mitochondria from N-plants were not affected (4), only in T-cytoplasm plants did the T-toxin uncouple phosphorylation from oxidation (12). Arntzen et al. (3) reported that an early effect of the HMT-toxin was the rapid closure of stomata, followed by a reduction in photosynthesis (CO_2 fixation) as well as in transpiration (Fig. 9–12). The action of the HMT-toxin involves inhibition of the active uptake of K^+ by guard cells in isolated epidermal strips. Thus, the toxin exerts a direct effect on stomatal function.

The inhibition of K^+ uptake by toxin-treated TMS corn roots may be directly related to membrane depolarization. Tipton et al. (215) suggest that the toxin arrests active transport through the plasma membrane of TMS (susceptible) root cells. Consequently, the cell is unable to maintain the necessary intracellular K^+ concentration. Preincubation of the microsomes from roots of T cytoplasm plants with the HMT-toxin strongly inhibited the K^+-stimulated ATPase activity. The same treatment of microsomes from N cytoplasm root cells had no effect (Table 9–8). All these results are consistent with the rapid electrolyte leakage detected after toxin treatment.

It is noteworthy that light is necessary for development of leaf necrosis caused by the toxin (14). Daly and Barna (31) demonstrated that very low concentrations of purified toxin inhibit dark CO_2 fixation as well as photosynthesis.

It would seem that the toxin exerts multiple effects on the host cell, but the receptor molecules have not yet been identified (236). Regarding the structure of the toxin, Kono and Daly (96) proposed a linear polyketol, active at 6.5×10^{-9} M. Its emperical formula is $C_{41}H_{68}O_{13}$, it has an MW of 768, and its structural formula is presented below.

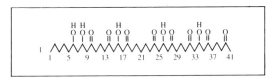

Helminthosporium carbonum toxin (HC-toxin). Race 1 of this fungus produces a toxin that affects corn genotypes that are natural hosts of the fungus. Thus, the toxin is a host-specific one (99). *H. carbonum* toxin can stimulate respiration, dark CO_2 fixation, nitrate uptake, and the uptake of some other solutes. This stimulatory effect on solute uptake is the earliest effect reported for the toxin (168) in addition to causing an increase in electrical potential across the plasmalemma. Resistant corn cultivars require a much higher concentration of HC-toxin to exhibit a similar stimulation of solute uptake. This toxin does not cause a general plasma membrane derangement in the sensitive host as do most other toxins. Its affects the plasmalemma more subtly, e.g. resulting in increased uptake of certain solutes by sensitive tissues that later become leaky and die.

The cyclic tetrapeptide *H. carbonum* toxin was isolated and crystallized (150). The toxin supports the growth of *H. victoriae* in corn, which is not a natural host for *H. victoriae* (24).

Periconia circinata toxin (PC-toxin). This fungus produces several closely related toxic polypeptides, each with a high degree of specificity for certain cultivars of grain sorghum (149). There are many parallels between the physiological actions of victorin and the *P. circinata* toxins. For example, they cause increased respiration, decreased growth, and decreased protein synthesis and disturb membrane function (46). The symptoms produced by this toxin in sensitive hosts are similar to those seen in plants inoculated with the fungus.

Phyllosticta maydis toxin (PM-toxin). *Phyllosticta maydis* is the causal agent of yellow leaf blight of corn. This fungus produces a host-specific toxin in culture that causes responses in the host similar to those caused by *Helminthosporium maydis* T-toxin (25, 236). In other words, *P. maydis* toxin is specific for corn containing the T-cytoplasm (Texas male sterile cytoplasm, Tcms). Indeed, the toxin has activity identical to that of *H. maydis* T-toxin, and the chemical properties of the

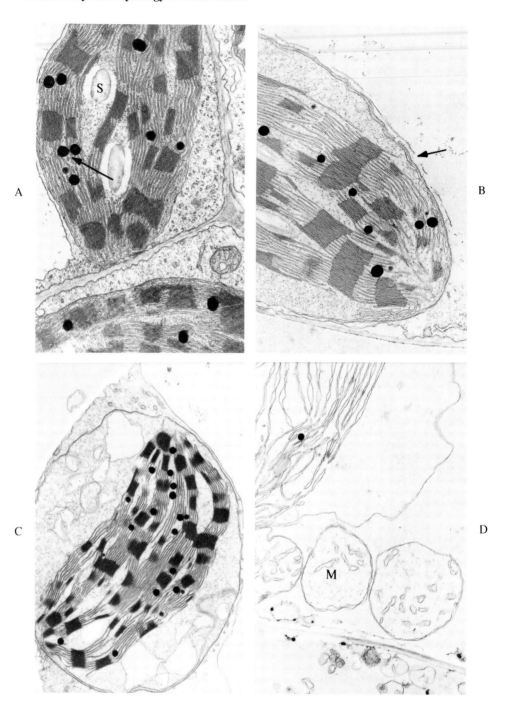

Fig. 9–11. (a) A portion of a mesophyll cell of corn (control) showing chloroplast with osmiophilic globules (arrow) and starch grains (S); dense cytoplasm, normal-appearing mitochondrion, plasmalemma, and tonoplast (× 21,000). (b) A portion of a mesophyll cell, 6 h after inoculation with *H. maydis*, showing broken tonoplast (arrow), slightly swollen chloroplast, and slightly less stroma than noted in control (×14,980). (c) A spherical chloroplast 12 h after inoculation. Note intense vesiculation in the stroma and separation of inner membrane from outer envelope (×10,140). (d) A portion of a mesophyll cell 24 h after inoculation revealing dispersed general lamellae, absence of stroma, and distorted chloroplast envelope. Mitochondria (M) lack ground material, and the plasmalemma is clearly discontinuous (×13,900) (after White et al., 231).

Table 9–7. Rates of leakage of electrolytes and carbohydrates from corn radicles and leaves containing either Texas male-sterile cytoplasm (T) or normal cytoplasm (N) in the presence and absence of toxin (after Halloin et al., 70).

Cytoplasm and tissue	Rate of leakage			
	Electrolyte		Total carbohydrate	
	μg/h/g		μg/h/g	
	Control	Toxin	Control	Toxin
N-roots	18.9	19.6	32	115
N-leaves	0.9	0.7	13	35
T-roots	20.7	26.1	97	309
T-leaves	1.1	2.6	15	56

two toxins seem to be identical or closely related. However, it is not known whether the same compound is produced by these fungi, because their chemical structures and primary sites of action have not been clarified.

The *P. maydis* toxin selectively inhibits seedling root growth, induces leaf chlorosis, and increases leakage of electrolytes in corn leaves containing the Tcms cytoplasm (Fig. 9–13). Mitochondria isolated from Tcms plants exhibited swelling, uncoupled oxidative phosphorylation, and altered O_2 uptake upon toxin treatment. Tissues with normal (N) cytoplasm were not affected except at very high concentrations of the toxin.

According to Comstock et al. (25), there is not a complete correlation between tissue sensitivity to the toxin and susceptibility to the fungal pathogen. Susceptibility of corn to infection by the fungus is under nuclear as well as cytoplasmic control (6), but sensitivity

Table 9–8. Effect of *H. maydis* race T toxin on K^+-stimulated ATPase activity of maize root microsomes (after Tipton et al., 215).

| Cytoplasm | Toxin | K^+-stimulated ATPase activity Pi, μ moles/h/g fresh tissue | | | | |
| | | Experiment number | | | | |
		1	2	3	4	5
N	−	2.60	2.41			2.41
N	+	2.60	2.30			2.28
T	−	3.08	2.00	2.43	2.41	2.50
T	+	0.00	0.73	0.27	0.39	0.00

Reaction mixtures contained 0.75 μ moles ATP (Tris salt, pH 6.8); 0.875 μ moles $MgCl_2$, when added; 12.5 μ moles KCl, when added; from 4.0 to 9.0 μ moles Tris-HCl buffer, pH 6.8 and enzyme, approximately 90 μg protein, in a final volume of 1.5 ml. Incubation was at 30 C for 30 min.

to this host-specific toxin is only cytoplasmically controlled. It is probable that some factor(s) in addition to this toxin is involved in symptom expression.

Host-specific Alternaria toxins. There are several reports from Japan on host-specific toxins produced by

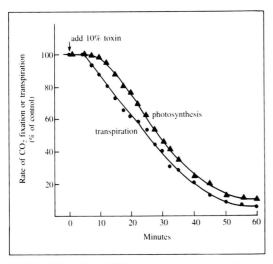

Fig. 9–12. The effect of HM-toxin on transpiration and photosynthesis in excised maize leaves (variety WF9[T-cms]x M 14). Control rates of transpiration and photosynthesis were 2.20 g $H_2O/dm^2/h$ and 34 mg $CO_2/dm^2/h$ (after Arntzen et al., 3).

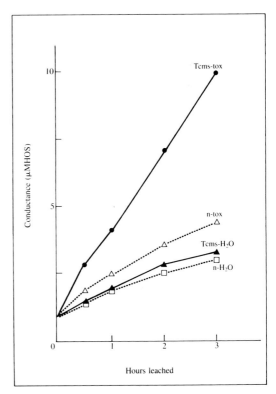

Fig. 9–13. Loss of electrolytes from Tcms and normal corn leaves treated with *Phyllosticta maydis* toxin (after Comstock et al., 25).

Table 9–9. Host-specific toxins (after Scheffer and Yoder, 168, modified).

Fungus	Host	Chemical nature	Selective toxicity reported by	Toxin isolated by
1. *Helminthosporium victoriae*	Oats (*Avena sativa*)	Unknown	Krupka (98)	Pringle and Braun (148)
2. *H. carbonum race 1*	Maize (*Zea mays*)	Cyclic tetrapeptide	Scheffer and Ullstrup (165)	Pringle and Scheffer (150)
3. *H. maydis* race T	Maize (*Zea mays*)	Linear polyketols	Smedegaard-Petersen and Nelson (182)	Kono and Daly (96)
4. *H. sacchari*	Sugarcane (*Saccharum officinarum*)	Sesquiterpene glycosides	Steiner and Byther (187)	Steiner and Strobel (188)
5. *Periconia circinata*	Grain sorghum (*Sorghum vulgare*)	Unknown	Leukel (103)	Pringle and Scheffer (149)
6. *Alternaria alternata* f. sp. *kikuchiana*	Japanese pear (*Pyrus serotina*)	Esters of epoxydecatrienoic acid	Tanaka (209)	Hiroe and Aoe (74)
7. *A. alternata* f. sp. *mali*	Apple (*Malus sylvestris*)	Cyclic tetrapeptides	Sawamura (164)	Sawamura (164)
8. *A. alternata* f. sp. *citri*	Emperor mandarin (*Citrus reticulata*) and rough lemon (*C. jambhiri*)	Hydrocarbon with carbonyl and hydroxyl groups	Pegg (143) and Kohmoto *et al.* (95)	Kohmoto *et al.* (95)
9. *A. alternata* f. sp. *lycopersici*	Tomato (*Lycopersicon esculentum*)	Esters of propanetricarboxylic acid and dimethyl-heptadecapentol	Gilchrist and Grogan (52)	Gilchrist and Grogan (52)
10. *Corynespora cassiicola*	Tomato (*L. esculentum*)	Unknown	Onesirosan *et al.* (131)	Onesirosan *et al.* (131)
11. *Phyllosticta maydis*	Maize (*Zea mays*)	Similar but not identical to T-toxin	Comstock *et al.* (25)	Comstock and Martinson (24)

different f. sp. of *Alternaria alternata* (*A. tenuis*) (52, 214). These are pathogenic to Japanese pear (*Pyrus serotina*), apple (*Malus sylvestris*), citrus (95) (*Citrus reticulata* and *C. jambhiri*) (95), and tomato (*Lycopersicon esculentum*) (199). Table 9–9 summarizes the literature on host-specific toxins.

Unsolved Problems

As regards the role of "host-specific" toxins in the specificity of pathogenicity or in pathogenesis, a number of questions remain unanswered. It seems logical that specificity of the pathogen is based on its ability to produce the host-specific toxin, and the susceptibility or resistance of the plant is determined by its sensitivity or insensitivity to the host-specific toxin. However, some facts seem to contradict this simple and apparently logical relationship, at least in the case of the *Helminthosporium* and other host-specific toxins.

(1) In spite of the toxin-insensitivity of cells of plant tissues resistant to the infecting microorganism, a few cells of the plant regularly die soon after natural infection (85). This phenomenon is known as the hypersen-sitive reaction, and at the same time the parasite's growth is stopped. The toxin-resistant (insensitive) cells, in this case, are killed by something other than the toxin. The pathogen is localized or killed in the necrotic tissue around the infection site by factor(s) other than host cell death, because this type of pathogen, being a necrotroph, grows vigorously in necrotic tissues of susceptible plants. Therefore, it would seem that the unsuccessful pathogenesis is related neither to plant insensitivity to the toxin nor to the hypersensitive tissue necrosis.

(2) Under conditions of natural infection only a few cells are killed in the hypersensitive necrotic spots of the resistant plant. The pathogen does not continue to kill plant cells as it does in the susceptible plant. The pathogen is stopped in the resistant plant by a mechanism obviously unrelated to the plant's insensitivity to the host-specific toxin.

(3) Avirulent strains of fungal pathogens that have lost the capacity to produce host-specific toxins still are capable of killing a few cells in the nonhost (hypersensitive) plants that are sensitive to the toxin. Here too, the

infecting fungus strains are confined to the infection sites. Thus, the course of unsuccessful pathogenesis seems to be similar with both the toxin-producing and the nontoxin-producing strains. Apparently, something other than the inability of the infecting strain to produce toxin is involved in the incompatible host-parasite relationship. The infected toxin-sensitive cells of the plant are killed even though the avirulent strain does not produce the host-specific toxin. At the same time the avirulent organism is prevented from growing and killing more host cells. One can deduce that the inability of the infecting strain to produce the host-specific toxin does not influence the specific nature of the host-parasite relation. It would seem that loss of the capacity to produce toxin and incompatibility are only correlative phenomena and not causally related.

(4) Susceptibility of maize to *Phyllosticta maydis* is under both nuclear and cytoplasmic gene control. On the other hand, sensitivity to the host-specific toxin of this fungus (PM-toxin) is controlled by extrachromosomal genes. Therefore, it was suggested (25) that factors in addition to the host-specific toxin are involved in the pathogenesis.

It is questionable that host-specific toxins are primary determinants of pathogenicity. This situation is perhaps better understood if one considers that host-specific toxins are involved only in symptom expression and that specificity in plant disease depends on other still unknown factors.

It perhaps helps to clarify the situation if one considers that in natural infections the first step in host-parasite incompatibility is the one that inhibits or kills the pathogen, after which the damaged pathogen releases an endotoxin(s), which could be a phytoalexin elicitor, and this causes the "hypersensitive" plant cell death, as was exemplified in the *Phytophthora*-potato complex (41, 94). Applying this hypothesis, one may presume that tissue necrosis is induced in a susceptible host-parasite combination by the host-specific "exotoxin." However, in a resistant combination, in which the plant is insensitive to the exotoxin, tissue necrosis (hypersensitive reaction) is induced by an "endotoxin" (elicitor) released by the infecting agent, which was "damaged" by the resistant host. In this way it is possible to explain the mechanism of necroses both in the susceptible host and in the plant that is insensitive to the host-specific toxin. This working hypothesis needs additional experimental verification.

Non-Host-Specific Toxins

Fusarial wilt toxins. The disease syndrome of fusarial wilts, which is characterized by epinasty, plugging and browning of xylem vessels (33), necrosis, and wilting, appears to be the result of a complex interaction of several toxins, enzymes, and hormones. Each of the toxins is able to produce only a portion of the disease syndrome. Wilt toxins are, however, nonspecific and cause some physiological changes not attributable to the pathogen. For example, fusaric acid, one of the most extensively studied wilt toxins, decreases respiratory

Fig. 9–14. Fusaric acid.

activity of tomato stem tissues, whereas tissues naturally infected with *Fusarium oxysporum* exhibit a marked increase in respiration (235).

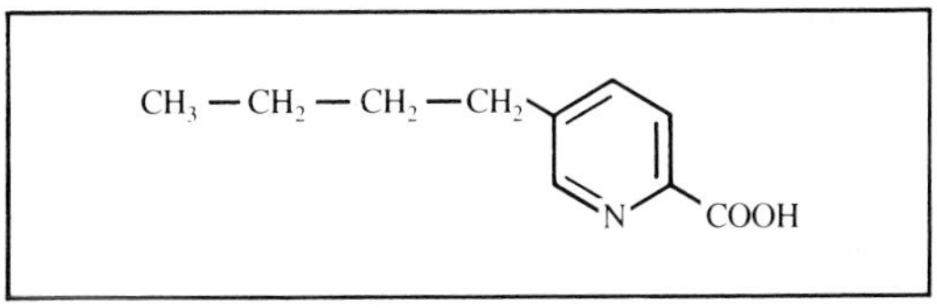

Fusaric acid is produced by many species of *Fusarium*. Chemically it is 5-*n*-butylpicolinic acid (Fig. 9–14). Metal ions are bound by fusaric acid through chelate formation, resulting in inhibition of Fe-porphyrin oxidases and the whole rate of respiration as well (49, 108, 206). The inhibition depends upon the presence of the carboxyl group in the toxin molecule. It can be reversed by addition of Fe^{+++}. Those enzymes that lack bound metals are not inhibited by fusaric acid.

The primary effect of fusaric acid produced in vivo seems to be on cell permeability. A consequence of this is the disruption of the balance of inorganic ions in diseased plants and leakage of electrolytes (50). Fusaric acid inhibits polyphenoloxidase competitively. The toxin inhibits this enzyme in lower concentration than it inhibits several other enzymes (15). The toxicity of fusaric acid in relation to the structure of the molecule has been extensively discussed by Kern (92).

Other low MW fusarial toxins are phytonivein (a steroid) from *Fusarium oxysporum* f. sp. *niveum*, phytolycopersin from f. sp. *lycopersici*, and lycomarasmin, which is produced by f. sp. *lycopersici*, f. sp. *vasinfectum*, and f. sp. *melonis* (92). Woolley et al. (233) proposed lycomarasmin to be a tripeptide (Fig. 9–15). It is a chelating agent; its toxicity is increased by the presence of iron, but it is detoxified by copper, which forms a stable chelate. The iron chelate is translocated into the leaves, where it is liberated, causing necrotic spots at the tips of the leaf blade. The equimolar iron-lycomarasmin chelate induces an early increase in tomato leaf respiration (120). This may be an explanation for the increased respiration of *Fusarium*-infected plants before the appearance of symptoms. In later stages of disease the respiration rate decreases, which may be due to the action of fusaric acid.

Other species of *Fusarium*, e.g. *F. solani* f. sp. *pisi*, produce toxic red pigments with a common *naphthazarine* structure (93). These naphthoquinones are toxic to bacteria, fungi, and tissues of higher plants. They inhibit anaerobic decarboxylation of pyruvate by

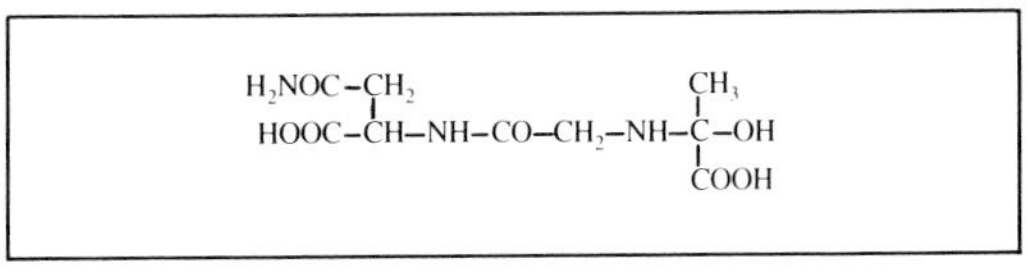

Fig. 9–15. Lycomarasmin.

Table 9–10. The effect of copper ion concentration in the nutrient solution on sensitivity to a naphthazarin toxin and susceptibility of peas to *Fusarium solani* f. sp. *pisi* (after Kern, 92).

Cu^{++} µg/ml	0	0.32	0.63	1.25	2.50	5.00
Toxin sensitivity[a]	2.9	2.6	2.5	2.4	1.5	0.8
Disease index[b]	3.3	3.0	2.6	2.1	2.0	1.5

a. Wilt index of pea cuttings with 50 µg isomarticin/g fresh weight.
b. Disease index of pea plants: 0 = no symptoms, 4 = total necrosis.

inhibiting thiamine pyrophosphate. Their toxicity for higher plants is reduced by copper ions. Copper probably chelates the toxin molecules and interferes with their transport through vessels and plasma membranes. The sensitivity of peas to one of these toxins and their susceptibility to the *Fusarium* pathogen decreased when plants were grown in nutrient solutions with increased copper content (Table 9–10). However, the inhibition of thiamine pyrophosphate by these toxins is not influenced by copper. In spite of all these results, the real role of fusarial toxins in disease symptom expression is uncertain.

Ophiobolin (cochliobolin). This toxin is produced by *Cochliobolus miyabeanus* (*Helminthosporium oryzae*), the causal agent of the rice leaf spot disease. *Ophiobolin* is able to produce at least a portion of the disease symptoms. In toxin-treated cells, just as in naturally infected epidermal cells of rice, polyphenols accumulate at an early stage of disease development (129). Similarly, an activated polyphenoloxidase has been demonstrated in diseased plants and as a consequence of treatment of ophiobolin (122). Changes in phenol metabolism seem to be an important factor in inducing disease symptoms caused by oxidation, and polymerization of phenolics produces the brown pigments characteristic of necotic spots (130). The toxin was isolated by Orsenigo (132) under the name *cochliobolin* and by Nakamura and Ishibashi (121), who called it *ophiobolin*. The toxin is a terpenoid having a novel structure (21, 127) (Fig. 9–16).

Helminthosporal. Another terpenoid toxin is produced by *Helminthosporium sativum* (*Cochliobolus sativus*), the causative agent of root rot disease of barley

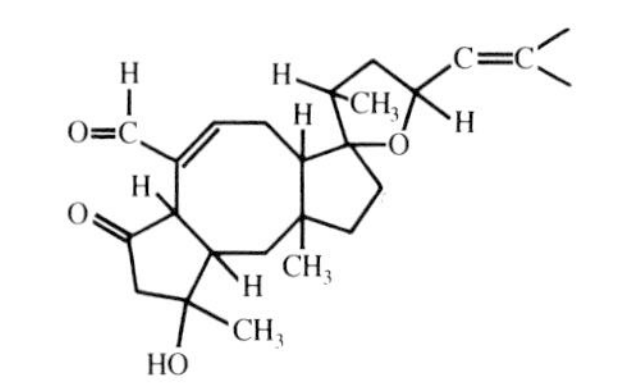

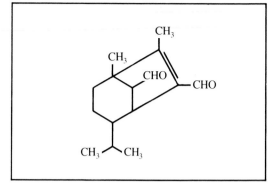

Fig. 9–17. Helminthosporal.

and wheat. Helminthosporal was isolated from the culture filtrate of the pathogenic fungus (114) and was found to be a sesquiterpenoid (Fig. 9–17). The biosynthesis of the toxin utilizes mevalonate (21).

This toxin inhibits electron transfer between flavoprotein dehydrogenase and cytochrome c. Oxidative phosphorylation in isolated mirochondria is also inhibited (210). A reduced analog of this toxin (helminthosporol) has a gibberellinlike growth-promoting action. The in vivo role of helminthosporal is uncertain at present.

Fusicoccin (FC). This is a diterpenoid glycoside produced by *Fusicoccum amygdali* via the mevalonic acid pathway (21). Fig. 9–18 shows the chemical structure of the toxin. Fusicoccin is probably involved in the wilting syndrome shown by almond and peach trees infected by the fungus *F. amydali* (65). It is not host-specific, and its precise site and mode of action are still obscure (20, 22). Recently it was suggested that fusicoccin specifically binds to a toxin receptor protein associated with the plasma membrane, in fact, with a membrane ATPase ion transport complex (191). The interaction with ATP-ase results in the stimulation of electrogenic H$^+$ extrusion (110, 111, 113). Activated ATPase may mediate the increase in H$^+$/K$^+$ exchange across the plasma membrane. One of the first symptoms

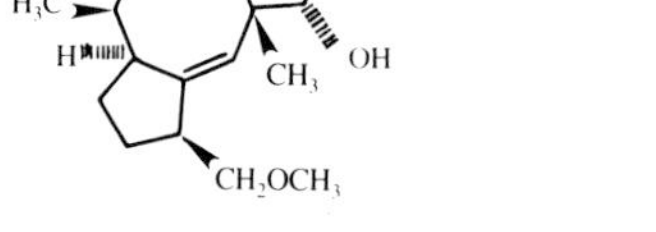

Fig. 9–16. Ophiobolin.

Fig. 9–18. Structure of fusicoccin (after Casinovi, 21).

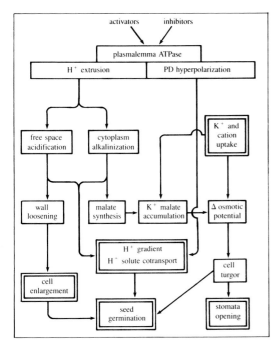

Fig. 9–19. The proposed relationship between FC-activated H$^+$ extrusion and some physiological processes. Final macroscopic responses are indicated by double rule (after Marrè, 111).

seems to be an increase in cell wall plasticity through an auxinlike mechanism (111, 112). This is probably causally related to H$^+$ extrusion. Cell wall loosening may lead to a disruption of the permeability of cell membranes. Fusicoccin also stimulates K$^+$ uptake, concomitant with hyperpolarization of the electrical potential across the plasma membrane. Furthermore, it enhances CO$_2$ fixation (20), respiration, water uptake, elongation of internodes, and cell enlargement (Fig. 9–19). The toxin increases the transpiration rate of cuttings by opening stomata (66, 221, 222). Fusicoccin is also very effective in inhibiting leaf abscission (44), as well as breaking seed dormancy, but the precise modes of these actions are not known (100).

One can hypothesize that the fundamental process activated by fusicoccin is the conversion of ~P energy into H$^+$ electrochemical gradient energy. All of the other effects of fusicoccin could be consequences of this primary action.

Pyricularin. The blast disease of rice is caused by *Pyricularia oryzae*, which produces an active material, pyricularin, that is nitrogen-containing (208). It is "semispecific" in that susceptible rice varieties are more susceptible to it than resistant ones.

High dilutions of pyricularin, such as 1:5,000,000, are able to increase respiratory activity of rice plants. Increases in respiration are similar for both resistant and susceptible varieties (207). This is in contrast with the finding that respiration of resistant *naturally* infected

tissue is less than that of infected susceptible plants. Apparently, the pathogen incites a different respiratory response than does pyricularin.

Pyricularin is toxic not only to higher plants but also to a variety of microorganisms. It is about ten times more toxic to *Pyricularia oryzae*, the fungus that produces the toxin, than it is to the host plant itself (128). Nevertheless, the fungus can produce large quantities of pyricularin in vitro. In culture fluids a pyricularin-binding protein has been found that inactivates pyricularin and destroys its antifungal property but does not alter its toxicity toward the rice plant. Consequently, the protein moiety of the toxic material controls its specificity. The occurrence of this protein has not been demonstrated in vivo. The fungus also produces in culture α-picolinic acid, which is similar in both structure and activity to fusaric acid.

Colletotin. Colletotin, the toxic agent of *Colletotrichum fuscum*, the fungus pathogen of *Digitalis lanata* and *D. purpurea*, may perhaps be considered a semihost-specific toxin. The symptoms caused by colletotin are similar to those elicited by the fungus on the host plant. In addition, *D. purpurea*, which is more resistant to *Colletotrichum*, is also more resistant to the toxin. On the other hand, the toxin is able to affect tomato plants that are not natural hosts of the pathogen, *C. fuscum* (55).

Colletotin disrupts cellular structure of foliar tissues, apparently by altering cell permeability. Crude colletotin contains both a polysaccharide and a peptide fraction, and it seems that integrity of the polysaccharide entity is essential for the maintenance of activity. The water-soaked translucent condition of the toxin-treated tissues is due to the passage of cell sap into the intercellular spaces, which is an early symptom of many necrotic diseases (104).

Alternaric acid. This is a hemiquinone derivative (Fig. 9–20) produced by *Alternaria solani* (early blight disease of tomato and potato). It induces severe wilt of seedlings of many plants but incites only a few of the symptoms incited by the pathogen (18).

Tentoxin. The toxin is produced in culture filtrate and by mycelia of *Alternaria tenuis* (*A. alternata*), the causal organism of seedling chlorosis in cotton and several other plant species. It is a cyclic tetrapeptide (Fig. 9–21) related chemically to the host-specific peptide toxins from *Helminthosporium victoriae*, *H. turcicum*, and *Periconia circinata*. Regarding the chemical nature of tentoxin, see Stoessl (190)

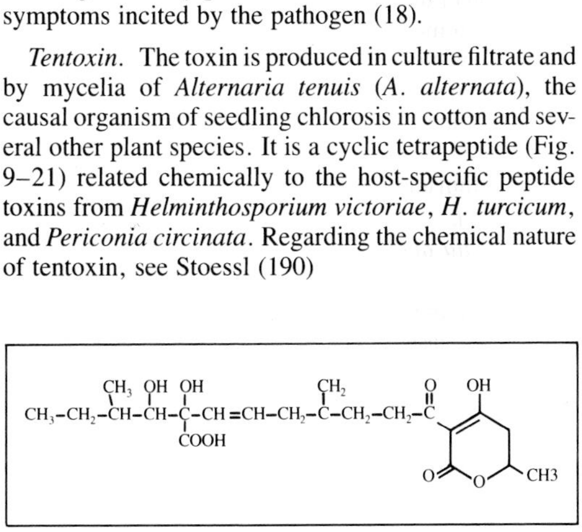

Fig. 9–20. Alternaric acid.

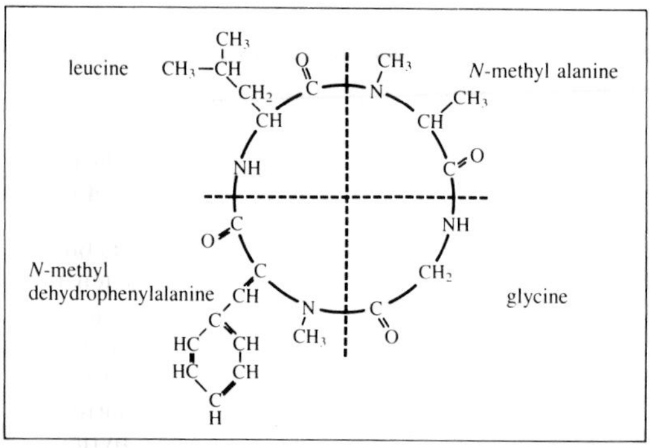

Fig. 9–21. Tentoxin.

It is believed that tentoxin participates in the infection process caused by *A. tenuis* by inhibiting chlorophyll accumulation and plastid membrane formation in germinating seedlings (69, 214). It disrupts normal plastid development rather than the synthesis of chlorophyll (69). Tentoxin binds with a very high affinity to chloroplast coupling factor (CF_1), a complex multiunit enzyme, and thereby inhibits photophosphorylation by this Ca^{++}-dependent ATPase (186). Spinach chloroplasts depleted in CF_1, when reconstituted with CF_1 from sensitive plants, became sensitive to the toxin; however, they were not sensitive when the CF_1 from resistant plants was used in reconstituting photophosphorylation activity (170). It was shown by Arntzen (2) that this toxin inhibits cyclic photophosphorylation but not reversible proton accumulation by isolated lettuce chloroplasts. At low concentrations, it also inhibits coupled electron flow in the presence of ADP and inorganic phosphate, but it does not affect basal electron flow (in the absence of ADP and phosphate) or uncoupled electron transprt. Tentoxin is an energy transfer inhibitor acting at the terminal steps of ATP synthesis.

Other phytotoxins (88, 199, 232, 236). Several toxic compounds were isolated from culture filtrates of pathogenic fungi, which may have some role in symptom expression of the disease. However, their role in vivo is uncertain in most cases; these include:

Zinniol is produced by *Alternaria zinniae*, *A. solani*, and *A. dauci*. It is a penta-substituted benzene (10).

Ascochitine is a hemiquinone derivative from *Ascochyta fabae*.

Diaporthin and skyrin are produced by *Endothia parasitica* (51).

Didymella applanata toxin affects young raspberry sprouts.

Myrothecium roridum toxin is perhaps involved in leaf spot disease of red clover.

Leptosphaerulina briosiana toxin causes leaf spots on alfalfa.

Alternaria tenuis phenolic toxins, other than tentoxin, may cause brown spot disease of tobacco.

Cercospora beticola toxin is involved in sugar beet leaf spot.

T-2 toxin is one of the toxins produced by *Fusarium tricinctum* and inhibits auxin-promoted elongation of hypocotyls (184).

Verticillium albo-atrum toxic proteins are involved in cotton wilt.

Phytophthora nicotianae var. *parasitica* toxin is involved in citrus disease.

Phytophthora megasperma var. *sojae* toxin contributes to symptom expression in root and stem rot of soybean.

Ceratocystis ulmi toxins. Peptidorhamnomannan induces wilt symptoms by physically interfering with water movement through the xylem. It is antigenic and has been demonstrated in host tissues by an immunoassay (166).

Two other toxins of *Ceratocystis*, a small protein and dextrans, probably contribute to symptom development in infected elm tissues (199).

Stemphylium botryosum toxins are causing necrosis on several host plants. Stemphylin is an aromatic glycoside, and stemphyloxin is a tricyclic compound possessing an unusual β-ketoaldehyde (11).

Pyrenophora teres toxins cause net blotch disease of barley. Two toxins are involved in symptom expression and in increasing respiration (8, 180, 181). One toxic compound appears to be N-(2-amino-2-carboxyethyl) aspartic acid. The second toxin is aspergillomarasmine A, which is associated with *Colletotrichum* and *Aspergillus*.

Rhynchosporosides. Several glycosides linked to 1,2-propanediol have been isolated from *Rhynchosporium secalis*, the causal organism of barley scald disease (199). A membrane receptor protein may have a role in sensitivity of some barley cultivars to one of these toxins.

Comparative Analysis of Disease Physiology— Toxins

A diverse group of metabolites, produced by plant pathogens, is capable of exerting a boad spectrum of adverse effects upon higher plants. These chemical substances include enzymes, plant growth regulators, and a number of unrelated organic substances that possess biological activity detrimental to plants. It is this latter group that we have termed *phytotoxins*.

Phytotoxins affect plant function at physiological concentrations, often displaying a certain subtlety of action. They are usually mobile in the sensitive host plant and are generally, but not exclusively, low in molecular weight. Since they are a diverse group of compounds, they do not have common structural features, and they include substances such as peptides, glycopeptides, phenolics, terpenoids, glycosides, and polysaccharides. In a number of cases, their structures have been determined satisfactorily.

There are about a dozen host-specific toxins; resistance and susceptibility of the host to the pathogen reflect insensitivity or sensitivity, respectively, to the host-specific toxin. The non-host-specific phytotoxins are able to reproduce at least a portion of the symptoms caused by the living pathogens in either host or nonhost plants.

Although viruses do not produce toxins, they may induce the accumulation of toxigenic substances in plants. Some of these appear to be involved in the production of the disease syndrome.

A number of sites of toxin action have been suggested. Some affect both cell permeability and transmembrane potential; as a consequence, ionic balance is disrupted, leakage of electrolytes occurs, enzymes are inhibited or stimulated, and increases in ethylene production and respiration have been recorded.

Sensitivity to the host-specific HS toxin appears highly correlated with the ability of host membrane proteins to bind toxin. Insensitivity of cultivars to host-specific toxins has been explained by inactivation of the toxin or a self-repair mechanism. There are a number of similarities in the physiological actions of the HV and PC toxins and, also, between the HMT and PM toxins. Multiple receptor sites have been postulated for HMT toxin; however, those receptor molecules have not been identified. Their receptivity or recognition of HMT toxin is indicated by changes in membrane permeability, stomatal closure, respiration, dark CO_2 fixation, and photosynthesis. The primary site of toxin action appears to be the mitochondrial membrane, but the plasma membrane may also be directly affected. The HC toxin probably affects the plasmalemma, which results in increased solute uptake and subtle damage to the plasmalemma, causing cell death.

Non-host-specific toxins, like fusaric acid, bind metal ions through chelate formation that results in inhibition of Fe-porphyrin oxidases. The toxins lycomarasimin and napthazarine also appear to exert their biological effects as chelating agents.

The toxin ophiobolin alters phenol metabolism of the host, and fusicoccin binds to a membrane receptor protein in which a part of the H^+/ATPase complex is resident. The interaction with ATPase results in the stimulation of electrogenic H^+ extrusion. This leads to an increase in cell wall plasticity, an auxinlike effect, and to other consequences.

Tentoxin is an energy transfer inhibitor, affecting the terminal steps of ATP synthesis in chloroplasts. It binds to a chloroplast coupling factor and, thus, selectively inhibits the light-dependent but not oxidative phosphorylation.

Among the toxins produced by pathogenic bacteria, the action of the wildfire toxin has been studied most intensively and is best understood. Called tabtoxin, it inhibits the action of glutamine synthetase, which interrupts the photorespiratory N-cycle, causing ammonia to accumulate in the tissues. It is the high level of ammonia that is responsible for disease symptom development. Phaseolotoxin induces dysfunction of the ornithine cycle by inhibiting ornithine carbamoyltransferase activity. The enzyme catalyzes citrulline formation from ornithine, hence toxicity has been linked to a failure of

affected plant cells to synthesize arginine and orotic acid. Rhizobitoxine blocks methione synthesis, whereas amylovorin, a wilt-inducing bacterial polysaccharide, causes a vessel blockage and possibly plasmolysis of the xylem parenchyma.

Additional Comments and Citations

1. A number of recent reports have further clarified the chemical nature of the active component of the *P. tabaci*-produced toxin. The mode of action of the toxin in causing chlorosis remains the inhibition of the enzyme glutamine synthetase (GS). However, the active compound is *not* tabtoxin, rather it is tabtoxinine-β-lactam (2-amino-4-(3-hydroxy-2-oxoazacyclobutan-3-yl)-butanoic acid) (TBL). Apparently TBL irreversibly inactivates GS and is responsible for the observed chlorosis. It is contended by the Durbin group that earlier preparations of tabtoxin also contained an enzyme that converted tabtoxin to TBL. Only purified TBL and not purified tabtoxin is capable of inhibiting GS. M. O. Thomas, P. J. Langston-Unkefer, T. F. Uchytil, and R. D. Durbin. 1983. Plant Physiol. 71:912–15.

2. R. Moore, W. Niemczura, O. Kwok, and S. Patil. 1984. Inhibitors of ornithine carbamoyltransferase from *Pseudomonas syringae* pv. *phaseolicola*. Revised structure of phaseolotoxin. Tetrahedron Lett. 25:3931–34. The precise structure of phaseolotoxin was revised in the aforementioned citation. In addition, the authors report that the major toxin octicidin is 20 times more active than phaseolotoxin in suppressing ornithine carbamoyltransferase.

3. Our knowledge pertaining to the host specific toxins (STs) has (as indicated by V. Macko. 1983. Structural aspects of toxins. In: Toxins and plant pathogenesis, J. M. Daly and F. J. Deverall (eds.). Academic Press. Australia, pp. 41–80) "taken a quantum leap." Specifically the chemistry and site of action of the *Helminthosporium sacchari* (HS) toxin and *H. maydis* or T-toxin are succinctly and authoritatively presented. In addition, basic facts concerning the chemistry of the non-specific fungal and bacterial toxins are also clearly described by Macko.

4. H. Schröter, A. Novacky, and V. Macko. 1985. Effect of *Helminthosporium sacchari*-toxin on cell membrane potential of susceptible sugarcane. Physiol. Plant Path. (in press). The authors recorded rapid membrane depolarization (plasmalemma) in 4-10 min. with the active component (a central bicarboxylic sesquiterpene linked to 2 residues of 5-0-(βgalactofuransyl) β-galactofuranoside (HS_4). The lower homologue with three galactose units (HS_3) is not only nontoxic but also appears to block the "receptor site" for HS_4.

5. Two important papers deal with the action of HMT toxin on mitochondrial membranes: M. J. Holden and H. Sze. 1984. *Helminthosporium maydis* T toxin in-

creased membrane permeability to Ca^{2+} in susceptible corn mitochondria. Plant Physiol. 75:235–37, and Kimber, A., and H. Sze. 1984. *Helminthosporium maydis* T toxin decreased calcium transport into mitochondria of susceptible corn. Plant Physiol. 74:804–9. Purified HMT toxin completely inhibited Ca^{2+} transport into the mitochondria of susceptible corn at 5 to 10 nanograms per ml, however transport into the mitochondria of resistant corn was decreased slightly by 100 nanograms per ml toxin. It would seem that mitochondrial and not microsomal membrane is a primary site of HMT toxin action. The toxin may decrease the energy-dependent ion transport by inhibiting the formation of or dissipating the electrochemical proton gradient generated by different redox substrates as well as ATP in mitochondria of the susceptible corn.

6. The binding of HS-toxin to membrane receptor(s) has been reinvestigated by M. S. Lesney, R. S. Livingston, and R. P. Scheffer. 1982. Effect of toxin from *Helminthosporium sacchari* on nongreen tissues and a reexamination of toxin binding. Phytopathology 72:844–49. There was no indication of toxin binding to membrane preparations from either susceptible or resistant sugarcane tissues. This is because a toxin preparation with much higher radioactivity than any used at present would be required. Although there must be a receptor for the toxin, it would seem that the site cannot be detected by the methods used by the present and the previous investigators. The site is not necessarily in the plasma membrane.

7. Another recent paper on the HS toxin by J. P. Duvick, J. M. Daly, Z. Kratky, V. Macko, W. Acklin and D. Arigoni. 1984. Biological activity of the isomeric forms of *Helminthosporium sacchari* toxin and of homologs produced in culture. Plant Physiol. 74:117–122, mainly deals with the function/structure relationships of the three isomeric forms of the toxin. The HS toxin consists of a central sesquiterpene aglycone linked to two residues of 5-*O*-(β-galctofuranosyl)-β-galactofuranoside. There was a lack of protection by D-galctose or α-D-galactopyranosides in protection of sugar cane tissue from symptoms induced by the toxin. Three early effects, namely depolarization, inhibition of dark CO_2 fixation, and ion leakage, seem to be associated with a principal action on the plasmalemma.

8. B. Barna, A. R. T. Sarhan, and Z. Király. 1985. The effect of age of tomato and maize leaves on resistance to a nonspecific and a host-specific toxin. Physiol. Plant Path. 26:153–58. Young tomato and maize leaf tissues whether produced naturally or as the result of delaying senescence by growing the plants with high levels of nitrate-nitrogen or by cytokinin treatments, have greater nonspecific resistance to fusaric acid and to HMT-toxin. This resistance seems to be associated with membranes containing higher phospholipid:sterol ratios than normal. Thus, the juvenile state of the host plant may affect reactions to non-specific as well as to specific toxins.

References

1. Anderson, E. S., and A. H. Rogers. 1963. Slime polysaccharides of the *Enterobacteriaceae*. Nature 198:714–15.

2. Arntzen, C. J. 1972. Inhibition of photophosphorylation by tentoxin, a cyclic tetrapeptide. Biochim. Biophys. Acta 283:539–42.

3. Arntzen, C.J., M. F. Haugh, and S. Bobick. 1973. Induction of stomatal closure by *Helminthosporium maydis* pathotoxin. Plant Physiol. 52:569–74.

4. Arntzen, C. J., C. Koeppe, R. F. Miller, and F. H. Peverly. 1973. The effect of pathotoxin from *Helminthosporium maydis* (race T) on energy-linked processes of corn seedlings. Physiol. Plant Path. 3:79–89.

5. Ayers, A. R., S. B. Ayers, and R. N. Goodman. 1979. Extracellular polysaccharide of *Erwinia amylovora*: A correlation with virulence. Appl. Environ. Microbiol. 38:659–66.

6. Ayers, J. E., R. R. Nelson, C. Koons, and G. L. Scheifele. 1970. Reactions of various maize inbreds and single crosses in normal and male-sterile cytoplasm to the yellow leaf blight organism (*Phyllosticta* sp.). Plant Dis. Reptr. 54:277–80.

7. Bablanian, R. 1975. Structural and functional alterations in cultured cells infected with cytocidal viruses. Prog. Med. Virol. 19:40–83

8. Bach, E., S. Christensen, L. Dalgaard, P. O. Larsen, C. I. Olsen, and V. Smedegaard-Petersen. 1979. Structures, properties and relationship to the aspergillomarasmines of toxins produced by *Pyrenophora teres*. Physiol. Plant Path. 14:41–46.

9. Backman, P A., and J. E. DeVay. 1971. Studies on the mode of action and biogenesis of the phytotoxin syringomycin. Physiol. Plant Path. 1:215–34.

10. Barash, I., H. Mor, D. Netzer, and Y. Kashman. 1981. Production of zinniol by *Alternaria dauci* and its phytotoxic effect on carrot. Physiol. Plant Path. 19:7–16.

11. Barash, I., G. Pupkin, D. Netzer, and Y. Kashman. 1982. A novel enolic β- ketoaldehyde phytotoxin produced by *Stemphylium botryosum* f. sp. *lycopersici*. Plant Physiol. 69:23–27.

12. Bednarski, M. A., S. Izawa, and R. P. Scheffer. 1977. Reversible effects of toxin from *Helminthosporium maydis* race T on oxidative phosphorylation by mitochondria from maize. Plant Physiol. 59:540–45.

13. Beier, R. C. 1980. Carbohydrate chemistry: Synthetic and structural investigation of the phytotoxins found in *Helminthosporium sacchari* and *Rhynchosporium secalis*. Ph.D. diss., Montana State Univ.

14. Bhullar, B. S., J. M. Daly, and D. W. Rehfeld. 1975. Inhibition of dark CO_2 fixation and photosynthesis in leaf discs of corn susceptible to the host-specific toxin produced by *Helminthosporium maydis*, race T. Plant Physiol. 56:1–7.

15. Bossi, R. 1959. Über die Wirkung der Fusarinsäure auf die Polypenoloxydase. Phytopath. Z. 37:273–316.

16. Braun, A. C. 1950. The mechanism of action of a bacterial toxin on plant cells. Proc. Nat. Acad. Sci. 36:423–27.

17. Braun, A. C. 1955. A study on the mode of action of the wildfire toxin. Phytopathology 45:659–64.

18. Brian, P. W., G. W. Elson, H. G. Hemming, and J. M. Wright. 1952. The phytotoxic properties of alternaric acid in relation to the etiology of plant diseases caused by *Alternaria solani* (Ell. et Mart.) Jones et Grout. Ann. Appl. Biol. 39:308–21.

19. Brown, J. H., M. J. Wille, L. S. Daley, and G. A. Strobel. 1982. Helminthosporoside inhibition of carbon fixation and apparent result of toxin mediated damage to membrane function. Plant Physiol. 69:143. (Suppl.)

20. Brown, P. H., and W. H. Outlow, Jr. 1982. Fusicoccin-

enhanced CO_2 fixation in guard cell protoplasts. Plant Physiol. 69:148. (Suppl.)

21. Casinovi, C. G. 1972. Chemistry of the terpenoid phytotoxins. In: Phytotoxins in plant diseases. R. K. S. Wood, A. Ballio, and A. Graniti (eds.). Academic Press, New York, pp. 105–25.

22. Chain, E. B., P. G. Mantle, and B. V. Milborrow. 1971. Further investigations of the toxicity of fusicoccins. Physiol. Plant Path. 1:495–514.

23. Clayton, E. E. 1936. Watersoaking of leaves in relation to the development of the wildfire disease of tobacco. J. Agr. Res. 52:239–69.

24. Comstock, J. C., and C. A. Martinson. 1974. Chromatographic comparison of *Phyllosticta maydis* and *Helminthosporium maydis* race T toxins. Proceed. Am. Phytopath. Soc. 1:34.(Abstr.)

25. Comstock J. C., C. A. Martinson, and B. G. Gengenbach. 1973. Host specificity of a toxin from *Phyllosticta maydis* for Texas cytoplasmically male-sterile maize. Phytopathology 63:1357–61.

26. Comstock, J. C., and R. P. Scheffer. 1973. Role of host-selective toxin in colonization of corn leaves by *Helminthosporium carbonum*. Phytopathology 63: 24–29.

27. Corey, R. R., and M. P. Starr. 1957. Colony types of *Xanthomonas phaseoli*. J. Bact. 74:137–45.

28. Crosse, J. E., R. N. Goodman, and W. H. Shaffer, Jr. 1972. Leaf damage as a predisposing factor in the infection of apple shoots by *Erwinia amylovora*. Phytopathology 62:176–82.

29. Cullimore, J. V., and A. P. Sims. 1980. An association between photorespiration and protein catabolism: Studies with *Chlamydomonas*. Planta 150:392–96.

30. Daly, J. M. 1981. Mechanisms of action. In: Toxins in plant disease. R. D. Durbin (ed.). Academic Press, New York, pp. 331–94.

31. Daly, J. M., and B. Barna. 1980. A differential effect of race T toxin on dark and photosynthetic CO_2 fixation by thin leaf slices from susceptible corn. Plant Physiol. 66:580–83.

32. DeVay, J. E., F. L. Lukezic, S. L. Sinden, H. English, and D. L. Coplin. 1968. A biocide produced by pathogenic isolates of *Pseudomonas syringae* and its possible role in the bacterial canker disease of peach trees. Phytopathology 58:95–101.

33. Dimond, A. E. 1970. Biophysics and biochemistry of the vascular wilt syndrome. Ann. Rev. Phytopathol. 8:301–22.

34. Dimond, A. E., and P. E. Waggoner. 1953. The cause of epinastic symptoms in *Fusarium* wilt of tomatoes. Phytopathology 43:663–69.

35. Dubos, R. J. 1946. The bacterial cell. Harvard Univ. Press, Cambridge.

36. Durbin, R. D. 1972. Bacterial phytotoxins: A survey of occurrence, mode of action and composition. In: Phytotoxins in plant diseases. R. K. S. Wood, A. Ballio, and A. Graniti (eds.). Academic Press, New York, pp.19–33.

37. Durbin, R. D. 1981. Toxins in plant disease. Academic Press, New York.

38. Erikson, D. 1945. Certain aspects of resistance of plum trees to bacteria canker. Part I. Some biochemical characteristics of *Pseudomonas mors-prunorum* (Wormald) and related phytopathogenic bacteria. Ann. Appl. Biol. 32: 44–52.

39. Erikson, D. 1945. Certain aspects of resistance of plum trees to bacterial canker. Part II. On the nature of the bacterial invasion of *Prunus* sp. by *Pseudomonas mors-prunorum* Wormald. Ann. Appl. Biol. 32:112–16.

40. Erikson, D., and H. B. S. Montogomery. 1945. Certain aspects of resistance of plum trees to bacterial canker. Part III. The action of cell-free filtrates of *Pseudomonas mors-prunorum* Wormald and related phytopathogenic bacteria on plum trees. Ann. Appl. Biol. 34:117–23.

41. Érsek, T., B. Barna, and Z. Király. 1973. Hypersensitivity and the resistance of potato tuber tissues to *Phytophthora infestans*. Acta Phytopath. Hung. 8:3–12.

42. Evans, M. L. 1973. Rapid stimulation of plant cell elongation by hormonal and non-hormonal factors. BioScience. 23:711–18.

43. Feder, W. A., and P. A. Ark. 1951. Wilt-inducing polysaccharides derived from crown gall, bean blight and soft rot bacteria. Phytopathology 41:804–8.

44. Feldman, A. W., A. Graniti, and L. Sparapano. 1971. Effect of fusicoccin on abscission, cellulase activity and ethylene production in citrus leaf explants. Physiol. Plant Path. 1:115–22.

45. Gardner, J. M., and C. I. Kado. 1972. Induction of the hypersensitive reaction in tobacco with specific high-molecular weight substances derived from the osmotic shock fluid of *Erwinia rubrifaciens*. Phytopathology 62:759.

46. Gardner, J. M., I. S. Mansour, and R. P. Scheffer. 1972. Effects of the host-specific toxin of *Periconia circinata* on some properties of sorghum plasma membranes. Physiol. Plant Path. 2:197–206.

47. Gardner, J. M., and R. P. Scheffer. 1973. Effects of cycloheximide and sulfhydril-binding compounds on sensitivity of oat tissues to *Helminthosporium victoriae* toxin. Physiol. Plant Path. 3:147–57.

48. Garibaldi, A., and D. F. Bateman. 1971. Pectic eyzymes produced by *Erwinia chrysanthemi* and their effects on plant tissue. Physiol. Plant Path. 1:25–40.

49. Gäumann, E. 1954. Toxin and plant diseases. Endeavour 13:198–204.

50. Gäumann, E. 1957. Fusaric acid as a wilt toxin. Phytopathology 47:342–57.

51. Gäumann, E., and W. Obrist. 1960. Über die Schädigung der Wasserpermeabilität pflanzlicher Protoplasten durch einige Welketoxine. Phytopath. Z. 37:145–58.

52. Gilchrist, D. G., and R. G. Grogan. 1976. Production and nature of a host-specific toxin from *Alternaria alternata* f. sp. *lycopersici*. Phytopathology 66:165–71.

53. Giovanelli, J., L. D. Owens, and S. H. Mudd. 1971. Mechanism of inhibition of spinach β-cystathionase by rhizobitoxine. Biochim. Biophys. Acta 227:671–84.

54. Giovanelli, J., L. D. Owens, and S. H. Mudd. 1972. β-Cystathionase in vivo inactivation by rhizobitoxine and role of the enzyme in methionine biosynthesis in corn seedlings. Plant Physiol. 51:492–503.

55. Goodman, R. N. 1959. Observations on the production, physiological activity and chemical nature of colletotin, a toxin from *Colletotrichum fuscum* Laub. Phytopath Z. 37:187–94.

56. Goodman, R. N. 1972. Phytotoxin-induced ultrastructural modifications of plant cells. In: Phytotoxins in plant diseases. R. K. S. Wood, A. Ballio, and A. Graniti (eds.). Academic Press, New York, pp. 311–29.

57. Goodman, R. N., J. S. Huang, and P.-Y. Huang. 1974. Host-specific phytotoxic polysaccharide from apple tissue infected by *Erwinia amylovora*. Science 183:1081–82.

58. Goodman, R. N., P. R. Stoffl, and S. B. Ayers. 1978. The utility of the fireblight toxin, amylovorin, for the detection of resistance of apple, pear and quince to *Erwinia amylovora*. Acta Hort. 86:51–56.

59. Goodman, R. N., and J. A. White. 1981. Xylem parenchyma plasmolysis and vessel disorientation are early signs of pathogenesis caused by *Erwinia amylovora*. Phytopathology 71:844–52.

60. Gorin, P. A. J., and J. F. T. Spencer. 1961. Structural relationships of extracellular polysaccharides from phytopathogenic *Xanthomonas* spp. Can. J. Chem. 39:2282–89.

61. Gorin, P. A. J., and J. F. T. Spencer. 1961. Extracellular acidic polysaccharides from *Corynebacterium insidiosum* and other *Corynebacterium* spp. Can. J. Chem. 39:2274–81.

62. Gorin, P. A. J., and J. F. T. Spencer. 1963. Structural relationship of extracellular polysaccharides from phytopathogenic *Xanthomonas* spp. Part II. *X. hyacinthi*, *X. transluscens* and *X. maculofolium-gardiniae*. Can. J. Chem. 41:2357–61.

63. Gottlieb, D. 1943. The presence of a toxin in tomato wilt. Phytopathology 33:126–35.

64. Gracen, V. E., C. O. Grogan, and M. F. Forster. 1972. Permeability changes induced by *Helminthosporium maydis*, race T, toxin. Can. J. Bot. 50:2167–70.

65. Graniti, A. 1964. The role of toxins in the pathogenesis of infections by *Fusicoccin amygdali* Del. on almond and peach. In: Host-parasite relations in plant pathology. Z. Király and G. Ubrizsy (eds.). Res. Inst. Plant Prot., Budapest, pp. 211–17.

66. Graniti, A., and N. D. Turner. 1970. Effect of fusicoccin on stomatal transporation in plants. Phytop. Medit. 9:160–67.

67. Gross, D. C., and J. E. DeVay. 1977. Population dynamics and pathogenesis of *Pseudomonas syringae* in maize and cowpea in relation to in vitro production of syringomycin. Phytopathology 67:475–83.

68. Gross, D. C., J. E. DeVay, and F. H. Stadtman. 1977. Chemical properties of syringomycin and syringotoxin: toxigenic peptides produced by *Pseudomonas syringae*. J. Appl. Bact. 43:453–63.

69. Halloin, J. M., G. A. deZoeten, G. Gaard, and J. C. Walker. 1970. The effects of tentoxin on chlorophyll synthesis and plastid structure in cucumber and cabbage. Plant Physiol. 45:310–14.

70. Halloin, J. M., J. C. Comstock, C. A. Martinson, and C. L. Tipton. 1973. Leakage from corn tissues induced by *Helminthosporium maydis* race T toxin. Phytopathology 63:640–42.

71. Hanchey, P., and H. Wheeler. 1979. The role of host cell membranes. In: Recognition and specificity in plant host-parasite interactions. J. M. Daly and I. Uritani (eds.). Jap. Sci. Soc. Press, Tokyo, pp. 193–210.

72. Hanchey, P., H. Wheeler, and H. H. Luke. 1968. Pathological changes in ultrastructure: Effect of victorin on oat roots. Amer. J. Bot. 55:53–61.

73. Hildebrand, E. M. 1939. Studies on fire-blight ooze. Phytopathology 29:142–57.

74. Hiroe, I., and S. Aoe. 1954. Phytopathological studies on black spot disease of Japanese pear, caused by *Alternaria kikuchiana* Tanaka (English Ser. 1) patho-chemical studies (1). On a new phytotoxin, phytoalternarin produced by the fungus. J. Fac. Agr. Tottori Univ. 2:1–26.

75. Hodgson, R., W. H. Peterson, and A. J. Riker. 1949. The toxicity of polysaccharides and other large molecules to tomato cuttings. Phytopathology 39:47–62.

76. Hodgson, R., A. J. Riker, and W. H. Peterson. 1945. Polysaccharide production by virulent and attenuated crown-gall bacteria. J. Biol. Chem. 158:89–100.

77. Hodgson, R., A. J. Riker, and W. H. Peterson. 1947. A wilt-inducing toxic substance from crown gall bacteria. Phytopathology 37:301–18.

78. Hoitink, H. A. J., R. L. Pelletier, and J. G. Coulson. 1966. Toxemia of halo blight of beans. Phytopathology 56:1062–65.

79. Hoitink, H. A. J., and S. L. Sinden. 1970. Partial purification and properties of chlorosis-inducing toxins of *Pseudomonas phaseolicola* and *Pseudomonas glycinea*. Phytopathology 60:1236–37.

80. Hooker, A. L., D. R. Smith, S. M. Lim, and J. B. Beckett. 1970. Reaction of corn seedling with male-sterile cytoplasm to *Helminthosporium maydis*. Plant Dis. Reptr. 54:708–12.

81. Huang, P.-Y. 1974. Ultrastructural modification by and pathogenicity of *Erwinia amylovora* in apple tissues. Ph.D. diss., Dept. Plant Pathology, Univ. of Missouri, Columbia.

82. Huang, P.-Y., and R. N. Goodman. 1976. Ultrastructural modifications in apple stems induced by *Erwinia amylovora* and the fire blight toxin. Phytopathology 66:269–76.

83. Husain, A., and A. Kelman. 1957. Presence of pectic and cellulolytic enzymes in tomato plants infected by *Pseudomonas solanacearum*. Phytopathology 47:111–12.

84. Husain, A., and A. Kelman. 1958. Relation of slime production to mechanism of wilting and pathogenicity of *Pseudomonas solanacearum*. Phytopathology 48:155–65.

85. Jennings, P., and A. J. Ullstrup. 1957. A histological study of three *Helminthosporium* leaf blights of corn. Phytopathology 47:707–14.

86. Jensen, J. H., and J. E. Livingston. 1944. Variation of symptoms produced by isolates of *Phytomonas medicaginis* var. *phaseolicola*. Phytopathology 34:471–80.

87. Jockusch, H., and B. Jockusch. 1968. Early cell death caused by TMV mutants with defective coat proteins. Mol. Gen. Genet. 102:204–9.

88. Kadis, S., A. Ciegler and S. J. Ajl (eds.). 1972. Microbial toxins. Vol. 8. Fungal toxins. Academic Press, New York.

89. Keck, R. W., and T. K. Hodges. 1973. Membrane permeability in plants: Changes induced by host-specific pathotoxins. Phytopathology 63:226–30.

90. Kenfield, D. S., and G. A. Strobel. 1981. Selective modulation of helminthosporoside binding by alpha-galactoside-binding proteins from sugar cane clones susceptible and resistant to *Helminthosporium sacchari*. Physiol. Plant Path. 19:145–52.

91. Kenfield, D. S., and G. A. Strobel. 1981. α-galactoside binding proteins from plant membranes: Isolation, characterization, and relation to helminthosporoside binding protein of sugarcane. Plant Physiol. 67:1174–80.

92. Kern, H. 1972. Phytotoxins produced by fusaria. In: Phytotoxins in plant diseases. R. K. S. Wood, A. Ballio, and A. Graniti (eds.). Academic Press, New York, pp. 35–44.

93. Kern, H., and S. Naef-Roth. 1967. Zwei neue, durch *Martiella*-Fusarien gebildete Naphthazarin-Derivate. Phytopath. Z. 60:316–24.

94. Király, Z., B. Barna, and T. Érsek. 1972. Hypersensitivity as a consequence, not the cause, of plant resistance to infection. Nature 239:456–58.

95. Kohmoto, K., R. P. Scheffer, and J. O. Whiteside. 1979. Host-selective toxins from *Alternaria citri*. Phytopathology 69:667–71.

96. Kono, Y., and J. M. Daly. 1979. Characterization of the host-specific pathotoxin produced by *Helminthosporium maydis*, race T, affecting corn with Texas male-sterile cytoplasm. Bioorg. Chem. 8:391–97.

97. Kono, Y., H. W. Knoche, and J. M. Daly. 1981. Structure: Fungal host specific. In: Toxins in plant disease. R. D. Durbin (ed.). Academic Press, New York, pp. 221–57.

98. Krupka, L. R. 1958. Increased ascorbic oxidase activity induced by the fungal toxin, victorin. Science 128:447–78.

99. Kuo, M. S., O. C. Yoder, and R. P. Scheffer. 1970. Comparative specificity of the toxins of *Helminthosporium carbonum* and *Helminthosporium victoriae*. Phytopathology 60:365–68.

100. Lado, P., A. Pennachion, F. R. Caldogno, S. Russi, and V. Silano. 1972. Comparison between some effects of fusicoccin and indole-3-acetic acid on cell enlargement in various plant materials. Physiol. Plant Path. 2:75–85.

101. Lee, D. L., and H. Rapoport. 1975. Synthesis of tabtoxinine-δ-lactam. J. Org. Chem. 40:3491–95.

102. Lesemann, D., and K. Rudolph. 1970. Die Bildung von Ribosomen-helices unter dem Einfluss des Toxins von *Pseudomo-*

nas phaseolicola in Blättern von Mangold (*Beta vulgaris* L.). Z. Pflanzenkr. Physiol. 62:108–15.

103. Leukel, R. W. 1948. *Periconia circinata* and its relation to milo disease. J. Agr. Res. 77:201–22.

104. Lewis, S., and R. N. Goodman. 1963. Morphological effects of colletotin on tomato and *Digitalis* foliage. Phytopathology 52:1273–76.

105. Luke, H. H., H. E. Warmke, and P. Hanchey. 1966. Effects of pathotoxin victorin on ultrastructure of root and leaf tissue of *Avena* species. Phytopathology 56:5l8–22.

106. Macko, V., K. Goodfriend, T. Wachs, J. A. A. Renwick, W. Acklin, and D. Arigoni. 1981. Characterization of the host-specific toxins produced by *Helminthosporium sacchari*, the causal organism of eyespot disease of sugarcane. Experientia 37:923–24.

107. Main C. E., and J. C. Walker. l97l. Physiological responses of susceptible and resistant cucumber to *Erwinia tracheiphila*. Phytopathology 6l:5l8–22.

108. Malini, S. 1966. Heavy metal chelates of fusaric acid: In vitro spectrophotometry. Phytopath. Z. 57:221–31.

109. Manning, J. M., S. Moore, W. B. Rowe, and A. Meister. 1969. Identification of L-methionine SR-sulfoximine as the distereoisomer of L-methionine S.-sulfoximine that inhibits glutamine synthetase. Biochemistry 8:2681–85.

110. Marrè, E. 1979. Fusicoccin: A tool in plant physiology. Ann. Rev. Plant Physiol. 30:273–88.

111. Marrè, E. 1980. Mechanism of action of phytotoxins affecting plasmalemma functions. Prog. Phytochem. 6:253–84.

112. Marrè, E., P. Lado, and F. R. Caldogno. 1972. On the relation between the growth promoting and the toxic action of fusicoccin. In:Phytotoxins in plant diseases. R. K. S. Wood, A. Ballio, and A. Graniti (eds.). Academic Press, New York, pp. 417–19.

113. Marrè, E., P. Lado, A. Ferroni, and A. B. Denti. 1974. Transmembrane potential increase induced by auxin, benzyladenine and fusicoccin. Correlation with proton extrusion and cell enlargement. Plant Sci. Lett. 2:257–65.

114. Mayo, P. de, E. Y. Spencer, and R. W. White. 1963. Terpenoids. IV. The structure and stereochemistry of helminthosporal. Can. J. Chem. 41:2996–3004.

115. McCammon, S. 1979. Prevention of the effects of victorin on susceptible oats. Ph. D. diss., Univ. of Kentucky, Lexington.

116. McIntire, F. C., W. H. Peterson, and A. J. Riker. 1942. A polysaccharide produced by the crown-gall organism. J. Biol. Chem. 143:491–96.

117. Mitchell, R. E. 1976. Isolation and structure of a chlorosis inducing toxin of *Pseudomonas phaseolicola*. Phytochemistry 15:1941–57.

118. Mitchell, R. E. 1981. Structure: Bacterial. In: Toxins in plant disease. R. D. Durbin (ed.). Academic Press, New York, pp. 259–93.

119. Mitchell, R. E., R. D. Durbin, and C. J. Lin 1978. *Pseudomonas tagetis* toxin: Purification and general biological and chemical characteristics. Phytopath. News 12:201.

120. Naef-Roth, S. 1959. Über die parasitogenen und toxigenen Veränderungen der Atmungsintensität bei Tomaten. Z. Pflkrankh. Pflpath. Pflschutz. 64:421–26.

121. Nakamura, M., and K. Ishibashi. 1958. On the new antibiotic "ophiobolin" produced by *Ophiobolus miyabeanus*. J. Agr. Chem. Soc. Jap. 32:732–44.

122. Nakamura, M., and H. Oku. 1960. Biochemical studies on *Cochliobolus miyabeanus*. IX. Detection of ophiobolin in the diseased rice leaves and its toxicity against higher plants. Ann. Takamine Lab. 12:266–71.

123. Nishimura, S., and R. P. Scheffer. 1965. Interactions between *Helminthosporium victoriae* spores and oat tissues. Phytopathology 55:629–34.

124. Novacky, A., and P. Hanchey. 1974. Depolarization of membrane potentials in oat roots treated with victorin. Physiol. Plant Path. 4:161–65.

125. Novacky, A., and A. L. Karr. 1977. Pathological alterations in cell membrane bioelectrical properties. In: Regulation of cell membrane activities in plants. E. Marrè and O. Cifferi (eds.). Elsevier North-Holland Biomedical Press, Amsterdam, pp. 137–44.

126. Novacky, A., and H. Wheeler. 1970. Victorin-induced changes in peroxidase isoenzymes in oats. Phytopathology 60:467–71.

127. Nozoe, S., M. Morisaki, K. Tsuda, Y. Iitaka, N. Takahashi, S. Tamura, K. Ishibashi, and M. Shirasaka. 1965. A structure of ophiobolin, C_{25}-terpenoid having a novel skeleton. J. Amer. Chem. Soc. 87:4968–70.

128. Ogasawara, N., K. Tamari, M. Suga, and K. Togashi. 1961. Biochemical studies on the blast disease of rice plant. Part XIX. Mechanism of the germination of spores of *Piricularia oryzae*: (1) Effects of piricularin and picolinic acid, the toxic substances produced by *Piricularia oryzae*, on the spore germination of this mould. J. Agr. Chem. Soc. Jap. 35:1315–23. (In Japanese.)

129. Oku, H. 1964. Host-parasite relation in *Helminthosporium* leaf spot disease of rice plant from the viewpoint of biochemical nature of the pathogen. In: Host-parasite relations in plant pathology. Z. Király and G. Ubrizsy (eds.). Res. Inst. Plant Prot., Budapest, pp. 183–91.

130. Oku, H. 1967. Role of parasite enzymes and toxins in development of characteristic symptoms in plant disease. In:The dynamic role of molecular constitutents in plant-parasite interaction. C. F. Mirocha and I. Uritani (eds.). Amer. Phytopath. Soc., St. Paul, MN, pp. 237–53.

131. Onesirosan, P., C. T. Mabuni, R. D. Durbin, R. B. Morin, D. H. Rich, and D. C. Arny. 1975. Toxin production by *Corynespora cassiicola*. Physiol. Plant Path. 5:289–95.

132. Orsenigo, M. 1957. Estrazione e purificazione della Cochliobolina, una tossina prodotta da *Helminthosporium oryzae*. Phytopath. Z. 29:189–96.

133. Owens, L. D. 1969. Toxins in plant disease: Structure and mode of action. Science 165:18–25.

134. Owens, L. D., J. F. Thompson, and P. V. Fennessey. 1972. Dihydrorhizobitoxine, a new ether amino-acid from *Rhizobium japonicum*. J. Chem. Soc. Chem. Commu. 715.

135. Owens, L. D., J. F. Thompson, R. G. Pitcher, and T. Williams. 1965. Rhizobial-induced chlorosis in soybeans: Isolation, production in nodules, and varietal specificity of the toxin. Plant Physiol. 40:927–30.

136. Owens, L. D., J. F. Thompson, R. G. Pitcher, and T. Williams. 1972. Structure of rhizobitoxine, an antimetabolic enolether amino-acid from *Rhizobium japonicum*. J. Chem. Soc. Chem. Commu. 714.

137. Patel, P. N., and J. C. Walker. 1963. Changes in free amino acids and amide content of resistant and susceptible beans after infection with the halo blight organism. Phytopathology 53:522–28.

138. Patil, S. S. 1972. Purification of the phytotoxin from *Pseudomonas phaseolicola*. Phytopathology 62:782.

139. Patil, S. S. 1974. Toxins produced by phytopathogenic bacteria. Ann. Rev. Phytopath. 12:259–79.

140. Patil, S. S., L. Q. Tam, and P. E. Kolattukudy. l972. Isolation and the mode of action of the toxin from *Pseudomonas phaseolicola*. In: Phytotoxins in plant diseases. R. K. S. Wood, A. Ballio, and A. Graniti (eds.). Academic Press, New York, pp. 365–72.

141. Patil, S. S., L. Q. Tam, and W. S. Sakai. 1972. Mode of action of the toxin from *Pseudomonas phaseolicola*. I. Toxin specificity, chlorosis, and ornithine accumulation. Plant Physiol. 49:803–7.

142. Payne, G., H. W. Knoche, Y. Kono, and J. M. Daly. 1980. Biological activity of host-specific pathotoxin produced by *Bipolaris* (*Helminthosporium maydis*) race T. Physiol. Plant Path. 16:227–39.

143. Pegg, K. G. 1966. Studies of a strain of *Alternaria citri* Pierce, the causal organism of brown spot of Emperor mandarin. Queensland J. Agr. Animal Sci. 23:15–28.

144. Pennypacker, B. W., P. E. Nelson, and R. S. Dickey. 1974. Histopathology of chrysanthemum stems artificially inoculated with *Erwinia chrysathemi*. Phytopathology 64:1344–53.

145. Person, A., and G. Beaud. 1980. Shut-off of host protein synthesis in vaccinia- virus-infected cells exposed to cordycepin. A study in vitro. Eur. J. Biochem. 103:85–93.

146. Person, A., F. Ben-Hamida, and G. Beaud. 1980. Inhibition of 40s-Met-tRNA ribosomal initiation complex formation by vaccinia virus. Nature 287: 355–57.

147. Politis, D. J., and R. N. Goodman. 1980. Fine structure of extracellular polysaccharide of *Erwinia amylovora*. Appl. Environ. Microbiol. 40:569–607.

148. Pringle, R. B., and A. C. Braun. 1957. The isolation of the toxin of *Helminthosporium victoriae*. Phytopathology 47:369–371.

149. Pringle, R. B., and R. P. Scheffer. 1967. Multiple host-specific toxins from *Periconia circinata*. Phytopathology. 57:530–32.

150. Pringle, R. B., and R. P. Scheffer. 1967. Isolation of the host-specific toxin and a related substance with nonspecific toxicity from *Helminthosporium carbonum*. Phytopathology 57:1169–72.

151. Raghow, R., and A. Granoff. 1979. Macromolecular synthesis in cells infected by frog virus. X. Inhibition of cellular protein synthesis by the inactivated virus. Virology 98:319–27.

152. Rai, P. V., and G. A. Strobel. 1969. Phytotoxic glycopeptides produced by *Corynebacterium michiganense* II. Biological properties. Phytopathology 59:53–57.

153. Ries, S. M., and G. A. Strobel. 1972. A phytotoxic glycopeptide from cultures of *Corynebacterium insidiosum*. Plant Physiol. 46:676–84.

154. Ries, S. M., and G. A. Strobel. 1972. Biological properties and pathological role of a phytotoxic glycopeptide from *Corynebacterium insidiosum*. Physiol. Plant Path. 2:133–42.

155. Roantree, R. J. 1971. The relationships of lipopolysaccharide structure to bacterial virulence. In: Microbiol toxins. Bacterial endotoxins. S. J. Ajl and S. Kadis (eds.). Academic Press, New York, 5:1–37.

156. Roberts, P. L., and K. R. Wood. 1981. Comparison of protein synthesis in tobacco systemically infected with a severe or a mild strain of cucumber mosaic virus. Phytopath. Z. 102:257–65.

157. Rose, J. K., H. Trachsel, K. Leong, and D. Baltimore. 1978. Inhibition of translation by poliovirus: Inactivation of a specific initiation factor. Proc. Nat. Acad. Sci. 75:2732–36.

158. Rudolph, K. 1969. Ein Phytotoxisches polysaccharid. Naturwiss. 56:569–70.

159. Rudolph, K., and M. A. Stahmann. 1966. The accumulation of L-ornithine in halo-blight infected bean plants (*Phaseolus vulgaris* L.) induced by the toxin of the pathogen *Pseudomonas phaseolicola* (Burkh.) Dowson. Phytopath. Z. 57:29–46.

160. Saftner, R. A. 1972. Promotion of cell elongation by host-specific fungal toxins. Master's thesis, Ohio State Univ., Columbus.

161. Sagot, J., and G. Beaud. 1979. Phosphorylation in vivo of a vaccinia-virus structural protein found associated with the ribosomes from infected cells. Eur. J. Biochem. 98:131–40.

162. Samaddar, K. R., and R. P. Scheffer. 1968. Effect of the specific toxin in *Helminthosporium victoriae* on host cell membranes. Plant Physiol. 43:21–28.

163. Samaddar, K. R., and R. P. Scheffer. 1970. Effects of *Helminthosporium victoriae* toxin on germination and aleurone secretion by resistant and susceptible seeds. Plant Physiol. 45:586–90.

164. Sawamura, K. 1966. Studies on the spotted disease of apples. VI. On the host-specific toxin of *Alternaria mali*. Ryall. Hort. Res. Sta. Jap. (Morioka), Series C, 4:43–59.

165. Scheffer, R., and A. J. Ullstrup. 1965. A host-specific toxic metabolite from *Helminthosporium carbonum*. Phytopathology 55:1037–38.

166. Scheffer, R. J., and D. M. Elgersma. 1981. Detection of a phytotoxic glycopeptide produced by *Ophiostoma ulmi* in elm by enzyme-linked immunospecific assay (ELISA). Physiol. Plant Path. 18:27–32.

167. Scheffer, R. P., and K. R. Samaddar. 1970. Host-specific toxins as determinants of pathogenicity. Rec. Adv. Phytochem. 3:123–42.

168. Scheffer, R. P., and O. C. Yoder. 1972. Host-specific toxins and selective toxicity. In: Phytotoxins in plant diseases. R. K. S. Wood, A. Ballio, and A. Graniti (eds.). Academic Press, New York, pp. 19–33.

169. Schrom, M., and R. Bablanian. 1979. Inhibition of protein synthesis by vaccinia virus. I. Characterization of an inhibited cell-free protein synthesizing system from infected cells. Virology 99:319–28.

170. Selman, B. R., and R. D. Durbin. 1978. Evidence for a catalytic function of the coupling factor 1 protein reconstituted with chloroplast thylakoid membranes. Biochim. Biophys. Acta 502:27–37.

171. Sequeira, L., and T. L. Graham. 1977. Agglutination of avirulent strains of *Pseudomonas solanacearum* by potato lectin. Physiol. Plant Path. 11:43–54.

172. Sijam, K., A. L. Karr, and R. N. Goodman. 1983. Comparision of the extracellular polysaccharides produced by *Erwinia amylovora* in apple tissue and culture medium. Physiol. Plant Path. 22:221–31.

173. Sijam, K., R. N. Goodman, and A. L. Karr. 1984. The effect of salts on the viscosity and wilt inducing capacity of the capsular polysaccharide of *Erwinia amylovora*. Physiol. Plant Path. 26:231–39.

174. Sinden, S. L., and J. E. DeVay. 1967. The nature of the wide-spectrum antibiotic produced by pathogenic strains of *Pseudomonas syringae* and its role in the bacterial canker disease of *Prunus persicae*. Phytopathology 57:102.

175. Sinden, S. L., J. E. DeVay, and P. A. Backman. 1971. Properties of syringomycin, a wide spectrum antibiotic and phytotoxin produced by *Pseudomonas syringae* and its role in bacterial canker disease of peach trees. Physiol. Plant Path. 1:199–213.

176. Sinden, S. L., and R. D. Durbin. 1968. Glutamine synthetase inhibition: Possible mode of action of wildfire toxin from *Pseudomonas tabaci*. Nature 219:379–80.

177. Sinden, S. L., and R. D. Durbin. 1969. Some comparisons of chlorosis-inducing pseudomonad toxins. Phytopathology 59:249–50.

178. Sinden, S. L., and R. D. Durbin. 1970. A comparision of the chlorosis-inducing toxin from *Pseudomonas coronafaciens* with wildfire toxin from *Pseudomonas tabaci*. Phytopathology 60:360–64.

179. Sloneker, J. H., and A. Jones. 1962. Extracellular bacterial polysaccharide from *Xanthomonas campestris* NRRL B-1459. Part 1. Constitution. Can. J. Chem. 40: 2066–71.

180. Smedegaard-Petersen, V. 1977. Isolation of two toxins

produced by *Pyrenophora teres* and their significance in disease development of net-spot blotch of barley. Physiol. Plant Path. 10:203–11.

181. Smedegaard-Petersen, V. 1977. Respiratory changes of barley leaves infected with *Pyrenophora teres* or affected by isolated toxins of this fungus. Physiol. Plant Path. 10:213–20.

182. Smedegaard-Petersen, V., and R. R. Nelson. 1969. The production of a host-specific pathotoxin by *Cochliobolus heterostrophus*. Can. J. Bot. 47:951–57.

183. Spencer, J. F. T., and P. A. J. Gorin. 1961. The occurrence in the host plant of physiologically active gums produced by *Cornebacterium insidiosum* and *Corynebacterium sepedonicum*. Can. J. Microbiol. 7:185–88.

184. Stahl, C., L. N. Vanderhoeff, N. Siegel, and J. P. Helgeson. 1973. *Fusarium tricinctum* T-s toxin inhibits auxin-promoted elongation in soybean hypocotyl. Plant Physiol. 52:663–66.

185. Staskawicz, B. J., and N. J. Panopoulos. l979. A rapid and sensitive microbiological assay for phaseolotoxin. Phytopathology 69:663–66.

186. Steele, J. A., T. F. Uchytil, R. D. Durbin, P. Bhatnagar, and D. H. Rich. 1976. Chloroplast coupling factor 1: A species-specific receptor for tentoxin. Proc. Nat. Acad. Sci. 73:2245–48.

187. Steiner, G. W., and R. S. Byther. 1971. Partial characterization and use of a host-specific toxin from *Helminthosporium sacchari* on sugarcane. Phytopathology 61:691–96.

188. Steiner, G. W., and G. A. Strobel. 1971. Helminthosporoside, a host-specific toxin from *Helminthosporium sacchari*. J. Biol. Chem. 246:4350–57.

189. Stewart, W. W. 1971. Isolation and proof of structure of wildfire toxin. Nature 229:174–78.

190. Stoessl, A. 1981. Structure and biogenetic relations: Fungal nonhost-specific. In: Toxins in plant disease. R. D. Durbin (ed.). Academic Press, New York, pp. 109–219.

191. Stout, R. G., and R. E. Cleland. 1980. Partial characterization of fusicoccin binding to receptor sites on oat root membranes. Plant Physiol. 66:353–59.

192. Strobel, G. A. 1967. Purification and properties of phytotoxic polysaccharide produced by *Corynebacterium sepedonicum*. Plant Physiol. 42:1433–41.

193. Strobel. G. A. 1970. A phytotoxic glycopeptide from potato plants infected with *Corynebacterium sepedonicum*. J. Biol. Chem. 345:32–38.

194. Strobel, G. A. 1971. Comparative biochemistry of the toxic glycopeptides produced by some plant pathogenic corynebacteria. In: Proc. Third Int. Conf. Plant Pathogenic Bact., H. P. Maas Geesteranus (ed.). Pudoc, Wageningen, Netherlands, pp. 357–65.

195. Strobel, G. A. 1973. The helminthosporoside-binding protein of sugarcane. J. Biol. Chem. 248:1321–28.

196. Strobel, G. A. 1973. Biochemical basis of the resistance of sugarcane to eyespot disease. Proc. Nat. Acad. Sci. 70:1693–96.

197. Strobel, G. A. 1974. The toxin-binding protein of sugarcane, its role in the plant and in disease development. Proc. Nat. Acad. Sci. 71:4232–36.

198. Strobel, G. A. 1979. The relationship between membrane ATPase activity in sugarcane and heat-induced resistance to helminthosporoside. Biochim. Biophys. Acta 554:460–68.

199. Strobel, G. A. 1982. Phytotoxins. Ann. Rev. Biochem. 51:309–33.

200. Strobel, G. A., and K. D. Hapner. 1975. Transfer of toxin susceptibility to plant protoplasts via the helminthosporoside binding protein of sugarcane. Biochem. Biophys. Res. Comm. 63:1151–56.

201. Strobel, G. A., and W. M. Hess. 1968. Biological activity of a phytotoxic glycopeptide produced by *Corynebacterium sepedonicum*. Plant Physiol. 43: 1673–88.

202. Strobel, G. A., and W. M. Hess. 1974. Evidence for the presence of the toxin- binding protein on the plasma membrane of sugarcane cells. Proc. Nat. Acad. Sci. 71:1413–17.

203. Strobel, G. A., W. M. Hess, and G. W. Steiner. 1972. Ultrastructure of cells in toxin-treated and *Helminthosporium sacchari*-infected sugarcane leaves. Phytopathology 62:339–45.

204. Suhayda, C. H., and R. N. Goodman. 1981. Early proliferation and migration and subsequent xylem occlusion by *Erwinia amylovora* and the fate of its extracellular polysaccharide (EPS) in apple shoots. Phytopathology 7l:697–707.

205. Tam, L. Q., and S. S. Patil. 1972. Mode of action of the toxin from *Pseudomonas phaseolicola*. II. Mechanism of inhibition of bean ornithine carbamolytransferase. Plant Physiol. 49:808–12.

206. Tamari, K., and J. Kaji. 1954. Studies on the mechanism of the plant growth inhibitory action of fusarinic acid, a toxin produced by *Gibberella fujikuoroi* WR., which is the causative mould of the bakanae disease of rice plant. Bull. Fac. Agr. Niigata Univ. 6:1–25.

207. Tamari, K., and J. Kaji. 1955. On the biochemical studies of the blast mould (*Piricularia oryzae*, Cavara), the causative mould of the blast disease of rice plant. Part 2. Studies on the physiological action of the piricularin, a toxin produced by the blast mould on rice plants. J. Agr. Chem. Soc. Jap. 29:85–190.

208. Tamari, K., N. Ogasawara, and J. Kaji. 196. Biochemical products of the metabolism of *Piricularia oryzae*. In: The rice blast disease. Johns Hopkins Univ. Press, Baltimore, pp. 35–68.

209. Tanaka, S. 1933. Studies on black spot disease of the Japanese pear (*Pyrus serotina* Rehd.). Mem. College Agr. Kyoto Imp. Univ. 28:1–31.

210. Taniguchi, E., and G. A. White. 1967. Site of action of the phytotoxin, helminthosporal. Biochem. Biophys. Res. Comm. 28:879–85.

211. Taylor, P. A., and R. D. Durbin. 1973. The production and properties of chlorosis-inducing toxins from a pseudomonad attacking timothy. Physiol. Plant Path. 3:9–17.

212. Taylor, P. A., D. P. Maxwell, and R. D. Durbin. 1971. Occurrence in Wisconsin of halo blight of timothy incited by a *Pseudomonas* species. Plant Dis. Reptr. 55: 361–62.

213. Taylor, P. A., H. K. Schnoes, and R. D. Durbin. 1972. Characterization of chlorosis-inducing toxins from a plant pathogenic *Pseudomonas* sp. Biochim. Biophys. Acta 286:107–17.

214. Templeton, G, E. 1972. *Alternaria* toxins related to pathogenesis in plants. In: Microbial toxins. Fungal toxins. S. Kadis, A. Ceigler, and S. J. Ajl (eds.). Academic Press, New York, 8:169–92.

215. Tipton, C. L., M. H. Mondal, and J. Uhlig. 1973. Inhibition of K+-stimulated ATPase of maize root microsomes by *Helminthosporium maydis* race T pathotoxin. Biochem. Biophys. Res. Comm. 51:725–28.

216. Trachsel, H., N. Sonenberg, A. J. Shatkin, J. K. Rose, K. Leong, J. E. Bergmann, J. Gordon, and D. Baltimore. 1980. Purification of a factor that restores translation of vesicular stomatitis virus m-RNA in extracts from poliovirus- infected Hela cells. Proc. Nat. Acad. Sci. 77:770–74.

217. Trimboli, R., P. C. Fahy, and K. F. Baker. 1978. Apical chlorosis and leaf spot of *Tagetis* spp. caused by *Pseudomonas tagetes* Hellmers. Aust. J. Agric. Res. 29:831–39.

218. Turner, J. G. 1981. Tabtoxin, produced by *Pseudomonas tabaci*, decreases *Nicotiana tabacum* glutamine synthetase in vivo and causes accumulation of ammonia. Physiol. Plant Path. l9:57–67.

219. Turner, J. G., and J. M. Debbage. 1982. Tabtoxin-induced symptoms are associated with the accumulation of ammonia formed during photorespiration. Physiol. Plant Path. 20:223–33.

220. Turner, M. T., and C. A. Martinson. 1972. Susceptibility

of corn lines to *Helminthosporium maydis* toxin. Plant Dis. Reptr. 56:29–32.

221. Turner, N. C. 1972. Fusicoccin—a phytotoxin that opens stomata. In: Phytotoxins in plant diseases. R. K. S. Wood, A. Ballio, and A. Graniti (eds.). Academic Press, New York, pp. 399–402.

222. Turner, N. C., and A. Graniti. 1969. Fusicoccin: A fungal toxin that opens stomata. Nature. 223:1070–71.

223. Vörös, J., and R. N. Goodman. 1965. Filamentous forms of *Erwinia amylovora*. Phytopathology 55:876–79.

224. Waitz, L., and W. Schwartz. 1956. Untersuchungen über die von *Pseudomonas phaseolicola* (Burkh.) Hervorgerufene Fettfleckenkrankheit der Bohne. Phytopath. Z. 26:297–312.

225. Wheeler, H. 1976. The role of phytotoxins in specificity. In: Specificity in plant diseases. R. K. S. Wood and A. Graniti (eds.). Plenum Press, New York, pp. 217–30.

226. Wheeler, H., and H. S. Black. 1963. Effects of *Helminthosporium victoriae* and victorin upon permeability. Amer. J. Bot. 50:686–93.

227. Wheeler, H., and E. Elbel. 1979. Time-course and antioxidant inhibition of ethylene production by victorin-treated oat leaves. Phytopathology 69:32–34.

228. Wheeler, H., and P. J. Hanchey. 1966. Respiratory control: Loss in mitochondria from diseased plants. Science 154:1569–71.

229. Wheeler, H., and H. H. Luke. 1963. Microbial toxins in plant diseases. Ann. Rev. Microbiol. 17:223–42.

230. Wheeler, H., A. S. Williams, and L. D. Young. 1971.

Helminthosporium maydis T-toxin as an indicator of resistance to southern leaf blight. Plant Dis. Reptr. 55:667–71.

231. White, J. A., O. H. Calvert, and M. F. Brown. 1973. Ultrastructural changes in corn leaves after inoculation with *Helminthosporium, maydis*, race T. Phytopatology 63:296–300.

232. Wood, R. K. S., A. Ballio, and A. Graniti (eds.). 1972. Phytotoxins in plant diseases. Academic Press, New York.

233. Woolley, D. W. 1948. Studies on the structure of lycomarasmin. J. Biol. Chem. 176:1291–98.

234. Woolley, D. W., R. B. Pringle, and A. C. Braun. 1952. Isolation of the phytopathogenic toxin of *Pseudomonas tabaci*, an antagonist of methionine. J. Biol. Chem. 197:409–17.

235. Wu, L. C., and R. P. Scheffer. 1962. Some effects of *Fusarium* infection of tomato on growth, oxidation and phosphorylation. Phytopathology 52:354–58.

236. Yoder, O. C. 1980. Toxins in plant pathogenesis. Ann. Rev. Phytopath. 18:103–29.

237. Yoder, O. C. 1981. Assay. In: Toxins in plant disease. R. D. Durbin (ed.). Academic Press, New York, pp. 45–78.

238. Zeller, W., and K. Rudolph. 1972. Einfluss des toxins von *Pseudomonas phaseolicola* auf die Permeabilität und den osmotischen wert von Mangoldblattern (*Beta vulgaris*, L.) Z. Pflanzenphysiol. 67:183–87.

239. Zimmermann, M. H., and J. McDonough. 1978. Dysfunction in the flow of food. In: Plant disease: An advanced treatise. J. S. Horsfall and E. B. Cowling (eds.). Academic Press, New York, 3:117–40.

10

Resistance to Infection

Introduction

Two major criteria control resistance, and, probably, each is mediated by more than one gene. Specifically, these are the host plants' pathogen-suppressing capacity and the pathogen's virulence. According to this hypothesis, each criterion is expressed as a range or in degrees, rather than as an "all-or-nothing" characteristic.

For example, *Helminthosporium victoriae* has, as its virulence determinant, the toxin victorin. Not only is virulence linked a priori to the production of the toxin, but isolates producing more toxin are more virulent (257). It has also been shown that sensitivity to the toxin is inherent in the oat cultivar Victoria and its progeny. Seedlings from resistant cultivars tolerated more than 4 $\times$ 10^5 times as much toxin (257). *Erwinia amylovora* has at least one known virulence determinant, its extracellular polysaccharide "amylovorin" (154), the production of which by an array of isolates appears to be proportional to their virulence in host tissue (20). The resistance features of the host plant, apple, are less well known. However, it would seem that phenolics (1) and a small protein may be involved (343, 344). Their levels in tissue and the rates at which they can be produced or released may determine just how resistant an apple cultivar is to *E. amylovora*. Although the virulence determinants for *Pseudomonas savastanoi* have been precisely assessed as its genes for IAA synthesis, the mode of resistance of the olive cultivars to the pathogen remains to be described (68, 69).

In any discussion of resistance the two reactants must be simultaneously considered, and it may be helpful to visualize the *interaction* as a totally new and single system. This system has been called by Loegering the *aegricorpus* (251) and is defined as "the single manifestation of specific genetic interactions in and between host and pathogen. Thus, Loegering's term *host reaction* type simultaneously expresses host resistance/susceptibility and the pathogens' avirulence/virulence as modified by the environment. Hence, in discussing host reaction we infer the genetic expressions of the two interactants.

Clearly, the coevolution of host and pathogen has permitted both to develop genetic characteristics that confer survival on each. Resistance, therefore, is an expression of the degree to which the host is able to restrict the pathogens' efforts to impair plant function.

Resistance to Viral Pathogens

The interaction between plant and virus can result in a wide diversity of responses, depending on the genome of both. Some examples are indicated in Table 10–1.

In a *susceptible* host, we might expect to observe characteristic symptoms and substantial systemic virus multiplication. *Resistance*, however, can take several forms. In many instances, virus remains restricted to cells at or adjacent to the site of infection and is unable to spread systemically. Cells at the center of this area may become necrotic, in which case the reaction is one of *hypersensitivity*. In other instances, however, systemic spread of virus may be restricted in the absence of necrosis, cells at the site of infection becoming chlorotic, or showing no sign of infection at all. In either case, although it may suffer localized damage, the plant is protected from the potentially severe consequences of systemic invasion and in that context may be said to be *resistant*. In addition, a plant is often said to be *resistant* when virus spreads systemically from the site of infection, but its accumulation is considerably less and symptoms generally milder than in those situationsresulting in susceptibility. This may be alternatively regarded as *low susceptibility* and, taken to its extreme, becomes *immunity*. *Tolerance*, too, may be regarded as a form of resistance, since there may be little visible expression of considerable virus multiplication. The plant may suffer little damage or loss of yield.

Resistance of various types, including cross-protection and localized and systemic resistance, can also be *induced* by previous virus infection or by treatment with various chemicals. Although induced resistance could be regarded as *immunity*, we use the term immunity here to signify extreme resistance, rather than resistance induced by previous infection or by other treatments of the plant.

Not all virus infections will fall clearly into the classes depicted in the table, which should be regarded as a guide to terminology. There is almost certainly a continuous spectrum of responses, with respect to both virus multiplication and symptom expression, controlled by equally diverse mechanisms. Additionally, both symptom expression and virus accumulation are dependent on environmental factors and plant age and may vary among different parts of the same plant. Many examples

Table 10–1. Host responses to virus infection (modified from Ragetli, 333).

	Response	Symptom severity	Virus multiplication
Susceptibility		+++	+++
Resistance:	immunity	−	−
	low susceptibility[a]	±	±
	tolerance	±	++
	hypersensitivity	+++[b]	+

a. May often be referred to as *resistance*.
b. Severe with respect to cells within the lesion.

have been documented elsewhere (44, 168, 232, 268, 269, 333).

The Hypersensitive Response

As indicated above, there are many virus/host combinations in which the plant responds to leaf inoculation with the formation of local lesions, groups of cells at the site of infection that display visible signs of metabolic change. If the reaction is one of hypersensitivity, these areas become necrotic, and virus is usually limited to the necrotic cells and the ones immediately surrounding them. However, localization need not necessarily accompany necrosis and hypersensitivity, and there are many instances (TRSV- and PVX-inoculated White Burley tobacco, for example), in which there is both localized necrosis and systemic infection. Conversely, necrosis need not be a prerequisite for localization. Lesions (e.g. those produced on *C. quinoa* or *C. amaranticolor* by CMV) may only become chlorotic, yet systemic invasion is prevented. In extreme cases, there may be no visible symptoms at all. Cucumber cotyledons inoculated with TMV are apparently symptomless, until cotyledons are decolorized and stained to reveal the presence of starch, with groups of cells in which starch has accumulated revealing the sites of infection. Nevertheless, virus particles are localized to cells in and adjacent to the starch lesions. It may be, therefore, that the mechanisms that lead to virus localization are different from those that lead to necrosis (e.g. 212). The former is not a necessary consequence of the latter. It is also of interest that some plants that react hypersensitively to leaf inoculation, and in which virus movement from the lesion area is prevented, may be systemically infected by other means (166, 446, 467). *N. glutinosa*, for example, becomes systemically infected with TMV when grafted to infected tobacco stock (446).

In addition to constituting a useful form of resistance, these responses provide a valuable means of plant virus bioassay (37, 269), since the number of lesions appearing on a leaf following virus inoculation is related to the number of infective particles in the inoculum.

The expression of hypersensitivity, though influenced by environmental factors such as temperature and

plant nutrition, depends primarily on the genome of both host and virus. In some cases, hypersensitivity may be controlled by a single dominant host gene. *Nicotiana glutinosa*, for example, responds hypersensitively to TMV, a response controlled by the N gene (184). This gene has been introduced into tobacco varieties producing Samsun NN and Xanthi nc, which are also hypersensitive to TMV at around 22 C. At higher temperatures, above about 28 C, lesions are chlorotic, rather than necrotic, and infection spreads systemically (352). In other cases, the situation may be a more complicated one (185). *Nicotiana sylvestris*, for example, carries the gene N′ and, while reacting hypersensitively to the U2 strain of TMV, becomes systemically infected with the U1 strain (379). Although most tobacco varieties become systemically infected with most strains of TMV, Havana 425 generally produces lesions. Conversely, while White Burley tobacco, said also to be N′ (423), becomes systemically infected with most strains of TMV, the *flavum* strain elicits a hypersensitive response (123). Cowpea (*Vigna sinensis*) is hypersensitive to most strains of CMV, though some (e.g. CMV–14) become systemic, and there are cowpea lines that become systemically infected by the majority of CMV strains. CMV-N produces lesions on most tobacco cultivars, while the majority of CMV strains infect systemically (416). Both the Tm-2 and Tm-2a genes, introduced into tomato cultivars from wild *Lycopersicon*, can confer hypersensitivity to TMV, especially when in heterozygous form. Generally, however, they cause effective suppression of TMV multiplication without necrosis, particularly when present in homozygous form or in combination with the Tm-1 gene. Unfortunately, we have not reached the point at which anything can be said about the gene products of either host or virus that mediate this response, or indeed any other type of symptom.

The time of appearance, rate of expansion and size of the lesion also depend on the virus and host, and on environmental influences. Some lesions appear quickly, TMV-induced lesions on Samsun NN tobacco (Fig. 10–1) appearing by ca. 28 h on plants maintained at 24 C, with the rate of virus synthesis slowing down progressively from 60 h (226); CMV lesions appear on cowpea within 36 h at 20 C. Alternatively, lesion development can take much longer, symptoms of PVS-inoculated *C. quinoa* leaves appearing as water-soaked lesions after 5 d, developing into well-defined, chlorotic lesions within 8 d at 27 C (378). While some lesions may have attained almost their final size when they appear, others continue to expand for a short period, several days, or indefinitely (112, 335). TMV lesions in the N′ gene-containing White Burley tobacco continue to grow, and size of lesions caused by different strains is correlated with the thermal stability of the coat protein. It is suggested that the properties of the coat protein subunit can influence the interaction with the localizing mechanism (127). Although the amount of virus within a lesion may be proportional to its size (243, 335, 403, 452), this need not be the case, particularly in sys-

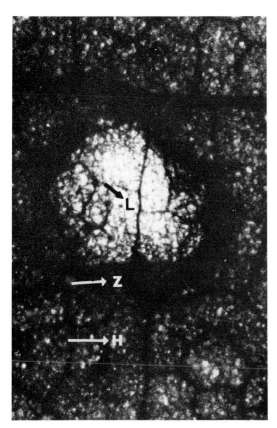

Fig. 10–1. Surface view of a TMV-induced local lesion in a portion of a mature leaf of *Nicotiana tabacum* L. var. Samsun NN, indicating healthy cells (H), darkly stained cells in a zone around the lesion (Z), and lightly stained necrotic cells of the lesion (L). Tissue fixed in glutaraldehyde and stained with OsO₄; calibration 1.0 mm (photographed by H.H. Lyon, from Israel and Ross, 196).

temically resistant tissue. Lesions induced by challenge inoculation of upper leaves of White Burley tobacco following primary inoculation of the lower leaves were smaller in size and fewer in number than on control leaves, though there was little reduction in virus multiplication (124).

The cytological changes in leaf tissues resulting from virus infection have been investigated in several virus/host combinations using both light microscopy and scanning and transmission electron microscopy, often with the ultimate objective of correlating biochemical with cytological changes and their consequences. Weintraub and Ragetli (448), for example, followed the progress of cells within TMV-induced lesions in *N. glutinosa* at intervals up to 78 h after inoculation, under conditions where lesions first became visible at about 33 h after inoculation.

In palisade and mesophyll cells, at the time of inoculation, chloroplasts contained a few small starch grains, and one of the first apparent changes, between 8 and 16

h, was an increase in size of the starch grains, which became surrounded by a swollen, opalescent vacuole, and a distortion of the lamellae. This appeared to be a transient phenomenon, with few enlarged starch grains present in infected cells between 28 and 32 h. At 24 h, cytoplasmic contents had moved from the cell wall toward the vacuole, and cytoplasmic membranes were beginning to rupture. The process of cytoplasmic membrane disruption was complete by 52 h, with cell contents becoming distributed throughout the cell space, and the appearance of small, vesiclelike inclusions. Also by this time, chloroplast membranes were almost completely disintegrated. Lamellae in the remaining parts of the chloroplasts were additionally distorted by large vacuoles, though many lamellae remained in stacks until the final stages of degeneration. Mitochondria appeared to increase in number between 22 and 44 h, persisting until 60 h; some were opaque, others beginning to disintegrate. From 60–78 h, discontinuities appeared in the membranes of some, and by 78 h some cells appeared to be devoid of recognizable mitochondria. It was apparent by this time that the infected cells were in their final stages of degeneration, with compression of the contents toward one end of the cell, where they became opaque; very little remained of the organelles. Finally, cells collapsed completely, becoming small, opaque, and compact masses, with cell walls breaking away from each other, producing large, intercellular spaces.

A similar picture of cell degeneration within the lesion area has been observed in TMV-inoculated Samsun NN tobacco (74, 196). At 24 C, the first cells become necrotic about 18 h after inoculation (402, 403), though at this stage lesions are not macroscopically visible. In experiments by Da Graca and Martin (73, 74), lesions first appeared 36 h after inoculation and became watersoaked, with collapsed areas of cells measuring approximately 0.3 mm in diameter at 40 h and by 72 h becoming slightly larger and necrotic. By both 40 and 72 h, cytoplasmic membranes in a large proportion of cells within the lesion were degenerating, organelles were being disrupted, and envelopes ruptured. The nuclei of a large portion of these disorganized cells were surprisingly intact, some becoming abnormally granular with enlarged nucleoli. Within mature (seven days old) lesions (196), palisade cells were collapsed and lacked vacuoles and most of their membrane-bound organelles and contained starch grains, ribosomes, recognizable chloroplast fragments, and, occasionally, intact nuclei.

The cellular damage within necrotic lesions induced in pinto bean by TMV has been graphically revealed by the scanning electron microscope (Fig. 10–2) (394). Here, with lesions first becoming visible 30 h after infection and reaching their final size after 72 h, there was little effect on mesophyll and lower epidermal cells, though the severe necrotic collapse of upper epidermal and palisade cells was clearly evident.

Comparable changes, broadly similar though differing in detail, have been reported for other viruses and hosts, including PVS-inoculated *C. quinoa* (378),

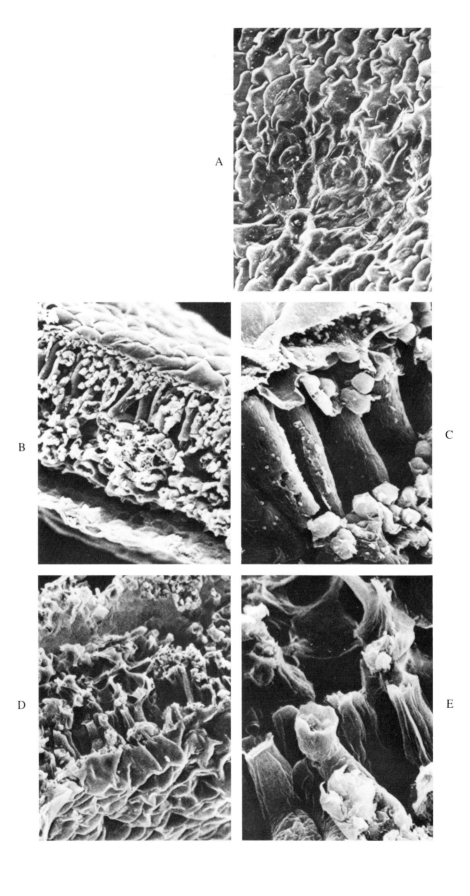

kidney bean infected with PVM (181, 418) and *G. globosa* inoculated with beet mosaic (349), PVX (3), or TBSV (11, 12, 322, 323).

Lesion Formation

The microscopic evidence suggests that one of the first changes in virus-infected cells destined to become necrotic is a loss of osmotic control, with disruption of the plasmalemma, tonoplast, and organelle membranes. Although there have been suggestions, the primary cause of these and subsequent changes remains obscure. The visual observations are supported by experiments designed to measure electrolyte leakage from virus-infected leaf tissue. There was stimulation of electrolyte leakage from TMV lesions in Xanthi nc tobacco (451) and CMV lesions in cowpea (205) several hours before lesion appearance, though not from leaves of TMV-inoculated Samsun tobacco, in which a systemic, rather than localized necrotic reaction develops. TMV inoculation of *N. glutinosa* also appears to cause alterations of cell membranes with concomitant leakage of cell constituents (308). It may be that increased permeability is a consequence of stimulated lipoxygenase activity (205) and is a primary event leading to cell disruption and necrosis, though it is not clear how it is mediated.

It is of interest that lesion formation seems to depend on the presence of the epidermis. No lesions developed following inoculation of TMV onto the exposed mesophyll cells of detached *N. glutinosa* or Xanthi nc tobacco leaves from which the lower epidermis had been removed (371); a few developed in *N. glutinosa*, though they appeared later than lesions on control leaves and were larger and paler in color (372). At the same time, the amount of virus produced, though not as great as in tissue with intact epidermis, was not negligible, indicating that the tissue had become infected, but necrosis had not, in the majority of cases, developed. Virus also multiplied in lower epidermis removed soon after inoculation with TMV, though again necrosis did not develop (see also 402). However, when epidermis was removed 8–10 h after inoculation, necrotic lesions developed on the epidermis-stripped tissue, whether or not the upper epidermis was also removed. Necrosis did not develop on stripped leaves when the epidermis was removed within 5 h of inoculation, even though virus was present.

Similarly, lesions failed to develop on the lower epidermis of CMV-inoculated cowpea leaves removed 2–3 h after infection, though there is doubt that virus multiplication had occurred. However, lesions did form on epidermis-free tissue when the epidermis was removed before lesions appeared (96). It would appear, therefore, that interaction between mesophyll and epidermal cells is required, at least in the early stages of lesion development, though the nature of this interaction remains unresolved.

In addition to appearing in normally hypersensitive hosts, necrotic lesions can be induced by various treatments in hosts, particularly tobacco, that a virus normally infects systemically. Such treatments include hot (50 C) water (122), ice-cold water (306), actinomycin D, chromomycin A3, or ultraviolet irradiation (307). Ultraviolet irradiation and dark treatment also induce lesion formation in *N. glutinosa*, Samsun NN, and Xanthi nc tobacco inoculated with TMV and maintained at 30 C, conditions under which a systemic infection would normally result. These induced lesions may differ in appearance from those in normally hypersensitive hosts and may also depend on the treatment selected (306) and the time of application in relation to inoculation (122). Application of one of these treatments may prove a convenient technique for enabling the initial sites of infection on a leaf surface to be identified. It remains to be determined, however, whether formation of these lesions is related to the normal mechanism of lesion production.

Symptom Limitation

When local lesions are formed, whether chlorotic or necrotic, there is clearly a limitation in the spread of symptoms, which is usually reflected in the cessation of lesion expansion. Symptom limitation may also be accompanied by a restriction in virus movement; the two may or may not be related.

Experiments by Shimomura (377), in which leaves of TMV-inoculated *N. glutinosa* plants kept at 30 C for 4–5 d and then returned to 22 C failed to develop lesions, suggest that active virus replication is required for the induction of the necrotic reaction, and that when virus replication is restricted or terminated, necrosis is not induced. Conversely, active virus replication need not lead to necrosis, since, for example, lesions can be considerably reduced in the presence of reducing agents, while at the same time virus replication is little affected (110, 388). There is also evidence for the presence of virus particles in cells beyond the lesion boundary, though in most cases it is not clear whether these cells would eventually become necrotic, or indeed, whether active virus replication is taking place.

Inoculation of a host that responds hypersensitively can lead to the development of resistance in tissue either adjacent to or distant from the lesions. Both symptom formation and virus movement may also be restricted in such locally or systemically resistant tissue. Lesions in resistant tissue may be smaller than those in primarily

Fig. 10–2. Scanning electron micrographs of healthy and TMV-inoculated pinto bean leaf. (a) Surface view of lesion, showing collapse of epidermal cells and stomata, ×360; (b) and (c), cross-section through healthy leaf showing intact epidermal, palisade, and mesophyll cells, ×360 and ×1440, respectively; (d) and (e), cross-section through a TMV lesion showing severe constrictions and distentions of palisade cells, ×360 and ×1440, respectively (from Stobbs et al., 394).

infected tissue, while at the same time containing the same amount of virus (250). It may be, therefore, that virus is restricted in resistant tissue by the same mechanism as operates in primarily inoculated leaves. Necrosis may be limited by a separate mechanism, which is stimulated in resistant tissue and may or may not be the same as operates in primarily infected tissue. There are suggestions that reduction in necrosis could be related to cytokinin stimulation. However, there is even less information on the limitation of necrosis than on its cause, and mechanisms remain purely speculative.

Visualization of Virus Particles in Lesions

Although virus particles were not detected in some early ultrastructural studies on local lesions formed by TMV in *N. glutinosa* (367, 448), subsequent observations on this virus/host combination by Milne (281), Hayashi and Matsui (170), and Weintraub et al. (449) revealed particles both within the necrotic zone and in cells around the periphery. In peripheral cells, particles were dispersed, but in necrotic cells, virus, apparent in only 2–5% of the cells observed 3 and 5 d after infection, appeared in discrete packets (281), a feature possibly resulting from water loss and shrinkage during the necrotization process. There were estimated to be 10^3 particles/cell, substantially less than in cells of hosts reacting systemically to TMV by at least two orders of magnitude.

Following these observations, and the detection of particles in TMV lesions in *Datura stramonium* (55), virus has been visualized both within the necrotic zone of the lesion and in cells at the immediate periphery in other combinations of virus and host, including Samsun NN tobacco (73, 196, 237) and pinto bean (389, 394) inoculated with TMV, *G. globosa* inoculated with TBSV (11, 12), and PVS-inoculated *C. quinoa* (378).

Virus Limitation

As we have seen, virus particles are in many instances, though not always, restricted to the local lesion area, and this needs an explanation. In contrast to the situation outlined in the preceding paragraphs, there are several possibilities, though none is entirely satisfactory.

Physical barriers: cell death. The severe cellular damage occurring in the necrotic zone of some types of lesions, resulting either from alterations in phenol metabolism or from another unidentified stimulus, suggested that virus particles were restricted to this area by death of the cells per se. It was proposed that metabolic changes extending radially from the lesion center led to the collapse of a ring of cells surrounding the center. Virus would be unable to pass freely into these cells; unable to support virus replication, the cells would form a barrier to further virus spread (196, 380).

This explanation is clearly inadequate. For example, when TMV lesions are allowed to develop in tobacco cultivars containing the N gene from *N. glutinosa* at around 20–22 C, and plants are then transferred to a temperature in excess of 28 C, a systemic infection results despite the formation of necrotic lesions. Similarly, limitation is not always a consequence of necrosis irrespective of temperature. Transfer of Samsun NN tobacco leaves systemically infected with TMV from 25 to 11 C results in the formation of lesions morphologically very similar to those induced in Xanthi nc tobacco at 25 C. There is clearly no correlation here between necrosis and virus localization (144, 145). While not ruling out the possibility, therefore, that cell death may be a contributory factor in some instances, we must look for other explanations.

Disruption of plasmodesmata and paramural bodies. Since passage of virus particles between cells takes place through plasmodesmata (103), disruption of these intercellular channels would be expected to interfere with virus spread. This may well result from cell death. Alternatively, in pinto bean inoculated with TMV, membrane-bound vesicular bodies (paramural bodies or plasmalemmasomes) were observed between the plasmalemma and cell wall of cells close to the margin of the lesions. They were found in a region beyond cells containing virus and, where they appeared in greatest number, plasmodesmata were apparently severed, providing a potential barrier to virus movement (389). However, paramural bodies, which also occur in healthy tissue (265), are very common in tissue systemically infected by several viruses and have only rarely been associated with lesion formation (266). It may be that the majority of structures observed by Spencer and Kimmins (389) were not paramural bodies, but callose (see below), and it is unlikely, therefore, that they contribute to virus limitation. Breakage of plasmodesmata could well be important, though this is, to date, a rare observation.

Deposition of callose, glycoproteins, and lignin. Deposition of polymers of various kinds has been found by many to be associated with lesion formation and postulated to be involved in virus limitation. However, to be convincing, appearance of such polymers needs to fulfill several criteria. Both the site and timing of formation are important. Polymers must be deposited around the lesion, in advance of virus spread from the lesion center, and formed in a situation where they are able to block virus movement in and around the plasmodesmata. With few exceptions, these criteria have not been met.

Callose, a β–1,3-glucan normally found in the wall of sieve elements, could fulfill this role. Its formation has often been observed to accompany lesion formation, and, if deposited in plasmodesmata connecting cells at the lesion periphery, could conceivably block virus spread (103). TMV-U1 forms small lesions in pinto bean leaves that normally reach their maximum size quickly and then stop enlarging. The necrotic zone is limited to cells of the upper epidermis and palisade and includes few lower mesophyll cells and none of the lower epidermis (394, 459; Fig. 10–3). Although Spencer and Kimmins (389) failed to find electron mi-

heat (458) or UV (Fig. 10–3) and dark (394) treatment, an increase that was correlated with a decrease in the rate of callose deposition at the lesion periphery. It was suggested that callose deposition at the lesion periphery, produced in response to necrosis of cells at the lesion center, was responsible for restriction of virus movement and that, following treatments that led to an increase in lesion size, callose was deposited too slowly to form an effective barrier. As a result, lesions continued to expand until callose was deposited in quantities sufficient to prevent further spread.

In *N. glutinosa*, too, lesion enlargement was correlated with lack of callose deposition. Only small quantities of callose were deposited following inoculation with TMV-U1, which produced lesions that continued to enlarge with time (458), while TMV-VM, which formed lesions of limited size, elicited significant callose deposition in walls of cells at the periphery of lesions on mature leaves (457). Transfer of VM-inoculated plants from 21 to 30 C within 24 h of inoculation resulted in large, expanding lesions, with a reduced callose deposition. Conversely, treatment of TMV-inoculated Samsun NN tobacco leaves with eosin Y, which stimulated callose formation, led to the formation of smaller and fewer lesions (376), again revealing a correlation between callose formation and lesion size. A similar correlation has been observed in TMV-infected French bean plants treated with boron (374) and in tissue with induced resistance (373). Inoculation of upper leaves of Samsun NN tobacco plants inoculated previously on their lower leaves with TMV leads to the formation of smaller lesions than on unprotected plants and to increased callose deposition. Callose was also observed, by both light and electron microscopy, in walls of cells at the margin of PVM lesions in kidney bean (181). In none of these investigations, however, was callose deposition observed prior to the appearance of visible necrosis, and, although careful light microscopy permits detection of cell-wall-associated callose,

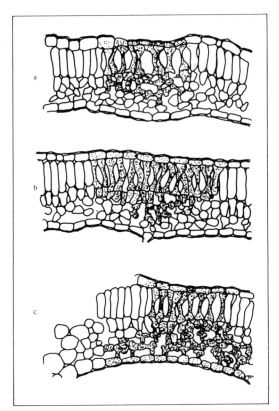

Fig. 10–3. Diagram of TMV-U lesion in pinto bean (a) and the shape following UV irradiation of upper (b) or lower (c) leaf surface (after Wu and Dimitman, 459).

croscopical evidence of callose deposition in sufficient quantities to account for lesion limitation, Wu found callose, identified by aniline blue staining, to be deposited in walls of nonnecrotic cells surrounding the necrotic zone (Fig. 10–4). Lesion size increased following

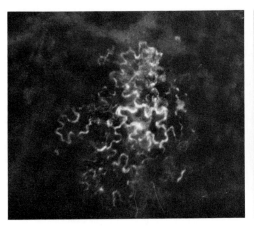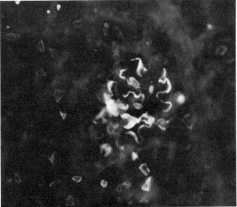

Fig. 10–4. Fluorescence of aniline blue-stained callose in epidermal cell walls in and adjacent to the site of infection. Left, TMV-inoculated *N. glutinosa* 20 h after inoculation; right, TMV-inoculated Samsun NN tobacco leaf 22 h after inoculation (after Shimomura and Dijkstra, 375).

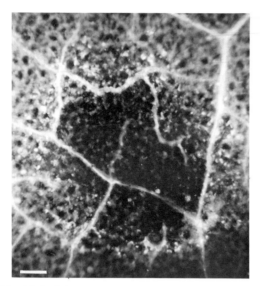

Fig. 10–5. Callose fluorescence in the peripheral zone of a TMV lesion in *N. glutinosa* 2 d after infection. Calibration, 100 μm (after Shimomura and Dijkstra, 375).

it does not permit a visualization of plasmodesmata, essential if restriction of cell-to-cell movement is to be demonstrated.

More recently, Shimomura and Dijkstra (376) have demonstrated histochemically the deposition of callose prior to necrosis in several virus/host combinations leading to lesion formation, including TMV-inoculated *N. glutinosa*, Samsun NN tobacco (Fig. 10–4) and pinto bean and TNV- and CPMV-inoculated pinto bean. As necrosis developed, deposition at the lesion center disappeared, probably as a result of degradation by β-1,3-glucan hydrolase (285); in the later stages deposition was apparent only at the lesion periphery (Fig. 10–5). Stobbs et al. (394) have also demonstrated the early deposition of callose in walls of cells in and adjacent to TMV lesions in pinto bean (Fig. 10–6), beyond the area of detectable virus. Faulkner and Kimmins have observed deposition in the walls of cells bordering fully expanded lesions in this host, which they too consider to be involved in virus localization (112).

Supportive electron microscope evidence for the involvement of callose has come from studies on TMV-inoculated *N. glutinosa*, in which Favali et al. (115) observed the occlusion of vascular bundles beyond the lesion periphery, with blockage of sieve plate pores and plasmodesmata connecting sieve tubes to companion cells by callose. Allison and Shalla (3) have also observed extensive deposition between wall and plasmalemma and in plasmodesmata connecting cells adjacent to the edge of PVX lesions in *G. globosa*. The plasmodesmata most severely distorted were those at the center, where callose deposition was greatest, Allison and Shalla suggesting that plasmodesmata were sealed by sacs of cytoplasm surrounded by callose. Callose deposition was extensive within the lesion but, im-

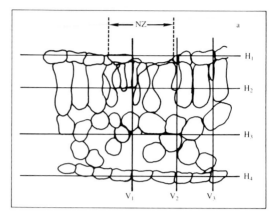

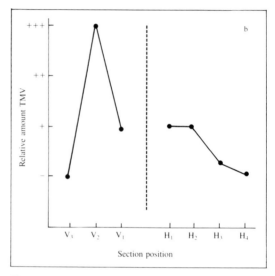

Fig. 10–6. Vertical (V) and horizontal (H) transects through a TMV lesion (NZ, necrotic zone) in pinto bean taken at the positions indicated (a) and virus particles visualized by electron microscopy. The relative amounts of virus at the transect sites are indicated in (b) (after Stobbs et al., 394).

portantly, was found at a greater distance from the lesion center than virus particles, and was therefore in a position to restrict virus movement. At variance with these observations, Pennazio et al. recently found callose deposition to occur later than the 2 d suggested by the earlier report and also found virus particles outside the zone of callose deposition (324).

Additionally, correlation between virus limitation and necrosis-associated callose deposition is not universal. For example, although Shimomura and Dijkstra (375) observed callose to be deposited in or around local lesions early in infection, when virus movement could be restricted, there was a comparable pattern of deposition in Samsun NN tobacco inoculated with TSWV (Fig. 10–7), TRV, and TBRV, in which systemic necrosis developed and virus was not limited. Similarly, callose deposition, in association with plasmodesmata,

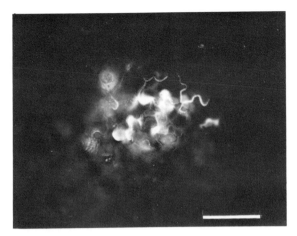

Fig. 10–7. Callose fluorescence in a TSWV-infected Samsun NN tobacco leaf. Calibration bar, 100 μm (after Shimomura and Dijkstra, 375).

failed to restrict movement of BWYV in beet (104, 105). Significantly, only in those cases where virus movement was restricted was callose deposition in veins observed (375; Fig. 10–8), leading Shimomura and Dijkstra to suggest that sealing of veins was more important in virus limitation.

Conversely, there are many instances when virus movement is curtailed in the absence of callose deposition at the appropriate time and place. Although PVX may elicit extensive callose formation in *G. Globosa*, for example, there is electron microscopic evidence that infection by TBSV, which is also restricted to the lesion area, results in very little callose formation, in a region in which it would be unlikely to be effective in limiting virus spread (11, 266). Also, several viruses that do not elicit a hypersensitive response in *C. quinoa* are localized in the absence of significant callose deposition (266), and, as we have already seen, TNV is localized to the areas of starch lesions in cucumber cotyledons, again in the absence of callose.

In evaluating the significance of these observations, a few cautionary notes are in order. First, callose is not easy to identify positively under the electron microscope and may, in some cases, be overlooked. Second, staining techniques used in light microscopic studies may detect substances other than callose. Third, only the electron microscope has the resolution to determine whether callose is deposited at sites where restriction of virus movement is possible.

It is conceivable, therefore, that in some cases callose is indeed making a contribution to virus limitation; in others, where virus is not restricted, it is not produced at all, or is produced either in insufficient quantity or at the wrong place or time to be effective. Its significance, however, remains circumstantial, and there is certainly not enough evidence to assign an unequivocal role to callose in virus limitation. It may simply be a non-specific response to necrosis.

Alternative polymers whose synthesis may be stimulated as part of a general wound reaction (457) and that

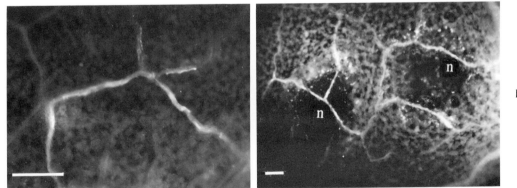

A

B

Fig. 10–8. Callose fluorescence in veins of (a) lower side of pinto bean leaf 40 h after inoculation and (b) Samsun NN leaf 48 h after inoculation with TMV (n, necrotic cells; calibration, 100 nm) (after Shimomura and Dijkstra, 375).

may contribute to secondary wall thickening (181, 418) include glycoproteins, lignins, and suberins. Kimmins and Brown (48, 209) have isolated two glycoproteins from pinto bean inoculated with either TMV or TNV, which were not apparent in healthy uninoculated leaves. Both contained acid-labile arabinose, mannose, and galactose, but while the polymer isolated between 24 and 36 h after inoculation contained β–1,4- and β–1,3-linked glucose residues, that isolated 5 d after inoculation contained only β–1,4-linked glucose. It was considered that deposition in cell walls in advance of infection could restrict the number of pathways for virus movement from developing lesions, a view unsupported by microscopical or other evidence.

Enhanced lignin deposition, probably resulting from a stimulation of precursors and enzymes involved in its formation (peroxidase and phenylalanine ammonia lyase, for example) may also be characteristic of the hypersensitive response. Lignin has been detected histochemically at the borders of both TMV and TNV lesions in bean (111) and was also considered to contribute to the restriction of virus movement (210). Appiano et al. (11), using both light and electron microscope techniques, found extensive lignin and suberin depositions in TBSV-inoculated *G. globosa*. Although formed at the periphery of the lesion area, they were deposited extracellularly and, like the small callose deposits that they also observed, were not considered capable of forming an effective barrier to virus movement. Similarly, Conti and colleagues (113–15) found extensive occlusion of tracheids in vascular bundles surrounding TMV lesions in *N. glutinosa*. The polymer became heavily labeled when ^{3}H-phenylalanine was supplied to the tissue, suggesting that it was lignin, though its electron opacity indicated it to be a material rich in phenolics rather than lignin itself. It was, however, formed too long after infection to be considered an effective barrier.

Finally, in cell walls surrounding TMV lesions in *N. glutinosa* and Conn. 423 tobacco, pectic acid has been reported to be replaced by insoluble calcium pectate (446). Although this calcium deposition has also been considered to play a role in virus limitation, there is no indication of precisely how the restriction would be achieved.

Clearly, the same comments made about callose apply to these other polymers. They may indeed, in some virus/host combinations, contribute to the formation of a physical barrier to virus movement. However, their association with hypersensitivity is no more than a correlation, and they have not yet been demonstrated to be produced either at a time or place to be effective.

Antiviral factors. The number of lesions that appear after mechanical inoculation of leaf tissue can, as indicated in Chap. 1, be influenced by a number of factors, including changes in the environment, previous virus infection, and treatment with various compounds, including hormones. Lesion size can also be influenced by treating the plant in various ways, usually after inocula-

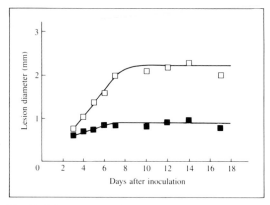

Fig. 10–9. Development of lesion size after TMV inoculation of young (□————□) or old (■————■) Xanthi nc tobacco leaves (after Weststeijn, 450).

tion. Leaf age at the time of inoculation is important, TMV inoculation of young leaves of Samsun NN (400, 402, 431) or Xanthi nc (Fig. 10–9) (450) tobacco producing larger lesions than inoculation of older leaves. Temperature too can influence symptom expression. TMV lesions appearing on *Nicotiana glutinosa* maintained at 28 C are larger than those that develop at lower temperatures, while around 35 C there is no necrosis and a systemic infection results (352). Brief shock treatment of inoculated leaves, with hot water, for example, can lead not only to formation of lesions on normally systemic hosts but also to increased lesion size, a 40 sec treatment of TMV-inoculated pinto bean leaves at 50 C 20 h after infection increasing lesion size 1.8 times compared to controls (462). However, though temperature-shock increases lesion size in several virus-host combinations (458, 462), in others (345), such as SBMV-inoculated pinto bean, lesion size decreases.

Shading plants increased the size of AMV lesions in bean (84), while detaching leaves after inoculation increased the size of TMV lesions (298). In some cases, the observed difference in lesion size has been correlated with biochemical changes, a decrease in rate of callose deposition accompanying increase in size of TMV lesions following heat treatment of bean, for example (458). Such correlations do not, however, constitute proof of a causal relationship, and the mechanisms underlying these differences remain to be determined.

Results of these and other treatments have led to the proposal of an alternative mechanism of virus limitation to the physical barriers outlined in the previous sections. First, it will be recalled that there are some instances, such as TMV-inoculated cucumber cotyledons, where virus is restricted in the absence of demonstrable physical barriers (67). Second, while postinoculation treatment with hormones (cytokinins, for example), generally led to a stimulation of RNA and protein synthesis and a reduction in lesion size, treatment with compounds that reduced RNA or protein synthesis had the reverse effect and resulted in an enhancement of lesion size. Treatment of pinto bean, *N. glutinosa*, or Samsun NN tobacco either a few hours before or up to two days after TMV inoculation with actinomycin D, an inhibitor of DNA-dependent RNA synthesis, enhanced lesion size. A stimulation in virus accumulation paralleled increase in lesion size, in accordance with observations on effect of leaf age (403) and detachment (298). Treatment of cucumber cotyledons 24 h after TMV inoculation stimulated virus accumulation, though the number of lesions was reduced; there was no information on lesion size (249). A similar response was observed following treatment of *N. glutinosa* or cucumber cotyledons 24 h after TMV inoculation with chloramphenicol (359), an inhibitor of protein synthesis on 70s ribosomes, or by exposure to short-wave UV irradiation (244). Ultraviolet irradiation affected lesion size on bean only when plants were maintained in darkness (244, 459).

Although a reduced rate of callose deposition accompanied the enhanced lesion size resulting from UV irradiation of TMV-inoculated pinto bean, indicating a correlation with a potential physical barrier to virus spread, no evidence of callose deposition could be found in TMV-inoculated cucumber cotyledons. This observation, coupled with the effects described in the preceding paragraph, led to the hypothesis (243, 246, 250) that virus multiplication in cells at the lesion periphery was inhibited by the induction in these cells of one or more factors whose biosynthesis was dependent upon an active RNA and protein synthesis. Prevention of the induction of such a compound, by inhibition either of host DNA-dependent RNA synthesis with actinomycin D or UV irradiation, or of protein synthesis by chloramphenicol, would permit virus replication in these cells, movement to adjacent cells, and hence the development of larger lesions. Clearly, and this applies particularly to UV treatment, it is important that the restriction mechanism be more susceptible to the treatment, at the time of its application, than virus replication. It also follows that, unless chloramphenicol was acting in some nonspecific manner, antiviral compound is synthesized in the organelles, on 70s ribosomes.

In many instances, the formation of lesions is accompanied not only by restriction of virus to cells in and adjacent to the lesion but also by the development of resistance to challenge inoculation, expressed either as fewer or smaller lesions, in uninfected tissue. It is not clear whether acquired (or induced) resistance represents an enhancement of mechanisms that normally lead to virus localization or whether, in some instances at least, additional factors such as inhibition of necrosis are involved. In either case, compounds that can inhibit virus replication have been isolated from both inoculated and systemically resistant leaves. They are described in some detail in the acquired resistance section of this chapter.

In the majority of cases, and particularly those in which necrosis does not occur, development of physical barriers seems inadequate to account for virus localization, and at the present time restriction of virus multiplication in cells around a lesion by a mechanism that involves the induction of "antiviral factors" is a likely possibility.

Immunity

Immunity, the complete absence of virus replication and symptom development following inoculation by mechanical or other means, may be the most common outcome of the interaction between virus and plant, and it is the objective of breeding for resistance to plant virus disease. Two types may be distinguished. In the first, all members of a particular species are immune to infection by a particular virus. Restriction may operate at any stage of the replication cycle, and the inability of the virus to infect is considered to be a result of its inability to utilize or interact with essential host cell components, such as translation initiation factors or RNA polymerases, which can be utilized by viruses that do infect the species. Such a species may be regarded as a "nonhost" for that particular virus. There are clearly many possible examples, such as the failure of tobacco or tomato to be infected by BCMV.

Immunity of the second type is controlled by one or more "resistance" genes, so that members of a species that do not contain the gene(s) are susceptible. Infection is blocked at an early stage, and its effects are undetectable. In potato, the Rx gene is said to confer immunity to all strains of PVX except X_{HB}, while Ry confers immunity to PVA and PVB (66, 199); it is possible that here "immunity" represents an efficient hypersensitivity (81, 82). Some soybean cultivars are apparently immune to beet western yellows (94), and some cowpea lines to the SB strain of CPMV (41). There are, however, very few examples of either type in which it can be stated with confidence that no cells become infected, since detailed electron microscopic studies have not been done.

In some cases at least, immunity, like hypersensitivity, is not maintained at the cell level. Of 55 cowpea lines immune to CPMV, all but one yielded susceptible protoplasts. Protoplasts from Arlington, however, accumulated virus to only ca. 1% of the level attained in susceptible protoplasts and were also highly resistant to inoculation by CPMV-RNA (40). CPMV will infect leaf protoplasts from tobacco, but not plants (195), and BMV, strain ATCC 66, will infect protoplasts of *Raphanus sativus*, a "nonhost" of this virus (136). In these instances, the cells are intrinsically capable of supporting virus replication. This would seem to suggest that in intact plants, either infection is blocked at a very early stage or some cells become infected but escape detection, and virus translocation to adjacent cells is restricted.

It would appear that immunity of tomato and bean plants to BMV, for example, is due to a deficiency in transport. Preinfection with TMV or DEMV, respectively, allows BMV to be transported from conducting tissue and, thence, to replicate in mesophyll cells (405). It would also seem that TMV is able to replicate in cowpea cells that become infected during mechanical inoculation but is unable to be transported to adjacent cells (398). It is also of interest that TNDV, normally limited to phloem tissue, can infect tobacco mesophyll protoplasts (227).

Tolerance

A plant that is tolerant to infection by a particular virus strain would, under the definition used here, support virus multiplication to levels approaching those of susceptible cultivars without exhibiting the symptoms characteristic of susceptibility. However, the concept of tolerance has been interpreted in many ways (43, 330) and is often used synonymously with resistance to include cases where virus multiplication is also suppressed, or indeed not measured at all. It has been taken to include instances where plants escape infection and to describe the situation in which a plant survives infection and sustains a low yield reduction irrespective of virus accumulation or symptom expression.

It may be that there are relatively few examples of tolerance that fall strictly within our definition and many of the published examples (e.g. 43, 187, 330) would be classed as resistant. Ambalema tobacco, for example, is said to be tolerant to TMV, with tolerance controlled by two recessive genes, together with one or more modifying genes (422). Virus accumulation is, however, substantially lower than in susceptible plants, particularly so in upper leaves (277), and it is possible that in Ambalema both virus multiplication and movement are restricted. The mechanisms controlling virus accumulation are temperature sensitive, plants containing the "tolerance" genes from Ambalema responding to TMV inoculation with severe symptoms and accumulation of moderately high virus concentrations when maintained at 28 C (24). In tobaccos T1 471 and T1 692, TMV accumulation does appear, on the basis of limited evidence, to approach that in susceptible cultivars without expression of overt symptoms (61).

Clearly, if a virus is approaching the same concentration in a tolerant cultivar as in a susceptible one without eliciting symptom expression, then either the tolerant plant does not respond to the stimuli responsible for symptom expression, it is able to withstand the consequences of these stimuli, or the stimuli are not produced. Until we have more information on the interaction between virus and host, and in particular on the biochemical changes expressed as symptoms, the basis for tolerance must remain undetermined.

Tolerance has proved a useful attribute in many crop-breeding programs, in control of beet yellows in sugar beet, for example (347). It has, however, disadvantages, not least of which is that it allows the development of a reservoir of virus that is a potential hazard to other plants. Mutation may also lead to the emergence of strains to which the plants are susceptible. Clearly, incorporation of genes that allow restriction of virus replication is to be preferred.

Resistance (Low Susceptibility)

Plant breeders have been striving to achieve resistance to virus infection in almost all important crop

plants for many years, and their achievements have been reviewed elsewhere (43, 186, 187, 268, 425). Such breeding programs have, of necessity, however, been conducted in the absence of any biochemical information on the action of the resistance genes involved. Here, we consider some selected examples that, on the one hand, indicate the progress being made in the study of mechanisms of resistance but, on the other, inevitably highlight the paucity of information currently available.

Cucumber/CMV. Several oriental cucumber cultivars are highly resistant to CMV infection and have been used extensively in breeding programs aimed at introducing resistance to CMV into commercial cultivars. Resistance in cultivars such as China, derived from Chinese Long (329), may be controlled by a single dominant gene (187), while that derived from Kyoto–3' seems to be controlled by at least two recessive genes. In cultivars containing resistance genes from these sources, and from Tokyo Long Green, both CMV accumulation (Fig. 10–10) and symptom severity are less than in susceptible cultivars (25, 35, 270, 280, 370, 444). Resistance can, however, be partially overcome by injection of actinomycin D into cotyledons of resistant cultivars near the time of virus inoculation (26, 35, 71, 296) or by ultraviolet irradiation (239). Considerably more infective virus is recovered from treated

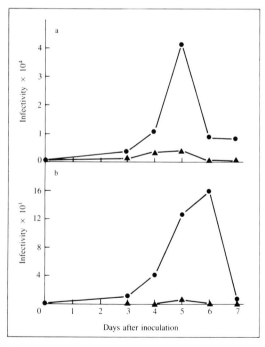

Fig. 10–10. Time course of development of CMV infection in the cotyledons of Ashley (●———●) and China (Kyoto) (▲———▲) seedlings measured in separate experiments as (a) extracted virus nucleoprotein and (b) extracted virus RNA; infectivity was recorded as mean lesion number/cowpea leaf/g cotyledon (after Maule et al., 270).

cotyledons than from untreated ones, suggesting a virus-induced host defense mechanism that requires DNA-dependent RNA synthesis.

This is not, however, the complete story. When protoplasts are prepared from a cultivar containing resistance genes from Kyoto–3′ and inoculated with CMV, the amount of virus produced per cell is less than in protoplasts from a susceptible cultivar, and the number of cells that become infected may also be less (71, 270, 271, 456). This partial maintenance of resistance at the single cell level, coupled with the observation that virus replication in "resistant" protoplasts is not stimulated by actinomycin D, suggests that resistance is at least partially mediated by mechanisms that are not induced. The observation that resistance is still maintained following inoculation with viral RNA puts its point of operation at a stage beyond uncoating of the virus particle.

By contrast, "resistance" conferred by hypersensitive genes is not maintained at the cellular level. Protoplasts from tobacco carrying the N gene, for example, support virus replication to the same extent as protoplasts from cultivars that do not have it (Fig. 10–11), although there are indications that virus yield may be influenced by hormones in the incubation medium. There is no cytological evidence of the necrosis that affects cells in infected leaf tissue or groups of cells (345). Virus replication would also appear to proceed

Table 10–2. Classification of tomato strains of TMV on the basis of symptom formation on tomatoes of different genotypes (after Pelham, 319).

	Strain			
Tomato genotype	0	1	2	1.2
Susceptible (+/+)	S	S	S	S
Tm-1	H	S	H	S
Tm-2	H	H	S	S
Tm-2²	H	H	H	H
Tm-1/Tm-2	H	H	H	H

S = susceptible; H = symptomless.

normally in cowpea protoplasts infected with CMV (149, 222), TMV, or TNV (455), combinations that lead to a hypersensitive response on leaf inoculation.

Study of the interaction between viruses and protoplasts may, therefore, allow a distinction to be made between resistance mechanisms that depend on groups of cells and that are, therefore, a tissue phenomenon, and those that do not. If, as in cucumber protoplasts, resistance is at least partially maintained at the cellular level, then protoplasts offer a potentially valuable system for the study of mechanisms of resistance.

Tomato/TMV. TMV is perhaps the most important virus infecting glasshouse-grown tomatoes, and wild *Lycopersicon* spp. have provided a source of several efficient "resistance" genes. Tm-l, Tm-2 and Tm-2² (= Tm-2ª) have been used most extensively (2, 119, 319).

Observation of the interaction between TMV strains and tomato plants incorporating these genes allows classification of the strains into four groups, on the basis of symptom expression (Table 10–2). Only plants containing Tm-2² are "resistant" to strains of all four groups.

The gene Tm-1 causes a suppression of both symptoms and virus multiplication, with 90–95% inhibition of virus accumulation in homozygous plants (Tm-1/Tm-1; GCR 237) and a 65–75% inhibition in heterozygous plants (Tm-1/+; J 484) (Fig. 10–12). It appears that one effect of the Tm–1 gene is to delay the appearance of infective virus, suggesting that the agent that limits virus replication is produced constitutively, rather than induced by virus infection (128). Also, the absence of an initial rise in infectivity following inoculation with a high concentration inoculum suggests that its action is at an early stage of virus replication, rather than on cell-to-cell spread. Virus multiplied in Tm-1/Tm-1 plants has a specific infectivity only ca. 10% of that recovered from susceptible plants, and it has been suggested that a reduced virus accumulation results from an effect on either assembly or stability of the virions (129–31).

It is of interest that symptoms are suppressed equally well in plants either homozygous or heterozygous for Tm-1, indicating that while suppression of virus multi-

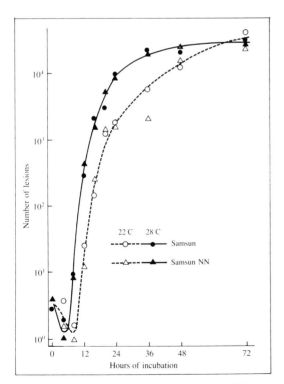

Fig. 10–11. Growth curve of TMV in protoplasts of Samsun and Samsun NN. Infectivity extractable from 10⁵ protoplasts plotted against time of incubation (after Otsuki et al., 311).

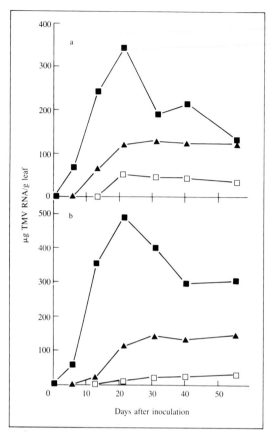

Fig. 10–12. Changes in TMV RNA concentration in TMV-infected tomato leaves. The host genotypes were: ■———■, +/+; ▲———▲, Tm-1/+; and □———□, Tm-1/Tm-1. (a) Changes in leaves that became infected by systemic spread of virus; (b) changes in leaves directly inoculated with strain O at 10 μg/ml (after Fraser and Loughlin, 128).

plication is gene-dosage dependent, symptom suppression is not. Additionally, curtailment of virus replication by the TM-1 gene in response to infection by TMV strain O is strongly temperature dependent, while suppression of symptom expression is not (130). Some TMV isolates of type 1 can also produce normal symptoms in plants containing the Tm-1 gene, though multiplication is still, to some degree, inhibited (128). These observations suggest that the effects of the Tm-1 gene on symptom formation and virus multiplication are at least partially separable.

The genes Tm-2 and Tm-2² confer hypersensitivity and reduce both symptom development and virus multiplication (162, 319), although prevention of symptom development can be temperature dependent (130). Protoplasts from plants containing these "resistance" genes are not resistant to infection. Protoplasts from plants homozygous for either gene are as susceptible to type O TMV, with respect both to the number of cells that become infected and to the virus yield per cell as those from the susceptible GCR 26 (286, 393). Similar-

ly, while plants of *Lycopersicon peruvianum* PI 128650 are "resistant" to TMV, protoplasts are susceptible. By contrast, protoplasts from plants homozygous for the Tm-1 gene do maintain resistance; GCR 237 cannot be infected with TMV-L (type O), or its RNA, although both leaf discs and protoplasts can be infected with TMV-CH2, a strain that overcomes the effects of the Tm-1 gene in intact plants. When plants are heterozygous for Tm–1, 1% of protoplasts become infected (compared to 61% of susceptible protoplasts) and produce 0.1% of the infectivity recovered from susceptible cells (286–88). These observations support the view that Tm–1 acts at an early stage of infection.

It is of interest that Tm–2 tomato plants become susceptible to TMV on pre-infection with a "helper" virus, PVX, while Tm–1 plants do not. It is suggested (405) that resistance is due to the restriction of virus transport and that PVX supplies a transport function. The nature of this function remains uncertain.

Cowpea/CCMV. The cowpea line Pl 186465 is resistant to CMV; although some virus can be recovered from inoculated primary leaves, plants remain symptomless, and virus is not recovered from trifoliate leaves. The following are the important features of the infection in resistant plants that differ from those in susceptible California Blackeye cowpeas (461).

First, the rate of virus accumulation is very much less than in susceptible plants and is continuous and linear throughout a 52-d period (Fig. 10–13). Second, CCMV is a multicomponent virus whose genome, consisting of four RNA species, is distributed among several different particles. When virus is purified from susceptible cowpeas and RNA is extracted and separated on polyacrylamide gels, the RNA profile obtained on scanning the gel is as indicated in Fig. 10–14. Although virus purified from resistant plants is similar in many respects to that from susceptible plants, its RNA profile indicates a 90% decrease in accumulation of RNA 3. This defi-

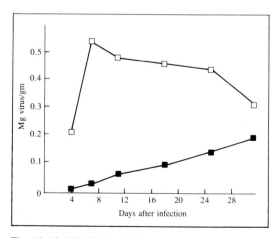

Fig. 10–13. CCMV (strain T) accumulation in susceptible California blackeye cowpeas (□———□) and resistant PI 186465 (■———■) (after Wyatt and Kuhn, 461).

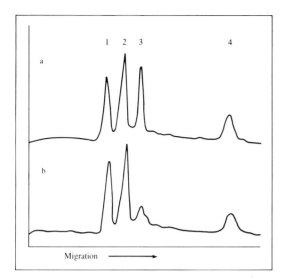

Fig. 10–14. Comparison of the RNA species (1–4) isolated from (a) CCMV produced in California Blackeye cowpeas and (b) PI 186465 cowpeas after electrophoresis in 2.7% polyacrylamide gels (after Wyatt and Kuhn, 461).

ciency in RNA 3 production does not, however, appear to affect the infectivity of the virus, and it is considered that the major contribution to resistance is a restriction on cell-to-cell spread, controlled by a single dominant host gene (230).

Perhaps not surprisingly, resistance, like the hypersensitive response, can be influenced by a number of factors. Cotton, for example, thought to be immune to TMV, in fact becomes subliminally infected when inoculated on the cotyledons, though very little virus can be recovered (63). It can be regarded as highly resistant. In susceptible plants, virus replication is usually greatest when plants are illuminated rather than kept in darkness. However, detachment of TMV-inoculated cotton cotyledons and maintenance in continuous darkness or treatment with ethylene reduces the expression of resistance and allows a substantially greater virus accumulation. The age of a leaf at the time of infection can also play a significant part in determining the severity of the ensuing disease (268, 400).

Although genetically determined resistance (or hypersensitivity) is by far the most satisfactory means of controlling virus diseases, susceptibility may, in some cases, be reduced by various treatments. To facilitate the assessment of their effect, or otherwise, tests are often performed on hypersensitive hosts, when a reduction in number or size of lesions formed on challenge inoculation might be expected. Perhaps of more practical importance, however, is the effect on virus replication and symptom expression in systemic hosts, where quantitation is more difficult. Cross-protection, or prior infection of plants with a mild strain of a virus with the objective of reducing the effect of subsequent natural infection by more virulent strains, has been used successfully in few cases. Inoculation of tomatoes at the

seedling stage wth TMV-MII–16, a nitrous acid mutant, has provided a useful form of protection. However, although symptom development on challenge inoculation is considerably reduced, replication is not by any means fully suppressed (59). There is also a promising mild strain of CaMV (412) that reduces symptom formation in Brussels sprouts (*Brassica oleracea* L. var. *gemmifera*) on challenge with a more severe strain. However, the same dangers inherent in the use of tolerant cultivars apply here (268).

In addition, there are several compounds that can interfere with virus replication. Treatment with virazole (or ribavirin; 1-β-D-ribofuranosyl–1,2,4-triazole–3-carboxamide) suppresses both virus replication and symptom expression in TSWV-infected tomato and tobacco (80) and multiplication of PVX, CMV, RCMV, and TMV in tobacco (50, 221), while DHT (2,4-dioxohexahydro–1,3,5-triazine) suppresses multiplication of TMV and PVX (50, 221).

Expression of resistance can also be found in some areas of leaves on plants that are otherwise heavily infected. In many quite severe mosaic diseases, including TYMV infection of Chinese cabbage (60) and TMV (15, 16, 268) or CMV (245) infection of tobacco, systemically infected leaves may contain an uneven distribution of virus. Dark green areas contain much less virus than yellow areas and appear not to support virus replication to any significant extent. Although adjacent to and apparently connected to yellow zones containing a high virus concentration, these green areas may remain almost virus-free for several weeks. They are also resistant to superinfection with strains of the same virus that infect systemically (16, 245). The basis for these resistance mechanisms remains to be determined. It is of interest, however, that in Xanthi nc tobacco inoculated with strains of CMV carrying the satellite CARNA–5, double-stranded RNA of this CMV- associated virus (ds-CARNA–5) accumulates in symptomless tissue from which virus cannot be recovered. Habili and Kaper (158, 159) suggest that this double-stranded RNA may be involved in resistance of the tissue to CMV.

There are, in addition, types of resistance that do not necessarily involve restriction of virus multiplication. Plants may, for example, be resistant to transmission by vectors. The CMV-resistant melon, PI 161375, is not only a poor source of virus for *Myzus persicae* and *Aphis gossypii* but is also resistant to transmission of CMV–14 by *M. persicae* and virtually completely resistant to transmission of this virus by *A. gossypii*, a type of resistance that supplements the restriction of virus multiplication in this cultivar (327). Resistance to transmission is controlled by a single dominant gene, which appears to be associated with nonpreference of plants for the vector. Transmission of virus through seed can also make a significant contribution to virus survival and dissemination, though factors influencing seed transmission are not well understood. Some plants, however, are more conducive to seed transmission than others. Though BSMV is often transmitted with high efficiency through barley seed, resistance to seed transmission can be conditioned by a single recessive gene (54).

Resistance, whether due to restricted replication of a virus that infects systemically or to a response that results in virus localization, is clearly important for survival of the plant, if not the virus. As we learn more about the intricacies of virus replication and host responses in susceptible virus/host combinations, the need for a greater understanding of resistance mechanisms is becoming more urgent.

Interference and Acquired Resistance

The phenomenon referred to as *interference* occurs when a virus is already present in cells of a host being inoculated with a second virus, or is added to an inoculum of a second virus; infection by the second virus may then be affected by the presence of the first. When, however, the first virus is not present either in the inoculum or the cells to be infected, but is restricted to other parts of the plant, infection by a second virus can still be suppressed, and in this case the phenomenon is known as *induced* or *acquired resistance*. Similar effects can also be produced, in some instances, by various macromolecules or by other microorganisms. Although it is probable that, in most instances, these effects are mediated at a stage later than the initial infection process, we include them at this point rather than in other chapters. They too have been the subject of several reviews (135, 167, 200, 242, 243).

Interference. Interference is most readily observed when one of the two viruses cause local lesions on the host to be infected; we have already seen, for example, that the number of lesions produced on *N. sylvestris* by the U2 strain of TMV can be drastically reduced by the presence in the inoculum of the mottle-forming strain, U1 (380). Though most commonly observed with related strains, interference of this sort also occurs between unrelated viruses (410), PVY interfering with lesion formation by cabbage black ringspot virus. Siegel's experiments (380) with TMV were suggested to indicate that U1 competed with U2, so that at high U1 concentrations, U2 was excluded from initiating infection. Though the evidence is not convincing, the fact that U1, when added only an hour after U2, fails to reduce subsequent lesion number may at least indicate interference at an early stage of virus replication. The reduction in lesion number produced by TMV-U1 on pinto bean by the U2 strain, to which the host was considered immune, has also been taken as evidence for exclusion, this time by a nonreplicating strain (460). However, it was subsequently observed that U2 did, in fact, replicate, producing microlesions (174). The size of the U1 lesion was also reduced in the presence of U2, indicating that events other than the establishment of infection were involved.

The VM strain of TMV causes smaller lesions on *N. glutinosa* than U1 and is more sensitive to ultraviolet irradiation. Addition of excess VM to U1 led to the formation of fewer lesions than appeared on inoculation by the U1 strain alone (336). Ultraviolet irradiation of leaves following inoculation by the mixture led to an increase in the number of U1 lesions, and it was inferred that the infection sites contained latent U1 particles that were only allowed to participate in infection when VM particles at the same site were inactivated. This would suggest that several particles may be present at an infection site, but only one participates in infection of that cell. However, other explanations are clearly possible (200). For example, UV irradiation soon after inoculation could affect susceptibility to infection and prevent creation of new infectible sites. Again, exclusion of one virus by another remains unproved.

There is no doubt, though, that two viruses, whether related or not, can infect and replicate in the same cell (88). Lauffer and Price (234) inoculated *N. langsdorfii* with a mixture of the aucuba and type strains of TMV and isolated both strains from some of the lesions that developed. This was interpreted to mean that more than one particle could participate in infection, though it is conceivable that the lesions containing both particles originated from closely adjacent cells. More convincingly, Goodman and Ross (150) inoculated tobacco with TMV and PVX and found 20% of cells in systemically infected leaves to be doubly infected. There are also many reports of phenotypic mixing following mixed infection (88, 201, 384, 404), particularly between strains of TMV.

Protoplasts, too, can be doubly infected in vitro by either related or unrelated viruses. Otsuki and Takebe (312) have observed the mixed infection of tobacco protoplasts by CMV and PVX and by the TMV strains OM and T (312), while Dawson and Watts (79) have achieved mixed infection with two strains of CCMV. Barker and Harrison have also achieved multiplication of two strains of RRV (29) or of RRV and TRV (28) in tobacco protoplasts.

It would appear, therefore, that the evidence for the exclusion of one strain by another in mixed inocula is not good. We must look for interference at later stages of virus replication, as we must in those instances where virus is already present in cells before inoculation with the challenge virus. This, again, is most readily observed when the challenge virus produces local lesions but also applies when both infect the host systemically. Indeed, the original observation of interference of this type, or "cross-protection," was the protection of tobacco against infection by a yellow strain of TMV by prior infection by a mild strain (276). More recently, the principle has been used on a commercial scale for the protection of tomatoes against the effects of severe TMV infection by spray inoculation with Rast's mild MII–16 strain (120).

Interference or protection is generally most pronounced between related strains and is not usually apparent between unrelated viruses. Consequently, it forms the basis of a test for relatedness, systemic infection by one virus significantly reducing either local lesion number or the development of systemic symptoms, following inoculation by a second. Systemic infection of *N. sylvestris*, for example, by TMV, inhibits local lesion formation by the aucuba strain (231). There are, however, many exceptions to the rule, and interference can be observed between unrelated viruses. Nitzany and Sela, for example (302), found that CMV reduces the number of TMV lesions on *N. glutinosa*. Alternatively, there may be an enhancement of infection by the challenge virus or, in some cases, no effect at all (200). The degree of protection is essentially assessed by change in symptom expression compared to controls, and there is rarely any attempt to assess the extent of multiplication of the challenge virus in a "protected" host. In tomatoes protected by the MII–16 strain of TMV, however, there is good evidence for suppression of multiplication of a more virulent challenge strain (59; Fraser, unpublished information).

Although several theories to account for cross-protection have been proposed, none is entirely satisfactory (e.g. 135, 163, 268). It has been suggested, for example, that infection by the first strain so alters the physiology of the plant that the efficiency of initial infection by the second is reduced, in an unspecified way. Alternative explanations, some of which could also apply to interference between viruses inoculated simultaneously, include the induction of a compound(s) by the protecting virus, which might interfere with replication by the second; adsorption of the challenge virus to aggregates of a related virus already present in the cell, so that it is unable to express its information; or "capture" of the genome of the challenge virus by viral capsid proteins of the first (86). This last possibility is, however, considered by some to be unlikely (468). It also seems unlikely that competition for essential metabolites is of general importance, in view of the often low concentrations reached by the protecting virus. Other suggestions include competition between the two viruses for essential "replication sites" within the cell, and recombination of the challenge virus genome with virus genomes already present. It has recently been observed (368) that the RNA of necrotizing strains of TMV is relatively more infectious on the yellow mosaic areas of *N. sylvestris* systemically infected with TMV than is intact virus. This would seem to suggest that prevention of uncoating of the challenge virus at least contributes to the resistance mechanism. Unfortunately, it is not yet possible to provide definitive answers.

Plants may also recover from infection (200) and be resistant to subsequent inoculation. Tobacco infected with TRSV, for example, exhibits severe symptoms on inoculated leaves, while leaves that develop later contain some virus, show only sporadic symptoms, and are resistant to reinfection. Similar characteristics are also shown by the "green islands" of tobacco leaves systemically infected with viruses causing a yellow mosaic.

Acquired resistance. Resistance to challenge inoculation can also be induced in cells that do not contain virus particles, either in the areas immediately surrounding necrotic lesions (localized-acquired resistance) or in tissue distant from that already inoculated (systemic acquired resistance). Again, resistance is most easily observed if the challenge virus also forms lesions, when lesions may be reduced in number, size, or both compared to control plants. Gilpatrick and Weintraub (148) reported in 1952 that inoculation of the upper leaves of *Dianthus barbatus* plants with CarMV, following inoculation of the lower leaves with the same virus, produced fewer lesions than did inoculation of control plants. This initial observation led to the confirmation of the phenomenon with other virus/host combinations, particularly by Loebenstein (242, 243), and to considerable investigation of the mechanisms involved. Localized resistance, expressed as the formation of both fewer and smaller lesions, following challenge with the same virus, was induced in tissue adjacent to TMV lesions on pinto bean leaves (463), and immediately surrounding TMV lesions on Samsun NN tobacco (345). Resistance was also systemically induced in half-leaves of Samsun NN tobacco following TMV inoculation of the adjacent half and in opposite leaves of pinto bean following inoculation of one of the primary leaves with TMV, AMV, or SBMV (Fig. 10–15) (346). In these cases, the principal effect was on lesion size; reduction in lesion numbers was generally less consistent and reproducible, only appearing after development of high levels of resistance in terms of lesion size.

Unlike the majority of examples of interference, induced resistance was found not to be generally virus-specific (Table 10–3). Primary infection of beans by TMV, for example, led to the production of smaller lesions on challenge of the opposite leaf with AMV or SBMV, and vice versa, and CMV induced resistance in opposite leaves of cowpea to AMV challenge (346). Localized or systemic resistance, affecting lesion size, number, or both can be induced by a variety of agents other than viruses (242, 243) including polyacrylic acid (143), yeast RNA (146), TMV protein (241), bacteria (399), fungi (228), chemicals such as $HgCl_2$ (399, 430), plant hormones (22, 434), and aspirin (10). Virus infection can also induce resistance to other pathogens (197, 198, 274). TMV inoculation of cucumber plants can also induce systemic resistance to TMV in the absence of necrosis (339).

Although systemic induced resistance is a widespread phenomenon, many of the more recent studies designed to clarify and explain it have concentrated on the infection with TMV of tobacco varieties containing the N gene derived from *N. glutinosa* (127). Inoculation of the lower leaves of Samsun NN (346, 429) or Xanthi nc (23, 202) plants induces resistance in upper leaves as assessed by reduction in lesion size (23, 132) and number (202, 340, 346, 430). This response is not, however, specifically controlled by products of the N gene, but is a response to necrosis. Systemic resistance can, for example, be induced in White Burley tobacco, which contains the N^1 rather than the N gene, or by the TMV strain *flavum*, which induces local lesion formation instead of the usual systemic infection (124).

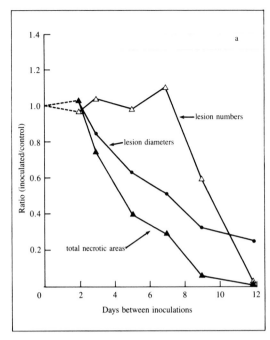
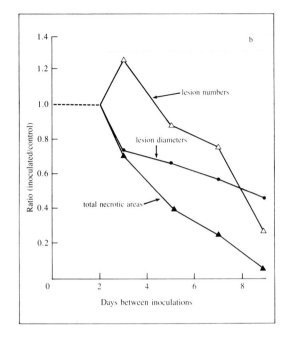

Fig. 10–15. Comparative data on lesions induced in primary leaves of pinto bean inoculated at various intervals following inoculation of the opposite leaves and in leaves opposite previously uninoculated ones. Data are expressed as ratios of lesion measurements in test leaves (opposite to leaves previously inoculated) to the corresponding measurements in control leaves (opposite to leaves rubbed with buffer). All first inoculations were on the same day and subsequent challenge inoculations on different days. Lesion area was calculated from lesion diameter and total necrotic area from lesion area and number. (a) SBMV used for both inoculations; (b) TMV used for first inoculation and SBMV for the second (after Ross, 346).

Lesion number seems most likely to be controlled by factors that affect the initial establishment of infection. It has been observed, for example, that tobacco leaves in which resistance has been induced by polyacrylic acid produce fewer lesions on challenge with TMV and are

Table 10–3. Relative size of lesions induced by various viruses in primary leaves of 'blackeye' cowpea inoculated 7 d following challenge inoculation of the opposite primary leaves with the same or other virus and in challenged leaves opposite previously uninoculated ones (after Ross, 346).

Viruses used		Diameter (in mm) of lesions in leaves opposite		Relative lesion size[a]
First inoculation	Second inoculation	Inoculated leaves	Uninoculated leaves	
AMV	AMV	0.47	0.94	50
CMV	AMV	0.54	1.13	48
CMV	CMV	0.29	0.55	53
TNV	CMV	0.25	0.55	45

a. Diameter of lesions in test leaves as percentage lesion diameter in controls.

less turgid than control leaves (58); as we have already seen, this might be expected to reduce their susceptibility to infection. Conversely, it would be expected that leaves with a greater turgidity would be easier to infect, producing more lesions than controls; indeed, this seems to be the case. Inoculation of *N. glutinosa* on the lower leaves induces *susceptibility* in upper leaves (132), leading to the formation of more (though slightly smaller) lesions on challenge inoculation with TMV than on control plants. This is correlated with a reduced abscisic acid concentration, which, by inference, would indicate a reduced water stress and increased turgidity in the upper leaves.

Lesion size, on the other hand, would be expected to be controlled by factors acting at later stages of lesion development, restricting virus multiplication, movement from initially infected cells, or the development of necrosis of infected cells. The evidence suggests that it is the extent of necrosis that is limited rather than virus multiplication. In leaves of Samsun NN tobacco in which resistance had been induced, Ross (346) found that a reduced lesion size/number was associated with reduced infectivity, but the results of more recent studies are at variance with these observations. The amount of virus recovered from upper leaves of Xanthi nc plants in which resistance was induced either by TMV inoculation (22) or HgCl$_2$ treatment (399) of lower leaves was little different from that obtained from control plants, in

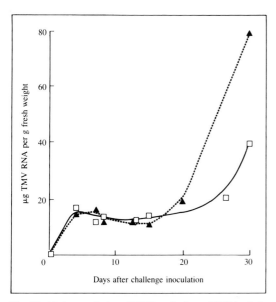

Fig. 10–16. Accumulation of TMV strain *flavum* RNA in challenge-inoculated leaves of White Burley tobacco. Upper leaves were challenge-inoculated with the local lesion-forming strain *flavum* at a dilution of $1:5 \times 10^2$. Ten days before the challenge-inoculation, the plants had been primarily inoculated with sterile phosphate buffer (controls) (□) or with TMV strain *flavum* at $1:10^2$ dilution (▲). Each point is the mean of three determinations (after Fraser, 124).

spite of reduced lesion size and number. These observations are supported by Fraser's studies with White Burley tobacco inoculated with the *flavum* strain of TMV (124), in which the accumulation of viral RNA was found to be the same in "resistant" as in control leaves (Fig. 10-16). In some cases it is possible that necrosis is reduced to such an extent that the reduction in lesion number apparent on visual observation is not a true reflection of the number of infection sites established. Balázs et al. (23), for example, have indications that inoculation of "resistant" Xanthi nc leaves with TMV produces several "microlesions" visible only on microscopic observation or by staining with I_2/KI for the accumulation of starch. In addition, when plants are inoculated, maintained at 32 C, and then returned to 24 C, the number of lesions that, subsequently, develop on resistant leaves is the same as that on controls. Visual observation of lesion number can, therefore, be misleading, unless coupled with microscopic observation of the leaf, which will allow detection of single infected cells. Unfortunately, not all the investigations attempting to explain resistance, some of which are referred to in the following paragraphs, have taken these considerations into account.

Enhanced peroxidase levels have been considered important in controlling lesion size (381, 382, 435), a higher peroxidase activity in "resistant" leaves allowing a more rapid necrosis, leading to the formation of smaller lesions. However, correlation between peroxidase activity and degree of resistance is not convincing (431), and its direct involvement in controlling lesion size seems unlikely.

Callose deposition around local lesions has also been considered important in limiting virus spread during the hypersensitive response, and Shimomura (373) considered callose an important factor in induced resistance, since its deposition was more extensive around the smaller lesions produced on TMV challenge inoculation of "resistant" Samsun NN leaves than around lesions on control plants. Clearly, this would only be a satisfactory explanation if the resistance mechanism were affecting the virus, either directly or indirectly, rather than necrosis.

It has also been suggested that the virus-induced synthesis of antiviral or interfering compounds may play an important role. In animal cells, various stimuli, including virus infection and treatment with double-stranded RNAs or polyanions, leads to the formation of host-specific glycosylated proteins, interferons, which, on interaction with susceptible cells, lead to the development in those cells of resistance to virus infection. The development of localized resistance, as assessed by reduction in lesion size or number, can be at least partially inhibited with actinomycin D, when administered near the time of primary inoculation (248), and it has been proposed that induced resistance is mediated through the induction of one or more compounds, possibly analogous to the interferons, that require transcription and are responsible for restriction of virus multiplication (rather than necrosis) in cells around the infection site. Since the size of lesions obtained on primary inoculation of a hypersensitive host can also be influenced by treatments that interfere with transcription (or translation), it has been suggested that virus localization is similarly mediated via compounds that restrict virus replication. Although it is possible that induced resistance, expressed either as fewer or smaller lesions, represents a potentiation of the same mechanisms that are responsible for virus limitation on primary inoculation, this may not necessarily be the case, and indeed mechanisms may differ among different host/virus combinations.

Nevertheless, attempts have been made to isolate antiviral or interfering factors from both systemically resistant and inoculated leaves. Loebenstein et al. (247) isolated a compound considered to be a protein from "resistant" leaves of *Datura stramonium* inoculated on the lower leaves with TMV, while Sela and his colleagues (362, 363) isolated an antiviral factor (AVF) from TMV-inoculated leaves of *N. glutinosa,* whose active component appeared to be RNA.

There were, however, two principal difficulties inherent in these early investigations: the purification of the putative agents and the estimation of their activity in a meaningful way. Antiviral or interfering activity was assessed by mixing virus with inocula before infection of a local lesion host, when a reduction in lesion number compared to that obtained with virus inocula mixed with appropriate control samples was obtained. Such a test does not, however, allow a distinction to be made between a compound that inhibits the establishment of infection and one that either directly or indirectly restricts virus replication.

Clearly, an assessment of the importance of such compounds rested on their introduction into cells in which their effect on virus replication could be tested. More recently, Loebenstein and Gera (246) reported the production of an inhibitor of virus replication (IVR) by TMV-infected protoplasts from Samsun NN, a tobacco cultivar incorporating the N gene from *N. glutinosa,* which confers hypersensitivity to TMV. The IVR is not produced by protoplasts from Samsun, which does not contain the N gene, and inhibits TMV replication in protoplasts from both cultivars. It also inhibits replication of TMV, CMV, and PVX in leaf disks from various hosts, indicating its lack of either virus or host specificity (140). Preliminary observations on partially purified preparations indicate the agent to consist of two biologically active components of MW 26 and 57,000.

Sela and his colleagues (137, 289, 360, 361, 364) have purified an AVF from TMV-inoculated *N. glutinosa* leaves, which they identify as a phosphorylated glycoprotein with an MW of ca. 22,000. It also appears to be produced in Samsun NN tobacco, and its formation is, therefore, considered to be dependent upon the presence of the N gene. It is suggested that a "pre-AVF" precursor is coded for by a "P" gene and is present in uninfected cells. The product of the N gene, or a modified form of it, participates enzymically in the activation of AVF, an activation that is completed by a phosphorylation step (Fig. 10-17). In vivo, induction of AVF is at least partially dependent upon both double-stranded RNA and cyclic nucleotides. It is postulated that the various stimuli that induce resistance act via the phosphorylation step of AVF activation.

AVF induction is associated with the N gene; it appears at the same time as virus localization and, when added to protoplasts, restricts virus replication at the picomole level, probably less than 1 molecule/cell. These observations, coupled with the evidence for pre-AVF activation (Fig. 10-17), support the view that AVF is stimulated in virus-infected cells by viral double-stranded RNA produced during its replication cycle, and is then responsible for restriction of virus replication. It is further suggested that virus replication is curtailed in neighboring cells by a hypothetical mechanism stimulated by AVF or an AVF-stimulated m-RNA, and which may involve the so-called "b" proteins (see below). It remains to be seen whether the AVF described by Sela and his colleagues is related to the IVR released by TMV-infected protoplasts.

Other proteins have also been considered to be involved both in virus localization (429, 433, 436) and in the development of induced resistance, expressed as a reduction in size and number of lesions that form on challenge inoculation (202, 429, 436). Inoculation of Samsun NN or Xanthi nc tobacco with TMV leads to the production, in both inoculated and uninoculated leaves, of up to ten protease resistant (325) soluble proteins that are not detectable in young, healthy plants (9, 430). Four are particularly prominent on polyacrylamide gel electrophoresis of tissue extracts; those from Samsun NN have been designated I-IV, and from Xanthi nc, b_4-b_1 (Figs. 10-18 and 10-19) in order of increasing electrophoretic mobility. Electrophoretic studies (9) have indicated three of these apparently distinct proteins (b_1-b_3) to be charge isomers, with an MW of ca. 14,200, while b_4 differs both in size and charge from b_1-b_3. Proteins II-IV from Samsun NN are similar in both charge and mobility on polyacrylamide gels to b_3-b_1. Though commonly referred to as "b"-proteins, they have, perhaps more explicitly, been termed *pathogenesis-related* (PR) proteins, or "stress-proteins," and an alternative nomenclature proposed (9), in which charge isomers are grouped together. Hence b_1, b_2, b_3 and b_4, or IV, III, II and I, become PRIa, Ib, Ic and II.

Their implication in resistance and localization arises from the fact that PR proteins appear in association with these phenomena in several hosts, including tobacco, bean (432), cowpea (70), cucumber (5), and *Gomphrena globosa* (141). They can be elicited by virus, bacterial or fungal infection of hosts that react hypersensitively (141), and in which resistance to a

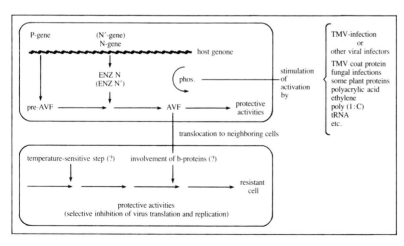

Fig. 10–17. A putative mechanism for the release of AVF activity (after Sela, 361).

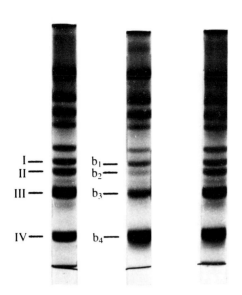

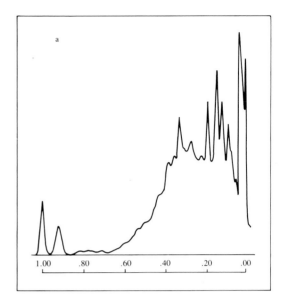

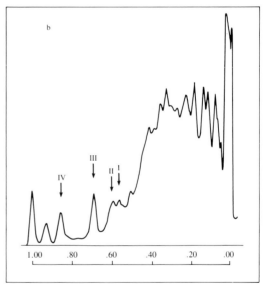

Fig. 10–18. Electrophoresis of extracts of TMV-infected leaves of Xanthi-nc and Samsun NN in polyacrylamide (10%) gels. (left) Crude protein solution from Samsun NN, (center) crude protein solution from Xanthi-nc, and (right) a mixture of crude protein from both cultivars (after Antoniw et al., 9).

Fig. 10–19. Densitometer tracings of electrophoretic patterns of centrifuged extracts from Samsun NN after electrophoresis in 10% acrylamide gels. (a) water- inoculated; (b) 7-d TMV-infected inoculated leaves (after van Loon and van Kammen, 436).

pathogen is induced. In addition, at least some are induced in treated and often also in untreated leaves by several compounds that also induce resistance, including polyacrylic acid (10, 143), benzoic acid, salicylic acid, aspirin (10, 454), and ethylene (432).

The appearance of PR proteins is not always, however, correlated with localization and induced resistance. Resistance can be induced in tobacco by $HgCl_2$ in their absence (429), and, conversely, PR proteins can be detected in the late stages of systemic infection. The four principal PR proteins have also been detected in uninfected Xanthi nc tobacco plants on flowering, where PRIa and Ib were shown to be identical to the corresponding proteins from virus-infected plants (125). PR proteins have also been found in uninfected Xanthi nc callus tissue (8), and PRAa is produced constitutively by the hybrid tobacco N. glutinosa x N. debneyii. Also, when lower leaves or inflorescences are removed from comparatively mature Xanthi nc plants, changing the PR protein content of the remaining leaves, then the number and size of lesions are totally unrelated to the concentration of PRIa in the tissue (125). Similarly, in TMV-infected Xanthi nc tobacco, there is no quantitative relationship between PR protein concentration and the degree of resistance induced, as assessed by reduction in either lesion number or size (126).

Although hypersensitivity to TMV, and hence virus localization, in both Samsun NN and Xanthi nc tobacco is determined by the presence of the N gene from N.

glutinosa, several observations suggest that the induction of PR proteins is not directly related to this gene. (1) The proteins from both Samsun NN and Xanthi nc, though similar to each other, are different from the corresponding proteins from N. glutinosa (429). (2) PR proteins can be induced by infecting tobacco cultivars not containing the N gene with viruses that elicit a hypersensitive response. The proteins induced by TNV infection of Samsun and Xanthi tobacco, for example, are apparently identical to those induced by TMV infection of Samsun NN and Xanthi nc (10). (3) Aspirin induces both resistance and PR proteins in White Burley, Xanthi, and Samsun tobacco, in addition to Samsun NN and Xanthi nc, again in the absence of the N gene (10). (4) Injection of polyacrylic acid at low concentration into Xanthi nc tobacco leaves induces resistance, expressed as fewer and smaller lesions on TMV inoculation, and also PRIa and Ib proteins. There is, however, no demonstrable induction of either resistance or PR proteins in Samsun NN, Samsun, or White Burley tobacco, indicating that PR induction by this compound is unrelated to the N gene (10). (5) Finally, inheritance of one of the tobacco proteins has been shown to occur independently of the N gene (141, 142).

The fact that PR proteins can be induced by several stimuli, including treatment with a variety of chemicals, and by infection with pathogens that induce necrosis suggests that there may be several pathways that lead ultimately to their induction, though they remain, for the moment, purely speculative. PR proteins are, for example, induced in tobacco by both ethylene and salicylic acid, though ethylene production is not stimulated by salicylic acid treatment. It has been suggested (432, 434) that ethylene is responsible for the induction of the synthesis of a compound, possibly analogous to salicylic acid, that in turn induces PR protein synthesis.

It is attractive, therefore, to assign a role for these PR proteins in the limitation of virus multiplication and spread, and hence in virus localization and induced resistance. Many parallels can be drawn between the appearance of PR proteins and interferon-mediated resistance to virus infection of animal cells (141). However, attempts to demonstrate an inhibitory effect on TMV replication in protoplasts have not proved successful (203) and, although PR proteins may be an important feature of a plant's response to stimuli, the evidence for any involvement in resistance remains no more than suggestive.

It may be that changes in hormone levels are of greater significance. We have already seen that decrease in lesion number is correlated with reduced abscisic acid concentration and that when resistance in Xanthi nc tobacco is induced by TMV infection, by application of leaf-spot inducing bacteria, or by $HgCl_2$ (399), reduced lesion number (without alteration in virus multiplication) is correlated with increased cytokinin content. The inference is that infection is initiated normally, but cytokinins prevent the development of necrosis, resulting in an apparent reduction in lesion number. Moreover, exogenously applied cytokinins can inhibit necrosis produced by $HgCl_2$, an observation supporting their putative role in suppressing virus-induced necrosis.

In conclusion, the evidence, based on relatively few examples, points to the fact that induced resistance results in an apparently reduced ability to form infection centers on challenge inoculation, an inhibition of the development of necrosis in cells surrounding the infection centers that do become established, or a combination of both these effects. On the basis of the studies completed so far, an effect on virus replication cannot be discounted. Clearly, much needs to be done to elucidate the biochemical mechanisms controlling these effects.

Resistance to Bacterial Pathogens

At present, the genetic expressions of resistance can, with few exceptions, be described *phenotypically* and have been categorized as morphological and biochemical. These have been further subdivided into *preformed* expressions, those characters of the host that develop naturally in the absence of any contact with the pathogen, and those that usually manifest themselves *postinfectionally*.

Morphological Expressions of Resistance

Preformed. Delivery of *Erwinia amylovora* cells by pollen and nectar-seeking insects is made difficult in the blossoms of some apple cultivars, according to Hildebrand (180), by virtue of their having tightly closed receptacles. Similarly, the blossoms of some apple and quince cultivars have smaller nectaries with fewer stomates and abundant hairs that impede passage of the inoculum to the highly susceptible nectarial stomate.

McLean (278) associated resistance of leaves of some *Citrus* species to the citrus canker bacterium *Pseudomonas citri* with variations in stomatal structure. Thus, *Citrus nobilis* var. *szinkum* (mandarin orange), which is resistant to *P. citri*, and Florida seedling grapefruit, *Citrus grandis*, which is susceptible, have stomates of the same size, general form, and mechanism of opening and closing. However, they differ strikingly in size of the cuticular ridge of the stomatal entrance (Fig. 10–20) which critically varies the opening in the cuticle—in the susceptible grapefruit the ridge of the entrance of the opening has its inner walls more nearly perpendicular to the leaf surface.

Bacteria can penetrate the stomatal opening only in a continuous film of water. When McLean studied the stomatal openings microscopically underwater, he observed air bubbles that extended across the aperture of the stomates or from one ridge of the entrance to the other. Thus, water films over both species of citrus leaf surfaces are held outside the stomatal opening by the ridges at the entrance. However, with changes in temperature, relative humidity, and wind-driven leaf movement, there is a greater or lesser tendency for water

Mandarin orange (resistant) Grapefruit (susceptible)

Fig. 10–20. Stomates of resistant and susceptible species of *Citrus* whose differing cuticular ridge structure determines, to a degree, intensity of infection by *Pseudomonas citri* (after McLean, 278).

covering the outer surface to be drawn into the stomatal opening. It is apparent that it will take less pressure to drive the bacteria-laden water film inward through the aperture of the grapefruit stomate than through the narrower aperture of the mandarin. This contention is supported by the generally accepted theory that the surface tension of water inhibits the transfer of liquid water through small natural openings, whereas gases, including water vapor, freely penetrate stomatal and lenticellate openings in either direction, according to the existing diffusion pressure of the gas. These minute apertures, therefore, seem to act as barriers to the movement of liquid water (with bacteria) into the substomatal chamber, since the openings are so small that a surface film of water bridges the two cutinized ridges of the entrance. Penetration is still further impeded when there is a gas on one side and liquid water on the other side of the opening. It requires tremendous pressures to force liquid water through these microscopic pores, which are, in addition, bordered by hard to wet surfaces. A detailed ultrastructural view of the stomatal opening, cuticular ridges, and passage into the substomatal cavity is seen in Fig. 1–17a.

An anatomical basis for resistance of sugar cane to ratoon stunting disease (RSD), caused by a *Corynebacterium*-like organism, has been advanced by Teakle et al. (408). Their measurements of water flow through healthy, single-node cuttings of a number of sugar cane clones indicated that a reduced flow rate of water (1.8 ml/min for an immune as opposed 16.8 ml/min for a susceptible clone) favored resistance. Microscopic observations revealed that resistant clones with slow flow rates had more profuse branching of the large metaxylem vessels in the nodes. For example, the large metaxylem vessel of an immune clone divides in the nodal region into 5 or 6 small branches, whereas in the susceptible one the large metaxylem vessel divides into 2 or 3 small vessels. The correlation of branching with reduced water flow is due to the smaller diameters of the branch vessels. Water flow through the vessels follows Poiseulles law that flow rate through a vessel is proportional to the fourth power of its radius. The question remains whether the relationship of reduced water flow to the observed immunity is causal or coincidental, although it would appear that the former is correct and that the reduced vessel size and water flow are accentuated by a bacterium-associated matrix, detected by Davis et al. (78), that appears to occlude vessels of infected plants. The resultant restriction at the node, due to both reduced vessel diameter and matrix accumulation, may serve to localize and limit the infection.

The stability of resistance in some varieties of alfalfa to *Corynebacterium insidiosum* prompted Cho et al. (65) to study the basis for it. Their investigations revealed that the resistant varieties had fewer vascular bundles, shorter vessel elements, and a thicker cortex in both stems and roots than the susceptible varieties. Movement of bacteria from cut-inoculated cotyledons was severely restricted, as compared to susceptible varieties. For example, the bacteria had traveled 10–25 mm

below the cotyledon in the susceptible plants in 48 h, it took 168 h for the bacteria to traverse that distance, and this in only one of 15 resistant seedlings. A previous study by Peltier and Schroeder (320) also suggested that the shorter and narrower vessel elements in the roots of resistant varieties restricted the movement of the bacteria. The data of Cho et al. (Table 10–4) revealed that the fewer vascular bundles, more deeply embedded in cor-

Table 10–4. Range and average number of vascular bundles or strands, length of vessel elements, and thickness of cortex in roots and stems of plants from alfalfa varieties resistant or susceptible to bacterial wilt[a] (after Cho et al., 65).

Variety	Roots Range	Roots Avg.	Stems Range	Stems Avg.
	Number of vascular bundles or strands/plant			
Resistant				
Ramsey	5–8	6.4 a	6–12	8.4 a
Ranger	5–8	6.4 a	6–10	8.2 a
Vernal	5–8	6.9 a	6–11	8.3 a
Susceptible				
African	6–9	7.7 b	6–11	8.8 b
Du Puits	6–10	7.8 b	6–12	9.4 c
Narragansett	6–10	7.2 b	6–12	9.1 bc
Rhizoma	6–10	7.7 b	6–12	9.5 c
	Length (μ) of vessel elements			
Resistant				
Ramsey	104–198	152 a	146–261	189 b
Ranger	94–219	153 a	157–230	183 b
Vernal	104–219	164 b	146–219	175 a
Susceptible				
African	136–251	184 d	157–282	215 d
Du Puits	125–240	175 c	146–261	203 c
Narragansett	125–261	187 d	157–271	214 d
Rhizoma	136–240	186 d	167–271	211 d
	Thickness (μ) of cortex			
Resistant				
Ramsey	136–324	229 a	63–125	99 ab
Ranger	173–324	239 a	63–135	104 a
Vernal	157–313	222 ab	73–135	96 b
Susceptible				
African	94–324	211 bc	63–125	86 c
Du Puits	94–292	200 c	63–94	80 cd
Narragansett	73–230	136 d	53–104	73 d
Rhizoma	115–313	195 c	63–125	86 c

a. Range and avg values based on 50 plants per variety. Length of vessel elements determined from three elements per plant. Differences in the average values were significant at the 1% level for the number of vascular bundles or strands and the length of vessel elements and at the 5% level for the thickness of the cortex. For each plant characteristic, data followed by the same letter are not significantly different according to Duncan's multiple range test.

tical tissue, were less likely to become infected than numerous bundles covered with a thin cortex. In addition, they conjectured that the bacteria would encounter more obstructions, at vessel ends, as they moved through the shorter vessels, which would account for their restricted mobility in resistant varieties. The observed restriction of bacterial movement may also reflect some physiochemical barriers that shall be subsequently discussed.

Postinfectional

Morphological aberrations developing as a consequence of infection appear to contribute to disease resistance. In the main, these structural impediments to the development of the pathogen take the form of barriers. Particularly dramatic, perhaps, is the development of tyloses, as detected in grape species by Mollenhauer and Hopkins (284). Their study of *Vitis vinifera*, *V. rotundifolia*, and *V. munsoniana* revealed the latter two to be much more resistant to this *Rickettsia*-like organism (RLO) than the *vinifera* (Thompson's Seedless). The basis for this difference was development of tyloses and gums in the vessels of *V. rotundifolia* and *V. munsoniana* at seven or eight times the frequency recorded in the *V. vinifera* cultivar. The tyloses and gums often encapsulated the bacteria and, occasionally, filled the infected elements of the xylem. Tyloses and gums were not commonly detected in the vessels of susceptible grapevines; however, they were detected in susceptible grapevines by Esau (102). Nevertheless, the authors suggest that gums and tyloses are responsible for localizing the bacteria in specific vessels (284).

It would seem that experiments designed to detect rates and intensities of tylose and gum formation would be most helpful in assessing their precise role in disease resistance. Wallis and Truter (440) have observed intense development of tyloses in tomatoes infected with *Pseudomonas solanacearum*; however, there is no record characterizing this phenomenon in tomato as a resistance reaction (see Chap. 8). If resistance in tomato to the Granville wilt pathogen were studied ultrastructurally, perhaps this relationship would be established. Attention is called, however, to the contention by Zimmermann and McDonough (471), citing earlier detailed histological studies, that tyloses develop when vessels are air filled after the cessation of water conduction. According to them, tyloses and gums are the result of the cessation of water conduction rather than a cause.

Still another barrier, apparently constructed by the plant as a consequence of the trauma associated with infection, is the suberization of surface layers of cells at or near wounds. A study by Maher and Kelman (262) has stimulated interest in the phenomenon of suberization as it pertains to resistance of potato tubers to degradation of the soft-rot pathogens *Erwinia carotovora* and *E. atroseptica*. The renewal of interest stems from observations by Lund (259), who noted that under conditions of ample oxygen supply soft-rot is readily arrested; however, when O_2 is limited, particularly by a water

Table 10–5. Effect of isopropyl N-(3-chlorophenyl) carbamate (CIPC) on suberin and periderm development by Katahdin potato slices held at 60 F, 14 d after treatment (selected data from Audia et al., 17).

CIPC (ppm)	Suberin[a]	Periderm[b]
0	3.9	4.2
5	4.1	3.6
25	4.3	1.3
50	3.7	0.8
100	3.9	0.1
500	3.7	0
1250	4.0	0

a. Values are ratings of intensity of tissue staining in ammoniacal gentian violet.
b. Values are ratings of intensity and numbers of cell layers dividing.

film, the soft-rotting *Erwinias* are particularly virulent. The apparent influence of water, surface or vacuum infiltrated, was speculated upon earlier (116, 157, 233) and has been discussed in detail in Chap. 4. Apparently, there is a cause-and-effect relationship between water and soft-rot, and it appears to be linked to the O_2 supply and its affect on the formation of suberin. There is some lack of clarity in the literature concerning the role of suberin and *wound-periderm* formation in retarding soft-rot development (13, 14). However, a report by Audia et al. (17) clearly distinguishes the roles that suberin and wound periderm play in this regard. Their data (Table 10–5) indicated that treatment of potato slices with the growth regulator CIPC, which completely suppressed periderm formation, did not sufficiently interfere with suberization to prevent it from inhibiting the development of *E. carotovora* in artificially inoculated potato slices.

The biochemical nature of suberin is described in some detail in Chap. 4. It should be mentioned, however, that the literature is "pointedly imprecise" in what is generally referred to as suberin (13). Kolattukudy and coworkers (106, 224, 225) have provided sufficient biochemical data to discriminate between suberin and cutin (see Fig. 4–12). These data should preclude the use of the phrase *cutin-suberinlike* in descriptions of these plant cell wall inclusions (14). In addition, Riley and Kolattukudy (338) have shown that suberin is covalently bound to ferulic acid, and the association of this substituted cinnamic acid with the hydroxylated fatty acid characterizes the latter as suberin rather than cutin (see structural features of suberin below).

The existence of phenolic acids as ferulic acid esters in plant cell walls of a number of plant species grasses has been reported by Hartley (169). Hence, the association of these ferulic acid esters with "ligninlike" material is discussed by Friend (133) in connection with postinfectional or induced resistance reactions. However, there is also evidence by Hartley, cited by Friend (133), that the ligninlike ferulic and *p*-coumaric acid esters in the cell walls of several species of grasses gave a negative phloroglucinol/HCl test for lignin. This suggests that suberin, rather than lignin, could be at least a part of the chemical barrier that forms, nonspecifically, as a consequence of infection or wounding (318, 337). This barrier cold block wall-degrading enzymes from their substrate, and, as suggested by Friend (133), this chemical barrier would necessitate a smaller total chemical change than the formation of a more extensive physical barrier. The question now arises, given the presence of ferulic acid in the wall of, for example, wounded or wound-inoculated potato tuber cells, of how this aromatic compound reacts with other wall compounds, e.g. carbohydrates and proteins. A partial answer relative to possible wall components that might react with ferulic acid has been supplied by Neukom and Markwalder (299). They suggested that in the presence of peroxidase and H_2O_2 the hemicellulosic arabinoxylans can become cross-linked by oxidative phenolic coupling of two ferulic acid residues. It is proposed that the dimer of ferulic acid is readily formed when conditions favor an oxidative redox potential. The reaction is presented in Fig. 10–21.

Cross-Linking of Arabinoxylan Chains

Thus, if cellulose microfibrils are ensheathed in hemicelluloses as suggested by Bauer (36), the blockage of reactive sites to hemicellulases may preclude their enzymatic degradation, as well as subsequent enzymatic cellulose depolymerization. Friend (133) has also suggested that ferulic acid could cross-link β-1–4

galactans by esterification through their C_2 or C_3 hydroxyls, thus preventing some galactanase activity.

Support for a suberin-based inhibition of wall-degrading enzyme activity is provided by the classic study of Zucker and Hankin (472). They postulated that the resistance of potato tuber slices to maceration by pectate lyases is due to the development of a "suberized layer" during a 24-h aging treatment following the wounding by slicing. Their model utilizes a purified pectate lyase, from the nonpathogen *Pseudomonas fluorescens*, and potato slices of the variety Kennebec. However, experiments with purified pectate lyase from *Erwinia carotovora* gave analogous results.

Resistance is based on the activity of phenylalanine ammonia lyase (PAL), which controls the production of chlorogenic acid. The data from pertinent experiments are presented in Table 10–6.

From experiment 1, it is apparent that treatment with cycloheximide suppressed PAL activity almost completely, and this precluded the development of almost all resistance to the purified pectate lyase.

Additional data (experiment 2) suggested that the removal of 0.3 mm of the "apparent" suberin layer resulted in 31% maceration (rather than 50%) of the aged potato slices, indicating a slight, but statistically significant, residual resistance to the lyase. In a corollary experiment, the authors reported that additional removal of 1.0-mm slices of tissue beneath the 0.3-mm suberin layer reduced still further the resistance of the potato slices to pectate lyase. These data were interpreted as indicating the existence of an intrinsic resistance factor other than suberin. However, in the light of more recent data (338), it is possible to suggest that the initially removed suberized layer represents a physical barrier containing a dense suberin deposition, whereas the cell walls of the deeper 1.0-mm layers of tissue may have

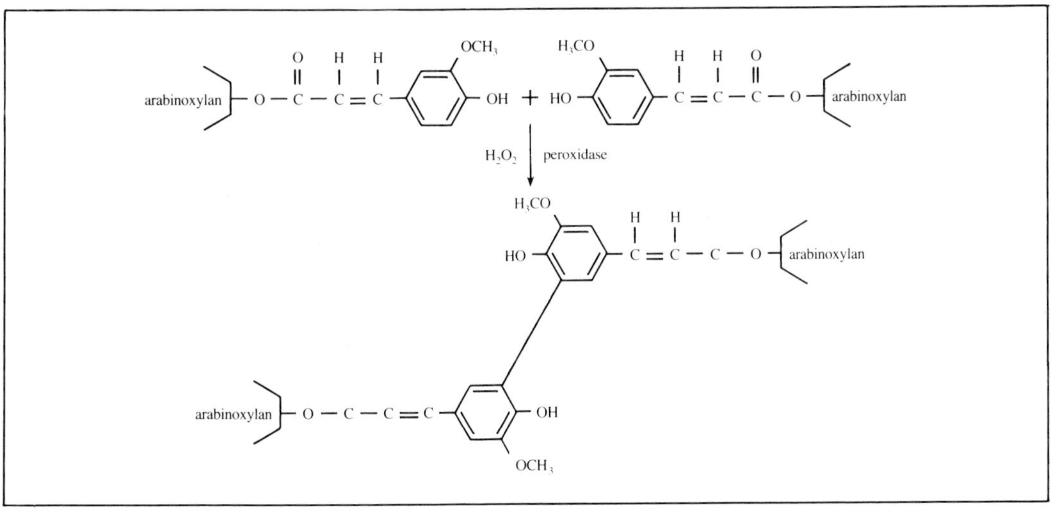

Fig. 10–21. Cross-linking of two arabinoxylan chains by oxidative coupling of two ferulic acid residues (after Neukom and Markwalder, 299).

Table 10–6. Factors affecting maceration of potato slices by purified pectate lyase produced by *Pseudomonas fluorescens* (after Zucker and Hankin, 472).

Treatment	% Maceration	PAL activity μg/g fresh wt.
	Experiment 1	
freshly cut	60	1
aged on water	10	66
aged on water + 10 μg/ml cycloheximide	60	1
	Experiment 2	
aged on water (suberin intact)	3	—
aged on water (suberin removed)	31	—
freshly cut	50	—

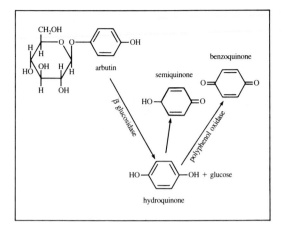

contained lesser amounts of, for example, chlorgenic acid-linked carbohydrate polymers.

Finally, it is important to relate the foregoing events to the conditions of oxygen paucity that favor the soft-rot of potato and other tuberous plant organs. Zucker and Hankin (472), like several previous investigators (116, 157), found water injection of tissue and vacuum infiltration of tissue with water inhibitory to the synthesis of PAL and potentiated susceptibility to soft rot. It appears reasonable that the exclusion of the oxygen by water could prevent oxidative coupling of phenolic acids, which as dimers cross-link a number of cell wall carbohydrate polymers, thus reducing their accessibility to enzyme degradation. It also seems plausible that reduced oxygen tension limits the activity of PAL and, hence, the synthesis of ferulic and other phenolic acids. The suppression of PAL by cycloheximide mimics the affect of low O_2 levels in infected tissues.

Biochemical Resistance Factors

Preformed. The question can legitimately be raised as to whether there are naturally occurring substances present in plants at sufficiently high concentrations to inhibit bacterial growth. A number of plant species contain rather high endogenous levels of phenolics, but are they antimicrobially effective in vivo? For example, apple and pear leaves and stems contain high levels of the glycosides phloridzin and arbutin, respectively. Young pear leaves may contain as much as 4–7% on a dry weight basis. A series of studies by Hildebrand and Schroth (178, 179) suggested that the glucoside of arbutin is hydrolyzed by β-glucosidase, removing glucose from the 4-position. The aglucone, dihydroquinone, then becomes antimicrobially active upon oxidation to the semiquinone.

Direct evidence for the antibacterial activity of the aglucone of arbutin is presented in Fig. 10–22. In 24 h, a clear zone of inhibition develops indicating the diffu-

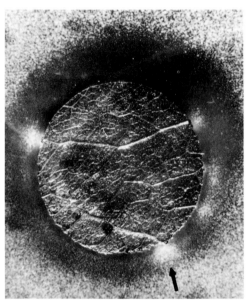

Fig. 10–22. The fate of the glycoside arbutin and its biological activity on release from a pear leaf disc on an agar plate seeded with *E. amylovora*. The zone of inhibition is attributable to one of the oxidized forms of hydroquinone. Growth of the bacterium at cut ends of major vessels (arrow) suggests the absence of β glucosidase in the presence of an oxidase inhibitor (courtesy of D. C. Hildebrand and M. N. Schroth).

sion of an antimicrobial substance from the disk. However, growth of the bacteria was not suppressed at a number of points along the margin of the leaf disk that coincided precisely with the positions of larger leaf veins. Aqueous extracts of vein and laminar tissue demonstrated similar amounts of arbutin; however, assays of leaf veins indicated that they were practically devoid of β-glucosidase. Similar data were presented by Smale and Keil (385) that antimicrobial activity and the attendant resistance in pears to *E. amylovora* probably require β-glucosidase modification of arbutin to the hydroquinone form. Later, Hildebrand (176) reexamined

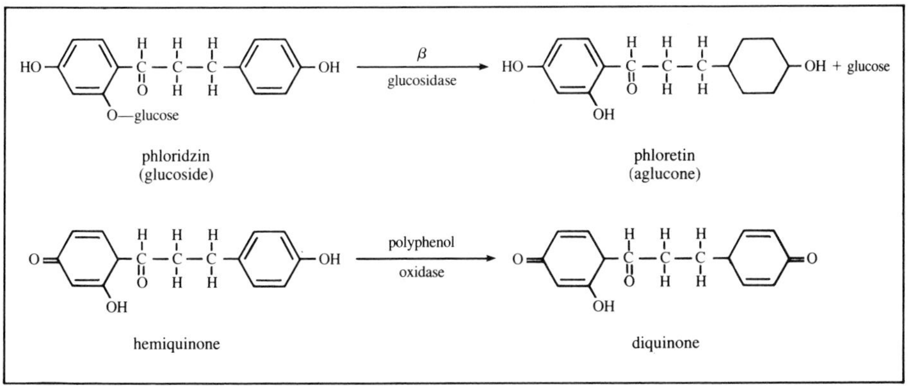

Fig. 10–23. Enzymatic conversion of the polyphenolic glucoside phloridzin to its antimicrobially active diquinone form.

this subject and found that in various tissues (young xylem, old xylem plus pith, bark, leaves, and veins), the disappearance of arbutin is accounted for as an increase in hydroquinone. The data also made it clear that neither young nor old xylem contains much endogenous arbutin, although β-glucosidase is present. This may explain why *E. amylovora* proliferates rapidly and preferentially, in the early stages of pathogenesis, in the xylem (395, 396). Hildebrand and Schroth (178, 179) also observed that the highly susceptible blossom nectaries in both susceptible and resistant pear cultivars contained little of the inhibitory hydroquinone. This was judged to be due to low amounts of β-glucosidase, rather than limiting amounts of arbutin.

It is conceivable that the presence of glucoside-hydrolyzing enzyme in pear tissue does not insure protection against successful bacterial invasion. The stability of the hydroquinone in its partially oxidized form, which, according to Geiger (139), is its most antimicrobially active state, may be of primary importance. Stability of the semiquinone is, in turn, controlled by the redox potential of the surrounding tissue and the degree of substitution of groups other than hydrogen in the phenyl ring. Stability of the hydroquinone in pear tissue was found to be impaired by the action of phenolase. In pear tissue, which had been frozen and thawed, antibiotic activity was reduced more than 50% in 1 h and to zero after 2 h by this enzyme (176). It is apparent, therefore, that the manipulation of tissue for the purpose of monitoring the levels of phenols and their redox state must be carefully controlled.

In analogous experiments, Goodman (unpublished data) sought to establish a relationship in apple between the glycoside phloridzin and its aglucone phloretin similar to that of arbutin and hydroquinone. Upon extraction, the phloretin levels in several varieties of apple were rather high, varying from 0.1–1.0% on a dry weight basis, in vegatative tissue of resistant as well as susceptible varieties. Although the level of phloretin in floral tissue was lower than detected in either leaf or stem tissue, comparable amounts were detected in the blossoms of both resistant and susceptible varieties.

Plurad (328), Goodman (unpublished data), and Addy and Goodman (1) determined the antibiotic activity of phloridzin, the predominant glucoside in apple, and its aglucone against *E. amylovora*. The enhanced antibacterial activity of the aglycone phloretin over the glucoside phloridzin was recorded (see Fig. 10–23). In this experiment, phloretin was produced by reacting phloridzin with β glucosidase.

From the foregoing, it is apparent that glycosides must be transformed in situ in order to exert an antimicrobial effect. Further, the aglucone must be maintained in its oxidized diquinone or semiquinone form for maximum resistance-inducing capability. The activity of a polyphenol-polyphenolase-quinone system is shown in Fig. 10–24.

Quinones. The antimicrobial activity of phenolic compounds is, as indicated above, linked with innate resistance, hypersensitive reaction-linked necrosis, failure of the pathogen to grow in discolored tissue, etc. These conditions have been offered as presumptive evidence of the antimicrobial potency of aromatic compounds in the tissue of higher plants. Assignment of numerical values (e.g., μg/ml) to the activity of these compounds either in vivo or in vitro and definition of their mode of action has rarely been provided.

A comprehensive study of the antibacterial action of quinones and hydroquinones (phenols) was conducted by Geiger (139). He observed that gram-positive bacteria are generally 10 to 20 times more sensitive to quinones than gram-negative ones. (Among the plant

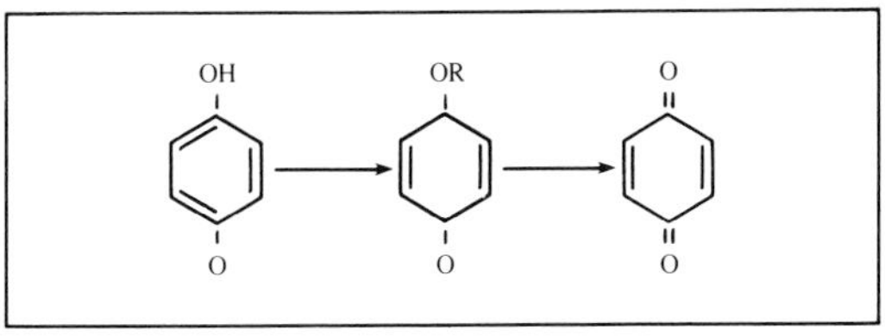

Fig. 10–24. Model system illustrating the conversion of hydroquinone to diquinone.

Table 10–7. The effect of monothioglycol on the antibacterial properties of quinones (after Geiger, 139).

Quinone	Sulfhydryl	Dilution units/ mg (DUM) (against *E. coli*)
2,5-dimethylquinone	none	12
	monothioglycol	1
2 methyl-5-hydroxy-1, 4-naphthoquinone	none	8
	monothioglycol	1

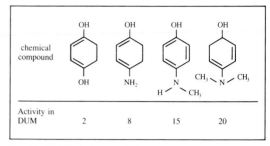

Fig. 10–26. The influence of nitrogen and methyl substitution on the microbial activity of hydroquinone against *E. coli* (after Geiger, 139).

pathogenic bacteria only some corynebacteria are gram positive.)

Geiger's study also revealed that the activity of quinones against Gram negative bacteria was influenced by the substitution on the benzene ring of methyl, hydroxyl, and sulfhydryl groups. Thus, derivatives of 1,4 benzoquinone in which the 2-, 3-, 5-, and 6- positions are occupied by substituents other than hydrogen showed a very low antimicrobial activity of less than one dilution-unit/mg (DUM) against *Escherichia coli*, whereas, 2,6-dimethyl quinone had an activity of 12 DUM. Similarly, derivatives of 1,4 naphthoquinone in which substitutions were made in the 2- and 3-positions were also inactive against gram negative bacteria. It is of interest that juglone (2-methyl-5-hydroxy-1,4-naphthoquinone), which is found in tissues of plant species of the genus *Juglans*, showed an activity of 12 DUM, whereas 2-methyl-3-hydroxy-1,4-naphthoquinone had an acitvity of 1 DUM.

These data suggested that the activity of quinones against Gram negative bacteria required the presence of a free position *ortho* to a *carbonyl* group. Geiger predicted that substitution of sulfhydryls in positions ortho to a carbonyl carbon would reduce the antimicrobial activity of quinones. The data supporting his contention are shown in Table 10–7. It is clear that the antimicrobial activity of the substituted quinone and naphthoquinone could be almost completely suppressed by the addition of 1 equivalent of a sulfhydryl-containing compound. The reaction is a 1,4 addition followed by rearrangement as shown in Fig. 10–25.

Hydroquinones. The activity of hydroquinones was, according to Geiger (139), much greater against gram-positive bacteria. However, compounds containing ni-

trogen were found to be more active against gram-negative bacteria than their oxygen analogs. The presence of methyl groups attached to the nitrogen atoms further potentiated their activity against *E. coli*, as shown in Fig. 10–26.

Mode of action of quinones and hydroquinones. Quinones and hydroquinones exert their affect against microorganisms by interrupting some vital enzymatic processes. This is exemplified by the fungicides chloronil and dichlone shown below.

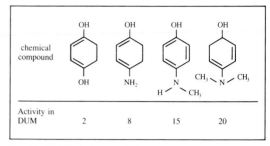

1.4-Benzoquinone   1.4-Naphthoquinone

McNew and Burchfield (279) and Owens (316) suggested that the fungitoxic activity of these compounds could be due to binding of the quinone nucleus to SH or NH_2 groups on the fungal (or bacterial) cell membrane or, for example, to disturbance of electron transport systems. According to Owens and Novotny (317), dichlone also inhibits phosphorylases, certain dehydrogenases, and carboxylases, as well as inactivating coenzyme A. It is also possible that they form stable cross-linkages between plant cell-wall carbohydrates and, thus, limit the activity of several wall-degrading enzymes (299).

The inhibitory effect of phenols on certain enzyme systems has been suppressed by cysteine. In fact, the data of Geiger shown in Table 10–7 concerning monothioglycol reveal how sulfhydryls limit the antibacterial activity of hydroquinones. It is apparent, therefore, that the efficacy of any aromatic compound as an endogenous disease resistance factor may be modified by reactants from a number of natural metabolic processes. Among these are oxidation reduction reactions; aminations, methylations, hydroxylations, and their reaction with mercaptans (SH groups).

Fig. 10–25. The inactivation of benzoquinone by a sulfhydryl-containing compound (after Geiger, 139).

It is apparent from the foregoing that oxidized aromatic compounds are effective antimicrobial agents; however, they are capable of reacting with reducing agents that negate their disease-inhibiting potential.

Importance of the redox (oxidation-reduction) potential. Maine (263) and Maine and Kelman (264) observed that tobacco stem tissue inoculated with *Pseudomonas solanacearum* exhibited a potentiated polyphenoloxidase activity that appeared to favor disease resistance. In order to test this premise, they designed an experiment to suppress polyphenoloxidase by feeding inoculated tissue SH groups via glutathione and hydrogen donors, such as ascorbic acid and sodium metabisulfite. The objective here was to keep the phenolics in their reduced state, rather than their antimicrobially active oxidized (quinone) configuration.

The influence of hydrogen donors on the innate resistance of tobacco plants was tested in the moderately resistant tobacco variety Coker-139. Potted plants were fed glutathione and ascorbic acid as soil drenches and then inoculated via stem puncture with *P. solanacearum.*

Disease indexes of 95 and 81, respectively, were registered for plants treated with glutathione and ascorbic acid, whereas the control rating was 38. When sodium sulfite was administered to the plants, the disease index was seven times greater than in the controls. It is possible that the hydrogen donors used in these experiments provided substrate for the dehydrogenases, i.e. ascorbic acid dehydrogenase and glutathione reductase, which interfere with the oxidation of phenols by the appropriate oxidases. It is also possible that the sulfhydryl groups were substituted into positions *ortho* to the carbonyl carbon of endogenous quinones, thus limiting their antimicrobial effectiveness.

Experiments by Leatham et al. (236) provide an additional modus operandi for the observations of Maine (263). In vitro experiments cross-linking the protein RNase by chlorogenic acid in the presence of peroxidase and H_2O_2 was suppressed by L-ascorbic acid. Leatham et al. (236) surmised the L-ascorbate inhibited the cross-linking of RNase by keeping the cross-linking chlorogenic acid in the reduced state.

The experiments of Lovrekovich et al. (253) suggest that in the process of infecting and degrading potato tuber tissue, *Erwinia carotovora* produces pectinases that, in ddition to macerating tissue, induce intensive polyphenol oxidation. Exposing potato tuber tissue to bacterial pectinase induced the production of a zone of dark oxidation products that became a barrier resisting further spread of the pathogen. This study also showed that *E. carotovora* produces cell-bound dehydrogenases that tend to keep the plant's phenols in a reduced and, hence, less antimicrobially active state. The equilibrium between host phenols and bacterial dehydrogenases was judged to be dependent on the number of bacterial cells present. According to Lovrekovich and Farkas (252), when the pectic enzymes produced by the bacteria diffuse out of the infected area "more rapidly than

the bacteria can follow," a barrier of oxidized phenols free of bacteria forms that serves to halt the progress of the pathogen. The likelihood that a comprehensive barrier may form is strengthened by the reports of Leatham et al. (236) and Neukom and Markwalder (299) describing cross-linking between diphenols and cell wall-associated carbohydrates and proteins.

Role of peroxidases. There is also evidence that, in addition to polyphenol-oxidase activity, heightened peroxidase activity also favors the development of disease resistance. Lovrekovich et al. (254) observed that injecting virulent cells of *Pseudomonas tabaci* into tobacco leaves increased peroxidase activity. Since peroxidase levels fluctuate markedly with age, experiments were performed on both young and old leaves (Fig. 10–27). Older leaves were uniformly high in peroxidase, whether infiltrated with 10^7 *P. tabaci* cells/ml or left untreated. However, young leaves showed a threefold increase in peroxidase after 3 d, and the increase was judged to be exclusively host-produced peroxidase. The injection of heat-killed *P. tabaci* (252) and TMV (255) also raised peroxidase levels in treated tissues approximately threefold. When tobacco leaves infiltrated with 10^9 cells of heat-killed *P. tabaci* were inoculated 24 h later with a virulent inoculum of 10^7 cells/ml of this pathogen, infection was suppressed. The positive correlation between increased peroxidase and suppression of infection was presumed to be cause and effect. This was further supported by experiments with bacterial cell-free extracts (protein) of *P. tabaci*, which, when infiltrated into tobacco leaf tissue, raised peroxidase levels 84% and protected the tissue against infection by a challenge inoculum of 10^7 cells/ml administered 24 h later. Injection of commercial RNA, DNA, or egg albumin was ineffective in this respect. Finally, infiltra-

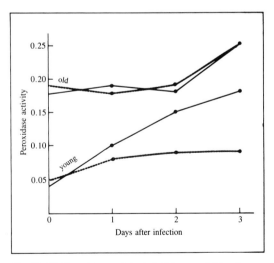

Fig. 10–27. Peroxidase activity in half-leaves of old and young tobacco tissues after infection with *P. tabaci.* Solid line: infected half-leaves. Dashed lie: untreated half-leaves (after Lovrekovich et al., 255).

tion of tobacco leaf tissue with commerical peroxidase increased the peroxidase level 100% and, similarly, protected the treated tissue against infection by a challenge inoculum. Boiled peroxidase was ineffective in providing protection.

It is apparent from the above that a potentiated level of peroxidase in tobacco leaf tissue favors the development of resistance to a bacterial pathogen. The manner in which peroxidase exerts its disease resistance affect is not well understood. However, some insight into a possible mechanism of action can be surmised from the studies of Urs and Dunleavy (420, 421) on the resistance fostered by peroxidase in soybean against *Xanthomonas phaseoli* var. *sojense*. These studies are described in some detail in Chap. 3. From their studies it was proposed that the soybean peroxidase in the presence of ascorbic acid and H_2O_2 was antibacterial. They envisioned the antibacterial activity being due to the production of H_2O_2 initially by the pathogen in host tissue. The level of peroxide is further elevated by the activity of ascorbic acid oxidase oxidizing ascorbic acid to dehydroascorbic acid and forming additional H_2O_2. The resistant cultivar (Clark 63) contains not only a larger amount of peroxidase but also four times as much ascorbic acid oxidase. Hence, the potential to oxidize endogenous phenol in Clark 63 could account for the recorded peroxidase-based resistance.

Postinfectional-Physiological

This category has, in recent years, become exceedingly complex. Where in 1967 we were content to consider the hypersensitive reaction (HR) as *the* inducible resistance response, there are now others. These include bacterial immobilization, in vivo agglutination, and plant immunization.

The hypersensitive reaction. The HR, as it was initially described by Klement et al. (219), has already been referred to peripherally in this chapter. It is characterized as an *active* defense or *resistance* reaction by the plant, and it permits visualization of the localization of an inoculum of an incompatible bacterial pathogen. It provides a plausible hypothesis for the inability of plant pathogenic bacteria to cause systemic infection in *nonhost* tissue. The concept of hypersensitivity proposes a specific molecular recognition reaction between host and pathogen, which initiates and then supports the development of resistance. As a consequence, the host is regarded as incompatible. There is good evidence (392) to support the contention that triggering the reaction requires actual contact between plant and bacterial cells. The lack of such molecular recognition between a compatible host and pathogen results in susceptibility, ostensibly because the active defense is not initiated (182, 183).

Visual evidence of HR is best discerned when an inoculum of $>5 \times 10^6$ cells ml of *Pseudomonas pisi*, or some other "incompatible" bacterial pathogen, is inoculated into tobacco leaf or other "nonhost" plant tissue (152). Such an inoculum dose causes confluent necrosis

exclusively in that region of the leaf that has been infiltrated with the bacteria. Bacterial cells are apparently free to move in the intercellular space, and they are able to multiply in the leaf tissue (Fig. 10–28), hence a reason for their immobility in tissues undergoing HR has been vigorously sought (218). However, neither the nature of the inducer molecule of HR nor the manner in which bacterial multiplication and movement is specifically suppressed is known. A comparison of growth of HR(incompatible)–inducing bacteria with compatible and saprophytic species is presented in Fig. 10–29.

It has been shown that *P. pisi* is localized in tobacco foliage intercellular space (153, 154), as is an avirulent isolate of *P. solanacearum* (366). A review of the sequence of events following injection of the bacteria should be helpful in visualizing the nature of this process. Within 1 h after infiltration, the fluid containing the bacteria, which causes a water-soaked appearance of the inoculated tissue, usually equilibrates with the atmosphere because of transpiration. After 6 h the infiltrated tissue is flaccid and electrolyte leakage from it is intense (220); after 18 h the tissue is completely desiccated and necrotic. This reaction is exceedingly rapid and, on close ultrastructural inspection, is a violent one. The initial study of Goodman and Plurad (156) revealed that by 6 h after infiltration with 5×10^4 *P. pisi* cells/cm^2 tobacco leaf tissue, all cellular membranes (plasmalemma, tonoplast, chloroplasts, mitochondria, and microbodies) were destroyed, seemingly at about the same time (Fig. 10–30). The induction time for HR, after which killing of the inoculum (in planta) with streptomycin can no longer suppress the reaction, is 1.5–2.0 h (218, 220).

The magnitude of the destructive power of HR must rank with the most toxic biological substances, e.g. the *Clostridium botulinum* type A toxin (147). We have calculated that HR in tobacco foliage caused by *P. pisi* reflects the affect of one bacterial cell ($\cong/\mu^3$) on one leaf parenchyma cell ($\cong 30,000 \ \mu^3$)(33). Our electron micrographs and calculations suggest that the reaction may occur at single points where a bacterial cell comes in contact with a plant cell (156). The inoculum level used by Klement et al. (219) (5×10^6 cells/ml) amplified cellular destruction so that it appears as confluent tissue necrosis (Fig. 10–30). Whereas, 5×10^6 bacterial cells/ml seems an overwhelming inoculum dose, it approximates a 1:1 ratio with tobacco leaf cells in one sq cm of tissue. Bacteria are rarely detected at this concentration in thin sections prepared for electron microscopy, thus making apparent the vast volume of tobacco leaf interellular space.

Two questions are clearly posed upon examination of HR: First, what is responsible for localizing the bacteria, and, second, what is the basis for the suppression of bacterial multiplication? Our initial ultrastructural study seemed to suggest that the answer to the first question was, simply, that the bacteria were trapped in intercellular space between the rapidly collapsing and desiccating parenchyma cells (153). This reflected the complete loss of membrane-selective permeability within 6

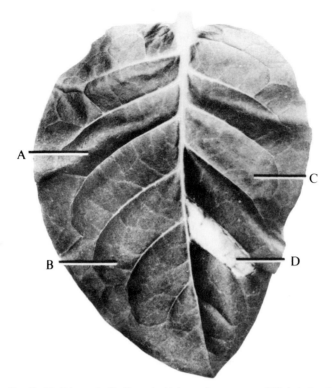

Fig. 10–28. Tobacco leaf infiltrated with increasing amounts of HR-inducing *P. pisi* cells 5×10^1 (A), 5×10^2 (B), 5×10^3 (C), and $5 \times 10^4/cm^2$ (D) wherein only "D" caused the macroscopic symptom of confluent necrosis (after Turner and Novacky, 419).

h after infiltration of the bacteria (155). The process of membrane damage appeared, from our membrane reconstitution experiments, to be detectable within 3 h (191, 192). These studies are described, in some detail, in Chap. 5 and suggest that, at least in the chloroplast, it is the protein component of the membrane that is denatured as a consequence of HR. The experiments of Huang and Goodman (190) indicated that HR may have its initial effects on membrane protein, and the compositional changes that occur are presented in Table 2–3. Experiments by Keen et al. (207) with blasticidin S revealed that this antibiotic, which inhibits protein synthesis at the translational level, completely inhibited the development of HR. Their data indicated that blasticidin S inhibited both plant cell necrosis and the accumulation of glyceollin, suggesting that both processes require the synthesis of protein. Evidence not quite as clear-cut, relating impaired protein synthesis to HR, was obtained earlier with actidione by Pinkas and Novacky (326). These experiments do not confirm membrane protein as the target for the bacterial elicitor of HR, but they do characterize HR as a process requiring specific protein synthesis. Loss of membrane integrity is the apparent cause of plant cell death. Hence, the final phase of HR and the most remarkable of its associated symptoms is the "rapid" death of the plant cell that

has come in contact with the incompatible pathogen. In fact, plant cell death caused by the incompatible bacterium may be four to five times faster than death caused by a compatible pathogen.

Subsequent experiments by Turner and Novacky (419) demonstrated that when low levels of incompatible (HR-inducing) bacteria, ca. 10^3 cells/ml were injected into tobacco leaf tissue, neither tissue collapse nor confluent necrosis developed. Nevertheless, the injected bacteria remained at the inoculation site, and microscopic observation revealed that single plant cells were killed by single bacteria (Fig. 10–28). These findings implied that in HR, bacteria were *not* necessarily localized by *collapsing tissue* and that necrosis proceeded at the cellular as well as the tissue level. Clearly, the earlier explanation for the localization of HR-bacteria by entrapment in collapsed tissue was no longer tenable. Obukowicz and Kennedy (305) used ultracytochemical methods to localize both phenols and polyphenoloxidase (PPO), primarily in the grana of tobacco leaf chloroplasts during HR development caused by the avirulent B1 strain of *Pseudomonas solanacearum*. They proposed that the activated PPO is part of the plant's *terminal* response during HR. It is during this period that subcellular organellar membranes are deranged.

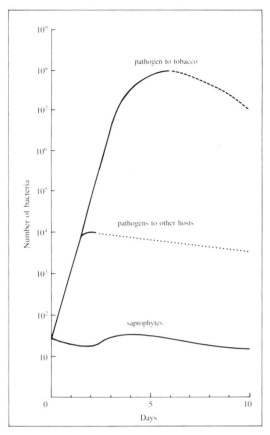

Fig. 10–29. Multiplication of phytopathogenic and saprophytic bacteria in tobacco leaf, idealized curves (after Klement et al., 219).

Nevertheless, the question regarding the *essentiality* of plant cell necrosis to inhibit growth of the incompatible bacterium remains unclarified. Király et al. (215) have suggested that the suppression of bacterial multiplication of *P. pisi* and *P. syringae* in tobacco leaf tissue is separate and distinct from tissue necrosis associated with HR. In their experiments, necrosis caused by either bacterium could be inhibited by prior infiltration of the tobacco with 1% albumin solution. This treatment did not influence the low rate of multiplication of the two incompatible species, as compared with the compatible species, *P. tabaci*. Hence, Király et al. (215) contend that necrosis accompanying HR is only a symptom associated with incompatibility (resistance) and is not causally related to the observed suppression of bacterial growth (Fig. 10–31).

Seemingly opposite results were obtained by Goodman (153; see Fig. 10–32) suggesting that, in the absence of visible necrosis, HR-inducing bacteria continued to grow in nonhost tissue. In this experiment, *Pseudomonas pisi* (10^8 cells/ml) was infiltrated into detached tobacco leaves and the injected water was allowed to equilibrate with the atmosphere for 2 h. Then one-half of the treated leaves were placed in a humid

chamber and the remaining half were left exposed to room atmosphere. Both groups of leaves had their petioles immersed in a reservoir of water. The population dynamics recorded in Fig. 10–32 clearly suggest that in leaves kept in the humid chamber, *which did not become necrotic* for at least 24 h, *P. pisi* grew to a level of 2×10^8 cells/gram of leaf tissue. During this period, leaves kept in room atmosphere supported growth of the inoculum from 4×10^6 to 2×10^7 in 12 h, after which the population fell to 5×10^6 cells/gram of leaf tissue. Hence, where necrosis ensued, the population of HR-inducing bacteria was eventually suppressed, but not so where necrosis failed to develop. Furthermore, in both treatments, membrane damage was detected and was massive after only 12 h. Thus, membrane-damage-associated HR occurred in both treatments, but only where necrosis developed was *P. pisi* suppressed. It is *unlikely* that contact of *P. pisi* with tobacco cell walls failed to occur, as the 2-h equilibration period probably insured movement of bacteria to the walls, a criterion for HR established by Stall and Cook (392). Furthermore, the intense membrane damage associated with HR developed in the leaves of both treatments (see Fig. 5–52). It would seem that in this experiment loss of water from membrane-damaged tissues fostered the development of necrosis. Where leaves were maintained in a humid environment, tissue collapse and desiccation-associated necrosis did not develop. This experiment suggests that two distinctly different types of necrosis may develop; several authors have described HR-necrosis in tobacco leaf tissue as being light in color (256, 365). Pathogenesis-related necrosis is frequently black or brown.

In relevant experiments, Holliday and Keen (182) were able to suppress glyceollin production in soybean with the herbicide glyphosate. Of particular interest is the fact that glyphosate, which blocks the shikimic acid pathway and, hence, aromatic amino acid synthesis, did not prevent plant cell necrosis from developing. Hence, these experiments were able to separate the process of cell death from aromatic and phytoalexin synthesis. According to Holliday and Keen, rather than phytoalexins causing plant cell death, host cell death may function as a signal for phytoalexin synthesis in soybean leaves. This study also noted that, although glyceollin synthesis was blocked, the incompatible race 1 of *P. glycinea* still grew much less well than the compatible race 2. Furthermore, the suppression of glyceollin production by glyphosate could be reversed by pretreating plants with phenylalanine and tyrosine, which restored the growth of race 1 to control levels. The herbicide glyphosate appears to be a useful tool with which to study the relationship of phytoalexin accumulation and host cell necrosis to bacterial population dynamics and disease resistance.

The positive relationship between HR induction in tobacco and identification of a bacterium as a plant pathogen appears less certain with the report of Azad and Kado (21) that HR$^-$ mutants of *Erwinia rubrifaciens* are, nevertheless, pathogenic in their host, wal-

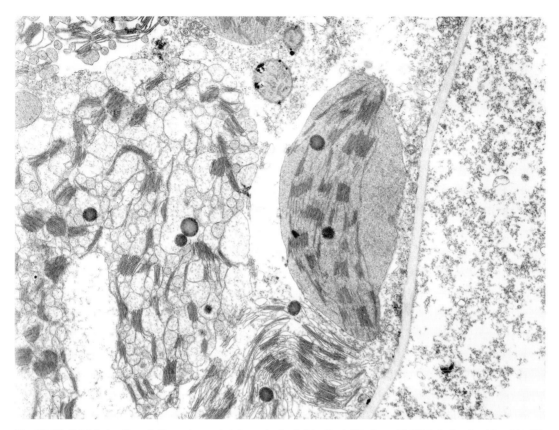

Fig. 10–30. Total destruction of all membranes in tobacco leaf cell 6 h after infiltration with HR-inducing *P. pisi* at 5×10^6 cells/ml. Ratio of 1 bacterial cell for each plant cell exposed to the inoculum (after Goodman and Plurad, 156).

nut. However, *E. rubrifaciens* is the only species with a known HR inducer. This inducer is reported to be pectic acid transeliminase (PATE). It is curious, however, that soft-rot-inducing bacteria such as *E. carotovora* (a PATE producer) do not cause HR in a number of test plant species.

Despite the intense research activity since its discovery in 1964 (219), several basic questions concerning HR remain. Specifically, we do not yet know the nature of either the inducer molecule or its receptor. How growth of HR-inducing bacteria is limited remains obscure, although phytoalexins, which will be discussed subsequently, appear to be partially effective in this regard (182). Finally, the basis for membrane disorientation caused by incompatible bacteria is also not known.

Immunization. Perhaps no research group has contributed more to the recent literature of immunization, from both the fundamental and practical points of view, than Joseph Kuć and his associates (38, 39, 56, 228, 229). Their contributions will be dealt with subsequently, but the pathway to our current status has been a long one, and earlier researches should be mentioned. Chester (64) details some of the early attempts at immunizing plants against infection. This review is suggested

not only as a history lesson but also for the ideas that one might gather or perhaps even to preclude "rediscovering the wheel."

Attention is called to the sagacity of Gäumann (138), who recognized that there were instances when the plant's survival of the primary infection sensitizes its disposition to resist infection so that it develops what he called "its full immunological efficiency," or the immune response.

One of the earliest substantive researches suggesting that plants could be immunized against infection was reported by Brown in 1923 (47), who injected heat-killed cells of *Agrobacterium tumefaciens* into daisy stem tissue and 24 h later injected additional killed cells into the *same* puncture. Immediately thereafter, she inoculated a virulent isolate, by needle prick, first in the internode just above the previous punctures, and then at the original puncture site. Galls formed on the controls and also on stems *above* the initial puncture, but not on those stems inoculated in the original puncture. From this study, it was clear that protection or immunity was afforded at the site of original puncture but did not extend to the internode above. The protective effect was not systemic. A much later study by Lippincott and Lippincott (240) indicated that competition at the

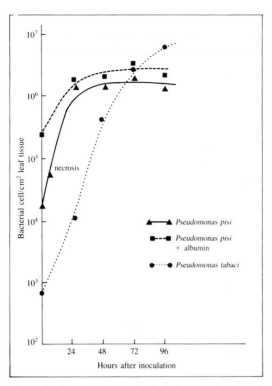

Fig. 10–31. Multiplication of *P. pisi* in Samsun tobacco leaf, in which the development of the hypersensitive necrosis was inhibited by albumin. The multiplication of *P. tabaci* in Samsun tobacco leaf (a compatible host-pathogen combination) is compared to that of *P. pisi*. Inoculum and albumin were applied to the half-leaves at 5×10^6 cells/ml *P. pisi* or 10^5 cells/ml for *P. tabaci*. Leaf discs (1 cm²) were cut at time intervals after inoculation and macerated in 0.1 ml sterile physiological saline per disc. Data reflect the bacterial cell number per cm² leaf tissue (after Király et al., 215).

wound site between heat-killed and virulent bacteria for attachment sites could account for the results obtained by Brown (47). Lovrekovich and Farkas (252) also used heat-killed cells of *P. tabaci* to inoculate half-leaves, and this treatment suppressed a challenge inoculum of living cells of the pathogen. Their study revealed that a time interval of 24–48 h between immunizing and challenge inocula was needed for the development of this "acquired" immunity. This suggested that the host plant plays an active role in the process and that young leaf tissue is more successfully immunized than older tissue. In addition, immunized tissue did not counter the effect of tabtoxin produced by *P. tabaci*. Hence, it appears that young metabolically active tissue has a greater capability of mounting an immune response than does older tissue, that the immunizing effect is directed against the bacteria per se, and that the phenomenon is not an antitoxin effect.

The study by Lippincott and Lippincott (240) indicated that pretreatment of bean leaves with an avirulent strain of *A. tumefaciens* need only precede the virulent challenge inoculum by 15 min to suppress infection. In addition, their data (Table 10–8) revealed that increasing the number of avirulent *A. tumefaciens* cells in an inoculum with a constant number of virulent cells resulted in a reduced number of tumors developing in inoculated leaves. Yet the report by Lovrekovich and Farkas clearly indicated that a time interval of 24–48 h was implicit for the development of the immune response in tobacco to *P. tabaci*. The first experiment suggests competition for infectible sites, whereas the second implies the need for the active synthesis of an immunizing substance. Because of the unique features associated with *A. tumefaciens* infection (427), i.e. insertion of a portion of the Ti plasmid into the host cell, it would seem that a successful resistance reaction only requires an efficient short-term blocking process at the wound site at the crucial moment of DNA passage into the host cell wound. Although pathogens like *P. tabaci* and *E. amylovora* might also be suppressed by com-

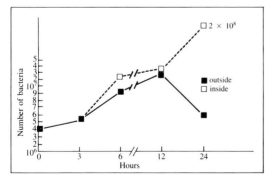

Fig. 10–32. Growth of an inoculum of 5×10^6 cells/ml of HR-inducing *P. pisi* in Samsun NN tobacco foliage; 2 h after infiltration the detached leaves with petioles immersed in water were placed in either a humid chamber (inside) or a normal room atmosphere (outside) (after Goodman, 153).

Table 10–8. With a constant number of virulent B6 *A. tumefaciens* cells, inhibition of tumor formation in wounded bean leaves is caused by an increasing number of avirulent 11BNV6 *A. tumefaciens* cells. The data suggest that 11BNV6 cells exclude B6 cells from some specific site within the wound (after Lippincott and Lippincott, 240).

Cells/ml of inoculum		11BNV6/B6 in inoculum	Mean of tumors/leaf	Percent of control
B6	11BNV6			
4.75×10^8	0	—	20	100
4.75×10^8	5.5×10^8	1.15	18	90
4.75×10^8	1.1×10^9	2.3	15.5	78
4.75×10^8	5.5×10^9	11.5	10	50
4.75×10^8	1.0×10^{10}	21.9	8.4	42

Table 10–9. Influence of the time between the first and second inoculations on the protective effect exerted by the initial inoculum (after Goodman, 151).

		No. healthy (H) and infected (I) shoots per time between inoculations							
		0.5 h		4 h		7 h		10 h	
First inoculum	Second inoculum	H	I	H	I	H	I	H	I
Yellow bacterium (35A)[a]	S-1[b]	10	0	3	6	8	4	9	1
Sterile water	S-1	1	8	0	9	0	10	0	10

a. Yellow bacterium is *Erwinia herbicola*.
b. S-1 is a wild-type virulent isolate of *E. amylovora*.

petitive inhibition at the wound site, their processes of pathogenesis probably require a more sustained resistance reaction in order to suppress continued multiplication and the effects of bacterial mass action.

A study by Goodman (151) reported protection of Jonathan apple against *E. amylovora* by an avirulent isolate of the pathogen, by *E. herbicola* a saprophyte, and by *P. tabaci*. The avirulent strain and *E. herbicola* were particularly effective in this respect. In a companion experiment, *E. herbicola* was used as the immunizing species (see Table 10–9). The data suggest that two types of suppression of the challenge inoculum were detected: a transient one that appeared within a half-hour and a second one that intensified over a 7- to 10-h interval between immunizing and challenge inoculations. The first may reflect the blockage of infectible sites, whereas the second suggests the synthesis of immunizing substances or a response to a sustained or intensifying stress. Later, in a similar experiment, McIntyre et al. (275) presented evidence that DNA from either virulent or avirulent *E. amylovora* cells provided systemic resistance to Bartlett pear against infection by virulent isolates of *E. amylovora*. Of interest was the fact that RNA from *E. amylovora* was ineffective in this regard. The role of DNA as an inducer of resistance has remained an enigma.

Experiments by Crosse (72) reporting the suppression of *Pseudomonas mors-prunorum*, by Averre and Kelman (18) with mixed inocula of virulent and avirulent isolates of *P. solanacearum* in tobacco, and by Hsieh and Buddenhagen (189) using *E. herbicola* to suppress *Xanthomonas oryzae* in rice provide additional evidence for the induction of an immune response in plants. The population dynamics in all these studies indicate a more favorable response if the immunizing isolate is present in significantly larger numbers than the virulent pathogen. Where mixed inocula were used (18, 72, 151), the data may be simply interpreted as indicating that induced immunity is more efficiently evoked by

larger populations of the inducing strain. The data also suggest that immunity must be evoked rapidly in order to counter the "eventually" superior growth potential of the pathogen in host tissue. Using a high titer of immunizing bacteria precludes the need for these bacteria to multiply in host tissue in order to evoke the immune response. In this regard, unpublished data from experiments by Nishimura, Ouchi, and Goodman infer that avirulent isolates of *E. amylovora* are able to immunize Jonathan apple better against isolates of the pathogen that are "slightly" less virulent than against isolates that are "maximally" virulent.

Five factors seem, therefore, to mediate the efficacy of the immune response: (1) comparative inoculum dose of inducer (immunizing strain) and pathogen, (2) degree of virulence of the pathogen, (3) the inherent resistance of the host, (4) the environmental conditions under which the interaction proceeds, and (5) time interval between immunizing and challenge inoculations.

It may be asked whether plants are not regularly immunized in nature by epiphytic species of incompatible bacteria that coincidentally enter host plant tissues at the same time that *comparatively few* cells of a compatible pathogen gain entry. It is tempting to presume that, in the field, the inoculum dose of the pathogen is more frequently low and, hence, is more manageable by the immunity-inducing species.

It is probably apparent to the reader, now, that immunization has been effected, experimentally, by heat-killed, avirulent, and incompatible bacteria. These are necessarily applied in high concentration in an effort to get a rapid titer rise of immune substances. What would the result be if immunization were attempted with only partially attenuated strains of the pathogen? Unpublished data of Nishimura et al. suggest that the higher the intrinsic virulence of the challenge inoculum, the more difficult it is to suppress. On a theoretical basis, at least, it would seem that protection efficiency would be enhanced if the immunizing strain was capable of limited growth in the host tissue. Ideally, the bacteria should be able to sustain just enough metabolic activity to fully trigger the development of the immune response.

We come finally to the question of practicality; can immunization be accomplished successfully in the field? The studies of Kuć et al. (229) have shown that plants can be systemically immunized against diseases caused by fungi, bacteria, and viruses by *restricted* infection with fungi, bacteria, or viruses. Immunization was follwed by a "booster" inoculation that protected the plant throughout the growing season. The host plant in these experiments was cucumber, and the pathogens were the tobacco necrosis viruses, *Pseudomonas lachrymans* and *Colletotrichum lagenarium*. The extent of immunization was found to be directly related to the number of lesions produced on the "inducer" leaf (first true leaf) until a saturation point (of either inoculum or lesions caused) is reached. Comparatively few lesions, as few as eight, caused by TNV on the "inducer" leaf immunized tissues above that leaf. The system employed by Bergstrom and Kuć (38) and Bergstrom et al.

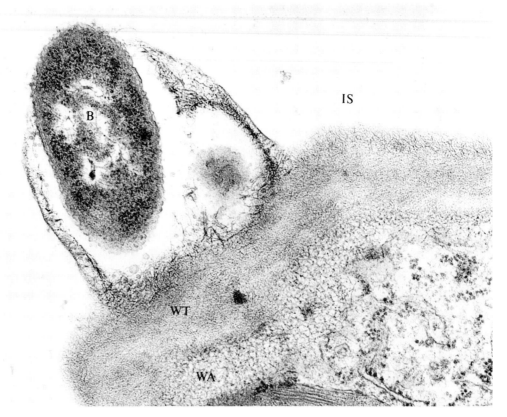

Fig. 10–33. Localization of *P. pisi* (B) by tobacco leaf cell at interface with intercellular space (Is) 3 h after inoculation; note wall thickening (WT) and wall apposition (WA), × 50,500 (Politis and Goodman, unpublished).

(39) was capable of immunizing cucumber plants against *E. tracheiphila* as well.

It is apparent that immunization of plants is possible and that the development of commercial methods and the selection of competent inducer strains may be accomplished in the not too distant future.

Bacterial envelopment and agglutination. In studying ultrastructural features of HR, Goodman (153) observed a *P. pisi* cell apparently immobilized by small granules of unknown origin. A subsequent time-course ultrastructural study revealed what seemed to be an active envelopment of HR-inducing bacteria by cuticular, fibrillar, and granular substances of "apparent" plant origin participating in the process (Fig. 10–33). Specifically, during a 6-h period, the *P. pisi* cells infiltrated into tobacco leaf intercellular space became increasingly enveloped and immobilized. Other investigators have since made similar observations in other plant-bacterial pathogen interactions (4, 57, 117, 155, 366). That the phenomenon of envelopment may be an artifact of transpiration and subsequent accumulation of intercellular debris around the bacteria was suggested (177). However, in the *P. pisi-* and *P. solanacearum-*tobacco

(155, 366) and *X. malvacearum-*cotton (4, 17) interactions, it seems illogical to characterize as artifacts such accompanying ultrastructural aberrations as wall thickening, plasmalemma vesiculation, or intense microfibril apposition precisely confined to points where plant cell walls are in close proximity or in contact with bacteria (see Fig. 10–34).

Other experiments that have revealed bacterial immobilization at the ultrastructural level are those of Huang et al. (194) that describe the agglutination of an unencapsulated avirulent isolate of *Erwinia amylovora* in apple petiole xylem vessels and that of Huang and Van Dyke (193) on the immobilization of *P. pisi* on the surface of tobacco callus cells. Both of these studies appear to strengthen the contention that incompatible bacteria are actively localized by plant cells. In the former case, the acapsular *E. amylovora* cells are aggregated by granules that appear to have their origin in neighboring xylem parenchyma cells (Fig. 10–35). In the latter, the incompatible *P. pisi,* a pathogen of pea, is localized on the surface of tobacco callus cells apparently by fibrils produced by the plant cells (Fig. 10–36).

The nature of the agglutinating granules and the manner in which they aggregate and immobilize *E.*

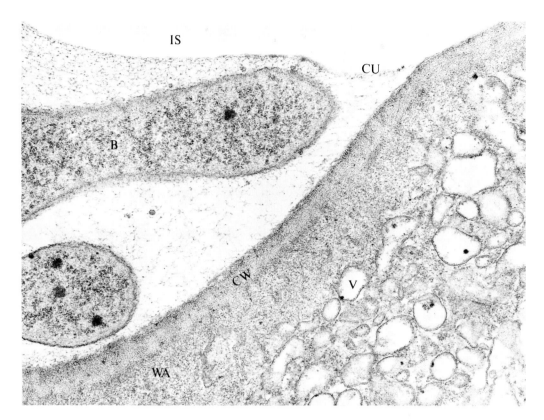

Fig. 10–34. Samsun NN tobacco leaf cell interfacing with intercellular space (IS) 3 h after infiltrating tissue with 10^7 cells/ml of *P. pisi*. Note bacteria (B) opposite site of apposition (WA) of cell wall-like material. Vesicles (V) appear to be carrying and depositing amorphous wall material. Note also fibrillar material seemingly originating at the cell wall (CW) surface forming the localizing cuticle (CU) (afer Politis and Goodman, unpublished).

amylovora in apple petiole xylem vessels have been postulated by Romeiro et al. (343, 344). First, it is imperative to recognize that early proliferation and long-distance transport of the pathogen take place in the xylem vessels, particularly the helical metaxylem (396, 397). Thus, the site of action of the granules is precisely where a resistance mechanism is needed to suppress the early proliferation of this pathogen (343).

The immobilizing granules appear to be synthesized in or released from xylem parenchyma into xylem vessels upon some stimulus emanating from the bacteria. The granules, approximately 60–80 nm in diameter, cause the avirulent, acapsular bacterial cells to agglutinate in situ. An agglutinin, possibly causally related to this phenomenon, has been extracted from apple seed, stem, and leaf tissue that has a dilution titer against the avirulent acapsular strain of 1024, and one of 32 against the capsular (virulent form) (343, 344). Hence, the virulent form of the pathogen is inherently less sensitive to this apparent resistance mechanism of the apple. In addition, the activity of the agglutinin can be completely suppressed by the extracellular polysaccharide (EPS) of the pathogen. It seems, therefore, that the pathogen has,

in its evolution, developed a survival mechanism capable of neutralizing a resistance mechanism of the host (Fig. 10–37). It is also possible that EPS diminishes the intensity or precludes activation of the host resistance mechanisms, one of these being the release of agglutinin into the xylem vessels.

It remains to be demonstrated, conclusively, whether the granules visualized in the electron microscope (194) and the agglutinin extracted from apple seed, stem, and leaf tissue (342, 343) are one and the same or are closely related proteins. Subsequent studies suggest that the apple agglutinin, named *malin* by Romeiro et al. (344) is a member of a group of small highly positively charged proteins (MW 12,600) called thionins that possess, in vitro, antibiotic activity (51). However, they seem widely distributed in plants, being most recently detected in tomato and melon seed (Addy, Karr, and Goodman, unpublished data), and may represent a previously unreported general plant disease resistance system.

A concluding comment concerning the divergence of opinion on how bacteria are immobilized in situ is warranted at this point. The procedure may be either an

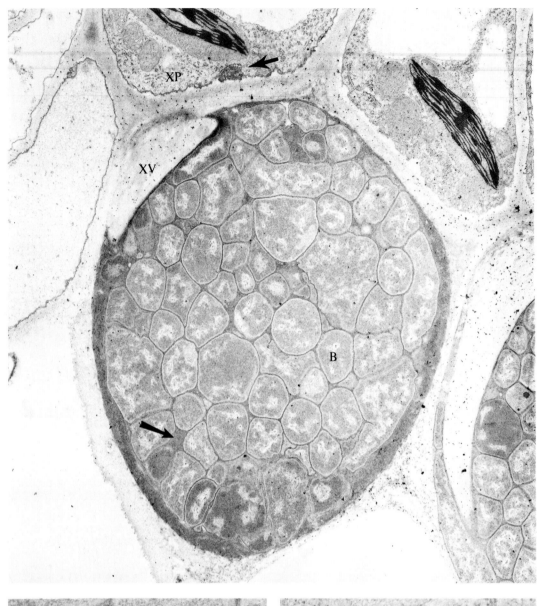

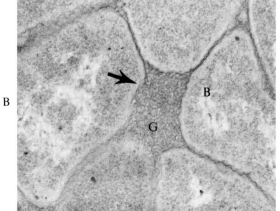

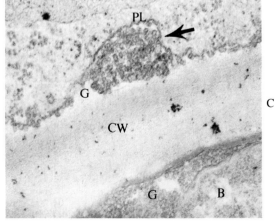

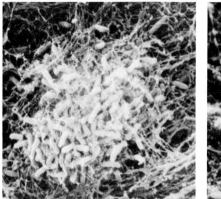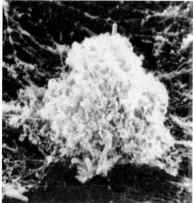

Fig. 10–36. Immobilization of incompatible *P. pisi* on the surface of tobacco callus cells by plant cell-elaborated fibrils. At left, 24 h after inoculation and, at right, after 48 h (after Huang and Van Dyke, 193).

important feature of resistance per se or an adjunct to HR.

Specifically, some contend (57, 156, 366) that the bacteria are enveloped by an active process that involves

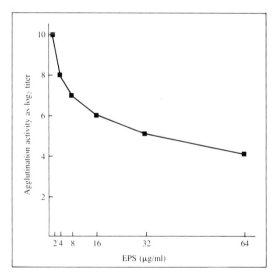

Fig. 10–37. The effect of EPS derived from cultures of strain E_9 on the agglutination of avirulent nonencapsulated strains of *E. amylovora*. Agglutination activity is given as $\log_2$ of the agglutination titer. The EPS concentration is displayed in $\mu g/ml$. No agglutination was observed when EPS concentrations of 128 $\mu g/ml$ or higher were used (after Romeiro et al., 344).

wall cuticle rupture and release of wall and cytoplasmic components that immobilize bacteria at the plant cell surface. The counter view holds that bacteria are trapped by plant cell surface debris that is loosened by water (inoculum) injected into intercellular space (177). As the injected water recedes, in response to transpiration, the debris is deposited on the bacteria in an "apparent" localizing process. The latter view depends extensively on the oft-observed entrapment of bacteria at the narrow junctures between plant cells and the apparent accumulation of debris at those sites in water-injected control tissue. However, this view does not explain how bacteria are enveloped along linear stretches of the cell wall surface (Fig. 10–38). The explanation appears not to acknowledge that the enveloped bacteria are usually opposite thickened plant cell wall or adjacent to areas of accumulating fiber-carrying vesicles or masses of granules that appear to be released from wall surface newly exposed by detachment of a cuticular layer (357) (Fig. 10–34).

It is possible, however, to explain the envelopment of the bacteria in yet another way. As the water is lost from intercellular space by transpiration, the receding film carries *fibers and granules back* toward the precise site of their release from the wall. The receding film of water is thus able to form with time (156) a thickening cuticular mantle around the incompatible bacteria, thereby immobilizing them. It would appear, from the study of Al-Mousawi et al. (4), however, that envelopment may be triggered by a number of agents and is nonspecific.

The infiltration of water only into intercellular space may redistribute wall material at the surfaces and

Fig. 10–35. a–c: A tangential section through Jonathan apple petiole tissue 24 h after inoculation with an avirulent mutant (E_8) of *E. amylovora*. The inoculum, 10^5 cells, was placed on the petiole tip immediately after leaf excision. (a) Xylem vessel (XV) is totally occluded by agglutinated bacteria ($\times$ 17,000). Arrows indicate sections magnified in band c. (b) magnification of agglutinated bacteria (B) in (a) by granules (G) $\times 41,000$. (c) Agglutinating granules between plasmalemma (PL) and cell wall (CW) separating xylem parenchyma cell (XP) and vessel lumen. Note granules either side of the cell wall; ($\times$ 41,000) (Goodman, unpublished).

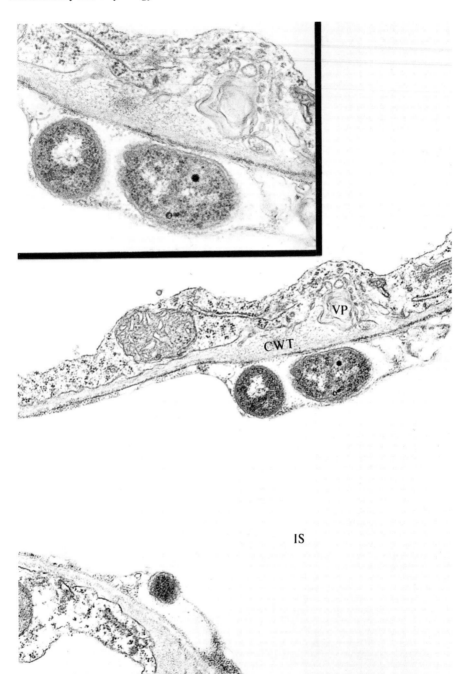

Fig. 10–38. Localization of *P. pisi* along linear stretches of tobacco plant cell wall 4 h after inoculation. Note cell wall thickening (CWT) and vesiculating plasmalemma (VP) opposite bacteria. Wall thickening is presumed to be a response of the plant wall to the bacterium and vesiculated plasmalemma possibly contributing to delivery of wall material; IS, intercellular space (after Goodman, et al., 155).

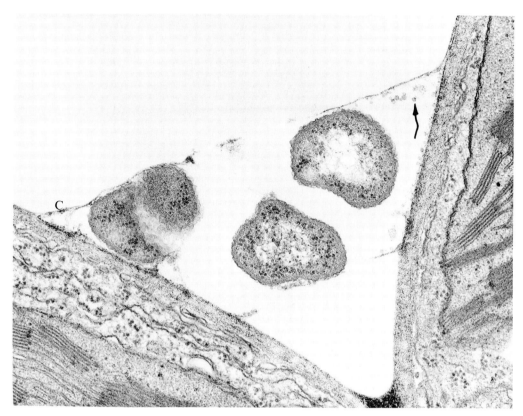

Fig. 10–39. A juncture between two tobacco leaf cells wherein *P. fluorescens* cells have been localized by an ephemeral wall cuticle (C) 6 h after inoculation with 10^9 cells/ml of *P. pisi*. Arrow designates localizing granules. Both tobacco and bacterial cells are unaffected (×42,800) (after Goodman et al., 155).

juncture of cells. However, this, too, remains to be conclusively established.

Postinfectional-Biochemical

Phytoalexins. Friend (133) has registered some doubts regarding phytoalexins as primary factors in resistance to fungal as well as bacterial pathogens. The basis for the uncertainty is the difficulty of demonstrating in vitro that phytoalexins are antibacterial at concentrations likely to be achieved in vivo. Nevertheless, a series of experiments by Keen and his associates (207, 208) present evidence that a number of pseudomonads that elicit HR in soybeans also cause the accumulation of glyceollin and related isoflavanoids in inoculated leaf tissues. The species that are effective, in this regard, include *Pseudomonas tomato*, *P. pisi*, *P. lachrymans*, *P. solanacearum*, *P. phaseolicola*, and *P. glycinea* (an incompatible race), whereas saprophytes *P. fluorescens* and *P. aeruginosa* and a compatible race of *P. glycinea* were unable to cause either HR or glyceollin accumulation.

The observation that saprophytic bacteria do not turn on phytoalexin synthesis and accumulation suggests that this feature of resistance is distinct from immobilization. Attention is called to the observation of Goodman et al. (155) and Sequeira et al. (366) that saprophytic species are immobilized in tobacco by a rather thin and ethereal cuticle (see Fig. 10–39). Although saprophytic bacteria are immobilized, the adjacent plant cells do not collapse, nor do they become necrotic. Instead, the infiltrated tissue slowly becomes chlorotic during the ensuing 7–10 d. It is of interest that tissue collapse and necrosis do occur in tissues infiltrated with HR-inducing bacteria. Hence, it would seem that plant cell death and phytoalexin production and accumulation are generally related if not causally linked. It is also apparent, however, that the complexity of the immobilizing cuticle developed in response to HR-inducing bacteria is much greater than that which envelops saprophytes.

In studies by Holliday et al. (183), shifting the temperature at which Harasoy soybean plants were held after inoculation with an incompatible race of *P. glycinea* from 22 C to 31 C clearly revealed that when HR is suppressed (Table 10–10), glyceollin does not accumulate beyond background levels. In addition, the bacterial population is 10-fold higher than in leaf tissue maintained at 22 C, which suggests a concomitant suppression of bacteriostasis. In related experiments, levels of glyceollin accumulation paralleled increases in bac-

Table 10–10. Symptom expression and glyceollin accumulation in soybean leaves infiltrated with *Pseudomonas syringae* pv. *glycinea* race 1, incubated at 31 C for various times and then incubated at 22 C for 48 h (after Holliday et al., 183).

| Incubation time (h) at 31 C | Symptoms | | Glyceollin (μg g^{-1} fresh weight)[a] |
	At time of temperature shift	48 h after temperature shift	
0[b]		N(H)	600
12	NS	N(H)	1,155
24	NS	N(H)	1,005
36	NS	N(H)	800
48	WS,C	WS,C,N(M)	450
72	WS,C	WS,C,N(M)	320
96	WS,C	WS,C,N(L)	360
96[c]			35
144[c]			80

NS: no symptoms; N: necrosis and browning (L, M, and H denote light, medium, and heavy); C: chlorosis; WS: watersoaking.
a. Glyceollin was extracted 48 h after shifting plants to 22 C; results represent mean of 3 repeated experiments corrected for absorption value of water-infiltrated control leaves (ca. 20 μg g^{-1}).
b. Continuous incubation at 22 C, glyceollin extracted 48 h after infiltration.
c. Continuous incubation at 31 C, glyceollin extracted 96 or 144 h after infiltration.

teriostasis. Hence, phytoalexin production seemed closely correlated with resistance, and glyceollin levels of 240 μg/g leaf tissue appeared necessary to support the development of bacteriostasis in the Harasoy-*P. glycinea* (race 1) combination.

Song, Karr, and Goodman (unpublished data) evaluated glyceollin production in fully imbibed, germinating, finely sliced Clark 63 soybean seeds. The sliced seeds were inoculated with ca. 10^6 cells of *P. glycinea* (compatible race 2) and incompatible species *P. syr-*

ingae or *P. tabaci*. Glyceollin production was monitored by HPLC at 24-h intervals, and the data concerning glyceollin accumulation and population levels achieved are presented in Table 10–11. It is apparent that the glyceollin accumulation from seeds, as shown earlier by Keen (206), is massive in incompatible combinations. This system revealed marked disparity in glyceollin production between compatible and the two incompatible species as the latter elicited 7 and 16 times as much glyceollin from soybean seeds. That *P. tabaci* quickly elicits exceedingly high glyceollin accumulation and, concomitantly, that the sliced soybean seeds support bacterial population increases of four orders of magnitude in 24 h, suggest that glyceollin may not directly suppress multiplication. How then is bacteriostasis imposed?

During the initial 24–48 h following inoculation, bacterial growth proceeds not just normally, as in pathogenesis, but explosively. It would seem, therefore, that in soybean HR fosters a period of hyperparasitism that is of finite duration and is followed by an intense synthesis and accumulation of the phytoalexin glyceollin. An alternative explanation is that wounding, by fine slicing of the imbibed soybean seed, provides an enormously rich substrate to support massive bacterial growth, which concomitantly elicits high levels of phytoalexin synthesis. The studies of Rahe and Arnold (334), and Keen and Holiday and Keen (182, 206) have suggested that plant cells exposed to HR-inducing pathogens *respond prior to or at their death by releasing elicitors of phytoalexin production*. These may be perceived as stress signals. It is, then, that the neighboring living cells actively synthesize phytoalexin and accumulation of phytoalexin occurs in the adjacent dead cells. Thus the dead cells become simultaneously a reservoir of antimicrobial substance and a chemical barrier or impediment restricting the development of the pathogen. It is possible that the intense, unrestricted, and sustained growth of the pathogen is necessary to insure adequate phytoalexin accumulation. It would also appear that the currently assessed threshold level for antibiosis is high.

The foregoing interpretations of the modus operandi

Table 10–11. Accumulation of glyceollin and growth of bacteria on sliced germination soybean seeds (after Song, Karr, and Goodman, unpublished data).

| Time after inoculation (h) | *P. glycinea* Race 2 | | *P. syringae* | | *P. tabaci* | |
	population	glyceollin	population	glyceollin	population	glyceollin
0	2.5×10^6	—	3.8×10^6	—	4.3×10^6	—
24	1.2×10^9	12.4	2.1×10^9	22.6	1.5×10^{10}	101.1
48	3.4×10^9	30.3	3.0×10^9	73.9	1.8×10^{10}	248.9
72	4.5×10^9	35.4	1.4×10^9	130.3	1.3×10^{10}	299.4
96	6.3×10^9	36.9	1.2×10^9	147.8	1.1×10^{10}	404.4
120	6.0×10^9	39.7	1.1×10^9	180.8	8.3×10^9	420.0
144	5.2×10^9	95.9	1.0×10^9	245.0	5.0×10^9	443.3

for phytoalexin suppression of bacterial pathogens should be considered a working hypothesis. It is not the purpose of this commentary to either validate or fault the phytoalexin concept. Time will be the final arbiter.

Resistance to Fungal Infections

The literature and terminology pertinent to the resistance of plants to fungal pathogens reflect the divergent interests (not goals) of researchers in genetics and plant breeding, epidemiology, disease control, plant physiology, and biochemistry. Plant breeders and geneticists are concerned with race specificity or nonspecificity, respectively, and vertical and horizontal or durable resistance. Plant pathologists are concerned with different forms of resistance based on the reaction types of the host plant. An example would be the disease scale of Stakman et al. (391) for infection of cereal leaves by rust fungi. The number of infection sites per unit leaf area and the size of the pustules is integral to an understanding of disease resistance, even if the host is "susceptible" according to the Stakman scale. It is also apparent that several susceptible plant cultivars permit sporulation of the fungal pathogen to develop slowly. Clearly, slowly sporulating types could also be useful in developing resistant types, in spite of their susceptibility as determined by the Stakman scale. Furthermore, plant pathologists have found, although rarely, that the yield of certain cultivars is not reduced by infection because they are tolerant of disease development. In this chapter an effort is made to describe resistance from biochemical and physiological points of view (212, 355, 365).

Preformed Resistance

Preformed chemical resistance factors exist in the host before infection, and they are toxic to the infecting agents. Many preexisting antimicrobial substances have been found in host plants, such as an array of phenolic compounds, phenolic glucosides, polyunsaturated compounds (wyerone), glucose esters (tuliposid) and benzoxazolinones (355, 356). According to Overeem (315), "It is difficult to discuss the role that these substances may play in resistance of plants to fungal or bacterial attack because so much has been suggested and so little has been convincingly proved." The hope that it would be easy to select resistant types of plants on the basis of preformed factors has proved overly optimistic. Preformed chemical factors may, nevertheless, be antimicrobially effective in resistant genotypes of susceptible species as well as in nonhost plants.

The classic examples of *preformed substances* are the phenolics governing resistance in colored onion scales against *Colletotrichum circinans* (439). Additional examples are cited in Chap. 6. The most thoroughly investigated cases are the tulip diseases caused by *Botrytis cinerea* and *Fusarium oxysporum* f. sp. *tulipae* and the tomato fruit disease caused by *Corticium rolfsii* (355).

Phenolic and other host substances can inhibit hydrolytic enzymes of the fungi, thereby restricting their activity. On the other hand, hydrolytic or "lysosomal" enzymes in the host vacuoles can inhibit fungi in resistant hosts.

The key role of host cell *membranes* in disease resistance and susceptibility has been treated in Chap. 1. Stable, unimpaired membrane integrity may also be considered a preformed resistance factor against facultative parasites.

It has been suggested that a primary or very early action of infection is on host plasma membranes. Membrane stability can protect plants against infection by facultative parasites (necrotrophs) and the stressful actions that tend to disorganize host cell membranes. Factors inducing juvenility in plants tend to maintain and stabilize membrane integrity. Cytokinins, for example, increase plant resistance to necroses whether they are caused by facultative parasites or by abiotic agents. Cytokinins can influence membrane synthesis and protein turnover and suppress ethylene production and a series of degradative enzyme activities; in addition, both membrane permeability and aromatic accumulation are suppressed. The juvenility effect of removal of the terminal bud on a shoot seems to be similar. Hardening plants in order to make tissues resistant to injury caused by freezing or chilling temperatures may also stabilize membranes to withstand these adverse actions. As a rule, phospholipid synthesis in membranes is increased and the phopholipid/sterol ratio increases. High nitrate-nitrogen nutrition may also induce juvenilitylike changes that cause plants to exhibit resistance to necroses caused by facultative parasites. High-nitrogen treatment of uninfected tomato plants caused an increase in phospholipid content of the membranes and a decrease in their free sterol content. In other words, the phopholipid/sterol ratio increased just as in the case of hardening. High N treatment also caused a sharp decrease in ion leakage from infected leaves, indicating that the changes in membrane composition were probably associated with membrane stability. Indeed, these plants were more resistant to infection caused by *Fusarium oxysporum* f. sp. *lycopersici*, to its toxin, fusaric acid, and to other toxic actions (33; see also Chap. 1). In summary, a high phospholipid/sterol ratio in the membranes appears to be associated with a stabilized membrane integrity, which, in turn, may be regarded as a preformed resistance factor to facultative parasites.

Amino acid-induced resistance, referred to in Chap. 5, may also be regarded as associated with preformed resistance. The excess of some *L*- and *DL*-amino acids probably disturbs normal host protein metabolism. Inhibitors of protein synthesis can reverse this type of resistance causing host plants to express their original infection type (32, 34). In fact, amino acids change the disposition of host plants to infection by altering the natural protein composition of plants before infection.

Preformed inaccessibility of several plants to the adverse actions of toxins can be the cause of "passive resistance." In this instance, the incompatible pathogen is neither inhibited nor suppressed (395, 465). Its toxin is incapable of inducing a pathological condition in the tissues. For example, the toxin of *Helminthosporium sacchari* cannot bind to a receptor protein on resistant plant membranes because this protein is different from the toxin- binding protein on susceptible sugarcane cell membranes. Consequently, the pathogen's toxin is not able to induce a proper environment for the fungus to develop. On the other hand, in susceptible sugarcane the toxin is bound by the receptor protein, thus the plant can be regarded as a host that has *preformed susceptibility*. After successful binding, the toxin can *induce* susceptibility (see below).

Active (Postinfectional) Resistance

This comprises all of the postinfectional metabolic processes that result in inhibition or killing of the infecting pathogen. One can distinguish two phases in active resistance: (1) the recognition of the pathogen ("determinative phase") and (2) the manifestation of inhibition or death of the pathogen that may or may not be connected to visible or other host reactions ("expressive phase").

The components of recognition on the surfaces of host and pathogen, posibily agglutinins or lectins, are present before infection in the host-pathogen complex, so they may also be regarded as "preformed" resistance factors. However, active processes of resistance may start only after the interaction between ligand and receptor (cognon and cognor). It would appear that recognition may be directly responsible for inhibition of the pathogen or it may trigger reactions in the pathogen or the host, thereby triggering the expressive phase of resistance. Furthermore, metabolic host responses during the expresssive phase may represent the mechanism of inhibition of the pathogen or may only be secondary biochemical alterations associated in a temporal sense with resistance but not causally related. Host response, in the latter case, is a sign associated with resistance by which one can easily determine the phenomenon.

The Gene-for-Gene Hypothesis

In his study of flax rust resistance, Flor (121) assumed that if one gene conditioned resistance in the host, there was a corresponding gene for virulence in the pathogen. If two or three genes determined resistance, two or three genes corresponded to avirulence in the pathogen. This is the complementary gene-for-gene hypothesis that is used to explain the coevolution of host-pathogen systems. The quadratic representation of host

and pathogen gene interaction designates genes for resistance (R) and avirulence (V) as dominant; those for susceptibility (r) and virulence (v) are recessive (97).

Three combinations are indicative of susceptibility (R-v, r-V, and r-v), and one (R-V) signifies resistance. The resistant combination requires an interaction between a gene for resistance in the host and a gene for avirulence in the pathogen. This has been interpreted to mean that the specific interaction between host and pathogen is for incompatibility. Susceptibility is regarded as a passive phenomenon due to the absence of genes for resistance and avirulence. Hence, resistance is regarded as an active process.

In reality, this is only a theoretical model and sheds little light on the biochemistry and physiology of infectious plant diseases. Unfortunately, the gene products of hosts and pathogens remain largely unidentified, thus one cannot transpose genetic ratios into molecular models. There are also difficulties in grouping different host reaction types in order to obtain well-defined resistant (R) and susceptible (S) host types. Quantitative measurements of R or S are frequently difficult because of our inability to measure pathogen growth in the host tissue accurately. In some instances, investigators studying the rust fungi have excluded the intermediate host reaction classes in order to obtain the suitable genetic ratios. Another difficulty is that the inheritance of virulence in several pathogens cannot be established because evidence for their sexual recombination has not been established.

It is noteworthy that in diseases where host-specific *toxins* have been described, the inheritance of host-pathogen interaction is inverse to the quadratic scheme. In this instance, the specific interaction is for compatibility (susceptibility). The gene product of the pathogen is the toxin, and the gene product of the host is the theoretical receptor.

Pathogen	Host	
	receptor	no receptor
toxin	+	−
no toxin	−	−

One can see that the host-gene product is only in the susceptible host with which the toxin interacts. On the other hand, the resistant plant lacks the crucial gene product. This would tend to support the hypothesis of active susceptibility, not active (induced) resistance. However, it is also conceivable that resistant tissues degrade or detoxify toxins or that resistant plants have a self-repairing mechanism against the toxic action. If this were the case, receptors for toxins would exist in both resistant and susceptible plants, and usual interpretation of the quadratic check would be permissible. Each of the above-mentioned instances would support the active resistance hypothesis (258). At present, gene products of a susceptible host (receptors), or those

Pathogen	Host	
	R	r
V	RV (−)	rV (+)
v	Rv (+)	rv (+)

of the resistant one (detoxification or self-repair), are not known in toxin-induced diseases. For that reason, there are limitations, at the present time at least, in utilizing either the gene-for-gene hypothesis or the quadratic check to define biochemical mechanisms of disease resistance.

Recognition versus Defense

HR has been regarded as an active defense mechanism against infection. "Defense" against pathogenic agents is an anthropomorphic aspect of the more general phenomenon of "recognition of foreignness" by any organism whether man, animal, or plant (49). Although pathogens may be inhibited from further growth or even killed in tissues of the resistant plant, one may regard resistance as being related to the phenomenon of recognition rather than to defense. The plant appears able to recognize "nonself" macromolecules of the pathogen by some primitive, as yet poorly defined, immune phenomenon. As a result of the recognition of foreignness, a positive rejection occurs and, in most cases, necrotic processes develop in the host and in the infecting, "foreign" agent.

Necrosis of host tissue is the most characteristic sign of HR. The theory of HR- induced resistance has been generally accepted as a logical explanation for the recognition of foreignness, according to which the plant "sacrifices" a few cells so that the rest can survive. The rapid death of the infected cells in the resistant plant reflects the hypersensitivity of the host. This explanation became more popular when it was discovered that in the hypersensitve necrotic tissues antimicrobial agents, phytoalexins, accumulated (291). HR, as well as the *phytoalexin hypothesis*, deals with the direct or indirect manifestation of the primary recognition phenomenon. This primary step in host resistance is certainly connected to its capacity to distinguish the foreign nonself from self (212, 365). The mechanism of recognition is the subject of much pathophysiological research. We will consider first the mechanism of hypersensitivity, then phytoalexins, and finally the mechanism of recognizing or distinguishing self from nonself.

Hypersensitivity. If the host or a nonhost plant is more than normally sensitive to an infecting agent and reacts with an early or rapid tissue necrosis, the plant is considered hypersensitive to the infecting agent. Hypersensitivity can exist between resistant host cultivars and cultivar-nonpathogenic parasites, or between nonhost species and species-nonpathogenic parasites. HR was initially described by Ward in 1902 (443) and later by Stakman in 1915 (390), who introduced the term *Hypersensitivity* into the literature. In its classical state, hypersensitivity takes the form of a small necrosis visible to the naked eye on different plant organs (Fig. 10–40). Thus, HR is regarded as synonymous with host cell or tissue necrosis. HR may effectively suppress obligate (biotrophs), as well as facultative parasites (necrotrophs). No quantitative data are available on the multiplication or growth of fungal pathogens in hypersen-

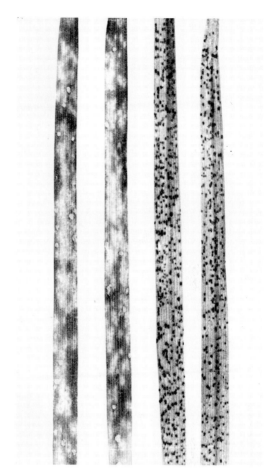

Fig. 10–40. Incompatible and compatible interaction between *Puccinia recondita* f. sp. *tritici* and wheat leaves. Right pair, susceptible reaction of the host. Left pair, hypersensitive (resistant) reaction of another host cultivar.

sitive plants, as compared to the susceptible ones. This is one reason the role of HR in disease resistance remains uncertain.

In some instances, disease resistance is not connected to the expression of the HR. Heath (171) has shown with rust disease that the HRs of host and nonhost plants in different incompatible plant-parasite pairs are different at the ultrastructural level. The extremely rapid host cell necrosis may or may not be associated with the simultaneous death of the rust haustorium. In some instances, growth of the rust may be rapidly restricted, even if necrosis does not take place. It is not certain whether one can regard this rapid response as a high degree of hypersensitivity without necrosis. If so, tissue necrosis would not be a crucial factor for hypersensitivity. The opposite view, that "without necrosis there is no hypersensitive reaction," may also be justified. Inhibition of rust fungi in resistant cultivars without host cell necrosis could be regarded simply as a distinct form of resistance that has nothing to do with hypersensitivity.

Hypersensitivity, as characterized by (rapid) cell necrosis and collapse, is a symptom of resistance to many fungal parasites. The mechanism of the HR in virus-induced local necroses and in the bacterial HR appears roughly similar, as well as for both biotrophic and necrotrophic fungi. Daly (76) and Farkas (109) have suggested that the biochemical lesions are very similar to those associated with stress, senescence, and mechanical injury (77). In biochemical events, membranes play a key role (453) and alterations in their structural and metabolic integrity can lead to decompartmentalization and then to uncontrolled mixing of enzymes and substrates that are normally isolated in subcellular organelles.

A respiratory increase is characteristic of many incompatible host-pathogen interactions, particularly when the fungus is a necrotroph. Smedegaard-Petersen and Stolen (387) reported that a barley variety, when infected by a particular race of powdery mildew (*Erysiphe graminis* f. sp. *hordei*), increased its respiration rate by 80% and reduced the grain yield, even though no symptoms become apparent. They contended that the unseen defense reaction had required energy and the greatly increased respiration deprived the resistant host of energy required for growth. According to Smedegaard-Petersen (386), nonpathogenic leaf surface fungi that are saprophytes may also induce increased respiration in nonhost plants. As a consequence, the grain yield is reduced (Fig. 10–41). Field experiments have shown that removal of fungal flora from leaf surfaces with fungicides increased the yield, in spite of the symptomlessness of the untreated barley plants.

The respiratory differences between hypersensitive and normosensitive (susceptible) hosts are sometimes large, but they may be small. Antonelli and Daly (7) compared the reactions of normosensitive and hypersensitive near-isogenic wheat lines toward race 56 of *Puccinia graminis* f. sp. *tritici*. They recorded no differences in the respiration rates for a long time after inoculation. Relatively high respiratory activity is believed to be necessary to furnish the energy and carbon units for several processes, including the synthesis of phenols, sterols, and terpenoid phytoalexins that accumulate in the hypersensitive necrotic tissues. These respiratory increases might also reflect other non-specific stresses (76). It is not known, however, whether the increased oxidation of phenols by phenolases or peroxidases is the cause of the breakdown in cellular structure, or whether the breakdown in cellular structure liberates the enzymes that can then oxidize the phenols.

The mechanism of rapid necrotization associated with HR to necrotrophic fungi was first investigated with the late blight disease of potato induced by *Phytophthora infestans*. In potato tissues treated with a narcotic, certain enzymatic processes and hypersensitive necroses are inhibited (292, 411). Poisons of a heavy metal enzyme, such as polyphenoloxidase, which is a typical Cu enzyme, also reduce the speed of the necrotic reaction (134). On the other hand, an increase in the

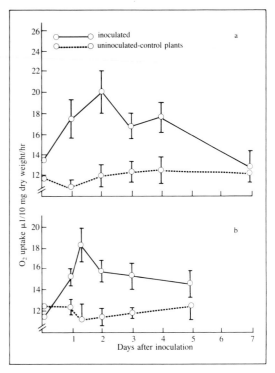

Fig. 10–41. (a) The effect of *Alternaria alternata* on respiration of barley plants. (b) The effect of *Cladosporium herbarum* on respiration of barley plants (after Smedegaard-Petersen, 386).

enzymatic- and nonenzymatic-reducing activity of the tissues results in a lowered probability for the appearance of necrosis, presumably by reducing the oxidized phenols. The appearance of necroses is usually correlated with the disappearance or lowering of the level of reducing compounds such as ascorbic acid. Reducing compounds given directly to the tissues reduce the degree of necrotization in rice infected with *Cochliobolus* and in rusted wheat (214, 309) and often the degree of resistance.

Respiratory enzyme inhibitors, such as sodium azide and 2,4-dinitrophenol, that inhibit ATP generation also suppress the development of hypersensitive necrosis in tuber tissues. It is of interest that this inhibition of necrotization can be reversed by the addition of ATP, but only in aged tubers. Otherwise, the aged tuber cells, if exposed to air for at least 20 h before inoculation, die very rapidly after inoculation, in comparison with fresh tuber tissues. Hence, it is once again apparent that senescence of tissues accelerates the process of necrotization (304). On the basis of these and other experiments (90) with SH reagents, Nozue et al. (304) suggest that cell membranes are damaged by the respiratory inhibitors and that ATP could reverse the alteration permitting unimpaired necrotization. Nozue et al. (303) also suggest that de novo protein synthesis is necessary for fresh potato tuber tissues to show necrosis. This was shown by using blasticidin S as a protein inhibitor. Clearly, hypersensitive necrosis is associated with an active metabolism in an incompatible host plant.

The situation seems dissimilar with the rust disease of wheat induced by *Puccinia graminis*, which is a true biotroph. Daly (77) has concluded that neither phenol nor hydroxyputrescine derivatives of cinnamic acid cause the hypersensitive necrosis. Furthermore, when an increase in peroxidase activity is induced by ethylene treatment, this does not induce the necrotic reaction. It would seem that peroxidase has no effect on hypersensitivity in the case of stem rust, e.g. it does not affect the compatible reaction and the hypersensitive response does not develop. The action of ethylene itself is quite unexpected, at least in a particular host-pathogen relationship in which resistance is determined by the *Sr6* gene in a temperature-sensitive reaction. Ethylene reverses the resistance reaction of *Sr6* plants at low temperatures to susceptibility, but hypersensitive necroses do not develop. Therefore, the action of ethylene on the development of tissue necrosis by rusts is dissimilar to necroses caused by viruses.' In addition, no similar effect of ethylene has been detected when the hypersensitive response of wheat to rust is controlled by the *Srll* gene. Detachment of leaves of some wheat types increases the hypersensitive reponse (273). This finding suggests that senescence favors the development of hypersensitive and normosensitive tissue necroses in diseases caused by viruses and facultative fungal parasites, respectively. The strange thing here is that detachment, which promotes senescence, also increases the susceptibility of wheat to rust. Increased hypersensitivity and increased susceptibility contradict the conventional theory of hypersensitive resistance in this instance.

Campbell and Deverall (53) also came to the conclusion that hypersensitive necrosis is not essential for expression of resistance in wheat to leaf rust. They inhibited hypersensitive necrosis by filtering out photosynthetically active wavelengths of light, as well as by the application of a photosynthesis inhibitor (DCMU). Despite the prevention of hypersensitivity, the incompatible isolate of *Puccinia recondita* was inhibited in the resistant host by a mechanism other than HR. They suggested that perhaps starvation of the pathogen leads to the inhibition of the fungus.

According to the classic view, the primary visible event in an incompatible host-parasite combination is the death of host cells, which precedes the cessation of fungal growth (261). However, some recent investigations on HR question the purpose and, therefore, the role of the hypersensitive response in disease resistance. HR necroses have been successfully induced in potato tuber, wheat leaf, and bean leaf tissues infected with compatible races of *Phytophthora infestans*, *Puccinia graminis*, and *Uromyces phaseoli*, respectively, when these fungi were inhibited or prevented from further growth by the application of chemicals and heat (31, 100, 213) or by a bacterial hyperparasite (175). The hypersensitive response, induced in originally compatible tissues, appears not to differ from that found in tissues of resistant cultivars infected with incompatible races. Even phytoalexins accumulated in the compatible tuber tissues when the compatible pathogen was inhibited. Leakage from dead *Phytophthora* mycelium, whether compatible or incompatible, after treatment by sonication or with chloroform, caused phytoalexin production and hypersensitive necrosis in tuber tissues. The chemical inhibitors and the compatible races separately did not cause phytoalexin accumulation, nor did they induce the hypersensitive response. In addition, Savel'eva and Rubin (354), Sato et al. (353), Varns and Kuć (437), and Bostock et al. (45) have reported that cell-free homogenates of the mycelium of *Phytophthora infestans* were able to induce HR in potato, whether the host was susceptible or resistant to the isolates of this fungus.

From the foregoing, it is possible to conclude that the first event in resistance is an unknown primary mechanism ("determinative phase," recognition) that influences, alters, or even inhibits the pathogen and that this adverse influence or inhibition then stimulates the fungus to release some material(s) that, in turn, causes the hypersensitive necrosis and accumulation of phytoalexin ("expressive phase"). In other words, HR and phytoalexins are the effect, not the cause, of resistance. In some instances, the adverse effects of the recognition phenomenon on the pathogen may be invisible, although still strong enough to cause "elicitors" to be released from the damaged pathogen, which may be responsible for hypersensitivity and phytoalexin synthesis. Érsek et al. (101) found that such "elicitor" compounds produced by *Phytophothora infestans* can influence phytoalexin production and tissue necrotization in potato tubers differently. It would seem that compounds of that origin cause secondary phenomena, e.g. tissue necrosis and phytoalexin production, and that there is no correlation between the two (45, 46).

Although it is difficult to separate in time the visible damage or inhibition of the host and parasite, a few recent data refer to the sequence of events during HR. Prusky et al. (331) compared the time course of the death of oat rust haustoria and that of host cells. The first microscopically observable event in the HR was the death of haustoria of *Puccinia coronata* f. sp. *avenae* in intact host cells. Host cell death occurred later. Similar results were noted when the HR was induced in a compatible host-pathogen combination with fungicide and heat treatment. It was concluded that in both natural and induced HR haustorial damage precedes host cell death. Induction of HR in the compatible combination with fungicide and heat treatment was similar to those instances mentioned above when HR was induced in plants compatible to *Puccinia*, *Uromyces*, and *Phytophthora* with chemicals or with a hyperparasite.

The growth of the pathogen is inhibited well before HR occurs in apple scab caused by *Venturia inaequalis* (301). It was demonstrated that the first microscopically observable signs of hypersensitivity, namely, granulation of cytoplasm and subsequent cell browning, occur at 20 and 40 h, respectively, after the inhibition of the fungus. The macroscopic symptoms are seen even later. The pathogen was inhibited in the hypersensitive hosts about 33 h earlier than an increase in phytoalexin content was detected. The authors concluded that neither phytoalexins nor hypersensitive necrosis represents the primary means of resistance to *Venturia inaequalis*.

Supporting this view, Mayama et al. (273), working with *Puccinia graminis*, concluded, "The hypersensitive response in rust disease is not a determinant for incompatible reactions." They claimed that hypersensitivity may be a stress symptom and is only incidental to resistance. It was mentioned earlier that when detached wheat leaves are inoculated with stem rust, both the susceptibility of the host and the intensity of the hypersensitive response are increased. This contradicts the view that the hypersensitive response is the cause of resistance.

Another report by Mayama et al. (272) brought the role of HR into question by showing that cell necrosis has no effect on the development of wheat stem rust. The hypersensitive response may be the consequence of a general stress that reflects an unknown mechanism of incompatibility.

The association of HR with resistance may be regarded as a means by which one can recognize visually the resistance of plants to nonpathogenic races of the parasite or to nonpathogens. According to Ward and Stoessl (442), the recognition phenomenon is the most fundamental aspect of specific resistance, and the hypersensitive response may or may not be a direct manifestation of this phenomenon. It would seem that the question of whether hypersensitivity is only a symptom associated with plant disease resistance or represents an integral part of the process is still a matter for debate.

According to the hypothesis of van der Plank (426), the hypersensitive host reponse, including the production of phytoalexins, occurs after the inhibition of the primary pathogen but well before the invasion of secondary pathogens. Therefore, hypersensitivity and phytoalexin would represent a preformed type of resistance against several secondary infections. This resistance (indeed, protection or acquired resistance) would be preformed only from the point of view of secondary pathogens.

The phytoalexin hypothesis. The phytoalexin hypothesis was first enunciated by Müller and Börger (293) in their studies on the acquired resistance of potato to *Phtophthora infestans*. The cut surface of tubers of an incompatible cultivar was inoculated with an incompatible race of the fungus. After some time, the tubers were reinoculated with a compatible race. As a result of the first infection, the tubers became resistant to the compatible race. Müller and Börger hypothesized that an antifungal principle, a phytoalexin, diffuses from the hypersensitive tissue and prevents the development of the compatible race of *Phytophthora* and of several other fungal species. The formation of phytoalexins in and around the hypersensitive cells of different plant

organs has been demonstrated in a comprehensive series of investigations.

According to van der Plank (426), the heart of the matter is that phytoalexins provide resistance to secondary, not primary, infections. He points out that all of the advocators of the phytoalexin theory of plant disease resistance forget that "what they showed was that a pretreated surface was resistant to secondary infection." There is no definitive evidence that phytoalexins can act similarly against primary infections. Accordingly, it would be necessary to reassess the experiments on the role of phytoalexins as well as of the hypersensitive reaction in plant disease resistance.

A sharp distinction must be made between primary and secondary infections. HR and the accumulation of phytoalexins are associated with the primary infection when the host and the infecting parasite are in an incompatible relationship. It has been inferred by many investigators that phytoalexins that can inhibit a subsequent infection must also be effective against the primary one. However, experimental data actually reflect an acquired resistance, which protects the plant against the second infection. The contention that phytoalexins can inhibit the primary parasite stems from too few reports (217, 383). The most interesting indirect evidence for the action of phytoalexins against primary fungal pathogens was based on the experiments with phytoalexin inhibitors (207, 283). Inhibition of PAL-activity with glyphosate or the phenylpropanoid pathway with L-2-aminooxy-3-phenylpropionic acid leads to the inhibition of glyceollin accumulation in soybeans and, at the same time, to partial suppression of resistance to *Phytophthora megasperma*. However, side effects of the inhibitors may disturb the whole picture.

The conclusion that host cell death and phytoalexin accumulation occur first, and that then the infecting pathogen is limited in growth, is widely accepted among phytopathologists. However, several data contradict this conclusion (46, 75, 213, 272, 273, 301). Some evidence suggests that inhibition of the pathogen is the primary event in some diseases and that hypersensitivity with phytoalexin accumulation is only a consequence. Additional investigations are needed to reveal the precise sequence of events in several host plants during infection by an incompatible primary pathogen. The exact time of cessation of fungal or bacterial growth and the time of hypersensitive cell death are exceedingly difficult to determine accurately in most cases, because the two events occur almost simultaneously, and because of the inadequate methods available at present.

It was recently pointed out that treating plant tissues with high MW cell wall components of infectious fungi can induce accumulation of phytoalexins and the hypersensitive response (6). This is the *elictor hypothesis*, which suggests that glucans, glycoproteins, or other products of cell walls of plant pathogens are responsible for the accumulation of phytoalexins. It has also been reported that, in some instances, low MW water-soluble glucans, *suppressors*, are released by compatible fungi, making the host susceptible. The suppressors inhibit the elicitation of phytoalexins and prevent their accumulation (89, 310). The inference here is that susceptibility lies in the binding of the suppressor rather specifically to the hypothetical receptor on the host cell membrane. However, the elicitors from *Phytophthora infestans* seem to be nonspecific. Ziegler and Pontzen (470) claim that another *Phytophthora* species produces suppressors of phytoalexins that are race-specific in relation to their soybean hosts. Hence, suppression, not the accumulation of phytoalexin, is induced specifically in the susceptible host. Their contention is that specific resistance depends on the lack of specific phytoalexin suppression.

Elicitors have been characterized from three fungal genera: *Phytophthora*, *Collectotrichum*, and *Fusarium* (6), and are, therefore, not universal and also not specific. Nevertheless, it seems logical that they are released only in incompatible host-pathogen combinations as a consequence of recognition. Hence, the production of elicitors implies that, in the expressive phase of resistance, incompatible pathogens are inhibited or damaged; therefore, they release substances that induce phytoalexin accumulation. Indeed, elicitors can be derived from fungal cultures near the end of growth and past the log-phase growth. At that time, degradation products accumulate in culture filtrates that are occasionally phytoalexin elicitors. However, to initiate this and other related metabolic events in the host and the pathogen, an early contact between germinating conidia of *Fusarium* and the host cells is necessary (160, 300). In the case of *Phytophthora megasperma*-soybean complex, a similar contact of the fungal pathogen with the host tissue is necessary to release elicitors of phytoalexins (466). Earlier investigations by Király et al. (213) and Érsek et al. (100) suggest that many races of fungal pathogens would be able to release elicitors in resistant hosts or in susceptible ones, provided that they were inhibited (damaged). However, no elicitation occurs in compatible host-race combinations because the fungus is neither inhibited nor damaged in host tissue. This hypothesis infers that, as a consequence of recognition, the incompatible pathogenic race is damaged or adversely affected in the resistant host, which is the basis for the release of the elicitor. The production of elicitors seems to be specific and occurs only in incompatible race-cultivar combinations. However, once isolated, the elicitors may induce phytoalexin, nonspecifically, to accumulate in other host species.

The correlation between phytoalexins and resistance seems not to be perfect in several instances. Bostock et al. found that, in contrast to stored tubers, freshly harvested tubers accumulate very low levels of phytoalexin after inoculation with an incompatible race of *Phytophthora infestans* (46). More importantly, phytoalexin accumulation was not different in fresh tubers inoculated with either compatible or incompatible races that, nevertheless, caused susceptible and resistant (hypersensitive) reactions, respectively (Fig. 10–42). Phy-

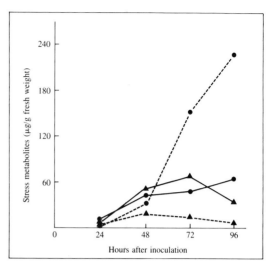

Fig. 10–42. Stress metabolite (terpenoid phytoalexin) accumulation in Kennebec potato tuber slices after inoculation with *P. infestans*. ● = inoculated with incompatible races; ▲ = inoculated with a compatible race. Solid lines are the averaged data from experiments using freshly harvested tubers. Broken lines are the averaged data using stored tubers (after Bostock et al., 46).

toalexin production appeared dependent on the physiological state of tuber tissues and was not correlated with either resistance or susceptibility. These results have raised doubts about the role of phytoalexins in resistance of potato tubers to *P. infestans*.

Lazarovits et al. (235) have shown that while phytoalexins decrease in soybean tissues with age, resistance increases. Furthermore, Ward and Lazarovits (441) demonstrated that soybean cultivars susceptible to *P. megasperma* remained susceptible, although the phytoalexin concentration increased to an inhibitory level after heat treatment. They increased the glyceollin level in infected soybean tissues by heat treatment and then returned the plants to low temperature. The high glyceollin concentration did not suppress the compatible pathogen in the susceptible cultivar. It was concluded that neither phytoalexins nor necrosis conditioned resistance to *P. megasperma*.

It was mentioned that *Venturia inaequalis* in a resistant apple cultivar was inhibited about 33 h prior to the detection of an increase in phytoalexin content in the host (301). Similarly, peas infected with an incompatible fungus, *Fusarium solani* f. sp. *phaseoli*, inhibited the fungus prior to the time that the synthesis of phytoalexin was detected (161) (Fig. 10–43). Ayers et al. (19) have shown that glucan elicitor from *P. megasperma* can protect soybean hypocotyls from infection by a compatible race if the elicitor is applied 10 h before inoculation. When the compatible race of the pathogen and the elicitor were applied simultaneously, the protection was unsuccessful. According to this experiment,

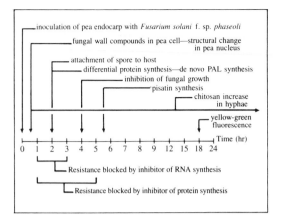

Fig. 10–43. Major events in the incompatible interaction between *F. solani* f. sp. *phaseoli* and pea endocarp tissue (after Hadwiger and Loschke, 161).

the phytoalexin accumulated after the application of the elicitor and, hence, is a preformed resistance factor that protects soybean plants against the compatible pathogen. Another glucan elicitor isolated from germ tubes of uredospores of *Uromyces phaseoli* (95, 188) protected bean leaf tissues from the fungus when it was infiltrated into the leaves at least 5 d before inoculation. In this instance, too, elicited phytoalexin had to be present in tissues before inoculation (as preformed resistance factor) in order to inhibit infection by a compatible pathogen.

These reports appear to support the hypothesis of van der Plank (426) that phytoalexins are produced after infection and, therefore, do not inhibit incompatible fungal races in resistant hosts. Rather, they may be resistance factors against subsequent inoculations, and, since the phytoalexins accumulate prior to the development of the second infection, they are in essence preformed. Van der Plank proposes a role for phytoalexins under natural conditions in inducing acquired resistance to secondary invaders.

According to some investigators, tolerance of several pathogens to phytoalexins can determine host susceptibility, and high sensitivity to phytoalexins could explain resistance. There are several proposals according to which successful degradation of phytoalexins by compatible pathogenic races and ineffective degradation by incompatible races would be the basis of suceptibility and resistance, respectively. Experimental data and conclusions are conflicting in this regard (409, 445). Nevertheless, Tegtmeier and Van Etten (409) have presented genetic evidence that the ability of *Nectria haematococca* to *degrade* and *tolerate* the phytoalexin pisatin correlates with its virulence in pea seedlings. A series of crosses were made between isolates that differed in sensitivity to pisatin. Of the 345 progeny analyzed, all of the moderately and highly virulent isolates were able to both tolerate and demethylate pisatin. The authors contended that their data support the hypothesis

that pisatin accumulation is an active mechanism of resistance in pea.

Other postinfectional reactions and compounds. It has been known that phenolic and other *aromatic compounds* are accumulated or released from glucosidic linkages in infected plants and that these componds are fungitoxic. Fungal glucosidases appear to have an important role in releasing those compounds from bound forms. Activation of enzymes that oxidize phenol compounds has also been implicated in disease resistance. However, an in planta role for these processes requires additional critically analyzed supportive data.

Another group of postinfectional "defense substances" was investigated by Lyr and Banasiak (260). These are the *volatile alkenales* and their corresponding alcohols and acids as well as their derivatives. The most common natural substance in many plant species is a hexenal whose structure is:

$$CH_3 - CH = CH - C = O$$
$$H$$

It is released by a membrane-bound lipoxygenase from unsaturated fatty acids (phospholipids). The enzyme is under the control of a cytokinin; under stress, senescence, or infection, the cytokinin level decreases in tissues and, as a consequence, lipoxygenase is activated. The released alkenale compounds appear to be effective against several fungi, including the ones that cause damping-off disease. The in planta role of these alkenales in active disease resistance remains to be shown.

Lignification could be a resistance factor against cultivar nonpathogens or species nonpathogenic fungi in response to infection. However, it is difficult to demonstrate that disease development is inhibited by the lignification in a resistant host, as it often occurs too late to appear effective (369, 414, 424). Nevertheless, some indication of resistance in wheat to *Septoria nodorum* is imparted by postinfectionally induced lignification (42); although lignification reduces penetration and fungal growth, "other mechanisms" appear to act to reduce fungal growth even further.

Papilla formation, or cell "wall apposition," after infection is also considered to be a host response that is involved in resistance (see Fig. 10–44). This, and a series of other host responses, coincide with the appearance of resistance. But it is difficult to distinguish between those responses that have a role in disease resistance and those that are only associated with it or are consequences of it (369, 413, 469).

Cell-wall polysaccharides from plants have been regarded as resistance factors against insects and microorganisms because their proteinases (with trypsin- or chymotrypsinlike activity) are inhibited by cell wall fragments produced after wounding, detachment, insect damage, or infection. The inhibitors do not usually inhibit proteolytic enzymes of the plant itself, and they have been regarded by Ryan (350) as wound hormones. Damage to the lower leaves induces proteinase inhibitor

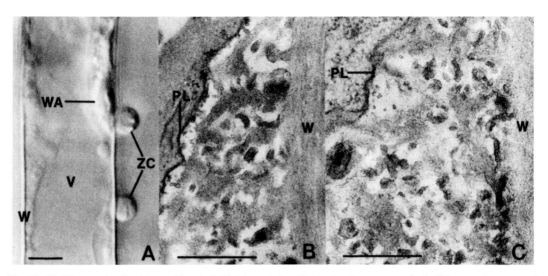

Fig. 10–44. Micrographs of wall appositions in kohlrabi root hairs. (A) A pair of cysts as seen in the living state by means of interference contrast optics. This photomicrograph, taken about 2 h after the inoculation of unshocked root hairs with zoospores of *O. brassicae*, shows the pair at the time of selection, before visible interactions had occurred. The root hair vacuole (V), root hair wall (W), mechanically induced wall apposition (WA), and zoospore cysts (ZC) are indicated. (B and C) Electron micrographs showing portions of heterogeneous wall appositions deposited between the plasmalemma (PL) and cell wall (W) by kohlrabi root hairs. Note the ultrastructural similarities between the mechanically induced (B) and the *O. brassicae*-induced (C) appositions. Scale bars, 0.5 μm (courtesy of Dr. James R. Aist, Cornell University).

activity in the upper leaves, hence the hypothetical induction signal appears to be systemic. The inhibitor accumulates in tomato plants resistant to *P. infestans* more intensively than in susceptible infected plants (321).

It was shown by Ryan et al. (351) that the proteinase inhibitor-inducing factor (PIIF) is a large pectic polysaccharide (rhamnogalacturonan I). This is released when host cell walls are degraded enzymatically as a result of infection, wounding, or insect attack. It is not known if PIIF itself is transported throughout the plant or if it produces a second message that is transported from the wound site.

It is remarkable that host cell wall fragmentation, as a result of infection or other stresses and senescence, can elicit chemical signals. One signal is localized and elicits phytoalexins, and a second one is systemic and hormonelike, initiating proteinase inhibitor synthesis. It is noteworthy that systemic-acquired resistance against secondary infection or injury, which develops in upper leaves of plants when the lower leaves are infected or injured, is associated with an increased level of cytokinin in the resistant leaves. Whether the PIIF action is related somehow to the action of cytokinin remains to be shown. (See below, on acquired resistance.)

Hydroxyproline-rich glycoproteins. It was shown in some instances of infection that the host cell walls accumulate hydroxyproline-rich proteins that are glycosylated with arabinose and galactose. For example, cell walls of melon seedlings infected with *Collectotrichum lagenarium* are enriched in hydroxyproline-rich

glycoproteins, and it was suggested (107) that this phenomenon is associated with resistance. The mechanism of action is not known. However, these glycoproteins may possess a lectinlike activity.

Protein inhibitors. Crude protein extracts of cell walls of several plants are capable of inhibiting the pathogen's polygalacturonase (118). However, they do not appear to have a role in the cultivar-specific resistance of plants to fungal infections. Perhaps they represent a general, nonspecific mechanism for resistance. This type of inhibitor can be regarded as a preformed resistance factor. However, there are data showing that they are activated after infection and earlier in resistant than in susceptible hosts (413).

Differential binding of a pathogen's pectolytic enzymes to host cell walls has been proposed as a mechanism of resistance (295, 332, 415). There are studies suggesting that cell walls of resistant plants have a greater capacity to bind endo-polygalacturonases of incompatible fungi than those of the compatible ones. Binding would be associated here with inhibition of the activity of enzymes. Another suggestion has been that binding would be essential for the pathogenicity of the infecting pathogens. The role of differential binding of wall-altering enzymes in resistance remains to be clarified.

Activated enzymes in diseased plants, like peroxidases, polyphenoloxidases, phenylalanine ammonia lyase (PAL), and ribonucleases have frequently been given a role in host-pathogen compatibility or incompatibility (see Chap. 5). These phenomena seem also to be related to injury and wound healing.

Gene-specific RNAs from resistant rust-infected wheats were thought to be involved in resistance, but the hypothesis (342) was withdrawn later (341) (see Chap. 5).

"Race-specific macromolecules" of pathogens were also claimed to be involved in plant resistance (438). Glycoproteins from incompatible races of *Phytophthora megasperma* f. sp. *glycinea* were thought to induce resistance in soybean to compatible races of the same pathogen by a mechanism that is different from the activation of phytoalexins. Recently, this hypothesis was withdrawn because of insufficiently sensitive bioassays applied in the experiments (83).

Common antigens (cross-reactive antigens, shared proteins) have been extensively treated in Chap. 5. In the incompatible host-pathogen relationship, common antigens are lacking or there are only a few proteins shared by the host plant and the parasite. The plant is resistant because it recognizes the foreign proteins. Thus, resistance depends upon unsuccessful molecular mimicry by the pathogen. When mimicry is successful, the host is susceptible, i.e., the host and its pathogen share common proteins. According to Barna et al. (30), DeVay et al. (85), and Charudattan and DeVay (62), common antigens are not involved in host cultivar/pathogenic race-specific resistance. Whether common antigens have any role in the relationship between host plant and the pathogen on a species level or in the establishment of the pathogen in the plant remains to be shown.

Toxin-binding proteins are believed to regulate susceptibility of sugarcane to the host-specific toxin of the fungus *Helminthosporium sacchari* (see also Chaps. 5 and 9). Resistant host cultivars contain a similar protein on the host plasma membrane; however, this preformed receptor does not bind the toxin (395). This hypothesis has been questioned by Lesney et al. (238), who claim that the methods are not sufficiently accurate to demonstrate the existence of a toxin-receptor protein on the surface of plant cells.

Postinfectionally induced RNAs and proteins were claimed by Tani et al. (406, 407) to have a role in specific resistance of oats to rust disease. Protein inhibitors were able to inhibit the expression of resistance, thus de novo synthesis of host proteins may have an active role in rust resistance. Whether the proteins are postinfectionally synthesized remains to be established (see also Chap. 5).

Agglutinins (lectins). The role of agglutinins and particularly lectins, from the point of view of resistance, has been treated in Chap. 5. These compounds are host membrane-bound or cell wall–bound agents, proteins or glycoproteins, that specifically bind to carbohydrates or to molecules containing carbohydrates on the surface of the pathogen. The binding causes agglutination or inhibition of cell division and growth of microorganisms. Wheat germ lectin and soybean lectin have been demonstrated to bind to hyphal tips and septa of different fungi, thereby inhibiting growth of hyphae and spore germination. Probably chitin synthesis is inhibited in the lectin-fungus interaction (27, 52, 108, 282). Thus, agglutinins may have a role in the essential process of recognition of pathogens, thereby initiating the "determinative phase" of postinfectional resistance. Only a few experiments with agglutinins have been performed that have some meaning from the point of view of plant pathology and disease resistance, and additional study appears warranted.

The most thoroughly investigated case is the binding of legume lectins to *Rhizobium* cell surfaces. This is presumed to be a self-recognition phenomenon: the lectin agglutinates the compatible strains, facilitating "infection" by the *Rhizobium* (see Chaps. 1 and 5). The agglutinin-fungus interactions seem to represent nonself recognitions.

It has been shown that a lectin in resistant soybean seeds may be a factor in resisance to *Phytophthora megasperma* var. *sojae* because it was bound by mycelial walls. However, direct evidence for involvement of the lectin has not been demonstrated.

Kojima et al. (223) purified a spore-agglutinating factor from sweet potato root. The factor is not a lectin, rather it is an agglutininlike pectic acid in host cell walls that contains galacturonic acid as its main component. It agglutinates differentially the germinated spores of several strains of *Ceratocystis fimbriata* in the presence of Ca^{2+} at pH 6.5. The phenomenon infers nonself recognition because it recognizes and inhibits the growth of the incompatible strains. The interaction between the surface of spores and the agglutinin molecule has not been defined. However, ionic interactions are important in the mechanism of agglutination. It is of interest that the pH value at which the differential agglutination takes place coincides with the pH of the sweet potato water-extract. The interaction may be a charge-charge phenomenon, rather than a classical lectin reaction that implies affinity for a specific molecular configuration (343). The germinated spores of the pathogenic fungal strain were highly insensitive to the agglutinin. On the other hand, all but one of the nonpathogenic strains were sensitive to it after germination (Table 10–12). Actually, the growth of the agglutinated spores was inhibited and the agglutinin seems to be a cell wall component. In plant infections caused by fungi, cell walls interact first with the infecting agets. Probably, host cell walls participate in the primary recognition process when plants are infected by fungi. It is possible that agglutinins, as cell wall components, are widely distributed in the plant kingdom and may have an important role in specific resistance (recognition).

The induced susceptibility hypothesis. When a specific toxin induces disease, it appears unlikely that an active resistance mechanism would be reponsible for the inhibition of the incompatible parasite and its toxin. According to some investigators the incompatible pathogen does not have a proper environment in the plant, therefore disease resistance need not involve an active inhibiting process. Toxins can induce a metabolic

Table 10–12. Agglutination of the germinated spores of various strains of *C. fimbriata* by spore-agglutinating factor of sweet potato (after Kojima et al., 223).

Concentration of spore-agglutinating factor, galacturonic acid/ml	Relative agglutination[a]						
	Sweet potato strain	Coffee strain	Prune strain	Cacao strain	Oak strain	Taro strain	Almond strain
μg							
0	—[b]	—	—	—	—	—	—
0.5	—	—	—	—	—	—	—
1.0	—	+1	—	+1	—	+1	—
1.5	—	+2	—	+1	+1	+1	—
2.0	—	+3	—	+2	+2	+2	—
2.5	—	+4	+1	+2	+3	+3	—
3.0	—	+5	+2	+2	+3	+4	—

a. Agglutination was observed in an assay mixture containing a high level (10 μg galacturonic acid/ml) of the spore-agglutinating factor.
b. − denotes no agglutination.

state in plants that facilitiate the growth of the pathogen, and this would be a case of induced susceptibility. Indeed, treating some nonhost plant tissues with toxins has rendered them susceptible to some species-non-pathogenic fungi (465). According to Daly (75), cereal rust fungi require an active switching off of normal metabolism of the susceptible host in order for parasitism to ensue. Each switch is probably triggered by changes in levels of growth hormones in the host such as indoleacetic acid, gibberellin, cytokinin, or ethylene. If a metabolic switch is not activated at a certain stage, the parasite cannot develop further in the plant. It would seem that with obligate parasites (biotrophs) susceptibility is an induced, active process and resistance is, perhaps, a passive phenomenon. At present, it is difficult to support this hypothesis, although the notion appears logical. With the current biological methods available, one cannot readily measure quantitatively the growth of the biotrophic fungi in either susceptible or resistant hosts, at least in the early stage of disease development. In other words, it is difficult to show that in a compatible host-pathogen complex the growth of the fungus is really stimulated by an activated host metabolism. If susceptibility is induced, fungus growth must be stimulated. It would also be necessary to demonstrate that the pathogenic development of the fungus remains at a low and constant level in resistant plants that are unable to provide the proper environment for the pathogen. This would be indicative of passive resistance. If the fungus is inhibited in growth, an induced and active resistance mechanism would be responsible. Evidence is lacking to either support or refute the hypotheses of induced susceptibility and passive resistance.

If the induced-susceptibility hypothesis is valid, sup-

pression of the hypothetical resistance mechanisms in the "expressive phase" would not increase the susceptibility of the plant. According to the experiments of Heath (172), this seems to be the case. When she applied heat shock and protein inhibitors before inoculation with rust, the resistant reaction of nonhost plants did not change, although the above-mentioned treatments inhibited the "resistance response." To a certain extent, suppression of phytoalexin production or other general signals of resistance would be regarded as induction of susceptibility. More data are needed to evaluate the role of suppression of the "resistance response" in the induced-susceptibility phenomenon.

Acquired Resistance to Secondary Infections

Acquired resistance develops after a primary infection and is effective against a second infection. This penomenon resembles acquired immunity in animals, at least superficially.

Local acquired resistance. There is both *local* and *systemic* acquired resistance in plants. The former was explained by Van der Plank (425, 426). He claims that phytoalexins and hypersensitivity develop between the inhibition of the primary pathogen and the invasion by the second pathogen. Under natural conditions, phytoalexins can provide protection against several second or subsequent infections. Since phytoalexins accumulate before the invasion of the second infection, these compounds could be regarded as preformed chemical resistance factors. According to Van der Plank, this type of acquired resistance is *localized* and *preformed* and constitutes a role in plant-disease resistance for phytoalexins and the hypersensitive reaction.

One of the most important early experiments on the

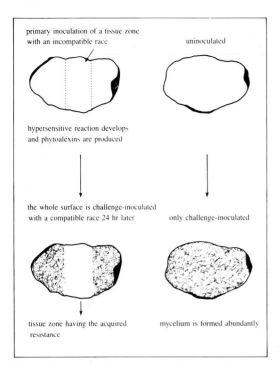

Fig. 10–45. Local acquired resistance to *P. infestans* in potato tuber tissue (modified from Müller and Börger, 293).

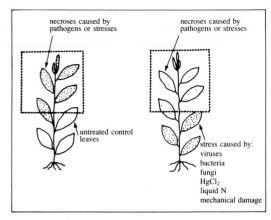

Fig. 10–46. Systemic-acquired resistance of the upper leaves of different plants to different pathogens or stresses caused by prior inoculation or treatment of the lower leaves with different pathogens or stresses, respectively.

acquired resistance was that by Müller and Börger (293) described above. This is a typical acquired resistance: as a consequence of a primary incompatible reaction, protection develops in the plant against subsequent secondary infection by a compatible pathogen (Fig. 10–45). Many similar experiments have been performed with other host cultivar-pathogenic race pairs and with several nonhost species and nonpathogenic fungus combinations. In fact, most experiments with phytoalexins have supported the existence of acquired resistance to secondary challenge infections. It was only inferred by Müller (290) and subsequent investigators that the fungitoxic phytoalexin inhibits both the secondary challenge inoculum (compatible pathogen) and the initial incompatible pathogen. However, the definitive evidence is still lacking.

When nonhost plants are inoculated with nonpathogenic fungi similar mechanisms are probably responsible for the acquired resistance subsequently exhibited to compatible pathogens (314). When wheat leaves were inoculated by the powdery mildew of barley (*Erysiphe graminis* f. sp. *hordei*) and then with the compatible wheat mildew (*Erysiphe graminis* f. sp. *tritici*), a resistance response developed that protected the wheat against challenge inoculum (99). The plants having the acquired resistance were resistant to compatible races of wheat mildew and wheat stem rust. The acquired resistance was not systemic. Acquired resistance associated with phytoalexins also appears to be localized. Many experiments concerning localized acquired re-

sistance have been conducted, but the underlying mechanism has been investigated sparingly (267).

Systemic-acquired resistance. It has been reported that the resistance induced by a primary infection can be distributed to other plant organs where it can inhibit secondary infections, thus reflecting a systemic-acquired resistance. For example, the infection of the lower leaves of tobacco and cucurbits by fungi, bacteria, or viruses was shown to induce systemic resistance in the upper leaves to fungi, bacteria, or viruses (23, 229, 346, 399). This phenomenon may be regarded as plant immunization (Fig. 10–46). Inducible systemic resistance of this type requires a defined time interval between induction and the challenge inoculum. Similarly, drops of a high concentration of $HgCl_2$, NaCl, or liquid N and mechanical abrasion applied to the lower leaves of cucumber can induce resistance in upper leaves to pathogens or to necroses caused by other stresses (91, 93, 211).

Resistance appears always to be induced by stress associated with necrosis. It would seem that a pathogen is not essential, only damage (stress), whether caused by pathogens or by chemical or mechanical injury. In summary, systemic acquired resistance is the result of stress and is effective against it. Necrotization of about 9% of one lower leaf surface of cucumber with different agents is sufficient to induce acquired resistance in the upper leaves. On the other hand, if the necrotization on the lower leaves damages too large a portion of the leaf surface and the leaves desiccate or die suddenly, acquired resistance does not develop (91). The signal for the systemic resistance in cucurbits is graft-transmissable from infected rootstock to scion (197). It is probable that resistance here simply means resistance to symptoms, i.e. necrotic spots. Doss and Hevesi (92) have demonstrated that in the upper leaves of cucumber that developed systemic-acquired resistance, only the necrotic symptoms of bacterial leaf spot caused by *Pseudo-*

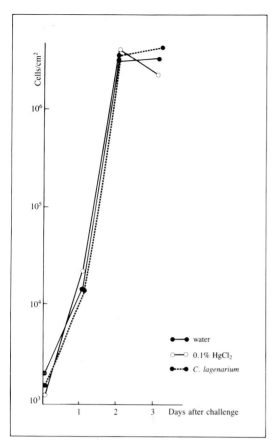

Fig. 10–47. Multiplication of *P. lachrymans* in tissues of upper cucumber leaves challenged 7 d after inoculation of the lower leaves with *C. lagenarium* or treatment with HgCl$_2$ (0.1%). Inoculum contained 3×10^5 cell ml^{-1} bacteria (after Doss and Hevesi, 92).

monas lachrymans were suppressed; multiplication of the bacteria was not affected (Fig. 10–47). Additional data concerning induction of systemic resistance against bacterial pathogens appears in the previous section of this chapter. Balázs et al. (23) also observed that the virus content of lesions in upper tobacco leaves experiencing induced resistance was not significantly altered, only the necroses were reduced. Regarding the growth and multiplication of fungi in the resistant leaves, the available data are contradictory and insufficient to assess the status of pathogenesis and of the pathogen in plants in which systemic resistance has been induced.

Several investigators have attempted to define the mechanism of systemic-acquired resistance. Elliston et al. (98) have shown that resistance is not due to the accumulation of phytoalexins in the upper leaves of systemically resistant bean plants. It was pointed out by Hammerschmidt et al. (165) that a threefold increase of peroxidase activity occurred in the resistant cucumber leaves after immunization well before the challenge inoculation with *Colletotrichum*. A booster inoculation

with *C. lagenarium* or with a necrotizing virus effectively enhanced systemic resistance and increased the activity of peroxidase. The timing of the appearance of enhanced enzyme activity and the expression of acquired resistance have shown that increased peroxidase acitivity developed later than the acquired resistance (Fig. 10–48). Furthermore, there was no correlation between peroxidase activity enhanced either by ethylene or senescence and the acquired resistance. Other careful investigations in several disease interactions have also shown that a direct relationship between the activity of peroxidase and resistance was not found (297, 358). Kuć and his associates have surmised that some peroxidase-requiring biochemical changes (e.g. lignification) could be responsible for systemic-acquired resistance (164). In resistant cucumber, lignifica-

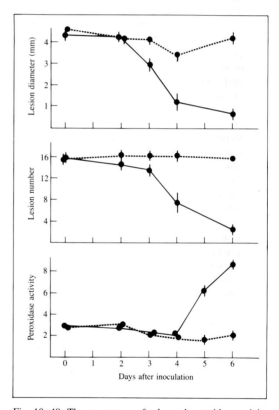

Fig. 10–48. The appearance of enhanced peroxidase activity and expression of induced resistance in leaf 2 of "SMR 58" cucumber as a function of time after the inducer inoculation of leaf 1. Solid lines represent plants in which systemic resistance was induced by infections of leaf 1 with *C. lagenarium*; dashed lines represent untreated controls. Vertical bars represent standard errors of the mean. Peroxidase activity is expressed as change in absorbance min^{-1} mg^{-1} protein using guaiacol as the hydrogen donor. Plants were challenged at the time leaf samples were taken to measure peroxidase activity. Symptoms (numbers of lesions and lesion size) were determined 1 w after challenge of leaf 2. Lesion size is the average of all necrotic lesions formed (after Hammerschmidt et al., 165).

tion occurs rapidly after penetration by infecting fungi, as compared to the susceptible cucumbers. Although the role of lignification in inhibiting the infecting fungi is not conclusively established in cucumbers, lignified host cell walls may increase their mechanical resistance, impair the degradative effect of fungal enzymes, or restrict the diffusion of toxins. The process may inhibit the pathogen by the action of phenolic lignin precursors or by lignification of the fungus. Direct evidence for this is still lacking.

Induced resistance in cucumber inoculated with *C. lagenarium* and tobacco with TNV has exposed changes in the soluble protein fraction in the leaves of these plants (5). These new proteins have been referred to as b-proteins, pathogenesis-related proteins, or stress proteins. Their possible role in acquired resistance is discussed in greater length in the virus section of this chapter.

It was shown that cytokinin increases in the resistant upper leaves of cucumber and tobacco. The first infection or stress applied to the lower leaves induced an increase in the cytokinin level in the lower as well as in the upper resistant leaves (23, 91, 93, 399). The development and disappearance of the acquired resistance coincided with the appearance or disappearance of increased cytokinin levels in the upper resistant leaves. The acquired resistance in cucumber correlated with the elevated cytokinin level whether resistance was induced by infection with *Colletotrichum lagenarium* or by treatment with mercuric chloride. It is interesting that the acquired resistance induced by the fungus reduced the necroses in the upper leaves that were caused by drops of mercuric chloride and vice versa. It is also noteworthy that the degradation of chlorophyll in the detached upper resistant leaves was slowed by the juvenility fostered by an increased cytokinin level. It has long been known that externally applied cytokinins reduced necrotic spots in tobacco (216).

In summary, the first stress applied to the lower leaves systematically induced an increase in the cytokinin hormone level in the upper leaves. The juvenility effect of the hormone could be involved in the suppression of necrotic symptoms caused by a second stress on the upper resistant leaves. In other words, stress in one organ (in the lower leaf) could induce synthesis or accumulation of an antistress hormone in another organ (upper leaf).

Acquired Susceptibility. It is known from several early papers (464) that preliminary inoculation of plant tissues with a compatible pathogen predisposes the infected cells to an originally incompatible race or to nonpathogens as secondary infections. For example, barley leaves primarily inoculated with a compatible race of powdery mildew (*Erysiphe graminis* f. sp. *hordei*) become susceptible to incompatible powdery mildew races or to the nonpathogenic *Sphaerotheca fuliginea* (314). It is noteworthy that, among 51 powdery mildew fungi collected from different plant species, 45 infected barley leaves that had been prior-inoculated with

Erysiphe graminis f. sp. *hordei* (417). This phenomenon has also been characterized as induced accessibility, because tissues are conditioned by the primary inoculation to general susceptibility to fungal infections (173). The phenomenon is connected to diseases that are caused by biotrophs (obligate parasites). Acquired susceptibility is localized in the area where the primary infection was effectively completed. It is assumed that cells having the acquired susceptibility cannot recognize a nonpathogen as a foreign entity because their metabolism is altered. However, the mechanism is unknown. It was proposed that the normal defense mechanisms of the plant are suppressed when the compatible relationship is established (314). Heat shock treatment prior to inoculation and experiments with inhibitors of protein and nucleic acid synthesis at a very early stage of host-parasite relationship by Ouchi and his associates have suggested that acquired susceptibility ("induced accessibility") is probably an active process. These findings are used to demonstrate the existence of the hypothetical induced susceptibility phenomenon (this has been treated in a previous paragraph). At present it is important to differentiate acquired susceptibility from induced susceptibility. The mechanisms may be common, but they probably refer to two phenomena. Induced susceptibility means that the pathogens (biotrophs) require an active change in the normal host metabolism to be able to parasitize the host plant. On the other hand, acquired susceptibility means that, as a result of a primary compatible host pathogen interaction, the altered metabolism renders the plant susceptible to a normally incompatible fungal race or to nonpathogen biotrophic species, as secondary infections.

Comparative Analysis of Disease Physiology— Resistance

There is an apparent lack of consensus in plant pathology literature in a number of specific areas, e.g. toxins, receptor sites, virulence factors, selective media, effective size of inoculum dose, site of virus capsid removal. However, no area is more differentially viewed or perceived with greater uncertainty than disease resistance. We, the authors, are also aware of a significant lack of consensus among ourselves. Our commentary reveals these differences of opinion by the emphasis, or lack thereof, we have placed on the various phenomena associated with disease resistance. The reader is urged to regard these differences as reflective of, primarily, our lack of precise information and to regard our uncertainty as a fair representation of the variance extant among researchers in plant pathology.

Physical barriers, complex and simpler ones (e.g. stomatal ridges, callose, and cutinized cell wall surfaces), that develop normally or in response to challenge by the three pathogen types have been observed to limit, diminish, or preclude infection. Barrier substances such as phenolics, suberin, callose, cutin, and lignin have been shown at microscopic and ultrastructural levels to impede infection. They constitute structures termed papillae, plasmalemmasomes, cuticular envelopments, and agglutinins that develop seemingly in juxtaposition to the pathogen and immobilize or physically limit further progress of the pathogen. There are instances in which these preformed or postinfectionally derived structures and substances constitute the first level of defense against the pathogen or become an adjunct to other biochemical events that, in concert, effectively limit the pathogen's advance. In some instances, these manifestations are feeble, appear too late, and are noneffective.

An analysis of the literature of resistance in plants to each of the major pathogen groups reveals some positive association with necrosis. Yet depending upon, for example, the virus/host combination and, importantly, microenvironmental conditions, the association may be equivocal. It is possible to establish: (1) virus localization (limitation of virus spread; reduced lesion size) with necrosis, (2) localization without necrosis, and (3) necrosis with systemic infection. An analogous spectrum of responses has been reported for bacterial and fungal pathogens in a variety of pathogen-host combinations. *Necrosis* that appears following inoculation, in what is designated a hypersensitive response, or local lesion development (a process that sorely needs definition with greater precision, regardless of pathogen type), results from rapid destruction of all organellar membrane systems. Unfortunately, the primary cause of membrane devastation remains unknown, as neither the inducer molecule nor the precise site of its action has been determined.

Aromatic substance accumulation, generally associated with necrosis and elevated levels of oxidizing enzymes, i.e. polyphenoloxidase and peroxidase, have long linked elevated aromatic synthesis to resistance. However, demonstrating a causal relationship to increased levels of preformed or pathogen-elicited (phytoalexin) substances has been difficult. That a system favoring oxidation of the aromatic compound fosters resistance is suggested by experiments with all pathogen groups wherein exposure of resistant cultivars to *reducing* conditions "breaks" resistance.

Clearly, gene control is implicit in the development of physical and biochemical resistance factors—both preformed and postinfectionally elaborated. However, microenvironmental conditions during the course of experimentation and the physiological condition of the plant material being used are decisive factors in the interaction between pathogen and host. Inherent virulence of the pathogen strain and size of the inoculum dose are critical elements in the evaluation of resistance factors.

Plant immunization has been attempted and moments of success have been recorded since the turn of the century; consequently, interest in this possibility has increased greatly. "Cross-protection" of plants against one virus pathogen by another, preferably a related strain, has been documented. Live-incompatible and heat-killed bacteria have protected plants against infection from a subsequent challenge inoculation with virulent compatible strains. Meager evidence suggests that weakly virulent bacterial strains are superior to avirulent or more aggressive isolates in this respect. Experiments have shown that a fungal pathogen will protect a host plant against virus or bacterium; virus against bacterium or fungus; and bacterium against virus or fungus. "Booster" or successive inoculations of cucumber plants with a bacterial pathogen have afforded protection for an entire fruiting season.

What, then, is the nature of the protecting substance? Does it provide localized protection, or can protection be systemic? Experiments with carnation and tomato and, earlier, with roses and daisy, leave the nature of the protecting substance unknown. It would seem that a greater research effort will be expended in exploring the possibilities of plant "immunization." It is also apparent that significant research effort will be expended in determining the linkages of necrosis and HR to resistance. Crucial, in these efforts, will be the development of methods to monitor the earliest metabolic and histologic signs of change in both host and pathogen as a consequence of their encounter.

Questions have also been raised that challenge the concept that resistance is the active process and susceptibility is the passive partner in the host-pathogen interaction.

Additional Comments and Citations

1. Recent information on pathogenesis-related proteins, and on the use of protoplasts in resistance studies, is provided in the following publications. L. C. van Loon (ed.). 1983. Workshop on pathogenesis-related (b) proteins. Netherlands J. Plant Path. 89:239–325, and B. D. Harrison, and M. A. Mayo. 1983. The use of protoplasts in plant virus research. In: Use of tissue culture and protoplasts in plant pathology. J. P. Helgeson and B. J. Deverall (eds.). Academic Press, Australia, pp. 69–137.

2. Control mechanisms and enzymes involved in phytoalexin biosynthesis are defined by J. Kimpel, and T. Kosuge. 1985. Metabolic regulation during glyceollin biosynthesis in green soybean hypocotyls. Plant Physiol. 77:1–7. Elicitor prepared from *Phytophthora megasperma* f. sp. *glycinea* provokes the accumulation of glyceollin, but clearly not in wounded control tissue. Levels are mediated by cotyledons and ethylene and controlled by enzymes PAL, CA4H, and pCL. There is an absolute correlation between increases in PAL and pCL and glyceollin accumulation. The described metabolic regulation of glyceollin is consistent with its putative role in an active, localized disease-resistance response.

3. J. P. Skou, H. J. Jörgensen, and U. Linholt. 1984. Comparative studies on callose formation in powdery mildew compatible and incompatible barley. Phytopath. Z. 109:147–68. The ml-o genes of barley for resistance are race-nonspecific and are responsible for a

durable type of resistance to powdery mildew (*Erysiphe graminis*). For barley lines and cultivars having the ml-o gene, early callose formation at a high rate is characteristic upon mildew infection.

References

1. Addy, S. K., and R. N. Goodman. 1974. Phenols in relation to pathogenesis induced by avirulent and virulent strains of *Erwinia amylovora*. Acta Phytopath. Hung. 9:227–86.

2. Alexander, L. J. Host-pathogen dynamics of tobacco mosaic virus on tomato. Phytopathology 61:611–17

3. Allison, A. V., and T. A. Shalla. 1974. The ultrastructure of local lesions induced by potato virus X: A sequence of cytological events in the course of infection. Phytopathology 64:784–93.

4. Al-Mousawi, H., P. E. Richardson, M. Essenberg, and W. M. Johnson. 1983. Specificity of the envelopment of bacteria and other particles in cotton cotyledons. Phytopathology 73:484–89.

5. Andebrhan, T., R. H. A. Coutts, E. E. Wagih, and R. K. S. Wood. 1980. Induced resistance and changes in the soluble protein fraction of cucumber leaves locally infected with *Colletotrichum lagenarium* or TNV. Phytopath. Z. 98:47–52.

6. Anderson, A. J. 1980. Studies on the structure and elicitor activity of fungal glucans. Can. J. Bot. 58:2343–48.

7. Antonelli, E., and J. M. Daly. 1966. Decarboxylation of indoleacetic acid by near isogenic lines of wheat resistant or susceptible to *Puccinia graminis* f. sp. *tritici*. Phytopathology 56:610–18.

8. Antoniw, J. F., J. S. H. Kueh, D. G. A. Walkey, and R. F. White. 1981. The presence of pathogenesis-related proteins in callus of Xanthi-nc tobacco. Phytopath. Z. 101:179–84.

9. Antoniw, J. F., C. E. Ritter, W. S. Pierpoint, and L. C. van Loon. 1980. Comparison of three pathogenesis related proteins from plants of two cultivars of tobacco infected with TMV. J. Gen. Virol. 47:79–87.

10. Antoniw, J. F., and R. F. White. 1980. The effect of aspirin and polyacrylic acid on soluble leaf proteins and resistance to virus infection in five cultivars of tobacco. Phytopath. Z. 98:331–41.

11. Appiano, A., S. Pennazio, G. D'Agostino, and P. Redolfi. 1977. Fine structure of necrotic local lesions induced by tomato bushy stunt virus in *Gomphrena globosa* leaves. Physiol. Plant Path. 11:327–32.

12. Appiano, A., S. Pennazio, and P. Redolfi. 1978. Cytological alterations in tissues of *Gomphrena globosa* plants systemically infected with tomato bushy stunt virus. J. Gen. Virol. 40:277–86.

13. Artschwager, E. 1927. Wound periderm formation in the potato as affected by temperature and humidity. J. Agr. Res. 35:995–1000.

14. Artschwager, E., and R. C. Starrett. 1933. Suberization and wound-cork formation in the sugar beet as affected by temperature and relative humidity. J. Agr. Res. 47:669–674.

15. Atkinson, P. H., and R. E. F. Matthews. 1967. Distribution of tobacco mosaic virus in systemically infected tobacco leaves. Virology 32:171–73.

16. Atkinson, P. H., and R. E. F. Matthews. 1970. On the origin of dark green tissue in tobacco leaves infected with tobacco mosaic virus. Virology 40:344–56.

17. Audia, W. V., W. L. Smith, Jr., and C. C. Craft. 1962. Effects of isopropyl N-(3- chlorophenyl) carbamate on suberin, periderm and decay development by Katahdin potato slices. Bot. Gaz. 123:255–58.

18. Averre, C. W. III, and A. Kelman. 1964. Severity of bacterial wilt as influenced by ratio of virulent to avirulent cells of *Pseudomonas solanacearum* in inoculum. Phytopathology 54:779–83.

19. Ayers, A. R., P. Albersheim, J. Ebel, and B. Valent. 1976. Host-pathogen interactions. X. Fractionation and biological activity of an elicitor isolated from the mycelial walls of *Phytophthora megasperma* var. *sojae*. Plant Physiol. 57:760–65.

20. Ayers, A. R., S. B. Ayers, and R. N. Goodman. 1979. Extracellular polysaccharide of *Erwinia amylovora*: A correlation with virulence. Appl. Environ. Microbiol. 38:659–66.

21. Azad, H. R., and C. I. Kado. 1984. Relation of tobacco hypersensitivity of pathogenecity of *Erwinia rubrifaciens*. Phytopathology 74:61–64.

22. Balázs, E., B. Barna, and Z. Király. 1976. Effect of kinetin on lesion development and infection sites in Xanthi-nc tobacco infected by TMV: Single-cell local lesions. Acta Phytopath. Hung. 11:1–9.

23. Balázs, E., I. Sziráki, and Z. Király. 1977. The role of cytokinins in systemic acquired resistance of tobacco hypersensitive to tobacco mosaic virus. Physiol. Plant Path. 11:29–37.

24. Bancroft, J. B., and G. S. Pound. 1954. Effect of air temperature on the multiplication of tobacco mosaic virus in tobacco. Phytopathology 44:481–82.

25. Barbara, D. J., and K. R. Wood. 1972. Virus multiplication, peroxidase and polyphenoloxidase activity in leaves of two cucumber (*Cucumis sativus* L.) cultivars inoculated with cucumber mosaic virus. Physiol. Plant Path. 2:167–73.

26. Barbara, D. J., and K. R. Wood. 1974. The influence of actinomycin D on cucumber mosaic virus (strain W) multiplication in cucumber cultivars. Physiol. Plant Path. 4:45–50.

27. Barkai-Golan, R., D. Mirelman, and N. Sharon. 1978. Studies on growth inhibition by lectins of penicillia and aspergilli. Arch. Microbiol. 116:119–24.

28. Barker, H., and B. D. Harrison. 1977. The interaction between raspberry ringspot and tobacco rattle viruses in doubly infected protoplasts. J. Gen. Virol. 35:135–48.

29. Barker, H., and B. D. Harrison. 1978. Double infection, interference and superinfection in protoplasts exposed to two strains of raspberry ringspot virus. J. Gen. Virol. 40:647–58.

30. Barna, B., E. Balázs, and Z. Király. 1978. Common antigens and rust diseases. Phytopath. Z. 93:336–41.

31. Barna, B., T. Érsek, and S. F. Mashaal. 1974. Hypersensitive reaction of rust-infected wheat in compatible host-parasite relationship. Acta Phytopath. Hung. 9:293–300.

32. Barna, B., and E. János. 1981. The role of protein metabolism of wheat in amino acid-induced resistance to rust. Acta Phytopath. Hung. 16:49–57.

33. Barna, B., A. R. T. Sarhan, and Z. Király. 1983. The influence of nitrogen nutrition on the sensitivity of tomato plants to fusaric acid and other toxic actions. Physiol. Plant Path. 23:257–63.

34. Barna, B., S. F. Mashaal, and Z. Király. 1977. Problems of involvement of protein metabolism in resistance of wheat to rust. In: Current topics in plant pathology. Z. Király (ed.). Akadémiai Kiadó, Budapest, pp. 35–40.

35. Barnett, A., and K. R. Wood. 1978. Influence of actinomycin D., ethephon, cycloheximide and chloramphenicol on the infection of a resistant and a susceptible cucumber cultivar with cucumber mosaic virus (Price's No. 6 strain). Physiol. Plant Path. 12:257–77.

36. Bauer, W. D., K. W. Talmadge, K. Keegstra, and P. Albersheim. 1973. The structure of plant cell walls. II. The hemicellulose of the walls of suspension- cultured sycamore cells. Plant Physiol. 51:174–87.

37. Bawden, F. C. 1964. Plant viruses and virus diseases. 4th ed. Ronald Press, New York.

38. Bergstrom, G. C., and J. Kuć. 1981. Induction of systemic resistance in cucumber to bacterial wilt and gummy stem blight. Phytopathology 73:203. (Abstr.)

39. Bergstrom, G. C., M. C. Johnson, and J. Kuć. 1982. Effects of local infection of cucumber by *Colletotrichum lagenarium*, *Pseudomonas lachrymans* or tobacco necrosis virus. Phytopathology 72:922–26.

40. Beier, H., G. Bruening, M. L. Russell, and C. L. Tucker. 1979. Replication of cowpea mosaic virus in protoplasts isolated from immune lines of cowpea. Virology 95:165–75.

41. Beier, H., D. J. Siler, M. L. Russell, and G. Bruening. 1977. Survey of susceptibility to cowpea mosaic virus among protoplasts and intact plants from *Vigna sinensis* lines. Phytopathology 67:917–21.

42. Bird, P. M., and J. P. Ride. 1981. The resistance of wheat to *Septoria nodorum*: Fungal development in relation to host lignification. Physiol. Plant Path. 19:289–99.

43. Bjorling, K. 1966. Virus resistance problems in plant breeding. Acta Agr. Scand. 16:119–36. (Suppl.)

44. Bos, L. 1970. Symptoms of virus diseases in plants. Pudoc, Wageningen.

45. Bostock, R. M., R. A. Laine, and J. A. Kuć. 1982. Factors affecting the elicitation of sesquiterpenoid phytoalexin accumulation by eicosapentaenoic acid and arachidonic acid in potato. Plant Physiol. 70:1417–21.

46. Bostock, R. M., E. Nuckles, J. W. Henfling, and J. A. Kuć. 1983. Effects of potato tuber age and storage on sesquiterpenoid stress metabolite accumulation, steroid glycoalkaloid accumulation and response to abscisic and arachidonic acids. Phytopathology 73:435–38.

47. Brown, N. A. 1923. Experiments with Paris daisy and rose to produce resistance to crown gall. Phytopathology 13:87–99.

48. Brown, R. G., and W. C. Kimmins. 1973. Hypersensitive resistance. Isolation and characterization of glycoproteins from plants with localized infections. Can. J. Bot. 51:1917–22.

49. Burnet, M. 1971. Self recognition: In colonial marine forms and flowering plants in relation to the evolution of immunity. Nature 232:230–35.

50. Byhan, O., and G. Schuster. 1980. RNA of virus and host plant as influenced by ribavirin and 2,4-dioxohexahydro-1,3,5-triazine. Tag.-Ber. Acad. Landwitsch-Wiss. DDR. 184:137–42.

51. Caleya, R. F., B. Gonzalez-Pascual, F. Garcia Olmedo, and P. Carbonero. 1972. Susceptibility of phytopathogenic bacteria to wheat purothionins in vitro. Appl. Microbiol. 23:998–1000.

52. Callow, J. A. 1977. Recognition, resistance and the role of plant lectins in host-parasite interactions. Adv. Bot. Res. 4:1–49.

53. Campbell, G. K., and B. J. Deverall. 1980. The effects of light and a photosynthetic inhibitor on the expression of the Lr20 gene for resistance to leaf rust in wheat. Physiol. Plant Path. 16:415–23.

54. Carroll, T. W., P. L. Gossel, and E. A. Hockett. 1979. Inheritance of resistance to seed transmission of barley stripe mosaic virus in barley. Phytopathology 69:431–33.

55. Carroll, T. W., and T. A. Shalla. 1965. Visualization of tobacco mosaic virus in local lesions of *Datura stramonium*. Phytopathology 55:928–29.

56. Caruso, F., and J. Kuć. 1979. Induced resistance of cucumber to anthracnose and angular leaf spot by *Pseudomonas lachrymans* and *Colletotrichum lagenerium*. Physiol. Plant Path. 14:191–201.

57. Cason, E. T., P. E. Richardson, M. K. Essenberg, L. A. Brinkerhoff, W. M. Johnson, and R. J. Venere. 1978. Ultrastructural cell wall alterations in immune cotton leaves inoculated with *Xanthomonas malvacearum*. Phytopathology 68:1015–21.

58. Cassells, A. C., A. Barnett, and M. Barlass. 1978. The effect of polyacrylic acid treatment on the susceptibility of *Nicotiana tabacum* cv. Xanthi-nc to tobacco mosaic virus. Physiol. Plant Path. 13:13–22.

59. Cassells, A. C., and C. C. Herrick. 1977. Cross protection between mild and severe strains of tobacco mosaic virus in doubly inoculated tomato plants. Virology 78:253–60.

60. Chalcroft, J. P., and R. E. F. Matthews. 1967. Role of virus strains and leaf ontogeny in the production of mosaic patterns by turnip yellow mosaic virus. Virology 33:659–73.

61. Chaplin, J. F., and G. V. Gooding. 1969. Reaction of diverse *Nicotiana tabacum* germplasm to tobacco mosaic virus. Tobacco Sci. 13:130–33.

62. Charudattan, R., and J. E. DeVay. 1981. Purification and partial characterization of an antigen from *Fusarium oxysporum* f. sp. *vasinfectum* that cross-reacts with antiserum to cotton (*Gossypium hirsutum*) root antigens. Physiol. Plant Path. 18:289–95.

63. Cheo, P. C., and J. S. Gerard. 1971. Differences in virus-replicating capacity among plant species inoculated with tobacco mosaic virus. Phytopathology 61:1010–12.

64. Chester, K. S. 1933. The problem of aquired immunity in plants. Q. Rev. Biol. 8: 129–54, 275–324.

65. Cho, Y. S., R. D. Wilcoxon, and F. I. Froheiser. 1973. Differences in anatomy, plant extracts, and movement of bacteria in plants of bacteria wilt resistant and susceptible varieties of alfalfa. Phytopathology 63:760–65.

66. Cockerham, G. 1970. Genetical studies on resistance to potato viruses X and Y. Heredity 25:309–48.

67. Cohen, J., and G. Loebenstein. 1975. An electron microscope study of starch lesions in cucumber cotyledons infected with tobacco mosaic virus. Phytopathology 65:32–39.

68. Comai, L., and T. Kosuge. 1980. Involvement of plasmid deoxyribonucleic acid in indoleacetic acid synthesis in *Pseudomonas savastanoi*. J. Bact. 143:950–57.

69. Comai, L., and T. Kosuge. 1983. Transposable element t at causes mutations in a plant pathogenic *Pseudomonas* sp. J. Bact. 154:1162–67.

70. Coutts, R. H. A. 1978. Alterations in the soluble protein patterns of tobacco and cowpea leaves following inoculation with tobacco necrosis virus. Plant Sci. Lett. 12:189–97.

71. Coutts, R. H. A., A. Barnett, and K. R. Wood. 1978. Aspects of the resistance of cucumber plants and protoplasts to cucumber mosaic virus. Ann. Appl. Biol. 89:336–99.

72. Crosse, J. E. 1965. Bacterial canker of stone fruits. VI. Inhibition of leaf scar infection of cherry by a saprophytic bacterium from leaf surfaces. Ann. Appl. Biol. 56:149–60.

73. da Graca, J. V., and M. M. Martin. 1975. Ultrastructural changes in tobacco mosaic virus-induced local lesions in *Nicotiana tabacum* L. cv. 'Samsun NN'. Physiol. Plant Path. 7:287–91.

74. da Graca, J. V., and M. M. Martin. 1976. An electron microscope study of hypersensitive tobacco infected with tobacco mosaic virus at 32°C. Physiol. Plant Path. 8:215–19.

75. Daly, J. M. 1972. The use of near-isogenic lines in biochemical studies of the resistance of wheat to stem rust. Phytopathology 62:392–400.

76. Daly, J. M. 1976. The carbon balance of diseased plants: Changes in respiration, photosynthesis, and translocation. In: Physiological plant pathology. R. Heitefuss and P. H. Williams (eds.). Springer-Verlag, Berlin, pp. 450–79.

77. Daly, J. M. 1976. Specific interactions involving hormonal and other changes. In: Specificity in plant diseases. R. K. S. Wood and A. Graniti (eds.). Plenum Press, New York, pp. 151–67.

78. Davis, M. J., A. G. Gillaspie, Jr., R. W. Harris, and R. H. Lawson. 1980. Ratoon stunting disease of sugarcane: Isolation of the causal organism. Science 240:1365–67.

79. Dawson, J. R. O., and J. W. Watts. 1979. Analysis of the products of mixed infection of tobacco protoplasts with two strains of cowpea chlorotic mottle virus. J. Gen. Virol. 45:133–37.

80. de Fazio, G., J. Caner, and M. Vicente. 1980. Effect of virazole (ribavirin) on tomato spotted wilt virus in two systemic hosts, tomato and tobacco. Arch. Virol. 63:305–9.

81. Delhey, R. 1974. Zur Natur der extremen Virusresistenz bie der Kartoffel. I. Das X-virus. Phytopath. Z. 80:97–119.

82. Delhey, R. 1975. Zur Natur der extremen Virusresistenz bie der Kartoffel. II. Das Y-virus. Phytopath. Z. 82:163–68.

83. Desjardins, A. E., L. M. Ross, M. W. Spellman, A. G. Darvill, and P. Albersheim. 1982. Host-pathogen interactions. XX. Biological variation in the protection of soybeans from infection by *Phytophthora megasperma* f. sp. *glycinea*. Plant Physiol. 69:1046–50.

84. Desjardins, P. R. 1969. Alfalfa mosaic virus-induced lesions on bean: Effect of light and temperature. Plant Dis. Reptr. 53:30–33.

85. DeVay, J. E., R. J. Wakeman, J. A. Kavanagh, and R. Charudattan. 1981. The tissue and cellular location of a major cross-reactive antigen shared by cotton and soil-borne fungal parasites. Physiol. Plant Path. 18:59–66.

86. de Zoeten, G. A., and R. W. Fulton. 1975. Understanding generates possibilities. Phytopathology 65:221–22.

87. Dimond, A. E. 1970. Biophysics and biochemistry of the vascular wilt syndrome. Ann. Rev. Phytopath. 8:301–22.

88. Dodds, J. A., and R. I. Hamilton. 1976. Structural interactions between viruses as a consequence of mixed virus infections. Ad. Virus Res. 20:33–86.

89. Doke, N., N. A. Garas, and J. Kuć. 1980. Effect on host hypersensitivity of suppressors released during the germination of *Phytophthora infestans* cystospores. Phytopathology 70:35–39.

90. Doke, N., and K. Tomiyama. 1978. Effect of sulfhydryl-binding compounds on hypersensitive death of potato tuber cells following infection with an incompatible race of *Phytophthora infestans*. Physiol. Plant Path. 12:133–39.

91. Doss, M. M. 1981. Mechanism of systemic acquired resistance in cucumber caused by pathogenic and non-pathogenic agents. Ph.D. diss., Res. Inst. Plant Protection Budapest, pp. 1–122.

92. Doss, M. M., and M. Hevesi. 1981. Systemic acquired resistance of cucumber to *Pseudomonas lachrymans* as expressed in suppression of symptoms, but not in multiplication of bacteria. Acta Phytopath. Hung. 16:269–72.

93. Doss, M. M., I. Sziraki, and Z. Király. 1980. Role of different stresses in inducing resistance to anthracnose by increasing the level of cytokinins in cucumber. Conf. New Endeavours in Plant Prot., Budapest, 2–5 Sept., p. 41. (Abstr.)

94. Duffus, J. E., and G. M. Milbrath. 1977. Susceptibility and immunity in soybean to beet western yellows virus. Phytopathology 67:269–72.

95. Ebrahim-Nesbat, F., H. H. Hoppe, and R. Heitefuss. 1982. Ultrastructural studies on the development of *Uromyces phaseoli* in bean leaves protected by elicitors of phytoalexin accumulation. Phytopath. Z. 103:261–71.

96. Ehara, Y., and T. Misawa. 1968. Studies on the infection of cucumber mosaic virus. V. The observation of epidermal cell in the local lesion. Tohoku J. Agr. Res. 19:166–72.

97. Ellingboe, A. H. 1982. Host resistance and host-parasite interactions: A perspective. In: Phytopathogenic prokaryotes II. M. S. Mount and G. H. Lacy (eds.). Academic Press, New York, pp. 103–17.

98. Elliston, J., J. Kuć, E. B. Williams, and J. E. Rahe. 1977. Relationship of phytoalexin accumulation to local and systemic protection of bean against anthracnose. Phytopath. Z. 88:114–30.

99. Érsek, T. 1973. Defense reaction induced by a primary inoculation with barley mildew on wheat seedlings. Acta Phytopath. Hung. 8:261–63.

100. Érsek, T., B. Barna, and Z. Király. 1973. Hypersensitivity

and the resistance of potato tuber tissues to *Phytophthora infestans*. Acta Phytopath. Hung. 8:3–12.

101. Érsek, T., Z. Király, and Z. Dobrovolszky. 1977. Lack of correlation between tissue necrosis and phytoalexin accumulation in tubers of potato cultivars. J. Food Saf. 1:77–85.

102. Esau, K. 1948. Anatomic effects of the viruses of Pierce's disease and phony peach. Hilgardia 18:423–82.

103. Esau, K. 1967. Anatomy of plant virus infections. Ann. Rev. Phytopath. 1:197–218.

104. Esau, K., and L. L. Hoefert. 1972. Development of infection with beet western yellows virus in the sugarbeet. Virology 48:724–38.

105. Esau, K., and L. L. Hoefert. 1972. Ultrastructure of sugarbeet leaves with beet western yellows virus. J. Ultrastruc. Res. 40:556–71.

106. Espelie, K. E., and P. E. Kolattukudy. 1979. Composition of aliphatic components of suberin of the endodermal fraction from the first internode of etiolated *Sorghum* seedlings. Plant Physiol. 63:433–35.

107. Esquerré-Tugayé, M.-T., C. Lafitte, D. Mazau, A. Toppan, and A. Touzé. 1979. Cell surfaces in plant-microorganism interactions. II. Evidence for the accumulation of hydroxyproline-rich glycoproteins in the cell wall of diseased plants as a defense mechanism. Plant Physiol. 64:320–26.

108. Etzler, M. E. 1981. Are lectins involved in plant-fungus interactions? Phytopathology 71:744–46.

109. Farkas, G. L. 1978. Senescence and plant disease. In: Plant disease. An advanced treatise. J. G. Horsfall and E. B. Cowling (eds.). Academic Press, New York, 3: 391–412.

110. Farkas, G. L., Z. Király, and F. Solymosy. 1960. Role of oxidative metabolism in the localization of plant viruses. Virology 12:408–21.

111. Faulkner, G., and W. C. Kimmins. 1975. Staining reactions of the tissue bordering lesions induced by wounding, tobacco mosaic virus, and tobacco necrosis virus in bean. Phytopathology 65:1396–1400.

112. Faulkner, G., and W. C. Kimmins. 1978. Fine structure of tissue bordering lesions induced by wounding and virus infection. Can. J. Bot. 56:2990–99.

113. Favali, M. A., M. Bassi, and G. G. Conti. 1974. Morphological, cytochemical and autoradiographic studies of local lesions induced by the U5 strain of tobacco mosaic virus in *Nicotiana glutinosa* L. Riv. Patol. Veg. 10:207–18.

114. Favali, M. A., G. G. Conti, and M. Bassi. 1977. Some observations on virus-induced local lesions by transmission and scanning electron microscopy. Acta Phytopath. Hung. 12:141–50.

115. Favali, M. A., G. G. Conti, and M. Bassi. 1978. Modifications of the vascular bundle ultrastructure in the "resistant zone" around necrotic lesions induced by tobacco mosaic virus. Physiol. Plant Path. 13:247–51.

116. Fernando, M., and G. Stevenson. 1952. Studies on the physiology of parasitism. XVI. The effect of the condition of potato tissue as modified by temperature and water content upon attack by certain organisms and their pectinase enzymes. Ann. Bot. 16:103–14.

117. Fett, W. F., and S. B. Jones. 1982. Role of bacterial immobilization of race-specific resistance of soybean to *Pseudomonas syringae* pv. *glycinea*. Phytopathology 72: 488–92.

118. Fisher, M. L., A. J. Anderson, and P. Albersheim. 1973. Host-pathogen interactions. VI. A single plant protein effectively inhibits endopolygalacturonases secreted by *Colletotrichum lindemuthianum* and *Aspergillus niger*. Plant Physiol. 51:489–91.

119. Fletcher, J. T. 1973. Tomato mosaic. In: The UK tomato manual. H. G. Kingham (ed.). Grower Books, London, pp. 196–208.

120. Fletcher, J. T., and J. M. Rowe. 1975. Observations and

experiments on the use of an avirulent mutant strain of tobacco mosaic virus as a means of controlling tomato mosaic. Ann. Appl. Biol. 81:171–79.

121. Flor, H. H. 1947. Inheritance of reaction to flax. J. Agr. Res. 74:241–62.

122. Foster, J. A., and A. F. Ross. 1975. The detection of symptomless virus-infected tissue in inoculated tobacco leaves. Phytopathology 65:600–10.

123. Fraser, R. S. S. 1969. Effects of two TMV strains on the synthesis and stability of chloroplast ribosomal RNA in tobacco leaves. Mol. Gen. Genet. 106:73–79.

124. Fraser, R. S. S. 1979. Systemic consequences of the local lesion reaction to tobacco mosaic virus in a tobacco variety lacking the N gene for hypersensitivity. Physiol. Plant Path. 14:383–94.

125. Fraser, R. S. S. 1981. Evidence for the occurrence of the "pathogenesis-related" proteins in leaves of healthy tobacco plants during flowering. Physiol. Plant Path. 19:69–76.

126. Fraser, R. S. S. 1982. Are "pathogenesis-related" proteins involved in acquired systemic resistance of tobacco plants to tobacco mosaic virus? J. Gen. Virol. 58:305–13.

127. Fraser, R. S. S. 1983. Varying effectiveness of the N′ gene for resistance to tobacco mosaic virus in tobacco infected with virus strains differing in coat protein properties. Physiol. Plant Path. 22:109–19.

128. Fraser, R. S. S., and S. A. R. Loughlin. 1980. Resistance to tobacco mosaic virus in tomato: Effects of the Tm-1 gene on virus multiplication. J. Gen. Virol. 48:87–96.

129. Fraser, R. S. S., and S. A. R. Loughlin. 1981. Biochemistry of resistance to tobacco mosaic virus in tomato. Vth Int. Virol. Cong., Strasbourg, p. 261. (Abstr.)

130. Fraser, R. S. S., and S. A. R. Loughlin. 1982. Effects of temperature on the Tm-1 gene for resistance to tobacco mosaic virus in tomato. Physiol. Plant Path. 20:109–17.

131. Fraser, R. S. S., S. A. R. Loughlin, and J. C. Connor. 1980. Resistance to tobacco mosaic virus in tomato: Effects of the Tm-1 gene on symptom formation and multiplication of virus strain 1. J. Gen. Virol. 50:221–24.

132. Fraser, R. S. S., S. A. R. Loughlin, and R. J. Whenham. 1979. Acquired systemic susceptibility to infection by tobacco mosaic virus in Nicotiana glutinosa L. J. Gen. Virol. 43:131–41.

133. Friend, J. 1976. Lignification of infected tissue. In: Biochemical aspects of plant-parasite relationships. J. Friend and D. R. Threlfall (eds.). Academic Press, London, pp. 201–303.

134. Fuchs, W. H., and E. Kotte. 1954. Zur Kenntnis der Resistenz von Solanum tuberosum gegen Phytophthora infestans de By. Naturwissenschaften 41:169–70.

135. Fulton, R. W. 1982. The protective effects of systemic virus infection. In: Active defense mechanisms in plants. NATO Advanced Study Institutes Series, vol. 37. R. K. S. Wood (ed.). Plenum Press, London, pp. 231–45.

136. Furusawa, I., and T. Okuno. 1978. Infection with BMV of mesophyll protoplasts from five plant species. J. Gen. Virol. 40:489–91.

137. Gat-Edelbaum, O., A. Altman, and I. Sela. 1983. Polyinosinic-polycytidylic acid in association with cyclic nucleotides activates the antiviral factor (AVF) in plant tissues. J. Gen. Virol. 64:211–14.

138. Gäumann, E. 1950. Principles of plant infection. (Trans. W. B. Breierley.) Crosley Lockwood and Son, London.

139. Geiger, W. B. 1946. The mechanism of antibacterial action of quinones and hydroquinones. Arch. Biochem. 11:23–32.

140. Gera, A., and G. Loebenstein. 1983. Further studies of an inhibitor of virus replication from tobacco mosaic virus-infected protoplasts of a local lesion responding tobacco cultivar. Phytopathology 73:111–15.

141. Gianinazzi, S. 1982. Antiviral agents and inducers of virus resistance: Analogies with interferon. In: Active Defense mechanisms in plants. NATO Advanced Study Institutes Series, vol. 37. R. K. S. Wood (ed.). Plenum Press, London, pp. 275–98.

142. Gianinazzi, S., P. Ahl, and A. Cornu. 1981. Genetic studies of b-proteins: A new approach to understanding the molecular basis of resistant mechanisms in Nicotiana. Vth Int. Virol. Cong., Strasbourg, p. 260. (Abstr.)

143. Gianinazzi, S., and B. Kassanis. 1974. Virus resistance induced in plants by polyacrylic acid. J. Gen. Virol. 23:1–9.

144. Gianinazzi, S., C. Martin, and J. C. Vallée. 1970. Hypersensibilité aux virus, temperature et protéines soluble chez le Nicotiana Xanthi nc. Apparition de nouvelles macromolécules lors de la répression de la synthese virale. Compte. Rend. Acad. Sci. 270:2383–86.

145. Gianinazzi, S., and C. Schneider. 1979. Differential reactions to tobacco mosaic virus infection in Samsun 'nn' tobacco plants. II. A comparative ultrastructural study of the virus induced necroses with those of Xanthi-nc. Phytopath. Z. 96:313–23.

146. Gicherman, G., and G. Loebenstein. 1968. Competitive inhibition by foreign nucleic acids and induced interference by yeast RNA with the infection by tobacco mosaic virus. Phytopathology 58:405–9.

147. Gill, M. D. 1982. Bacterial toxins: A table of lethal amounts. Microbiol. Rev. 46:86–94.

148. Gilpatrick, J. D., and M. Weintraub. 1952. An unusual type of protection with the carnation mosaic virus. Science 115:701–2.

149. Gonda, T. J., and R. H. Symons. 1979. Cucumber mosaic virus replication in cowpea protoplasts: Time course of virus, coat protein and RNA synthesis. J. Gen. Virol. 45:723–36.

150. Goodman, R. M., and A. F. Ross. 1974. Enhancement of potato virus X synthesis in doubly infected tobacco occurs in doubly infected cells. Virology 58:16–24.

151. Goodman, R. N. 1967. Protection of apple stem tissue against Erwinia amylovora infection by avirulent strains and three other bacterial species. Phytopathology 57:22–24.

152. Goodman, R. N. 1968. The hypersensitive reaction in tobacco: A reflection of changes in host cell permeability. Phytopathology 58:872–73.

153. Goodman, R. N. 1972. Electrolyte leakage and membrane damage in relation to bacterial population, pH, and ammonia production in tobacco leaf tissue inoculated with Pseudomonas pisi. Phytopathology 62:1327–31.

154. Goodman, R. N., J. S. Huang, and P.-Y. Huang. 1974. Host-specific phytotoxic polysaccharide from apple tissue infected with Erwinia amylovora. Science 183:1108–82.

155. Goodman, R. N., P.-Y. Huang, and J. A. White. 1976. Ultrastructural evidence for immobilization of an incompatible bacterium, Pseudomonas pisi, in tobacco leaf tissue. Phytopathology 66:754–64.

156. Goodman, R. N., and S. B. Plurad. 1971. Ultrastructural changes in tobacco undergoing the hypersensitive reaction caused by plant pathogenic bacteria. Physiol. Plant Path. 1:11–15.

157. Gregg, M. 1952. Studies on the physiology of parasitism. XVII. Enzyme secretion by strains of Bacterium carotovorum and other pathogens in relation to parasitic vigor. Ann. Bot. 16:235–50.

158. Habili, N., and J. M. Kaper. 1981. Cucumber mosaic virus-associated RNA 5. VII. Double-stranded form accumulation and disease attenuation in tobacco. Virology 112:250–61.

159. Habili, N., and J. M. Kaper. 1982. Is accumulation of the ds form of cucumber mosaic virus associated RNA 5 in infected tobacco plants part of a defence mechanism against virus infection? Vth Int. Virol. Cong., Strasbourg, p. 261. (Abstr.)

160. Hadwiger, L. A., J. M. Beckman, and M. J. Adams. 1981. Localization of fungal components in the pea-Fusarium interaction

detected immunochemically with antichitosan and antifungal cell wall antisera. Plant Physiol. 67:170–75.

161. Hadwiger, L. A., and D. C. Loschke. 1981. Molecular communication in host-parasite interactions: Hexosamine polymers (chitosan) as regulator compounds in race-specific and other interactions. Phytopathology 71:756–62.

162. Hall, T. J. 1980. Resistance at the Tm-2 locus in the tomato to tomato mosaic virus. Euphytica 29:189–97.

163. Hamilton, R. I. 1980. Defences triggered by previous invaders: Viruses. In: Plant disease. An advanced treatise. J. G. Horsfall and E. B. Cowling (eds.). Academic Press, New York, 5:279–303.

164. Hammerschmidt, R., and J. Kuć. 1982. Lignification as a mechanism for induced systemic resistance in cucumber. Physiol. Plant Path. 20:61–71.

165. Hammerschmidt, R., E. M. Nuckles, and J. Kuć. 1982. Association of enhanced peroxidase activity with induced systemic resistance of cucumber to *Colletotrichum lagenarium*. Physiol. Plant Path. 20:73–82.

166. Hansen, A. J. 1974. Systemic tobacco mosaic virus infection of a 'resistant' N-gene-carrying tobacco hybrid raised from infected callus culture. Virology 57:387–91.

167. Harrison, B. D. 1982. Active resistance of plants to viruses. In: Active defense mechanisms in plants. NATO Advanced Study Institutes Series, vol. 37. R. K. S. Wood (ed.). Plenum Press, London, pp. 193–209.

168. Harrison, B. D., and A. F. Murant. CMI/AAB. Descriptions of plant viruses. Commonw. Mycol. Inst. and Assn. Appl. Biol., Kew.

169. Hartley, R. D. 1973. Carbohydrate esters of ferulic acid as components of cell walls of *Lolium multiflorum*. Phytochemistry 12:661–65.

170. Hayashi, T., and C. Matsui. 1965. Fine structure of lesion periphery produced by tobacco mosaic virus. Phytopathology 55:387–92.

171. Heath, M. C. 1976. Hypersensitivity, the cause or the consequence of rust resistance? Phytopathology 66:935–36.

172. Heath, M. C. 1979. Effects of heat shock, actinomycin D, cycloheximide and blasticidin S on non-host interactions with rust fungi. Physiol. Plant Path. 15:211–18.

173. Heath, M. C. 1980. Effects of infection by compatible species or injection of tissue extracts on the susceptibility of non-host plants to rust fungi. Phytopathology 70: 356–60.

174. Helms, K. 1965. Interference between two strains of tobacco mosaic virus in leaves of pinto bean. Virology 27:346–50.

175. Hevesi, M., and S. F. Mashaal. 1975. Contribution to the mechanism of infection of *Erwinia uredovora*, a parasite of rust fungi. Acta Phytopath. Hung. 10:275–80.

176. Hildebrand, D. C. 1970. Fireblight resistance in *Pyrus*: Hydroquinone formation as related to antibiotic activity. Can. J. Bot. 48:177–81.

177. Hildebrand, D. C., M. C. Alosi, and M. N. Schroth. 1980. Physical entrapment of pseudomonads in bean leaves by films formed at air water interfaces. Phytopathology 70:98–109.

178. Hildebrand, D. C., and M. N. Schroth. 1963. Relation of arbutin-hydroquinone in pear blossoms to invasion by *Erwinia amylovora*. Nature 197:513.

179. Hildebrand, D. C., and M. N. Schroth. 1964. Fireblight resistance in *Pyrus*: Localization of arbutin and beta glucosidase. Phytopathology 59:1534–39.

180. Hildebrand, E. M. 1954. The blossom-blight phase of fireblight and methods of control. Cornell Univ. Agr. Exp. Sta. Mem. 207.

181. Hiruki, C., and J. C. Tu. 1972. Light and electron microscopy of potato virus M lesions and marginal tissue in red kidney bean. Phytopathology 62:77–85.

182. Holliday, M. J., and N. T. Keen. 1982. The role of phytoalexins in the resistance of soybean leaves to bacteria: Effects of glyphosate on glyceollin accumulation. Phytopathology 72:1470–74.

183. Holliday, M. J., M. Long, and N. T. Keen. 1981. Manipulation of the temperature-sensitive interaction between soybean leaves and *Pseudomonas syringae* pv. *glycinea*—implications on the nature of determinative events modulating hypersensitive resistance. Physiol. Plant Path. 19:209–16.

184. Holmes, F. O. 1938. Inheritance of resistance to tobaccomosaic disease in tobacco. Phytopathology 28:553–61.

185. Holmes, F. O. 1954. Inheritance of resistance to viral diseases in plants. Adv. Virus Res. 2:1–30.

186. Holmes, F. O. 1960. Inheritance in tobacco of an improved resistance to infection by tobacco mosaic virus. Virology 12:59–67.

187. Holmes, F. O. 1965. Genetics of pathogenicity in viruses and of resistance in host plants. Adv. Virus Res. 11:139–61.

188. Hoppe, H. H., B. Humme, and R. Heitefuss. 1980. Elicitor induced accumulation of phytoalexins in healthy and rust infected leaves of *Phaseolus vulgaris*. Phytopath. Z. 97:85–88.

189. Hsieh, S. P. Y., and I. W. Buddenhagen. 1974. Suppressing effects of *Erwinia herbicola* on infection by *Xanthomonas oryzae* and on symptom development in rice. Phytopathology 64:1182–85.

190. Huang, J. S., and R. N. Goodman. 1972. Alterations in structural proteins from chloroplast membranes of bacterially induced hypersensitive tobacco leaves. Phytopathology 62:1428–34.

191. Huang, J. S., P.-Y. Huang, and R. N. Goodman. 1973. Reconstitution of a membrane-like structure with structural proteins and lipids isolated from tobacco thylakoid membranes. Amer. J. Bot. 60:80–85.

192. Huang, J. S., P.-Y. Huang, and R. N. Goodman. 1974. Ultrastructural changes in tobacco thylakoid membrane protein caused by a bacterially induced hypersensitive reaction. Physiol. Plant Path. 4:93–97.

193. Huang, J. S., and C. G. Van Dyke. 1978. Interaction of tobacco callus tissue with *Pseudomonas tabaci*, *P. pisi* and *P. fluorescens*. Physiol. Plant Path. 13:65–72.

194. Huang, P.-Y., J. S. Huang, and R. N. Goodman. 1975. Resistance mechanisms of apple shoots to an avirulent strain of *Erwinia amylovora*. Physiol. Plant Path. 6: 283–87.

195. Huber, R., G. Rezelman, T. Hibi, and A. van Kammen. 1977. Cowpea mosaic virus infection of protoplasts from Samsun tobacco leaves. J. Gen. Virol. 34:315–23.

196. Israel, H. W., and A. F. Ross. 1967. Fine structure of local lesions induced by tobacco mosaic virus in tobacco. Virology 33:272–86.

197. Jenns, A. E., and J. Kuć. 1979. Graft transmission of systemic resistance of cucumber to anthracnose induced by *Colletotrichum lagenarium* and tobacco necrosis virus. Phytopathology 69:753–56.

198. Jenns, A. E. and J. Kuć. 1980. Characteristics of anthracnose resistance induced by localized infection of cucumber with tobacco necrosis virus. Physiol. Plant Path. 17:81–91.

199. Jones, R. A. C. 1981. The ecology of viruses infecting wild and cultivated potatoes in the Andean region of South America. In: Pests, pathogens and vegetation. J. H. Thresh (ed.). Pitman, London, pp. 89–107.

200. Kassanis, B. 1963. Interactions of viruses in plants. Adv. Virus Res. 10:219–55.

201. Kassanis, B., and C. Bastow. 1971. Phenotypic mixing between strains of tobacco mosaic virus. J. Gen. Virol. 11:171–76.

202. Kassanis, B., S. Gianinazzi, and R. F. White. 1974. A

possible explanation of the resistance of virus-infected tobacco plants to second infection. J. Gen. Virol. 23: 11–16

203. Kassanis, B., and R. F. White. 1978. Effect of polyacrylic acid and b proteins on TMV multiplication in tobacco protoplasts. Phytopath. Z. 91:269–72.

204. Kato, S. 1976. Ethylene production during lipid peroxidation in cowpea leaves infected with cucumber mosaic virus. Ann. Phytopath. Soc. Jap. 43:587–89.

205. Kato, S., and T. Misawa. 1976. Lipid peroxidation during the appearance of hypersensitive reaction in cowpea leaves infected with cucumber mosaic virus. Ann. Phytopath. Soc. Jap. 42:472–80.

206. Keen, N. T. 1981. Evaluation of the role of phytoalexins. In: Plant disease control. Resistance and susceptibility. R. C. Staples and G. H. Toenniessen (eds.). Wiley-Interscience, New York, pp. 155–77.

207. Keen, N. T., T. Érsek, M. Long, B. Bruegger, and M. Holliday. 1981. Inhibition of the hypersensitive reaction of soybean leaves to incompatible Pseudomonas sp. by blasticidin S, streptomycin or elevated temperature. Physiol. Plant Path. 18:325–37.

208. Keen, N. T., and M. J. Holliday. 1982. Recognition of bacterial pathogens by plants. In: Phytopathogenic prokaryotes II. M. S. Mount and G. H. Lacy (eds.). Academic Press, New York, pp. 179–217.

209. Kimmins, W. C., and R. G. Brown. 1973. Hypersensitive resistance. The role of cell wall glycoproteins in virus localization. Can. J. Bot. 51:1923–26.

210. Kimmins, W. C., and D. Wuddah. 1977. Hypersensitive resistance: Determination of lignin in leaves with a localized virus infection. Phytopathology 67:1012–16.

211. Király, Z. 1979. Betydning af den erhvervede resistens mod plante-sygdomme. Ugeskrift for Jordbrug (Copenhangen), 49:1201–3.

212. Király, Z. 1980. Defenses triggered by the invader: Hypersensitivity. In: Plant disease. An advanced treatise. J. G. Horsfall and E. B. Cowling (eds.). Academic Press, New York, 5:201–24.

213. Király, Z., B. Barna, and T. Érsek. 1972. Hypersensitivity as a consequence, not the cause, of plant resistance to infection. Nature 239:456–58.

214. Király, Z., and G. L. Farkas. 1962. Relation between phenol metabolism and stem rust resistance in wheat. Phytopathology 52:657–64.

215. Király, Z., M. Hevesi, and Z. Klement. 1977. Inhibition of bacterial multiplication in incompatible host-parasite relationships in the absence of the hypersensitive necrosis. Acta Phytopath. Hung. 12:247–56.

216. Király, Z., and J. Szirmai. 1964. The influence of kinetin on tobacco mosaic virus production in Nicotiana glutinosa leaf disks. Virology 23:286–88.

217. Kitazawa, K., and K. Tomiyama. 1969. Microscopic observations of potato cells by compatible and incompatible races of Phytophthora infestans. Phytopath. Z. 66: 317–24.

218. Klement, Z. 1982. Hypersensitivity. In: Phytopathogenic prokaryotes II. M. S. Mount and G. H. Lacy (eds.). Academic Press, New York, pp. 149–77.

219. Klement, Z., G. L. Farkas, and L. Lovrekovich. 1964. Hypersensitive reaction induced by phytopathogenic bacteria in the tobacco leaf. Phytopathology 54:474–77.

220. Klement, Z., and R. N. Goodman. 1967. The hypersensitive reaction to infection by bacterial plant pathogens. Ann. Rev. Phytopath. 5:17–44.

221. Kluge, S. 1980. On the effects of virazole, dioxohexahydrotriazine and polyacrylic acid on virus multiplication and host plant proteins. Tag.-Ber. Acad. Landwirtsch.-Wiss. DDR 184:143–58.

222. Koike, M., T. Hibi, and K. Yora. 1977. Infection of cowpea mesophyll protoplasts with cucumber mosaic virus. Virology 83:413–16.

223. Kojima, M., K. Kawakita, and I. Uritani. 1982. Studies on a factor in sweet potato root which agglutinates spores of Ceratocystis fimbriata, black rot fungus. Plant Physiol. 69:474–78.

224. Kolattukudy, P. E., and V. P. Agrawal. 1974. Structure and composition of aliphatic constituents of potato tuber skin (suberin). Lipids 9:682–91.

225. Kolattukudy, P. E., K. Kronman, and A. J. Poulose. 1975. Determination of structure and composition of suberin from the roots of carrot, parsnip, rutabaga, turnip, red beet and sweet potato by combined liquid-gas chromatography and mass spectrometry. Plant Physiol. 55:567–73.

226. Konate, G., M. Kopp, and B. Fritig. 1982. Rates of tobacco mosaic virus multiplication in systemic and necrotic responding tobacco varieties: Biochemical approach to the study of hypersensitive resistance to viruses. Phytopath. Z. 105:214–25.

227. Kubo, S., and Y. Takanami. 1979. Infection of tobacco mesophyll protoplasts with tobacco necrotic dwarf virus, a phloem-limited virus. J. Gen. Virol. 42:387–98.

228. Kuć, J. 1981. Multiple mechanisms, reaction rates and induced resistance in plants. In: Plant disease control. R. C. Staples and G. H. Toenniessen (eds.). Wiley, New York, pp. 259–72.

229. Kuć, J. 1982. Plant immunization-mechanisms and practical implications. In: Active defense mechanisms in plants. NATO Advanced Study Institutes Series, vol. 37. R. K. S. Wood (ed.). Plenum Press, New York and London, pp. 157–78.

230. Kuhn, C. W., S. D. Wyatt, and B. B. Brantley. 1981. Genetic control of symptoms, movement and virus accumulation in cowpea plants infected with cowpea chlorotic mottle virus. Phytopathology 71:1310–15.

231. Kunkel, L. O. 1934. Studies on acquired immunity with tobacco and aucuba mosaics. Phytopathology 24:437–66.

232. Kurstak, E. (ed.). 1981. Handbook of plant virus infections. Comparative diagnosis. Elsevier/North Holland, Amsterdam.

233. Lapwood, D. H. 1957. Studies in the physiology of parasitism. XXIII. On the parasitic vigor of certain bacteria in relation to their capacity to secrete pectolytic enzymes. Ann. Bot. 21:167–84.

234. Lauffer, M. A., and W. C. Price. 1945. Infection by viruses. Arch. Biochem. 8:449–68.

235. Lazarovits, G., P. Stossel, and E. W. B. Ward. 1981. Age-related changes in specificity and glyceollin production in the hypocotyl reaction of soybeans to Phytophthora megasperma var. sojae. Phytopathology 71:94–97.

236. Leatham, G. F., V. King, and M. A. Stahmann. 1980. In vitro protein polymerization by quinones or free radicals generated by plant or fungal oxidative enzymes. Phytopathology 70:1134–40.

237. Legrand, M., B. Fritig, and L. Hirth. 1976. Enzymes of the phenyl-propanoid pathway and the necrotic reaction of hypersensitive tobacco to tobacco mosaic virus. Phytochemistry 15:1353–59.

238. Lesney, M. S., R. S. Livingston, and R. P. Scheffer. 1982. Effects of toxin from Helminthosporium sacchari on nongreen tissues and a reexamination of toxin binding. Phytopathology 72:844–49.

239. Levy, A., G. Lobenstein, M. Smookler, and T. Drori. 1974. Partial suppression by UV irradiation of the mechanism of resistance to cucumber mosaic virus in a resistant cucumber cultivar. Virology 60:37–44.

240. Lippincott, B. B., and J. A. Lippincott. 1969. Bacterial attachment to a specific wound site as an essential stage in tumor initiation by Agrobacterium tumefaciens. J. Bact. 97:620–28.

241. Loebenstein, G. 1962. Inducing partial protection in the host plant with native virus protein. Virology 17:574–81.

242. Loebenstein, G. 1972. Inhibition, interference and acquired resistance during infection. In: Principles and techniques in plant virology. C. I. Kado and H. O. Agrawal (eds.). Van Nostrand/Rheinhold, New York, pp. 32–61.

243. Loebenstein, G. 1972. Localization and induced resistance in virus-infected plants. Ann. Rev. Phytopath. 10:177–206.

244. Loebenstein, G., R. Chazan, and M. Eisenberg. 1970. Partial suppression of the localizing mechanism to tobacco mosaic virus by UV irradiation. Virology 41:373–76.

245. Loebenstein, G., J. Cohen, S. Shabtai, R. H. A. Coutts, and K. R. Wood. 1977. Distribution of cucumber mosaic virus in systemically infected tobacco leaves. Virology 81:117–25.

246. Loebenstein, G., and A. Gera. 1981. Inhibitor of virus replication released from tobacco mosaic virus infected protoplasts of a local lesion-responding tobacco cultivar. Virology 114:132–39.

247. Loebenstein, G., S. Rabina, and T. van Praagh. 1966. Induced interference phenomena in virus infections. In: Viruses of plants. A. B. R. Beemster and J. Dijkstra (eds.). Elsevier/North Holland Medical Press, Amsterdam, pp. 151–57.

248. Loebenstein, G., S. Rabina, and T. van Praagh. 1968. Sensitivity of induced localized acquired resistance to actinomycin D. Virology 34:264–68.

249. Loebenstein, G., B. Sela, and T. van Praagh. 1969. Increase of tobacco mosaic local lesion size and virus multiplication in hypersensitive hosts in the presence of actinomycin D. Virology 37:42–48.

250. Loebenstein, G., S. Spiegel, and A. Gera. 1982. Localized resistance and barrier substances. In: Active defense mechanisms in plants. NATO Advanced Study Institutes Series, vol. 37. R. K. S. Wood (ed.). Plenum Press, London, pp. 211–30.

251. Loegering, W. Q. 1966. The relationship between host and pathogen in stem rust of wheat. Proc. 2nd. Int. Genetics Symp., Lund 1963. Hereditas 2:167–77. (Suppl.)

252. Lovrekovich, L., and G. L. Farkas. 1965. Induced protection against wildfire disease in tobacco leaves treated with heat-killed bacteria. Nature 205:823–24.

253. Lovrekovich, L., H. Lovrekovich, and M. A. Stahmann. 1967. Inhibition of phenol oxidation by *Erwinia carotovora* in potato tuber tissue and its significance in disease resistance. Phytopathology 57:737–42.

254. Lovrekovich, L., H. Lovrekovich, and M. A. Stahmann. 1968. The importance of peroxidase in the wildfire disease. Phytopathology 58:193–98.

255. Lovrekovich, L., H. Lovrekovich, and M. A. Stahmann. 1968. Tobacco mosaic-virus induced resistance to *Pseudomonas tabaci* in tobacco. Phytopathology 58:1034–35.

256. Lozano, J. C., and L. Sequeira. 1970. Differentiation of races of *Pseudomonas solanacearum* by a leaf infiltration technique. Phytopathology 60:834–40.

257. Luke, H. H., and H. E. Wheeler. 1955. Toxin production by *Helminthosporium victoriae*. Phytopathology 45:453–58.

258. Luke, H. H., and H. E. Wheeler. 1964. An intermediate reaction to victorin. Phytopathology 54:1492–93.

259. Lund, B. M. 1973. The effect of certain bacteria on ethylene production by plant tissue. In: Fungal pathogenicity and the plant's response. R. J. W. Byrde and C. V. Cutting (eds.). Academic Press, New York, pp. 69–86.

260. Lyr, H., and L. Banasiak. 1983. Alkenales, volatile defence substances in plants as models for new plant protection agents. Acta Phytopath. Hung. 18:3–12.

261. Maclean, D. J., J. A. Sargent, I. C. Tommerup, and D. S. Ingram. 1974. Hypersensitivity as the primary event in resistance to fungal parasites. Nature 249:186–87.

262. Maher, E. A., and A. Kelman. 1983. Oxygen status of potato tuber tissue in relation to maceration by pectic enzymes of *Erwinia carotovora*. Phytopathology 73:537–39.

263. Maine, E. C. 1960. Physiological responses in the tobacco plant to pathogenesis by *Pseudomonas solanacearum* causal agent of bacterial wilt. Master's thesis, Dept. Plant Pathology, North Carolina State Univ., Raleigh.

264. Maine, E. C., and A. Kelman. 1961. The influence of reducing substances on resistance to bacterial wilt in tobacco. Phytopathology 51:491–92.

265. Marchand, R., and A. W. Robards. 1968. Membrane systems associated with the plasmalemma of plant cells. Ann. Bot. 32:457–71.

266. Martelli, G. P. 1980. Ultrastructural aspects of possible defence reactions in virus-infected plant cells. Microbiologica 3:369–91.

267. Matta, A. 1982. Mechanisms in non-host resistance. In: Active defense mechanisms in plants. NATO Advanced Study Institutes Series, vol. 37. R. K. S. Wood (ed.). Plenum Press, London, pp. 119–41.

268. Matthews, R. E. F. 1980. Host plant responses to virus infection. In: Comprehensive virology. H. H. Fraenkel-Conrat and R. R. Wagner (eds). Plenum Press, New York, 16:297–359.

269. Matthews, R. E. F. 1981. Plant virology. 2d ed. Academic Press, New York.

270. Maule, A. J., M. I. Boulton, and K. R. Wood. 1980. Resistance of cucumber protoplasts to cucumber mosaic virus: A comparative study. J. Gen. Virol. 51:271–79.

271. Maule, A. J., M. I. Boulton, and K. R. Wood. 1982. Use of protoplasts in virus resistance studies. In: Active defense mechanisms in plants. NATO Advanced Study Institutes Series, vol. 37. R. K. S. Wood (ed.). Plenum Press, London, pp. 342–43.

272. Mayama, S., J. M. Daly, D. W. Rehfeld, and R. C. Daly. 1975. Hypersensitive response of near-isogenic wheat carrying the temperature-sensitive Sr6 allele for resistance to stem rust. Physiol. Plant Path. 7:35–47.

273. Mayama, S., D. W. Rehfeld, and J. M. Daly. 1975. The effect of detachment on the development of rust disease and the hypersensitive response of wheat leaves infected with *Puccinia graminis tritici*. Phytopathology 65:1139–42.

274. McIntyre, J. L., J. A. Dodds, and J. D. Hare. 1981. Effects of localized infections of *Nicotiana tabacum* by tobacco mosaic virus on systemic resistance against diverse pathogens and an insect. Phytopathology 71:297–301.

275. McIntyre, J. L., J. Kuć, and E. B. Williams. 1975. Protection of Bartlett plear against fireblight with deoxyribonucleic acid from virulent and avirulent *Erwinia amylovora*. Physiol. Plant Path. 7:153–70.

276. McKinney, H. H. 1929. Mosaic diseases in the Canary Islands, West Africa and Gilbraltar. J. Agr. Res. 39:557–78.

277. McKinney, H. H. 1943. Studies on the genotypes of tobacco resistant to the common-mosaic virus. Phytopathology 33:300.

278. McLean, F. T. 1921. A study of the structure of the stomata of two species of citrus in relation to citrus canker. Torrey Bot. Club Bull. 48:101–6.

279. McNew, G. L., and H. P. Burchfield. 1951. Fungitoxicity and biological activity of quinones. Contrib. Boyce Thompson Inst. 16:357–74.

280. Menke, G. H., and J. C. Walker. 1963. Metabolism of resistant and susceptible cucumber varieties infected with cucumber mosaic virus. Phytopathology 53:1349–55.

281. Milne, R. G. 1966. Electron microscopy of tobacco mosaic virus in leaves of *Nicotiana glutinosa*. Virology 28:527–32.

282. Mirelman, D., E. Galun, N. Sharon, and R. Lotan. 1975. Inhibition of fungal growth by wheat germ agglutinin. Nature 256:414–16.

283. Moesta, P., and H. Grisebach. 1982. L-2-aminooxy-3-phenylpropionic acid inhibits phytoalexin accumulation in soybean with concomitant loss of resistance against Phytophthora megasperma f. sp. glycinea. Physiol. Plant Path. 21:65–70.

284. Mollenhauer, H. H., and D. L. Hopkins. 1976. Xylem morphology of Pierce's disease-infected grapevines with different levels of tolerance. Physiol. Plant Path. 9:95–100.

285. Moore, A. E., and B. A. Stone. 1972. Effect of infection with TMV and other viruses on the level of a α-1,3-glucan hydrolase in leaves of Nicotiana glutinosa. Virology 50:791–98.

286. Motoyoshi, F., and N. Oshima. 1977. Expression of genetically controlled resistance to tobacco mosaic virus infection in isolated tomato leaf protoplasts. J. Gen. Virol. 34:499–506.

287. Motoyoshi, F., and N. Oshima. 1978. Resistance and susceptibility of tomato protoplasts to tobacco mosaic virus infection. IVth Int. Virol. Cong., The Hague, p. 263. (Abstr.)

288. Motoyoshi, F., and N. Oshima. 1979. Standardization in inoculation procedure and effect of a resistance gene on infection of tomato protoplasts with tobacco mosaic virus RNA. J. Gen. Virol. 44:801–6.

289. Mozes, R., Y. Antignus, I. Sela, and I. Harpaz. 1978. The chemical nature of an antiviral factor (AVF) from virus-infected plants. J. Gen. Virol. 38:241–49.

290. Müller, K. O. 1956. Einige einfache Versuche zum Nachweis von Phytoalexinen. Phytopath. Z. 27:237–54.

291. Müller, K. O. 1959. Hypersensitivity. In: Plant pathology: An advanced treatise. J. G. Horsfall and A. E. Dimond (eds.). Academic Press, New York, 1: 469–519.

292. Müller, K. O., and A. Behr. 1949. Mechanism of Phytophthora resistance of potatoes. Nature 163:498–99.

293. Müller, K. O., and H. Börger. 1941. Experimentelle Untersuchungen über die Phytophthora-Resistenz der Kartoffel. Arb. Biol. Reichsanst. Land-Forstw. 23:189–231.

294. Murant, A. F., and R. K. S. Wood. 1957. Factors affecting the pathogenicity of bacteria to potato tubers. II. Ann. Appl. Biol. 45:650–63.

295. Mussell, H. W., and L. L. Strand. 1977. Pectic enzymes: Involvement in pathogenesis and possible relevance to tolerance and specificity. In: Cell wall biochemistry related to specificity in host-pathogen interactions. B. Solheim and J. Raa (eds.). Universitetsforlaget, Oslo, pp. 31–70.

296. Nachman, I., G. Loebenstein, T. van Praagh, and A. Zelcer. 1971. Increased multiplication of cucumber mosaic virus in a resistant cucumber cultivar caused by actinomycin D. Physiol. Plant Path. 1:67–72.

297. Nadolny, L., and L. Sequeira. 1980. Increases in peroxidase are not directly involved in induced resistance in tobacco. Physiol. Plant Path. 16:1–8.

298. Nakagaki, Y., and T. Hirai. 1971. Effect of detached leaf treatment on tobacco mosaic virus multiplication in tobacco and bean leaves. Phytopathology 61:22–27.

299. Neukom, H., and H. U. Markwalder. 1978. Oxidative gelation of wheat and flour pentosans: A new way of cross-linking polymers. Cereal Foods World 23:374–76.

300. Nichols, J., J. M. Beckman, and M. A. Hadwiger. 1980. Glycosidic enzyme activity in peas tissue and pea-Fusarium solani interactions. Plant Physiol. 66:199–204.

301. Nicholson, R. L., S. van Soyc, E. B. William, and J. Kuć. 1977. Host-pathogen interactions preceding the hypersensitive reaction of Malus sp. to Venturia inaequalis. Phytopathology 67:108–14.

302. Nitzany, F. E., and I. Sela. 1962. Interference between cucumber mosaic virus and tobacco mosaic virus on different hosts. Virology 17:549–53.

303. Nozue, M., K. Tomiyama, and N. Doke. 1977. Effect of blasticidin S on development of potential of potato tuber cell to react hypersensitively to infection by Phytophthora infestans. Physiol. Plant Path. 10:181–89.

304. Nozue, M., K. Tomiyama, and N. Doke. 1978. Effect of adenosine 5'-triphosphate on hypersensitive death of potato tuber cells infected by Phytophthora infestans. Phytopathology 68:873–76.

305. Obukowicz, M., and G. S. Kennedy. 1981. Phenolic ultracytochemistry of tobacco cells undergoing the hypersensitive reaction to Pseudomonas solanacearum. Physiol. Plant Path. 18:339–44.

306. Ohashi, Y., and T. Shimomura. 1971. Induction of local lesion formation on leaves systemically infected with virus by a brief heat or cold treatment. Ann. Phytopath. Soc. Jap. 37:211–14.

307. Ohashi, Y., and T. Shimomura. 1979. Induction of necrotic lesions by ultraviolet irradiation on leaves systemically infected with tobacco mosaic virus. Ann. Phytopath. Soc. Jap. 45:247–54.

308. Ohashi, Y., and T. Shimomura. 1982. Modification of cell membranes of leaves systemically infected with tobacco mosaic virus. Physiol. Plant Path. 20:125–28.

309. Oku, H. 1960. Biochemical studies on Cochliobolus miyabeanus. VI. Breakdown of disease resistance of rice plant by treatment with reducing agents. Ann. Phytopath. Soc. Jap. 25:92–98.

310. Oku, H., T. Shiraishi, and S. Ouchi. 1977. Suppression of induction of phytoalexin, pisatin by low-molecular-weight substances from spore germination fluid of pea pathogen, Mycosphaerella pinodes. Naturwiss. 64:643.

311. Otsuki, Y., T. Shimomura, and I. Takebe. 1972. Tobacco mosaic virus multiplication and expression of the N gene in necrotic responding tobacco varieties. Virology 50:45–50.

312. Otsuki, Y., and I. Takebe. 1976. Double infection of isolated tobacco mesophyll protoplasts by unrelated plant viruses. J. Gen. Virol. 30:309–16.

313. Otsuki, Y., and I. Takebe. 1976. Double infection of isolated tobacco leaf protoplasts by two strains of tobacco mosaic virus. In: Biochemistry and cytology of plant-parasite interaction. K. Tomiyama, J. M. Daly, I. Uritani, H. Oku, and S. Ouchi (eds.). Kodansha, Tokyo, pp. 213–22.

314. Ouchi, S., C. Hibino, H. Oku, M. Fujiwara, and H. Nakabayashi. 1979. The induction of resistance or susceptibility. In: Recognition and specificity in plant host-parasite interactions. J. M. Daly and I. Uritani (eds.). Jap. Sci. Soc. Press, Tokyo, pp. 49–65.

315. Overeem, J. C. 1976. Pre-existing antimicrobial substances in plants and their role in disease resistance. In: Biochemical aspects of plant-parasite relationships. J. Friend and D. R. Threlfall (eds.). Academic Press, New York, pp. 195–206.

316. Owens, R. G. 1953. Studies on the nature of fungicidal action. I. Inhibition of sulfhydryl-, amino-, iron- and copper dependent in vitro by fungicides and related compounds. Contrib. Boyce Thompson Inst. 17:221–42.

317. Owens, R. G., and H. M. Novotny. 1958. Mechanism of action of the fungicide dichlone (2,3-dichloro-1,4-naphthoquinone). Contrib. Boyce Thompson Inst. 19: 463–82.

318. Pearce, R. B., and J. P. Ride. 1982. Chitin and related compounds as elicitors of the lignification response in wounded wheat leaves. Physiol. Plant Path. 20:119–23.

319. Pelham, J. 1972. Strain-genotype interaction of tobacco mosaic virus in tomato. Ann. Appl. Biol. 72:219–28.

320. Peltier, G. L., and F. R. Schroeder. 1932. The nature of

resistance in alfalfa to wilt (*Aplanobacter insidiosum* L. Mc.). Nebraska Agr. Exp. Sta. Res. Bull. 63, 28 p.

321. Peng, F. H., and L. L. Black. 1977. Increased proteinase inhibitor activity in response to infection of resistant tomato plants by *Phytophthora infestans*. Phytopathology 66:958–63.

322. Pennazio, S., A. Appiano, and P. Redolfi. 1979. Changes occurring in *Gomphrena globosa* leaves in advance of the appearance of tomato bushy stunt virus necrotic local lesions. Physiol. Plant Path. 15:177–82.

323. Pennazio, S., G. D'Agostino, A. Appiano, and P. Redolfi. 1978. Ultrastructure and histochemistry of the resistant tissue surrounding lesions to tomato bushy stunt virus in *Gomphrena globosa* leaves. Physiol. Plant Path. 13:165–71.

324. Pennazio, S., P. Redolfi, and C. Sapetti. 1981. Callose formation and permeability changes during the partly localized reaction of *Gomphrena globosa* to potato virus X. Phytopath. Z. 100:172–81.

325. Pierpoint, W. S., N. P. Robinson, and M. B. Leason. 1981. The pathogenesis-related proteins of tobacco: Their induction by viruses in intact plants and their induction by chemicals in detached leaves. Physiol. Plant Path. 19:85–97.

326. Pinkas, J., and A. Novacky. 1971. The differentiation between bacterial hypersensitive reaction and pathogenesis by the use of cycloheximide. Phytopathology 61:906.

327. Pitrat, M., and H. Lecoq. 1980. Inheritance of resistance to cucumber mosaic virus transmission by *Aphis gossypii* in *Cucumis melo*. Phytopathology 70:958–61.

328. Plurad, S. B. 1967. Apple aphids and the etiology of fireblight disease. Ph.D. diss., Dept. of Plant Pathology, Univ. of Missouri, Columbia.

329. Porter, R. H. 1929. Reaction of Chinese cucumbers to mosaic. Phytopathology 19:85–86.

330. Posnette, A. F. 1969. Tolerance of virus infection in crop plants. Rev. Appl. Mycol. 48:113–18.

331. Prusky, D., A. Dinoor, and B. Jacoby. 1980. The sequence of death of haustoria and host cells during the hypersensitive reaction of oat to crown rust. Physiol. Plant Path. 17:33–40.

332. Raa, J., B. Robertson, B. Solheim, and A. Tromso. 1977. Cell surface biochemistry related to specificity of pathogenesis and virulence of microorganisms. In: Cell wall biochemistry related to specificity in host-pathogen interactions. B. Solheim and J. Raa (eds.). Universitetsforlaget, Oslo, pp. 11–30.

333. Ragetli, H. W. J. 1967. Virus-host interactions, with emphasis on certain cytopathic phenomena. Can. J. Bot. 45:1221–34.

334. Rahe, J. E., and R. M. Arnold. 1975. Injury-related phaseollin accumulation and its implications with regard to host-parasite interaction. Can. J. Bot. 921–28.

335. Rappaport, I., and S. G. Wildman. 1957. A kinetic study of local lesion growth on *Nicotiana glutinosa* resulting from tobacco mosaic virus infection. Virology 4:265–74.

336. Rappaport, I., and J. H. Wu. 1962. Release of inhibited virus infection following irradiation with ultraviolet light. Virology 17:411–19.

337. Ride, J. P., and R. B. Pearce. 1979. Lignification and papilla formation at sites of attempted penetration of wheat leaves by non-pathogenic fungi. Physiol. Plant Path. 15:79–92.

338. Riley, R. G., and P. E. Kolattukudy. 1975. Evidence for covalently attached *p*- coumaric acid and ferulic acids in cutins and suberins. Plant Physiol. 56:650–54.

339. Roberts, D. A. 1982. Systemic acquired resistance induced in hypersensitive plants by nonnecrotic localized virus infections. Virology 122:207–9.

340. Rohloff, H., and B. Lerch. 1977. Soluble leaf proteins in virus infected plants and acquired resistance. 1. Investigations on *Nicotiana tabacum* cvs. 'Xanthi-nc' and 'Samsun'. Phytopath. Z. 89:306–16.

341. Rohringer, R. 1976. Evidence for direct involvement of nucleic acids in host-parasite specificity. In: Specificity in plant diseases. R. K. S. Wood and A. Graniti (eds.). Plenum Press, New York and London, pp. 185–98.

342. Rohringer, R., N. K. Howes, W. K. Kim, and D. J. Samborski. 1974. Evidence for a gene-specific RNA determining resistance in wheat to stem rust. Nature 249:585–88.

343. Romeiro, R., A. Karr, and R. N. Goodman. 1981. *Erwinia amylovora* cell wall receptor for apple agglutinin. Physiol. Plant Path. 19:383–90.

344. Romeiro, R., A. Karr, and R. N. Goodman. 1981. Isolation of a factor from apple that agglutinates *Erwinia amylovora*. Plant Physiol. 68:772–77.

345. Ross, A. F. 1961. Localized acquired resistance to plant virus infection in hypersensitive hosts. Virology 14:329–39.

346. Ross, A. F. 1966. Systemic effects of local lesion formation. In: Viruses of plants. A. B. R. Beemster and J. Dijkstra (eds.). North Holland, Amsterdam, pp. 127–50.

347. Russell, G. E. 1978. Plant breeding for pest and disease resistance. Butterworth, London.

348. Russell, T. E., and R. S. Halliwell. 1974. Response of cultured cells of systemic and local lesion tobacco hosts to microinjection with TMV. Phytopathology 64:1520–26.

349. Russo, M., and G. P. Martelli. 1969. Cytology of *Gomphrena globosa* L. plants infected by beet mosaic virus (BMV). Phytopath. Medit. 8:65–82.

350. Ryan, C. A. 1978. Proteinase inhibitors in plant leaves: A biochemical model for pest-induced natural plant protection. Trends Biochem. Sci. July:148–50.

351. Ryan, C. A., P. Bishop, G. Pearce, A. G. Darvill, M. McNeil, and P. Albersheim. 1981. A sycamore cell wall polysaccharide and a chemically related tomato leaf polysaccharide possess similar proteinase inhibitor-inducing activities. Plant Physiol. 68:616–18.

352. Samuel, G. 1931. Some experiments on inoculating methods with plant viruses, and on local lesions. Ann. Appl. Biol. 18:494–507.

353. Sato, N., K. Tomiyama, N. Katsui, and T. Masamune. 1968. Isolation of rishitin from tuber of interspecific potato varieties containing different late blight resistance genes. Ann. Phytopath. Soc. Jap. 34:140–42.

354. Savel'eva, O. N., and B. A. Rubin. 1963. On the nature of the physiologically active toxin of the fungus *Phytophthora infestans*. Fiziol. Rast. 10:189–94.

355. Schlösser, E. W. 1980. Preformed internal chemical defenses. In: Plant disease: An advanced treatise. J. G. Horsfall and E. B. Cowling (eds.). Academic Press, New York, 5:161–77.

356. Schönbeck, F., and E. W. Schlösser. 1976. Preformed substances as potential protectants. In: Physiologial plant pathology. R. Heitefuss and P. H. Williams (eds.). Springer-Verlag, Berlin, 4:653–78.

357. Scott, F. M. 1964. Lipid deposition in intercellular space. Nature 203:164–65.

358. Seevers, P. M., and J. M. Daly. 1970. Studies on wheat stem rust resistance controlled at the Sr6 locus. II. Peroxidase activities. Phytopathology 60:1642–47.

359. Sela, B., G. Loebenstein, and T. van Praagh. 1969. Increase of tobacco mosaic virus multiplication and lesion size in hypersensitive hosts in the presence of chloramphenicol. Virology 39:260–64.

360. Sela, I. 1981. Antiviral factors from virus-infected plants. Trends Biochem. Sci. 6:31–33.

361. Sela, I. 1981. Plant-virus interactions related to resistance and localization of viral infections. Adv. Virus Res. 26:201–37.

362. Sela, I., and S. W. Applebaum. 1962. Occurrence of an antiviral factor in virus-infected plants. Virology 17:543–48.

363. Sela, I., I. Harpaz, and Y. Birk. 1966. Identification of the active component of an antiviral factor isolated from virus infected plants. Virology 28:71–78.

364. Sela, I., A. Hauschner, and R. Mozes. 1978. The mechanism of stimulation of the antiviral factor (AVF) in *Nicotiana* leaves. The involvement of phosphorylation and the role of the N-gene. Virology 89:1–6.

365. Sequeira, L. 1980. Recognition. In: Plant disease: An advanced treatise. J. G. Horsfall and E. B. Cowling (eds.). Academic Press, New York, 5:179–200.

366. Sequeira, L., G. Gaard, and G. A. De Zoeten. 1977. Interaction of bacteria and host cell walls: Its relation to mechanisms of induced resistance. Physiol. Plant Path. 10:43–50.

367. Shalla, T. A. 1959. Relations of tobacco mosaic virus and barley stripe mosaic virus to their host cells as revealed by ultrathin tissue-sectioning for the electron microscope. Virology 7:193–219.

368. Sherwood, J. L., and R. W. Fulton. 1982. The specific involvement of coat protein in tobacco mosaic virus cross protection. Virology 119:150–58.

369. Sherwood, R. T., and C. P. Vance. 1982. Initial events in the epidermal layer during penetration. In: Plant infection: The physiological and biochemical basis. Y. Asada, W. Bushnell, S. Ouchi, and C. Vance (eds.). Jap. Sci. Soc. Press, Tokyo, pp. 27–44.

370. Shifriss, O., C. H. Myers, and C. Chupp. 1942. Resistance to mosaic virus in the cucumber. Phytopathology 32:773–84.

371. Shimomura, T. 1971. Necrosis and localization of infection in local lesion hosts. Phytopath. Z. 70:185–96.

372. Shimomura, T. 1977. The role of epidermis in local lesion formation on *Nicotiana glutinosa* leaves caused by tobacco mosaic virus. Ann. Phytopath. Soc. Jap. 43:159–66.

373. Shimomura, T. 1979. Stimulation of callose synthesis in the leaves of Samsun NN tobacco showing systemic acquired resistance to tobacco mosaic virus. Ann. Phytopath. Soc. Jap. 45:299–304.

374. Shimomura, T. 1982. Effects of boron on the formation of local lesions and accumulation of callose in French bean and Samsun NN tobacco leaves inoculated with tobacco mosaic virus. Physiol. Plant Path. 20:257–61.

375. Shimomura, T., and J. Dijkstra. 1975. The occurrence of callose during the process of local lesion formation. Neth. J. Plant Path. 81:107–21.

376. Shimomura, T., and J. Dijkstra. 1976. Effect of eosin Y on the formation of local lesions and on the accumulation of callose in 'Samsun NN' tobacco and French bean leaves inoculated with tobacco mosaic virus. Neth. J. Plant Path. 82:109–18.

377. Shimomura, T., and Y. Ohashi. 1977. Tobacco mosaic virus multiplication and necrotization in local lesion hosts. Ann. Phytopath. Soc. Jap. 43:224–27.

378. Shukla, P., and C. Hiruki. 1975. Ultrastructural changes in leaf cells of *Chenopodium quinoa* infected with potato virus S. Physiol. Plant Path. 7:189–94.

379. Siegel, A., and S. G. Wildman. 1954. Some natural relationships among strains of tobacco mosaic virus. Phytopathology 44:277–82.

380. Siegel, A., and M. Zaitlin. 1964. Infection process in plant virus diseases. Ann. Rev. Phytopath. 2:179–202.

381. Simons, T. J., and A. F. Ross. 1971. Changes in phenol metabolism associated with induced systemic resistance to tobacco mosaic virus in Samsun NN tobacco. Phytopathology 61:1261–65.

382. Simons, T. J., and A. F. Ross. 1971. Metabolic changes associated with systemic induced resistance to tobacco mosaic virus in Samsun NN tobacco. Phytopathology 61:293–300.

383. Skipp, R. A., and B. J. Deverall. 1972. Relationship between fungal growth and host changes visible by light microscopy during infection of bean hypocotyls (*Phaseolus vulgaris*) susceptible and resistant to physiologic races of *Colletotrichum lindemuthianum*. Physiol. Plant Path. 2:357–74.

384. Skotnicki, A., P. D. Scotti, and A. Gibbs. 1977. Particles produced during a mixed infection by two tobamoviruses contain coat proteins of both viruses. Intervirology 8:60–64.

385. Smale, B., and H. Keil. 1966. A biochemical study of the intervarietal resistance of *Pyrus communis* to fireblight. Phytochemistry 5:1113–20.

386. Smedegaard-Petersen, V. 1982. The effect of defense reactions on the energy balance and yield of resistant plants. In: Active defense mechanisms in plants. R. K. S. Wood (ed.). Plenum Press, London, pp. 299–315.

387. Smedegaard-Petersen, V., and O. Stolen. 1981. Effect of energy-requiring defense reactions on yield and grain quality in a powdery mildew-resistant barley cultivar. Phytopathology 71:396–99.

388. Solymosy, F., G. L. Farkas, and Z. Király. 1959. Biochemical mechanism of lesion formation in virus-infected plant tissues. Nature 184:706–7.

389. Spencer, D. F., and W. C. Kimmins. 1971. Ultrastructure of tobacco mosaic virus lesions and surrounding tissue in *Phaseolus vulgaris* var. Pinto. Can. J. Bot. 49:417–21.

390. Stakman, E. C. 1915. Relation between *Puccinia graminis* and plants highly resistant to its attack. J. Agr. Res. 4:139–200.

391. Stakman, E. C., M. N. Levine, and W. Q. Loegering. 1944. Identification of physiologic races of *Puccinia graminis tritici*. U.S. Dept. Agric. Bur. Entom. Plant Quarant. E. 617.

392. Stall, R. E., and A. A. Cook. 1979. Evidence that bacterial contact with the plant cell is necessary for the hypersensitive reaction but not the susceptible reaction. Physiol. Plant Path. 14:77–84.

393. Stobbs, L. W., and B. H. Macneill. 1980. Response to tobacco mosaic virus of a tomato cultivar homozygous for gene Tm-2. Can. J. Plant Path. 2:5–11.

394. Stobbs, L. W., M. S. Manocha, and H. F. Dias. 1977. Histological changes associated with virus localization in TMV-infected pinto bean leaves. Physiol. Plant Path. 11:87–94.

395. Strobel, G. A. 1982. Phytotoxins. Ann. Rev. Biochem. 51:309–33.

396. Suhayda, C. H., and R. N. Goodman. 1981. Early proliferation and migration of *Erwinia amylovora* and subsequent xylem occlusion. Phytopathology 71:697–707.

397. Suhayda, C. H., and R. N. Goodman. 1981. Infection courts and systemic movement of ^{32}P-labeled *Erwinia amylovora* in apple petioles and stems. Phytopathology 71:656–60.

398. Sulzinski, M. A., and M. Zaitlin. 1982. Tobacco mosaic virus replication in resistant and susceptible plants: In some resistant species, virus is confined to a small number of initially infected cells. Virology 121:12–19.

399. Sziráki, I., E. Balázs, and Z. Király. 1980. Role of different stresses in inducing systemic acquired resistance to TMV and increasing cytokinin level in tobacco. Physiol. Plant Path. 16:277–84.

400. Takahashi, T. 1971. Studies on viral pathogenesis in plant hosts. I. Relation between host leaf age and the formation of systemic symptoms induced by tobacco mosaic virus. Phytopath. Z. 71:275–84.

401. Takahashi, T. 1972. Studies on viral pathogenesis in plant hosts. III. Leaf age dependent susceptibility to tobacco mosaic virus infection in 'Samsun NN' and 'Samsun' tobacco plants. Phytopath. 75:140–55.

402. Takahashi, T. 1973. Studies on viral pathogenesis in plant hosts. IV. Comparison of early process of tobacco mosaic virus

infection in the leaves of 'Samsun NN' and 'Samsun' tobacco plants. Phytopath. Z. 77:157–68.

403. Takahashi, T. 1974. Studies on viral pathogenesis in plant hosts. VI. The rate of primary lesion growth in the leaves of 'Samsun NN' tobacco to tobacco mosaic virus. Phytopath. Z. 79:53–66.

404. Taliansky, M. E., T. I. Atabekova, and J. G. Atabekov. 1977. The formation of phenotypically mixed particles upon mixed assembly of some tobacco mosaic virus (TMV) strains. Virology 76:701–8.

405. Taliansky, M. E., S. I. Malyshenko, E. S. Pshennikova, and J. G. Atabekov. 1982. Plant virus-specific transport function. II. A factor controlling virus host range. Virology 122:327–31.

406. Tani, T., and H. Yamamoto. 1979. RNA and protein synthesis and enzyme changes during infection. In: Recognition and specificity in plant host-parasite interactions. J. M. Daly and I. Uritani (eds.). Jap. Sci. Soc. Press, Tokyo, pp. 273–87.

407. Tani, T., H. Yamamoto, G. Kadota, and N. Naito. 1977. Induction of susceptibility in oat leaves to avirulent rust fungi by blasticidin S, a protein synthesis inhibitor. In: Current topics in plant pathology. Z. Király (ed.). Akadémiai Kiadó, Budapest, pp. 25–34.

408. Teakle, D. S., J. M. Appleton, and D. R. L. Steindl. 1978. An anatomical basis for resistance of sugar cane to ratoon stunting disease. Physiol. Plant Path. 12:83–91.

409. Tegtmeier, K. J., and H. D. Van Etten. 1982. The role of pisatin tolerance and degradation in the virulence of *Nectria haematococca* on peas: A genetic analysis. Phytopathology 72:608–12.

410. Thomson, A. D. 1960. Possible occurrence of interfering systems in virus-infected plants. Nature 187:761–62.

411. Tomiyama, K., R. Sakai, N. Takase, and M. Takakuwa. 1957. Physiological studies on the defence reaction of potato plant to the infection by *Phytophthora infestans*. IV. The influence of preinfectional ethanol narcosis upon the physiological reaction of potato tuber to the infection by *P. infestans*. Ann. Phytopath. Soc. Jap. 21:153–58.

412. Tomlinson, J. A., and R. J. Shepherd. 1978. Studies on mutagenesis and cross-protection of cauliflower mosaic virus. Ann. Appl. Biol. 90:223–31.

413. Touzé, A., and M.-T. Esquerré-Tugayé. 1982. Defense mechanisms of plants against varietal non-specific pathogens. In: Active defense mechanisms in plants. NATO Advanced Study Institutes Series, vol. 37. R. K. S. Wood (ed.). Plenum Press, London, pp. 103–17.

414. Touzé, A., and M. Rossignol. 1977. Lignification and the onset of premunition in muskmelon plants. In: Cell wall biochemistry related to specificity in host-pathogen interactions. B. Solheim and J. Raa (eds.). Universitetsforlaget, Oslo, pp. 289–93.

415. Tronsomo, A., and A. M. Tronsomo. 1977. Biochemistry of pathogenic growth of *Botrytis cinerea* in apple fruit. In: Cell wall biochemistry related to specificity in host-plant pathogen interactions. B. Solheim and J. Raa (eds.). Universitetsforlaget, Oslo, pp. 263–266.

416. Troutman, J. L., and R. W. Fulton. 1958. Resistance in tobacco to cucumber mosaic virus. Virology 6:303–16.

417. Tsuchiya, K., and K. Hirata. 1973. Growth of various powdery mildew fungi on the barley leaves infected preliminarily with the barley powdery mildew fungus. Ann. Phytopath. Soc. Jap. 39:396–403.

418. Tu, J. C., and C. Hiruki. 1971. Electron microscopy of cell wall thickening in local lesions of potato virus-M infected red kidney bean. Phytopathology 61:862–68.

419. Turner, J. B., and A. Novacky. 1974. The quantitative relation between plant and bacterial cells involved in the hypersensitive reaction. Phytopathology 64:885–90.

420. Urs, N. V. R., and J. M. Dunleavy. 1974. Bactericidal activity of horseradish peroxidase on *Xanthomonas phaseoli* var. *sojensis*. Phytopathology 64:542–45.

421. Urs, N. V. R., and J. M. Dunleavy. 1974. Function of peroxidase in resistance of soybean to bacterial pustule. Crop Sci. 14:740–44.

422. Valleau, W. D. 1952. Breeding tobacco for disease resistance. Econ. Bot. 6:69–102.

423. Valleau, W. D., and E. M. Johnson. 1943. An outbreak of plantago virus in Burley tobacco. Phytopathology 33:210–19.

424. Vance, C. P., T. K. Kirk, and R. T. Sherwood. 1980. Lignification as a mechanism of disease resistance. Ann. Rev. Phytopath. 18:259–88.

425. Van der Plank, J. E. 1968. Disease resistance in plants. Academic Press, London.

426. Van der Plank, J. E. 1975. Principles of plant infection. Academic Press, New York.

427. Van Larbecke, N., G. Engler, M. Holsters, S. Van den Elsacker, I. Zaenen, R. A. Schilperoort, and J. Schell. 1974. Large plasmid in *Agrobacterium tumefaciens* essential for crown gall-inducing activity. Nature 252:169–70.

428. van Loon, L. C. 1975. Polyacrylamide disc electrophoresis of the soluble leaf proteins from *Nicotiana tabacum* var. 'Samsun' and 'Samsun NN'. Physiol. Plant Path. 6:289–300.

429. van Loon, L. C. 1975. Polyacrylamide disc electrophoresis of the soluble leaf proteins from *Nicotiana tabacum* var. 'Samsun' and 'Samsun NN'. IV. Similarity of qualitative changes of specific proteins after infection with different viruses and their relationship to acquired resistance. Virology 67:566–75.

430. van Loon, L. C. 1976. Specific soluble leaf proteins in virus-infected tobacco plants are not normal constituents. J. Gen. Virol. 30:375–79.

431. van Loon, L. C. 1976. Systemic acquired resistance, peroxidase activity and lesion size in tobacco reacting hypersensitively to tobacco mosaic virus. Physiol. Plant Path. 8:231–42.

432. van Loon, L. C. 1981. The induction of pathogenesis-related proteins in virus- infected tobacco and bean plants. Vth Int. Virol. Cong., Strasbourg, p. 260. (Abstr.)

433. van Loon, L. C. 1982. Regulation of changes in proteins and enzymes associated with active defence against virus infection. In: Active defense mechanisms in plants. NATO Advanced Study Institutes Series, vol. 37. R. K. S. Wood (ed.). Plenum Press, London, pp. 247–73.

434. van Loon, L. C., and J. F. Antoniw. 1982. Comparison of the effects of salicylic acid and ethephon with virus-induced hypersensitivity and acquired resistance in tobacco. Neth. J. Plant Path. 88:237–56.

435. van Loon, L. C., and J. L. M. C. Geelen. 1971. The relationship of polyphenoloxidase to symptom expression in tobacco var. 'Samsun NN' after infection with tobacco mosaic virus. Acta Phytopath. Hung. 6:9–20.

436. van Loon, L. C., and A. van Kammen. 1970. Polyacrylamide disc electrophoresis of the soluble leaf proteins from *Nicotiana tabacum* var. 'Samsun' and 'Samsun NN'. II. Changes in protein constitution after infection with tobacco mosaic virus. Virology 40:199–211.

437. Varns, J. L., and J. Kuć. 1971. Suppression of rishitin and phytuberin accumulation and hypersensitive response in potato by compatible races of *Phytophthora infestans*. Phytopatholgy 61:178–81.

438. Wade, M., and P. Albersheim. 1979. Race-specific molecules that protect soybeans from *Phytophthora megasperma* var. *sojae*. Proc. Nat. Acad. Sci. 76: 4433–37.

439. Walker, J. C., and M. A. Stahmann. 1955. Chemical nature of disease resistance in plants. Ann. Rev. Plant Physiol. 6:351–66.

440. Wallis, F. M., and S. J. Truter. 1978. Histopathology of

tomato plants infected with *Pseudomonas solanacearum*, with emphasis on ultrastructure. Physiol. Plant Path. 13:307–17.

441. Ward, E. W. B., and G. Lazarovits. 1982. Temperature-induced changes in specificity in the interaction of soybeans with *Phytophthora megasperma* f. sp. *glycinea*. Phytopathology 72:826–30.

442. Ward, E. W. B., and A. Stoessl. 1976. On the question of "elicitors" or "inducers" in incompatible interactions between plants and fungal pathogens. Phytopathology 66:940–41.

443. Ward, H. M. 1902. On the relations between host and parasite in the bromes and their brown rust *Puccinia dispersa* Erikss. Ann. Bot. 16:223–315.

444. Wasuwat, S. L., and J. C. Walker. 1961. Relative concentration of cucumber mosaic virus in a resistant and a susceptible cucumber variety. Phytopathology 51:614–16.

445. Weinstein, L. T., M. G. Hahn, and P. Albersheim. 1981. Host-pathogen interactions. XVIII. Isolation and biological activity of glycinol, a pterocarpan phytoalexin synthesized by soybeans. Plant Physiol. 68:358–63.

446. Weintraub, M., W. G. Kemp, and H. W. J. Ragetli. 1961. Some observations on hypersensitivity to plant viruses. Phytopathology 51:290–93.

447. Weintraub, M., and H. W. J. Ragetli. 1961. Cell wall composition of leaves with a localized virus infection. Phytopathology 51:215–19.

448. Weintraub, M., and H. W. J. Ragetli. 1964. An electron microscope study of tobacco mosaic virus lesions in *Nicotiana glutinosa* L. J. Cell Biol. 23:499–509.

449. Weintraub, M., H. W. J. Ragetli, and V. T. John. 1967. Some conditions affecting the intracellular arrangement and concentration of tobacco mosaic virus particles in local lesions. J. Cell Biol. 35:183–92.

450. Weststeijn, E. A. 1976. Peroxidase activity in leaves of *Nicotiana tabacum* var. Xanthi nc. before and after infection with tobacco mosaic virus. Physiol. Plant Path. 8:63–71.

451. Weststeijn, E. A. 1978. Permeability changes in the hypersensitive reaction of *Nicotiana tabacum* cv. Xanthi nc. after infection with tobacco mosaic virus. Physiol. Plant Path. 13:253–58.

452. Weststeijn, E. A. 1981. Lesion growth and virus localization in leaves of *Nicotiana tabacum* cv. Xanthi nc. after inoculation with tobacco mosaic virus and incubation alternately at 22°C and 32°C. Physiol. Plant Path. 18:357–68.

453. Wheeler, H. E. 1976. Permeability alterations in diseased plants. In: Physiological plant pathology. R. Heitefuss and P. H. Williams (eds.). Springer- Verlag, Berlin and New York, pp. 413–29.

454. White, R. F. 1979. Acetylsalicylic acid (aspirin) induces resistance to tobacco mosaic virus in tobacco. Virology 99:410–12.

455. Wieringa-Brants, D. H., F. A. Timmer, and M. H. C. Rouweler. 1978. Infection of cowpea mesophyll protoplasts by tobacco mosaic virus and tobacco necrosis virus. Neth. J. Plant Path. 84:239–40.

456. Wood, K. R., M. I. Boulton, and A. J. Maule. 1980. Application of protoplasts in plant virus research. In: Plant cell cultures: Results and perspectives. F. Sala, B. Parisi, R. Cella, and O. Cifferri (eds.). Elsevier, Amsterdam, pp. 405–10.

457. Wu, J. H. 1973. Wound-healing as a factor in limiting the size of lesions in *Nicotiana glutinosa* leaves infected by the very mild strain of tobacco mosaic virus (TMV-VM). Virology 51:474–84.

458. Wu, J. H., L. M. Blakely, and J. E. Dimitman. 1969. Inactivation of a host resistance mechanism as an explanation for heat activation of TMV-infected bean leaves. Virology 37:658–66.

459. Wu, J. H., and J. E. Dimitman. 1970. Leaf structure and callose formation as determinants of TMV movement in bean leaves as revealed by UV irradiation studies. Virology 40:820–27.

460. Wu, J. H., and I. Rappaport. 1961. An analysis of the interference between two strains of tobacco mosaic virus on *Phaseolus vulgaris* L. Virology 14:259–63.

461. Wyatt, S. D., and C. W. Kuhn. 1979. Replication and properties of cowpea chlorotic mottle virus in resistant cowpeas. Phytopathology 69:125–29.

462. Yarwood, C. E. 1958. Heat activation of virus infections. Phytopathology 48:39–46.

463. Yarwood, C. E. 1960. Localised acquired resistance to tobacco mosaic virus. Phytopathology 50:741–44.

464. Yarwood, C. E. 1965. Predisposition to mildew by rust infection, heat, abrasion, and pressure. Phytopathology 55:1372.

465. Yoder, O. C. 1980. Toxins in pathogenesis. Ann. Rev. Phytopath. 18:103–29.

466. Yoshikawa, M., M. Matama, and H. Masago. 1981. Release of a soluble phytoalexin elicitor from mycelial walls of *Phytophthora megasperma* var. *soyae* by soybean tissues. Plant Physiol. 67:1032–35.

467. Zaitlin, M. 1962. Graft transmissibility of a systemic virus infection to a hypersensitive host: An interpretation. Phytopathology 51:1222–23.

468. Zaitlin, M. 1976. Viral cross protection: More understanding is needed. Phyopathology 66:382–83.

469. Zeyen, R. J., and W. R. Bushnell. 1979. Papilla response of barley epidermal cells caused by *Erysiphe graminis*: Rate and methods of deposition determined by microcinematography and transmission electron microscopy. Can. J. Bot. 57:898–913.

470. Ziegler, E., and R. Pontzen. 1982. Specific inhibition of glucan-elicited glyceollin accumulation in soybeans by an extracellular mannan-glycoprotein of *Phytophthora megasperma* f. sp. *glycinea*. Physiol. Plant Path. 20:321–31.

471. Zimmermann, M. H., and J. McDonough. 1978. Dysfunction in the flow of food. In: Plant disease: An advanced treatise. J. G. Horsfall and E. B. Cowling (eds.). Academic Press, New York, 3:117–40.

472. Zucker, M., and L. Hankin. 1970. Physiological basis for a cycloheximide-induced soft rot of potatoes by *Pseudomonas fluorescens*. Ann. Bot. 34:1047–62.

Index

Professor Robert N. Goodman (*left*) has been on the faculty of the University of Missouri since 1952—first in the Department of Horticulture, then joining the newly formed Department of Plant Pathology in 1967 and serving as its Chairman from 1968 to 1979. His professional career has been in phytobacteriology with continuing research on *Erwinia amylovora,* initially on control with antibiotics, later on ultrastructural features of pathogenesis and resistance linked to hypersensitivity. Recently, epidemiology and pathogenesis of *Agrobacterium tumefaciens* in grapes and ultrastructural facets of plasmid insertion into host cells have received his attention. Honors include his College's Jr. and Sr. Faculty Research awards and Fellowship in the American Phytopathological Society. He was awarded Guggenheim and Laor Foundations fellowships to study with Professor Ernst Gäumann in Switzerland, an NIH special fellowship to work with Professor R. D. Preston in Leeds (U.K.), and was appointed Visiting Sr. Scientist to work on lectins with Professor N. Sharon at the Weizmann Institute in Israel. In 1985, he was awarded a University Development Leave to conduct polysaccharide polymer research at Unilever Laboratories, Sharnbrook, U.K. Dr. Goodman has coauthored two books and several chapters and, with his students and associates, more than 140 journal articles.

Professor Dr. Zoltán Király (*center*) has been on the staff of the Research Institute for Plant Protection of the Hungarian Academy of Sciences since 1955 and has been its Director since 1980. His research has been broadly based in the biochemical aspects of plant disease. With his associates and students, he has made the Budapest laboratories an internationally renowned "School" of pathophysiology. In earlier years, Dr. Király's research was concerned with the synthesis of aromatic compounds in plants and their possible role in resistance. More recently, his studies have focused on pathologically redirected plant metabolites and their relationship to resistance and susceptibility. Dr. Király has been Adjunct Professor at the Agricultural Universities at Godolo and Szeged. Hungary has awarded him its highest honor in electing him to full membership in its National Academy of Sciences. He has conducted research for periods of one year at the University of Missouri (twice) and at the Agricultural University in Copenhagen. He is editor of *Acta Phytopathologica* of the Hungarian Academy of Sciences and has written several books, chapters, and, with his associates, more than 200 journal articles.

Dr. K. Roger Wood (*right*) is Senior Lecturer in the Department of Microbiology, University of Birmingham (U.K.). His D.Phil. degree was awarded in 1966 from Oxford, where he studied and conducted research in virology. He was awarded a postdoctoral fellowship at the National Cancer Institute, Bethesda, where he worked on the

Polyoma virus. He joined the faculty at Birmingham University initially as Research Fellow to Professor Harry Smith, FRS, in 1967 and in the intervening years rose to his Senior Lecturer position. His research, over the years, has been concerned with the bases for resistance in plants to virus infection and symptom expression. More recently, these studies have focused on gene function, with special attention given to virus-specific protein in relation to symptom expression. Dr. Wood has been for many years on the editorial board of *Physiological Plant Pathology*. Together with his students and associates, he has written 50 journal articles. His teaching responsibilities are unique, presenting courses regularly in Plant Virology, Animal Virology, and Molecular Biology.